FULL QUANTUM EFFECTS IN CONDENSED MATTER PHYSICS

As physical science advances, theoretical simulations become increasingly reflective of realistic systems, and experimental observations become more precise and refined. Thus, going beyond the Born–Oppenheimer approximation is inevitable. This book bases its discussion of condensed matter physics on the Schrödinger equation, considering both nuclear and electronic degrees of freedom. Particular attention is given to two types of phenomena: those, such as nuclear quantum effects, for which the Born–Oppenheimer approximation, although applicable in principle, is progressively weakened in practice, and those that cannot be applied at all, such as phenomena exhibiting nonadiabatic effects. In practical systems, the full quantum nature of condensed matter, as emphasized in this book, cannot be overlooked when performing accurate simulations or measurements of material properties and phenomena. This book offers state-of-the-art quantum theoretical and experimental methods, valuable for undergraduates, graduates, researchers, and industry professionals in fields such as physics, chemistry, materials science, energy, and environmental science.

ENGE WANG is President Emeritus of Peking University, Honorary Director of Kavli Institute of Theoretical Sciences, and a Professor of Physics. He is renowned for his research on full quantum effects in light-element materials. A member of the Chinese Academy of Sciences and Fellow of the American Physical Society, Wang has been honored with a series of awards, including the Tan Kah Kee Science Award in Mathematics and Physics, the Humboldt Research Award, and the TWAS Award in Physics. He has coauthored over 350 papers, edited and been a member of editorial boards of over 10 international journals, and coauthored two books: *Computer Simulations of Molecules and Condensed Matter: From Electronic Structures to Molecular Dynamics* (World Scientific, 2018) and *Water: Basic Science* (Springer, 2023).

FULL QUANTUM EFFECTS IN CONDENSED MATTER PHYSICS

Beyond the Born–Oppenheimer Approximation

ENGE WANG

Peking University

Tsientang Institute for Advanced Study

CAMBRIDGE UNIVERSITY PRESS

SCIENCE PRESS

CAMBRIDGE
UNIVERSITY PRESS

Shaftesbury Road, Cambridge CB2 8EA, United Kingdom

One Liberty Plaza, 20th Floor, New York, NY 10006, USA

477 Williamstown Road, Port Melbourne, VIC 3207, Australia

314–321, 3rd Floor, Plot 3, Splendor Forum, Jasola District Centre, New Delhi – 110025, India

Cambridge University Press is part of Cambridge University Press & Assessment,
a department of the University of Cambridge.

We share the University's mission to contribute to society through the pursuit
of education, learning and research at the highest international levels of excellence.

www.cambridge.org
Information on this title: www.cambridge.org/9781009563918
DOI: 10.1017/9781009563901

When citing this work, please include a reference to the DOI 10.1017/9781009563901

First published 2026

Cover image: Atomic image credit: Enge Wang & Peking Univ. Team;
Earth image: fotograzia / Moment / Getty Images

A catalogue record for this publication is available from the British Library

A Cataloging-in-Publication data record for this book is available from the Library of Congress

ISBN 978-1-009-56391-8 Hardback

To Charles, Milo, and Lan
and
To the memory of my grandmother

Contents

Preface

The importance of taking full quantum effects into account in condensed matter physics became apparent to me over the course of the last three decades, while my students and I were researching materials containing light elements such as hydrogen, lithium, boron, carbon, nitrogen, and oxygen. These elements combine to form a large number of interesting materials, including liquid and solid water, diamond, graphene, boronene, carbon nitride, carbon nanocones, boron–carbon-nitride nanotubes, and crystalline and porous silicon carbide. It has long been assumed that the physical properties of heavy-element materials (e.g., transition metals) are more complex than those of light-element materials. Although this seems intuitively reasonable, physical reality does not always conform to human intuition. An increasing number of studies have found that light-element materials exhibit full quantum effects (especially when nuclear quantum effects are non-negligible), resulting in novel and complex physical phenomena.

The concept of full quantum effects arises naturally from the Schrödinger equation. In this book, the term *full quantum effects* denotes the physical and chemical properties predicted by the full quantum theory with nuclear and electronic degrees of freedom included. By "full quantum theory," we mean those theories and computational approaches that not only consider the quantized electrons of a macromolecular or condensed matter system in the Schrödinger equation but also account for the quantized nuclei as well as the coupled motion of these quantum particles. In quantum chemistry, it is preferable to solve the Schrödinger equation directly in terms of the wavefunctions of the electrons and nuclei. However, such studies are generally only possible for small-scale systems containing at most a few atoms. For condensed matter systems with a large number of atoms, the customary first step is to apply the Born–Oppenheimer approximation from computational physics; then, various further approximations are used to solve the Schrödinger equation. For example, the nucleus is normally considered to be a classical point-like particle, because its mass is much larger than that of the electron; in most cases, only the electrons are strictly described by wavefunctions. Even if the Born–Oppenheimer approximation holds, this approach neglects the quantum coupling between the motions of the electrons and the nuclei. Treating the nuclei and the electrons separately and simply summing the solutions automatically implies the adiabatic approximation.

In this book, the discussion of condensed matter physics is based on the Schrödinger equation considering all nuclear and electronic degrees of freedom. Particular attention is paid to two types of phenomena: those (such as nuclear quantum effects) for which the Born–Oppenheimer approximation, although applicable in principle, is constantly weakened in practice and those to which it cannot be applied at all (i.e., which exhibit nonadiabatic effects). To avoid terminological or conceptual misunderstanding, we begin by reviewing some common concepts in condensed matter physics and explore their natural relationship with full quantum effects.

Several different theoretical concepts have been introduced over the years to explain the discoveries of experimental condensed matter research. For example, zero-point energy, zero-point vibration, quantum fluctuations, quantum tunneling, and electron–phonon coupling have all been widely accepted as explanations for a wide range of physical phenomena: hydrogen symmetrization and metallization; proton tunneling; superliquids; supersolids; spin liquids; quantum critical behavior; quantum phase transitions; quantum paraelectricity; ultra-low thermal conductivity; negative thermal expansion; superconductivity; phonon-assisted optical absorption; and isotope engineering of energy bands. These concepts are, in fact, all related to the nuclear component of full quantum effects. They can only be understood if the nucleus is considered to be a quantized particle, which is fundamentally different from considering it to be a classical one. However, only a few researchers have unified these physical concepts from a fully quantum-mechanical perspective.

In the original formulation of the Born–Oppenheimer approximation, the nuclear quantum effect was described incompletely but not entirely neglected. However, this approximation has been repeatedly weakened in applications over the last century (for instance, in the early adiabatic "ball-and-stick" model of electronic states). Therefore, it has become common practice to use harmonic or quasi-harmonic approximations from perturbation theory to describe what are referred to here as full quantum effects. For instance, phonons are constructed based on the vibrations of the Born–Oppenheimer potential energy surface. When a nonadiabatic process such as a phonon-assisted electron transition is considered, the system is typically modeled using various Born–Oppenheimer potential energy surfaces to construct the corresponding phonons, and the electron–phonon coupling process is described accordingly. In fact, with such an approach, researchers are unconsciously moving beyond the scope of the Born–Oppenheimer approximation and into the realm of full quantization, as described by the Born–Huang expansion.

Notably, these calculations assume that the harmonic or quasi-harmonic approximations hold. The full quantum physics of condensed matter discussed in this book includes not only cases in which perturbation theory holds for the solid-state system but also ones in which it does not. For example, we consider the case in which the solid-state system is in a nonequilibrium state, and the atomic nuclear configuration is far from the most stable structure of the electron ground state. In addition, we address liquid condensed matter systems and even other complex situations.

Including full quantum effects is also essential for studying the nonadiabatic effects in electron–nucleus coupled dynamics, which are entirely neglected in the original Born–Oppenheimer approximation. Nonadiabatic effects have become increasingly important in the study of non-equilibrium physical phenomena such as spectral broadening, intensity changes, and peak-position shifts caused by changes in excited-state occupancy during electron transitions. These effects also govern phonon-vibration-frequency variation with electron or hole doping concentration in graphene for example, as well as optically induced ultrafast charge transfer, topological phase transitions and charge-density waves (charge order), the molecular Berry phase, surface catalysis, and photolysis, among other physical processes.

Although it is possible to explain and predict many experimental outcomes in condensed matter physics from first principles, it is important not to confuse the physical concepts behind each success. Otherwise, theoretical results that should not have been compared with experimental data may be used in publications, without a correct understanding of the underlying approximations. This is problematic, because, without such understanding, it may frequently be impossible to discern whether the agreement or disagreement between computational and experimental results comes from the error offsets at the computational level, over-interpretation of the experimental results, or limitations of the experimental techniques. In many cases, there is no way for the reader or even the author to know. One of the purposes of this book is to point out the circumstances under which the Born–Oppenheimer approximation is acceptable and those under which nonadiabatic phenomena become non-negligible.

At the Born–Oppenheimer level, this book will explain when nuclear quantum effects, which are often ignored in current condensed-matter computational physics, can have detrimental consequences for the description of certain physical properties. Such considerations are crucial in many branches of current research in physics, chemistry, and materials science, but they have rarely been discussed systematically and clearly in traditional condensed matter physics. For example, nuclear quantum effects and nonadiabatic effects are excluded from the most frequently used method in condensed-matter computational physics: first-principles Born–Oppenheimer molecular-dynamics simulation. Therefore, adopting a fully quantum-mechanical point of view toward condensed matter physics is necessary. Solving the many-body Schrödinger equation for a real condensed matter system with the Born–Huang expansion remains a difficult task even today. However, this difficulty does not mean that further advances are impossible, nor does it justify adopting incomplete physical pictures of phenomena.

Depending on the aim of the study, the evolution of the slow nuclear part of the system relative to the fast electronic part is generally treated adiabatically. However, its quantum particle behavior is retained so that the nuclear quantum effects can be subsequently investigated. Furthermore, the evolution of the slow nuclear component can also be considered classically to preserve its nonadiabatic properties by including electron–nucleus coupling, and study nonadiabatic effects. In fact, serious works that consider both nuclear-quantum and nonadiabatic effects have also been thriving in recent years. Some of these efforts will also be covered in this book.

While most of the examples discussed in this book involve light elements, heavier elements (including heavy elements) can also exhibit full quantum effects. For example, the quantum paraelectric mechanism in ferroelectric materials and the ultralow and glass-like thermal conductivity of chalcogenide single crystals demonstrate that nuclear quantum effects are not exclusive to light elements. Moreover, in some physical processes involving coordinate uncertainty, the quantum fluctuations may be more significant than is generally believed, because nuclear quantum behavior and nucleus–electron quantum coupling exist in many-body condensed matter systems of all elements, according to the Schrödinger equation. Nuclear quantum effects are presently best studied using light-element materials, in which such effects are more pronounced. Nonadiabatic effects can be observed during electron transitions, when the nonequilibrium process is dominated by the nucleus–electron quantum coupling. Ultrafast processes, such as the quantum coupling of induced electron transition to the atomic motion on lattice, are currently the most studied examples of nonadiabatic effects; related research involves laser-induced charge density waves and superconducting states in layered titanium diselenide. For both the nuclear quantum effects and the nonadiabatic effects, no deterministic relationship has been established with the atomic masses of the elements in the materials.

In real-world condensed matter systems, nuclear quantum effects and nonadiabatic effects coexist and influence each other. A few studies on full quantum effects have now begun to take both types into account; these studies have not been limited to a particular class of elements.

With these considerations in mind, this book begins with an introduction to the full quantum effects in condensed matter, providing the reader with an overview of the relevant physical and chemical problems, followed by a discussion of full quantum theory. Next, the various experimental methods of studying full quantum effects in depth are discussed, along with their limitations. It is impossible to exclude full quantum effects from an experimental observation. The challenges encountered in previous full quantum-effects research were either caused by a lack of experimental precision or by the problem being too complex to be explained without resorting to different approximations. Considering this situation, the reader can easily understand why the Born–Huang expansion is discussed first, and newer theoretical approaches are developed from this foundation. The aim is to establish a clear theoretical framework for the full quantum effects.

Obviously, it is insufficient to study the full quantum effects from a theoretical perspective alone; the full quantum states must also be rigorously observed and measured. This requires precise characterization tools, novel experimental design, and careful sample preparation (e.g., atomic-level isotopic-purity control). In recent years, advances in ultra-high spatial-resolution imaging and in ultra-high energy-resolution spectroscopy have facilitated the detection of nuclear quantum effects, while the development of ultrafast time-resolved laser technology and its incorporation into current spectroscopic methods have aided in the detection of nonadiabatic effects. The establishment of extreme external-field conditions also allows us to manifest or suppress the full quantum effects artificially during an experiment, allowing precise control. Most of this experimental research has been conducted within the last decade.

In the study of full quantum physics in condensed matter, close coordination of theoretical and experimental work is important. For example, nuclear quantum effects can be partially characterized by experimental measurements via isotope substitution. The result can then be compared with a first-principles path-integral molecular dynamics simulation.

In the study of hydrogen-bonding network structures and phase transitions in liquid and solid water, such a combination of theory and experiment has been fully exploited by my group, enabling us to conduct original research on the full quantum effects.

Although ubiquitous in condensed matter physics and significant across a broad range of topics in many frontier fields, full quantum effects did not attract considerable research interest until recently because of limitations in theoretical understanding and experimental techniques. After a brief overview of the historical development and status quo of the field, this book first introduces physical (Chapter 2) and then chemical (Chapter 3) problems in full quantum effects in condensed matter, providing the reader with a general conceptual foundation. Chapters 4 and 5 thoroughly discuss the theoretical basis and calculational methods of full-quantum-effect research. The development of experimental techniques and external-field extreme conditions are explained in Chapter 6. Chapters 7 and 8 focus on the full quantum effects in hydrogen, hydrides, and other typical element and compound materials, using the methods introduced earlier. In Chapters 9–11, existing or potential applications of full quantum effects in energy, environmental, and device research are introduced. Chapters 1–8 of this book present well-established topics and can be used as a university textbook. However, some of the research topics brought up in Chapters 9–11 are preliminary and require further investigation.

As the book approaches completion, I have a sense that the advances both in theory and in experiment over the past several years have made it possible for us to return to the original Schrödinger equation of quantum mechanics and study full quantum effects in condensed matter physics from a new perspective. The big picture will become simple and clear if we see things this way. To make the concept of full quantum effects more accessible to people, two issues concerning the current way of thinking must be addressed. First, condensed matter theory based on the second quantization of a model Hamiltonian must be considered in conjunction with first-principles full quantum simulation approaches, such as first-principles electron–phonon coupling theory. Second, a general consensus needs to be established between theoretical physicists and chemists, who use entirely different languages reflecting their different research backgrounds and focuses. Moreover, the challenges of this emerging field of condensed matter physics must be addressed in a coordinated way.

After completing this heavy undertaking, I feel quite happy and relaxed. Let us now return to the two main purposes of the book. On the one hand, it is meant to provide the readers with a clear entry point into the fundamental physics of condensed matter, especially for problems beyond the adiabatic ball-and-stick model of electronic states. On the other hand, it is intended to facilitate the development of this newly emerging research field and steer the discussion in the right direction. Most of the problems in this field can be described as "many-body" and "quantum" in nature; these are the two central features of condensed matter systems. I hope that the readers of this book some years from now will regard it as an early comprehensive review that contributed to the establishment of full quantum physics of condensed matter. In the present era, I hope that my work will spark interest in this research field and encourage its readers to participate in the study of full quantum effects in condensed matter physics.

As physical science develops, theoretical simulations will become increasingly reflective of realistic systems, and experimental observations will become more precise and refined. Therefore, going beyond the Born–Oppenheimer approximation is inevitable. In practical systems, the full quantum physics of condensed matter emphasized in this book cannot be overlooked when performing accurate simulations or measurements of material properties. I believe that, with the rapid development of the theoretical approach and the substantial improvement of experimental techniques, the Born–Huang expansion will become the dominant paradigm in condensed matter physics in the coming years, especially for descriptions of the physical properties related to nuclear quantum effects and nonadiabatic effects. In contrast to the traditional paradigm of condensed matter physics research built in the framework of the Born–Oppenheimer approximation, the new paradigm presented below is meant to go beyond the Born–Oppenheimer approximation to include the nuclear quantum and nonadiabatic effects in a complete manner. Therefore, as far as fundamental research is concerned, studying the full quantum effects will improve our understanding of the real world considerably.

Although condensed matter physics is the primary focus of this book, some of the results presented here are also relevant to research in chemistry, materials science, and other related fields (such as energy science and environmental science). Thus, the results of this basic research will not only advance frontier science but also create opportunities for potential applications.

It has been about five years since I had the idea for this book. From the first draft, the book has benefitted from the discussions with my colleagues and students, including Professors Wenlong Wang, Sheng Meng, Guangyu Zhang, Jimin Zhao, and Xingjiang Zhou from the Institute of Physics, Chinese Academy of Sciences, and Professors Xinzheng Li, Ji Chen, Ying Jiang, Xi Lin, Shuang Jia, Peng Gao, and Lei Liu from Peking University. The two institutions have influenced my research of the subject in a variety of ways. Professor Niu Qian (China University of Science and Technology) has also provided valuable insight into defining the concept of full quantum effects. Dr. Luhao Zhang kindly read the manuscript, and Dr. Chang Liu carefully formatted the figures. More friends participated in reading and revising the first draft than can be listed here, but I would like to thank them all for their valuable contributions and cooperative spirit and apologize for not naming them individually. I am also thankful to the Tsientang Institute for Advanced Study, a place that made this task pleasant in the final stage.

I hope that the final product proves worthy of the effort and will contribute to the enhancement of condensed matter physics.

1
Overview of Full Quantum Effects in Condensed Matter

1.1 Schrödinger Equation and Its Approximate Solution

Quantum mechanics is one of the greatest scientific achievements of the twentieth century, and development of the theory took nearly three decades. In the late nineteenth century, physicists were confronted with the challenge of understanding the "ultraviolet divergence" in blackbody radiation. Heat can be transferred in three ways: conduction, convection, and radiation. The ability of any surface to emit electromagnetic radiation is directly proportional to its ability to absorb incident electromagnetic radiation. The term "blackbody" refers to the physical body that exhibits the strongest ability to absorb, and thus also to emit, electromagnetic radiation at a given temperature (see Figure 1.1). Research on blackbody radiation involves investigating the electromagnetic radiation emitted from the blackbody surface at a certain temperature; in other words, it examines the relationship between the blackbody radiation energy density and frequency. Some experimental measurements of blackbody radiation energy density at different frequencies are shown in Figure 1.2 [Lu G.1997]. It is observed that for a blackbody at a given temperature, the energy density of the emitted electromagnetic field peaks at some point as the frequency varies, while falling toward zero on both sides of the peak.

The seemingly simple relationship shown in Figure 1.2 was confusing to nineteenth-century physicists; although many attempted to solve the problem, they failed to think outside the regime of classical physics. For example, in 1893, W. Wien discovered that the frequency corresponding to the maximum energy is proportional to the blackbody temperature [Wien1893] and derived a distribution formula relating the radiated energy to frequency. This formula agreed well with the experimental results at most frequencies but deviated significantly from the experiment in the low-frequency region (see the dash-dotted curve in Figure 1.2). The approach adopted by Lord Rayleigh and J. H. Jeans in 1899 was based on electrodynamics and statistical physics [Rayleigh1899]. Their theory agreed well with the experiment in the low-frequency range but showed non-negligible discrepancies and even divergence in the high-frequency region (see the dashed curve in Figure 1.2). Because these higher frequencies often lie in the ultraviolet band, this problem is called the "ultraviolet catastrophe."

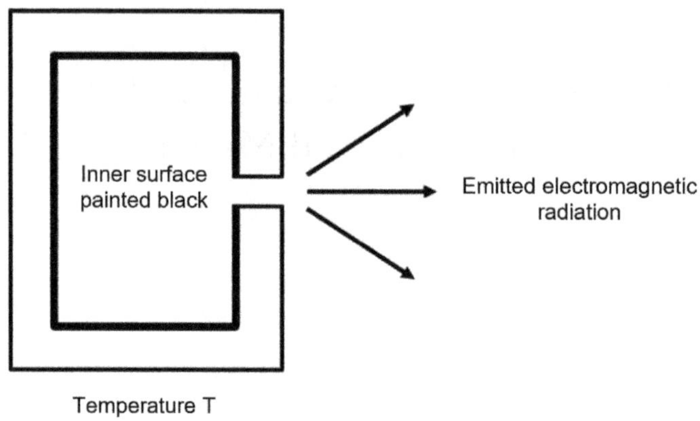

Figure 1.1 Schematic of a blackbody radiator.

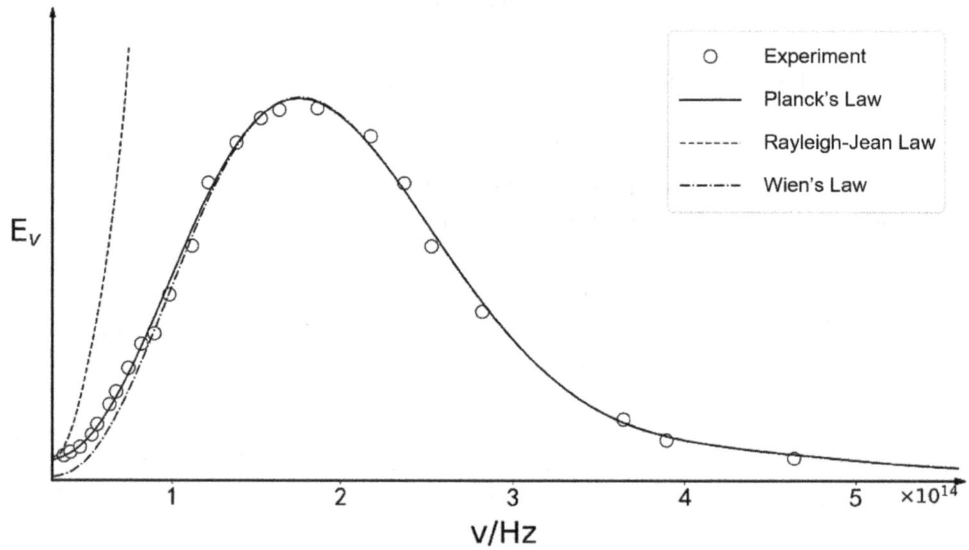

Figure 1.2 Blackbody energy density as a function of frequency ν. Reprinted from Lu G.[1997].

In 1894, M. Planck began to investigate the problem of blackbody radiation. After years of research, Planck published two papers in 1900 [Planck1900a,b] and presented his finding on the quantum problem at a German Physical Society meeting in Berlin on December 14. He proposed the famous quantum hypothesis: Electromagnetic radiation energy at frequency ν can only be a multiple of $h\nu$, where the constant h (now called Planck's constant) must be used as the magnitude of the energy quantum.

In physics, some fundamental constants are obtained experimentally (e.g., the speed of light c or the electric charge e), while others are derived from various laws (i.e., Newton's gravitational constant G or Boltzmann's constant k_B). Planck's constant, however, was discovered by individual creative intelligence alone. Using the quantum hypothesis, he utilized electrodynamics and statistical physics methods to derive a new formula that related radiation energy to frequency known as Planck's law, which agreed perfectly with the experimental results available at that time (see the solid curve in Figure 1.2). Planck was awarded the Nobel Prize in Physics in 1918 for his important contribution to the discovery of energy quanta. Afterward, Paul A. M. Dirac commented on this important discovery, saying that Planck ushered in a new quantum era in the history of science.

However, it is not clear why the radiation energy is quantized. For people who are accustomed to solving problems with classical physics, this was incredibly perplexing. It was not until 1913 that N. Bohr first combined Planck's quantum hypothesis with Rutherford's concept of the atomic nucleus and proposed a quantum theory based on the atomic model, thus cleverly answering the question [Bohr1913]. According to this model, electrons are confined to specific orbitals outside the nucleus, and energy must be absorbed or emitted for electrons to jump between them (as shown in Figure 1.3). Since the orbits are not contiguous, the radiation energy cannot be continuous. Although the Bohr model has subsequently undergone multiple modifications and refinements, we should acknowledge that this explanation of the origin of quantum effects is the most straightforward and simplest. R. P. Feynman once stated that if the history of human science were to be summarized into a single sentence, it would be: "Everything is made of atoms," indicating the significance of the atomic model for the development of science. Bohr was awarded the Nobel Prize in Physics in 1922 for his work on atomic structure and quantum radiation.

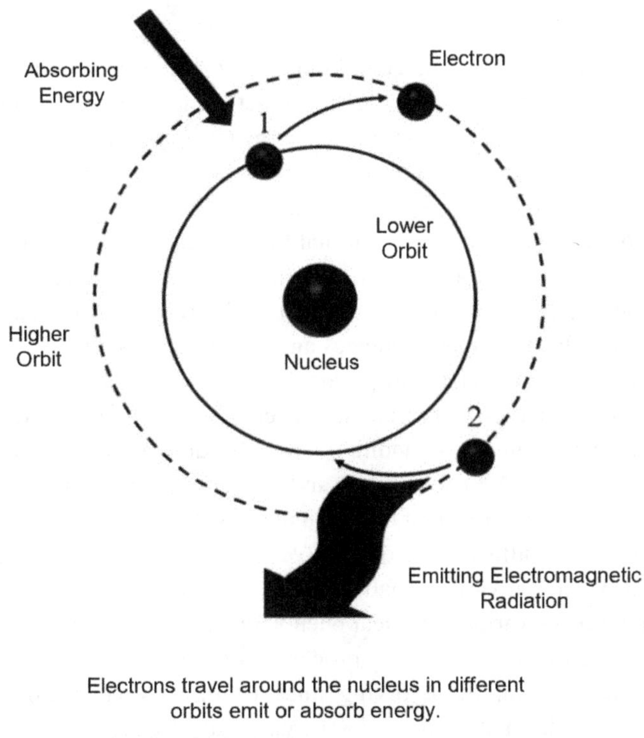

Figure 1.3 Bohr's model of atomic structure.

The establishment of quantum theory posed a great challenge to classical physics, but it also provided opportunities to build an entirely new scientific framework. The Schrödinger equation marked the transition from early quantum theory to modern quantum mechanics [Schrödinger1926]. Since its formulation by E. Schrödinger in 1926, the Schrödinger equation has unquestionably become the fundamental tool for describing the microscopic world consisting of electrons and nuclei in the nonrelativistic limit. The role of the Schrödinger equation in quantum mechanics is comparable to that of Newton's equations in classical mechanics. One of the major challenges in physics, chemistry, and materials science research is to use Schrödinger's equation to solve physical, chemical, and materials problems in real world that are closely related to our daily lives [Andzelm1992, Georges1996, Giannozzi2009]. Schrödinger won the Nobel Prize in Physics in 1933.

The general expression of the Schrödinger equation is as follows:

$$\hat{H}\Psi\left(\vec{r},\vec{R},t\right)=i\hbar\frac{\partial}{\partial t}\Psi\left(\vec{r},\vec{R},t\right), \tag{1.1}$$

where $\hbar = h/(2\pi)$ is the reduced Planck constant, from which it can also be seen that Planck not only introduced the concept of quantum but also laid the foundation for the establishment and development of quantum physics in general. Equation (1.1) corresponds to the Hamiltonian, for simplicity in the nonrelativistic regime, which is expressed as

$$\hat{H}=-\sum_{i=1}^{N}\frac{1}{2}\nabla_i^2+\frac{1}{2}\sum_{i\neq i'}V\left(\vec{r}_i-\vec{r}_{i'}\right)-\sum_{j=1}^{M}\frac{1}{2M_j}\nabla_j^2+\frac{1}{2}\sum_{j\neq j'}V\left(\vec{R}_j-\vec{R}_{j'}\right)+\sum_{i,j}V\left(\vec{r}_i-\vec{R}_j\right), \tag{1.2}$$

It follows from Eq. (1.2) that the Hamiltonian of a real system without external fields contains five key terms of action: the electron kinetic energy term, the electron–electron Coulombic repulsion term, the nuclear kinetic energy term, the nucleus–nucleus Coulombic repulsion term, and the electron–nucleus Coulombic attraction term [Li X.Z.2014]. These five terms are the most fundamental components of the Hamiltonian of a polyatomic system (molecules or condensed matter). External field terms can, of course, be added at any time. Note that we adopted the atomic unit system, with electron mass $m_0 = 1$ and $\hbar = 1$. M_j in Eq. (1.2) is the relative mass of the jth nucleus with respect to the electron, satisfying $M_j \gg 1$. $\vec{R}$ and $\vec{r}$ are the positions of nucleus and electron, respectively.

In the twentieth century, a standard protocol was developed for investigating the fundamental behavior of a matter at the quantum level. The quantum properties of the system were researched experimentally; in addition, the Schrödinger equation with the appropriate Hamiltonian was solved by theoretical and numerical approximation to predict new phenomena or to explain the relevant experimental results. Combining the theoretical and experimental approaches allowed the qualitative and quantitative description of a diverse range of physical systems at various spatial and temporal scales, providing a more comprehensive understanding of various physical phenomena in the objective world.

Knowledge of elementary excitation processes in atomic, molecular, and condensed matter systems has been rapidly evolving since 1930. This has not only enhanced our knowledge of the world at the microscopic scale but also promoted the development and application of new materials and devices, thus playing an essential role in society's transition from the electric to the electronic era. Over time, secondary subjects such as atomic and molecular physics, condensed matter physics, and quantum optics were established under the primary subject of physics [Maiman1960, Anderson1972]; each subject thrived in the mid twentieth century. From the viewpoint of research methodology, the development of these subjects depended on successfully applying basic quantum mechanics to various material systems, leading to the discovery of novel phenomena and new physics. The new material fabrication techniques and quantum device designs inspired by these advances have undoubtedly increased the scientific and technological development of modern society and will continue to play a new and even more indispensable role in the future.

Most attempts to solve the Schrödinger equation (Eqs (1.1) and (1.2)) for a condensed matter system (i.e., a system containing multiple atoms and electrons) begin by assuming the validity of the Born–Oppenheimer approximation, first proposed by M. Born and J. R. Oppenheimer in 1927, 1 year after the Schrödinger equation, in a theoretical publication studying the vibrational and rotational quantum states of small molecules [Born1927]. The Born–Oppenheimer approximation relies on the significant difference in mass between electrons and nuclei. Given that the mass of an electron is much smaller than that of the nucleus, when the electron and nucleus exhibit comparable kinetic energies at a given temperature, the electron's speed is much higher than that of the nucleus. Consequently, the characteristic relaxation time scale of electrons is considerably shorter than that of the nucleus at the dynamic level. This temporal distinction allows for the separate consideration of electrons and nuclei, permitting the use of the adiabatic approximation in their motion analysis. Due to the shorter relaxation times of electrons, they can rapidly follow the nucleus's movements and return to their eigenstates. Given the characteristic differences in speed between electrons (light and quick) and atomic nuclei (heavy and slow), the adiabatic approximation has become widely accepted. Since its introduction, the Born–Oppenheimer approximation has served as a fundamental principle in modern computational condensed physics and chemistry. (X. Z. Li's illustration in Figure 1.4 visually depicts the impact of the difference in characteristic time scales).

Applying this assumption to the terms in Eq. (1.2), it would be reasonable to ignore the nuclear kinetic-energy term from the Hamiltonian and the corresponding coupling of the electron–nucleus interactions at the quantum-mechanical level. In other words, when describing the electronic structure, we can treat the nuclei as a set of point-charged particles fixed in space, and these nuclei only act as an external field to the many-body quantum system composed of electrons. Elcctrons in this interacting many-body quantum system assume a series of discrete energy levels for the particular nuclear configuration or external potential field. The energies of these energy levels, together with the classical Coulomb potential between the nuclei, make up the total energy of the polyatomic molecular or condensed matter system for a given nuclear configuration, in which these nuclei are treated as classical particles. In this case, we can construct a series of potential energy surfaces in the higher dimensional space of nuclear configurations ($3N - 3$, where N corresponds to the number of nuclei). An example of such a Born–Oppenheimer potential energy surface for a water molecule is shown in Figure 1.5. In principle, a nucleus is a wave packet on the

Figure 1.4 Principle of the Born–Oppenheimer approximation. Because of their different masses, electrons and nucleus have different characteristic relaxation time scales, those of the electrons being much shorter. For example, the relaxation of an electron is much faster compared to the nucleus. Thus, during every step of the nuclear movement, electrons can relax to their original eigenstates. This is analogous to a slow boy who accidentally disturbs a wasps' nest. As the wasps fly much more quickly than the boy runs, they can follow the boy while remaining in their cloud-like "eigenstate." Reprinted from Li X.Z.[2014]. Reproduced with permission from Peking University Press, Beijing.

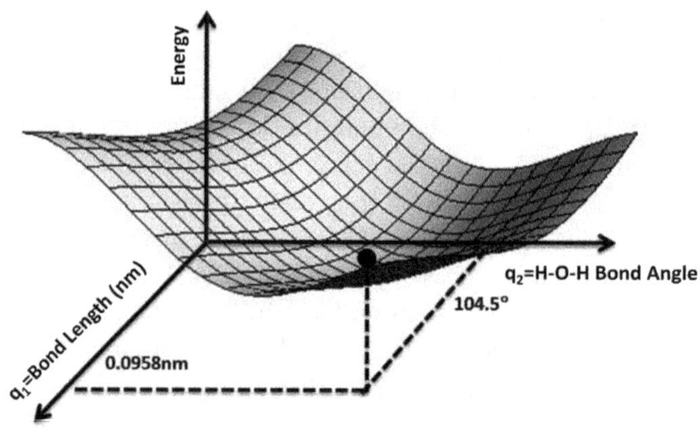

Figure 1.5 Electronic ground-state Born–Oppenheimer potential energy surface of a water molecule. The dimension of the conformational space inside the molecule is 6 (because $3N - 3 = 6$, where N is the number of atoms). The figure uses the hydrogen bond length and the H–O–H bond angle as axes. As can be seen from the figure, when the nuclei of the hydrogen and oxygen atoms are treated as classical particles, the equilibrium bond length is 0.958 Å and the bond angle is 104.5°, corresponding to the minimum of the surface.

Born–Oppenheimer potential energy surface; its vibrational and rotational quantum states can be described using this wave packet.

Within the framework of the Born–Oppenheimer approximation, the overall wavefunction is the product of the nuclear wave packet and the electronic wavefunction, with the nuclear positions serving as parameters for the determination of the electronic wavefunction. The whole system is free to move on one potential energy surface but cannot jump between surfaces corresponding to different electronic states. From a mathematical perspective, the separation-of-variables method can handle the wavefunction that contains the electronic and nuclear degrees of freedom: by treating the nuclei and electrons separately and simply summing the solutions. From a physics viewpoint, this approximation is an adiabatic treatment due to the difference in relaxation time scales of the two particles (electrons and nuclei). For this reason, the Born–Oppenheimer approximation is also referred to as the adiabatic approximation. (This formulation of the Born–Oppenheimer approximation is also called the adiabatic formulation in later studies when the nonadiabatic effects are concerned.) Based on this formulation, it is possible to take the nuclear kinetic term and the nucleus–electron nonadiabatic interaction term between the nucleus and the electron into account at the theoretical level, that is, to include the quantum description of nuclei, which is a recent development in physical science and is the focus of this book.

The Born–Oppenheimer approximation, along with Schrödinger's equation, constituted a traditional paradigm of modern physical science research. For theoreticians, it is the starting point for the vast majority of simulations; for experimentalists, it is the theoretical tool on which most researchers rely when interpreting or predicting observations. However, it should be noted that although this approximation does not purposely ignore the wave-packet property of the nucleus, people tend to emphasize the importance of "treating the nucleus as a point-charged particle fixed in space" when describing the electronic structure in practice, thus reducing the quantized "wave packet" to a classical "point-charged particle." Moreover, when describing the nuclear motion, they are likely to introduce two more approximations into their calculation without realizing them.

One of these approximations treats the nucleus as a classical particle, as in the ball-and-stick model frequently used to describe crystal and molecule structures. (Figure 1.6 shows the ball-and-stick model of a methane molecule.) In this model, the quantum glue of the valence-electron-cloud distribution in a specific nuclear configuration is described

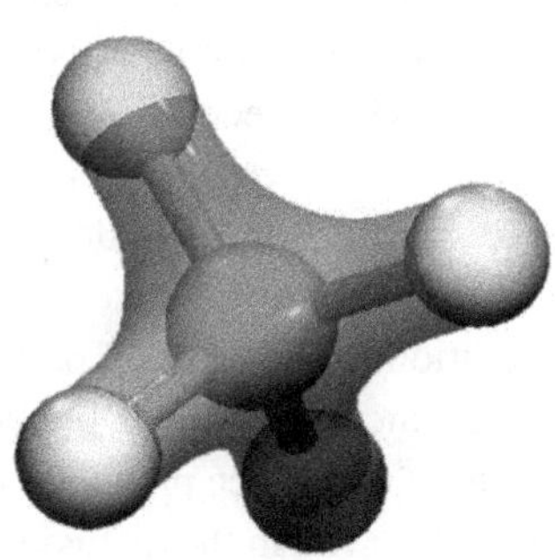

Figure 1.6 Ball-and-stick model of a methane molecule. The small white balls represent the hydrogen atomic nuclei, and the large grey ball represents the carbon atomic nucleus. Electrons act as cloud-like quantum glue, forming covalent bonds to bind the nuclei in their equilibrium positions; these bonds are represented by sticks between the balls. The ball-and-stick model is a simple and effective way to understand the structure of molecular and condensed matter systems. Reprinted figure with permission from Mendoza[2014]. Copyright 2024 by the American Physical Society.

as a chemical bond between the nuclei and is represented by a stick. Similarly, the nucleus (along with the core electrons) is represented by a ball. By using a series of simple balls and sticks, one can then understand the crystal structure in a general way. Subsequently, consideration is given to nuclear movement around the equilibrium position. The simple ball-and-stick model fails to reflect the essential fact that the nucleus is also a wave packet. Using the language of first-principles molecular dynamics (an approach developed in the 1980s), first-principles Born–Oppenheimer molecular dynamics simulations do not treat the atomic nuclei according to the Born–Oppenheimer approximation rigorously because of the underlying assumption that each nucleus (with its core electrons) can be regarded as a classical particle.

Some literature incorrectly states that the description of nuclear motion in Born–Oppenheimer molecular dynamics is equivalent to the Born–Oppenheimer approximation. As a result of this misunderstanding, people tend to overlook the fact that the nucleus is a quantum particle in their theoretical descriptions of many molecular or condensed matter systems. It should be noted that, this fact is not missing in the original formulation of the Born–Oppenheimer approximation, which considers the nuclear ground state of a system on the Born–Oppenheimer potential energy surface, and is therefore incomplete. However, this part is currently missing when people talk about first principles Born–Oppenheimer molecular dynamics simulations. The discussion of this issue has often led to misunderstandings, which this book seeks to clarify.

The second most favored approach takes the wave-packet properties of the nucleus into account (e.g., when describing phonon vibrations). However, it is customary for physicists to place the system near the minimum of the Born–Oppenheimer potential energy surface (corresponding to a stable structure) and apply the perturbation expansion. The second-order expansion corresponds to the commonly used harmonic approximation (HA). The higher-order expansion, on the other hand, can include certain anharmonic effects. However, it must be noted that approximations such as perturbation theory are inefficient when dealing with systems that are far from stable structures (e.g., liquid systems or the breaking and recombining of chemical bonds in reactions).

Anharmonic nuclear quantum effects have recently become a key frontier research topic in physical science. Using liquid water as an example, a condensed matter that is an integral part of our lives, the isotopic substitution of hydrogen and deuterium can result in differences in physical properties that cannot be described by classical thermodynamic statistics and kinetics, nor by traditional methods of perturbation expansion. A simple analysis reveals that the perturbation method is more likely to fail in describing systems composed of light elements. In the field of surface physics and catalytic reactions, quantum tunneling (Figure 1.7) is commonly observed in the diffusion of light-element atoms at low temperatures, accompanied by bond breaking and formation. This type of phenomenon also cannot be approximated by the aforementioned perturbation expansion. These conceptual nuances based on perturbation theory in describing quantum properties of nuclei, which result in discrepancies in understanding condensed matter physical properties and even lead to the absence of certain important properties, are also among the main focuses of this book.

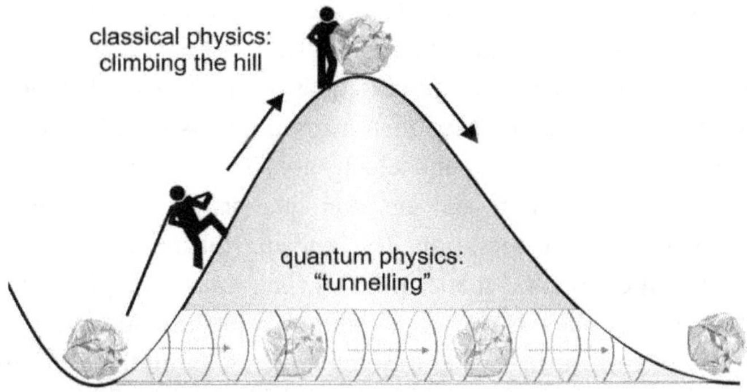

Figure 1.7 Quantum tunneling. It shows the transport behavior of microscopic particles crossing the potential barrier from one side to the other, even when their kinetic energy is lower than the height of the potential. When the classical thermal motion is not enough to overcome the system's potential barrier, quantum tunneling can complement and help with the process. Note that tunneling can be achieved via many paths, each with a certain associated probability; the final result is the integral of these paths. This effect is often more pronounced at low temperatures. At high temperatures, its contribution in comparison with that of thermal excitation approaches zero, and the quantum-mechanical description of the system returns to the classical limit.

Even if the Born–Oppenheimer approximation is strictly followed, some important physical concepts are still missing from the description of real materials; these will often be mentioned in this book. One of them is the nonadiabatic effect, which often appears in the study of nonequilibrium dynamics. This effect involves the characteristic relaxation time scales of electronic and nuclear motion. Its true physical counterpart is that in some specific cases, characteristic scales of relaxation times for electron and nucleus motions do not differ by several orders of magnitude; consequently, the influence of the excited electron state on the nuclear motion and the influence of the nuclear motion on the electron eigenstate transition can no longer be ignored [Krüger2022].

Thus, in theory, nonadiabatic effects associated with the electron eigenstate transition must be considered in addition to the existing nuclear quantum effects (see Section 1.4 for the definitions of "nuclear quantum effect" and "nonadiabatic effect"). In practice, for problems in which nonadiabatic effects predominate, the description of nuclear motion can usually be restricted to the classical regime for simplicity.

Furthermore, the Born–Oppenheimer adiabatic approximation fails to consider other physical phenomena associated with the full quantum effect, such as the Berry phase. It has been shown that during the motion of the nucleus, the electronic ground state wavefunction evolves and accumulates a geometric phase [Mead1979]. Recently, Q. Niu et al. discovered that this geometric phase influences the equations of nucleus motion through a nonlocal effective magnetic field, the molecular Berry phase, and thus modifies the vibrational modes. As a result, the degeneracy of the optical branches at the center of the Brillouin zone is removed from the phonon spectrum of some systems with time reversal symmetry breaking [Saparov2022]. In recent years, with the invention (particularly in physical chemistry, etc.) of precision measurement tools such as ultrafast spectroscopy, nonadiabatic effects have been attracting more attention as they play a crucial role in the design and development of energy and environmental materials.

From the introduction of the Schrödinger equation to the establishment of the Born–Oppenheimer approximation, condensed matter research has mainly been concerned with the quantum properties of electrons and electron–electron many-body correlation. The quantum properties of the nucleus and electron–nucleus quantum effects, on the other hand, were excluded from consideration. By ignoring the nuclear quantum effects, the mainstream methods of condensed matter physics (i.e., first-principles electronic structure calculation and first-principles Born–Oppenheimer molecular dynamics) only partially quantize the many-body electron–nucleus system when solving the Schrödinger equation. In other words, only the electrons are treated as quantum particles, while the nuclei are treated as classical particles.

Therefore, this book focuses on full-quantum condensed matter physics where both the electrons and the nuclei are treated as quantum particles, and while the electron–electron many-body correlation effect are considered, nucleus-nucleus many-body correlation effects are also taken into account, as well as the nonadiabatic coupling of electron–nucleus quantum motion. In this way, in addition to a more rigorous theoretical extension of the Born–Oppenheimer approximation can be established, the theoretical description of many physical and chemical properties and the understanding of experimental phenomena beyond the scope of the Born–Oppenheimer approximation are the focus of the discussion in this book.

Among these discussions, the Born–Huang expansion proposed by M. Born and K. Huang in the 1950s provides a simple and effective theoretical framework [Born1954]. Because of its unique status, the Born–Huang expansion will often be our starting point in this book when solving the Schrödinger equation for polyatomic systems in all degrees of freedom. Based on this expansion, we will discuss when the Born–Oppenheimer approximation is acceptable and when nonadiabatic effects beyond its scope are relevant. At the level of the Born–Oppenheimer approximation, we will also explain when nuclear quantum effects can significantly impact on the physical description of a real system. All these are crucial concepts in current physics, chemistry, and materials science research, but they have rarely been discussed or systematically clarified. As a result, the issues embedded in this way of thinking, as well as the newly emerging full-quantum condensed-matter physics research, have not been given sufficient attention. Appropriately acknowledging and understanding these issues is the key to the content and approach of this book.

To help readers understand the primary focus of this book, the core problem of full-quantum condensed-matter physics can be summarized as the phenomena that are neglected by the adiabatic ball-and-stick model of the electronic states when describing the physical properties of the condensed matter system. Based on the Born–Huang expansion, these phenomena can be categorized as nuclear quantum effects (NQEs) and nonadiabatic effects (NAEs). These two ideas are naturally coupled when we use the Born–Huang expansion to rigorously treat actual systems. Its derivation and other details will be discussed later.

The most common NQEs are the zero-point energy, zero-point vibrations, quantum fluctuations, quantum tunneling, and the accompanying quantum delocalization. These effects are more prominent in low-temperature experiments, where, for example, the drift diffusion of nuclei is often dominated by quantum tunneling. Similarly, when analyzing the stability of some structures, the zero-point energy provides a correction to the internal energy even when the temperature is close to absolute zero. Terms such as quantum tunneling, zero-point energy, or quantum fluctuation are often used by experimentalists to describe their findings without sufficient awareness that these concepts are connected to NQEs. All of these can be truly understood only when the atomic nucleus is treated as a quantum particle, which fundamentally differs from treating it as a classical particle.

In the study of NQEs, researchers often reveal and compare physical properties using isotopic substitution. To the best of my knowledge, this idea of isotope was first introduced in 1913 by F. Soddy, who showed that it is possible for the same element to have two or more different types of atoms with the same number of electrons and protons but different numbers of neutrons [Soddy1913]. In condensed matter, isotopes of the same element may have different physical properties; this is known as the isotope effect [Vértes2011]. It is more pronounced in light elements because of the large relative differences in isotopic atomic masses (e.g., deuterium atoms are 100% heavier than normal hydrogen atoms).

NAEs are more commonly observed when overall properties of a condensed matter system are influenced by the coupling of mutually superimposed electron–nucleus wavefunctions. At the kinetic level, the entire system evolves as electrons transition from the ground state to excited states, affecting future electron and nuclear kinetic processes on a time scale ranging from tens of femtoseconds to a few picoseconds. Essentially, NAEs target the quantum coupling of the electron–nucleus motion. Because of recent advances in ultrafast spectroscopy, the study of NAEs has become a frontier research area in physical science; in particular, considerable efforts focus on optically induced quantum-mechanical phenomena, such as topological phase transitions and charge-density waves. Many-body correlation, which is a key factor in condensed matter physics and the isotope effect, can play important roles in the study of nonadiabatic processes as well.

Finally, while this book covers full-quantum physics research, this does not preclude other undiscussed research directions from being considered as quantum physics. On the contrary, quantum mechanics has supported many recent research directions in physics, chemistry, and materials science, such as superconductivity, topology, quantum information, quantum computing, and cold-atom physics [Leuenberger2001, Hasan2010, Qi X.L.2011]. These are beyond the scope of our discussion and are thus not explored in detail in this book. The main objective of this book is to summarize the description of quantum-mechanical phenomena important for physical science research, based on the Born–Oppenheimer approximation and the Born–Huang expansion. Psychologically, it aims to correct some biases in current understanding and point out concepts often overlooked when conducting condensed matter physics research, highlighting the potential consequences of such biases. We will use NQEs and NAEs, which are often ignored in most current research, as entry points and discuss the coupling between the quantum motion of the electron and the nucleus. We will make it clear which areas and problems can be further studied more systematically and rigorously from the perspective of full quantum physics. By introducing the concepts of full quantum physics, we hope to provide a fresh perspective on new areas of condensed matter physics, chemistry, and materials science that are important but are currently not attracting enough research attention.

1.2 Introduction to Full Quantum Effects

Section 1.1 covered the origins of quantum mechanics, the establishment of Schrödinger's equation, the approximate solution of Schrödinger's equation for molecular or condensed matter systems, and the rigorous consideration of full-quantum-physics problems. With this background, we can now precisely define full quantum effects, starting with Schrödinger's equation. ***"Full quantum effects" refers to the physical and chemical properties based on a full quantum-mechanical theory of the nuclear as well as electronic degrees of freedom.*** The full quantum theory is a theoretical and computational approach to quantize all nuclei and electrons in a polyatomic macromolecule or condensed matter system starting from Schrödinger's equation (Eqs (1.1) and (1.2)), while taking the coupling motion of these quantum particles into consideration.

Historically, the formulation of the full quantum effect occurred almost simultaneously with the development of quantum mechanics. In addition to the Copenhagen school led by N. Bohr, scholars from the University of Göttingen played a crucial role. Among them were M. Born (professor of theoretical physics at Göttingen who supervised W. K. Heisenberg, W. E. Pauli, E. P. Jordan, and J. R. Oppenheimer, among others), J. Franck (professor of experimental physics at Göttingen who shared the 1925 Nobel Prize in Physics with G. L. Hertz for discovering the law of electron–atom collisions; his work on photon–molecule interaction inspired the early discussion of full quantum effects), and D. Hilbert (professor of mathematics at Göttingen, who developed many fundamental ideas in linear algebra; he was the mentor of M. Born and later collaborated with E. Noether, E. P. Wigner, J. von Neumann, and others). The Göttingen school developed a broad range of ideas and theories, mainly in the 1920s, laying the mathematical foundations for understanding the quantum mechanics. Finally, in 1926, E. Schrödinger from Vienna presented his landmark equation, signifying the establishment of quantum mechanics. At this stage, research still focused mainly on electronic states.

Born realized the importance of Schrödinger's equation in explaining the microworld physics problems and soon applied it to small molecules. In 1927, he and J. R. Oppenheimer proposed the Born–Oppenheimer approximation, which decouples the electronic motion from the nuclear motion. At the same time, Franck, Born's colleague at the University of Göttingen and lifelong friend, reported experimental findings on light absorption, together with fluorescence selection rules that predicted the existence of quantum effects beyond the Born–Oppenheimer approximation [Franck1926]. Some scholars were aware of these phenomena at the time but lacked the proper language to describe them. Before Franck reported his results in 1926 [Franck1926], spectroscopy was limited to the study of the optical properties associated with electronic states; the effect of nuclear motion on the spectrum was not under consideration. The key quantity in the description of the atomic spectrum, which has been gradually established by early quantum mechanics theory, concerns only the electronic state. Although the nuclear quantum state was reported by P. J. W. Debye, A. Einstein, and others in their study of specific heat capacity of solids [Einstein1911, Debye1912], this vibrational state was not coupled to the electron quantum state. The study of the coupling between nuclear vibrational quantum states and electronic quantum states did not commence until Franck's work in 1926 [Franck1926].

We begin by describing Franck's basic observations in modern language. The quantum numbers describing the electronic degrees of freedom must be considered simultaneously with those describing the nuclear degrees of freedom of the initial and final quantum states of the system to explain the absorption selection rule. Figure 1.8 further clarifies this issue. During light absorption, the initial state of the system generally consists of electronic and nuclear vibrational ground states; the final state corresponds to certain nuclear vibrational state in the electronic excited state. When light absorption occurs, if the dependence of the electron transition dipole matrix element on the nuclear position is neglected (also known as the Franck–Condon approximation), the selection rule requires that the final nuclear vibrational state must have the greatest overlap with the initial nuclear vibrational state in real space. This was the basic light-absorption selection rule reported by Franck.

A similar mechanism occurs in the fluorescence experiment. Before fluorescence, the vibrational excited state has relaxed to the vibrational ground state, while the electronic state remains excited. During the fluorescence process, the initial state of the system is thus described by the nuclear vibrational ground state in the electron excited state, and the final state corresponds to some nuclear vibration state in the electronic ground state. Under the Franck–Condon approximation, the selection rule requires that the final nuclear vibrational state have the greatest overlap with the initial nuclear vibrational state in real space. If the dependence of the electron transition dipole matrix element on the nuclear position

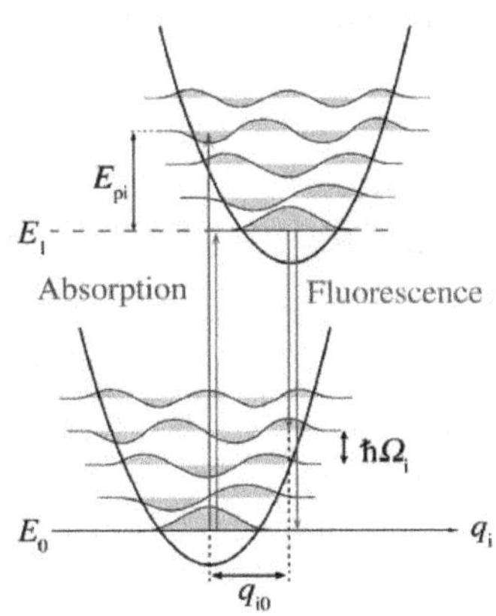

Figure 1.8 Franck–Condon principle. The lower and upper curves represent the electronic ground-state and excited-state potential energy surfaces, respectively. The waves on each potential energy surface correspond to the nuclear vibrational states.

is considered (i.e., if we go beyond the Franck–Condon approximation), this selection rule needs to be corrected accordingly. In the case where this correction is negligible, the selection rule (the Franck–Condon Principle) plays an indispensable role in the current theoretical description of molecular spectra.

To describe absorption and fluorescence, a simple, clear, and specific scientific language has been used. In fact, the Born–Huang expansion supports this description; it was formally introduced in 1954 [Born1954]. Franck and others had already emphasized the importance of going beyond the Born–Oppenheimer approximation and adopting a full quantum treatment when describing the physical properties of real materials. However, the simple language associated with the full quantum effect did not exist until the 1950s. The theoretical explanation of the Franck experiment provided by E. Condon (using the analogy of overlapping pointers in a pendulum model) was essentially phenomenological [Condon1928]. Because Condon's formulation was the first theoretical explanation of Franck's experiment, the selection rule is known as the Franck–Condon Principle. It is worth noting that Condon's phenomenological language is outdated, and it takes considerable time to formulate a complete modern description using the Born–Huang expansion.

After Condon's work, the concepts of the polaron (Figure 1.9), electron–phonon coupling, and polariton theory were proposed to understand the quantum-mechanical coupling of electronic and nuclear motions in the framework of perturbation theory [Landau1933, Frohlich1950, Bardeen1955]. These efforts not only laid a solid theoretical foundation for the emergence of the Born–Huang expansion but also formed the theoretical basis for the most classical theory, that is, the Bardeen–Cooper–Schrieffer (BCS) theory of superconductivity [Bardeen1957]. These works have played a crucial role in the development of modern physics.

Parallel to these advances, the lattice dynamics theory ushered in a golden age of development in the 1930s and 1940s; the fundamental ideas of lattice vibration were refined, significantly contributing to solid-state physics. These advances, together with polaron, electron–phonon coupling, and polariton theory, established a solid theoretical foundation for the Born–Huang expansion, which we will investigate later. The proposal of

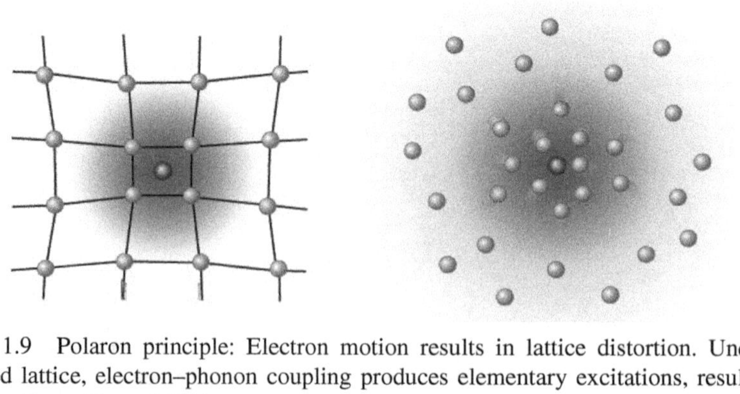

Figure 1.9 Polaron principle: Electron motion results in lattice distortion. Under the distorted lattice, electron–phonon coupling produces elementary excitations, resulting in polarons. From Chevy[2016].

the Born–Huang expansion was made primarily in two papers. The first paper, by Born, was included in the 1951 issue of the Journal of the University of Göttingen [Born1951]. Because it was written in German, which after the Second World War was largely replaced by English as the leading scientific language, this article has often been overlooked. The paper was titled "*Die Gultigkeitsgrenze der Theorie der idealen Kristalle und Ihre Uberwindung*," which translates to "The Validity Limits of the Ideal Crystal Theory and How to Overcome Them." In this work, Born presented a form of wavefunction expansion for a many-body system consisting of multiple electrons and atomic nuclei. Evidently, he addressed the problem using the Born–Oppenheimer approximation (narrowly understood) to study crystals, which tends to treat the nuclei as a set of classical particles with a perfectly periodic lattice configuration. In this paper, he attempted to generate a set of electronic eigenstates that was complete in Hilbert space using the perfect lattice configurations and attempted to expand the total electron-nuclear wavefunction using these eigenstates as a basis (Eq (1.3)). To some extent, Born's method retained the characteristics of the Born–Oppenheimer approximation, since the system wavefunction still consisted of products of the electronic-state wavefunction and the nuclear wavefunction before summing the electronic state,

$$\Psi^j\left(\vec{r},\vec{R},t\right)=\sum_{n=1}^{\mathrm{el}}\chi_n^j\left(\vec{R},t\right)\Phi_n\left(\vec{r},\vec{R}_0\right), \tag{1.3}$$

where $\vec{R}_0$ represents the configuration of the crystal equilibrium position. Because the potential energy surfaces of the electron states are summed up, the impact of the Born–Oppenheimer approximation is weakened when dealing the separation of the variables in the overall wavefunction expansion; the nonadiabatic processes in the nuclear and electron motion can be partially included in calculations. This form of expansion provides new ways to describe many-body wavefunctions of a quantum system containing electrons and nuclei, though the quantum motions of nuclei and electrons in this description are not fully coupled in an *on-the-fly* manner.

Equation (1.3) was further refined in Appendix VIII of the classic textbook *The Dynamical Theory of Crystal Lattices* by M. Born and K. Huang, which was published in 1954. Electronic eigenstates in the refined equation depend directly on the immediate nuclear configuration $\vec{R}$, not on the crystal equilibrium position $\vec{R}_0$:

$$\Psi^j\left(\vec{r},\vec{R},t\right)=\sum_{n=1}^{\mathrm{el}}\chi_n^j\left(\vec{R},t\right)\Phi_n\left(\vec{r},\vec{R}\right). \tag{1.4}$$

The main difference between Eqs (1.3) and (1.4) is that the electronic wavefunction of the latter no longer depends on the crystal equilibrium position $\vec{R}_0$. Rather, it depends on the immediate atomic nuclear configuration $\vec{R}$. This formula couples the nuclear and the electronic quantum motions, so that it can describe nonadiabatic processes such as electron–phonon coupling more accurately. Equation (1.4) is often referred to as the Born–Huang expansion, and it is the basis of the full-quantum simulations (that is, simulations of practical systems where electrons and nuclei are treated as quantized particles) discussed in this book.

The Born–Huang scheme did not immediately become the main theoretical tool for studying the nature of electron–phonon quantum interaction in condensed matter. This was because first-principles calculation of electronic structure and related physical properties was far from mature at the time. Instead of the Born–Huang expansion, an approach based on the model Hamiltonian and the second quantization formulation was commonly used to study electron–phonon coupling in solid systems. In this approach, which was mentioned in many solid-state theory courses, the electronic state of the system is often described by a known Bloch state. From 1930s to 1970s, the Bloch state in a simple metal system often corresponded to certain plane wave (PW) of the free electron gas. Phonons (collective vibrational excitations of atomic nuclei) played a key role in this approach. The overall state of the system was characterized by the nuclear vibrational states and the electronic states. Because of the electron–phonon interaction, the basis was described by second quantization. After the diagonalization of the Hamiltonian containing the electron–phonon interaction, the system eigenstate became a linear superposition of electronic and nuclear eigenstate products. This formulation worked well for most model systems (and indeed, the physics behind the method was in fact equivalent to the Born–Huang expansion). As a result, electron–phonon coupling in solids has been described using the language of second quantization since the 1930s. (Figure 1.10 illustrates Cooper-pair generation in the BCS theory.) L. D. Landau, H. Fröhlich, J. Bardeen, D. Pines, L. N. Cooper, J. R. Schrieffer, and H. Y. Fan all made significant early contributions to the second-quantized approach [Landau1933, Fröhlich1950, Fan H.Y.1951, Bardeen1955, 1957]. The theory was further refined by the outstanding work of W. L. McMillan, R. C. Dynes, P. B. Allen, V. Heine, and M. Cardona [McMillan1968, Dynes1972, Allen1976, 1981].

In a similar way that first-principles electronic-structure calculations developed, theoretical chemists have made relatively systematic use of the Born–Huang expansion since the 1970s, compared with physicists studying solid-state theory. Considerable progress has been made in quantum mechanically explaining simple gas-phase reactions. Recently, researchers in condensed matter physics have also started to explore the

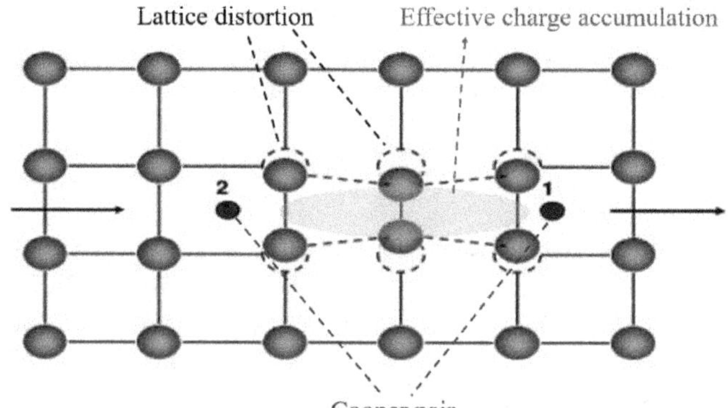

Figure 1.10 Cooper-pair generation according to the Bardeen–Cooper–Schrieffer theory. The lattice distortion produces an effective attraction between electrons, numbered 1 and 2 in the figure, to form a Cooper pair.

first-principles method for electron–phonon coupling, using the early electron–phonon coupling formulations mentioned above. The development of computational methods has allowed the computation of phonon spectra using first-principles electronic structure calculations. It has enabled electron–phonon coupling calculations for real materials, based on electronic Bloch states in Kohn–Sham orbitals. These methods have undoubtedly pushed the boundaries of computational condensed matter physics [Chang K.J.1985, Baroni2001, Guistino2017]. In Section 1.3, we shall summarize these areas of research interest in physics, chemistry, and materials science, and systematically review this long development process from the theoretical and experimental perspectives.

1.3 Development of Computational Methods and Measurement Techniques

In theoretical research, the challenges we face often require us to focus on the development of both theoretical approaches and computational techniques. Because we are using the Born–Huang expansion of the Schrödinger equation as the theoretical starting point for discussing full quantum effects in condensed matter systems, important advances in theoretical research from different periods of time will be presented here with special emphasis on numerical computation. In experimental research, we will focus on the observable details reflecting the full quantum effect and look at how various experimental techniques were developed and pushed to the limits of resolution. Finally, we combine both approaches and show how an external-field modulation platform can uncover the special physical and chemical properties that are dominated by full quantum effects.

1.3.1 Computational Methods

The wavefunction method is one of the earliest computational methods in physical science and has played an irreplaceable part in the development of condensed matter computational techniques (particularly in first-principles electronic structure calculation and dynamics methods). From the historical point of view, the Born–Huang expansion was proposed in 1954, but theoretical research on condensed matter systems at the time was primarily concerned with understanding and developing different theoretical models; little to no further work on the study of the wave function form of condensed matter systems was carried out using the Born–Huang expansion after its emergence. This was because numerical computation was not considered as an independent research tool for condensed matter physics; moreover, the computer capacity at the time was limited. When studying specific problems, people relied on ideal models, perturbation theory, and special analytic functions. The writing style of *The Dynamical Theory of Crystal Lattices*, the textbook by Born and Huang, is itself reflective of this trend. The application of theoretical progress to real-world systems remained limited. Furthermore, the Born–Huang expansion lacked the analytical solutions available with approaches based on model Hamiltonians.

Until the 1950s, computational simulation was not a formal research tool in physics, materials science, and theoretical chemistry. Figure 1.11 shows the principle of calculating

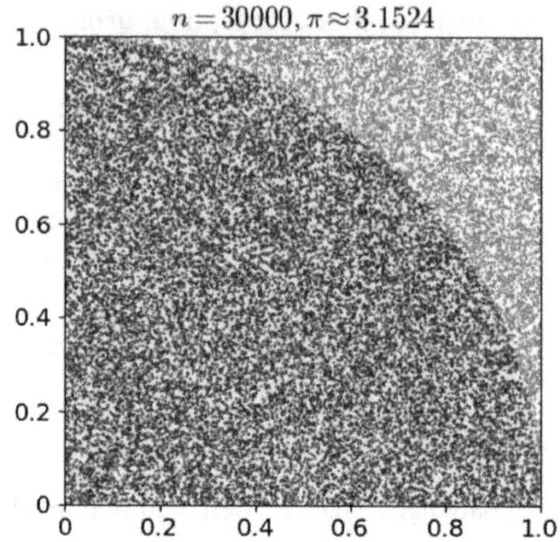

Figure 1.11 Estimating π using Monte Carlo sampling. The deep grey dots are the random number inside the circle and the light grey ones are outside. The value of π can be calculated based on the results.

π based on Monte Carlo sampling, signifying the development of computer technology and algorithms. Numerous ground-breaking works then emerged sequentially. Two of the most notable papers were the work on Monte Carlo sampling by N. Metropolis, A. W. Rosenbluth, M. N. Rosenbluth, A. H. Teller, and E. Teller, and the Los Alamos National Laboratory (LANL) molecular dynamics simulation technical report written by E. Fermi, J. Pasta, and S. Ulam. B. J. Alder, A. Rahman, and other scholars not only conducted pioneering research in the field [Alder1959, Rahman1964] but also actively promoted the development of similar numerical methods in later years.

In the molecular dynamics or Monte Carlo studies conducted in the mid twentieth century, the pair-potential model was conventionally used to describe the nuclear interactions in condensed matter systems. One of its advantages was the low computational cost; this enabled large-scale simulations with longer simulation times and more data, bringing this method closer to the concept of "computational research." Typical examples of successful interaction-potential applications include the Lennard-Jones potential [Lennard-Jones1924] and Buckingham potential [Buckingham1938]. A molecular dynamics simulation configuration based on the Lennard-Jones potential is illustrated in Figure 1.12.

In systems with relatively simple chemical environments, such as molecular or ionic crystals, using interaction potentials in analytic form allows accurate descriptions of nuclear interactions. By analyzing the nuclear configuration through molecular dynamics or Monte Carlo methods, statistical information about the nucleus as a classical particle at finite temperatures can be obtained via numerical simulations in specific materials. However, in systems with relatively complex chemical environments, such as those with bond breakage during chemical reactions, the electronic structure changes so drastically under different nuclear configurations that the nuclear interaction potential can no longer

Figure 1.12 Example of a molecular dynamics simulation based on the Lennard-Jones potential

be expressed analytically. It is therefore necessary to develop a sampling method that calculates the electronic structure in different nuclear configurations immediately after the nucleus has been displaced. This method is typically based on electronic structure calculations in a particular atomic nuclear configuration. Since the electrons are interacting with a many-body quantum system, the solution to this problem is rather complicated; even with today's computational methods, there is still considerable room for refinement. We will briefly review the evolution of electronic structure description chronologically, then look at how the electronic structure calculation methods were combined with molecular dynamics and Monte Carlo sampling in the 1980s.

The significance of electronic structure calculations for a particular nuclear configuration can be well summarized by a sentence from *Electronic Structure: Basic Theory and Practical Methods*, by R. M. Martin at the University of Illinois at Urbana-Champaign (UIUC): "Since the discovery of electron as a particle in 1896–1897, the theory of electrons in matter has ranked among the great challenges in theoretical physics. The fundamental basis for understanding materials and phenomena ultimately rests upon understanding electronic structure" [Martin2004]. Martin alludes to the discovery of the electron by J. J. Thomson, which was well before the formulation of Schrödinger's equation. The Schrödinger equation and the Born–Oppenheimer approximation provide us with a basic equation that many-electron wavefunction should satisfy for a given configuration of atomic nuclei. Despite the use of classical approximations to the nucleus, it is still very difficult to determine how to solve this equation for a many-electron problem. The challenge posed by the Schrödinger equation lies in its many-body nature. The methods of solving this problem can be roughly divided into three categories: the wavefunction method, the density functional theory, and the quantum Monte Carlo technique. All these electronic structure calculation methods have undergone independent development while remaining integrated and mutually reinforcing.

One of the earliest key concepts was the self-consistent field method. The idea was introduced by R. B. Lindsay in 1924 in an attempt to solve the electronic structure of alkali metals using the Bohr model [Lindsay1924]. D. R. Hartree realized that the same approach could be applied using Schrödinger's equation [Hartree1928]. The Hartree method, which treats the many-body electron wavefunction as a product of single-electron wavefunctions, creatively combines Schrödinger's equation, the variational principle, and the self-consistent field method, but its many-electron wavefunction fails to satisfy exchange antisymmetry, making it impractical for real-world applications. Utilizing the determinant introduced by J. C. Slater in 1929 [Slater1929] to describe the wavefunction antisymmetry due to electron exchange [Fock1930], V. A. Fock and Slater generalized the Hartree method. The modified Hartree method retains the characteristics of Schrödinger's equation, the variational principle, and the self-consistent field method, while using the Slater determinant to describe the wavefunction antisymmetry due to electron exchange [Fock1930]. In this way, the inter-electron exchange interactions can be characterized, but higher-order correlation effects are still missing. To include them, the overall wavefunction can be written as a linear combination of Slater determinants, each corresponding to a different electronic-state configuration. Such post-Hartree–Fock methods, widely used in quantum chemistry, can be broadly categorized into configuration-interaction methods, Moller–Plesset perturbation theory, and coupled-cluster methods. One advantage of these methods is their systematically improved accuracy.

However, since the Hartree–Fock method treats each Slater determinant separately, its computational cost increases as N^4, where N is the number of electrons. Taking a linear combination of determinants to solve the higher-order problem is even more computationally expensive (for full configuration-interaction methods, the computational cost even grows exponentially). As a result, the applicability of Hartree–Fock methods to real materials is limited.

The development of density functional theory made a decisive contribution to condensed matter physics, theoretical chemistry, and computational materials science in terms of accuracy and predictability. Density functional theory has its root in the Thomas–Fermi model of the electronic states of atoms, developed in the 1930s. In this variational model, the density rather than the many-body wavefunction was the focus of attention in the total energy calculations. The density for any complex many-electron system can be described as a function with respect to the three variables x, y, and z, whereas the number of variables in the many-body wavefunction is $3N$ (N being the number of electrons). Thus, using the density rather than the wavefunction clearly lowers the numerical computation cost. The total energy of the electrons in the Thomas–Fermi model consists of the electron kinetic energy, the potential energy of electron–nucleus interaction, and an electron interaction term. The kinetic energy term is estimated from the electron density distribution, assuming that electrons at each position are surrounded by a uniform electron gas. This is the predecessor of the local density approximation.

A closer look at the three terms in the total energy reveals that for a system state that is far from being a uniform electron gas, there is a significant difference of the kinetic energy because the concept of orbitals is missing. For this reason, the application of the Thomas–Fermi model in practical systems was severely limited. In 1951, that is, after the establishment of the Thomas–Fermi model but before the formal establishment of density functional theory, Slater proposed a density functional description of the Hartree–Fock method using density as the basic variable. Slater's modification significantly reduced the computational cost of the Hartree–Fock method and was considered a major algorithmic advance at the time. However, the most exciting developments took place between 1964 and 1965.

In 1964, P. Hohenberg and W. Kohn formulated two theorems that laid the theoretical groundwork for the density functional theory [Hohenberg1964]. The first theorem stated that in a many-electron system, there are one-to-one correspondences among the ground-state many-body wavefunction, the external potential, and the ground-state electron-density distribution. The second theorem stated that the total ground-state energy and the ground-state electron-density distribution can be calculated by minimizing a certain function of the density distribution. With these two theorems, the total energy of a many-electron system can be computed using the electron density distribution function as a functional, and the system can be solved by combining the functional with the variational principle. In 1965, W. Kohn and L. J. Sham introduced the idea of orbitals and established a relatively accurate scheme for calculating the electron kinetic energy [Kohn1965]. Based on these works, modern density functional theory was formally established and has evolved rapidly during the next nearly 60 years [Parr1989, Koch2001, Cohen2012]. Figure 1.13 shows the electron-density distribution of a nitroglycerine molecule.

The traditional wavefunction approach and the density functional theory have significantly transformed computational condensed matter physics, theoretical chemistry, and computational materials science in the last 60 years, and the main contributors to make important progress in these fields were awarded the 1998 Nobel Prize in Chemistry. However, the computational cost of the former grows rapidly as the system size

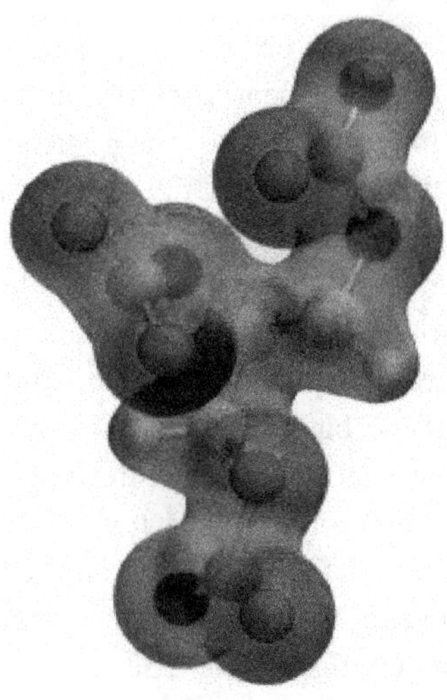

Figure 1.13 Example of an electron-density distribution, the key idea in density functional theory. The image represents a nitroglycerine molecule.

increases (making it unsuitable for condensed matter systems); the latter lacks a complete treatment of many-body electron correlation effects, and systematic improvement of the theory has remained challenging. These limitations make it difficult to perform high-precision electronic structure calculations in condensed matter systems that include many-body electron-correlation effects. The quantum Monte Carlo (QMC) method provides a good alternative for electronic structure calculation. This method is particularly suitable for large-scale parallel computing. The supercomputing advances in the last 20 years have allowed first-principles QMC method to become part of the triumvirate for real-material electron-structure calculation, along with the wavefunction method and the density functional theory.

Due to the limitations of the early first-principles calculations, comparisons between the result and the experimental data must be conducted to precisely evaluate the Born–Oppenheimer potential energy surfaces, and the accuracy of the electronic structure calculation can thus be determined. However, making such comparisons can be difficult when the experimental data may not exist, or when the experimental results may contain inaccuracies due to measurement errors. It is also possible that the experimental quantities can be measured but cannot be rigorously simulated. Therefore, improving the electronic structure calculation accuracy in full quantum computing is pertinent. Fortunately, after years of research and development, we now have versatile calculation methods for theoretically verifying the accuracy of the Born–Oppenheimer potential energy surface, thus providing an alternative to experimental comparison. This also means that the full quantum effect calculations are closer to producing exact solutions within the framework of the Born–Huang expansion, which provides a reliable reference and theoretical basis for future experimental measurements. Our discussion here is restricted to the basic framework of nonrelativistic quantum mechanics, as specified at the beginning of this book. Detailed discussion of how these electronic structure calculation methods can theoretically approach the exact solution will be explained in Chapter 4.

The three methods mentioned above provide a realistic solution to the electronic structure calculation of a specific static nuclear configuration under the Born–Oppenheimer approximation. However, our understanding of the nuclear motion, an idea closely linked to electronic structure calculation, remains at the level of molecular dynamics or Monte Carlo simulations based on the classical description of the nuclei. Integrating first-principles electronic structure calculation with nuclear motion is a challenging task that necessitates a range of theoretical and computational tools.

The simplest approach is to rely on the Born–Oppenheimer approximation and consider the nucleus as a classical particle. A basic molecular dynamics calculation typically begins by computing the self-consistent electronic structure for a specific nuclear configuration. Then, it calculates the forces on the nucleus in this specific configuration according to the Hellmann–Feynman Theorem, generates new configurations by moving the nucleus in a specific time step according to Newton's equation, and computes the self-consistent electronic structure for the new configuration. The force analysis is performed for every updated nuclear configuration, and the statistical sampling is also repeated. Evidently, if the electronic structure calculation method lacks an algorithm to calculate the nuclear forces for a particular configuration according to the Hellmann–Feynman Theorem, we can only resort to Monte Carlo methods instead of molecular dynamics for sampling. Because the Born–Oppenheimer approximation is the fundamental approximation within this theoretical framework, this molecular dynamic approach is often called the "first-principles Born–Oppenheimer molecular dynamics" approach. As mentioned before, the Born–Oppenheimer approximation here is only a restricted version of the original Born–Oppenheimer approximation; therefore, the molecular dynamics method is still limited despite all the improvements.

Even with these limitations, the realization of the molecular dynamic method was not simple. In the past (especially decades ago), it was extremely time-consuming to perform self-consistent electronic structure calculations for a particular nuclear configuration. In 1985, R. Car and M. Parrinello suggested that it was possible to bypass many of these computational conundrums by using an extended Lagrangian [Car1985] such that the nuclear configuration change near the Born–Oppenheimer potential energy surface could be described more efficiently. This first-principles Car–Parrinello molecular dynamics approach flourished for about a decade. In the late 1990s and early 2000s, as self-consistent algorithms for electronic structures were refined and computer capacity was improved, it was gradually replaced by the first-principles Born–Oppenheimer molecular dynamics method, which is now one of the most widely used methods for describing the statistical and dynamical properties of a real system.

Based on such these methods, at the theoretical level, one can basically visualize the substances of the objective world as follows. Each nucleus can be regarded as a ball surrounded by a cloud of electrons, with the electron–nucleus interactions taking the form of quantum glue. The interactions alter the nuclear configuration in real time and simultaneously change the state of the quantum glue. If we use classical mechanics to describe the changes in the nuclear configuration, then the existence, kinetic properties, and chemical and biological reactions of most materials can be understood using this image. Figure 1.14 shows a schematic of the proton tunneling in liquid water based on first-principles molecular dynamics simulation. The first-principles Born–Oppenheimer molecular dynamics method agrees with this picture.

As self-consistent electron-structure calculations are required for each configuration, the simulation time scale is typically restricted to be within a few hundred picoseconds at most. This is much shorter than the time scale required for normal chemical-bond breakage. To examine events with low probabilities, additional sampling enhancement is generally required for computational simulations based on similar molecular dynamics methods. For this reason, molecular simulation with enhanced sampling targeting complex systems with high degrees of freedom has become a frontier research topic in theoretical chemistry since the 1990s [Roux1995, Bartels1998, Laio2002, Gao Y.Q.2008].

Most of the enhanced sampling methods mentioned above focus on the equilibrium statistical properties of complex condensed matter systems. The study of kinetic properties has also attracted the research interest of theoretical chemists since the mid 1980s. Within the framework of the ball-and-stick model, the Nose–Hoover thermostat method with a slowly varying heat bath was proposed by S. Nose, W. G. Hoover, and

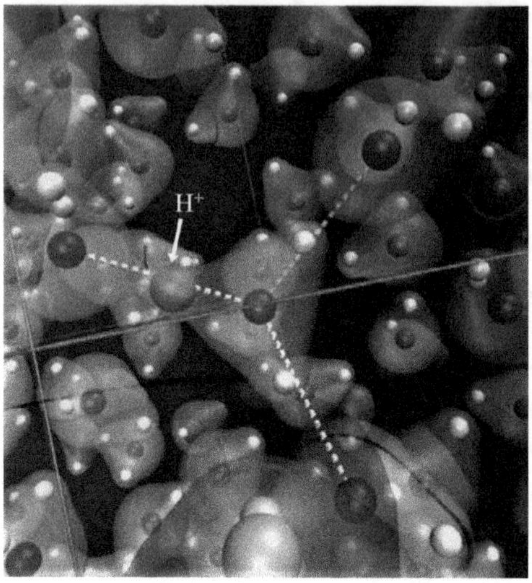

Figure 1.14 Proton tunneling in liquid water based on first-principles Born–Oppenheimer molecular dynamics simulation. The silver balls represent hydrogen nuclei, the deep grey balls indicate oxygen nuclei, and the centered large light grey ball corresponds to the protons (hydrogen ions, H^+) in water. The electron-density distribution in real time (on-the-fly) is described by the contour plane.

others [Nose1984a,b, 1986, Hoover1985, 1986, Martyna1992]. Semiclassical quantum dynamics beyond the ball-and-stick model has increased in popularity since the 1970s owing to the pioneering work of W. H. Miller at UC Berkeley [Miller1974, Tromp1986, Voth1989, Shao J.S.1999, Shi Q.2003, Liu J.2006]. Even today, computational methods similar to molecular dynamics simulation remain an active research area in theoretical chemistry. Along with the development of semiclassical quantum dynamics, full quantum gas-phase reaction dynamics has also made rapid progress in the last 50 years owing to advances in triatomic molecular-scattering theory, first-principles electronic structure calculations, fast Fourier transforms, and other methods. The Born–Huang expansion has gradually become a standard method and has broad application in these studies.

Compared to the gas-phase reaction, for real condensed matter systems, the study of NQEs and NAEs has also become an important topic since the 1980s, with numerical path integrals playing an especially crucial role. As previously mentioned, the computational cost of quantum-dynamics methods grows exponentially with the system size when solving the Schrödinger equation with nuclear motion, making such methods problematic when applied to realistic systems, especially condensed matter systems. Feynman's path integral method, one of the nine standard formalisms of quantum mechanics, provides an alternative [Feynman1949, 1953a,b,c, 1965].

The path-integral formalism was developed by R. P. Feynman in the late 1940s and early 1950s. This profound formalism connects quantum mechanics with methods from classical mechanics. In other words, the solution of quantum-mechanical problems transformed into generalized classical scheme, such that the solution infinitely approaches the real quantum world. However, numerical simulation was not widely accepted as a research tool in physics, theoretical chemistry, and materials science at that time. After the introduction of numerical methods such as molecular dynamics in the years 1950–1970, D. Chandler and P. G. Wolynes realized that it was possible to combine imaginary-time path-integral methods at finite temperature with molecular dynamics methods, thus ushering in a new era of NQE simulation in condensed matter systems [Chandler1981]. Building on this work, B. J. Berne et al. extended the methods to more realistic systems [Berne1986]. D. M. Ceperley and E. L. Pollock combined path integrals with Monte Carlo methods in the mid 1980s and performed atomic-level simulations of superfluid ^{4}He at low temperatures, attracting considerable attention in condensed matter physics research [Pollock1987, Ceperley1995]. In both methods, modeling the temperature with imaginary time ensured an accurate description of statistical properties. Nevertheless, the force field of interatomic interactions significantly restricted the methods' application to processes related to chemical-bond breakage. To address this issue, M. E. Tuckerman, D. Marx, M. Parrinello, and M. L. Klein combined the first-principles Car–Parrinello molecular dynamics method with path integral molecular dynamics (PIMD) [Tuckerman1996, Marx1996, Tuckerman2002].

In the twenty-first century, researchers attempted to combine first-principles Born–Oppenheimer molecular dynamics/Monte Carlo sampling with path integral to simulate Born–Oppenheimer potential energy surfaces [Zhang Q.F.2008, Li X.Z.2010, 2011, Ceriotti2012]. Results of a first-principles PIMD simulation for oxalic acid crystals are shown in Figure 1.15. The methods of electronic structure calculation have also evolved from density functional theory to a generalized theory that includes a variety of techniques, such as the QMC method [Morales2010]. Recently, these methods have also

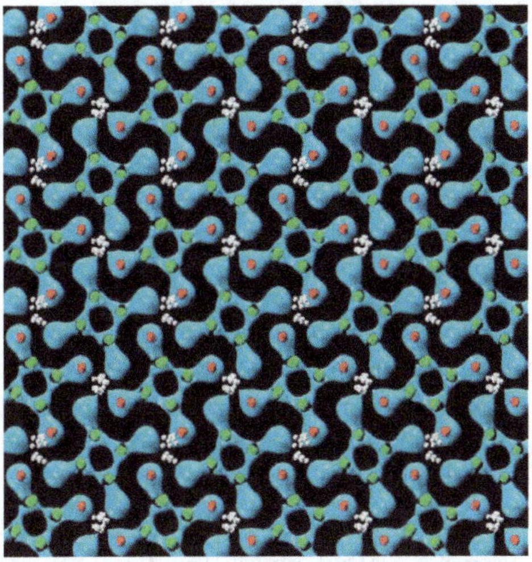

Figure 1.15 Simulation of oxalic acid crystals using the first-principles path-integral molecular dynamics method. The white spheres represent the hydrogen atom paths at finite temperature, the red spheres are the paths of oxygen, the green spheres indicate carbon-atom paths, and the blue cloud corresponds to the electron density distribution.

been used in conjunction with enhanced sampling, for example, thermodynamic integration [Pérez2011, Feng Y.X.2015], which further increases the applicability of path integral numerical methods.

Although the study of electron–phonon coupling based on model Hamiltonians began in the 1930s, similar work based on first-principles electronic structure calculations only dates to the 1980s. They have been continuously improved and refined in recent years, laying a solid theoretical foundation for full-quantum theoretical approaches and numerical simulations of condensed matter. Chapters 4 and 5 of this book are devoted to the derivation and development of such computational methods, and to a large extent cover the frontier research work performed in this field.

1.3.2 Measurement Techniques

From the experimental perspective, nuclear quantum and nonadiabatic effects are naturally included in the observations of physical and chemical properties of real materials. While there are many theoretical methods for dealing with full quantum effects, there is no specific experimental approach that can detect them systematically and effectively. Nevertheless, there are many conventional methods to detect the full quantum effect, such as infrared spectroscopy, Raman spectroscopy, Brillouin spectroscopy, sum-frequency generation (SFG) spectroscopy, ultrafast spectroscopy, angle-resolved photoemission spectroscopy (ARPES), nuclear magnetic resonance (NMR), single crystal X-ray diffraction (XRD), electron energy-loss spectroscopy (EELS), neutron spectroscopy, electron diffraction, and other imaging techniques. These technologies are coupled with various external fields to form a complex interweaving of experiments that reflect the progress of states-of-the-art science and technology. On the other hand, these advances in study of full quantum effects place additional demand on the development of future technology.

One simple and common method to identify the full quantum effects is replacing normal atoms with their isotopes. By measuring the changes in the physical properties of the material before and after isotopic substitution, the nuclear quantum effect can be uncovered and compared with theoretical calculations. However, conventional diffraction and spectroscopy techniques have low spatial resolution, making it difficult to collect full-quantum-effect information at the atomic scale. One of the important experimental breakthroughs in this field was the hydrogen-sensitive qPlus-based scanning probe microscopy (SPM) developed by Y. Jiang's group at Peking University [Guo J.2022]. Another relevant technology is atomically resolved EELS using a scanning transmission electron microscope (STEM), developed by P. Gao's group at Peking University [Li N.2023]. These techniques push the problem of studying full quantum effects to the atomic limit. In recent years, E. G. Wang, Y. Jiang, X. Z. Li, L. M. Xu, J. Chen, and Q. F. Zhang from Peking University have made extensive observations of full quantum effects, including phase transitions in liquid and solid water [Meng X.Z.2015, Guo J.2016a, Peng J.B.2018a,b, Tian Y.2022], and proton tunneling on surface and in bulk materials [Zhang Q.F.2008, Guo J.Q.2022]. This team and P. Gao's and L. Liu's groups have also studied energy-band isotopic engineering in hexagonal boron-nitride-isotope junctions by EELS in TEM (Transmission Electron Microscope) [Li N.2023].

M. Cardona et al. at the Max Planck Institute for Solid State Physics became interested in solid-state electron–phonon coupling as early as the 1970s [Allen1981]. Starting from the late 1990s, NMR was used to study proton tunneling. By measuring the variation in the spin-lattice relaxation time as a function of magnetic field and temperature, the coordinated proton-tunneling process was quantified [Brougham1999]. Subsequently, single-crystal XRD was adopted to investigate phase transitions in ice at high pressures and explore the delocalization of hydrogen nuclei [Loubeyre1999]. X-ray and neutron scattering methods have been used to obtain the momentum distribution and corresponding radial-distribution functions of OO/OH/HH pairs during particle scattering in water; the results were directly compared with the first-principles PIMD prediction (i.e., the full quantum simulation). These methods have found broad applications in researching nuclear quantum effects in bulk-phase water and other complex systems (see Figure 1.16). In 2017, A. Nilsson's group at Stockholm University discovered that nuclear quantum effects can influence the thermodynamic response and the correlation length of supercooled water. For water and heavy water, the correlation lengths show maxima at different temperatures, implying a significant isotope effect [Kim K.H.2017].

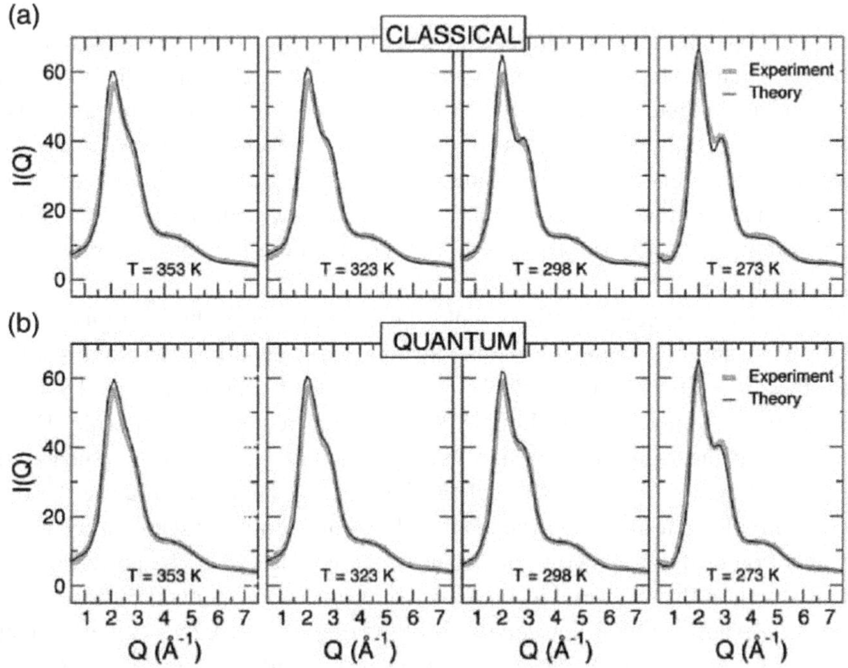

Figure 1.16 Measured and calculated X-ray scattering spectra for liquid water at four different temperatures. The black curve indicates the experimental data; the grey curve shows the theoretical calculation. Panel (a) shows the classical nuclear simulation result, and panel (b) gives the results of the simulation that takes into account the nuclear quantum effect after making a quantum treatment of the nucleus. Reprinted with permission from Paesani[2009]. Copyright 2024 American Chemical Society.

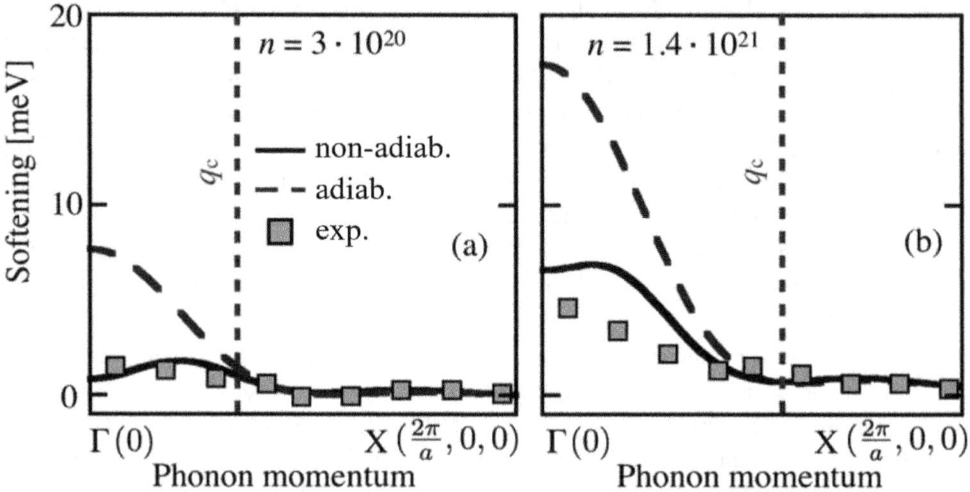

Figure 1.17 Renormalized phonon vibration frequencies in a heavily boron-doped diamond, for doping concentrations of (a) 3×10^{20} cm^{-3} and (b) 1.4×10^{21} cm^{-3}. The experimental results (squares), adiabatic-theory prediction (dashed curve) and nonadiabatic-theory prediction (solid curve) are labeled. Reprinted figure with permission from Caruso[2017]. Copyright 2024 by the American Physical Society.

These methods can also be applied to studying nonadiabatic processes. For instance, Caruso et al. obtained high-precision phonon spectra of heavily doped diamond and studied the associated nonadiabatic effects. Figure 1.17 shows the renormalized results of the longitudinal optical phonon vibration frequencies in the heavily doped diamond. When the experimental observations were compared to the full-quantum-effect simulation, the presence of an NAE in the heavily boron-doped diamond was clearly indicated. For a doping concentration of 1.4×10^{21}cm^{-3}, the NAE changed the phonon frequency by nearly 15 meV [Caruso2017].

Nuclear quantum effects are very complex because they occur both within and between the molecules, and the competing relationships between different vibrational modes increase the difficulty of quantitatively detecting and distinguishing NQEs in experiments. The recently developed technique of deep inelastic neutron scattering (DINS) can measure the nuclear quantum motion of hydrogen and heavier elements (such as D or O atoms). By comparing the average nuclear kinetic energy measured along the different directions of molecular axes with the results of full quantum calculations, the anisotropy and competition of the momentum distribution can be analyzed. The corresponding theoretical research in this area provides a basis for direct experimental verification of competing NQEs [Andreani2005, Pietropaolo2008] (Figure 1.18). DINS can also be applied to the study of water in critical states, including supercooled water, pre-melting of ice, and confined water. In short, it provides new ways to assess the singular physical properties and NQEs of water under critical conditions. For example, C. Andreani et al. (University of Rome, Italy) found that hydrogen atoms in supercooled water exhibit excessive average kinetic energy, indicating quantum delocalization [Pietropaolo2008] (Figure 1.19).

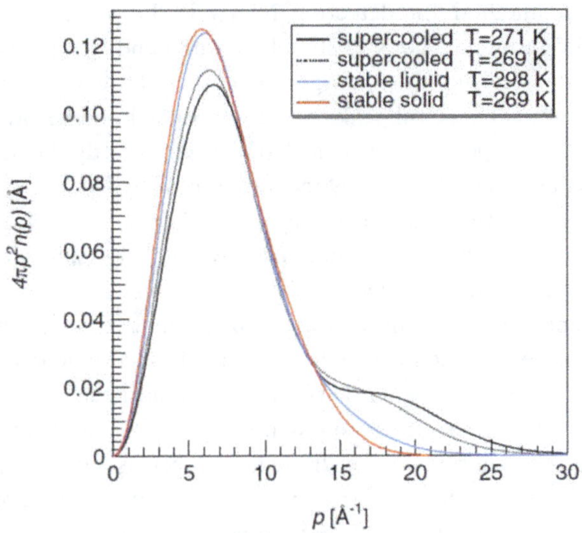

Figure 1.18 Proton-momentum distribution for water under different conditions. Results were obtained using deep inelastic neutron scattering spectroscopy; the plot reflects the hydrogen nuclear motion. The curves correspond to solid ice at 269 K (red line), liquid water at 298 K (blue line), supercooled water at 269 K (dotted line), respectively. Reprinted figure with permission from Pietropaolo[2008]. Copyright 2024 by the American Physical Society.

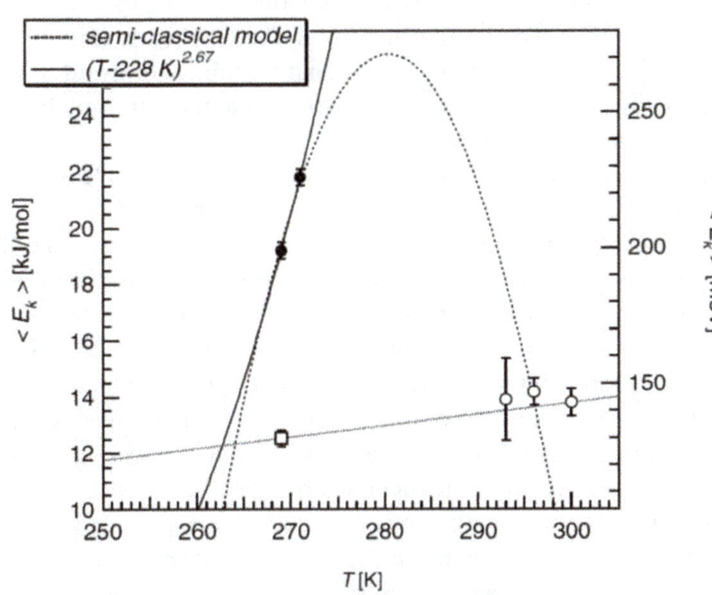

Figure 1.19 Hydrogen-nuclear kinetic energy (measured with deep inelastic neutron scattering spectroscopy) as a function of temperature under different conditions: ice (hollow squares), liquid water (hollow dots), and supercooled water (solid dots). The hydrogen nuclei in supercooled water exhibit excessive average kinetic energy, indicating quantum delocalization. Reprinted figure with permission from Pietropaolo[2008]. Copyright 2024 by the American Physical Society.

Another accurate method for detecting NQEs is the vibrational spectroscopy of hydrogen bonds (in particular the X-H stretching and bending modes). In experiments, the nuclear quantum effects on the hydrogen-bond interaction can easily be reflected by comparing the vibrational spectral peaks of the hydrogen bond before and after isotopic substitution. Vibrational spectroscopy can also be used to study the quantum fluctuation of hydrogen atoms and how this affects the strength of hydrogen bonds [Hirsch1986]. The sum-frequency vibrational spectroscopy (SFVS) invented by Y. R. Shen et al, for example, is a type of nonlinear laser spectroscopy with high surface resolution that is capable of obtaining vibrational spectra of the liquid water surface (interface). SFVS utilizes a fundamental principle in nonlinear optics, that is, when two beams of light hit a material, both the refracted and reflected wave have frequencies equal to the sum of the two incident light frequencies. This quadratic nonlinear optical process can only be observed in systems where inversion symmetry is broken, such as at a surface or interface of a material. Therefore, sum-frequency spectroscopy is a particularly effective experimental tool for obtaining molecular structural information about such surfaces and interfaces. It has been applied to a variety of *in situ* environments, such as the interfaces between water and other gases, liquids, or solids. Moreover, it imposes little restriction on experimental parameters such as temperature and pressure. SFVS has opened up many new prospects in surface science and inspired original studies of different interfacial structures in water. For example, in 2012, M. Bonn et al. at the Max Planck Institute for Polymer Research reported that NQEs can affect the bond orientation of water at the water–vapor interface. The OH bonds tend to point upward into the vapor phase, whereas the OD bonds tend to point downward toward the bulk water, which is a direct manifestation of the nuclear quantum effect on hydrogen bonds [Nagata2012] (shown in Figure 1.20).

Diffraction and spectroscopy are mainly used to examine the nuclear quantum effects on light-element systems in the ground state. By contrast, ultrafast laser pump–probe techniques can detect not only the excited energy states and NQEs in ultrafast dynamics processes, but also their associated NAEs. This method adjusts the time difference between the pump and probe light pulses reaching the sample. When the probe light pulse has a different delay time from the pump light pulse, the light intensity change after the light passes through the sample is recorded. The most common ultrafast pump–probe technique is the ultrafast time-resolved transient absorption spectroscopy, which is often used to monitor the absorbance change of water molecules in the excited state and to record the time evolution of the particle number distribution at each excited energy level. The data can be further fitted to obtain the potential energy function $V(r, R)$ of the O–H···O interaction between water molecules, where r denotes the O–H bond length and R denotes the O–O spacing. The anharmonic characteristic of the potential energy enables detection of vibrational excited states and consequently implies quantum delocalization of the O–H···O hydrogen bond. This method has enhanced the understanding of various quantum-mechanical problems in proton transfer and water decomposition at room temperature [Bakker2002]. Experimental studies of water decomposition using ultrafast spectroscopy have also significantly contributed to the theory of nonadiabatic effects. S. Meng

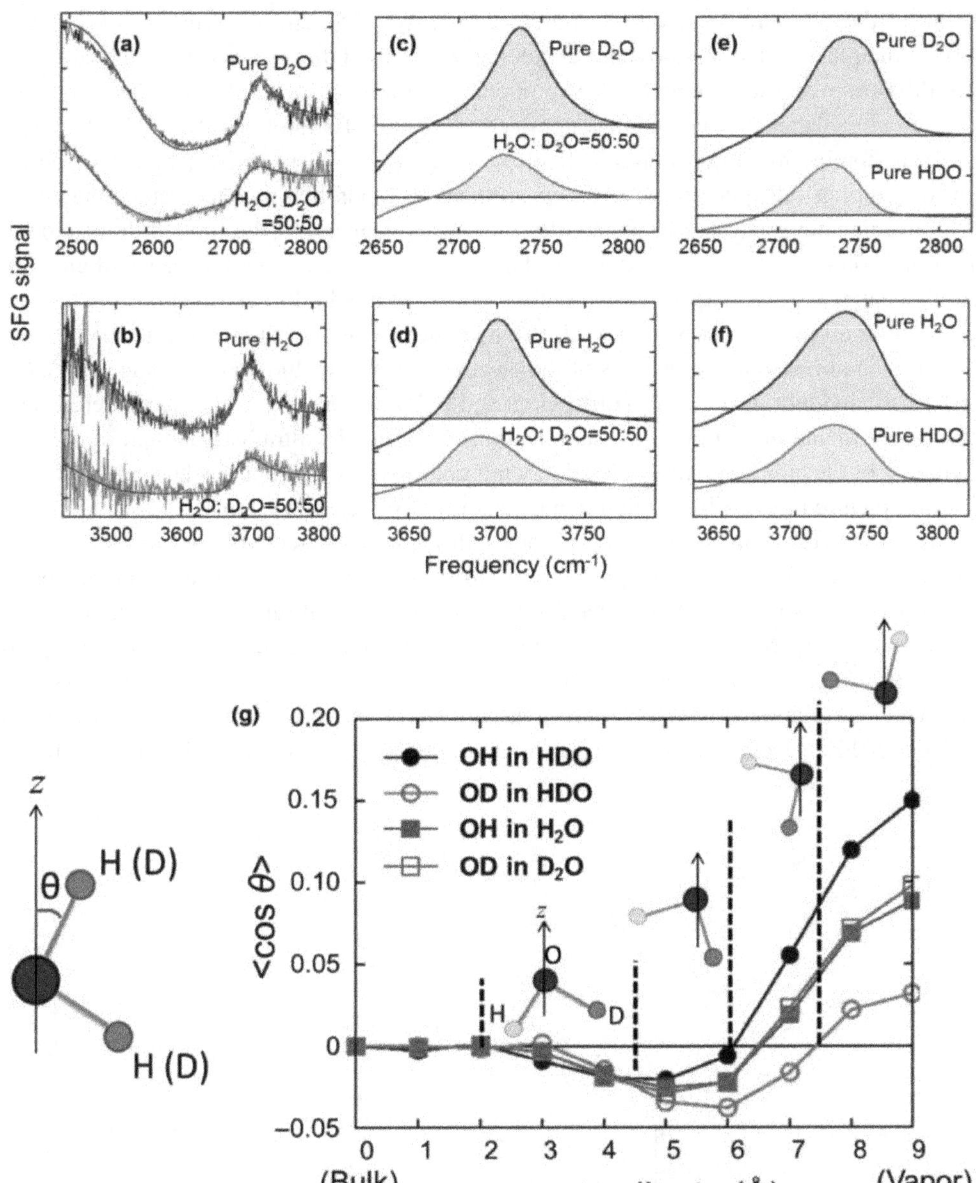

Figure 1.20 (a) and (b) Experimentally measured sum-frequency vibrational spectra of a liquid water surface. (c) and (d) Fitted sum-frequency vibrational spectral peaks of liquid water surface. (e) and (f) Sum-frequency vibrational peaks predicted by a molecular dynamics simulation. (g) Surface-orientation distribution of OH/OD bonds in water obtained by the molecular dynamics simulation. Because of the nuclear quantum effect, OH bonds on the surface have a stronger tendency to point along the direction out of the surface compared to OD bonds. In contrast, the OH and OD bonds on the bottom point to other water molecules. Reprinted figure with permission from Nagata[2012]. Copyright 2024 by the American Physical Society.

et al. at the Institute of Physics, Chinese Academy of Sciences, recently developed a first-principles excited-state-dynamics calculation method that effectively deals with non-equilibrium processes [Meng S.2008a]. In collaboration with E. G. Wang et al., they also studied nonadiabatic dynamics in water decomposition of photolysis [You P.W.2021a].

Thus, combining ultrafast laser technology with traditional spectroscopy shows promise for developing full-quantum-effect measurement methods that can accurately characterize nonadiabatic processes in nonequilibrium states. Generally, the time scale of atomic motion is on the order of picoseconds or femtoseconds, whereas the time scale of electron motion can reach the order of attoseconds. On the one hand, ultrafast spectroscopy and its derivatives allow the observation of electronic structures within the sample and are able to capture the nonadiabatic dynamics of the electron system (including electronic spins). On the other hand, ultrafast electron diffraction and XRD can observe the lattice-ordered state and lattice quantum motion of the sample, thus revealing the ultrafast quantum-dynamical behavior of the atomic system. Effectively integrating these two approaches will provide a holistic picture of the full quantum dynamics of the condensed system.

In 2012, Hellmann et al. incorporated ultrafast laser technology into time- and angle-resolved photoemission spectroscopy to study nonadiabatic dynamic processes in several typical states of charge density waves driven by ultrafast lasers (Figure 1.21) [Hellmann2012]. Madeo et al. used ultrafast laser technology in conjunction with photoelectron microscopy to directly observe the nonadiabatic process of phonon-assisted momentum-forbidden dark excitons in atomically thin semiconductors (Figure 1.22) [Madeo2020]. Recently, Zinchenko et al. integrated attosecond laser technology with soft

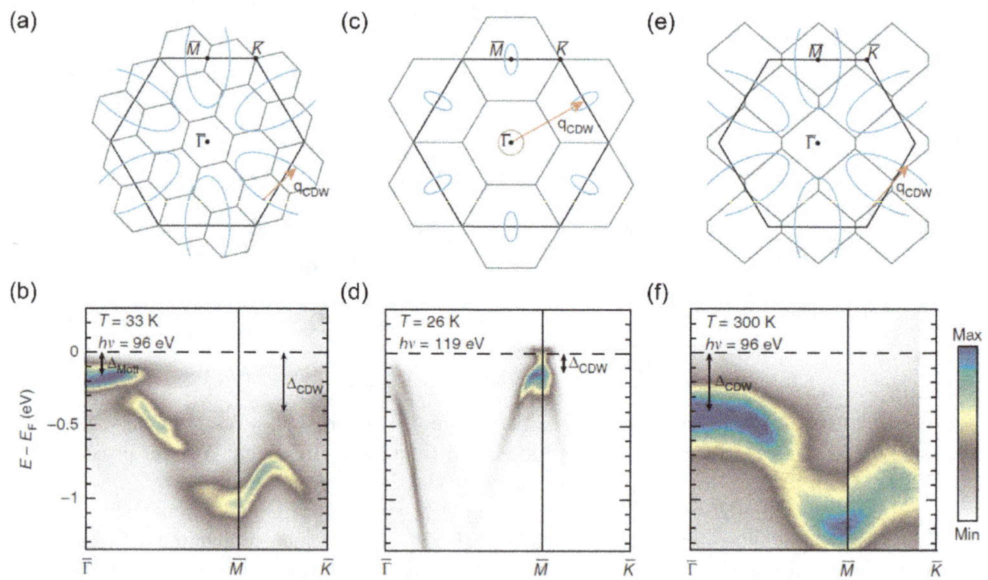

Figure 1.21 Observation of the three most arguably intriguing charge-density-wave phases driven by ultrafast laser in layered transition-metal dichalcogenides. From Hellmann[2012]. Reproduced with permission from SNCSC.

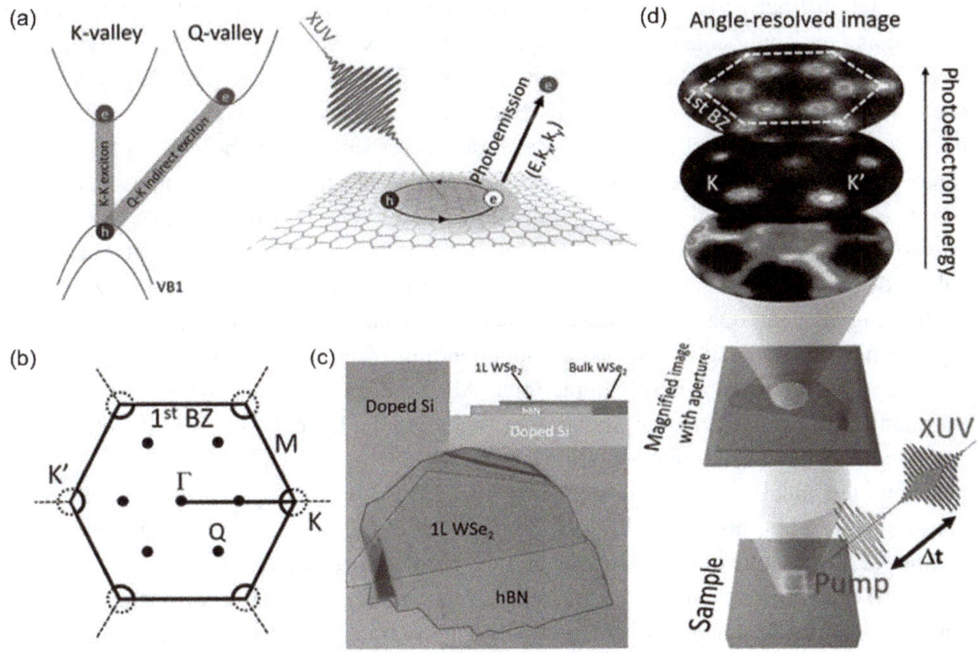

Figure 1.22 Dark exciton measurements and experimental setup in a photoexcited WSe_2 monolayer. From Madeo[2020]. Reprinted with permission from AAAS.

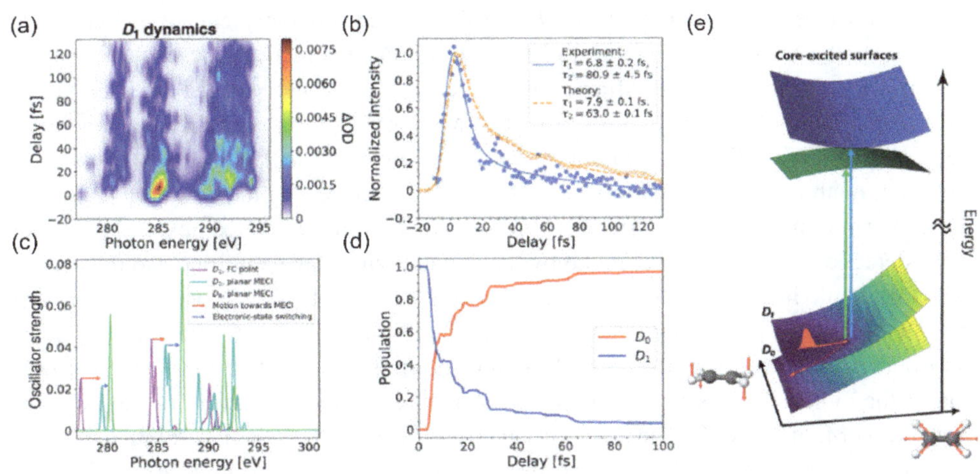

Figure 1.23 Attosecond soft X-ray transient absorption measurement of nonadiabatic processes in electronically excited $C_2H_4^+$ molecules. MECI: minimum-energy conical intersection. (a) and (c–d) are theoretical calculation. (b) contains both theoretical and experimental results. From Zinchenko[2021]. Reprinted with permission from AAAS.

X-ray transient absorption to study the nonadiabatic evolution of the electronic states in excited $C_2H_4^+$ molecule (Figure 1.23(E), D_0 and D_1 state); their findings clearly demonstrate an electronic-state switching on ultrafast time scales [Zinchenko2021].

Although diffraction and spectroscopy are widely used methods in full-quantum-effect research and are applicable to complex systems, their spatial resolution is severely limited (usually on the order of hundreds of nanometers to tens of micrometers). Moreover, the information obtained by these methods often suffers from the average-treatment effect resulting from atomic and molecular ensemble. For example, in the study of NQEs in water, the interaction between the nuclear quantum state and the environment can lead to severe line broadening because the quantum state of the hydrogen nucleus is susceptible to environmental influence. The limitations of the existing experimental methods make it difficult to characterize NQEs accurately and quantitatively. Therefore, techniques for probing nuclear quantum states at the single-bond/single-atom precision level are urgently needed to grasp the physical understanding of the nuclear quantum effects and how it affects electron–nucleus interactions. All these problems fall within the scope of full-quantum condensed-matter physics that is beyond the Born–Oppenheimer approximation (especially its narrow interpretation).

Recent technical advances in scanning tunneling microscopy (STM), atomic force microscopy (AFM), and other high-resolution probe techniques show promise in overcoming the limitations of spectroscopy and diffraction methods, allowing for *in situ* real-space full-quantum-state probing at the single-bond/single-atom level. After years of continuous effort, Y. Jiang et al. at Peking University obtained significant results that have sparked widespread interest in this field [Guo J.2014, 2016a, 2017, Meng X.Z.2015, Peng J.B.2018a,b, Ma R.Z.2020, Tian Y.2022].

STM imaging is based on the concept of electron tunneling and offers very high spatial resolution. Since its development in 1982, STM has considerably enhanced our understanding of the microscopic world and started an era of direct atomic observation and manipulation. Its inventors, G. Binnig and H. Rohrer, were awarded the Nobel Prize for Physics in 1986. Following the invention of STM, a series of scanning probe technologies was developed with different functions, and AFM, invented in 1986, is one of the most important methods. While STM converts the tunneling current to the detection signal and can only work on conductive surfaces, AFM uses force or force-related physical quantities as the detection signal; it is not limited by the conductivity of the substrate and thus can be used for atomic imaging to characterize insulating materials. In addition, STM measures the electronic state information near the Fermi energy level in the sample and can image the frontier orbital of the molecules. However, its tip is not sensitive to the nuclear position of the molecule, making it difficult to image the chemical structure of the molecule directly. AFM, on the other hand, is expected to achieve higher spatial resolution in short-range imaging because the Pauli repulsion is more sensitive to atomic positions. In 1998, F. J. Giessibl at University of Regensburg invented the qPlus sensor using a quartz tuning fork. The qPlus sensor replaces the conventional silicon cantilever in the AFM probe and substantially improves the detection of short-range forces between the tip and the atoms. The tip of the qPlus sensor can also be made using conductive metals to build a system that allows both AFM and STM operation.

Since the 1990s, STM has become the most popular method for studying water on solid surfaces. Groups led by K. Morgenstern at the University of Bochum and M. Salmeron at UC Berkeley have conducted notable research on hydrogen bonds in water clusters (Figure 1.24, Figure 1.25) adsorbed on different metal surfaces [Michaelides2007, Maier2015]. K. Thürmer's team at Sandia National Laboratories used STM to probe the wetting behavior of Pt (111) and other metal surfaces (Figure 1.26) [Nie S.2010]. Although these studies have provided a relatively clear physical picture of water adsorption and infiltration on solid surfaces, it is still not possible to resolve the internal structure of the water molecules and the hydrogen bond orientation.

Studying the surface diffusion of a single hydrogen atom or its isotopic atom can uncover the quantum tunneling mechanism induced by zero-point energy at low temperatures. In 2000, W. Ho et al. at UC Irvine achieved breakthrough results in the atomic-scale

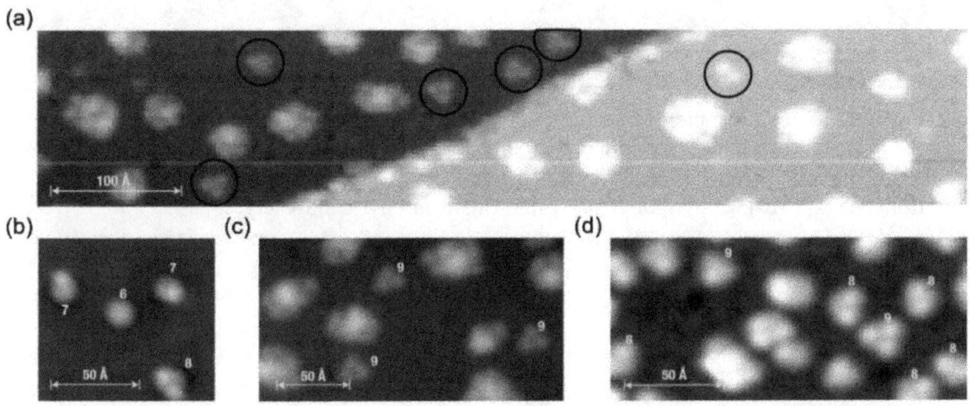

Figure 1.24 Scanning tunneling micrographs of water clusters on Cu (111) and Ag (111) surfaces: (a), (c), and (d) H_2O clusters on the Cu (111) surface; (b) D_2O clusters on the Ag (111). The number of water molecules included in each cluster is labeled in the figures. From Michaelides[2007]. Reproduced with permission from SNCSC.

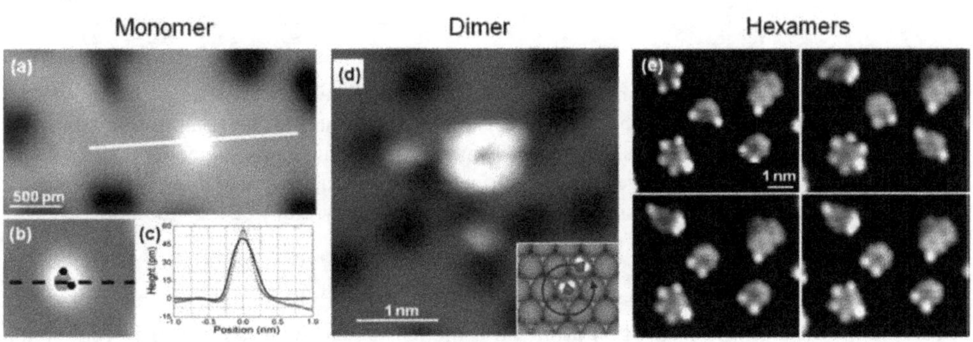

Figure 1.25 Scanning tunneling micrographs (STMs) of surface water structure on Ru (0001): (a) 4 K image of isolated water-monomer adsorption; (b) simulated STM; (c) theoretical (black solid curve) and experimental (grey curve) water-molecule-adsorption height profiles; (d) water-dimer adsorption, with a schematic on the bottom-right; (e) images of four water hexamers and a larger water polymer (top right) on Ru (0001) at 50 K, acquired at 1 min interval. Reprinted with permission from Maier[2015]. Copyright 2024 American Chemical Society.

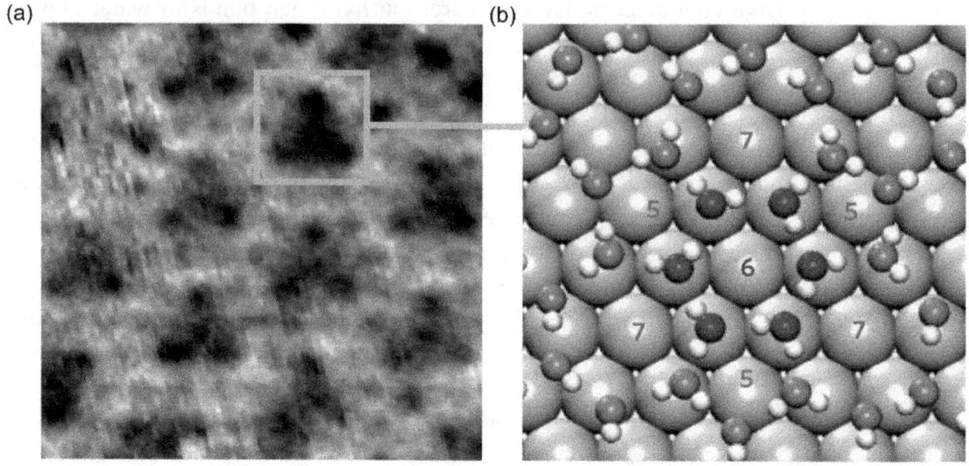

Figure 1.26 (a) Scanning tunneling micrograph of a √39×√39 water layer deposited on Pt (111) and (b) the corresponding atomic model. The numbers in (b) indicate the number of neighboring water molecules around each location. Reprinted figure with permission from Nie S.[2010]. Copyright 2024 by the American Physical Society.

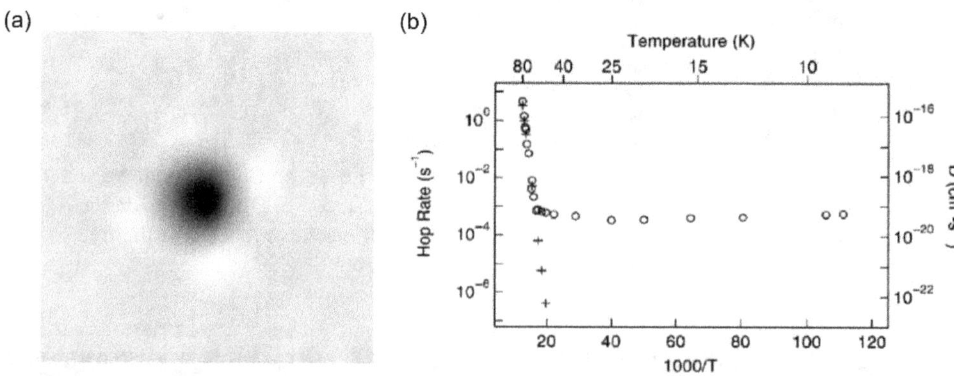

Figure 1.27 (a) Scanning tunneling micrograph of a hydrogen atom adsorbed on Cu(001) at 9 K. (b) Hydrogen diffusion rate as a function of temperature. The H atoms are represented by dots and the D atoms by crosses. The hydrogen diffusion is temperature independent at low temperatures, suggesting that surface hydrogen quantum tunneling is involved. Reprinted figure with permission from Lauhon[2000]. Copyright 2024 by the American Physical Society.

study of hydrogen-adsorption systems. For the first time, STM imaging and vibrational spectroscopy of a single hydrogen atom on a metal surface were achieved in real space; hydrogen and deuterium atoms were detected individually. The individual hydrogen diffusion rate at different temperatures was subsequently measured with STM; it was independent of temperature below 60 K, suggesting that quantum tunneling of hydrogen atoms occurs on metal surfaces in low temperature region [Lauhon2000] (as shown in Figure 1.27). Additionally, STM can also be used to study proton transfer by quantum tunneling in hydrogen bond systems. H. Okuyama's group at Kyoto University studied NQEs in water surfaces/interfaces, finding that the molecules in surface–interface water dimers shift their roles as hydrogen-bond donor and acceptor, and that this rearrangement is related to quantum tunneling. Furthermore, they found that the dynamics of the OH group and its adsorption orientation on the Cu (110) surface are also results of quantum tunneling by the hydrogen nuclei [Kumagai2008, 2009].

The quantum tunneling effect described in the preceding research works occurs only with a single hydrogen nucleus. However, the hydrogen nuclei in a hydrogen-bond network are not independent of each other, but rather form a strongly correlated system. Therefore, proton transfer in hydrogen-bonded systems can involve many-body quantum behavior. Kumagai and his coworkers discovered that the H_2O-$(OH)_n$ $(n = 2-4)$ hydrogen chain on a Cu(110) surface [Kumagai2011] and the porphycene molecules on an Ag(110) surface undergo stepwise many-particle tunneling [Koch2017] (Figure 1.28). In addition, they successfully performed real-space observation of carbon atom tunneling [Lin C.2019].

Based on electron tunneling, the STM study on nuclear quantum effects is limited. The tunneling currents are sensitive to the outer electrons of atoms, which easily shield the tunneling electrons and reduce their interactions with the nucleus. One solution for this is to increase the STM sensitivity to its limit, so that it detects the nuclear quantum state indirectly by probing the weak coupling between the nuclear quantum state and the outer electron state. Another approach is to extend STM to inelastic electron tunneling spectroscopy (IETS), which probes the coupling between the outer electrons and the molecular vibrations and identifies the vibrational modes of individual molecules, thus enabling NQE observation at the single-bond/single-atom scale.

Since 2012, Y. Jiang's group at Peking University has made remarkable progress in both directions. Through precisely tunable tip–water and substrate–water coupling, they

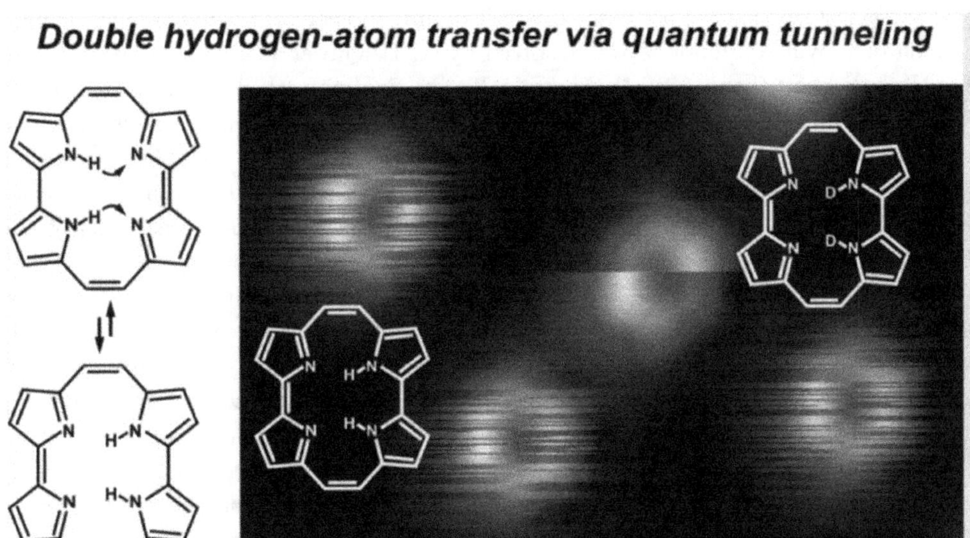

Figure 1.28 Double hydrogen-atom tunneling within a porphycene molecule on an Ag(110) surface, observed by scanning tunneling microscopy. Analysis reveals that the hydrogen tunneling occurs in a stepwise fashion. Figure created by Koch[2017], DOI: 10.1021/jacs.7b06905 licensed under CC BY.

significantly enhanced the distribution density of the frontier molecular orbitals of water near the Fermi level and obtained sub-molecular-orbital images with extremely high signal-to-noise ratios. Their results allow for precise localization of hydrogen nuclei in water monomers and tetramers on NaCl(001) film, as well as hydrogen-bond directionality (chirality) identification in real space. The many-body quantum states and coordinated quantum-tunneling dynamics of hydrogen nuclei in hydrogen-bond networks are also reported in their subsequent works [Guo J.2014, Peng J.B.2018a,b]. Using tip-enhanced IETS, the first high-resolution vibrational spectrum of a single water molecule was obtained, and the hydrogen-bond strength was thus determined. Isotopic substitution experiments combined with quantum simulations reveal that quantum fluctuation of the hydrogen bond can be even similar to its thermal effect at room temperature. This implies that the nuclear quantum effect of hydrogen is more than a simple modification of classical interactions, and that it is significant enough to alter the structure and properties of water. Further analysis shows that the anharmonic quantum fluctuations of hydrogen nuclei weaken the weaker hydrogen bonds and strengthen the stronger bonds. This result may be applicable to many other hydrogen-bonded systems, and it clarifies some of the long-standing debates about the quantum nature of hydrogen bonds [Guo J.2016a].

The insensitivity of STM to nuclei and core electrons can be compensated by qPlus-AFM. The qPlus-AFM probe can operate stably with sub-Ångstrom resolution (<100 pm). The tip can be placed very close to the sample surface, substantially increasing detection sensitivity to short-range forces between the tip and the atoms and enabling direct atomic-level imaging of chemical bonds. Using qPlus noncontact atomic-force microscopy (NC-AFM), L. Gross et al. at IBM-Zurich Research Center and M. Crommie's research group at UC Berkeley imaged the backbone structure and bond orders of organic molecules, and successfully differentiated the products in surface-chemistry reactions [Gross2012, Oteyza2013] (Figure 1.29). The intermolecular hydrogen-bond configuration of 8-hydroxyquinoline was resolved by X. H. Qiu et al. at the National Center for Nanoscience [Zhang J.2013] (Figure 1.30). Y. Sugimoto et al. at the University of Tokyo applied qPlus-AFM to the study of surface water and obtained hydrogen-bond configurations by imaging one- and two-dimensional water clusters on a Cu(110) surface [Shiotari2017] (Figure 1.31). Recently, Y. Jiang's research team produced another groundbreaking work, developing noninvasive AFM imaging based on higher-order electrostatic force. They developed the qPlus sensor to improve the force detection sensitivity to the order of fN. Based on this technique, they obtained an electric quadrupole moment tip by modulating the charge distribution of the tip with atomic precision, detecting the extremely weak high-order electrostatic force signal between the tip and the water molecule, thus enabling the imaging and localization of hydrogen nuclei in the water molecule [Peng J.B.2018a,b]. AFM has also led to the study of hydrogen-bond symmetry in 2D ice resulting from quantum delocalization [Tian Y.2022]. Thus, AFM provides a unique and powerful experimental tool for studying NQEs in weakly interacting light-element systems.

Because scanning probe microscopy is a surface-sensitive technique, using it to obtain information at buried interfaces can be difficult. Transmission electron microscopy (TEM), however, provides atomic-level spatial resolution that can overcome this difficulty. In principle, TEM can probe NQEs and NAEs at surfaces and interfaces. The

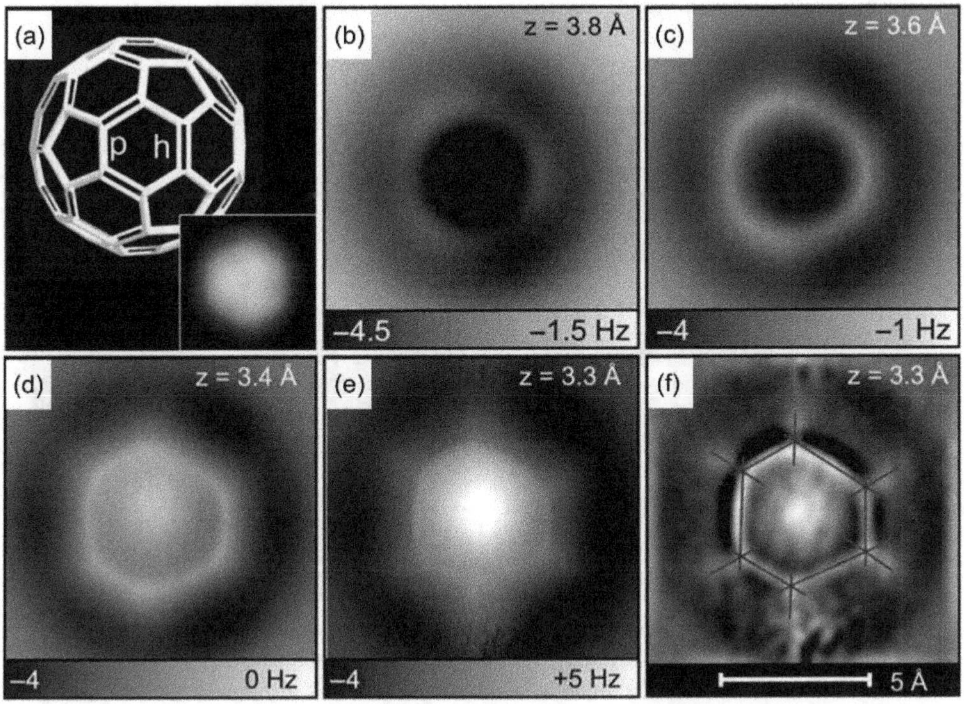

Figure 1.29 Chemical structure and bond orders of C_{60} molecules adsorbed on Cu(111) surface imaged with qPlus noncontact atomic force microscopy (NC-AFM). (a) C_{60} atomic model in the middle; STM image on bottom-right inserted. (b)–(d) AFM images of C_{60} molecules at different tip heights. (f) Laplace-filtered and flattened image of (e), used to differentiate single and double bonds. From Gross[2012]. Reprinted with permission from AAAS.

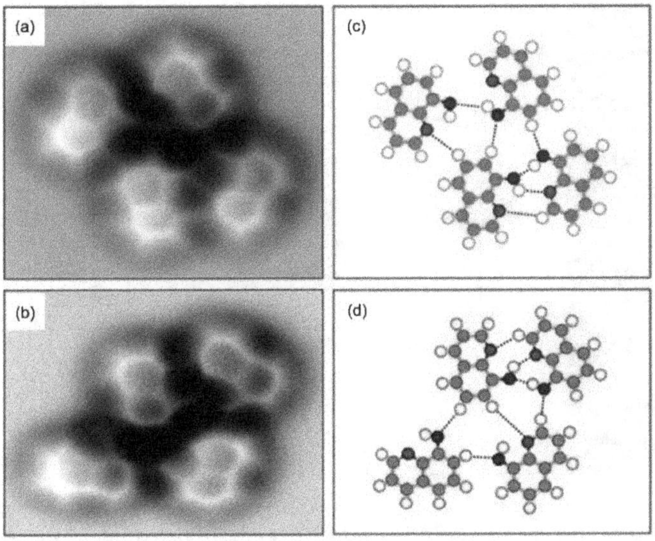

Figure 1.30 Intermolecular-bond formation observed using qPlus noncontact atomic force microscopy (NC-AFM). (a) and (b) AFM images of two configurations of 8-hydroxyquinoline tetramer adsorbed on a Cu(111) surface. (c) and (d) Corresponding chemical-structure models. From Zhang J.[2013]. Reprinted with permission from AAAS.

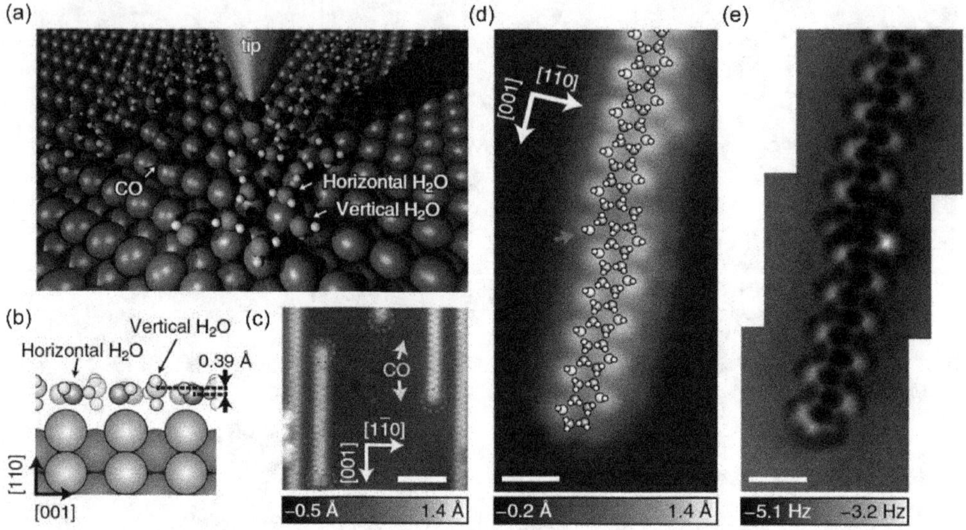

Figure 1.31 1D water-chain structure on a Cu(110) surface measured by qPlus noncontact atomic force microscopy (NC-AFM). (a) Schematic representation of AFM measurement with a CO-terminated tip. (b) Simulated atomic configuration of water chain on Cu(110) surface. (c) STM image that includes a larger area of the water chains. (d) and (e) Respective STM and AFM images of a special water chain with a kink and a terminal. The atomic structure of the water chain in (d) is superimposed on the STM image. Figure created by Shiotari[2017] licensed under CC BY 4.0.

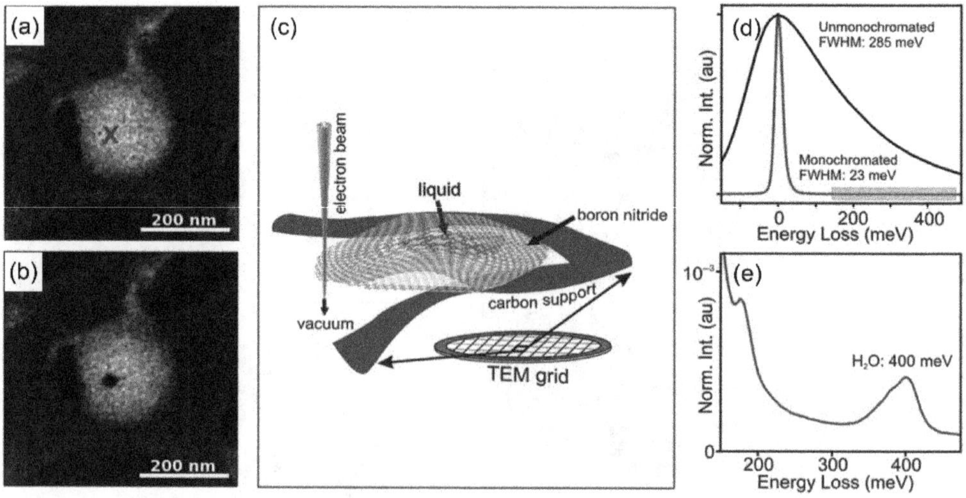

Figure 1.32 Local vibrational spectra measured by TEM-based electron-energy-loss spectroscopy (EELS). (a) and (b) TEM images of BN-encapsulated liquid water layer before and after electron-beam irradiation. (c) Illustration of test sample and EELS measurement process. (d) EELS spectrum measured by the unmonochromated (upper curve) and monochromated (lower curve) electron beams. (e) Measured vibrational spectrum of the water layer. Figure created by Shiotari[2017] licensed under CC BY 4.0.

recently developed monochromator can correct spherical aberration in TEM, improving both ultrahigh energy and spatial resolution. When combined with TEM, EELS becomes another powerful tool that allows simultaneous probing of the isotope spatial distributions and their vibrational spectra. For example, J. R. Jokisaari et al. adopted this approach and measured the vibrational spectrum of liquid water with high resolution; by encapsulating water between boron nitride sheets, the differences in vibrational spectra of H_2O and D_2O could be identified [Jokisaari2018] (Figure 1.32).

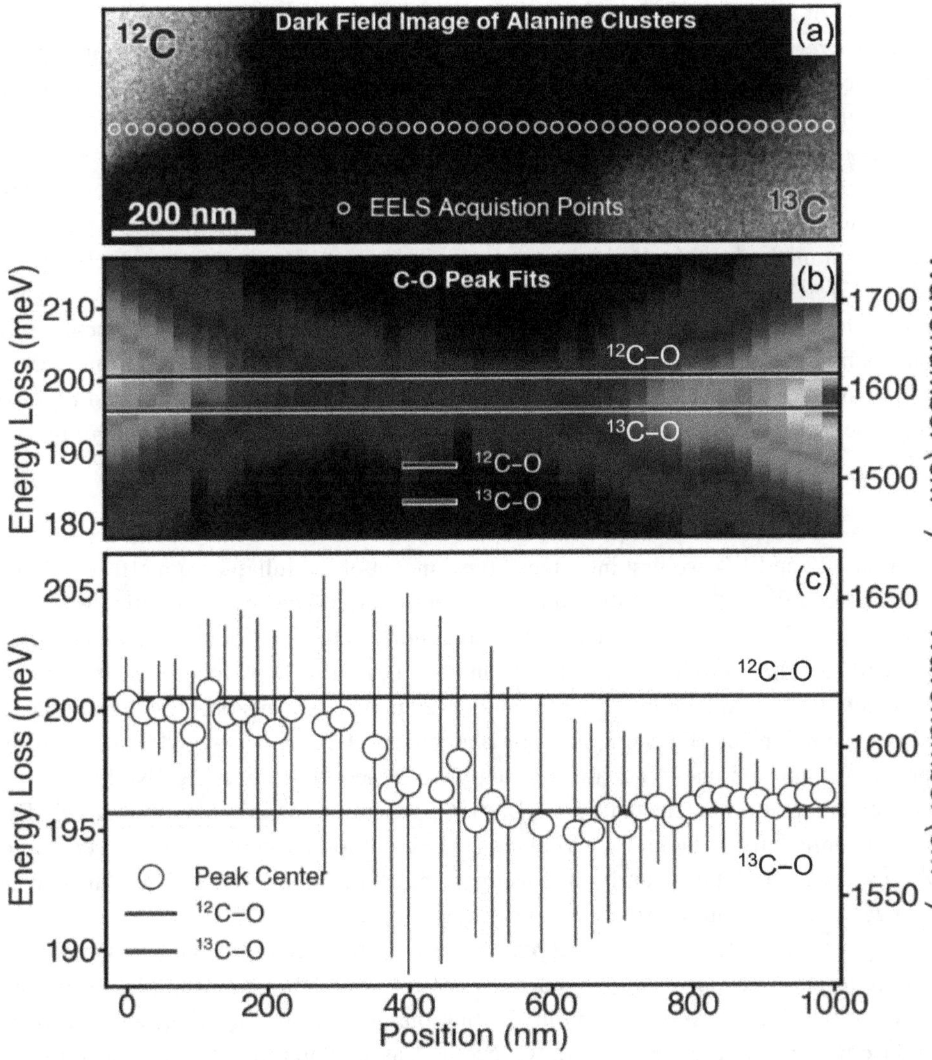

Figure 1.33 Spatially resolved local vibrational spectrum of polyalanine. (a) TEM image of polyalanine; the electron-energy-loss spectrum (EELS) acquisition points are marked by circles. (b) Corresponding local vibrational spectral measurement. (c) Standard deviation of the peak plotted against spatial position of the vibrational spectrum. ^{13}C and ^{12}C, isotopes of carbon, have different C–O vibrational peaks. From Hachtel[2019]. Reprinted with permission from AAAS.

J. A. Hachtel et al. resolved bond formation of ^{13}C and ^{12}C to O in isotope-labeled alanine with nanoscale precision. Compared to ^{12}C-labeled alanine, a 4.8 meV redshift in the vibrational energy of the C–O asymmetric stretching mode was observed for ^{13}C-labeled alanine [Hachtel2019] (Figure 1.33). Very recently, P. Gao et al. at Peking University have studied the atomic-scale isotope effects at the h-^{10}BN/h-^{11}BN interface using a four-dimensional EELS based on TEM [Li N.2022]. These TEM-based spectroscopic methods, with their ultrahigh energy resolution and spatial resolution, particularly their ability of atomic-scale resolving, open up new routes for reliable experimental observation of the full quantum effect at interfaces.

Taking everything into account, it is reasonable to argue that the development of computational methods in condensed matter physics since the 1950s (in particular the PIMD method based on first-principles electronic structure calculations developed in the 1980s, also known as the first-principles path integral molecular dynamics method or simply the PIMD method), along with the recent advances in various experimental measurement techniques for extreme condition measurements in external fields (in particular the detection methods with ultrahigh spatial resolution and ultrafast temporal resolution), have served as the foundation for studies of full quantum effects in condensed matter physics. These methods, in addition to improving our understanding of the physical mechanisms behind a myriad of complex processes, provide the means to move from qualitative to quantitative characterizations of new phenomena in physical science and to move beyond the Born–Oppenheimer approximation (especially its narrow understanding).

1.4 Challenges in Studying Full Quantum Effects

In Sections 1.2 and 1.3, we first introduced the concept of the full quantum effect and then discussed the current status of full-quantum-effect research. The key information is that the full quantum effect is naturally included in quantum mechanics (for clarity, we are referring specifically to Schrödinger's Eqs (1.1) and (1.2)). As this book focuses on the physics and chemistry of full quantum effects in condensed matter systems obeying Schrödinger's equation, we will place heavy emphasis on phenomena that can be studied with the Born–Oppenheimer approximation but are constantly weakened in practice (e.g., NQEs), as well as phenomena that are neglected by the Born–Oppenheimer approximation (e.g., NAEs). To avoid confusion, we begin by reviewing some common concepts in condensed matter physics and explore their natural relationship with full quantum effects. Then, the current challenges of full-quantum-effect research will be discussed.

A large variety of theoretical concepts have been introduced over the years to explain the discoveries of experimental condensed matter research. For example, zero-point energy, zero-point vibration, quantum fluctuations, quantum tunneling, and electron–phonon coupling have all been widely accepted as explanations for a wide range of physical phenomena: hydrogen symmetrization and metallization; proton tunneling; spin liquids; superliquids; supersolids; quantum critical behavior; quantum phase transitions; quantum paraelectricity; ultralow thermal conductivity; negative thermal expansion; superconductivity; phonon-assisted optical absorption; and isotope engineering of energy bands. These concepts are, in fact, all related to the nuclear component of the full quantum effect, that is, they can only be understood when the nucleus is treated as a quantized particle, which is fundamentally different from treating it as a classical one. However, only a few researchers have unified these physical concepts from a fully quantum-mechanical perspective.

In the original formulation of the Born–Oppenheimer approximation, the nuclear quantum effect was described incompletely but not entirely neglected. However, this approximation has been repeatedly weakened in applications over the last century (for instance, in the early adiabatic "ball-and-stick" model of electronic states). Therefore, it has become common practice to use harmonic or quasi-harmonic approximations from perturbation theory to describe what is referred to here as full quantum effects. For instance, phonons are constructed based on the vibrations of the Born–Oppenheimer potential energy surface. When a nonadiabatic process such as the phonon-assisted electron transition is considered, the system is typically modeled using different Born–Oppenheimer potential energy surfaces to construct the corresponding phonons, and the electron–phonon coupling is described accordingly. In fact, with such an approach, researchers are subconsciously moving beyond the scope of the Born–Oppenheimer approximation and into the realm of full quantization as described by the Born–Huang expansion. However, these calculations assume that the harmonic or quasi-harmonic approximations hold.

In other words, our understanding of the problem has not changed since 1951, when Born proposed the expansion treatment of static structure in solids in terms of $\vec{R}_0$, with the electronic and nuclear states not interacting on-the-fly. The problems of full-quantum condensed-matter physics discussed in this book include not only cases in which perturbation theory holds for the solid-state system but also ones in which it does not. For example, we consider the case in which the solid-state system is in a nonequilibrium state, and the atomic nuclear configuration is far from the most stable structures of electron ground-state. We also address liquid condensed matter systems and other complex situations.

Including full quantum effects is also essential for examining the nonadiabatic effects in electron–nucleus coupled dynamics, which are entirely neglected in the original Born–Oppenheimer approximation.[1] Nonadiabatic effects have become increasingly important in the study of nonequilibrium phenomena such as spectral broadening, intensity changes, and peak-position shifts caused by changes in excited-state occupancy during an electron transition. These effects also govern phonon-vibration-frequency variation with electron or hole doping concentration in graphene, as well as ultrafast charge transfer, optically induced topological phase transitions, optically induced charge-density waves (charge order), the molecular Berry phase, surface catalysis, and photolysis, among other physical processes.

In other words, full-quantum condensed-matter physics aims to describe the physical and chemical phenomena that are beyond the Born–Oppenheimer approximation.

[1] It is possible to treat electron–phonon coupling with the harmonic approximation, based on the Born–Oppenheimer potential energy surfaces of different electronic states, as mentioned earlier. However, doing so requires going beyond the scope of the Born–Oppenheimer approximation and into the Born–Huang regime. We must also note that in systems where nonadiabatic processes cannot be neglected, the nuclear configuration will significantly differ from its most stable form of the electronic ground state, and the harmonic approximation will no longer be valid.

This includes problems that can be considered by the Born–Oppenheimer approximation but have been ignored in the past and problems where the approximation no longer holds. Some of these problems may have been discussed from different perspectives in previous research, but there is no general consensus on the fundamental physics behind them. Therefore, adopting a fully quantum-mechanical point of view toward condensed matter physics is necessary. Solving the many-body Schrödinger equation for a real condensed matter system with the Born–Huang expansion remains a difficult task even today. However, this difficulty does not mean that further advances are impossible nor does it justify adopting incomplete physical pictures of phenomena.

While the ball-and-stick model provides a simple and clear description of crystal structures, our theoretical starting point in understanding the real physical and chemical properties of molecules or condensed matter systems (many-body systems consisting of electrons and nuclei) should always be the Born–Huang expansion described in Eq. (1.4). The reason for this is that the Born–Huang expansion naturally includes all quantum information about electrons, nuclei, and electron–nucleus interactions. However unfortunately, except for a few gas-phase reaction problems in theoretical chemistry, we are currently unable to perform rigorous full quantum simulations of real systems, which are systems containing multiple nuclei and electrons, based on the Born–Huang expansion alone.

One relatively feasible computational solution to this problem is to describe the electronic structure using a first-principles approach, after which path-integral methods, for example, can be used to either determine the statistical properties of the nucleus quantum mechanically or to describe its quantum dynamical property semiclassically. To evaluate the nonadiabatic effects implied by Eq. (1.4), it is necessary to make multiple other approximations. In practice, if we ignore the nuclear wavefunction in Eq. (1.4) and treat the nuclear motion as classical orbital motion, then NQEs are neglected. Similarly, if we fail to consider the correlation between the electronic and nuclear states and ignore the summation to the electronic potential energy surfaces, then NAEs are excluded from our calculation. The first-principles Born–Oppenheimer molecular dynamics method – the most commonly used calculation method in condensed matter physics – fails to consider both.

Therefore, studies on full quantum effects in condensed matter systems can be naturally divided into two parts: one focusing on NQEs and the other on NAEs (Figure 1.34). It is worth noting that the integration of the two types of effects is also an active current research topic [Wang E. G.2016, 2022]. This book will start by discussing NQEs and NAEs, and introducing the current theoretical research status, the challenges we face, and the available experimental methods. The detailed discussion of these topics can be found in Chapters 4–6.

As shown in Figure 1.34, the basis for full-quantum-effect studies is electronic structure calculation. After nearly a century of development, the simulation of electronic states in condensed matter systems has been proven quite successful, but the study of NQEs and NAEs due to electron–nucleus quantum coupling has only been of interest in the last three to four decades.

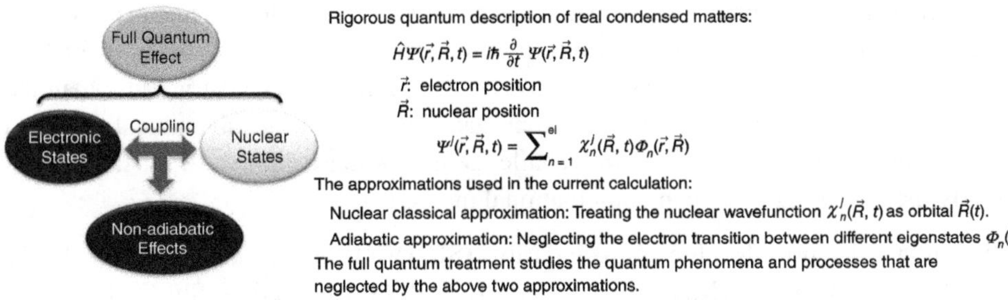

Figure 1.34 Full-quantum-effect research in condensed matter physics. The full quantum effects include nuclear and electronic states, as well as nonadiabatic effects, while the Born–Oppenheimer approximation in narrow interpretation focuses only on electronic states.

From a theoretical point of view, NQEs refer to the quantum-mechanical properties of molecules or condensed matter systems after expressing the nuclei and the electrons as wavefunctions. NQE research can be roughly divided into two parts: statistics and dynamics. The imaginary-time path integral based on the density matrix allows the rigorous description of NQEs at finite temperature and characterization of the physical and chemical properties of the system from a statistical perspective. This line of research began in the early 1980s with the work of D. Chandler, P. G. Wolynes, B. Berne, D. M. Ceperley, and E. L. Pollock. Later research focused on combining numerical methods with first-principles electronic structure calculations, determining the numerical sampling validity, or improving the ergodicity of the results [Tuckerman1996, Marx1996, Zhang Q.F.2008, Ceriotti2009a,b]. Recently, combining statistical methods with other molecular simulation tools (e.g., thermodynamic integration and enhanced sampling) [Perez2011, Feng Y.X.2015] has attracted considerable interest, as well as nuclear-configuration sampling based on a more accurate electronic structure theory [Morales2010, Chen Y.L.2023]. We have briefly discussed these topics above, and a detailed discussion will be provided in Chapter 5.

A more serious challenge for condensed matter NQE research lies in the simulation of its dynamical properties. Dynamics simulation usually begins with the quantum correlation function, defined as

$$C_{\mathrm{AB}}(\mathrm{t}) = \frac{1}{Z}\mathrm{Tr}\left[\hat{A}^{\beta} e^{i\hat{H}t/\hbar}\hat{B}e^{-i\hat{H}t/\hbar}\right], \tag{1.5}$$

where $\hat{H}$ denotes the Hamiltonian of the system, $\hat{A}$ and $\hat{B}$ are operators corresponding to physical observables, and $Z = \mathrm{Tr}\left[e^{-\beta\hat{H}}\right]$ is the partition function, such that $\hat{A}^{\beta} = e^{-\beta\hat{H}}\hat{A}$. Different dynamic properties can be derived from different quantum correlation functions: For instance, chemical reaction rates are related to flow correlation functions, and infrared spectra are related to dipole moment correlation functions.

In Eq. (1.5), the NQE can originate from either the Boltzmann distribution operator $e^{-\beta\hat{H}}$ or the time-evolution operator $e^{-i\hat{H}t/\hbar}$. The Boltzmann operator contains quantum statistical information that can be treated by the imaginary-time method, similar to how temperature was treated using an imaginary-time operator in our earlier discussions. The time-evolution operator contains quantum dynamical information. Because of its sign, using it for large molecules and condensed matter systems involves a computational cost not less than that of the traditional wave-packet-dynamics approach. For this reason, dynamical evolution operators have posed the main research challenge in this field since the 1970s. There are currently the following five main computational methods in use:

(1) Semiclassical initial-value representation (SC-IVR), developed by W. H. Miller et al. at UC Berkeley [Miller1970, Wang H.B.1998, Liu J.2009, Shi Q.2003];
(2) Forward–backward semiclassical dynamics (FBSD), developed by N. Makri and J. S. Shao [Shao J.S.1999];
(3) Centroid molecular dynamics (CMD), proposed by J. S. Cao at the University of Utah (currently at MIT) and G. A. Voth at the University of Chicago [Cao J.S.1993];
(4) Ring-polymer molecular dynamics (RPMD), proposed by D. E. Manolopoulos et al. at the University of Oxford [Craig I.R.2004, 2005a,b];
(5) Path-integral Liouville dynamics, developed by J. Liu at Peking University [Liu J.2014].

Despite this progress, these methods are still incapable of accounting for quantum coherence effects. As a result, developing a dynamical method for describing quantum coherence in condensed matter systems is a significant unsolved problem.

In addition to NQEs, the theory of full-quantum condensed-matter physics also considers NAEs based on the Born–Huang expansion. These are processes resulting from the coordinated evolution of electrons and nuclei in a real molecular or condensed matter system. With the recent advances in time-dependent density functional theory (TDDFT) [Runge1984, van Leeuwen1996, Meng S.2008a,b] and nonadiabatic dynamics [Tully1990], atomic-scale nonadiabatic dynamical processes in condensed matter systems are gradually becoming understandable at the electronic structure level. TDDFT has been successfully applied to linear and nonlinear responses of many-electron systems in external fields, as well as dynamics simulations of collective excitation formation and decay. However, many challenges remain, including the quantitative description of the electron excitations and dynamics in real systems, qualitative understanding of electron–phonon interaction, and the explanation of new phenomena induced by electron–electron interactions. The experimental photolysis of water at a TiO_2 electrode provides a good example [Fujishima1972]. Further, no theoretical method of accurately calculating the TiO_2 electrode surface structure and the molecular structure on the water interface has gained universal acceptance [Bikondoa2006, Bandura2008, Liu L.M.2010]. Even less is known about the dynamics of processes such as light absorption, excitation, or charge transfer, and there is little understanding of the mechanisms of decomposition and reaction dynamics in excited water molecules. There also exist many novel physical phenomena induced by electron excitation under an external field, such as the transition from solid to liquid, ferroelectric to antiferroelectric, or ferromagnetic to antiferromagnetic, all of which harbor rich interactions and physical processes. These processes are related to electron excitation and nonadiabatic effects; developing corresponding theoretical simulation methods is undoubtedly the key to understanding them.

To study electron excitation, we will focus on its initial step: optical absorption. Conventional theoretical methods utilize classical electromagnetic simulations and the macroscopic dielectric parameters of materials. However, these methods fail to consider microscopic quantum properties such as electron–hole excitation, collective electron-motion dynamics, and hot electron transport and transfer. On the other hand, the traditional wavefunction and local basis-vector methods are inefficient in capturing the periodicity of surface systems. Therefore, the most sensible option, both in terms of rigor and computational cost, is to continue using (time-dependent) density functional theory to investigate the physiochemical processes triggered by electron excitation in real macromolecular and condensed matter systems.

Recent experimental studies have benefited from the development of various high-resolution measurement schemes and a platform for extreme external field conditions. When dealing with different condensed matter systems, it is of paramount significance to consider the effect of nuclear motion on the physical properties of quantum mechanics. Specifically, this involves comparing the measurements with the influence of nuclear quantum motion on electronic structures and other elementary processes predicted by the first-principles approach. Ultrafast spectroscopy with high temporal resolution and STM as well as AFM probe images with ultrahigh spatial resolution have played key roles in the development of experimental techniques for full-quantum-effect measurements [Guo J.2017, 2022].

Traditional experimental tools for the study of NQEs in hydrogen-bonded systems are spectroscopic methods include infrared spectroscopy, Raman spectroscopy, SFVS, NMR, XRD, neutron scattering, etc. Combining these methods with isotope substitution allows us to identify the nuclear quantum effects on hydrogen bond strength; this result can be compared with a full quantum calculation for confirmation. Nonetheless, one major limitation of these research tools is their spatial resolution, which ranges from hundreds of nanometers to tens of micrometers. Moreover, the information obtained often suffers from the average treatment effect because the signals of many hydrogen bonds are superimposed, making it impossible to learn about the intrinsic properties of hydrogen bonds and individual hydrogen-bond configurations at the atomic level. Furthermore, the nuclear quantum states become more susceptible to the local environment when confined at the interface. This can cause severe spectral line broadening, preventing accurate and quantitative characterization of nuclear quantum effects. High-resolution atomic-level detection of nuclear quantum states is therefore required to uncover the theoretical details of NQEs, their interaction with hydrogen bonds, and associated isotope effects. All of these require the development of more precise experimental techniques under extreme conditions. In addition to the difficulty of detecting atoms at the material surface, measuring within the material or at the interface can present additional technical challenges to the experiments. Y. Jiang et al. [Guo J.2014, 2016a,b, Meng X.Z.2015, Peng J.B.2018a,b, Tian Y.2022] and P. Gao et al. [Li N.2022] at Peking University have made some progress in this direction [Wang E.G.2022]. However, until now, the study of NQEs at the atomic level has been limited to hydrogen and a few light-element atoms, such as boron and carbon. It would be interesting to explore how to extend this study to other elements.

One of the key research directions in experimental studies of condensed matter full quantum effects involves effectively incorporating external-field platforms with measurement techniques such as ultrafast time-resolved spectroscopy, ultrahigh spatial-resolution scanning probes, and ultrahigh energy- and spatial-resolution EELS in STEM. Advanced large-scale experimental facilities (such as attosecond light sources, neutron-scattering sources, or synchrotron radiation sources) and comprehensive external-field platforms (such as extremely low temperatures, ultrahigh pressures, strong magnetic fields, or ultrafast strong light fields) provide new possibilities for research in this area. These methods open a new avenue for in-depth experimental studies in full-quantum condensed-matter physics.

The key computational challenges of full-quantum-effect research in condensed matter physics are as follows: (1) development of simulation methods for nuclear quantum effects, in particular first-principles theoretical methods that focus on nuclear quantum dynamics; (2) development of simulation methods for nonadiabatic effects, especially the establishment of theoretical foundations and algorithms for nonequilibrium processes caused by electron excitation under external fields; (3) development of simulation methods combining nuclear quantum effects with nonadiabatic effects, including quantum-semiclassical methods.

The NQE simulation methods need to be improved both in terms of statistics and dynamics. At the statistics level, not only must path-integral numerical methods be combined with more accurate electronic structure calculations, but molecular simulation techniques (such as thermodynamic integration and enhanced sampling) must be incorporated. In terms of theoretical dynamics, it is necessary to develop a methodology for describing the coherence of condensed matter that brings together the benefits of both the wavefunction and the path-integral approaches. For the nonadiabatic dynamics method in real macromolecular or condensed matter systems, the study of microscopic physiochemical processes induced by electron excitation using first-principles methods based on density functional theory and Time-Dependent Density Functional Theory (TDDFT) should be continued, as these are the most rigorous approaches and have the best computational efficiency. However, these techniques should be combined with molecular dynamics methods that consider not only electron but also nuclear dynamics, and that apply to realistic material systems. Finally, the integration of the nonadiabatic effect and the nuclear quantum effect within the framework of first-principles methods and its application in electronic structure calculation and optical-property simulation are also the key in the development of full-quantum-effect simulation methods.

It is worth noting that all three research directions require an accurate description of electronic structure. Therefore, the continuous optimization and development of first-principles electronic structure algorithms (e.g., quantum Monte Carlo methods) remains a top priority. With a solid foundation of full-quantum-physics theory, the goal of gradually extending the full-quantum-effect discussion to more complex processes in chemistry, materials science, energy science, environmental science, and even life science will no longer be a dream.

Parallel to the development of computational methods, condensed matter theoretical research has addressed the full-quantum-effect problems since the 1930s. Earlier in Section 1.2, we highlighted the work of Landau, Fröhlich, Bardeen, Pines, Cooper, Schrieffer, H. Y. Fan, McMillan, Dynes, Allen, Heine, Cardona, and others who made important contributions to modeling methods [Landau1993, Frohlich1937, 1950, Fan H.Y.1951, Bardeen1955, 1957, McMillan1968, Dynes1972, Allen1976, 1981]. Early theoretical studies were mainly concerned with electronic structure models: People used the free electron gas to describe the electronic states of metals or used the tight-binding approximation to describe the electronic states in semiconductors [Chadi1975], semimetals [Xu J.H.1993], or insulators [Harrison1980]. Many effective modeling methods have thus been proposed based on the tight-binding approach. Along with the emergence of first-principles electronic structure calculation methods in the 1980s, accurate first-principles simulation for electron–phonon coupling in real systems has become a research topic in modern condensed matter physics [Chang K.J.1985, Baroni2001, Guistino2017]. For this reason, we need to pay special attention to the combination of first-principles condensed-matter computational methods with existing theories based on model Hamiltonians and described by second quantization (e.g., first-principles electron–phonon coupling methods), in addition to the three key challenges outlined earlier, which will open up novel research opportunities in the field.

The goal of recent experimental work has been to develop precise characterization methods, such as ultrafast time-resolved spectroscopy and ultrahigh spatial resolution scanning probes, and effective combinations of the two. For these methods to succeed, both the techniques and the environment must meet rigorous experimental standards under extreme conditions, and the material samples themselves create new design and preparation challenges due to high quality of single crystallization, isotope purity, and precisely atomic-level controlled preparation of defect concentration and location. Quantum statistical effects are ubiquitous in real materials. Therefore, the characterization techniques and the material quality need to be optimized in special environments to allow effective experimental observations of the full quantum effect and its statistical and dynamical properties at the single-atom/defect or single-bond level going beyond the Born–Oppenheimer approximation.

Moreover, during full-quantum-effect characterization and measurement, powerful means of manipulation are required to enhance or reduce NQEs and NAEs in experiments. However, research in this direction is still limited. It is possible to actively enhance or suppress NQEs by combining atomic-precision material design with surface and interface engineering, isotope purification and substitution, and controlled doping of light element ions; the development of extreme-condition platforms under various external fields (such as ultrahigh pressure, extremely low temperature, strong magnetic field, ultrafast ultra-intense light field, etc.) can also help. Through these experiments, new ways of manipulating physical properties in a multitude of condensed matter systems beyond the traditional Born–Oppenheimer approximation could be explored, potentially leading to the design of devices based on full-quantum effect.

As in all development stages of condensed matter physics, integrating theory and experiment is crucial in studying the full quantum effect. Theoretical and experimental studies complement each other, leading to higher technical demands. The results will have a significant and far-reaching impact.

1.5 Example: Full Quantum Effects in Water

Light-element materials represent a new frontier in condensed matter physics and materials science. It has long been assumed that the physical properties of heavy elements (e.g., transition metals) are more complex than those of light-element materials. Although this seems intuitively reasonable, physical reality does not always conform to human intuition. An increasing number of studies have found that light-element materials exhibit full quantum effects (especially when nuclear quantum effects are non-negligible), resulting in novel and complex physical phenomena. For example, many complex physical puzzles associated with elemental condensed systems composed of simple hydrogen (H) or helium (He) cannot be answered with certainty despite more than a century of effort. We will address these issues in Chapters 7–11 of this book. In order to provide the reader with an example of how to approach a realistic condensed matter system that everyone is familiar within terms of the full quantum effect from the start, we will first introduce the full quantum effect problems for liquid and solid water, consisting of light elements hydrogen and oxygen. Through this example, we will explain the importance of full quantum effects in understanding the anomalous behavior of water.

Note that while most of the examples discussed in this book involve light elements, heavier elements (including heavy elements) can also exhibit full quantum effects. According to Eqs (1.1) and (1.2), nuclear quantum behavior and nuclear–electron quantum coupling are present in any many-body condensed system, regardless of composition. For example, the quantum paraelectricity mechanism in ferroelectric materials and the ultralow thermal conductivity of high-quality chalcogenide sulfide compounds demonstrate that nuclear quantum effects are not exclusive to light elements. Moreover, in some physical processes involving coordinate uncertainty, the quantum fluctuations may be more significant than is generally believed at low temperature, as exemplified by life processes. Nuclear quantum effects are presently best studied using light-element materials, in which such effects are more pronounced. Nonadiabatic effects can be observed during electronic transitions between ground state and excited states and nucleus evolution among different potential energy surfaces, when the nonequilibrium process is dominated by the nucleus–electron quantum coupling. Ultrafast processes such as the atomic motion evolution process induced by quantum coupling of the electron transition and the lattice vibration are currently the most studied examples of NAEs; related research involves laser-induced quantum-mechanical phenomena and superconducting states in layered titanium diselenide. For these studies related to both NQEs and NAEs, no deterministic relationship has been established with the mass of the elements. In real-world condensed matter systems, both effects coexist and influence each other. A few studies on full quantum effects have now begun to take both NQEs and NAEs into account, which are applied to any kind of elements.

The light elements here refer specifically to elements with small atomic masses, particularly those in the first and second rows of the periodic table. Their atomic numbers range from 1 to 10. Because of their low nuclear mass, a simple analogy is that when the electrons move, so do the nuclei (as shown in Figure 1.35(b)). Compared to other

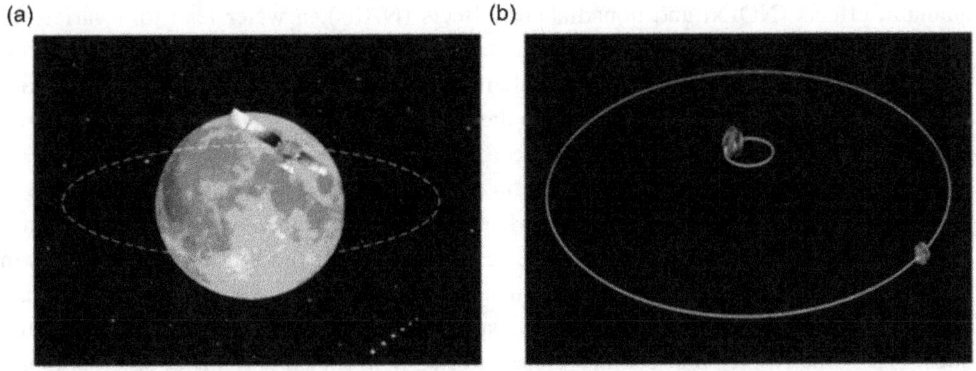

Figure 1.35 Analogies of electron motion and relative nuclear motion in (a) a heavy-element system and (b) a light-element system

elements with heavier nuclear masses (Figure 1.35(a)), the electron–nucleus quantum coupling in light elements is stronger and cannot be neglected. This is precisely why the full quantum effects are more pronounced in condensed matter systems composed of light-element atoms. To obtain comprehensive information on full quantum effects in light-element condensed matter systems, it is necessary to begin with the original Schrödinger equation and use the Born–Huang expansion. In other words, we must consider the wavefunctions of the electron and nuclei, as well as the quantum coupling between them.

Starting from the mid 1990s, E. G. Wang et al. explored the synthesis and characterization of light-elemental materials both experimentally and theoretically. The elements studied by his group included hydrogen (H), lithium (Li), boron (B), carbon (C), nitrogen (N), and oxygen (O). These elements combine to form different materials, such as liquid and solid water, diamond, graphene, boronene, carbon nitride, carbon nanocone, boron–carbon-nitride nanotubes, and crystalline and porous silicon carbide [Wang E.G.1997, Ma X.C.1999, Bai X.D.2000, Wu K.H.2001, Meng S.2002, Zhang G.Y.2003, Yang J.J.2004, Wang W.L.2006, Pan D.2008, Xu Z.2008, Zhang Q.F.2008, Liu K.H.2009, 2012, 2014, Watkins2011, Sun Z.R.2012, Chen J.2013, 2014a,b, Shi Z.W.2014, Guo J.2014, 2016a,b, 2017, Meng X.Z.2015, Xu X.Z.2017, Peng J.B.2018a,b, Wang L.2019, Ma R.Z.2020, Li N.2021, 2023, You P.W.2021a, Tian Y.2022, Huang X.2023, Hong J.N.2024, Wu D.2024]. These materials and their related interesting problems have become hot topics in condensed matter physics, chemistry, material science, and other multidisciplinary frontier areas including energy and environmental science. In 2010, the first PhD student in the study of full quantum effects from E. G. Wang's group, Q. F. Zhang, completed his thesis on *A study of full quantum effects of proton diffusion in* $BaZrO_3$ *by path integral molecular dynamics* at the Institute of Physics, Chinese Academy of Sciences [Zhang Q.F.2010]. Over the past 20 years or so, E. G. Wang's group has made significant progress in both NQE and NAE studies, theoretically and experimentally.

Composed of the light elements, namely hydrogen and oxygen, water is one of the most intricate condensed matter systems. Hydrogen bonding interactions through nuclear quantum effects (NQEs) and nonadiabatic effects (NAEs) in water lead to a variety of rich physics and anomalous behavior, which provides an ideal platform to study the full quantum effects beyond the Born–Oppenheimer approximation. Leveraging the cumulative research of E. G. Wang's group over the past few years, I shall first take water as an example to discuss these full quantum effects.

Water is the most abundant and common condensed matter in nature, indispensable to human life, health, and daily activities. Perhaps because people are too familiar with water, it is often assumed that there are no more scientific questions to be studied. However, in fact, many fundamental scientific questions involving water remain unsolved to this day. Indeed, "What is the structure of water?" was listed as one of the 125 pressing scientific questions of the twenty-first century presented in *Science*. Just take the simple ice (solid water) as an example – its phase diagram is proving to be far more complex than previously imagined. Some scientists have even referred to water as an "enduring mystery" (Ball, 2008, Bartels-Rausch, 2013, Meng, S., 2023].

Water molecules form a variety of complex and variable structures in the liquid or solid state through hydrogen bond networks. Along with metallic bonds, covalent bonds, ionic bonds, and van der Waals forces, hydrogen bonds constitute one of the five fundamental types of interaction in condensed matter. Due to the relatively weak strength of hydrogen bonds, there has been no effective experimental technique to directly measure them. In the theoretical treatments of hydrogen bonds in condensed matter systems, people have always been under the narrow understanding of the Born–Oppenheimer approximation – treating hydrogen nuclei as classical particles. In other words, most of the work in the current water science still only stays at the level of electron quantization but ignores the aspect of the quantum properties of nuclei. As early as the 1950s, it had been proposed that considering nuclear quantum effects, that is, the full quantum treatment of nuclei and electrons, might affect the strength of hydrogen bonds and the dynamic processes of hydrogen bond networks, as well as macroscopic physicochemical properties. But this prediction has always been unable to provide definite experimental and theoretical evidence scientifically.

Research in this field faces two major challenges. Firstly, on the experimental side, we need to develop atomic-level manipulation and detection techniques, precisely assemble water clusters of various sizes. At the same time, we need to push the detection techniques to the limit, which includes obtaining atomic imaging within water molecule and the hydrogen bonds between water molecules, quantitatively measuring the strength of a single hydrogen bond, and studying the quantum fluctuations of a single hydrogen bond caused by full quantum effects. In addition, on the theoretical side, we need to break through the narrow understanding of the Born–Oppenheimer approximation and develop a set of first-principles quantum-mechanical simulation methods that are effective for condensed matter systems. After considering the zero-point vibrations and quantum tunneling of atomic nuclei introduced by full quantum effects, we calculate their impact on the strength of hydrogen bonds and the dynamic structure of hydrogen bond networks, as well as the resulting changes in the macroscopic physical and chemical properties of water. However, the research difficulties in both of these aspects are extremely great.

In response to this fundamental problem in physical science, Y. Jiang, X. Z. Li, and others at Peking University have conducted systematic and in-depth research from both experimental and theoretical perspectives [Guo J.2016a]. Firstly, using single-molecule manipulation technique, they skillfully built a single-hydrogen-bonding model between a water molecule and a Cl atom on the Au-supported NaCl (001) surface. Then, they developed a unique tip-enhanced inelastic electron tunneling spectroscopy (TE-IETS), based on the ultrahigh-resolution scanning tunneling microscope (STM) system (illustrated on the left in Figure 1.36). These breakthroughs in molecular manipulation and spectroscopy enabled them to significantly enhance the signal-to-noise ratio, and successfully pushed the study of vibrational spectroscopy of water to the limit of a single hydrogen bond, achieving precise measurements of the single hydrogen bond strength. The results revealed that NQEs contribute up to 14% to bond strength (illustrated on the right in Figure 1.36). This finding proved that the influence of NQEs is similar to the contribution of thermal effects at room temperature. Moreover, they were able to quantitatively determine the change of quantum fluctuation in single-hydrogen-bond through isotope substitution. To gain a deeper understanding of the physical mechanism of this microscopic phenomenon, they executed full quantum simulations on the strength of a single hydrogen bond. They used the first-principles path integral molecular dynamics (PIMD) software package, which has been collaboratively developed since 2005 by the research group of E. G. Wang and researchers from Sweden and

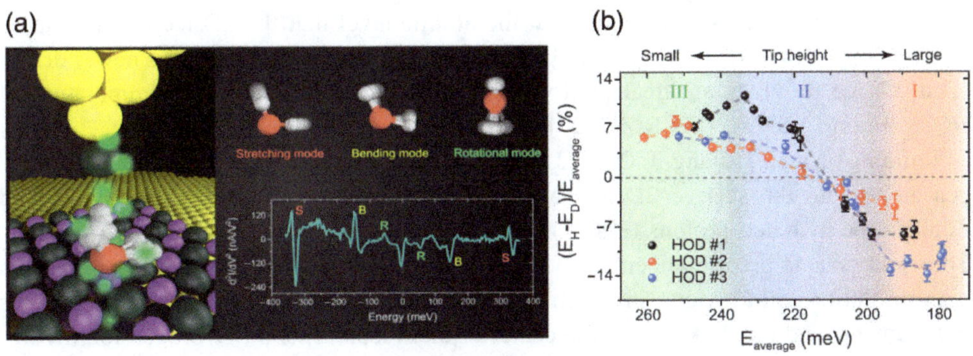

Figure 1.36 Experimental and theoretical studies of a single hydrogen bond in water. (a) shows the experimental setup. The vibrational spectrum of the water molecules is obtained at the single-bond level using inelastic electron tunneling spectroscopy. The bond strength is subsequently determined from the redshift in the O–H bond stretching frequency. Finally, the nuclear quantum effects on the hydrogen-bond strength are characterized by isotope substitution. This experimental approach has a very high (single-bond level) spatial resolution compared to traditional spectroscopic methods. Graph in (b) shows the difference between the OH and OD bond energies as the tip height varies. The calculation is carried out by first-principles path-integral methods and compared with experimental data. From Guo J.[2016a]. Reprinted with permission from AAAS.

the United Kingdom (interested readers can refer to Appendix A for more details). Through comparison, they established a unified law about the influence of NQEs on the properties of hydrogen bonds: The quantum fluctuation induced by the anharmonic zero-point motion of the nucleus strengthens strong hydrogen bonds and weakens weak ones, thereby causing changes in the configuration of the hydrogen bond network. This is the first time that the influence of nuclear quantum effects on the strength of hydrogen bonds has been rigorously proven both experimentally and theoretically. These results clarified the full quantum nature of hydrogen bonds and networks that had long been debated in academia, and unveiled the origins of anomalies of water/ice. This result clearly tells us that the full quantum effect cannot be neglected in the study of single-hydrogen-bond vibration, as well as various physical and chemical properties related to the breaking and recombination of hydrogen bonds in condensed matter systems.

Despite increasing research indicating that nuclear quantum effects have significant impact on systems composed of light elements, most studies in this field, so far, are still limited to detection of this existing effect. Whether the nuclear quantum effect can be used as a new physical parameter to realize new states and properties through its manipulation is still a direction to be explored. To answer this question, Y. Jiang's research group developed a non-invasive high-order electrostatic force imaging technique based on a qPlus-type atomic force microscope (AFM), which could identify the location of hydrogen nuclei in water with atomic precision [Peng J.B.2018a,b]. Utilizing this technique breakthrough, they successfully grew the first 2D ice structure on a hydrophobic metal surface by precise control on the growth kinetic parameters. Combined with theoretical simulations, they determined at the atomic level that the 2D ice has a "bilayer interlocking" structure, corresponding to a unique growth mechanism. This newly developed 2D ice provides an ideal platform for the realization of active manipulation of NQEs (Figure 1.37(a)) [Ma R.Z.2020].

Building on this, Y. Jiang, J. Chen, J. Guo, and others further doped high-density hydrogen atoms into the surface 2D ice. A novel ice structure consisting of self-assembled Zundel-type hydrated protons ($H_5O_2^+$) and water molecules is obtained (Figure 1.37(b)). They were able to precisely identify the Zundel configuration of hydrated protons through high-resolution AFM images, in which the oxygen atoms appear as pairs of bright spheres in the image, and the dark area in the center of them represents the hydrogen ion (hydrogen atomic nucleus H^+) shared between two water molecules. This study found that by manipulating nuclear quantum effects, in the high-density hydrogen-doped 2D ice, the hydrogen ions between oxygen atoms appear a clear tendency toward symmetrization, thus causing the difference between OH covalent bonds and O–H hydrogen bonds to completely disappear. They also conducted first-principles path integral molecular dynamics calculations and found that the high-density doping of hydrogen atoms significantly reduced the barrier of quantum tunnelling, thereby promoting the quantum delocalization of the hydrogen nucleus between water molecules. This is the first time that people used the nuclear quantum effect as a new physical parameter, and through manipulating it, a new matter state is obtained – a two-dimensional ice with metallization tendencies which is

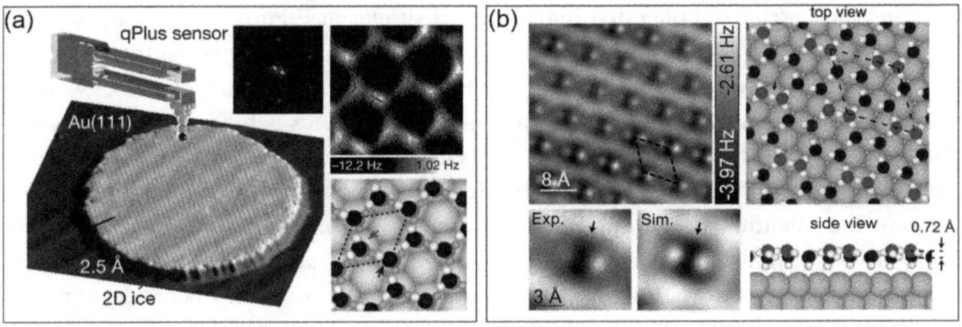

Figure 1.37 Two-dimensional ice with metallization tendencies and symmetrical hydrogen-bond configuration at atmospheric pressure. (a) Experimental setup, sub-molecular AFM images, and atomic model of bilayer 2D ice on Au(111) surface. (b) 2D hexagonal ice with symmetric hydrogen-bond configuration, obtained by doping hydrogen atoms on the Au(111) surface. The light grey spheres in the atomic model correspond to the oxygen atoms in the Zundel-type hydrated protons ($H_5O_2^+$); the enlarged AFM image shows the Zundel-type hydrated protons, and the grey arrows inside the diagram indicate the positions of the hydrogen ions in the symmetric hydrogen bonds. Figure 1.37(a) from Ma R. Z. [2020]. Reproduced with permission from SNCSC; Figure 1.37(b) from Tian Y.[2022]. Reprinted with permission from AAAS.

stable under conditions of room temperature and normal pressure [Tian Y.2022]. It has broken the "ice rules" that have been widely adopted for more than 90 years under normal pressure, and obtained a new matter state with a symmetrized hydrogen bond configuration. The growth and control method of the two-dimensional ice may be applicable to the study of other low-dimensional ice, which was subsequently confirmed by other experiments [Bartelt2022]. These research results tell us that the full quantum effect is not only indispensable when studying physical problems related to hydrogen bond vibration and the breaking and recombination of hydrogen bonds, but it is also very likely to drive the hydrogen bond network to undergo structural phase transitions and induce new states beyond traditional theoretical predictions.

Another significant impact of the full quantum effect in water is the quantum tunneling of hydrogen nuclei H^+ (protons). As the protons in the hydrogen bond network of water usually have strong correlation, in addition to the relatively well-studied phenomenon of stepwise quantum tunneling of protons, theory predicts that protons may also undergo concerted quantum tunneling. However, definitive data proving the existence of this concerted quantum tunneling was not available until 2015, when Y. Jiang, J. Chen, and X. Z. Li combined the submolecular STM technique with a real-time detection method to track proton transfer within individual water tetramer cluster on the NaCl(001) surface. They directly observed the quantum tunneling dynamics process of protons within the water tetramer cluster, and confirmed that this tunneling process is completed in concert by four protons under the influence of full quantum effects (as shown in Figure 1.38(a)). They carried out first-principles path integral molecular dynamics calculations and proved that nuclear quantum effects would greatly reduce the potential barrier of concerted tunneling (as shown in Figure 1.38(b)), thus significantly increasing the probability of proton transfer in water clusters. At the same time, they achieved controlled flipping of the water tetramer chirality between clockwise and anticlockwise states by adjusting the tip height. The physics behind is the full quantum effects change the flipping batters. To further prove the universality of the concerted tunneling, they conducted isotope substitution experiments and found that only by considering the quantum properties of nuclei can the phenomenon of concerted proton transport in the hydrogen bond network be accurately described. This is the first time that people have confirmed, on the microscopic scale, the existence of correlated nuclear quantum effects, proving that the configuration of water's hydrogen bonds is not static, but is constantly undergoing breaking and recombination processes; also, information and energy are often transmitted back and forth in the hydrogen bond network through concerted proton tunneling. This work clarified the debate in the community over more than the past 20 years about whether concerted tunneling exists, and provided new insights into the understanding of phase transitions and zero-point entropy of ice [Meng

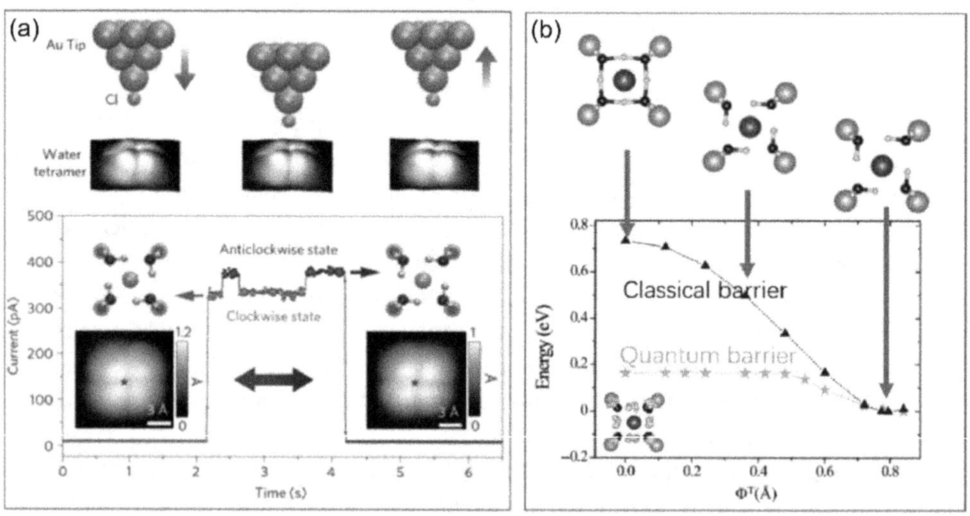

Figure 1.38 Schematic of concerted hydrogen tunneling in a water tetramer. (a) Experimental setup. The switching dynamics of water-tetramer chirality is monitored by the tunneling current, whose intensity can be easily adjusted by varying the tip height. In this way, the switching rate between the two chirality states in the water tetramer can be determined experimentally. (b) Theoretical simulation. When the nucleus is treated classically, the calculation yields a relatively high transition barrier (dashed line with triangular markers); when treated quantum mechanically, the potential barrier height is significantly reduced due to quantum tunneling. The nuclear quantum effect more effectively reduces the potential barrier during concerted tunneling (solid line with pentagram markers) than during stepwise tunneling. This implies that the experimentally observed proton-transfer process is accomplished through concerted tunneling. Figure 1.38(a) from Meng X. Z. [2015]. Reproduced with permission from SNCSC. Figure 1.38(b) is reproduced from Feng Y. X.[2018b], with the permission of AIP Publishing. The material is modified.

X.Z.2015]. Figure 1.38 illustrates the process of concerted quantum tunneling of protons in water clusters.

For the study of nonadiabatic dynamics in water, S. Meng et al. at the Institute of Physics, Chinese Academy of Sciences, have developed a first-principles excited-state dynamical method that can effectively analyze nonequilibrium processes (see Appendix B). It considers the NAEs associated with electronic transitions in condensed matter system such as water, thus allowing the description of the atomic transition on different potential energy surfaces. Through real-time simulations of the quantum process during water photolysis, S. Meng and E. G. Wang et al. found that photoexcited gold clusters can generate hot electrons in water. These hot electrons are subsequently injected into the high-energy orbitals of water molecules, inducing rapid proton transfer; observation of water photolysis on the 100 fs time scale was thus realized [Yan L.2018, You P.W.2021a]. These results cannot be obtained using the Born–Oppenheimer approximation because the excited electrons will make transitions between different orbitals until they reach a specific water molecule; a Born–Oppenheimer molecular dynamics simulation can only force the electrons to remain in ground state orbitals, and water photolysis will not occur. More surprisingly, hydrogen atoms from different water molecules were observed colliding and combining to form hydrogen molecules. The theoretical computation predicted a quantum efficiency of 0.06% for H_2 production, which is surprisingly consistent with experiment (0.05%). This suggests that the photolysis of water follows a chain-reaction mechanism: plasmon excitation $\rightarrow$ decay of equipartition plasmon $\rightarrow$ photoelectron generation $\rightarrow$ photoelectron colliding with the first water molecule $\rightarrow$ hydrogen atom detaching from water and colliding with a nearby water molecule $\rightarrow$ detached hydrogen atoms colliding with each other to form a hydrogen molecule $\rightarrow$ H_2 release (Figure 1.39) [Yan L.2018]. This was the first time that hydrogen production in water photolysis was described at the atomic level from the perspective of nonadiabatic quantum dynamics. The ultrafast electron–ion process of plasmon-induced water photolysis was demonstrated theoretically, revealing completely the chain reaction mechanism of the hydrogen production.

These examples involving water, a system consisting of the light elements H and O, highlight the significance of full quantum effects in condensed matter physics. They also reflect the current status of full-quantum condensed-matter physics, which will be discussed in detail in Chapters 2–3. We have summarized some of the reliable theoretical tools for going beyond the adiabatic Born–Oppenheimer approximation, as well as precise experimental techniques, and have shown with specific examples how successfully these can be combined, demonstrating a new research paradigm of full-quantum condensed-matter physics, with applications to specific examples.

In fact, an important goal of physical science research is to provide an accurate, microscopic-level description of the physical and chemical phenomena occurring in real systems with a quantum many-body interaction. The development of such descriptions allows us to precisely adjust their properties, thereby contributing to novel device design, assembly, and applications. Thus, the challenges we face in condensed matter physics, chemistry, and materials science are fundamentally the same: We strive

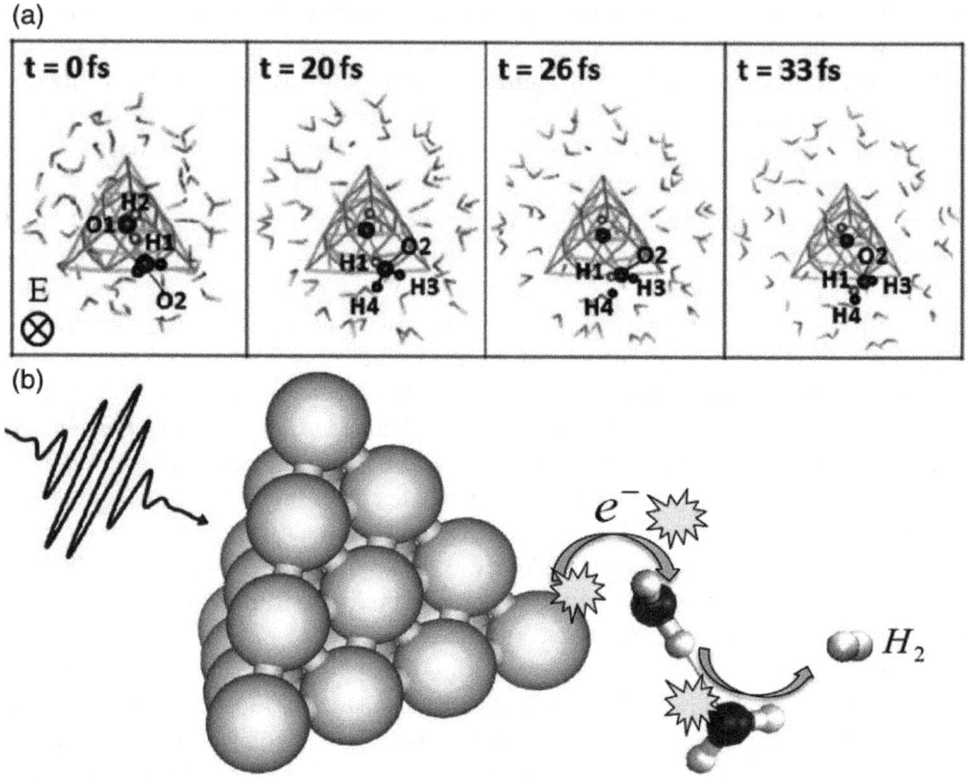

Figure 1.39 Nonadiabatic-dynamics simulation of water photolysis: (a) photosplitting of water molecules around a gold cluster, producing hydrogen gas; (b) illustration of a chain-reaction-like mechanism in which excited electrons cause hydrogen atoms to collide at high speeds with surrounding water molecules. Reprinted with permission from Yan L.[2018]. Copyright 2024 American Chemical Society.

to understand the novel physical and chemical properties arising from the many-body quantum mechanics described by the Schrödinger's equation. This book discusses these physical and chemical problems from the full quantum perspective. Despite significant progress in theory and computational methods of electronic structure since the advent of quantum mechanics, performing full quantum calculations of electrons and nuclei in realistic systems remains a challenge. In terms of experimental research, although some observations naturally include NQEs and NAEs, effective and precise measurement techniques that specifically target these in electron–nucleus quantum coupling are still limited. To this day, identifying the contributions of NQEs and NAEs in experiments remain a difficult task. We have considerable work ahead of us in terms of understanding the observed physical and chemical phenomena. Addressing these issues is the main focus of full-quantum-effects study.

Discovering new physical states and new manipulation techniques requires an accurate physical understanding of real and complex condensed systems. The full-quantum condensed-matter physics emphasized in this book is indispensable for the accurate simulation and measurement of physical properties in real systems. Although the primary focus of the book is basic science, that is, improving our understanding of the objective world through the study of full quantum effects in condensed matter physics, some of the results presented here are also relevant to research in applied fields (including energy and environmental science). Thus, this fundamental research work will not only advance frontier science but also create opportunities for interdisciplinary and practical applications, contributing to the sustainable development of science, technology, and socioeconomics for humankind.

2

Full Quantum Effects in Condensed Matter Physics

As indicated by the historical review in Chapter 1, it has taken a long time for quantum mechanics to be fully integrated into the study of realistic materials. Before introducing the theoretical and experimental full-quantum-effect methods in detail, we provide an overview in Chapters 2 and 3 of the phenomena in physics and chemistry, respectively, for the understanding of which integration of full quantum effects is indispensable.

One of the primary objectives of condensed matter physics is uncovering and understanding the fundamental properties of materials, including mechanical, thermal, acoustic, optical, electrical, and magnetic properties. Condensed matter is composed of atoms, and their physical properties are the macroscopic manifestations of microscopic processes involving electrons and nuclei, which are governed by full quantum effects. Thus, to reveal the physical nature of a many-body system, it is necessary to have a full quantum-mechanical description of the interacting electrons and nuclei. Full quantum effects are ubiquitous in condensed matter; however, they are more pronounced in some materials than in others.

Superfluidity and supersolidity, along with superconductivity, have received considerable attention. They are fundamentally resulted from the full quantum effect: They would not exist if only the electrons were quantized, without considering the quantum properties of the nuclear motion. Similarly, phonons are essentially quanta of nuclei vibrating around their equilibrium positions. However, in traditional descriptions of the optical and electrical properties of solids, nuclei are treated classically (e.g., the adiabatic ball-and-stick model discussed in Chapter 1) and the nucleus near equilibrium are loosely approximated using perturbation theory that only considers harmonic vibrations. When phonon-assisted electron transitions are considered, for example, in the study of optical properties, the coupling of the electron and the nuclear quantum motion is even more vital for an accurate conclusion. To understand the properties of real materials, it is necessary to fully understand electron–nucleus interactions, which requires expressing nuclear quantum behavior in terms of wavefunctions. As condensed matter physics progresses, consideration of full quantum effects will be the key to the discovery of new quantum materials, the fabrication of novel devices with special functions, and other applications.

Full quantum effects have been previously defined as the physical and chemical properties predicted by the full quantum theory, based on the nuclear and electronic degrees of freedom – more specifically, the properties that are neglected in the adiabatic ball-and-stick model of electronic states. They can be divided into two categories – nuclear quantum effects (NQEs) and nonadiabatic effects (NAEs). However, these two effects often coexist and interact with each other.

The ball-and-stick model depicts the nuclei as classical particles; only electron wavefunctions are considered when studying the quantum properties of condensed systems, and electron–nucleus quantum coupling is ignored. This approach is simple and valid in most scenarios, but not in all. In fact, nuclear fluctuations always affect the electron motion and the electron–nucleus interaction, even if only slightly. In extreme cases, such as light-element systems or systems with very close potential energy minima, the nuclei are even capable of quantum tunneling and of quantum fluctuations at absolute zero temperature, resulting in phenomena that cannot be explained by classical physics. Moreover, the quantum nature of the nuclei means that zero-point energy is always present in a real system: The ground state of the system always has non-zero energy irrespective of how low the temperature is, which is the reason why quantum liquids and solids exist at low temperatures. Using the ball-and-stick model based on adiabatic electronic states to describe these properties of condensed matter systems gives rise to non-negligible errors.

Neglect of NQEs is not the only weakness of the traditional approaches. During the dynamical evolution of condensed matter, electrons and nuclei evolve simultaneously, and they do so quantum mechanically. The evolution can be considered as electron transitions between different energy levels among different Born–Oppenheimer potential energy surfaces determined by the nuclear configuration. These processes manifest as NAEs and have a significant impact on the forces acting on the nuclei and on nuclear trajectories. They also influence the density distribution and dynamical evolution of the electrons, resulting in a cascade of physical phenomena and processes that cannot be described in the adiabatic approximation.

In Sections 2.1–2.2, we discuss the use of recently developed theoretical and experimental techniques for investigating the structure and physical properties of condensed matter by considering the NQEs, NAEs, and both together and comparing the results to those obtained without consideration of full quantum effects. The comparative discussion of chemical properties is deferred until Chapter 3. The purpose of the present chapter is to provide readers who are encountering full quantum effects for the first time with a comprehensive conceptual understanding of the condensed matter physics problems, and to prepare them for further exploration of the intrinsic connection between specific macroscopic phenomena and their underlying microscopic mechanisms. The explanation to some of these problems will inevitably involve specific research examples. However, in Chapters 2 and 3, we focus on the introduction of the full-quantum phenomena rather than their explanations. A systematic study of the full quantum effects in specific problems or typical systems is provided in Chapters 7 and 8, following the introduction of available theoretical and experimental tools.

2.1 Nuclear Quantum Effects

Nuclear quantum effects (NQEs) are the properties exhibited by molecules or condensed matter systems that depend on both the nuclear and electronic wavefunctions. The discussion of NQEs can be rigorous and include nonadiabatic effects, or it can be approximate within an adiabatic framework. However, the theoretical debate over the origin of NQEs can be fully resolved only when the nucleus is treated as a quantum rather than as a classical particle.

2.1.1 Atomic Structure

First and foremost, when considering the impact of quantum effects on nuclei, we must distinguish the atomic structure of condensed matter – the distribution and arrangement of atoms or molecules in solid, liquid, or gas – when NQEs are included from the structure predicted by classical ball-and-stick models. For some physical properties, the discrepancies between classical and quantum predictions are not significant; this has led researchers to separate the electronic degrees of freedom from the nuclear degrees of freedom for convenience and to treat the nuclei classically. Nevertheless, in some cases, especially in the regime of low temperature and high pressure, the full quantum effects are non-negligible and bring about evident changes in observed physical properties, even leading to the emergence of new physics.

A typical example is the phase transition problem of helium, which has been studied for over a century. At normal atmospheric pressure, it is impossible to produce solid helium from liquid helium even if the temperature is infinitely close to absolute zero (see the phase diagram in Figure 2.1). This is because of helium's low mass and large zero-point energy. Helium is an inert gas, so its electron-shell structure is saturated, allowing only weak interactions between helium atoms. As a result, even when the temperature approaches absolute zero, the interactions between helium atoms at atmospheric pressure are not strong enough to overcome the zero-point energy, preventing the entire system from entering the solid state. This is also why low-temperature conditions in experiments are generally realized using liquid helium. Increasing the atmospheric pressure of liquid helium at low temperatures can effectively reduce the helium-atom spacing and enhance their interactions, which is a feasible way to overcome the zero-point energy effect and produce solid helium. In fact, liquid helium has been observed to freeze at very low temperatures, but only with pressures of at least 25 atm. Full quantum effects are primarily responsible for this important property.

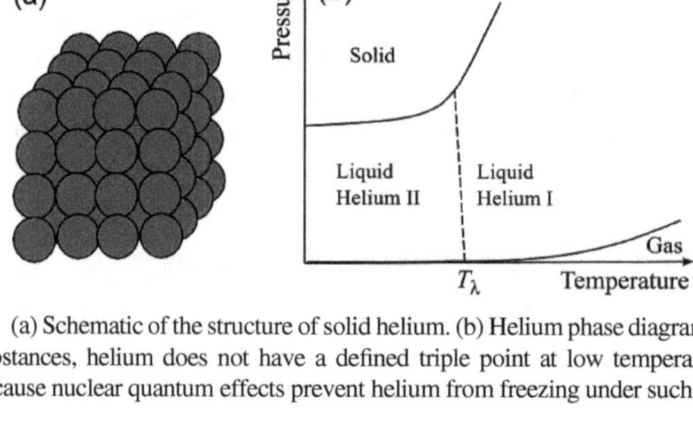

Figure 2.1 (a) Schematic of the structure of solid helium. (b) Helium phase diagram. In contrast to most substances, helium does not have a defined triple point at low temperature and low pressure because nuclear quantum effects prevent helium from freezing under such conditions.

It is well-known that liquid helium can be superfluid. In fact, the solid form of helium might exhibit even more bizarrely non-classical behavior than the liquid and gaseous forms because it could be a supersolid. A helium-4 (^{4}He) atom contains two protons, two neutrons, and two electrons. Thus, it is a boson with an even number of spins. This means that at sufficiently low temperatures, all ^{4}He atoms in a system can remain in the same ground state: The exchange of positions between two helium atoms is not impeded in any way. For this reason, atoms in solid ^{4}He flow without resistance, a phenomenon known as supersolidity. Although helium-3 (^{3}He) has an odd number of spins and behaves similar to a fermion, two ^{3}He atoms can pair up to form a bosonic particle; therefore, ^{3}He systems might also exhibit supersolid properties. Supersolids are arguably one of the most prominent macroscopic manifestations of nuclear quantum fluctuations. Some recent theoretical and experimental studies of supersolids are discussed in later sections.

Another phase-transition process that is affected by full quantum effects occurs in the condensed system composed purely of the hydrogen element. Similar to solid helium, molecular solid hydrogen can be produced directly from hydrogen gas in the molecular state; however, this phase transition occurs at atmospheric pressure when the temperature falls below 14 K. If we ignore the slow conversion between ortho-hydrogen and para-hydrogen and assume that the solid hydrogen is in the ground state, the hydrogen molecules in the solid can rotate freely when the pressure is relatively low (phase I), which can be also called supersolid. This is a direct result of the competition between the zero-point energy of molecular rotation and the anisotropic intermolecular-interaction strength. At low pressures, hydrogen molecules rotate freely. This is because the special interactions between them cannot eliminate the minimum energy associated with their rotation (zero-point energy). As the pressure rises, the distance between molecules decreases, and the anisotropy of intermolecular interactions increases; this restricts the molecular rotation, and the solid hydrogen transitions into phase II.

Using isotope substitution to replace hydrogen with deuterium significantly lowers the boundary pressure between phases I and II (see Figure 2.2). At a pressure of 80 GPa and a temperature of 50 K, the angular distribution of molecular solid deuterium shows two peaks (Figure 2.2(b)), indicating that the rotation of the deuterium molecule is restricted. By contrast, no angular distribution peaks are observed for molecular solid hydrogen under the same conditions (Figure 2.2(c)), suggesting a rotationally free state. Deuterium molecules have a lower zero-point rotation energy, so the anisotropic intermolecular interaction does not need to be very strong to restrict the rotation. Consequently, the boundary between phases I and II occurs at lower pressure. The essential reason is that the rotational zero-point energy of deuterium molecules in solids is smaller. At the same temperature, it is not sufficient to destroy the anisotropic characteristics between molecules. At this low temperature, the corresponding thermal energy is also very low. The combined result of these two effects cannot exceed the energy controlling the anisotropic tendency of the orientation of deuterium molecules (as shown in Figure 2.2(b)). It has been found that the molecular rotation of solid deuterium is free only when the temperature is raised to 150 K, so no characteristic peaks in the angular distribution of deuterium molecules can be observed. This physical picture was further confirmed by X.Z. Li, A. Michaelides, and

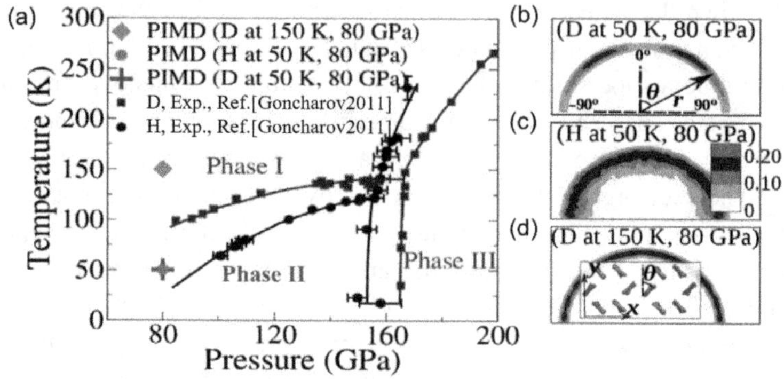

Figure 2.2 (a) Phase diagram of solid hydrogen. The black curve with circle markers and the grey curve with square markers represent the boundaries of phase I (top left), phase II (bottom left), and phase III (right) for hydrogen and deuterium, respectively. Reprinted from Goncharov[2011], with the permission of AIP Publishing. The isotope dependence of the phase I/II transition has been explained by path-integral molecular-dynamics simulations [Li X.Z.2013]. (b) At a temperature of 50 K and a pressure of 80 GPa, the rotation of the deuterium molecule is restricted (two peaks in the angular-distribution plot). (c) Hydrogen molecules under the same conditions rotate completely freely (no characteristic peak is found); this corresponds to the supersolid state. (d) The molecular rotation of the molecular solid deuterium is less restricted if the pressure remains at 80 GPa and only the temperature increases from 50 K to 150 K. This is a result of the competition between the zero-point energy and thermal energies that trigger the molecular rotation process and the anisotropic intermolecular interaction strength, the interacting competitive relationship can be controlled by temperature and pressure. Used with permission of IOP Publishing, from Li[2013], Copyright 2024; permission conveyed through Copyright Clearance Center, Inc.

others in 2013 through path-integral molecular-dynamics simulations [Li X.Z.2013]. The specific phase-transition process is depicted in Figure 2.2.

Hydrogen is the first element in the periodic table. It belongs to the same group as the alkali metals but differs from them in many ways. Under normal conditions, two hydrogen atoms combine to form an H_2 molecule, which is similar to the behavior of chlorine atoms from the halogen group and oxygen atoms from the chalcogen group. In 1935, Wigner and Huntington predicted that hydrogen would become a solid under high pressure, similarly to alkali metals. The valence electrons bound within the molecule can be pressurized to form itinerant electrons, and the free movement of these electrons gives rise to conductivity in solid hydrogen, transforming it into a metal [Wigner1935]. In 1968, Ashcroft suggested that metallic hydrogen is a superconductor with a T_c possibly above room temperature because of the strong electron–phonon coupling [Ashcroft1968]. Producing metallic hydrogen has been one of the Holy Grails of condensed matter physics and materials science. Attempts to produce it at ultrahigh pressure and ultralow temperature have been made in the past [Mao H.K.1994]. However, no conclusive evidence for metallic-hydrogen production has been obtained thus far. Theoretical calculations predict that hydrogen metallization occurs at around 380 GPa [Monacelli2021]. Several preliminary experimenters have reported light-absorption/reflection and resistance measurements indicating hydrogen metallization at a pressure between 360 and 490 GPa.

If the pressure continues to increase, the condensed hydrogen would theoretically become a softer metal similar to alkali metals. This is caused by the underlying change in chemical bonding – the weakened anisotropy due to the interatomic interactions. At low temperatures, because of large quantum fluctuations, the metallic hydrogen may become a low-temperature liquid due to NQEs. In 2004, E. Babaev and N. W. Ashcroft used the Ginsberg-Landau theory to explore the possibility of an NQE-induced superconducting and superfluid phase transition in molecular solid hydrogen at high pressure [Babaev2004]. The melting curve of solid hydrogen at low pressure was calculated by S. A. Bonev et al. in the same issue of *Nature*; the slope (dT/dP) of the curve was negative, which indicates that a low-temperature liquid phase exists at high pressures [Bonev2004].

Using this idea, J. Chen, X. Z. Li, and E. G. Wang from Peking University, collaborating with Michaelides and Pickard from University College London (both currently professors at the University of Cambridge) and Needs from the University of Cambridge, performed first-principles molecular-dynamics and path-integral molecular-dynamics simulations of hydrogen from 500 to 1,200 GPa [Chen J.2013, 2014]; the only difference between the two methods was that NQEs were included in the latter. According to their results (Figure 2.3), when NQEs were considered, the melting curve of hydrogen from 500 to 900 GPa had a negative slope. With the lowest simulation temperature set to 50 K, the system still

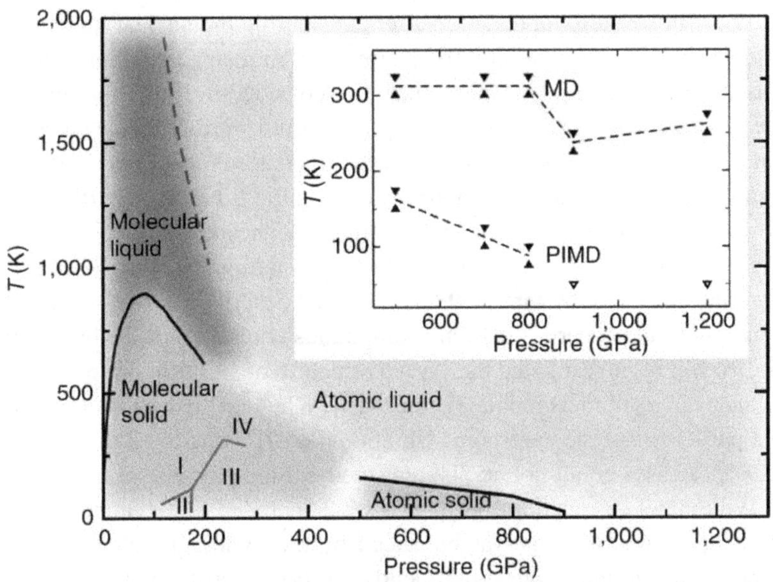

Figure 2.3 Calculated phase diagram of hydrogen at high pressures; the simulation extends to much higher pressures than those in Figure 2.2. From 500 to 1,200 GPa, the melting curve predicted by the path-integral molecular-dynamics simulations (lower curve in the inset) has a negative slope, while the melting curve given by the classical molecular-dynamics simulations (upper curve in the inset) does not follow this trend. Thus, a low-temperature liquid phase may exist in the region between the two curves, due to nuclear quantum effects. Figure created by Chen J.[2013] licensed under CC BY 3.0.

behaved similarly to a liquid at 900 and 1,200 GPa. By contrast, the approach that treated the nucleus classically did not produce a notable negative slope of the melting curve. Thus, in the region between the grey and black dashed curves in the inset of Figure 2.3, the system was in a low-temperature liquid state caused by NQEs. This provides a convenient way to study the liquid–liquid transition of hydrogen (see Section 7.1 for a detailed derivation).

Lithium is the third element in the periodic table, following hydrogen and helium. Therefore, it is reasonable to expect significant NQEs of lithium in its phase diagram. As an alkali metal, solid lithium has a body-centered cubic (bcc) crystal structure with a coordination number of 8. In 1947, after applying pressure to solid lithium at 77 K, C. Barrett discovered the formation of a new phase: the fcc phase of ^{7}Li [Barrett1947]. At zero pressure (0 GPa), lithium transforms into a close-packed structure (later believed to be the 9R crystal structure) below 77 K. In 2017, after considering NQEs, G. Ackland et al. found that this close-packed structure is thermodynamically unstable and is not the 9R structure; this was shown experimentally using synchrotron XRD in diamond anvil cells and theoretically using multiscale simulations with density functional theory and molecular dynamics [Ackland2017].

The crystal structure of lithium has been an intriguing research topic for decades. Why does such a simple metal have so many phases at high pressure? What is the mechanism of its phase transition? In order to answer this question, it has been thoroughly studied by F. Gorelli et al. They have shown that the experimentally observed oC88 (*C2mb*) phase near 70 GPa can only be properly understood after considering the nuclear quantum properties of the atoms in solid lithium [Gorelli2012]. However, in their calculations, the description of NQEs is not very accurate (in particular, the consideration of anharmonic effects is incomplete); therefore, the lithium phase diagram should be re-computed at a higher accuracy. It can be argued that if the pressure is further increased (above 80 GPa), the influence of full quantum effects on the phase diagram of metallic lithium (in particular, on phases such as oC40 and oC24) will become even more pronounced.

NQEs can alter the atomic structures and corresponding states not only for pure condensed materials made of the three lightest elements described above but also for atomic structure and physical properties of some compounds composed of the hydrogen element. In 2016, I. Errea et al. investigated the crystal structure of sulfur hydride at high pressures using the genetic algorithm technique [Errea2016]. When the hydrogen atoms were treated as classical particles and the zero-point vibrational energy due to NQEs was ignored, the $Im\bar{3}m$ and $R3m$ phases could not be distinguished at high pressure in the simulated XRD spectrum. When nuclear quantum properties were included, the symmetry of the chemical bond between two S atoms was enhanced by the hydrogen delocalization resulting from zero-point nuclear vibration; consequently, the crystal structure was dominated by the $Im\bar{3}m$ phase. Similarly, in a study of high-pressure LaH_{10}, Errea et al. found that if NQEs were excluded from consideration, multiple minima appeared on the Born–Oppenheimer energy surface, corresponding to different crystal structures (see Figure 2.4(b) on the left) [Errea2020]. However, when NQEs were included, all classical atomic structures collapsed into a full-quantum crystal configuration, the $Fm\bar{3}m$ structure. The potential energy surface was smoothed so that the entire system had only one minimum (see Figure 2.4(b) on

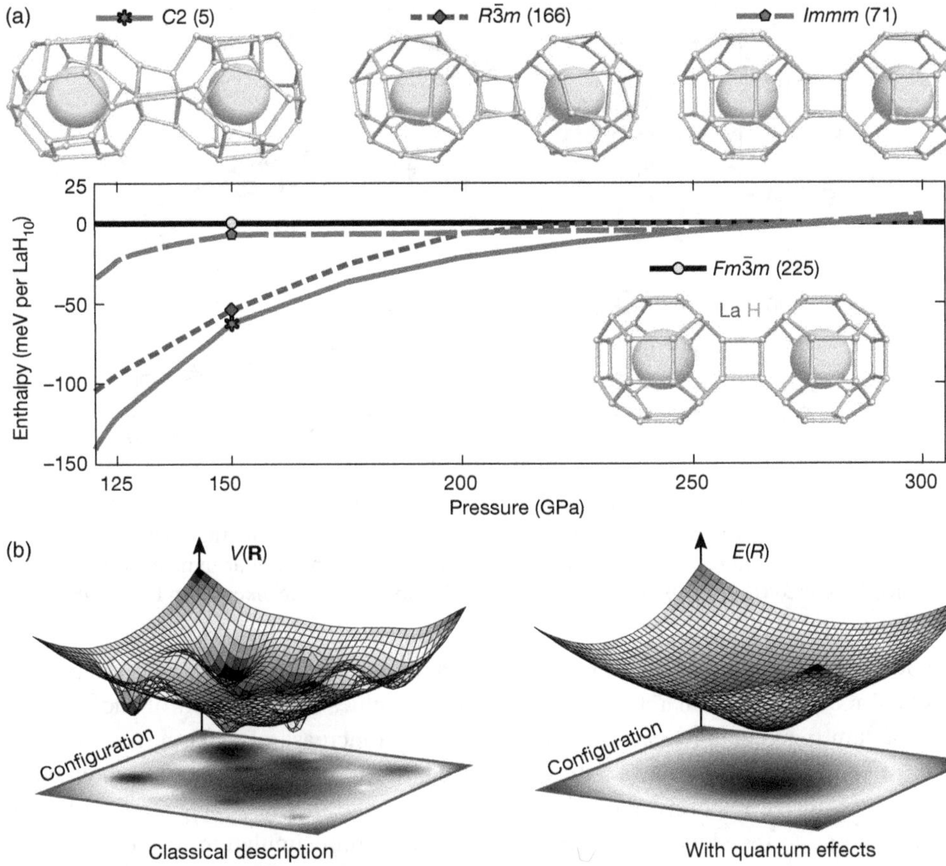

Figure 2.4 (a) Different crystal structures of LaH_{10} under high pressure and their corresponding enthalpies of formation (neglecting nuclear quantum effects). (b) Potential energy surfaces without (left) and with (right) consideration of nuclear quantum effects. Neglecting nuclear quantum effects gives rise to multiple minima on the energy surface (left panel in figure (b)); when nuclear quantum effects are included, all classical atomic configurations collapse to the most stable ($Fm\overline{3}m$) structure, resulting in a potential energy surface with a single minimum (right panel in figure (b)). From Errea[2020]. Reproduced with permission from SNCSC.

the right). These examples confirm the decisive role of full quantum effects in determining the crystal structure of given compounds.

The evidence presented above demonstrates how full quantum effects can alter the crystal structure of hydrogen-rich materials. In addition to solid materials, common liquid structures are also sensitive to full quantum effects. For example, B. Cheng et al. calculated water density as a function of temperature in different phases [Cheng B.2019]. Figure 2.5 shows the density–temperature curves for cubic ice (I_c), hexagonal ice (I_h), and liquid water (L), considering both the classical (CL) and quantum (Q) treatments of the nuclei, respectively. The numerical simulations were subsequently compared with experimental results (reported by Hare1987). According to Figure 2.5, the theoretical calculations that

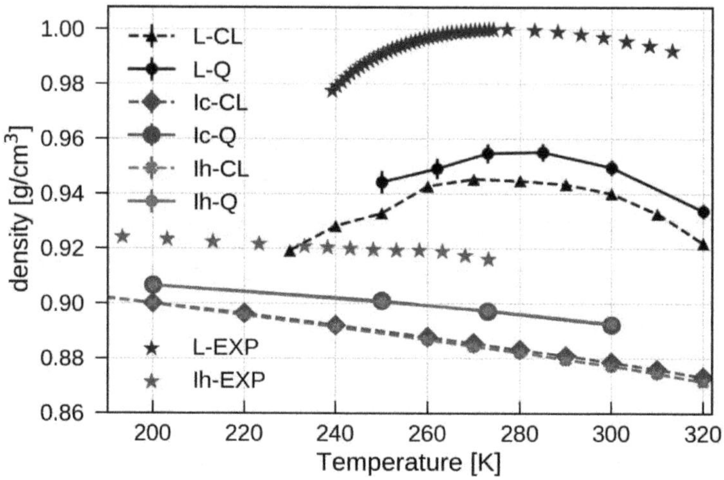

Figure 2.5 Density–temperature curves of cubic ice (I_c), hexagonal ice (I_h), and liquid water (L) predicted by classical (CL) and quantum (Q) treatments of the nuclei, respectively. The quantum results were obtained using path-integral molecular-dynamics simulations. Reprinted from Cheng B.[2019]. The experimental data were taken from Hare[1987].

included NQEs agreed well with the experimental data, and the temperature of maximum density for liquid water matched the experimental values. Comparison of the classical and quantum treatments reveals that NQEs lead to an increase of ~1% in the density of the three phases of water. The densities predicted by full-quantum simulation show better agreement with the experimental values. In terms of the thermal-expansion coefficient (corresponding to the slope of the curve), the full quantum simulation results also agree better with the experimental data. This indicates that treating the nuclei quantum mechanically allows us to obtain a more accurate thermal expansion coefficient than treating them classically.

The radial distribution functions (RDFs) between different atoms of liquid water (O–O, O–H, H–H) are depicted in Figure 2.6. We see that the NQEs decrease the magnitude of all three RDFs, which is consistent with the experimental observations. Moreover, the quantum simulation results for the first peaks of the O–H and H–H RDFs differ significantly from the classical-approximation results and are more consistent with experiment. These studies substantiate the importance of full quantum effects for the atomic structure of liquids.

2.1.2 Electronic Structure

The effect of nuclear quantum motion on electronic structure has long been an active research area in solid state physics. During the development of solid state quantum theory in the mid twentieth century, Bardeen and Pines et al. recognized the importance of quantizing the nuclear motion and established various theoretical methods to evaluate the effects of electron–phonon coupling on the conductivity of metals. Similar theoretical

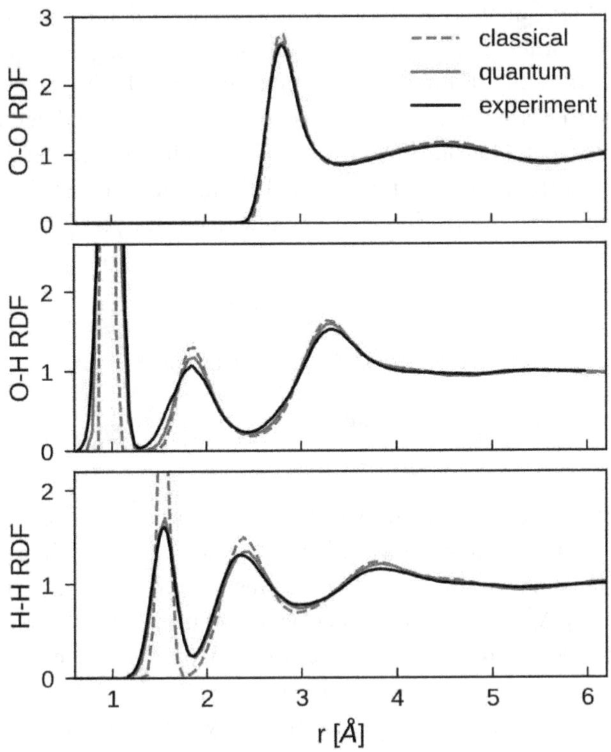

Figure 2.6 Oxygen–oxygen (O–O), oxygen–hydrogen (O–H), and hydrogen–hydrogen (H–H) in liquid water radial distribution functions (RDFs) as a function of bond length. The dashed, grey solid, and black curves show results of classical treatment of the nuclei, quantum simulation of the nuclei, and experiment (data from [Skinner2014, Soper2000, Chen W.]). Reprinted from Cheng B.[2019].

models have also been extended to semiconductor physics. Bardeen, Pines, and others developed several methods to correct the energy-band characteristics in semiconducting materials based on considering the nuclear motion. In the 1950s, H.Y Fan and Migdal et al. proposed theoretical models for phonon renormalized electronic structures based on perturbation theory [Fan H.Y.1951, Migdal1958]. From the 1970s to 1980s, Allen, Heine, and Cardona unified these early theoretical models into the "AH theory," which describes electronic energy-band structure at finite temperatures [Allen1976, 1981]. These theoretical works were concerned with developing models that better explain experimental data. Although NQEs such as energy-gap renormalization introduced by the zero-point vibration were naturally included in these methods, this was never explicitly stated. Rather, the electron–phonon coupling was explained from the perspective of thermal excitation. As a result, these models were only effective at characterizing specific materials and did not provide a complete physical picture. More recent research on first-principles electron–phonon coupling theory will be introduced in Section 5.2. For the historical development of the first-principles method, the interested reader can consult [Ziman1960, Giustino2017].

Studies of nuclear quantum motion and electronic structure renormalization made significant progress after 2010. In particular, S. G. Louie, M. L. Cohen, and other researchers at UC Berkeley [Giustino2010] creatively combined early electron–phonon coupling theory with first-principles methods and successfully realized quantitative phonon renormalization calculations. Their method also allowed methodical discussion of the contribution of different phonon modes and excitations. Generally, the most important phonon-induced renormalization terms in the electron energy-band structure can be written as

$$\Delta\epsilon_{\mathrm{nk}} = \Delta^{SE}\epsilon_{nk} + \Delta^{DW} \int_{nk} , \tag{2.1}$$

where $\Delta^{SE}\epsilon_{nk}$ is the phonon-induced self-energy correction

$$\Delta^{SE}\epsilon_{\boldsymbol{nk}} = \sum_{m\neq n,v} \int \frac{d\boldsymbol{q}}{\Omega_{\mathrm{BZ}}} \frac{2n_{\boldsymbol{q}v}+1}{\int_{\boldsymbol{nk}} - \int_{m,\boldsymbol{k}+\boldsymbol{q}}} \left| g_{mn,v}\left(\boldsymbol{k},\boldsymbol{q}\right)\right|^2 . \tag{2.2}$$

Here, n is the energy-band index; the vectors k and q are the wave vectors of the electron and phonon wavefunctions, respectively; v indicates the phonon branch; n_{qv} is the Bose–Einstein occupation factor; Ω_{BZ} is the volume of the Brillouin zone; and $g_{mn,v}(k,q) = \langle m, k+q | \Delta_{qv} V | nk \rangle$ is the electron–phonon matrix element. In fact, $\Delta^{SE} \in_{nk}$ includes the first-order correction due to single-phonon-induced electron scattering. Its physical significance is represented by the SE term of the Feynman diagram in Figure 2.7(a).

The second term of Eq. (2.1), $\Delta^{DW} \in_{nk}$, is the Debye–Waller correction term

$$\Delta^{DW}\epsilon_{nk} = \sum_{m\neq n,v} \int \frac{d\boldsymbol{q}}{\Omega_{\mathrm{BZ}}} \frac{2n_{\boldsymbol{q}v}+1}{\epsilon_{\boldsymbol{nk}} - \epsilon_{\boldsymbol{m},\boldsymbol{k}}} \left| g^{DW}_{mn,v}\left(\boldsymbol{k},\boldsymbol{q}\right)\right|^2 . \tag{2.3}$$

It can be considered a second-order correction. The electron–phonon coupling matrix element $g^{DW}_{mn,v}(k,q)$ is very complicated; the detailed derivation can be found in the literature [Giustino2010]. This term represents the interaction of the electron with two phonons, corresponding to the DW term shown in Figure 2.7(a).

Using density functional theory, F. Giustino et al. calculated the temperature dependence of the direct band gap of diamond. Their result shows a significant zero-point renormalization correction (Figure 2.7(b)) with a magnitude of 0.6 eV, which agrees with the experimental value. Figure 2.7(c) shows the SE and DW contributions to the band-gap renormalization: The SE term dominates at low temperature, but the DW term becomes more important as the temperature increases. Figure 2.7(d) shows that absorption-band edge broadening becomes more prominent as the temperature rises, which is in good agreement with the experimental results. The energy-level broadening due to zero-point vibrations was ~0.2 eV. The full-quantum-effect correction to the electron structure of diamond is covered in more detail in subsection 8.4.1.

Nuclear quantum effects (NQEs) (including zero-point vibrations) on electron energy-band renormalization can also be observed in many materials besides diamond. For example, K. Ishii et al. recently calculated the temperature dependence of the band gap for different semiconductor materials (Figure 2.8) [Ishii2021]. Among the tetrahedral

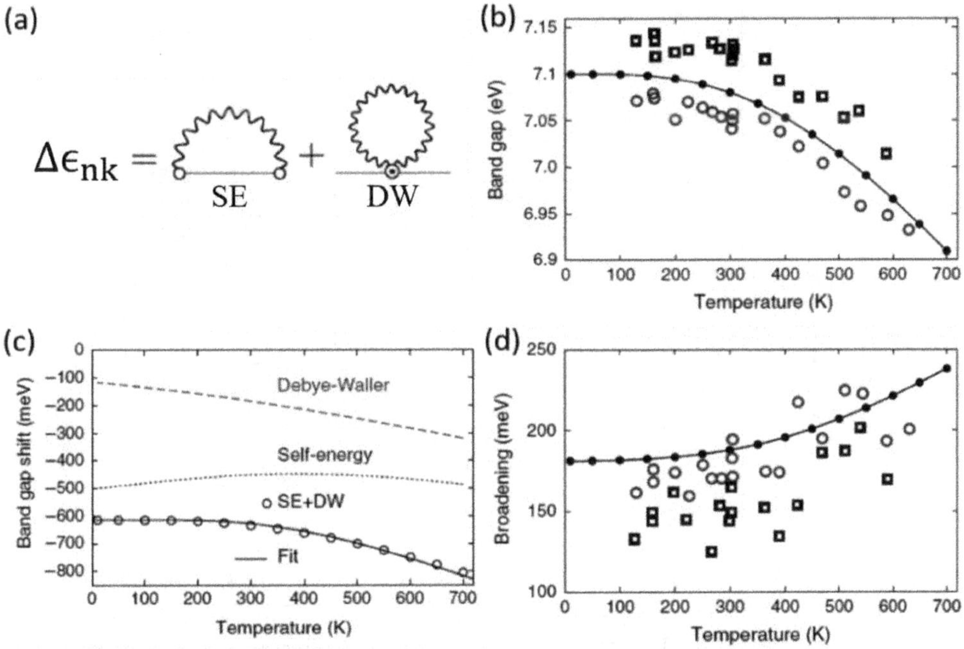

Figure 2.7 (a) Feynman diagram of electron–phonon interaction process. Solid and curved lines represent the electron and phonon Green's functions, respectively. (b) Temperature dependence of the calculated direct band gap (black curve) of diamond; the circular and square dots are the experimental data obtained by different methods described in [Logothetidis1992]. (c) Band-gap renormalization correction as a function of temperature, reflecting the contributions of different physical processes. (d) Temperature-dependent broadening of the band-edge energy levels with the same notation as (b). Reprinted figure with permission from Giustino[2010]. Copyright 2024 by the American Physical Society. (SE: self-energy; DW: Debye–Waller.)

semiconductor materials discussed in the paper, diamond had the most significant zero-point vibrational renormalization correction, followed by BN, SiC, BP, Si, and Ge. This indicates that the magnitude of the renormalization effect is inversely proportional to the mass of the element; as is typical of NQEs, lighter elements exhibit greater renormalization corrections to the electronic structure. Therefore, it is expected that for materials containing lighter elements, the electronic structure is more significantly affected by the renormalization of the full quantum effects.

In addition to the total renormalization effect, more interesting differences in physical effects can be found if the contributions of the two terms in Eq. (2.1) are comparable to each other. In the materials composed of relatively lighter elements, C and BN, the first-order process by single-phonon-induced electron scattering (FM term in Figure 2.8) is significantly more important, while the second-order DW process by two phonons is relatively less important. For materials with heavy elements, the second-order DW process dominates. In BP, the contributions of first-order and second-order processes are approximately the same. In SiC, Si, and Ge, the electronic structure renormalization is completely

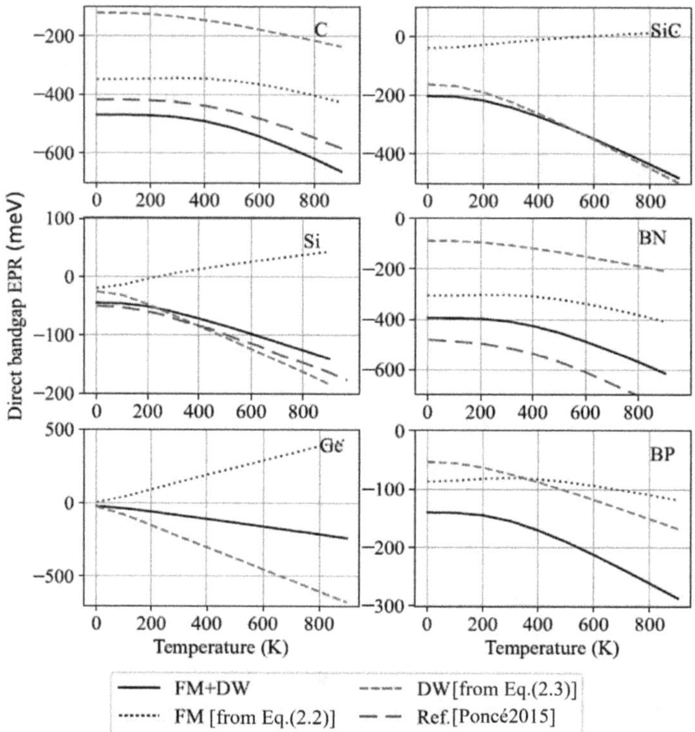

Figure 2.8 Temperature dependence of the direct band gap for a variety of semiconductors. The FM term represents the first-order phonon process; DW is a second-order phonon process. Figure created by Ishii[2021] licensed under CC BY 4.0.

dominated by the second-order phonon process. In particular, the first-order process in Ge behaves in the opposite way from the second-order process; this causes the two terms to cancel out, weakening the total renormalization effect. Therefore, although electronic structure renormalization corrections may not be evident in some materials, this does not imply that NQEs are unimportant, a point constantly emphasized in this book.

2.1.3 Thermodynamics

The thermal properties of condensed matter involve many fundamental observables, for example, the specific heat, thermal expansion coefficient, and thermal conductivity. Some of the most evident macroscopic phenomena caused by heat are the phase transitions among the solid, liquid, and gaseous states of matter. Many of the phenomena associated with light-element phase transitions are related to NQEs. The temperature-induced various solid–liquid phase transitions of hydrogen and helium, as well as the production of metallic hydrogen at high pressure, are all good examples (see the discussion in subsection 2.1.1.). In recent years, NQEs on thermal conductivity have been explored in many condensed matter systems (especially in light-element materials).

Experimental evidence shows that NQEs can significantly affect the thermal expansion properties of solids. The thermal expansion coefficient can be calculated from the equation of state of the system. The free energy of a crystal can be written as

$$F = U - kT\ln Z = U + \sum_i \left[\frac{1}{2}\hbar\omega_i + kT\ln\left(1-\exp\left(\frac{-\hbar\omega_i}{kT}\right)\right)\right], \tag{2.4}$$

where U is the ground state energy of the crystal. The second term is the free energy of the lattice vibration in the harmonic approximation. The pressure is given by

$$P = -\frac{\partial}{\partial V}\left(U + \sum_i \frac{1}{2}\hbar\omega_i\right) - \sum_i \left(\frac{1}{\exp\left(\frac{\hbar\omega_i}{kT}\right)-1}\right)\frac{\partial}{\partial V}(\hbar\omega_i). \tag{2.5}$$

Taking anharmonic effects into account, the equilibrium distance between atoms will be affected by the temperature, and the vibrational frequency will depend on the change in crystal volume. Considering Eq. (2.5), E. Grüneisen introduced the parameter $\gamma = -\frac{d\ln\omega_i}{d\ln V}$, which takes a positive value (between 1 and 2) for most crystalline materials. As the temperature increases at constant pressure, the vibrational energy E increases and the crystal expands. Using the Grüneisen parameter γ, the thermal expansion coefficient of a crystal can be expressed as

$$\alpha = \frac{\gamma}{B}\frac{d}{dT}\left(\frac{E}{V}\right) = \frac{\gamma}{B}C_v\ , \tag{2.6}$$

This is called the Grüneisen relationship, where B is the isothermal bulk modulus. The relationship suggests that the thermal expansion coefficient is proportional to the specific heat per unit volume C_v.

However, at low temperatures, negative thermal expansion coefficients have been observed in germanium, silicon, arsenic, and other element materials. The thermal expansion coefficient of silicon as a function of temperature is plotted in Figure 2.9. When the temperature is between 10 K and 125 K, α is negative. The origin of this behavior is a long-standing puzzle. The authors of early studies based on a quasi-harmonic approximation in the framework of perturbation theory believed that low-energy phonons decrease in frequency with volume contraction. Recently, Kim et al. provided a new explanation based on a fully anharmonic study including NQEs [Kim D.S.2018]: The atoms undergo an additional vibration because of the presence of zero-point energy at low temperatures. This zero-point vibration, caused by NQEs, interacts with higher-energy phonon modes, and consequently affects the thermal-expansion properties of the material.

The zero-point energy of anharmonic modes is a factor that must be considered in explaining the negative thermal expansion coefficient. NQEs can result in anharmonic coupling between all phonon modes, including those that are not excited at low temperature. This changes the self-energy of the excited low-energy phonons, altering the relationship between the free energy and the crystal volume. Figure 2.9 compares the thermal expansion coefficient of single-crystal silicon obtained by anharmonic stochastic time-dependent

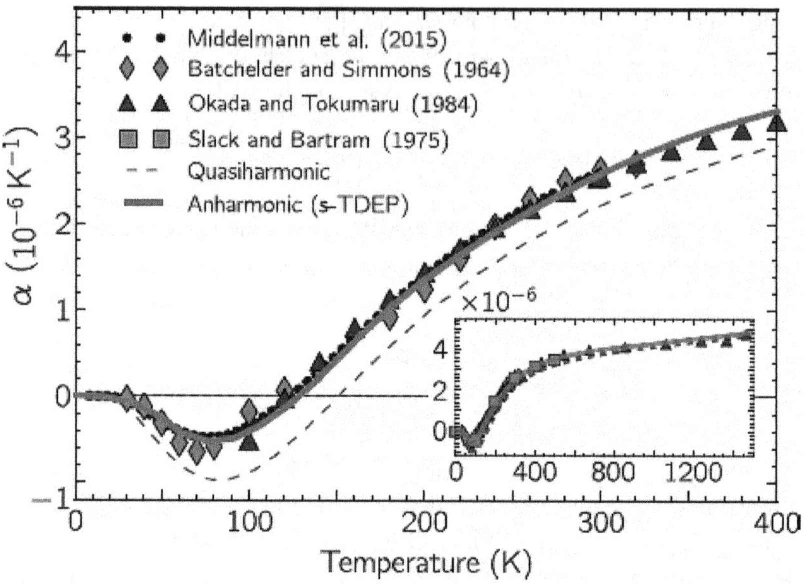

Figure 2.9 Calculated and experimental thermal coefficients of silicon as a function of temperature. The dashed curve represents the results of the quasi-harmonic approximation, and the solid curve corresponds to the anharmonic calculation; the latter is in good agreement with experiment. Reprinted from Kim[2018].

effective potential (s-TDEP) calculations with the experimental values, taking NQEs into consideration. The calculated coefficients are in good agreement with experiment. Although the quasi-harmonic approximation (the dashed curve in the figure) also reflects the anomaly of the coefficient, its prediction clearly deviates from the experimental results. Note that diamond, too, exhibits a negative thermal expansion coefficient; the anomaly in diamond is more prominent than in silicon because of lighter carbon nucleus.

It is worth noting that NQEs affect thermal properties not only at low temperatures but also at high temperatures near the material's melting point.

In addition, it has been recently shown that NQEs can influence the thermal conductivity of solid materials. As shown in Figure 2.10(a), single crystal $BaTiS_3$, a hexagonal single-crystal perovskite chalcogenide, has a glass-like ultralow thermal conductivity that decreases as a function of temperature. Because $BaTiS_3$ has a perfect crystal structure and low defect density, conventional perturbation theory and phonon scattering caused by crystal defects are insufficient to explain the conductivity decrease. It is possible that this anomaly is associated with full quantum effects. Neutron diffraction experiments have shown that with the decrease in temperature, the large displacement of titanium atoms is caused by quantum tunneling between two corresponding crystal configurations; these two structures' shallow potential wells connect to form a W-shaped potential well (as shown in Figure 2.10(b)). Inelastic neutron and X-ray scattering experiments have confirmed the existence of this W-shaped potential well with a quantum tunneling frequency lower than THz (as shown in Figure 2.10(c)). The tunneling process caused by the displacement of

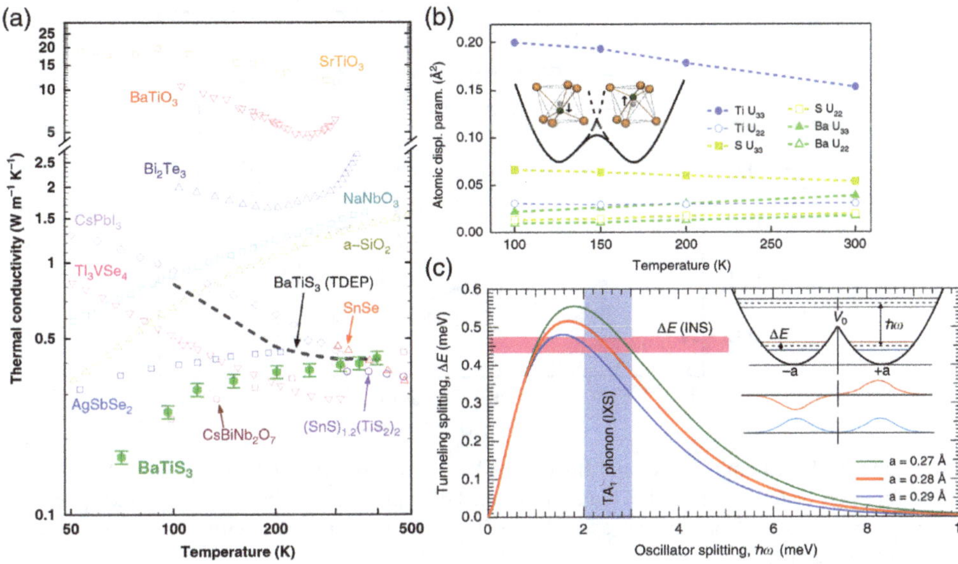

Figure 2.10 Hexagonal single-crystal $BaTiS_3$ with glass-like ultralow thermal conductivity, induced by nuclear quantum tunneling. (a) Measured thermal conductivity versus temperature for $BaTiS_3$ (green solid dots). (b) Atomic displacement versus temperature. The inset shows the structural model of titanium atoms in shallow potential wells. (c) Nuclear-quantum-tunneling splitting energy versus phonon energy for different double-well potential models. By comparing the tunneling splitting energy (red region) with the TA_1 phonon energy (blue region), the potential-well separation distance was calculated to be 0.28 Å. Figure created by Sun B.[2020] licensed under CC BY 4.0.

titanium atoms can effectively scatter heat-carrying phonons, thereby significantly reducing the thermal conductivity [Sun B.2020]. This study is also the first to experimentally discover that atoms with larger mass (such as titanium atoms) can undergo high-frequency quantum tunneling in the sub-terahertz range under the influence of NQEs. Therefore, it successfully explains the ultralow thermal conductivity phenomenon of high-quality single crystal $BaTiS_3$ under conditions far higher than the low-temperature region, and the new mechanism that produces this abnormal thermal insulation phenomenon. The glass-like ultralow thermal conductivity at a low temperature is interesting in thermoelectrics, especially for defect-free crystals.

Other examples of how full quantum effects change the thermal properties of materials include the phase transition between liquid and solid water and the thermodynamic stability of ice structures [Cheng B.2019]. Despite the simple chemical structure of water, the thermodynamic properties of its liquid and solid phases have always been a contentious research topic. The main challenge (largely disregarded until a few years ago) lies in how to describe the NQEs accurately. In general, when lighter atoms are replaced by heavier atoms of the same element, the volume of the material decreases. However, replacing the hydrogen in water with deuterium results in a greater ice volume. Another peculiar aspect of this effect is that it becomes stronger as the temperature rises. This anomaly was recently

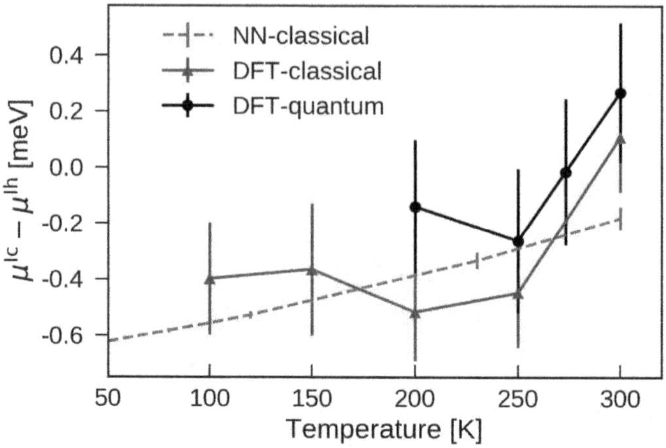

Figure 2.11 Chemical-potential difference between cubic (I_c) and hexagonal (I_h) ice as a function of temperature. DFT-classical and DFT-quantum indicate classical and quantum simulations of nuclei, respectively. DFT, density functional theory. Reprinted from Cheng B.[2019].

investigated by B. Pamuk et al., who have included NQEs in their theory [Pamuk2012, Ganeshan2013].

Another complication in water structure and phase-transition theory is the stability of cubic and hexagonal ice. The presence of zero-point entropy and proton disorder, as well as the small energy differences between the two ice phases caused by harmonic vibrations below 1 meV/H_2O, make it difficult to determine which structure is more stable under given conditions. Therefore, NQEs that include anharmonic motion are critical to understanding this issue. Figure 2.11 shows the chemical potentials of two ice phases (cubic and hexagonal). The "NN (Neural Network) -classical" line in the figure corresponds to the chemical-potential difference when the nucleus is treated as a classical particle. In the classical case, the chemical potential of the cubic ice is lower at low temperature, indicating higher stability. However, after considering NQEs (by comparing the "DFT-classical" and "DFT-quantum" lines), it was found that the hexagonal phase becomes more stable near room temperature (300 K). Here "DFT-classical" and "DFT-quantum" indicate classical and quantum simulations of nuclei, respectively. If the heat of the phase transition from cubic to hexagonal ice is constant for temperatures in the 200–300 K range, the formation enthalpy of the phase transition is 1.0 ± 0.5 meV/H_2O (based on the chemical potential difference); this is close to the experimental result of 0.1–1.7 meV/H_2O [Carr2014].

Figure 2.12 compares calculated and observed values of the chemical-potential difference between liquid water and hexagonal ice [Cheng B.2019]. Table 2.1 summarizes the melting temperature (T_m) and the melting heat (H_f) of ice. After including NQEs, the calculated T_m is only 2% off the experimental value. Comparing quantum and classical process, we find that NQEs reduce the T_m of H_2O by 8 K. If H is replaced by its isotope D, the melting temperature T_m increases by (7 ± 2) K, which is close to the results obtained by other theoretical methods. Summarizing the above discussion, in light of these results, we see that the calculated thermodynamic entropy and enthalpy are in better agreement with

Table 2.1 *Comparison of the calculated melting temperature (T_m) and melting heat (H_f) of the solid–liquid phase transition of water with the experimental values. The figures in parentheses represent statistical errors in the last digit.*

Model	Melting temperature (K)	Melting heat (meV/H_2O)
NN-classical	279.6(4)	67.8(2)
DFT-classical	275(2)	58(2)
DFT-H_2O quantum	267(2)	52(3)
DFT-D_2O quantum	274(2)	58(2)
experiment(H_2O)	273.15	62.3
experiment(D_2O)	276.97	64.5

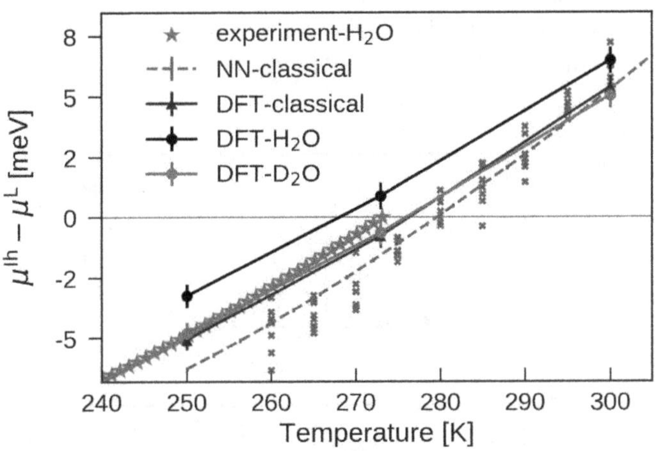

Figure 2.12 Chemical-potential difference between liquid water and hexagonal ice as a function of temperature. The results are calculated by different methods. The DFT-H_2O and DFT-D_2O indicate quantum simulations of nuclei, while NN-classical and DFT-classical indicate classical simulations of nuclei. Neural Network (NN) and density functional theory (DFT) are used to model the Born–Oppenheimer potential energy surface. The experimental data (experiment-H_2O) are labeled by light grey pentagram stars. Reprinted from Cheng B.[2019].

the experimental results after considering NQEs. In addition, the simulations that include NQEs also result in better agreement between the calculated structural parameters and the experimental values. This suggests that NQEs can have a significant impact on the thermodynamic properties of both liquid and solid phases of water.

2.1.4 Optical Properties

As discussed in Sections 2.1.1–2.1.3, full quantum effects can significantly influence both the atomic and electronic structures of condensed matter systems. Therefore, it is reasonable to assume that they can also influence the optical properties of these systems. In this section, several typical examples of NQEs on the optical properties of molecules and condensed matter are discussed. These examples involve materials such as hydrazine, lithium clusters, diamond-like clusters, and diamond.

In general, the peak width of the optical absorption spectrum increases when NQEs are considered. The hydrazine molecule, H_2N-NH_2, provides an example. Its optical properties were investigated by A. Kaczmarek et al. in 2009 using the ring-polymer molecular dynamics (RPMD) method, an improved version of PIMD that describes the statistical properties and relevant dynamical properties of the condensed matter system [Kaczmarek2009]. When taking NQEs into account, the calculated light absorption peak was found to be broader; this can be explained by the quantum fluctuation of chemical bond length (Figure 2.13). Kaczmarek et al. noticed that the hydrazine spectra measured by J. A. Syage et al. [Syage1992] had three major broad peaks at 195, 171, and 130 nm, and three weak shoulders at 218, 178, and 150 nm (Figure 2.13(c)). Overall, the RPMD calculations (quantum) had better agreement with the experimental data than calculations (classical) that did not include NQEs.

Figure 2.14 shows the single-photon nonresonant ionization spectrum, which can be measured using a method similar to that used for the deep UV spectrum. The experimental

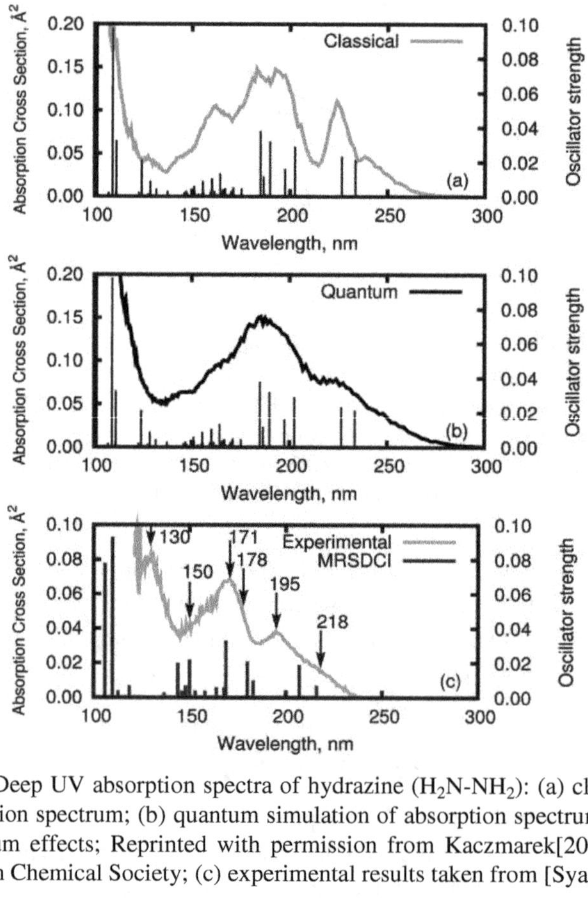

Figure 2.13 Deep UV absorption spectra of hydrazine (H_2N-NH_2): (a) classical simulation of absorption spectrum; (b) quantum simulation of absorption spectrum that includes nuclear quantum effects; Reprinted with permission from Kaczmarek[2009]. Copyright 2024 American Chemical Society; (c) experimental results taken from [Syage1992].

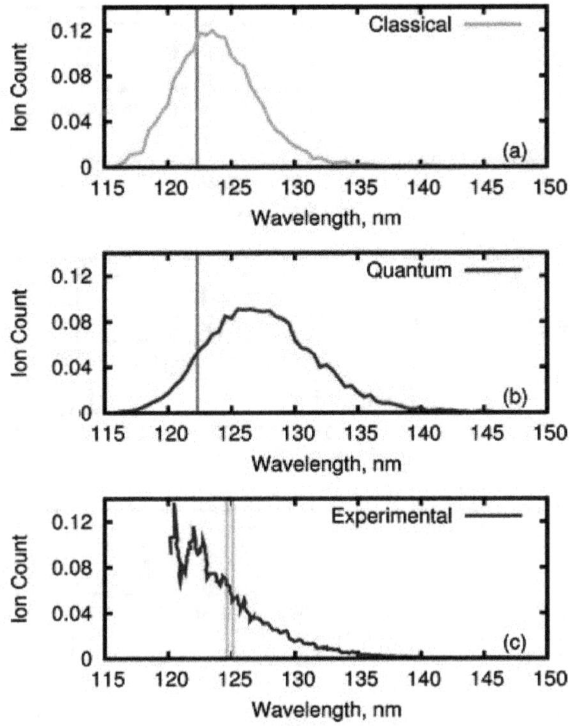

Figure 2.14 Single-photon ionization efficiency of hydrazine molecules. (a) Simulated absorption spectrum using the classical approximation. (b) Simulated absorption spectrum including nuclear quantum effects. (c) Experimental results taken from [Syage1992], where the two vertical grey lines indicate the first ionization potentials given by the experiment. Reprinted with permission from Kaczmarek[2009]. Copyright 2024 American Chemical Society.

ionization energy has been reported to be 124.6 nm or 125.1 nm. Comparing the ionization energy spectra calculated with and without NQEs, we see that, while the quantum calculations show a redshift with respect to the experimental value, the slope at low frequencies and the curve pattern (indicated by the vertical lines of the experimental data) are consistent with observation. This phenomenon can be explained by the fact that the electronic structure of the ground state hydrazine molecule corresponds to the structure of the dissociated molecule with distortion. The ionization energy of the ionized molecular structure is smaller than the ionization energy of the static equilibrium structure, so the ionized molecular structure is more energetically stable. The calculations considering NQEs show a shift in the average ionization energy relative to the equilibrium values of molecular structures. This shift is caused by quantum properties of the nucleus and cannot be obtained by treating the nuclei classically.

Figure 2.15 shows the infrared (IR) spectrum of the hydrazine molecule in the gas phase at 300 K. In the classical simulation (Figure 2.15(a)), the signal with maximum intensity occurs at 966 cm^{-1}, which corresponds to the asymmetric NH_2 wagging oscillation.

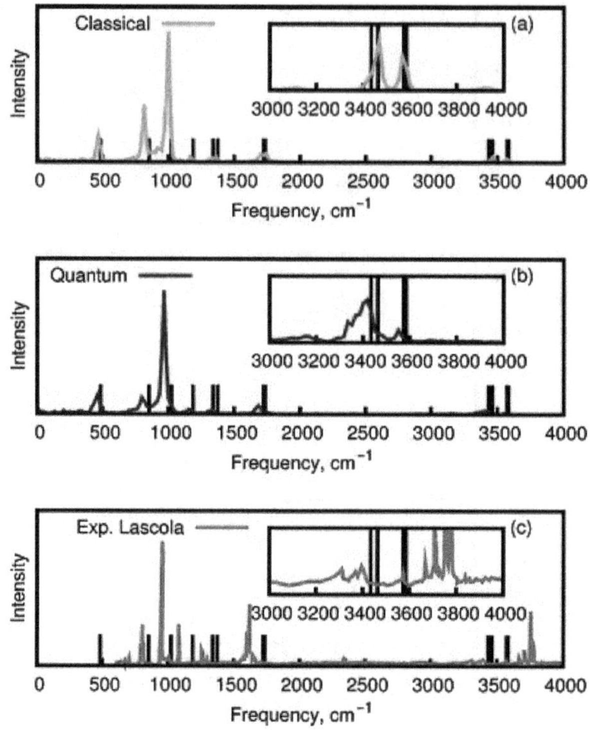

Figure 2.15 Infrared absorption spectra of hydrazine molecules. (a) Calculated absorption spectra using the classical treatment of nuclei. (b) Calculated absorption spectra using the quantum treatment of nuclei. (c) Experimental results taken from [Lascola1988]. Reprinted with permission from Kaczmarek[2009]. Copyright 2024 American Chemical Society.

Including the anharmonic vibrations causes a redshift in this peak, with a magnitude ranging from several wavenumbers up to 40 cm^{-1}. Taking NQEs into account, the vibrational frequencies exhibit an additional redshift of up to 50 cm^{-1}. When compared to other vibrational modes, the N–H stretching mode was observed to differ the most from the experimental results. Further, the use of the RPMD method incorporating NQEs is a significant improvement over classical molecular dynamics. The calculated deep-UV absorption, ionization, and IR spectra are only consistent with experiments when NQEs are taken into account. This corroborates the importance of NQEs in understanding the deep-UV continuum in the absorption spectrum, the tail excitation of the single-photon ionization spectrum, and the anharmonic IR spectral shift.

NQEs also affect the optical properties of lithium clusters (e.g., Li_8) [Sala2004]. Due to recent advances in experimental technique, in-depth study of the dynamics and photoexcitation of metal clusters at temperatures below 50 K is now possible, providing more systematic insights into the relationship between nuclear quantum behavior and light absorption. F. Della Sala et al. computed the optical absorption spectra of Li_8 clusters at 10 K and 50 K using the path-integral molecular-dynamics method (Figure 2.16). Their simulation utilized a molecular configuration with C_{3v} symmetry, where one Li atom is surrounded by seven others; this leads to a "core/shell" structure (see Figure 8.9(f)). The theoretical optical absorption spectrum is characterized by two main peaks centered at 2.65 eV and 2.85 eV, respectively. A weaker resonance peak can be observed at 2.3 eV. Further comparison of the classical to the quantum absorption spectrum reveals many differences. In the first place, the classical absorption spectrum varies significantly with temperature, whereas the quantum spectrum is less temperature dependent. Secondly, at temperatures of 10 K and 50 K, the quantum fluctuations wash out the spectral characteristics of bimodal peaks, leading to a broad, featureless, and unimodal spectrum. This phenomenon cannot occur without considering NQEs. In particular, at 50 K, quantum fluctuations lead to a slower decay of the main peak in the high-energy regime, whereas the curve evidently increases on the low-energy side (see Figure 2.16(c)). The experimental results show a similar featureless broad unimodal peak, with a slower peak decay on the high-energy side (although the position of the measured peak is located at an energy that is lower than the calculated value). The calculated quantum absorption spectrum at 50 K comes closest to the experimental data in terms of general spectral characteristics.

To explore NQEs and the limits of classical simulations further, Della Sala et al. modified the classical approximation and increased the Lorentzian broadening factors in order

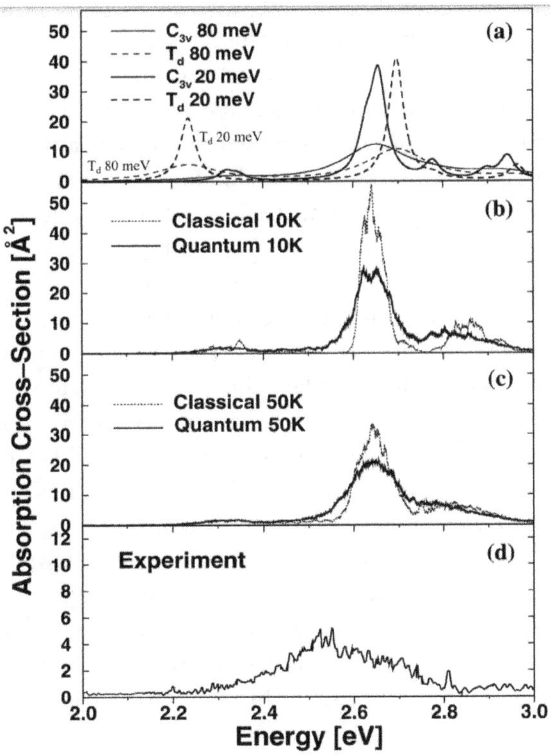

Figure 2.16 Absorption spectra of Li_8 clusters calculated by classical and quantum methods Reprinted figure with permission from Della Sala[2004]. Copyright 2024 by the American Physical Society. The experimental data were taken from [Blanc1992].

to simulate the experimentally observed broad peak. However, this did not produce a spectrum that matches the experimental result in terms of peak position, shape, and intensity. Closer analysis reveals that, because of quantum fluctuations of the Li-Li bond length, the bond lengths in the core and shell layers overlap significantly. By contrast, when only classical thermal fluctuations are considered, there is a clear gap in the bond length between the two groups of bonds. Therefore, the potential energy profile predicted by including NQEs can exhibit interatomic attraction over a wide interval, which is clearly reflected in the absorption spectrum [Della Sala2004]. The study of lithium clusters thus once again substantiates the importance of NQEs (zero-point vibrations) for interpreting experimental observation of low-temperature optical spectra of light-element systems.

NQEs influence the spectra not only of materials composed of lighter elements as discussed above, but also of carbon-containing systems. In 2013, F. Giustino et al. from the University of Oxford (currently at the University of Texas, Austin) investigated NQEs on light-absorption properties of clusters of diamondoid, a class of nanomaterials with diamond-like structure [Patrick2013]. Adamantane ($C_{10}H_{16}$) is the smallest diamondoid, and other diamondoids can be regarded as compounds made of adamantane cage structures. Diamondoids are hard, biocompatible, and they are specially characterized by emitting monochromatic light, making them suitable for use as fluorescent markers and single-photon sources. In fundamental research, diamondoids are ideal systems for testing theories of the electronic structure of carbon nanomaterials, especially the interactions between electronic and nuclear dynamics. The highest occupied states of diamondoid are localized on C–C bonds; their energies are sensitive to the variation of bond lengths arising from nuclear motion. On the other hand, the lowest unoccupied states are distributed on the diamondoid surface and are also sensitive to the collective vibration of adsorbed hydrogen atoms.

If the quantum characteristics of the nucleus are not taken into account and the nuclei are fixed at their equilibrium positions, the absorption spectra calculated by density functional theory (DFT) and the more accurate time-dependent DFT are essentially identical and are indicated by the grey thin lines shown in Figure 2.17(a). The calculated absorption peaks are sharp, whereas the experimental spectra are broad and featureless. This noticeable discrepancy suggests that standard computational methods, which ignore NQEs and assume the atoms are at rest in the lattice, do not reflect the actual situation. To take NQEs into account, Patrick et al. adopted the semiclassical Franck–Condon approximation [Lax1952] to calculate the absorption cross section. However, this approach still used the adiabatic approximation, which separates the electron motion from the nuclear motion, and the absorption cross section was expressed as a weighted average of the final nuclear states based on Monte Carlo sampling. The absorption spectra calculated using this method (with quantum treatment of nuclei) reproduce the experimental spectra with excellent accuracy at the electron-volt scale (see the grey-shaded regions of Figure 2.17(b)). The results demonstrate that NQEs induce a spectral line broadening of up to 1 eV and a redshift of the absorption threshold as large as 0.6 eV.

It is worth noting that a semiclassical approach cannot explain sharp peaks of adamantane and diamantane near the spectrum edge because more accurate calculations are required to account for finer selection rules due to nuclear quantum vibrations. To address

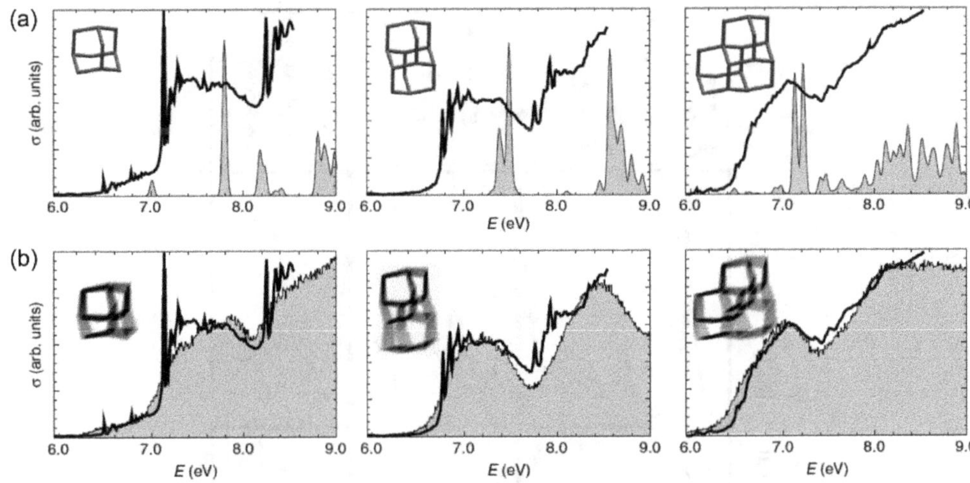

Figure 2.17 Absorption spectra of adamantane, diamantane, and triamantane (left to right). (a) The light grey area shows the spectra calculated with nuclei fixed at their static equilibrium positions. The corresponding atomic-structure model is provided in the inset. (b) The light grey area shows the calculated spectra after considering nuclear quantum effects; the fuzzy atomic configuration in the inset represents the quantum nature of the nuclei. The black curves correspond to the experimental results. From Patrick[2013]. Licensed under CC BY 4.0.

this issue, Giustino et al. included the coupling contributions of nuclear vibrational quantum states based on the Born–Huang expansion in their calculation of the electric dipolar transition. The calculated absorption spectra are clearly in better agreement with the fine structure observed experimentally (Figure 2.18) and provide more insight into the physical nature of several of the characteristic absorption peaks. For example, the 149.4 meV vibrational mode corresponds to the A_1 stretching mode of the C atom in the cage, and the two satellite peaks correspond to the A_1 deformation modes of the C–C–C bridge.

Nuclear quantum vibrations can also affect the optical properties of diamond. Taking a first-principles approach, Spanish and Italian researchers have collaboratively explained the effect of zero-point vibrations on the optical spectrum of diamond in terms of electron–phonon coupling [Cannuccia2011]. Because of NQEs, the vibrational ground state of an atom in the absolute zero-temperature limit still has a non-vanishing zero-point vibrational energy; this results in electronic structure renormalization via electron–phonon coupling. The coherent wave packet (electron–phonon entangled state) formed by this coupling is called a "polaron." A condensed matter system may be conventionally described by a continuous system of polarons, and its optical-absorption spectrum may be obtained by solving the spectral function of this polaronic system. Studies using this quantum dynamical approach have shown that full-quantum effect affects optical properties and the dynamics of electronic structure. Figure 2.19 shows the experimental and theoretical curves of the second derivatives of the dielectric function of diamond (convolution of the spectral function of the conduction band and the valence band); the spectral function of the conduction band is shown in the inset. The results of the quasiparticle approximation (dashed curve)

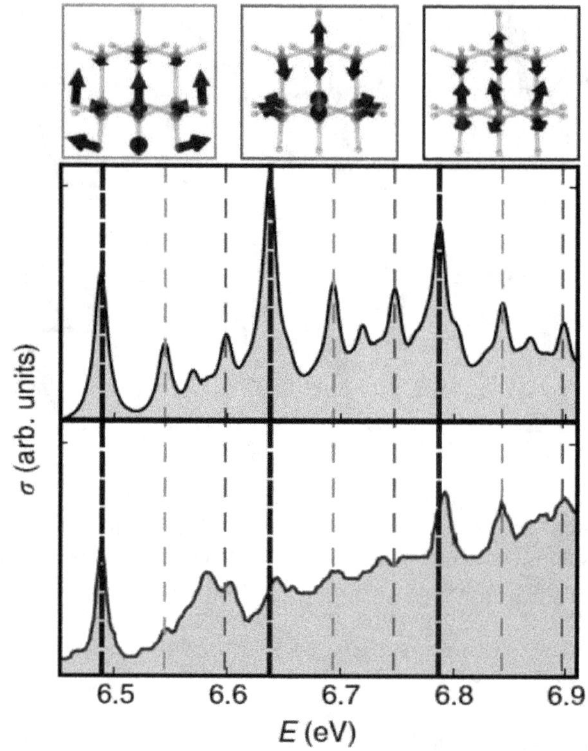

Figure 2.18 Fine structure in the optical spectrum of adamantane near the onset of absorption. The solid line in the lower panel is the experimental spectrum [Landt2009] and the solid line in the center panel is based on the Born–Huang expansion. The vertical dashed lines correspond to the vibrational modes discussed in the text. The thick black dashed lines represent the A_1 mode with energy 149.4 meV. From Patrick[2013]. Licensed under CC BY 4.0.

were compared with quantum dynamical theory (solid curve), and some important features that reflect the dynamical effects of the electron–phonon coupling were observed in the spectral lines. The theoretical results including NQEs agreed well with the experiments [Logothetidis1992], whereas the conventional harmonic and quasi-harmonic approximations based on the static structure failed to explain the sub-bandgap (<7.18 eV) absorption peaks at 6.71 eV and 6.94 eV. This is a good indication that the full quantum effects induce strong dynamical effects on the optical spectrum of diamond.

In addition, Giustino et al. calculated the light-absorption coefficient of diamond at different temperatures [Zacharias2016]; a detailed discussion is provided in subsection 8.4.1, where we describe the optical properties of carbon materials. In brief, Giustino et al. compared calculations that included full quantum effects to methods that only considered the static nuclear configuration (completely ignoring the contribution of the nuclear quantum vibrations). They found that light-absorption coefficients were more accurately predicted at different photon energies when electron–phonon coupling was considered, and that the calculated absorption spectra were consistent with experimental observation.

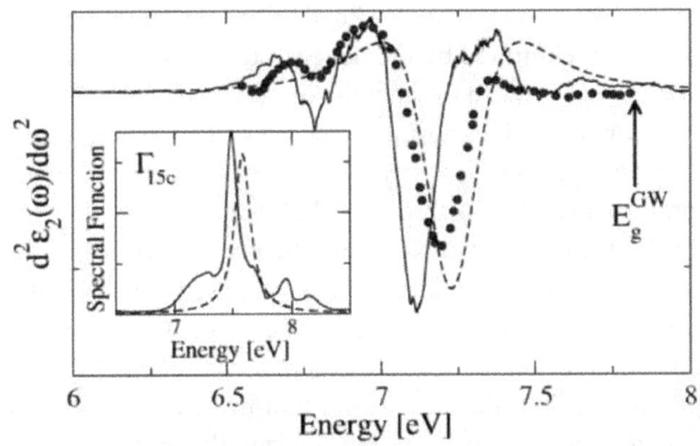

Figure 2.19 Experimental [Logothetidis1992] second derivative of the dielectric function of diamond (dots), compared with results calculated with the quasi-harmonic approximation (dashed line) and quantum dynamical theory (solid line). The inset shows the spectral function of the conduction band. Reprinted figure with permission from Cannuccia[2011]. Copyright 2024 by the American Physical Society.

This study demonstrated that the full quantum effects can reduce the threshold energy for light absorption from above 7 eV to below 5.5 eV – a significant change in a material's optical properties. These examples clearly demonstrate that not only the zero-point vibration renormalization correction of the bandgap but also that the phonon-assisted indirect transitions in semiconductors can be understood using a first-principles approach based on non-perturbative treatment of electron–phonon coupling. The reader can refer to subsection 5.2.2 for the full theoretical derivation.

2.1.5 Quantum Paraelectric Phenomenon

Ferroelectricity, and the related quantum paraelectricity, also provide good examples of the role of NQEs. Ferroelectric materials possess spontaneous polarization that is reversible under external fields at finite temperatures. The polarization of materials arises from the separation of positive and negative charge centers, which is attributed to the stable polarized structure of the crystal on the Born–Oppenheimer potential surface. Because polarization of ferroelectricity is due to the polarization characteristic of the stable structure of such crystals, ferroelectrics tend to exhibit long-range magnetic-ordered polarization when the temperature is low. At high temperatures, the polarization is overwhelmed by thermal fluctuations.

However, in the 1960s, people discovered a class of materials represented by strontium titanate ($SrTiO_3$) that exhibited the so-called quantum paraelectric phenomenon (quantum paraelectricity). This once again profoundly revealed the complexity of the microscopic quantum world. In many materials, as the temperature decreases and thermal fluctuations are suppressed, the dielectric constant increases, signaling the ferroelectric transition. However, this is not the case for a quantum paraelectric material: Even when the temperature approaches absolute zero, no ferroelectric transition is observed. This is because of the presence of strong quantum fluctuations, or more precisely, because of the quantum critical behavior exhibited by the degenerate states associated with lattice vibrations. Quantum fluctuations replace thermal fluctuations as the temperature drops, preventing the formation of ferroelectric long-range ordering. This is a temperature-independent NQE, that is, a paraelectric phenomenon caused by quantum control. Quantum paraelectric materials thus represent a vivid manifestation of NQEs.

In 1952, J. H. Barrett investigated NQEs on the paraelectric–ferroelectric transition at low temperatures. Even before this, J. C. Slater proposed a model of ferroelectricity in $BaTiO_3$ [Slater1950]. This model addressed the large deviations of barium atoms from their equilibrium positions observed at low temperatures; treating the barium atoms as simple harmonic oscillators, it introduced the quantum-mechanical treatment of polarization theory. Building upon Slater's theory, Barrett realized that not only the barium atoms but all the atoms must be treated quantum mechanically [Barrett1952]. In other words, the calculation must consider the behavior of barium, titanium, and oxygen nuclei as they vibrate around their equilibrium positions. This led to Barrett's famous formula

$$\chi = \frac{M}{(T_1/2)\coth\left(\frac{T_1}{2T}\right) - T_c}, \tag{2.7}$$

where χ is susceptibility, M is a constant, T_1 is a temperature related to the ground state energy of the quantum oscillator, and T_c is the quantum critical temperature. It can be seen that the NQEs vanish (i.e., the Barrett formula reduces to the classical Curie–Weiss formula) when temperature $T \gg T_1$. By fitting the results from earlier experiments, Barrett predicted that the polarization rate of strontium titanate would plateau at low temperatures.

As low-temperature experimental techniques advanced, experimental physics moved into the liquid-helium temperature regime, allowing for more precise measurements of physical phenomena. One of the most famous examples was the experimental study of the dielectric properties of strontium titanate by Swiss scientists K. A. Müller and H. Burkard in 1979 (Figure 2.20). As the temperature decreased, the dielectric constant of the material increased rapidly, which is consistent with the classical Curie–Weiss law. However, when the temperature fell below 10 K, the dielectric constant no longer varied with temperature. This trend continued at least until 0.3 K. This experiment demonstrated the existence of the quantum paraelectricity predicted by Barrett. These works advanced the development of low-temperature physics experiments, and in 1987 K. A. Müller was awarded the Nobel Prize in Physics, along with J.G. Bednorz, for the discovery of high-temperature superconductivity.

With the development of computational methods such as the pseudopotential techniques in the 1990s, first-principles calculations were widely used in simulations of physical properties of real materials. At the theoretical level, instead of resorting to empirical or semi-empirical modeling, researchers began to understand quantum paraelectricity on a new level. Vanderbilt et al. developed a first-principles effective-Hamiltonian method to describe the atomic structure and polarization of ferroelectric materials at finite temperatures [Zhong W.1995]. This unique approach, which marked a new direction in the study of NQEs in strontium titanate,

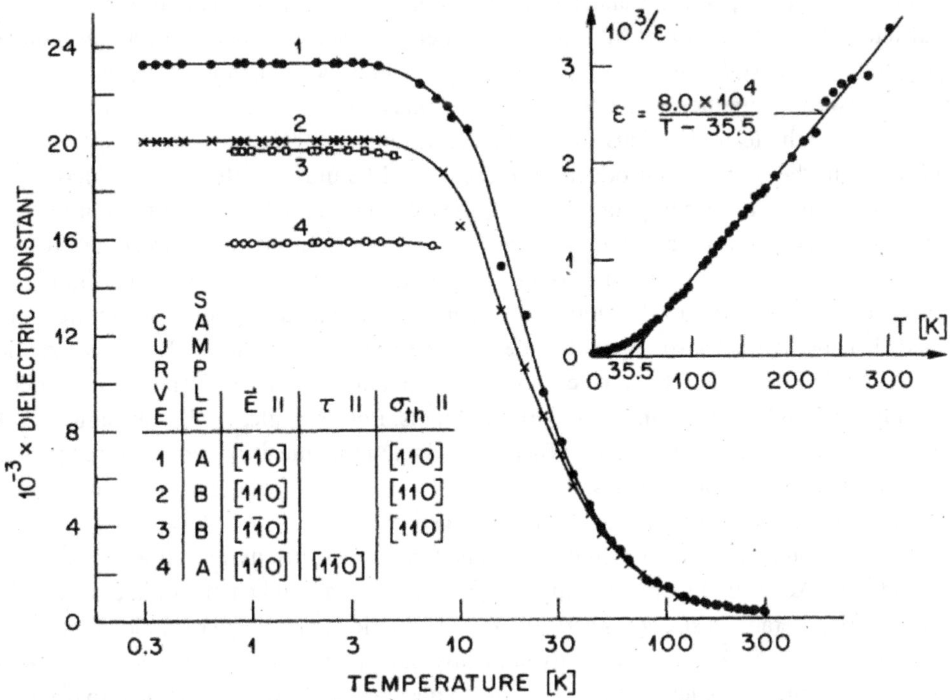

Figure 2.20 Dielectric constant of $SrTiO_3$ as a function of temperature. In the temperature range 0.3–10 K, the dielectric constant does not vary with temperature, revealing quantum paraelectricity caused by nuclear quantum effects. The inset shows the classical Curie–Weiss fit, which agrees with the experiments at high temperatures but deviates significantly at low temperatures. Reprinted figure with permission from Müller[1979]. Copyright 2024 by the American Physical Society.

no longer assumed the traditional view of atoms. Instead, it examined the system from the perspective of phonon quasiparticles, reflecting the collective atomic motion. Vanderbilt and coworkers believed that the quantum effects associated with the uncertainty principle are not only influenced by the nuclear wavefunction but also by the uncertainty of the corresponding particle positions in space. Consequently, the zero-point energy satisfies

$$\Delta E \geq \frac{\hbar^2}{8m\left(\Delta q\right)^2} \sim k_B T_{QM} \tag{2.8}$$

where m is the nuclear mass, Δq denotes the uncertainty of the spatial coordinates, and T_{QM} is the characteristic temperature of the quantum effects. The ferroelectric transition is induced by the softening of the ferroelectric phonon. From the quasiparticle point of view, the uncertainty in spatial coordinates arises from the degeneracy of opposite polarization states. Therefore, Δq should be taken as the difference of phonon coordinates in different polarization states. Although the various atomic masses m in strontium titanate are large, resulting in a large effective phonon mass, the small q value makes the NQEs more pronounced than expected. Vanderbilt et al. combined the model Hamiltonian with the path-integral Monte Carlo (PIMC) method to perform calculations on strontium titanate. They found that NQEs not only quantitatively change the temperature of the structural phase transition from the cubic to the tetragonal phase (by about 35–50 K), but also qualitatively affect the ferroelectricity of the system [Zhong W.1996].

Similar to the previous introduction of the glass-like ultralow thermal conductivity of the chalcogenide sulfur compound BaTiS3, the quantum paraelectricity of strontium titanate shows once again that NQEs are not exclusive to light elements. In some physical processes involving coordinate uncertainty, quantum fluctuations may be quite significant. Moreover, it is possible to develop new quantum materials at higher temperatures, provided that the characteristic spatial scales can be tuned and reduced. Extensive studies [Müller1979, Zhong W.1996] have shown that quantum paraelectric phenomena occur not only in $SrTiO_3$ but also in other similar systems. For example, NQEs can significantly suppress spontaneous polarization in $BaTiO_3$, reducing the transition temperature by about 35–50 K. A more detailed discussion is provided in Chapter 8, where we introduce typical full quantum effects in systems containing oxygen.

Using full quantum effects to modulate quantum paraelectricity has attracted considerable interest. According to a new study, an applied external field can modulate the relative ferroelectric interaction strengths and quantum fluctuations, revealing the hidden ferroelectric phases. K. A. Nelson et al. have dynamically induced the hidden ferroelectric phase in undoped quantum-paraelectric $SrTiO_3$ using an ultrafast terahertz electric field to drive the directional ion displacements. They observed a significant change in the phonon-excitation spectrum after the external electric field was applied, demonstrating the existence of a new ferroelectric phase [Li X.2019].

Another approach involves the use of element doping to modulate the nuclear quantum fluctuations, allowing observation of their effects on material ferroelectricity. Several studies suggest that doping with elements such as Ba or Pb can weaken the quantum fluctuations in $SrTiO_3$, thus inducing the ferroelectric phase transition. In $Ba_xSr_{1-x}TiO_3$ (BSTO), changing the Ba concentration causes the material to transition from quantum paraelectric to quantum ferroelectric, and finally to a classical ferroelectric [Wu H.2006].

Isotope substitution is another way to modulate NQEs. It can have a significant effect on ferroelectricity and can be used to modify and optimize ferroelectric devices. The literature shows that increasing nuclear mass effectively suppresses quantum fluctuations [Kvyatkovskiĭ2001]. The influence of this isotope effect is especially pronounced in systems containing light element atoms. A related discussion can be found in Section 11.5.

In recent years, an approach similar to Vanderbilt's has been used by X. Z. Li's group at Peking University to study the ferroelectric–paraelectric transition [Ye Q.J.2018]. Building on this, X.F. Zhang et al. found that $BaFe_{12}O_{19}$ also shows quantum paraelectricity [Zhang X.F.2020]. Their results are presented in Figure 2.21. The same group also discovered that NQEs can induce a Goldstone-like quantum state in $BaMnO_3$ [Zhang X.F.2021].

Other topics closely related to the above discussion include quantum magnetism and quantum spin liquids, both of which are currently very active research topics in condensed matter physics. As the temperature approaches zero, quantum fluctuations dominate and

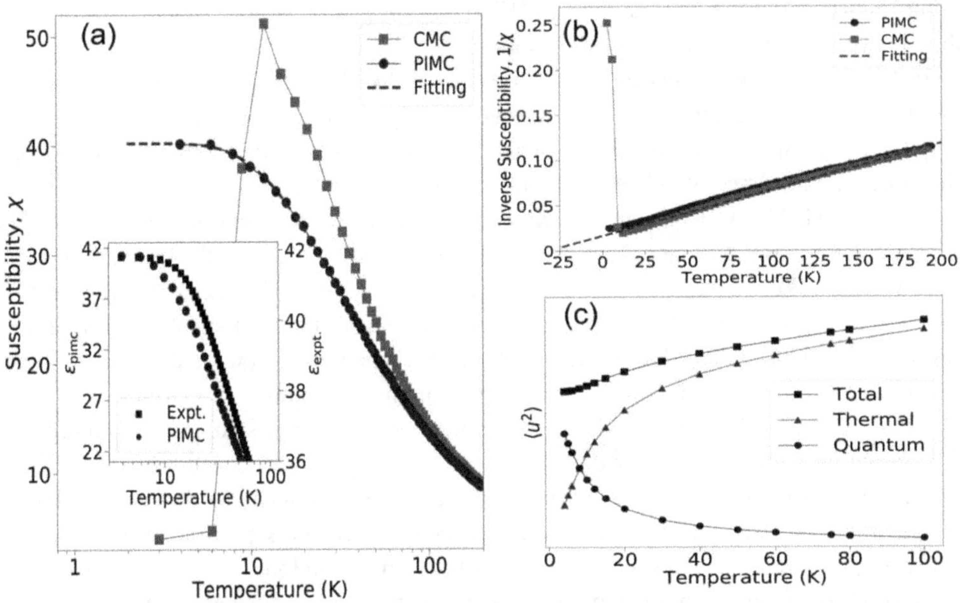

Figure 2.21 Quantum paraelectricity of $BaFe_{12}O_{19}$. (a) Susceptibility vs. temperature. The experimental data (black squares in the inset figure) were obtained by Y. Sun's group at the Institute of Physics, Chinese Academy of Sciences [Shen S.P.2016]. The grey squares are from a classical Monte Carlo (CMC) simulation that treated the nuclei classically; the black dots represent a path-integral Monte Carlo (PIMC) simulation that treated them quantum mechanically. The comparison shows that the low-temperature plateau is caused by nuclear quantum effects rather than by steric hindrance effects as previously believed. This is more evident in the inverse susceptibility shown in (b). Relationship between the inverse of the electrode polarization rate and temperature (c). Quantum fluctuations dominate at low temperatures. Reprinted figure with permission from Zhang X. F.[2020]. Copyright 2024 by the American Physical Society.

long-range magnetic ordering is suppressed. The corresponding condensed matter state is a quantum spin liquid with an infinite number of highly entangled superimposed spin configurations. Because the microscopic detail is difficult to understand and previous model-based theoretical studies often couple multiple degrees of freedom, it is difficult to explain all the experimental data satisfactorily. A first-principles many-body calculation method based on full quantum considerations has the potential to overcome such constraints. To realize quantum spin liquids experimentally and characterize them theoretically, it is essential to develop suitable preparation and control techniques for quantum magnetic materials.

2.1.6 Superfluids, Supersolids, and Quantum Defects

In the first section of this chapter, we discussed the crystal structure and phase transitions of helium. Superfluidity (first discovered in liquid ^{4}He) and supersolidity are important physical properties closely related to nuclear quantum properties. Theoretical work on superfluidity is already well-developed [Kleinert1989], so it is not discussed at length in this section. Our focus will be on supersolidity.

In 1939, M. Wolfke suggested that superfluidity-like behavior may also exist in solids [Wolfke1939]. Since the 1970s, the concept of supersolidity has received considerable attention. The theoretical studies conducted by A. J. Leggett and others inspired multiple searches for experimental evidence of supersolids.

In 2004, M. H. W. Chan and E. Kim et al. measured the rotational inertia of solid ^{4}He and reported experimental evidence for supersolidity [Kim E.2004a,b, 2008, Chan M.H.W.2008, West2009]. In 2007, J. Beamish et al. measured the shear modulus of solid ^{4}He and found its temperature dependence very similar to that of the rotational inertia of solid ^{4}He [Day2007]. This has led to debates about whether the torsional-oscillator signal is caused by the supersolid state or by changes in shear modulus. To explore this issue, M. H. W. Chan et al. redesigned their experiment with a harder sample cell and reported the absence of supersolid ^{4}He with an accuracy of 4 parts per million [Kim D.Y.2014]. Thereafter, supersolid-detection methods involving unidirectional mass flow received widespread attention. R. B. Hallock et al. placed a solid between two superfluid reservoirs. After applying pressure to one reservoir, they observed an anomalous mass flow in solid ^{4}He [Ray2011]. A similar phenomenon was also observed by J. Beamish and Z. G. Cheng et al. By mechanically squeezing solid helium in one direction, they observed a mass flow that could not have been caused by elastic effects [Cheng Z.G.2015, 2016]. They also conducted controlled experiments on solid ^{3}He and observed behavior different from that of solid ^{4}He. Their results support the existence of superfluidity in solid ^{4}He [Cheng Z.G.2018a].

Because shear-modulus measurements of solid ^{4}He may affect the interpretation of rotational inertia experiments [Kim E.2004a,b, Day2007], accurate determinations of defect motion and plastic deformation are necessary. In solid ^{4}He, the strong zero-point motion leads to a large wavefunction overlap between helium atoms. As noted by Leggett and other theorists, this is one of the reasons that full quantum effects must be taken into consideration [Andreev1969; Leggett1970]: The atoms no longer exhibit characteristics of classical particles and become indistinguishable. In other words, atomic exchange occurs frequently, making defects in solid helium highly mobile. Helium impurities and vacancies behave as quasi-particles, and dislocation lines can be regarded as free vibrating strings. In the elastic region, the energy loss due to defect motion is negligible; however, in the plastic region, the interactions between defects are enhanced and energy loss is evident. Therefore, studying the plastic-deformation process in a quantum solid with frequent particle exchange allows us to simulate the defect-evolution process of conventional materials in a short period of time, providing a reference for related research in materials science and physical metallurgy. However, observation of plastic deformation in quantum solids remains a significant technical challenge.

In 2018, Beamish and Z. G. Cheng studied the plastic properties of solid ^{4}He and observed sudden stress release caused by lattice slips. They suggested that the slip events were caused by dislocation avalanches under large shear stresses. They employed piezoelectric actuators to deform solid helium at low temperatures and measured its plastic properties. By combining multiple actuators together, a maximum shear deformation of up to

0.4% was realized. Plastic deformation occurred beyond the yield point. It was found that the dislocation avalanche traveled at speeds close to that of sound and that energy was released simultaneously in the form of phonons. This was the first time that acoustic emission was observed in a quantum solid [Cheng Z.G.2018b].

2.1.7 Superconductivity

Toward the end of our discussion of the physical properties of NQEs, we briefly introduce the mechanism of superconductivity. Electron–phonon coupling is naturally related to the quantum-mechanical formulation of nuclear motion, as phonons are quasiparticles that describe nuclear quantum properties. Evidently, one cannot discuss the electron–phonon coupling without considering nonadiabatic processes. In general, problems involving many-body interactions between electrons and atoms must be studied in a fully quantum-mechanical framework based on the Born–Huang expansion. Consequently, the analysis of full quantum effects can be quite complex: We cannot simply categorize one physical problem as an NQE problem and another as an NAE problem. For organizational purposes, the role of the nuclear quantum state in electron–phonon coupling will be discussed first, and that of the electronic quantum state will not be introduced until the following (NAE) section.

Electron–phonon coupling is extensively discussed in traditional textbooks of solid-state physics. It is customary to use nonrelativistic quantum field theory (the language that describes quantum many-body systems) and in particular the second-quantization representation to define the electronic and nuclear vibrational basis states, introducing creation and annihilation operators. One can then construct and diagonalize the Hamiltonian (which contains the electron–phonon interaction) using the defined basis. The eigenstates of the system are linear combinations of the product of the electronic and nuclear eigenstates [Fan H.Y.1951, Bardeen1955, 1957]. In a physical sense, this linear combination is consistent with the Born–Huang expansion. The only difference is that, in the quantum-field-theory approach to a solid, the nuclear position is described by the equilibrium position R_0 of the lattice. In the Born–Huang expansion, on the other hand, the nuclear position is described as the *on-the-fly* position R.

From the advent of quantum mechanics in the 1920s to the proposal of first-principles electronic structure calculations in the 1980s, many people have used the language of second quantization when studying the elementary excitation problems caused by electron–phonon coupling (including superconductivity) in simple condensed matter systems [Landau1933, Frohlich1950, Fan H.Y.1951, Bardeen1955, 1957, McMillan1968, Dynes1972, Allen1976, 1981]. In metals, the electrons have generally been treated as constituting a free (or nearly free) electron gas, and the eigenstates of nuclear motion have often been described in terms of simple harmonic oscillations. In insulators and semiconductors, electron states have usually been described in the tight-binding approximation. The reason for this is very simple: When computer capacity was limited, the second-quantization language provided the most pragmatic and straightforward way to discuss physical problems.

Superconductivity physics is a vast field of study; in this book, we do not intend to pursue it in depth. Rather, we focus on the discovery of new superconductors in light-element systems at high pressures, with particular emphasis on the role of electron–phonon coupling, an interaction closely associated with NQEs.

As early as 1968, Ashcroft predicted based on the BCS theory that, if hydrogen becomes metallic at high pressure, strong electron–phonon coupling could turn it into a high-temperature or even room-temperature superconductor [Ashcroft1968]. At the time, knowledge of crystal structure was mostly derived from X-ray analyses, and research based on Ashcroft's idea was therefore limited. In the late 1990s and early 2000s, developments in first-principles structure searching significantly facilitated the advancement of this field. The key to finding superconducting materials is to search for structures that can break the strong covalent bonds between hydrogen molecules; the electronic states of hydrogen can then move to near the Fermi surface of the system, resulting in effective scattering of phonons.

In recent years, theoreticians have predicted many high-pressure hydrogen-rich compounds or hydrides that can superconduct at high temperatures. For instance, Y. M. Ma's group at Jilin University predicted a superconducting-transition temperature T_c of 80 K for H_2S at 160 GPa [Li Y.W.2014]; T. Cui's group predicted that T_c for H_3S can reach 204 K at 200 GPa [Duan D.2014]. Such predictions have drawn the attention of experimentalists to high-temperature superconductivity in sulfur hydride.

In 2015, M. I. Eremets et al. experimentally reported superconductivity in sulfur hydride at pressures above 150 GPa [Drozdov2015]: The sample resistance dropped abruptly at 203 K, suggesting a superconducting transition, which was in agreement with the theoretical calculations of T. Cui's group. I. Errea et al. conducted a theoretical study of this system and discovered the important contribution of NQEs: The zero-point energy drives H_2S into $Im\overline{3}m$ H_3S around 150 GPa [Errea2015]. The superconducting properties of sulfur hydride were also calculated at different pressures. If the hydrogen atoms were treated as classical particles and the zero-point vibrational energy was neglected, the system remained in the $R3m$ phase without hydrogen-bond symmetrization [Errea2016]. However, the superconducting transition temperature of the $R3m$ phase did not agree with experimental observation. Only when NQEs were taken into account was the crystal in the $Im\overline{3}m$ phase with hydrogen-bond symmetrization, exhibiting the superconducting transition observed in experiments. The calculated transition temperature also agreed with the experimental measurement.

The general picture of the superconductivity properties of sulfur hydride under pressure has become clear: Below 150 GPa, the system is in the $P\overline{1}$ -H_2S phase with a superconducting-transition temperature below 80 K; above 150 GPa, it is in the $Im\overline{3}m$ -H_3S phase with a transition temperature of up to nearly 200 K. (The two crystal configurations are illustrated in Figure 2.22.) Given the importance of superconductivity in condensed matter physics, many hidden physical details remain to be investigated. For example, the variation in the electron density of states (DOS) near the Fermi surface was not included in early Eliashberg equation calculations [Li Y.W.2014, Duan D.2014]. In 2015, D. A. Papaconstantopoulos et al. pointed out that there is a distinct van Hove singularity in the electron DOS near the Fermi surface of $Im\overline{3}m$ -H_3S. This singularity does not exist in the $P\overline{1}$ -H_2S phase [Papaconstantopoulos2015].

In 2016, W. Sano et al. conducted a systematic study on $Im\overline{3}m$ -H_3S at 250 GPa and $P\overline{1}$ -H_2S at 140 GPa using first-principles calculation. They found a difference in the electronic DOS for the two structures (see Figure 2.23) [Sano2016]. Based on the more accurate Migdal–Eliashberg equation, the effects of the following issues on T_c were investigated

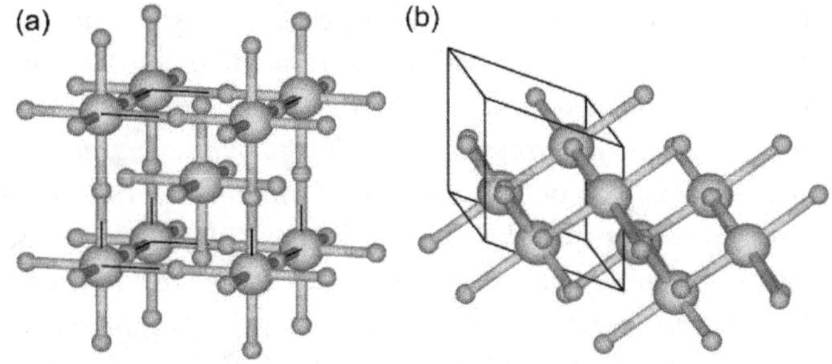

Figure 2.22 Two typical structures of sulfur hydrides: (a) $Im\overline{3}m$-H_3S and (b) $P\overline{1}$-H_2S. The large balls in the figure correspond to S and the small balls to H.

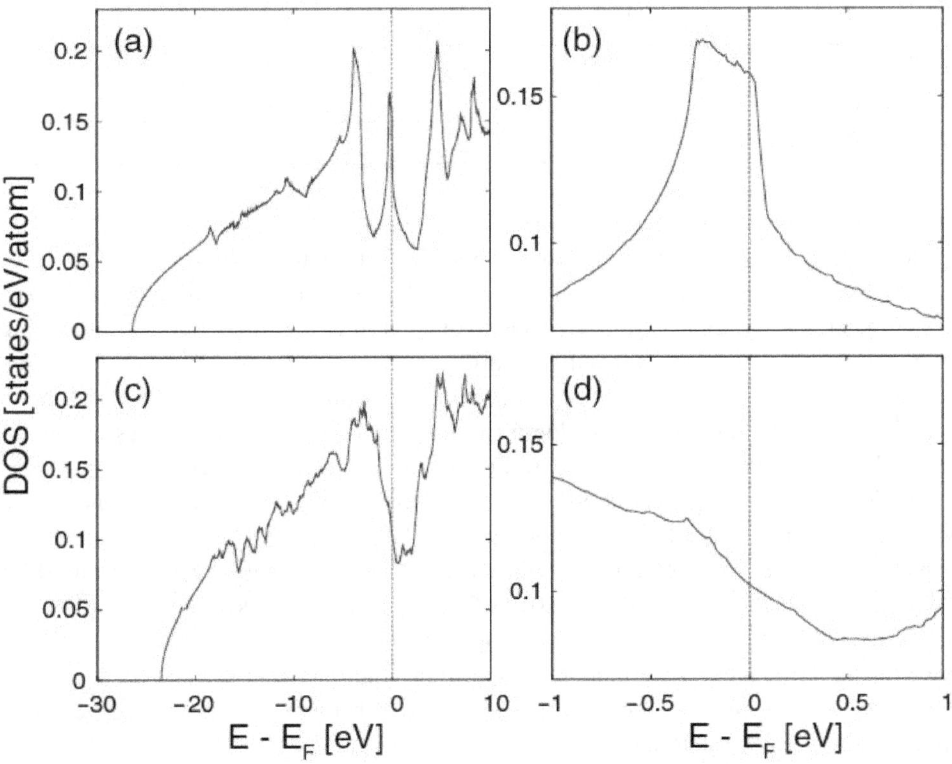

Figure 2.23 Density of states (DOS) of (a) $Im\overline{3}m$-H_3S calculated at 250 GPa, and (c) $P\overline{1}$-H_2S at 140 GPa. (b) and (d) are enlarged views of (a) and (c), respectively, around the Fermi energy level E_F. Reprinted figure with permission from Sano[2016]. Copyright 2024 by the American Physical Society.

Table 2.2 *Superconducting-transition temperature of $Im\bar{3}m$-H_3S at 250 GPa and $P\bar{1}$-H_2S at 140 GPa. The first row (const. DOS) shows the calculated result based on the Eliashberg equation without considering the variation in the electron density of states at the Fermi level when calculating the electron–phonon coupling coefficient. The second row (DOS) represents the result from the more accurate Migdal–Eliashberg equation, which includes the variation in the electronic density of states and the self-consistent calculation of the self-energy. The third row (DOS + ZPR) denotes the result after considering the zero-point renormalization of the electron dispersion based on the second row; the fourth row (DOS + ZPR + AH) further includes the anharmonic term of the phonon spectrum. Upon comparing these results, we note that last row of data agrees well with the experimental measurements. The differences in these results reflect the effects of different physical processes on the superconducting transition temperature T_c. Nevertheless, all the results predict a high T_c for $Im\bar{3}m$-H_3S at 250 GPa. This indicates that these details, which play an important role in the conventional calculation, do not qualitatively change the high-temperature physical picture of superconductivity in the hydrogen sulfide system. [Sano2016].*

Density of state	$Im\bar{3}m$-H_3S (K)	$P\bar{1}$-H_2S (K)
const DOS	225	56
DOS	193	63
DOS + ZPR	202	44
DOS + ZPR + AH	181	34

further: (1) the variation in the electronic DOS; (2) the effect of zero-point renormalization (ZPR) on the electronic DOS; and (3) anharmonic effects on the phonon (Table 2.2). These issues have been considered crucial in conventional studies of the superconducting properties of materials, and they are indeed capable of correcting the superconducting-transition temperature in the high-pressure sulfur hydride system to some extent; nevertheless, they do not change the physical picture mentioned earlier. In short, the strong electron–phonon coupling (corresponding to NQEs) dominates the superconducting properties in this system making it fundamentally different from conventional superconductors.

Studies of lanthanum hydride compounds (LaH_{10}) were carried out independently by M. I. Eremets and R. J. Hemley's group in 2019. They found that the superconducting transition temperature Tc of LaH_{10} could reach 260 K at pressures ranging from 180 to 200 GPa [Drozdov2019, Somayazulu2019]. Overall, these findings amply demonstrate the importance of NQEs in understanding the superconductivity of hydride compounds at high pressures, as well as the superconductivity of other light-element systems such as doped diamonds. These issues are thoroughly discussed in Chapters 7 and 8, where we discuss specific light-element materials.

At present, the high-temperature superconductivity exhibited by these light-element compounds at high pressure falls within the framework of traditional BCS theory. If the Debye temperature and electron–phonon coupling coefficient are increased further, the superconducting transition temperature can theoretically reach or exceed room temperature.

The high-pressure conditions in these studies were chosen to enhance the electron–phonon coupling. Nonetheless, strong electron–phonon coupling can cause lattice instability, which is why the superconducting-transition temperature predicted by early BCS theory has an upper limit, namely the McMillan limit. Realizing high-temperature superconductivity via electron–phonon coupling at atmospheric pressure remains a technical challenge. For unconventional superconductivity, the mechanism of which is still a subject of debate, it may be helpful to consider the possibility of pairing electrons in the strongly correlated system from the perspective of full quantum effects.

After considering the NQEs, Y. Jiang, J. Chen, and E. G. Wang recently compared the pressure-induced delocalization of hydrogen atoms in bulk ice and 2D ice. They found that both low-dimensional structures and hydrogen doping facilitate the quantum delocalization of hydrogen atoms at low (or even atmospheric) pressures, leading to symmetric hydrogen-bonding configurations. The O–H covalent bonds and hydrogen bonds in ice become indistinguishable, leading to metallization. This work has provided a promising new route for exploring the metallization and superconductivity of 2D ice at atmospheric pressure [Tian Y.2022]. It is reasonable to expect that light-element superconductors with higher transition temperatures and lower transition pressures can be developed through synergic modulation of pressure, dimensionality, and doping.

2.2 Nonadiabatic Effects

The nonadiabatic effect, a physical phenomenon occurring in real processes, represents the concurrent evolution of electrons and nuclei in a molecular or condensed matter system according to the laws of quantum mechanics, as opposed to the simple superposition of electron and nuclear systems evolving separately over time under the Born–Oppenheimer approximation. The electrons and nuclei in a real condensed system are constantly interacting with each other, and thus their dynamic evolutions are tightly coupled and inseparable. In the past, it has been a common practice to adopt the adiabatic approximation, that is, to treat the electrons dynamically in a static configuration of nuclei, but the assumption of adiabaticity is not valid in many cases. NAEs can be incorporated by treating the nuclei as either quantum or classical particles; the latter approach yields only approximations of the rigorous solutions provided by the former.

Many important physical problems require a nonadiabatic treatment. For example, NAEs influence the vibrations and collisions of atoms that lead to the breaking and reformation of chemical bonds. These processes inevitably affect ultrafast interfacial electron transfer, the strength of electron–phonon coupling, the formation of charge density waves, and the Berry phase. They are also heavily involved in affecting superconductivity transition temperature and inducing structural phase transitions. For example, in the study of the superconductivity of magnesium diboride, it was found that nonadiabatic effects can cause the movement and splitting of energy bands near the Fermi surface, thereby changing the distribution of electrons and causing changes in the strength of the electron–phonon coupling.

The study of nonadiabatic phenomena in condensed matter systems has now become popular, and many research groups around the world have devoted considerable efforts to it. Notably, S. Meng's group at the Institute of Physics, Chinese Academy of Sciences, proposed a first-principles nonadiabatic quantum-dynamics method, with which they have extensively explored NAEs in a variety of condensed matter systems [Meng S.2008a, Lakhotia H.2020, Lian C.2020, You P.W.2021b, Guan M.X.2021, Xu J.Y.2022, Guan M.X.2022]. A detailed description of their method is provided in Appendix B.

As early as in 1928, M. Born and V. Fock proposed the adiabatic theorem. The theorem states that if the external time-dependent perturbations change slowly with time, and there is an energy gap between the eigenstate under consideration and other eigenstates of the system, or the eigenvalue under study is far from the rest eigenvalues of the same Hamiltonian, the system will remain in its instantaneous eigenstate [Born1928]. In other words, if the external perturbations change slowly, the system can adapt to changes in the environment and remain in equilibrium. This allows describing the entire system in terms of a single eigenstate rather than a superposition of eigenstates.

A nonadiabatic process is the one in which external perturbations cause the coupling between different eigenstates. From the definition, it is clear that the physical quantities used to distinguish between adiabatic and nonadiabatic processes are the characteristic time of the perturbation change and the characteristic energy of the system. More specifically, we can consider a system with an intrinsic relaxation time of $2\pi/\omega^{\mathrm{in}}$ that is perturbed with frequency ω^{ex}. Suppose that the system is in an eigenstate with energy e_{i} and that its other eigenenergies are represented by e_{j}, with $\mathrm{i} \neq \mathrm{j}$. Then the adiabatic theorem is valid only if $\omega^{\mathrm{in}} \gg \omega^{\mathrm{ex}}$ and $\left|e_{\mathrm{i}} - e_{\mathrm{j}}\right| \gg 0$. A nonadiabatic system violates one or both of these conditions. The following paragraphs discuss some typical cases of nonadiabatic processes.

One important case is the time-dependent evolution of the electron degrees of freedom in a system in which an external field is present (e.g., the tunnel ionization of a single molecule or atom in a strong laser field). The system can be considered nonadiabatic if the external field changes rapidly. The characteristic time/energy scale of the system can be described by the Keldysh parameter $\gamma \equiv \frac{\sqrt{2I_{\mathrm{P}}}}{E_{\mathrm{max}}}\omega$, where I_{P} is the ionization potential, E_{max} is the electric field amplitude, and ω is the laser frequency.

The other case concerns electrons and nuclei in the absence of external fields. In this case, the change in the electric field of the electron subsystem is caused by the motion of the positively charged nuclei. The adiabatic theorem holds when $\omega^{\mathrm{ele}} \gg \omega^{\mathrm{nuc}}$ and $\Delta E \gg 0$, where E refers to the energies of the electron subsystem. Then, the motion of electrons and nuclei can be treated separately, and the evolution of the system can be described by solving the time-dependent Schrödinger equation for a series of fixed nuclear configurations, which is equivalent to evolving the system adiabatically on a single potential energy surface. These conditions justify invoking the Born–Oppenheimer approximation. Conversely, the Born–Oppenheimer approximation fails when the time scales of the electronic and nuclear motions are comparable and the energy difference between potential energy surfaces is small; these are the conditions under which nonadiabatic effects must be considered.

In subsections 2.2.1–2.2.6, we cover some typical systems to elucidate NAEs in the full quantization process. We will discuss nonadiabatic effects on the nuclear motion and the electronic motion separately.

2.2.1 Atomic Collisions

When an atom collides with a solid surface, it generates a time-dependent electric field with a characteristic frequency $\leq 10^{13}$ Hz. Because the mass of an electron is significantly less than that of an atom, the frequency of electron vibration inside the solid will be one or two orders of magnitude greater than the frequency of the electric field induced by the atomic collision. Using this difference in time scale, it is possible to study the nonadiabatic effects during collisions between atoms and a solid surface.

The usual approach is to compare the experimental results of inelastic scattering of hydrogen atoms colliding with insulator, metal, and semiconductor surfaces with the theoretical results of Born–Oppenheimer molecular-dynamics simulation [Bünermann2021]. Past research has shown that the overall scattering process of hydrogen colliding with insulator surfaces largely satisfies the description given by the Born–Oppenheimer approximation. For example, collisions between hydrogen atoms and xenon can be described by the Born–Oppenheimer molecular dynamics simulation. In the case of a metal surface, the high-energy hydrogen atoms excite electron–hole pairs during the process, but the atomic scattering can still be treated by the weakly coupled action in the electronic-friction approximation (an approach similar to the perturbation correction). At this point, the Born–Oppenheimer approximation deviates from experiment, but only slightly.

The study of atomic collisions with semiconductor surfaces provides a good example of how the limit of the Born–Oppenheimer approximation can be further tested. High-speed incident atoms not only excite electrons in semiconductors and help them jump from the valence band to the conduction band but also cause significant energy transfer during the dynamical process of electron–nucleus evolution. All these effects indicate the existence of full-quantum physical phenomena beyond the Born–Oppenheimer approximation.

Recently, K. Krüger et al. found that in the Born–Oppenheimer approximation, it is difficult to understand the scattering of high-speed hydrogen atoms off the semiconductor surface Ge(111)$c(2 \times 8)$ [Krüger2022]. In their experiments, they collected the energy-loss and angular-distribution spectra of the scattered hydrogen atoms. As the energy of the incident hydrogen atoms increased, the energy-loss spectrum changed from single-peaked to double-peaked (Figure 2.24).

When the incident-atom energy was smaller than the Ge surface bandgap of 0.49 eV, the energy-loss spectrum of the scattered hydrogen atoms agreed well with the Born–Oppenheimer molecular-dynamics simulation result (Figure 2.24(a)). This indicates that the scattering process corresponds to the adiabatic channel. In other words, although the hydrogen atoms may be excited to produce phonons in the solid during dynamical evolution, the energy loss is small enough that the scattering can still be described by the Born–Oppenheimer approximation. This is why the first experimental peak on the left in Figure 2.24 agrees almost perfectly with the Born–Oppenheimer theoretical prediction. However, as the energies of the incident hydrogen atoms gradually increased and became larger than the bandgap of the Ge surface, a second peak began to emerge in the energy-loss spectrum (the peak on the right in Figure 2.24(b–d)). The Born–Oppenheimer result (solid curve in Figure 2.24(b–d)) only predicts the first peak and completely fails to predict the second peak. Moreover, further study revealed that the ratio of the number of the scattered H atoms through the second channel increases with increasing incident hydrogen-atom energy (from 1.0:1.3 at $E_i = 0.99$ eV to 1.0:9.8 at $E_i = 6.17$ eV). Consequently, this experimentally discovered new channel cannot be explained by the Born–Oppenheimer molecular-dynamics simulations, and the adiabatic approximation fails. The second peak is associated with the nonadiabatic channel. It is worth noting that it appears at an energy value immediately above the Ge surface bandgap, indicating that the minimum energy loss in it is equal to the semiconductor-surface bandgap value. This means that the electron transition from the valence band (VB) to the conduction band (CB) of the Ge semiconductor is caused by the hydrogen-atom scattering. This dynamical process is realized through a nonadiabatic VB-CB channel, and energy transfer occurs. Interestingly, as the energy of the incident hydrogen atoms increases, the ratio of the number of the scattered H atoms through the nonadiabatic channel increases, which clearly indicates the increasing importance of the nonadiabatic process.

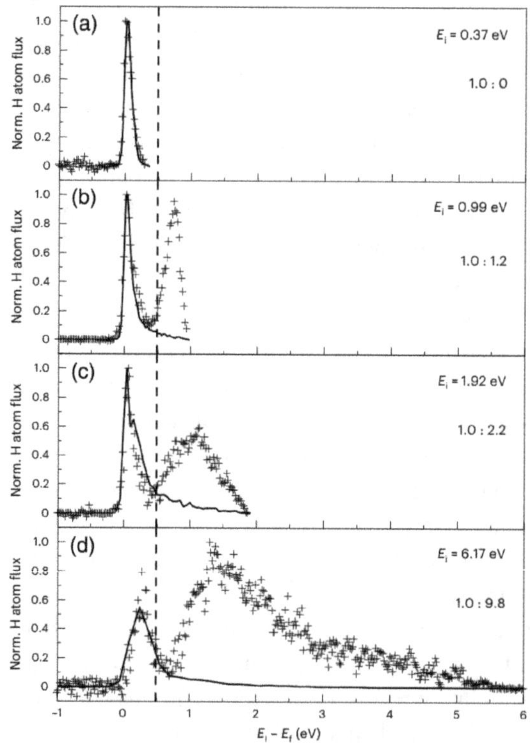

Figure 2.24 Energy-loss spectrum of scattered hydrogen atoms on germanium surface (Ge(111)c(2 × 8)). The incident hydrogen atoms follow the [110] direction, the incident and scattering angles are both 45°, and the Ge surface temperature is 300 K. (a)–(d) Energy-loss spectra at different incident energies E_i, including experimental results (+) and molecular dynamics simulations (black solid curve): (a) $E_i = 0.37$ eV, (b) $E_i = 0.99$ eV, (c) $E_i = 1.92$ eV, and (d) $E_i = 6.17$ eV. The Ge surface bandgap (0.49 eV) is indicated by a vertical dashed line. The ratio in the figure indicates the ratio of the number of the scattered H atoms through adiabatic valence-band to conduction-band channels to that of nonadiabatic valence-band to conduction-band channels measured experimentally. All experimental curves are normalized to the peak intensity. Figure created by Krüger[2022] licensed under CC BY 4.0.

The hydrogen-atom scattering results on the semiconductor Ge surface are also closely linked to the incidence angle. With the incident-atom energy fixed at 0.99 eV, the scattered hydrogen-atom energy was measured experimentally as the incidence angle varied (30°, 45°, and 60°). A polar plot (Figure 2.25) shows that the angular distribution of the nonadiabatic VB-CB channel was narrower than that of the adiabatic channel, exhibiting maximum scattering flux near the specular scattering angle (indicated by the arrow in the figure). On the other hand, when the incidence angle was fixed, increasing the energy of the scattered hydrogen atoms led to the broadening of angular distributions in both the adiabatic and nonadiabatic VB-CB channels, but more noticeably in the latter, which is clear as the theoretical results in Figure 2.25(d–f) cannot reproduce the experimental polar plot within the dotted line in Figure 2.25(a–c). As in the previous discussion, the Born–Oppenheimer molecular-dynamics simulation results only agree with the experimental results in the adiabatic channel. This indicates that, when the new physical processes resulting from full quantum effects dominate the system, the Born–Oppenheimer approximation is no longer valid.

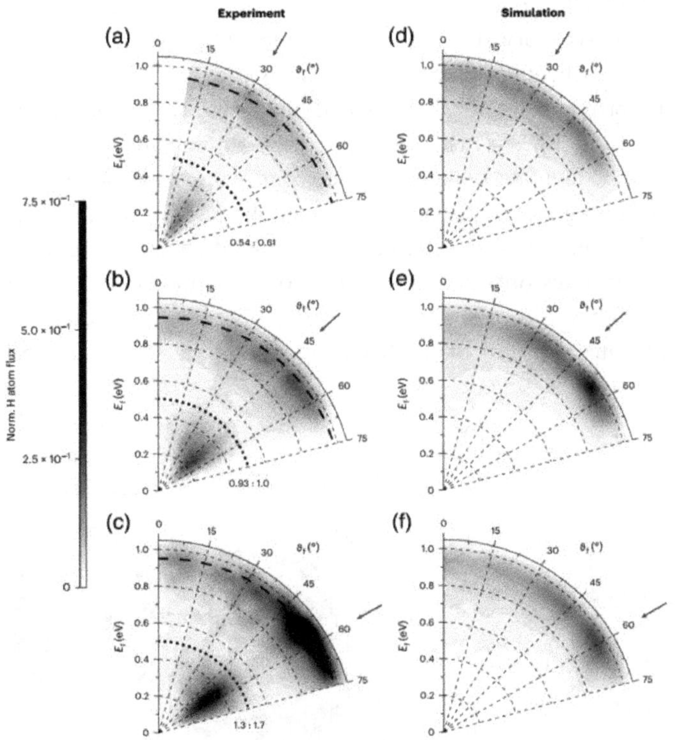

Figure 2.25 Hydrogen-atom scattering off a semiconductor surface (Ge(111)$c(2 \times 8)$) for different incidence angles at a fixed incidence energy of 0.99 eV. (a)–(c) Experimental results of hydrogen atom scattering at different incidence angles: (a) 30°, (b) 45°, and (c) 60°. The ratio of the number of H atoms scattered through the nonadiabatic VB-CB channel to those scattered through the adiabatic channel is also indicated for each angle. (d)–(f) Born–Oppenheimer molecular-dynamics results: (d) 30°, (e) 45°, and (f) 60°. The arrows refer to the specular scattering angle of hydrogen atoms. Figure created by Krüger[2022] licensed under CC BY 4.0.

The collision of N_2^+ ions and Ar atoms is also a well-known process exhibiting nonadiabatic effects. Experimental results by S. Schlemmer et al. revealed anomalous behavior caused by the rotational relaxation of N_2^+ ions [Schlemmer1999]. With a relatively large binding energy (1.109 eV), N_2^+-Ar collisions at low temperatures produce a long-lived bound state; this is generally presumed to induce rapid redistribution of the rotational states. Nevertheless, Schlemmer et al. reported a very small vibrational relaxation-rate coefficient $k_J(90\text{ K}) = (1.4 \pm 0.4) \times 10^{-11}\text{cm}^3\text{ s}^{-1}$, which contradicts the theoretical prediction. If collisions occur at the Langevin frequency $k_L = 7.4 \times 10^{-10}\text{ cm}^3\text{ s}^{-1}$, this suggests that the N_2^+ ion can remain stable after approximately 50 collisions with Ar atoms. (According to the Langevin model, collisional complexes are formed when the collision energy is sufficient to overcome the rotational potential barrier.) Schlemmer et al. attributed this unexpected result to hidden motions or additional potential barriers. They suggested that an accurate first-principles calculation or a quantum-mechanical treatment with values on the order of few meV would be required to provide a conclusive answer.

To address this problem, O. T. Unke et al. constructed the (N^{2+}-Ar) potential energy surface at the UCCSD(T)-F12a/aug-cc-VTZ level of accuracy using reproducing-kernel Hilbert space (RKHS) theory [Unke2016] and obtained good agreement with the experimental results. For verification, more accurate calculations at the CASSCF/MRCI+Q level were carried out, proving that the accuracy of the previous one was sufficient. Based on these results, they further performed quantum close-coupling and quasi-classical-trajectory calculations. As the calculated potential energy surface was relatively accurate, and the quasi-classical result agreed with that of the quantum calculation, the difference between the experimental observation and theoretical computation may originate in NAEs. The rotational relaxation rates obtained using the above methods ranged from ~5.34×10^{-10} $\text{cm}^3\text{ s}^{-1}$ to $6.96 \times 10^{-10}\text{ cm}^3\text{ s}^{-1}$. The results were consistent with the expected redistribution of the rotational states: 70% of the collisions resulted in a change in the ionic rotational

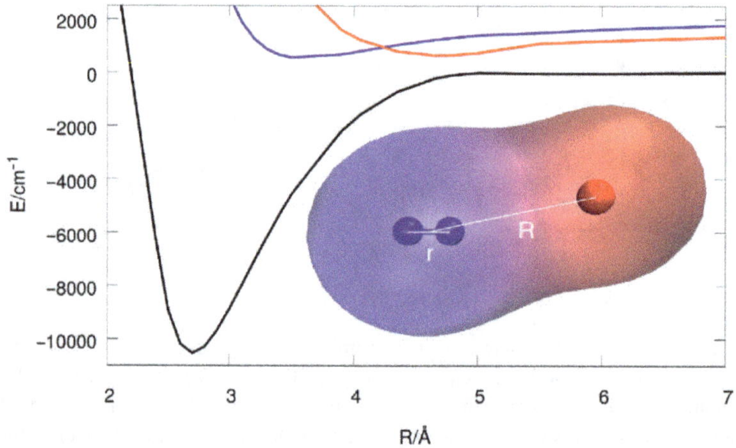

Figure 2.26 Interaction energy as a function of collision distance (R); the internal interaction energy turning point (r = 1.072 Å) of the three lowest electron states of (N^{2+}-Ar) at v = 0 vibration is shown. The black curve corresponds to the ground state and the red/blue curves to the lowest excited state. Figure created by Bircher[2017] licensed under CC BY.

quantum state. In fact, the difference in the interaction energy also supports this claim. Figure 2.26 depicts the calculated interaction energy at different distances. It can be seen that the difference between the ground state and the excited state is only 700 cm^{-1} at a relatively long distance, indicating that NAEs must be considered in this problem.

2.2.2 Atomic Vibrations

Another very powerful tool for detecting nonadiabatic effects is vibrational spectroscopy. NAEs in linear absorption spectra include the decrease in the occupancy of electronic excited states due to internal conversion or inter-system crossing, resulting spectral line broadening, spectral intensity changing, and spectral-peak position shifting. In nonlinear pump–probe modes, it is also possible to probe the physical and chemical properties of the materials directly at specific time scales, such as the time-resolved excited-state spectroscopy.

Theoretically, nonadiabatic vibrations can be studied either by semiclassical methods based on the quantum mechanics or by hybrid quantum and classical methods [Bircher2017]. In Chapter 5, we will introduce several theoretical approaches to the study of NAEs. Figure 2.27 shows the time-resolved stimulated-emission spectrum of the pyrazine S2 state calculated by different models (i.e., with and without NAEs). A semiclassical method based on the multiple-surface-dephasing representation was adopted by Zimmermann et al. to study the spectra containing nonadiabatic effects [Zimmermann2014]. This is very similar to using MSDR to estimate adiabatic properties, except that when nonadiabatic effects are included, the dipole autocorrelation function instead of the degree of overlap is calculated semiclassically, using the Wigner phase-space formalism of quantum mechanics. The nonadiabatic trajectory can be obtained using either fewest-switches surface hopping or the local mean-field dynamics method, which is equivalent to extending Ehrenfest dynamics method from a single trajectory to multiple trajectories. A comparison of the spectra calculated by these two methods reveals that the signal decay on the time scale of tens of femtoseconds is due to NAEs, especially the decay of the S2 state to S1 state.

The vibrational states in hydrogen-bond systems also exhibit pronounced NAEs. When two molecules are dimerized via hydrogen bonding, the intramolecular and intermolecular vibrational couplings are often strong, with pronounced quantum tunneling effects. As a result, an energy-level crossing, also known as a conical intersection of the vibration spectrum, is easily formed. The conical intersection in the electron-energy-band structure determines the ultrafast photophysical and photochemical properties of the system. It is natural to wonder whether the conical intersection of the vibration spectrum has similar properties. In general, the conical intersection of the vibration spectrum determines the ultrafast relaxation processes (usually on the order of 100 fs) in hydrogen-bonded systems. The corresponding vibrational characteristics of the system are extremely complex. Furthermore, the typical hydrogen bond dissociation energy and intramolecular vibrational frequencies involved in OH bond vibrations are of approximately the same order of magnitude. Both are often associated with hydrogen bond breakage.

P. Hamm et al. recently studied adiabatic processes involving the conical intersection of the vibration spectrum in the HCO_2-H_2O system [Hamm2015]. Some relevant properties under gas-phase conditions had been previously studied by Q. Xue et al. [Xue Q.2004].

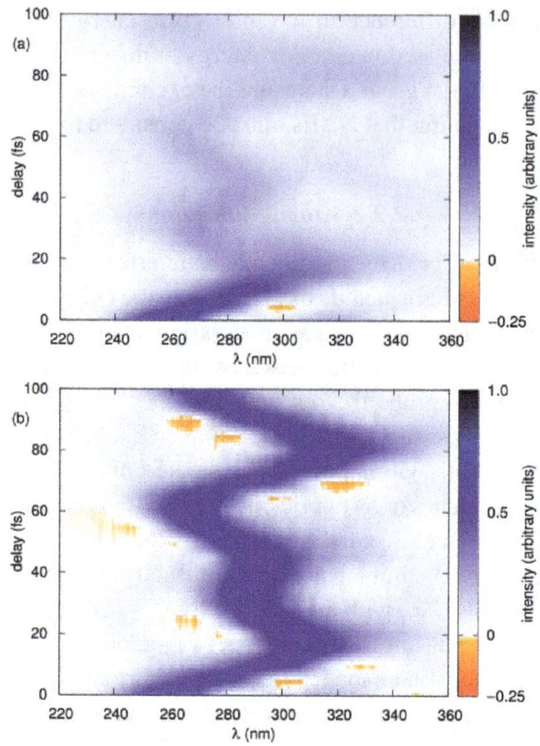

Figure 2.27 Nonadiabatic effects on the time-resolved stimulated emission spectrum of pyrazine, calculated (a) with and (b) without nonadiabatic coupling. In (a), the signal decays with increasing delay time τ due to nonadiabatic effects; (b) shows that this decay does not occur when nonadiabatic effects are neglected. Both spectra are calculated within the framework of the multiple-surface-dephasing representation method combined with the fewest-switches surface-hopping method. Reprinted from Zimmermann[2014], with the permission of AIP Publishing.

Figure 2.28(a) shows the diabatic potential energy surfaces of stretching and bending vibrations of two local oxy–hydrogen bonds with respect to the Q1 vibrational mode of water. It can be seen that the stretching vibrations (labeled as |1,0,0> and |0,1,0>) and bending vibrational modes (|0,0,2>) of the two bonds have many intersections, which evolve into conical intersections of phonon spectra when intermolecular vibrational degrees of freedom are considered. With these potential energy surfaces, it is possible to derive the relaxation pathway of the vibrational energy (with nonadiabatic coupling implied). Surprisingly, the population in the HOH bending vibrational state |0,0,2> grows no faster than that in |0,0,1> (Figure 2.28(b)), although the latter does not have any conical intersection with the excited O–H stretching mode. Thus, unlike the electronic states, the conical intersection of the vibration spectrum cannot efficiently facilitate energy-flow pathways. This suggests that the adiabatic approximation is far from accurate and that nonadiabatic effects affecting these vibrational modes are crucial.

NAEs are also evident in the phonon vibrations of solids. Examples of nonadiabatic effects on atomic vibrations in graphene will be introduced later when we discuss the full

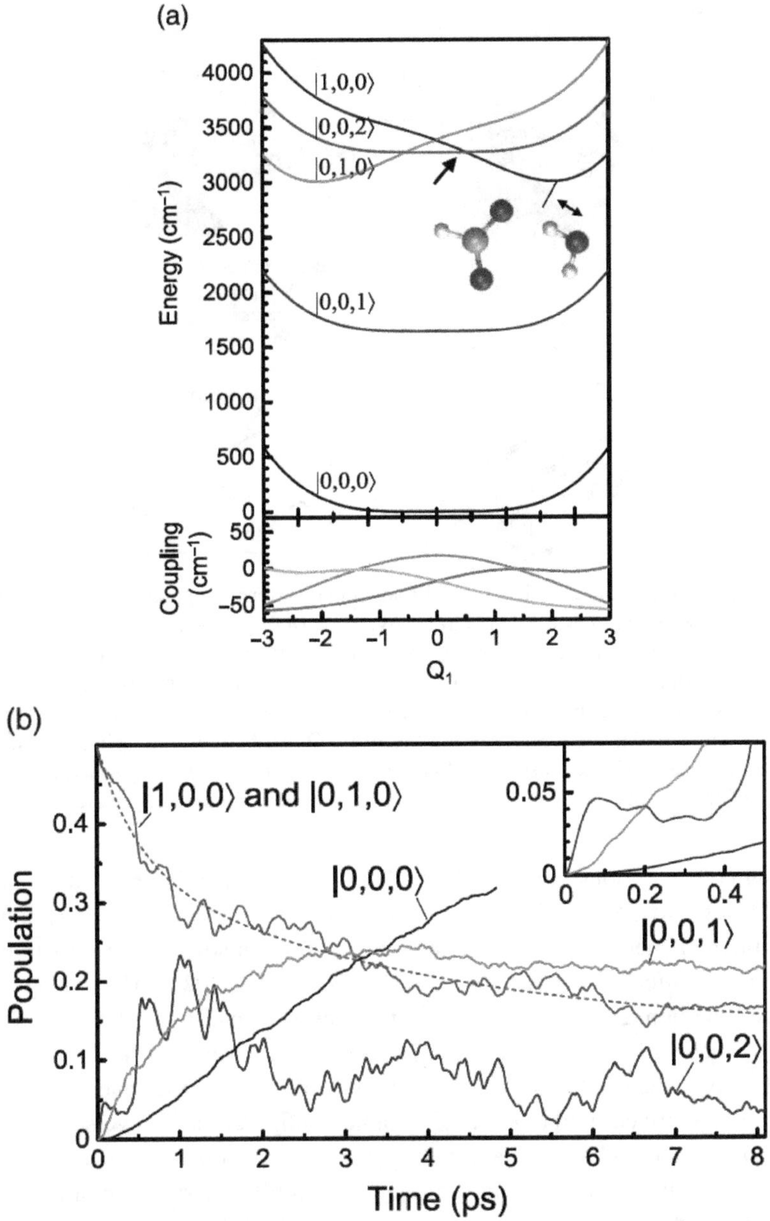

Figure 2.28 (a) Nonadiabatic potential energy surfaces of two local O–H stretching vibrations in water. (b) Vibrational relaxation of the O–H stretching mode under "vertical" excitation. Reprinted from Hamm[2015], with the permission of AIP Publishing.

quantum effect in spectroscopy (Chapter 6). The linear Dirac structure of its electronic energy band makes nonadiabatic characteristic more evident in graphene than in other semiconductor materials. S. Pisana et al. found that the phonon vibration frequency (G-peak) of bilayer graphene at room temperature deviates from the eigenfrequency predicted by the adiabatic approximation and the deviation significantly increases as the electron/

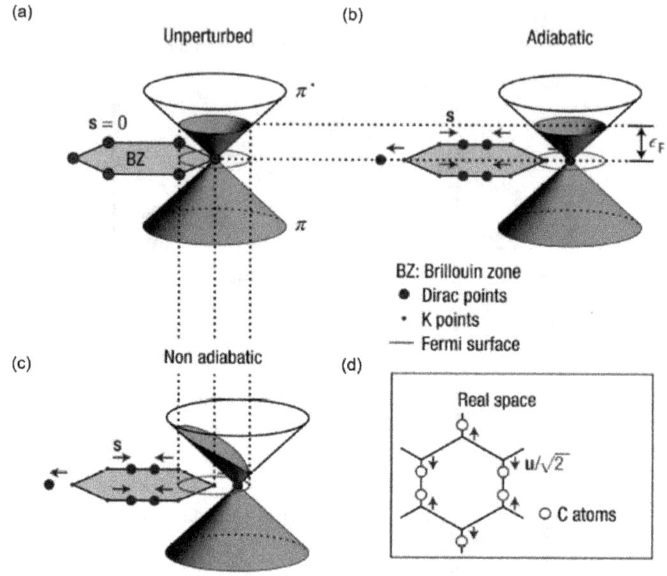

Figure 2.29 Schematic of adiabatic and nonadiabatic Born–Oppenheimer approximation in graphene, obtained by Raman spectroscopy: (a) perfect crystal at rest, with Dirac points lying at the K points of the momentum space, and electron concentration controllable by voltage; (b) adiabatic process obeying the Born–Oppenheimer approximation, with E_{2g} phonons causing lattice distortion, and the center of the circular Fermi surface at the Dirac points; (c) nonadiabatic process beyond the Born–Oppenheimer approximation, with the E_{2g} phonons causing lattice distortion, but the center of the circular Fermi surface deviating from the Dirac points; (d) schematic of vibrational modes of atoms in graphene.

hole doping concentrations increase [Pisana2007]. G.Y. Zhang's group at the Institute of Physics, Chinese Academy of Sciences, reported that when the Fermi energy is nearly half of the E_{2g} phonon energy at low temperatures, E_{2g} phonon softening occurs in monolayer graphene, and its lifetime is more than doubled [Zhao Y.C.2020]. They also successfully corroborated the nonadiabatic electron–phonon coupling and logarithmic Kohn anomaly predicted by theory. To understand the experimental observations, they noted that, during atomic vibrations, the electrons do not follow the atom's dynamics and do not stay in the ground state determined by the *on-the-fly* atomic configuration. Conversely, the electrons exhibit a certain "inertia" and remain in their original electronic state because of the presence of a gapless linear energy band. Consequently, a tilted Dirac cone of occupation distribution is formed, which is a direct manifestation of the NAEs. Figure 2.29 illustrates the adiabatic and nonadiabatic processes corresponding to the atomic vibrations of graphene.

2.2.3 Ultrafast Charge Transfer

As quantum many-body systems consisting of electrons and nuclei, the coupling between the two is critical to the properties of condensed matter. NAEs influence not only the state of nuclear motions (i.e., atomic vibrations and collisions) mentioned in the previous sections but also electronic motions. This results in behaviors that cannot be described by the conventional adiabatic approximation. There are many existing examples, such as NAE-induced ultrafast charge transfer at interfaces, charge-order reconstruction, and optically induced phase transitions.

The electronic excited behaviors of two-dimensional materials have recently attracted considerable interest. For example, X. P. Hong et al. found that when the MoS_2/WS_2 heterojunction interface was excited by a 670nm light source, only electrons in MoS_2 layer were excited at first, but holes were transferred to the WS_2 layer within 50 fs, realizing ultrafast spatial separation of oppositely charged carriers [Hong X.P.2014]. This phenomenon could be used in high-efficiency light detectors and solar cells. Employing a time-dependent density-functional method to simulate nonadiabatic electron–nuclear dynamics, J. Zhang et al. investigated the influence of the interfacial atomic composition, stacking configuration, twisting angle, defects, and other factors on the heterojunction excitation process [Zhang J.2017]. They found that different stacking configurations can alter the rate of optically induced ultrafast charge transfer by up to a factor of ten. Additionally, the efficiency of excited-electron transfer in heterojunctions differs among materials. For example, electron–hole separation occurs within 20 fs in $MoSe_2/WSe_2$ heterojunctions. Interestingly enough, they found that the ultrafast-charge-transfer rate does not have a quadratic dependence on the electron-coupling matrix elements, as predicted by Fermi's golden rule, but rather shows an exponential dependence (Figure 2.30). Further analysis showed that the interlayer interaction of two-dimensional materials exhibits strong electron coupling and nonadiabatic properties that cannot be described by perturbation theory alone. Fermi's golden rule is based on time-dependent perturbation theory; the light-excited two-dimensional heterojunctions leading to ultrafast charge transfer by a nonadiabatic process defies this rule because it goes beyond the scope of perturbation theory. J. Zhang et al. also discovered that the charge-transfer efficiency is extremely sensitive to the interlayer electron-coupling strength in van der Waals heterostructures: A small change in coupling strength can cause a significant increase in charge-transfer efficiency. This result could potentially lead to the development of novel high-efficiency optoelectronic and energy devices. J. Zhang et al. found that nonadiabatic processes are highly dependent on electron wavefunction coupling and that strong nonadiabatic processes offer the prospect of constructing ultrasensitive electronic devices.

Recently, several research groups have conducted extensive theoretical and experimental studies on the nonadiabatic process of photoelectric conversion in solar cells. S. Meng et al. studied electron–hole separation dynamics on dye-absorbed nanowires due to photoexcitation [Meng S. 2008a,b]. Rapid electron injection of cyanin natural dyes occurs at the titanium-dioxide nanowire surface, and the calculated time scale agrees well with the experimental measurements. Furthermore, they found that the adsorption state of dye molecules on the oxide surface greatly affected the electron-injection time scale. These results demonstrate that electron–hole separation can be realized within 200 fs, ensuring the proper operation of the dye-sensitized solar device.

In addition, the electron–hole separation in this system is significantly influenced by the molecule type, size, adsorption configuration, and surface defects. Related studies have

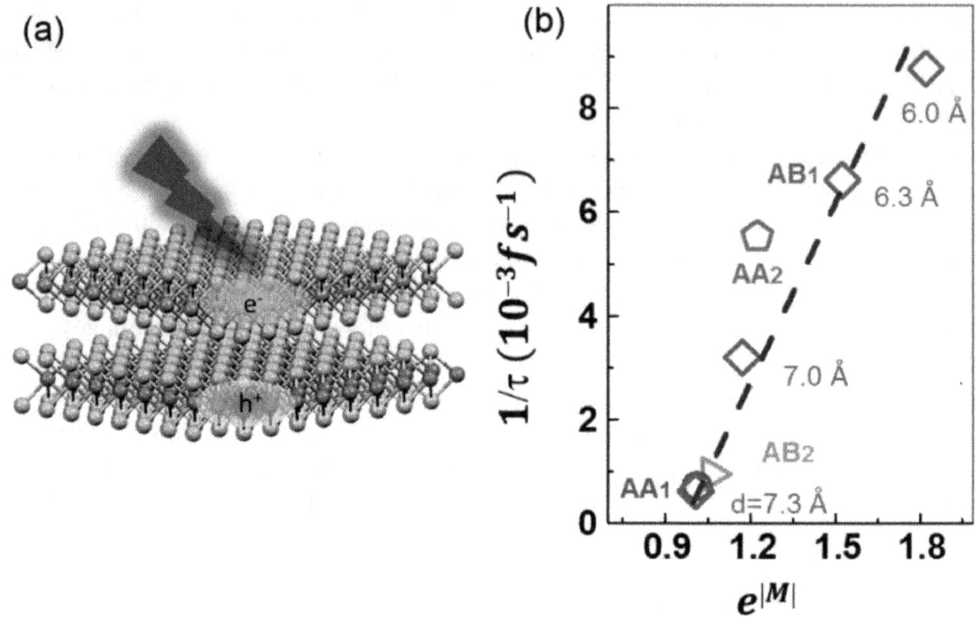

Figure 2.30 Nonadiabatic dynamics of interlayer ultrafast charge transfer in van der Waals heterojunctions. (a) Schematic diagram of ultrafast charge transfer. (b) Exponential dependence of charge transfer rate on interlayer electron coupling matrix element M in different configurations, with evident nonadiabatic characteristics that cannot be explained by Fermi's golden rule. AA_1, AA_2, AB_1, and AB_2 denote different stacking configurations of two layers of 2D material: in AA, the lattice orientations are the same, with identical or staggered atoms in the upper and lower layers; in AB, the orientations are opposite. Figure created by Zhang J.[2017] licensed under CC BY 4.0.

established a theoretical basis for understanding accurately the process of electron–hole dynamics microscopically and thus for optimizing the photovoltaic conversion efficiency of solar cells [Meng S.2010].

To prolong the interfacial electron–hole recombination time and to control external conditions, W. Ma et al. developed a real-time method to simulate the recombination dynamics of photoexcited electrons and holes within the framework of time-dependent density functional theory. By studying the initial recombination process and averaging over multiple trajectories, charge recombination at an interface was simulated [Ma W.2013]. The electron-injection time scales of two structurally similar model molecules, N1 and N2, were found to be comparable (150–160 fs). However, compared to N1, N2 had an extra phenyl ring in the bridge. Its electron-recombination process was therefore four or five times slower: The recombination time scales for N1 and N2 were 6 and 23 ps, respectively. This means that the electron collection and energy conversion efficiency could be improved by a factor of ~4. This calculation is in agreement with the experimental observations of Haid et al.; they found that the recombination process was five times slower, with efficiencies of 1.24% and 8.21% for N1 and N2, respectively [Haid2012].

Using a first-principles approach based on time-dependent density functional theory, W. Ma et al. succeeded in predicting the energy-conversion efficiency of dye-sensitized solar cells [Ma W.2014]. This was achieved by calculating the precise absorption spectrum of the dye molecules, the electronic structure at the interface, and the lifetime of electron injection and recombination. Taking two groups of pure organic dyes as examples, the short-circuit currents of the relevant cells were obtained accurately by this method. By calculating the potential difference between the Fermi energy level of the electron in the titanium dioxide conduction band and the redox couple in electrolyte, the open-circuit voltage of the cell could be accurately acquired. The calculated photovoltaic conversion efficiency of the solar cell agreed well with the experiment, with an error of 1%–2%.

As this example shows, photoexcitation promotes electrons to specific excited states, where they undergo nonadiabatic processes such as electron relaxation and charge transfer. In most scenarios, these processes no longer follow the *on-the-fly* atomic configuration and vibrational modes [Meng S.2008a,b]. Instead, they are mainly determined by the previous electronic state and the strength of the electron state coupling, exhibiting strong nonadiabatic effects. The motion of the excited electrons is very sensitive to a variety of microscopic factors, such as electron coupling strength; effectively modulating it has the potential to improve the efficiency of photodetector and energy devices.

2.2.4 Electron–Phonon Coupling

Electron–phonon coupling in solids is one of the fundamental problems in condensed matter physics; it expresses the basic forms of interactions of a multibody system. From the previous examples, we know that it plays a crucial role in superconductivity [Tian Y.C.2016, Wu Q.2020]; it is also related to fundamental physical properties such as ultrahigh thermal conductivity [Tian Z.Y.2022]. When introducing superconductivity in subsection 2.1.7, we focused on NQEs on superconductivity in terms of the nuclear quantum state. In the present discussion of NAEs, we are more interested in how the electronic state affects the lattice vibrational state through electron–phonon coupling, especially in the presence of optical fields. Please refer to Chapter 5 for a detailed description of theoretical methods, including the first-principles calculation of electron–phonon coupling.

Recently, it has been realized that in materials, such as metals, heavily doped semiconductors, and narrow-bandgap semiconductors, the electron–phonon coupling is significantly influenced by NAEs. The electron–phonon couplings given by calculations based on the adiabatic Born–Oppenheimer approximation are often very different from the real situation, the electron–phonon coupling strength in metals and narrow-bandgap semiconductors is underestimated by ~10%–40% [Hu S.Q.2022]. For a wide-gap semiconductor with a doping concentration of 1.4×10^{21} cm^{-3}, F. Caruso et al. found that when the phonon frequency is less than the Kohn anomaly point the electron–phonon coupling in heavily boron-doped diamond crystals induces significant phonon-mode renormalization. According to the adiabatic approximation, the change in phonon-frequency softening value is 22 meV, whereas nonadiabatic theory yields a phonon-frequency softening value change of 7 meV. The nonadiabatic calculation agrees well with the X-ray scattering experimental measurement (5.3 meV) [Caruso2017]. Because of its inaccurate description of phonon frequencies, the adiabatic approximation overestimates the electron–phonon coupling constant by 10% and the superconducting-transition temperature is even overestimated by 50%. Evidently, proper consideration of NAEs is required to account for electron–phonon couplings accurately.

As another example, E. Cappelluti et al. suggested that NAEs open new channels for σ electron–phonon coupling in systems with small Fermi pockets, forming nonadiabatic Cooper pairs. This successfully explains the high transition temperature (39 K) of MgB_2 superconductors [Cappelluti2002]. Similarly, M. Lazzeri et al. found that, taking NAEs into consideration, charge doping in graphene leads to significant changes in the E_{2g} phonon frequency at the Γ center of the Brillouin zone; in the adiabatic approximation, on the other hand, this frequency only weakly depends on the amount of charge doping [Lazzeri2006]. The nonadiabatic effect can also lead to a sharp change in phonon linewidth for certain charge-carrier concentrations. Pronounced nonadiabatic effects on electron–phonon coupling properties can even be observed in insulators subjected to ultrafast laser excitation. All these findings indicate that NAEs cannot be neglected when discussing electron–phonon coupling effects [Hu S.Q.2022].

Nonadiabatic optical modulation of the electron–phonon coupling has therefore received considerable attention with the advancements in ultrafast laser technology. Such coupling contributes to many physical phenomena, including light-enhanced superconductivity [Mankowsky2014], polariton generation [Peng S.2021], and hidden charge order [Lian C.2020]. It also plays a vital role in energy conversion, energy transfer, solar cells, and the design of functional quantum materials. Despite various experimental studies, the properties and mechanisms of light-modulated electron–phonon coupling have not been fully understood because of the complex interactions involved and the dynamical coupling between multiple degrees of freedom. In particular, the effects of optical excitation on electron–phonon coupling and its dynamic evolution remain unknown. Because of these theoretical limitations, only the ground state has been considered in the literature, even for the cases with ultrafast laser excitation. Therefore, establishing an effective description of nonequilibrium electron–phonon coupling, understanding its physical mechanism, and tracking dynamic modulation by optical excitation remain critical theoretical issues to be addressed.

Because of its unique energy-band structure, graphene has become an ideal platform for studying light-modulated electron–phonon coupling dynamics. The gapless Dirac cone not only gives rise to a broad frequency band and ultrafast optical responses but also provides effective resonant electron–phonon coupling channels between Dirac fermions and phonons. Recently, S. Q. Hu et al. investigated the dynamic mechanism of optically induced electron–phonon coupling enhancement in graphene away from equilibrium using real-time density-functional molecular dynamics simulations [Hu S.Q.2022]. They found that electron–phonon coupling accelerates photocarrier relaxation in photoexcited graphene and induces a Fano resonance in the phonon spectrum. Based on this discovery, they reconstructed the electron–phonon coupling strength and effective electron–phonon coupling-matrix elements for the nonequilibrium case by analyzing the energy-transfer rate between photocarriers and phonons, proposing a method for characterizing and tracking the excited-state dynamics in real time. The results obtained by their method show good agreement with experimental results. By tracking the dynamic changes in total strength of electron–phonon coupling λ and its matrix elements, optically enhanced electron–phonon coupling was observed during the photocarrier relaxation process in the photoexcited graphene. The coupling strength in the excited states could be enhanced by several times up to being more than 10 times stronger than that in the ground state (see Figure 2.31). Analysis of the energy-resolved electron–phonon coupling revealed that this enhancement originates from the nonequilibrium dynamic distribution of photocarriers, which provides additional channels for electron–phonon scattering and is essential for modulating the electron–phonon coupling strength.

Nonadiabatic excited-state electron–phonon coupling depends on the photocarrier states. This dependence can be used not only for real-time electron–phonon coupling measurements but also for precise manipulation of the coupling strength via optical techniques. More interestingly, since electron–phonon coupling is associated with carrier dynamics, the topological properties of photocarriers (e.g., chirality and valley selectivity) may introduce new physical discoveries and technological applications.

Shockley–Read–Hall (SRH) non-radiative recombination of carriers in semiconductors is also dominated by nonadiabatic electron–phonon coupling. During SRH recombination, electrons in the conduction band and holes in the valence band are captured by mid-gap defective levels through phonon scattering, resulting in electron–hole recombination. In photoelectronic devices such as light-emitting diodes and solar cells, this can reduce the carrier lifetime and consequently lower the energy conversion efficiency. Reliable first-principles study was not developed until the last decade.

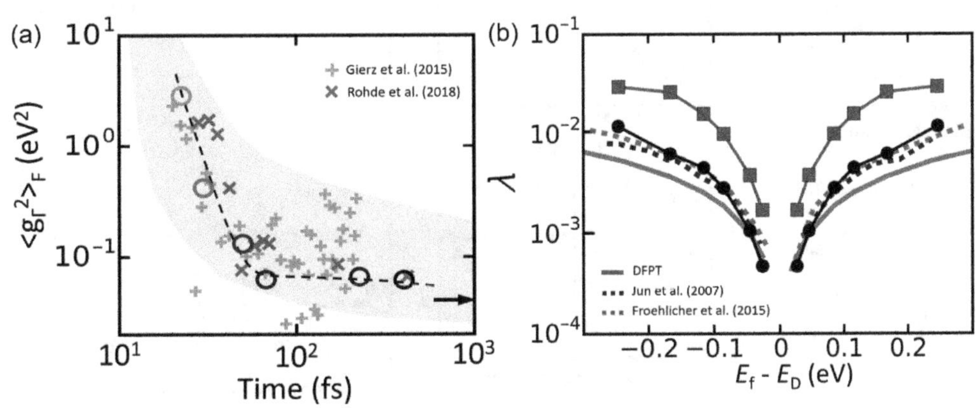

Figure 2.31 (a) Non-equilibrium electron–phonon coupling strength as a function of time in photoexcited graphene. (b) Optically induced enhancement of electron–phonon coupling strength λ (curve with square markers: 50 fs of photoexcitation; curve with circle markers: ground state). DFPT stands for Density Functional Perturbation Theory. The inset in (a) shows schematically the coupling of photocarrier and phonon after photoexcitation. Figure created by Hu S.Q.[2022] licensed under CC BY 4.0.

Within the framework of perturbation theory, L. Shi et al. proposed a novel variational equation to calculate the electron–phonon coupling-matrix elements in real space, thereby substantially reducing the computational cost. They calculated the defect-state capture coefficients due to zinc impurities and nitrogen vacancies in GaN under the adiabatic and Condon approximations [Shi L.2012]. The results were significantly lower than the experimental values, because nonadiabatic effects were excluded from the calculation. (Further, the changes in ground-state and excited-state phonon frequencies and vibrational modes were neglected in the Condon approximation.) Later, A. Alkauskas et al. adopted the static model to consider NAEs (at least partially), focusing on a particular phonon mode only. By this assumption, they came up with a simpler formula, yielding theoretical results closer to the experimental values [Alkauskas2014]. Inspired by Alkauskas's work, L. Shi et al. combined the static approximation with their original variational equation, greatly improving the accuracy of theoretical prediction [Shi L.2015].

In addition, J. Zhao's group at the University of Science and Technology of China recently used nonadiabatic dynamics to investigate carrier recombination induced by defect states in titanium dioxide, two-dimensional black phosphorus, and organic–inorganic perovskite systems [Zhang L.2018, 2019, Chu W.2020]. Their method utilized trajectory surface hopping for a better description of NAEs. The improved calculation accuracy of this method allowed real-time observation of carrier recombination (Figure 2.32), contributing to a deeper understanding of the underlying physical mechanisms of NAEs. This work revealed that the key parameters determining electron–hole

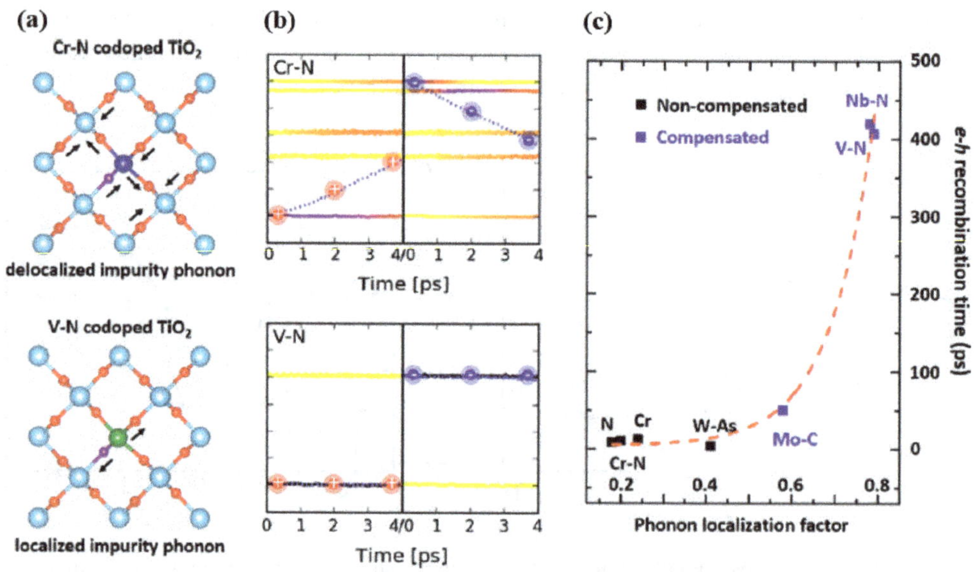

Figure 2.32 (a) Delocalized phonon modes of Cr-N-doped TiO_2 and localized phonon modes of V-N-doped TiO_2. (b) Carrier recombination process of Cr-N- and V-N-doped TiO_2. (c) Electron–hole recombination times with different dopants. (Red dashed curve indicates an exponential fit.) Reprinted with permission from Zhang L.[2018]. Copyright 2024 American Chemical Society.

recombination depend on whether it is neutrally doping in titanium dioxide and whether the impurity phonon modes are localized [Zhang L.2018].

2.2.5 Optically Induced Phase Transitions

Optically induced transitions between different structural phases enable modulation of a material's various properties at ultrahigh response rates. They have become a major research topic in the field of quantum materials and ultrafast science. During the phase transition, optically induced electron hopping alters the number of occupancies of the excited-state charge carrier and directly affects the force on the atoms. As a result, the electron and phonon systems can be coherently excited. The energy dissipation of the system is a typical nonadiabatic process in which fermions (e.g., electrons) and bosonic quasiparticles (e.g., phonons or magnons) follow complex relaxation pathways involving scattering between multiple electron and phonon modes. Establishing theoretical models that can depict the interactions between different degrees of freedom separately will not only further deepen our understanding of nonadiabatic effects in experimentally complex phase transitions but will also serve as a foundation for the design of materials and devices with special properties.

For monolayer transition-metal chalcogenide compounds, the crystal structure is determined by the number of electrons occupying the d orbitals in the transition metal, as well as the energy splitting due to the ligand field. Among a number of transition metal chalcogenides, monolayer $MoTe_2$ has one of the smallest values of the energy difference between the semiconducting $2H$ phase and the metallic $1T'$ phase (~42 meV per single cell). Therefore, multi-phase modulation can be achieved by temperature change, stress, and electron doping. In 2015, for the first time, S. Cho et al. experimentally used laser modulation to realize the $2H$-to-$1T'$ phase transition [Cho2015]. However, the understanding of ultrafast optically induced ordered phase transitions is still limited to the ground-state study.

In 2022, M.X. Guan et al. systematically investigated the nonadiabatic ultrafast structural phase transition of monolayer $MoTe_2$ from the semiconducting to metallic states under a laser field [Guan M.X.2022]. They found that only two Raman-enhanced optical phonon modes – the out-of-plane A'_1 mode and the in-plane E' mode – were excited under weak-field excitation, and the atoms only oscillated harmonically around their equilibrium positions. These two phonon modes correspond to the system with the strongest electron–phonon coupling, and the induced atomic motion will generate asymmetric change in the potential energy surface. Therefore, the structural change of $MoTe_2$ is directional under photoexcitation.

The phase transition of monolayer $MoTe_2$ under strong-field excitation was divided into three stages, and the phonon-excitation amplitudes of the A'_1 and E' modes were found to be about three times larger than those with weak-field excitation. Decoherence occurred after 0.4 ps. The amplitudes decreased gradually, accompanied by transitions to other phonon modes that led to anharmonic motion. The acoustic phonon mode at the M point of the reciprocal space [LA(M)] and the in-plane E'' modes emerged sequentially, resulting in ordered structural deformation. Within 1 ps, the semiconducting $2H$ phase had turned to a nonequilibrium metallic $2H^*$ phase (see Figure 2.33). Extending the excitation time allowed the $MoTe_2$ to reach thermal equilibrium and transform into the $1T'$

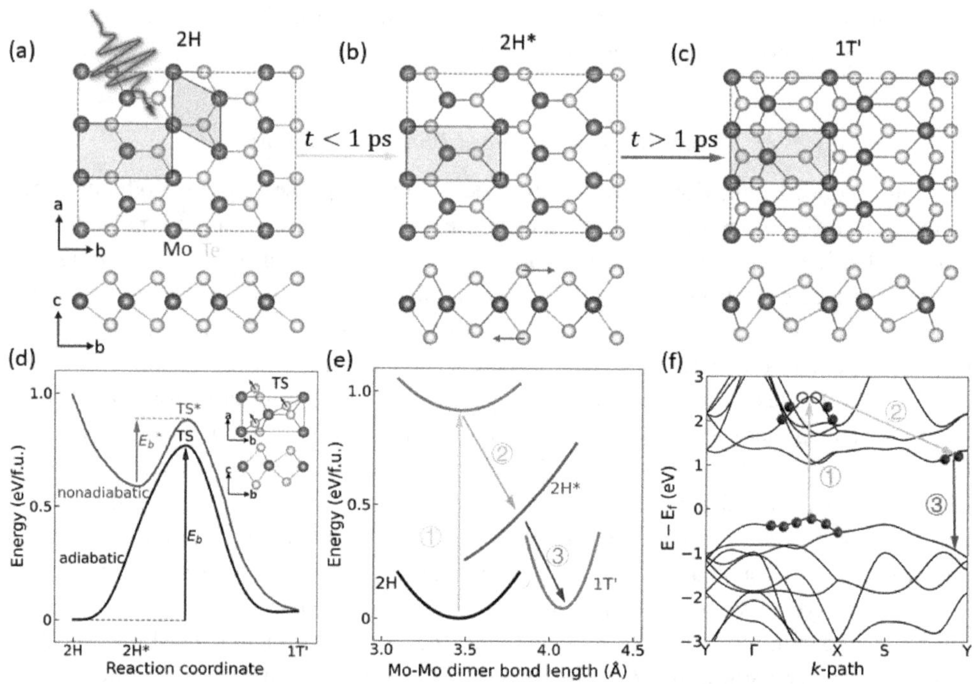

Figure 2.33 (a)–(c) Schematic of the laser-induced phase transition from the semiconducting $2H$ phase to the metallic $1T'$ phase in monolayer $MoTe_2$. (d) Energy profile along the phase-transition pathway during adiabatic and nonadiabatic phase transitions. (e)–(f) Potential energy surfaces and carrier distribution changes in the three stages of the phase transition. Reprinted figure with permission from Guan M. X.[2022]. Copyright 2024 by the American Physical Society.

phase. Photoexcitation effectively lowers the phase-transition energy barrier, and ordered phonon excitation leads to a phase-transition process that significantly differs from the thermal-equilibrium transition at high temperatures.

By analyzing the energy distribution of carriers, M. X. Guan et al. found that this phase-transition process can be attributed to the electron–phonon interactions induced by photocarriers. The forces applied to the nuclei evidently differ when the carriers occupy conduction bands in different energy ranges. For example, when the electrons are excited to high-energy conduction bands, the forces induce atomic displacement along the direction of the A'_1 and E' vibration modes. When electrons are excited to the bottoms of conduction bands, the nuclear forces along the A'_1 and E' vibrational vectors are significantly reduced and the forces responsible for Mo-Mo bond shrinkage become dominant. As a result, the structural distortion pathway can be manipulated by adjusting the intensity and frequency of the laser. These theoretical results provide a clear parameter range for realizing the semiconductor–metal phase transition. The $2H$ phase is the most common ground-state structure in monolayer transition-metal chalcogenides, as well as the basic building block of various multilayer and bulk materials. This work provides new ways of studying phonon scattering in transition-metal sulfides and designing different nonequilibrium phase-transition processes.

Recently, theoretical and experimental studies of topological quantum states and topological quantum materials have been increasing and have become an important frontier in condensed matter physics. As a new concept of material classification, topological order has become an important parameter in the study of condensed matter as well as their local symmetry. In crystalline materials, the topological order is generally connected to the Berry phase of the system. A deep understanding of topological and geometric phases is relevant to many fundamental problems in condensed matter physics, such as the classification of quantum phases, phase transitions, and multiple zero-gap meta-excitations. In topological materials, electrons and phonons exhibit many novel physical properties, such as chiral anomalies and chiral Landau levels, due to their prominent Berry curvature or Berry phase effects and the couplings between electrons, phonons, and spins. Photoexcitation can be used to distinguish different interactions and manipulate states of matter to extract information about fundamental physical properties of materials, structural phase transitions, and related quantum states. An in-depth understanding of the optically induced correlation between the macroscopic behavior of topological materials and their microscopic atomic and electronic structures has become the primary goal of current condensed matter research.

The photoelectric response of topological materials is closely related to their microscopic electronic structure. Especially for chiral topological semimetals, the carriers near the energy-band crossing have divergent Berry curvatures. Therefore, when the Fermi surface is located near a band crossing, the photoelectric response of the system is significant at low frequencies and exhibits low-energy divergence. This provides a new theoretical basis for developing efficient IR and terahertz photoelectric detectors [Liu J.2020].

Theoretical studies of nonlinear optical phenomena in topological materials can usually be carried out by combining the calculation of ground-state properties with symmetry analysis. However, this method has several limitations. First, the real-time dynamical information of the excited carriers in momentum space and real space is missing, making direct comparison with the time-resolved experimental results impossible. Secondly, electron–phonon and photon–phonon couplings are not included in this approach – especially nonadiabatic couplings, which are crucial for the occurrence of certain phase transition processes. Moreover, perturbation-based theoretical analysis cannot describe physical processes under strong optical fields.

First-principles time-dependent density-functional molecular dynamics simulations, on the other hand, can describe the electron–lattice correlations without relying on the adiabatic approximation. A description of the dynamical behavior of electron-lattice correlations is given. M.X. Guan et al. systematically studied electronic structure during topological phase transitions in the type-II Weyl semimetal WTe_2 under photoexcitation (see Figure 2.34). They reported a selective excitation of carriers near the Weyl points, which are determined by the atomic orbital symmetry and the selection rules. In contrast to the regular spin selection rule for chiral excitation, the excitation pathway could be controlled by altering the polarization direction and photon energy of the linearly polarized light source [Guan M.X.2021].

Asymmetric carrier excitation will induce photocurrents in different directions in real space, thus affecting the direction and symmetry of the interlayer displacement in

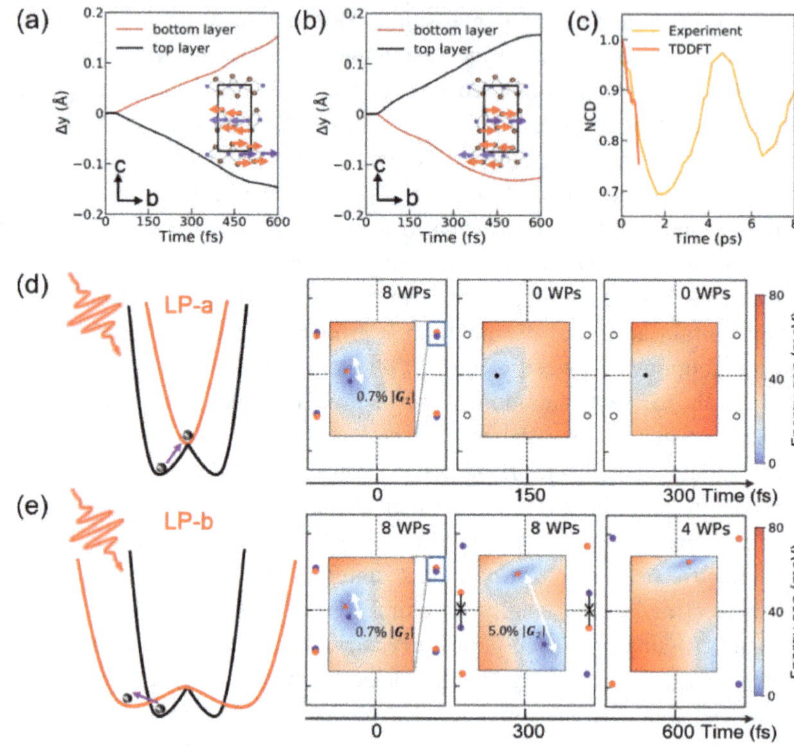

Figure 2.34 (a) and (b) Relative interlayer motion along the a-axis and b-axis of a crystal in the polarization direction of a linearly polarized source; the insets show the corresponding motion modes. (c) Comparison of theoretical simulations with experimental observations. (d) and (e) Symmetry evolution of the system. The position, number, and degree of separation of the two nearest Weyl points in the $k_z = 0$ plane are shown. Figure created by Guan M.X.[2021] licensed under CC BY 4.0.

the system (as shown in Figure 2.34(a) and (b)). The topological properties of WTe_2 are closely associated with the number of Weyl points and their degree of separation in momentum space. Therefore, asymmetric carrier excitation will induce changes in Weyl quasiparticle behavior in momentum space and thus alter the topological properties of the system. M. X. Guan et al. provide a clear phase diagram for the photoinduced phase transition (Figure 2.35). The carrier excitation near the Weyl points is influenced not only by its chirality but also by the atomic-orbital characterizations of the wavefunctions. The two factors result in similar effects but have distinctly different mechanisms. The study conducted by M. X. Guan et al. lays a theoretical foundation for a better understanding of the Weyl point singularity.

2.2.6 Optically Induced Charge Density Waves

In complex condensed matter systems, the study of the coupling between particles and quasiparticles (e.g., electrons, phonons, magnons, and plasmons) plays a key role in

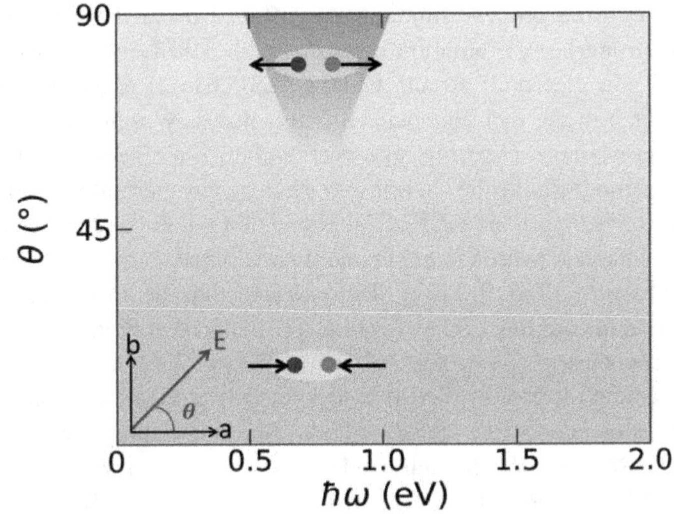

Figure 2.35 Phase diagram of light-driven T_d-WTe_2 topological phase transition depending on the linearly polarized photon energy ($\hbar\omega$) and polarization direction (θ). Figure created by Guan M. X.[2021] licensed under CC BY 4.0.

understanding and manipulating physical properties. The interaction of electrons with other collective excitations causes periodic distortions in a perfect lattice, resulting in the formation of a charge density wave (CDW), also known as a charge order. CDWs are closely associated with quantum phenomena such as superconductivity and many-body correlations. Their formation mechanism is still an active research topic in condensed matter physics: They may arise from Fermi surface nesting of electrons, electron–phonon coupling, or other (perhaps unknown) factors.

Laser excitation is an effective tool for studying and manipulating complex and entangled particle or quasiparticle interactions. However, this method inevitably induces nonadiabatic motion of electrons and atomic nuclei. Thus, lasers can not only help in understanding the ground-state and excited-state properties of CDW materials but can also introduce new physical states through nonadiabatic dynamic processes. Under laser irradiation, the collective oscillation modes in the CDW phase are excited, and the collective oscillations occur in both the lattice and electron systems. Ultrashort laser irradiation is capable of closing the charge-order band gap (due to crystal structure distortion) and the Mott gap (due to strong electron–electron correlation) in the CDW phase. Observation shows that CDWs are not disrupted at low temperatures – the Mott gap of electrons closes without significant changes in the atomic configuration, indicating the presence of a nonadiabatic process. In general, photoexcited CDWs have more complex dynamics than do CDWs in their ground state.

In 2019, J. Zhang et al. carried out theoretical simulations of ultrafast excitation in $1T$-TaS_2, a typical CDW material [Zhang J.2019]. The study demonstrated that photoexcitation could induce a new collective oscillation mode, accompanied by a periodic emergence of a transient structure with metallic properties, the M state. This photoexcitation-induced process significantly differs from a thermally induced phase transition. The strong laser irradiation created a high density of electron–hole pairs in the CDW phase. The electronic structure of $1T$-TaS_2 changes drastically in less than 50 fs, and the bandgap in the two-dimensional plane closes. Within a few hundred femtoseconds, electron–phonon scattering increases the lattice temperature, producing a new lattice-oscillation mode unlike the phonon mode induced by thermal excitation. According to J. Zhang et al., the laser-induced CDW phase in $1T$-TaS_2 undergoes a non-thermal insulator–metal phase transition. This is mainly attributed to the photo-induced charge density wave phase of $1T$-TaS_2. They demonstrated that electron–electron correlation and electron–phonon coupling effects in the CDW material are vital in the laser-induced phase transition process.

In layered titanium diselenide ($TiSe_2$), lasers not only generate exciton-induced CDWs but also induce superconducting states by controlling the doping concentration and the stress magnitude. Recently, this has attracted much attention as a way to provide a test bed to study the CDW–superconductor transition and as a new platform for Bose–Einstein condensation. Because of the complexity of exciton–phonon coupling, existing research tools cannot easily distinguish the roles of the two quasiparticles in the formation of CDWs. It has been recently discovered that lasers can be used to “animate” the excitons and phonons, so that they can be studied separately with ultrafast techniques. The response time of excitons is short and that of phonons is long, which makes it possible to distinguish their roles in the time domain. However, spectroscopy is unable to provide simultaneously ultrafast time-resolved and ultrahigh space-resolved information in regular experiments, leading to much controversy in this research area.

To resolve this issue, C. Lian et al. have developed theoretical methods that can deal with the ultrafast dynamics of laser-excited quantum materials. They developed a first-principles approach with both ultrafast time-resolution and ultrahigh spatial resolution to study the dynamics of laser-excited titanium diselenide. By considering NAEs, they uncovered a new CDW-formation mechanism [Lian C.2020]. The study revealed that the charge order was disturbed by the laser excitation instantly (<10 fs), causing ions to move in the opposite direction from the periodic lattice distortion. A complete lattice-distortion inversion could be achieved under strong laser irradiation (as shown in Figure 2.36). This finding confirms that excitons play an important role in CDW stability. Nevertheless, CDWs cannot be created by excitons only: The laser can only suppress 20% of the charge order, not 100%.

Further, first-principles dynamics simulations have revealed that a self-amplified mechanism of electron–phonon coupling occurs during photoexcitation, producing a stable CDW. With an external heat bath for slow annealing of the nuclear motion, the periodic distortion can be completely frozen even at ultralow annealing rates. This indicates that the periodic lattice distortion dynamics depend sensitively on the initial nuclear motion: After this weak motion is dissipated by the heat bath, subsequent periodic lattice distortion will not occur. The recovery and inversion of the lattice distortion is realized through a self-amplified process: “charge-order disturbance → nuclear motion → further charge-order disturbance → even greater nuclear motion.” By calculating the

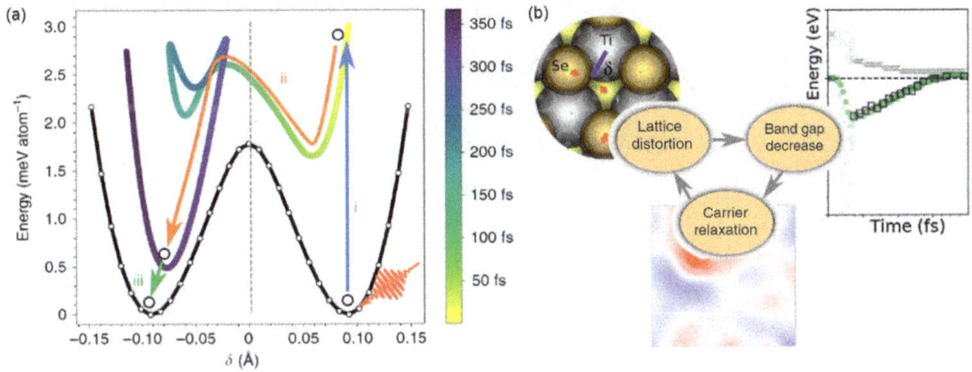

Figure 2.36 (a) Nonadiabatic electron–ion dynamic trajectories in photoexcited TiSe2. The vertical axis indicates the instantaneous potential energy of the system and the horizontal axis represents the displacement of the Ti atom from the high-symmetry equilibrium position. The colors correspond to time after electron excitation. (b) Illustration of the self-amplified electron–phonon mechanism in the charge-ordering process, clearly showing the nonadiabatic effect of electron-lattice coupling. Figure created by Lian C.[2020] licensed under CC BY 4.0.

time-dependence of the energy band structure, the time scale of this process has been found to be ~100 fs. The theoretical calculations agree well with the experimental results measured by time-dependent angle-resolved electron spectroscopy.

Through these numerous examples, we have introduced a variety of physical phenomena caused by nonadiabatic effects in condensed matter. They clearly demonstrate the vital role of NAEs in the quantum coupling of electronic and nuclear motion. The novel and rich physical processes revealed by these experiments cannot be explained by theoretical methods based on the Born–Oppenheimer adiabatic approximation. Studying electronic and nuclear NAEs in-depth provides new insights into much-debated problems in condensed matter physics. Only when full quantum effects are considered can problems involving charge-density waves, electron–phonon coupling, the molecular Berry phase, or superconductivity be adequately clarified. This line of research, already at the cutting edge, will become even more important in the future.

3

Full Quantum Effects in Chemistry

Although the concept of full quantum effects originated in physics research, it has had a broad and profound impact on chemistry. The main purpose of chemistry is to understand and control chemical reactions, which can be regarded as the continuous migration of atoms (i.e., constant change of atomic positions) due to the breaking of old bonds and formation of new bonds. Full quantum effects enable certain chemical reactions that do not occur if classical approximations held exactly. In particular, when light-element atoms (such as hydrogen and helium) or molecules (such as H_2, O_2, and CH_4) are involved, the quantum tunneling and quantum vibration of the nuclei as well as the potential energy surface splitting and dynamic evolution induced by electron–nucleus quantum coupling can have a decisive impact on the reaction process.

As early as 1927, when discussing the tautomerism phenomena in chiral molecules, physicist F. Hund predicted that quantum mechanisms (often referred to at the time as "quantum effects of nuclei") would play a crucial role in chemical reactions. In 1932, another physicist, E. Wigner, introduced the concept of "quantum tunneling" into chemistry. Three years later, H. Eyring, M. G. Evans, and M. Polanyi proposed their famous transition-state theory (TST) of chemical reactions based on statistical thermodynamics and quantum mechanics. It provided a concise picture of chemical kinetics within the classical framework and established a theoretical foundation for understanding chemical reactions. In the original paper by Eyring, Evans, and Polanyi, the importance of quantum tunneling in certain chemical reactions was highlighted, despite it being outside the framework of classical theory; however, most chemists ignored this. Only by the 1970s did work begin to systematically address the effects of quantum tunneling and quantum vibration of nuclei on chemical reactions.

In Chapters 1 and 2, we discussed full quantum effects predominantly using the language of physics, investigating particular physical phenomena. This chapter focuses on the implications of full quantum effects for chemistry. Therefore, in the initial section of this chapter, we reintroduce some basic concepts using chemical language so that both physicists and chemists can benefit from this book.

In chemical reactions, the microscopic motion of atoms and molecules is governed by quantum mechanics. However, even for the simplest chemical reaction problems, solving the time-dependent Schrödinger equation based on basic quantum-mechanical principles and accurately calculating the dynamics of the entire reaction is almost impossible. Therefore, various numerical approximations are used instead. Because electrons are considerably lighter than nuclei, they move much faster, and their spatial distribution can be adjusted to accommodate position changes of the related nuclei. It is common to treat nuclear and electronic motions separately; this method is called the Born–Oppenheimer approximation or adiabatic approximation. It is assumed that the electrons are in adiabatic states (i.e., in simple cases, they cannot exchange heat with the fixed nuclei via kinetic energy while in continuous motion). Thus, solving the time-dependent Schrödinger equation can be simplified to solving an instantaneous eigenequation.

Quantum chemistry studies the electronic structure of molecules, and dynamics describes the relative motion of atoms. For a specific molecular structure or chemical reaction, the Born–Oppenheimer approximation is used to determine the spatial positions of the nuclei in one of the possible configurations, and then the energy of the electronic states in that configuration is calculated using quantum-mechanical methods. Thus, the energy of the electronic states can be calculated for every possible atomic configuration and expressed as a function of the nuclear coordinate. The resulting Born–Oppenheimer potential energy surface is the basis for all discussions of physical and chemical processes under the adiabatic approximation, as well as the starting point for studying chemical dynamics. When nuclei are treated classically, chemical processes can be regarded as cases of atomic motion on the potential energy surface.

Under normal circumstances, when the Born–Oppenheimer approximation is valid, the potential energy surfaces of the different electronic states are separated from one another. The approximation fails when the potential energy surfaces cross or are very close to each other, and dynamic processes governed by nonadiabatic effects occur. For example, in photochemical reactions of polyatomic molecules, photon absorption induces an electron transition from the ground state to excited states with higher energy, which triggers nonadiabatic electron–nucleus coupling. In this case, the quantum motion of electrons and nuclei and the quantum coupling between the two must be considered. Most current nonadiabatic molecular dynamics methods are based on a hybrid quantum–classical theoretical framework. In other words, the electrons are treated as quantum particles moving in accordance with quantum mechanics, whereas the nuclei are treated as classical particles moving in accordance with Newtonian mechanics. However, in the reaction dynamics of light elements, especially those involving hydrogen atom transfer, quantum tunneling and vibration of the hydrogen nuclei are important factors that cannot be ignored. In such cases, the nucleus cannot be treated as a classical particle, and full quantum effects must be considered when solving the Schrödinger equation for electron–nucleus quantum coupling. Our discussion of the above problems begins with the Born–Huang expansion (Eq. (1.4)).

The presence or absence of nuclear quantum-tunneling effects in chemical reactions is often inferred experimentally by investigating the dependence of the reaction rate on temperature. For a chemical reaction driven by thermal fluctuations, the reaction rate predicted by the classical TST depends on the free-energy potential (activation energy) and pre-exponential factor. The temperature dependence of the reaction rate constant satisfies the Arrhenius equation (see Eq. (3.1)). In contrast, for a reaction with pronounced

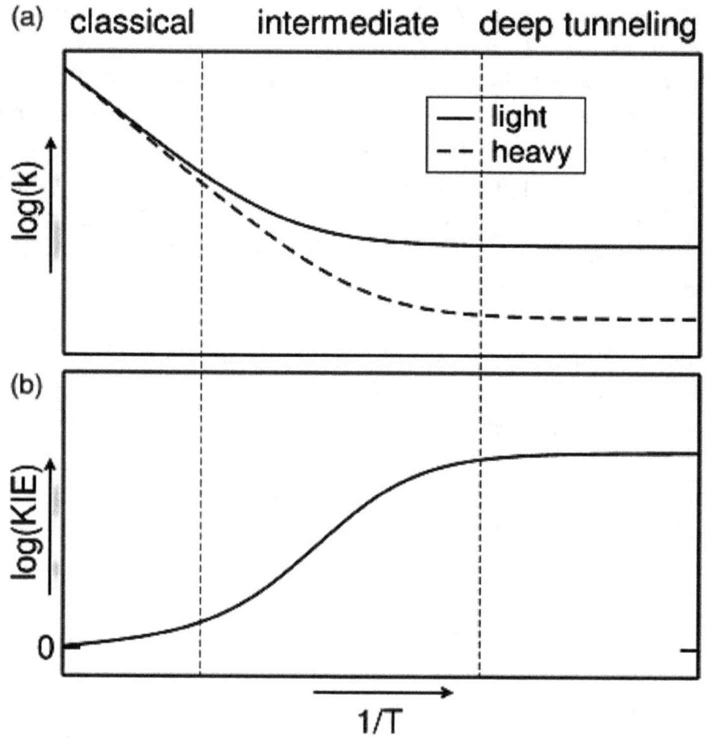

Figure 3.1 Temperature dependence of isotope chemical-reaction rate (a) and quantified kinetic isotope effect (b), considering full quantum effects.

quantum tunneling, the logarithm of the reaction rate no longer depends linearly on the reciprocal of temperature. At low temperatures, the logarithmic reaction rate does not vary with temperature (see Figure 3.1); this cannot be understood if the nucleus is treated classically.

The kinetic isotope effect (KIE) is one of the most direct methods for verifying the presence of quantum tunneling in chemical reactions. In the KIE experiment, the reactant atom is replaced with its isotope during the chemical reaction. Reactions with and without this substitution are performed separately. By comparing the two reaction rates, information about the full quantum effects of the chemical reaction is obtained. As illustrated in the upper panel of Figure 3.1, if the reaction involves breaking the bonds between the reactant atom and the atoms of other elements, the reaction rate should differ when the reactant atom is replaced with its isotope. Because chemical bonds are less likely to break for heavier atoms, replacing labeled reactants with heavier isotopes leads to slower reaction rates in most cases. For example, if the atoms being studied are H and D, the reaction rate k_H involving a C–H bond is usually several times higher than the reaction rate k_D involving a C–D bond when excluding full quantum effects. However, if quantum tunneling occurs, the ratio k_H/k_D is much greater than 10. This issue will be discussed in detail in subsections 3.2.3 and 3.2.4. Note that as the KIE may be affected by other complex factors, such as diffusion limitation, NQEs on the reaction process cannot be ignored completely, even if the experimentally measured KIE values are not very large [Hama2015].

Nuclear quantum tunneling affects not only the reaction rate but also the reaction products, allowing for the product selectivity to be controlled. In a "parallel reaction" (a relatively common phenomenon in organic chemistry), two or more reaction pathways simultaneously yield two or more products, typically determined via thermodynamic or kinetic control. The selective regulation of reaction products is based on the following principle: If the products form before reaching chemical equilibrium, the mechanism using the production rate difference to control the product distribution is called "kinetic reaction control," and the major products are called "kinetic products." The mechanism using the production rate difference depends on the heights of the potential barriers – for example, TS (1) or TS (2) in Figure 3.2. If the reaction system is allowed to reach equilibrium before product separation, the mechanism using the thermal energy difference of the various products to control the product distribution is referred to as "thermodynamic reaction control," and the main products are called "thermodynamic products." The mechanism for using the thermal energy difference depends on the depths of the potential wells – for example, see the lowest energy points at C or B in Figure 3.2.

The situation is very different for chemical reactions affected by nuclear quantum tunneling. From a simplistic perspective, tunneling effects may appear to be a special case of kinetic control when the energy of the potential barrier is zero. However, this assumption is naïve. For reaction pathways through potential barriers with relatively high energies, quantum tunneling occurs when the width of the barrier is sufficiently narrow. In this case, thermodynamic products may be obtained even under kinetic control conditions. Therefore, quantum tunneling is an independent process for chemical reaction products. In addition to

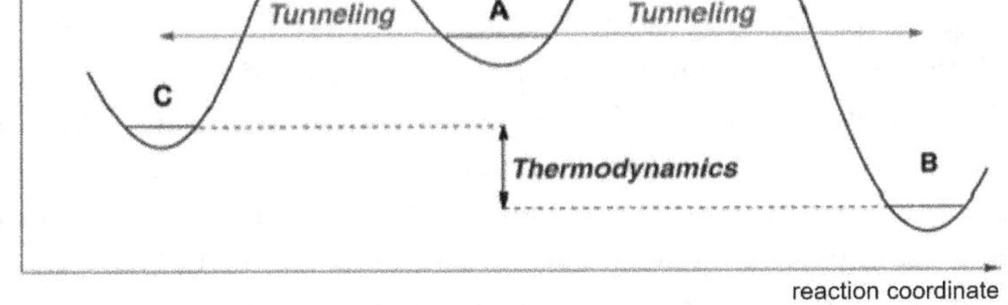

Figure 3.2 Illustration of nuclear quantum-tunneling effects on the product selectivity of a complex chemical reaction. In addition to the traditional concepts of thermodynamic and kinetic reaction control, tunneling control must be considered as an important factor in modulating the reaction products. Reprinted with permission from Schreiner[2017]. Copyright 2024 American Chemical Society.

traditional kinetic and thermodynamic control processes, it is necessary to consider the tunneling control process (indicated by the horizontal arrow in Figure 3.2) [Schreiner2017]. For thermodynamic reaction control, the product selectivity is higher for the product with a lower free energy (such as energy well B in Figure 3.2). For kinetic reaction control, the product is generated faster through the reaction pathway with a lower-energy potential barrier (such as the TS (1) barrier in Figure 3.2). For tunneling reaction control, product selectivity depends on a reaction pathway with a narrower potential barrier (such as the barrier between energy wells A and C in Figure 3.2).

If the nucleus is treated classically, chemical reactions through a reaction potential barrier with relatively high energy are unlikely to occur when thermal fluctuations are not sufficient to drive the reactants across the potential barrier. However, full quantum effects, specifically nuclear tunneling, can violate this classical law. In interstellar chemical reactions, hydrogen, helium, and a large number of formaldehyde molecules can react through nuclear tunneling in low-temperature environments such as outer space, breaking the traditional chemical reaction limit. Consequently, the synthesis of numerous organic molecules from simple inorganic elements is possible in interstellar chemistry. It is, therefore, reasonable to infer that such tunneling is a key mechanism in the formation of complex organic compounds required for early life or even in the origin of life itself; in addition, it may be accompanied by nonadiabatic processes resulting from many-body quantum coupling under radiation.

3.1 Transition-State Theory of Chemical Reactions

The establishment of TST is a major milestone in the history of chemistry. This simple, yet profound and universal way of visualizing the mechanisms and basic kinetic processes of chemical reactions has enabled a wide range of applications. The development of TST can be traced back to the introduction of the concepts of the potential energy surface and saddle point into chemical reaction theory by H. Pelzer and E. Wigner in 1932. In 1935, H. Eyring, M. G. Evans, and M. Polanyi formulated TST in its standard form based on statistical thermodynamics and quantum mechanics. In 1938, Wigner presented a rigorous argument to justify the theory.

In a chemical reaction, the transformation of reactant molecules into product molecules is accompanied by the breaking of old bonds and the formation of new ones. As shown in Figure 3.3, the reacting system passes through a short-lived transition state (also called an activated or transition-state complex), in which the old chemical bonds of the reactant molecules are on the verge of breaking, and new bonds in the product molecules are about to be formed. The total reaction rate is determined by the conversion rate of the reactant changing from its initial state to the transition-state complex. In 1889, based on numerous experiments, the Swedish chemist S. A. Arrhenius proposed a basic expression for reaction rate k. This is expressed as

$$k = Ae^{-\frac{E}{k_B T}}, \tag{3.1}$$

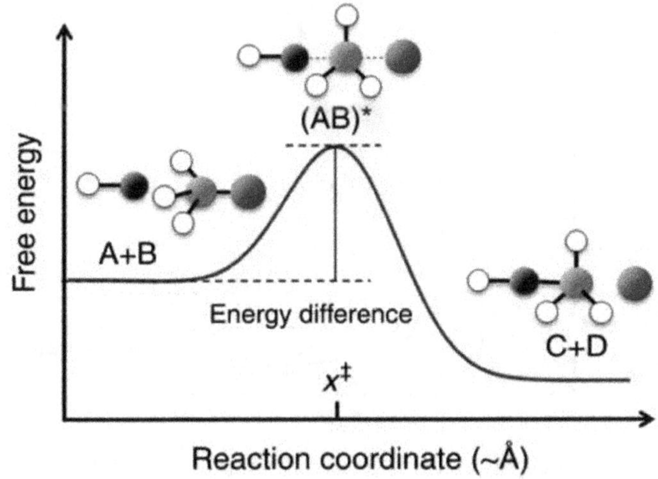

Figure 3.3 Schematic of the potential energy profile of classical transition-state theory. A + B, (AB)*, and C + D correspond to the reactant, transition-state complex (also called a transition state), and generated product, respectively. The energy difference between the reactant and the transition-state complex is the transition potential barrier. Figure created by Vestergaard[2016] licensed under CC BY 4.0.

where E is the height of the potential barrier along the reaction coordinate (i.e., the activation energy for the reaction), and A is the pre-exponential factor. This equation is known as the Arrhenius equation. Eyring later derived an expression for the pre-exponential factor in terms of the partition functions of the reactant and transition-state complex. The Arrhenius equation expresses the reaction in a simple form corresponding to a clear physical picture, that is, chemical reactions must overcome the activation energy, follow the principle of minimal energy pathway.

Early TST operated within an intrinsically classical framework; it assumed that the momentum and position along the reaction coordinates could be determined simultaneously, contrary to the uncertainty principle in quantum mechanics.

The dividing surface is an important concept in TST. As shown in Figure 3.4, the potential energy surface can be divided into reactant, product, and transition-state regions. The transition-state region is the region of configuration space near the dividing surface, XY, of the reaction. In Figure 3.4, the reactant region is below the dividing surface and transition-state region (see the arrow AB + C), while the product region is above the dividing surface and transition-state region (see the arrow A + BC). Thus, the potential energy surface provides the stage for the reaction process to occur, and the easiest pathway that connects reactants and products on the potential energy surface is the trajectory the entire chemical reaction takes. Reaction rates are determined by calculating the flux integral of the trajectory through the dividing surface.

The two basic assumptions adopted by classical TST are the local equilibrium assumption and the no-recrossing assumption [Luo Y.R.1983]. The former implies that the

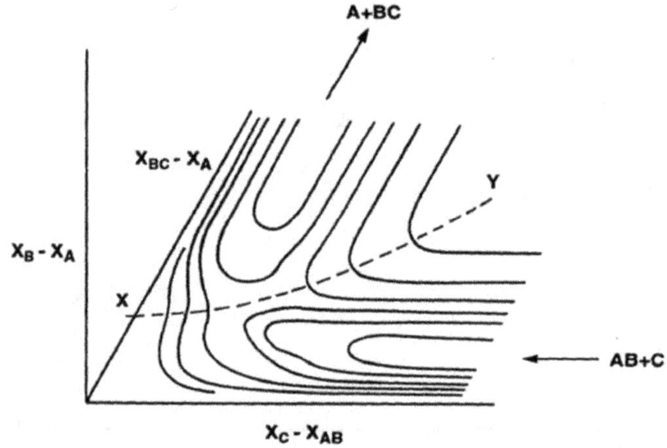

Figure 3.4 Potential energy contours of the chemical reaction AB + C → A + BC in the collinear case. The dashed line XY represents the dividing surface of the reaction. Reprinted figure with permission from Marcus[1993]. Copyright 2024 by the American Physical Society.

Born–Oppenheimer approximation holds near the dividing surface and that the variables of the Hamiltonian are separable; this is also known as the adiabatic condition. The second assumption is that the reaction trajectory crosses the dividing surface only once and does not return to the reactant region. In TST calculations, the atomic position most closely linked to the breaking or formation of bonds is generally chosen as the reaction coordinate, and the configuration corresponding to the highest free energy along the reaction trajectory (i.e., the saddle point) is the transition state. However, it is unclear how an appropriate dividing surface should be determined. This is a key issue in the study of chemical reactions. In early TST, the dividing surface was typically chosen to pass through the saddle point. However, this is no longer the only method used.

Since the 1970s, the assumptions underlying TST have been further examined, and quantum effects have been included, leading to the development of a generalized (or variational) TST. In the generalized TST, the dividing surface is determined by the variational method. Numerous review articles have been published on these improvements; interested readers can refer to references [Luo Y.R.1983, Truhlar1983] for further reading. In this section, we focus on new developments in TST related to full quantum effects.

In actual chemical reactions, full quantum effects cannot be neglected. In particular, the role of full quantum effects is decisive at low temperatures or when light elements are involved. The quantum energy levels of the reaction system and nuclear quantum tunneling are the most significant full quantum effects to observe in the reaction process. The zero-point energy and discrete energy levels of the reactant, transition-state complex, and product affect the potential barriers and, thus, the conversion rate and pathway of the overall reaction. The reactant occupies the minimum potential energy surface and exhibits discrete quantum energy levels. The quantized transition state with discrete energy levels

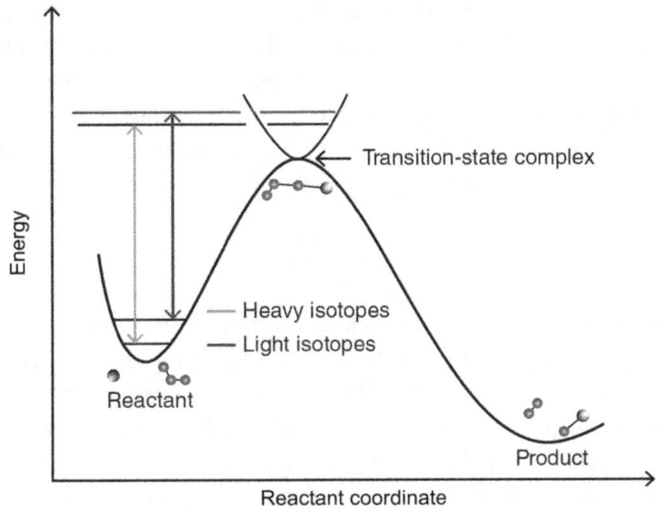

Figure 3.5 Zero-point energies of the reactant and the transition-state complex. Figure created by Yang S.[2016] licensed under CC BY 4.0.

also occupies the minimum energy surface along the direction of the vertical reaction coordinate. The transition potential barrier and interaction between the reactant and transition states depend on these two sets of discrete energy levels (see Figure 3.5). The quantum tunneling effect, on the other hand, belongs to the category of kinetics and allows the occurrence of reactions that are impossible in the classical framework.

In TST, the role of full quantum effects in chemical reactions can be studied either by quantum mechanically correcting the classical TST or by developing a completely new theory that is entirely within the framework of quantum mechanics. Let's start to briefly describe the first approach.

It has long been known that the classical TST only provides an upper limit for the reaction rate. The true reaction rate is determined by multiplying this upper limit by the system-dependent transmission coefficient, γ. Therefore, when making quantitative corrections to TST, the corrected reaction rate can be expressed as the classical rate multiplied by the correction factor given by

$$k^{\mathrm{qm}} = \gamma k^{\mathrm{cl}}, \tag{3.2}$$

where the superscripts "qm" and "cl" indicate quantum and classical reaction rates, respectively.

One simple correction method incorporates the quantum energy-level effect in TST by introducing a zero-point energy correction and replacing the classical potential with a quantum or adiabatic vibrational potential [Luo Y.R.1983]. This approximation is applicable at low temperatures or low energies, where the reaction system remains mostly in the ground state. Garrett et al. used variational TST to derive the reaction rate after considering the quantum energy levels. The transmission coefficient γ corresponds to the adiabatic vibrational ground state of the system [Garrett1980].

A further correction along this line treats the reaction coordinate classically but quantizes the internal degrees of freedom. The classical partition function for the internal binding degrees of freedom of the reactant and transition state is replaced with a quantum partition function [Hu X.G.1991]. Based on the harmonic approximation of the potential energy surface, γ in Eq. (3.2) can be written as the ratio of the quantum partition function to the classical partition function [Song K.2016]:

$$\gamma = \frac{Z_{\mathrm{qm}}^{\neq} Z_{\mathrm{cl}}^{\mathrm{R}}}{Z_{\mathrm{cl}}^{\neq} Z_{\mathrm{qm}}^{\mathrm{R}}}, \tag{3.3}$$

where $Z_{\mathrm{qm}}^{\neq}$ and $Z_{\mathrm{cl}}^{\neq}$ denote the quantum and classical partition functions of the transition state, respectively, and $Z_{\mathrm{qm}}^{\mathrm{R}}$ and $Z_{\mathrm{cl}}^{\mathrm{R}}$ represent the quantum and classical partition functions of the reactant state, respectively. Note that the transition state is a saddle point on the potential energy surface, and there is an imaginary frequency associated with it when expanded to the second order. However, the calculation of the transition-state partition function does not include the contribution of the imaginary frequency.

Traditional TST fails to describe quantum tunneling. This can cause significant errors, particularly when analyzing reactions at low temperatures. The inclusion of tunneling is equivalent to multiplying the classical reaction rate by a correction factor [Song K.2016], which is referred to as the tunneling transmission coefficient. Within classical TST, this transmission probability is 0 for energies below the potential barrier and 1 for energies above it. The quantum tunneling effect makes this probability nonzero for energies below the potential barrier and less than 1 for energies above the potential barrier (nonclassical reflection). The tunneling transmission coefficient includes the quantum tunneling effect below the potential barrier and the nonclassical reflection effect above the potential barrier [Truhlar1996]. The quantum tunneling pathway may be completely different from the classical pathway that requires the minimum energy and may constitute a "shortcut" [Luo Y.R.1983]. At this time, the reaction may occur through a potential energy path close to the minimum energy path (MEP), which is called Small Curvature Tunneling (SCT). For this situation, starting from the MEP to find the SCT is the most effective method. Of course, the reaction may also occur through an energy path far from the MEP, that is, Large Curvature Tunneling (LCT). For instance, when studying the exchange reaction of hydrogen atoms and molecules $H + H_2 \rightarrow H_2 + H$, R. A. Marcus and M. E. Coltrin found that the tunneling reaction path connecting the vibrational amplitude extrema of the given vibrational states (the dashed line between P and P′ in Figure 3.6) [Marcus1977] was much shorter than the path of steepest ascent calculated according to classical TST. This corresponded to a narrow potential barrier. The calculated reaction rates using this shortcut were one to two orders of magnitude greater than those obtained using the standard path. Another instance is the reaction process of carbon atoms in organic molecules (see subsection 8.4.3).

An alternative to correcting the classical theory is to develop a rigorous TST based on the basic principles of quantum mechanics. However, as stated previously, quantum TST cannot be derived from classical TST: The two basic assumptions of the classical theory imply that momentum and position in the reaction coordinates can be measured at the same time, contradicting the uncertainty principle. The quantum TST must consider quantized

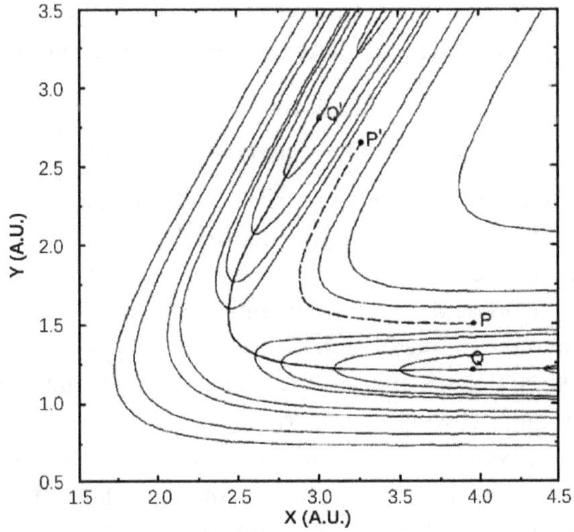

Figure 3.6 $H + H_2 \rightarrow H_2 + H$ reaction potential energy surface. The solid curve through Q and Q′ is the path of steepest ascent, and the dashed curve through P and P' is the quantum-tunneling pathway connecting the maximum vibrational amplitudes in the given vibrational states (here the zero-point states). Reprinted from Marcus[1977], with the permission of AIP Publishing.

nuclei. A groundbreaking advance occurred when W. H. Miller et al. derived a rigorous expression for the quantum reaction rate, which can be expressed as the time integral of the quantum-mechanical current–current correlation function [Miller1983]. Because the calculation of time-dependent current functions requires solving the time-dependent Schrödinger equation and a rigorous reaction-rate formulation requires evaluating the time integral at an infinite time, further approximations are needed for practical calculations [Truhlar1996, Song K.2016]. However, this time-dependent approach naturally includes all of the dynamic effects.

Another class of widely used methods is path-integral quantum TST [Voth1993]. This approach considers all the quantum-mechanical effects of the system and can also include the anharmonic effects of the potential energy surface. The details of such methods are discussed in Chapter 5, where we cover the dynamic properties of full quantum effects.

In recent years, both modified classical approaches and entirely quantum-mechanical approaches to TST have been developed and have found many important applications. However, there is no one-size-fits-all solution for all chemical reaction problems. Many substantial advances in TST over the past few decades relate to NQEs and quantum dynamics. We believe that a more precise treatment of full quantum effects will result in a more comprehensive TST in the future.

3.2 Nuclear Quantum Effects

Full quantum effects can affect the chemical properties of condensed matter. In fact, in the study of materials at the microscopic level, it is difficult to distinguish between problems belonging to physics and chemistry. Here, we simply follow this convention and focus on the full quantum effects of condensed matter in chemical reactions at the molecular level.

3.2.1 Surface Catalytic Reactions

Developing theories that can accurately predict chemical reaction rates is critical, not only for basic theoretical chemistry but also for industrial catalytic applications. Theoretical studies of real catalytic reactions often involve complex calculations, making it difficult to directly compare theoretical and experimental results. One possible solution to this problem is to choose only simple chemical reactions as model systems. Unfortunately, even for the simplest model system, very few comparative studies have been reported. This is because determining accurate experimental measurement of the reaction rate is incredibly challenging, even if the most basic reactions in surface chemistry are selected.

The process by which hydrogen atoms combine into diatomic molecules on transition-metal surfaces was studied recently by D. Borodin et al. [Borodin2022], using H atoms as well as heavier D atoms as reactants. The products obtained were H_2, HD, and D_2. The isotope effects on reaction rates can be used to study NQEs. It is generally believed that NQEs are only pronounced at low temperatures and that a classical description is sufficiently accurate at high temperatures. For example, for gas-phase reactions, it is commonly assumed that NQEs do not need to be considered above 500 K. However, the group from Germany measured the reaction rate of H atoms combining to form H_2 molecules on a Pt surface in the temperature range of 250–1,000 K. By comparing these results with model calculations, they found that NQEs were critical to the reaction process, even at temperatures greater than 500 K. The electron spin degrees of freedom were also found to be crucial. Neglecting the NQEs and electron spin degeneracy would cause an overestimation of the reaction rate by one to three orders of magnitude.

Borodin et al. used pulsed molecular beams of an H_2/D_2 mixture to irradiate Pt (111) and (332) surfaces. Velocity-resolved kinetic measurements were performed to obtain an accurate instantaneous reaction rate. Figure 3.7 shows the experimentally measured and theoretically simulated H atom recombination rates, as well as their temperature and time dependence. It can be seen that the theoretical model agreed well with the experimental data over a wide temperature range, which supports the validity of the theoretical analysis. They constructed a quantum rate model (QRM) that considered isotope effects on the reaction rate. The specific temperature-dependent reaction rate (e.g., for the formation of H_2 molecules) can be expressed as follows:

$$k_{H_2}(T) = \left\langle S_0^{H_2} \right\rangle(T) \sqrt{\frac{k_B T}{2\pi m_{H_2}}} \frac{Q_{H_2}/V}{\left(Q_{H^*}/A\right)^2} \exp\left(-\frac{E_0^{H_2}}{k_B T}\right), \tag{3.4}$$

where $\left\langle S_0 \right\rangle(T)$ is the thermal sticking probability, E_0 is the absorption energy, Q_{H^*} is the reactant partition function, and Q_{H_2} is the product partition function. V and A represent the reference volume and area for partition function evaluation, respectively. ${E_0}^{H_2,HD,D_2}$ and $\left\langle S_0^{H_2,HD,D_2} \right\rangle(T)$ can be obtained from experiments, while the value of Q_{H_2} for the gas

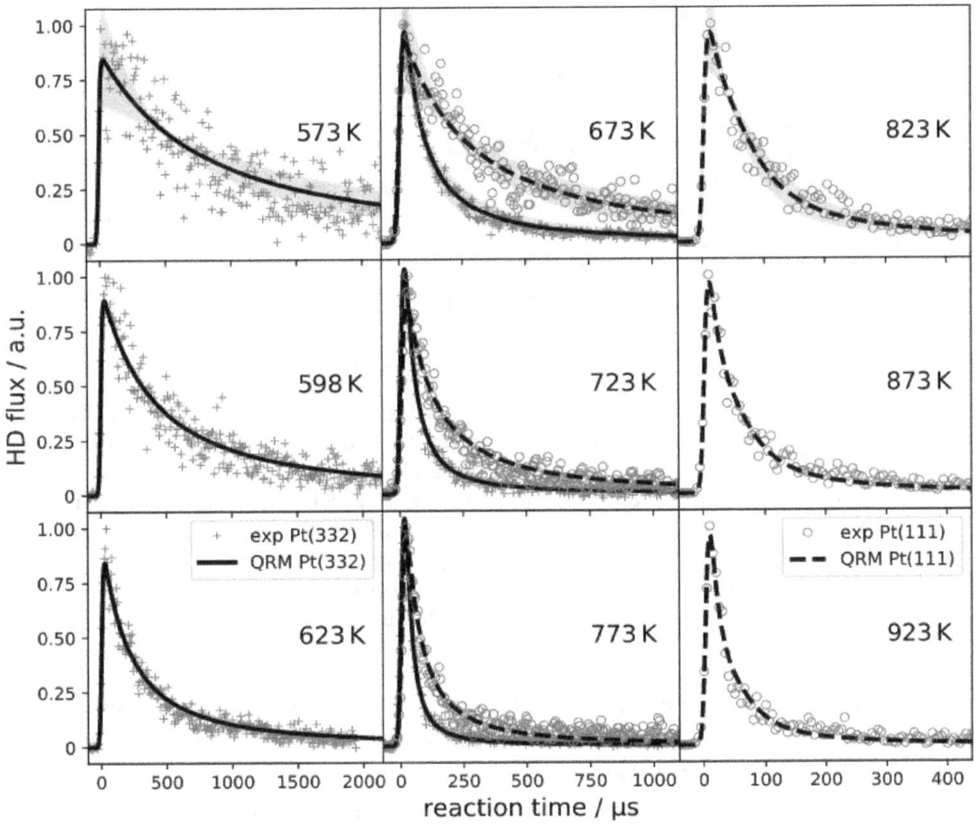

Figure 3.7 Velocity-resolved-kinetics measurements of H atom recombination into HD molecules at the Pt (111) (∘) and (332) (+) surfaces compared with quantum rate model (QRM) simulation results (dashed ((111) surfaces)) and solid curves ((332) surfaces)) for the different surfaces. From Borodin[2022]. Reprinted with permission from AAAS.

phase has also been obtained through extensive experimental measurements. However, the reactant partition functions $Q_{H^*,D}$ must be determined for the reaction rate calculation to be accurate. Borodin et al. used quantum potential energy sampling to compute $Q_{H^*,D}$. The partition functions included contributions from both nuclei and electrons. More specifically, corresponding to the contribution of the nucleus part, the quantum energy levels and nuclear wavefunctions were obtained by solving the Schrödinger equation, and the nuclear partition function was calculated based on these results. For the electronic contribution, as each electron orbital can be occupied by two electrons with opposite spins, the partition function was calculated in a subspace with twofold spin degeneracy.

Figure 3.8 shows the results of different experiments. It can be seen that older experimental data have large uncertainties, whereas those obtained by Borodin et al. are much more accurate. The prediction results of different theoretical models, including QRM, harmonic transition-state theory (hTST), the classical rate model (CRM), and QRM without the contribution of electron spin, are also shown in Figure 3.8(a). Evidently, the QRM

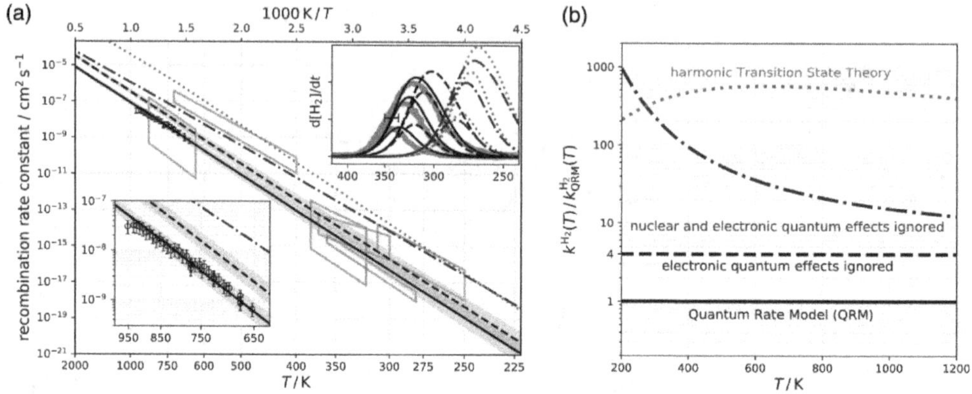

Figure 3.8 Reaction rates of H atoms combining to form H_2 molecules on the Pt (111) surface. (a) Experimental data measured by Borodin et al. (denoted by ∘) are compared with the results of the quantum rate model (QRM) (black solid line), the harmonic transition-state theory (upper light gray dotted line), the classical rate model (gray dashed–dotted line), and QRM neglecting electron spin (deep gray dashed line). The gray boxes around the lines indicate the rate uncertainties. The inset graph at the top right shows the results from previous work measured by different groups. (b) Comparison of the reaction rates predicted by four theoretical models. From Borodin[2022]. Reprinted with permission from AAAS.

results, which considered both NQEs and electron spin, are in excellent agreement with the experiment over a wide temperature range, while the other models deviate from the experimental results to some extent.

In order to compare the different models more clearly, the ratios of the reaction rates predicted by the different models are compared with the QRM rate in Figure 3.8(b). The worst prediction is given by hTST, which overestimates the reaction rate by two to three orders of magnitude over a wide temperature range of 200–1,200 K. The overestimation results from the use of harmonic oscillators without interactions to approximate the description of the potential energy surface and the lack of a recrossing correction from the transition state back to the reactant state. Although CRM outperforms hTST, the model overestimates the rate constant by a factor of ~20, even at high temperatures (>1,000 K). The CRM partition function is computed using complete potential energy surface sampling, a method that neglects the quantum effects of nuclei and electrons. However, CRM yields much better predictions than hTST, especially at high temperatures, because it accounts for anharmonic effects. This is because the difference between the quantum and classical descriptions of the nucleus becomes less pronounced at high temperatures, and the anharmonic effects from thermal fluctuations are enhanced. Overall, in these methods above, the QRM simulation results agree best with the experimental values. Based on the QRM result, if the contribution of the electron spin to the partition function is ignored, the reaction rate given is overestimated by a factor of four. This is because two H atoms can combine to form four states (spin singlet and triplet), and only the singlet state is the ground state of the product. If spin degeneracy is not considered, then all four states can be products, and the estimated reaction rate will be four times larger than the actual rate.

By comparing theoretical models that account for different physical effects with the experimental results, Borodin et al. demonstrated the importance of considering the full quantum effects on both nuclear quantum state and electron spin state to accurately describe catalytic reactions on metallic surfaces.

3.2.2 Ultracold Chemical Reactions

Controlling the quantum states of molecules has long been considered desirable because it allows chemists to break chemical bonds at will and form new ones by connecting with any desired atoms. Preparing ultracold molecules with designated quantum states to study their interaction laws is the first step toward making quantum state control a reality. As early as 1917, A. Einstein published "On the quantum theory of radiation" [Einstein1917]. According to his theory, when a traveling atom is irradiated by an incident light of the correct frequency, not only is the photon energy absorbed, triggering an electronic transition outside the nucleus, but the momentum of the atom also decreases slightly with each photon collision until its value is reduced to a minimum; thereby, an atom irradiated by a laser can be cooled and trapped. The Doppler effect present in the laser cooling and trapping of atoms is illustrated in Figure 3.9. Using this principle, S. Chu et al. reported the cooling of Na atomic gas to a temperature of 240 μK via laser cooling in 1985. Bose–Einstein condensation, a macroscopic quantum phenomenon, was realized in 1995 based on a combination of laser cooling and trapping techniques and the evaporative cooling technique. This work has opened a new chapter in physics in which the preparation of ultracold atoms and molecules with defined quantum states is possible.

A molecule is the smallest unit of matter that retains all of its chemical properties. Molecules serve as the bridge between quantum mechanics, which describes the microscopic properties of matter, and chemistry, which studies the laws of atom–molecule interactions. In particular, accurately measuring a Bose–Einstein condensate or degenerate Fermi gas of molecules allows us to obtain quantum information regarding ultracold chemical reactions and discover new states of matter. It has profound implications for

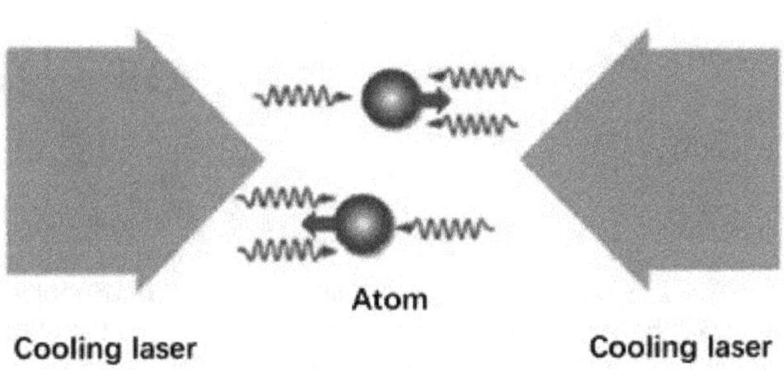

Figure 3.9 Schematic of the Doppler effect in the laser cooling and trapping of atoms. Compared to the atomic transition frequency in the laboratory frame, the cooling-laser frequency is redshifted. Atomic deceleration can be achieved when atoms interact with incident photons. Figure created by Wang B.Y.[2016] licensed under CC BY 4.0.

finding quantum-mechanical applications in data storage and transmission and for understanding and controlling chemical reactions, which is the fundamental task of chemistry research. However, diatomic molecules are much more complicated than single atoms [Schawlow1981]. Ultracold molecules not only have multiple internal energy levels but also rotate and vibrate, making the study of ultracold chemistry complex and challenging. Although realizing complete quantum-state control of an ultracold molecular gas remains a challenge, rapid advances in laser cooling technology and ultracold physics over the last decade have led to more efficient and stable ultracold-molecule preparation methods [Wu J.Z.2018]. These methods are highly flexible, maneuverable, and precise, opening new avenues in ultracold molecule and ultracold chemistry research. The chemical reactions of ultracold atoms and molecules are dominated by quantum mechanics, exhibiting ubiquitous NQEs. Studying these reactions can enrich our understanding of the quantum world, which, in turn, can further improve our knowledge of ultracold atom–molecule interactions and ultimately lead to the development of effective applications.

One of the most important recent advances in the exploration of ultracold chemical reactions is the preparation of fermionic diatomic ultracold molecules, along with the production of a quantum degenerate Fermi gas. The polar potassium–rubidium molecule (KRb) is one of the most studied examples of such fermionic diatomic ultracold molecules [Ni K.K.2008]. Polar molecules have both long-range interactions and tunable dipole–dipole interactions. Their permanent electric dipole moment adds an additional electric-field "control" that can have a significant impact over the course of a reaction at ultracold temperatures. Because K is a fermion (with a half-odd integer total spin) and Rb is a boson (with an integer total spin), molecules made of the two elements exhibit fermionic characteristics. According to the Pauli exclusion principle, when the nuclear spins of all molecules in the lowest rotational-vibrational state are equal, quantum degeneracy requires antisymmetry in the other degrees of freedom for the wavefunctions of fermions. This antisymmetry can stabilize the ultracold KRb molecular system and suppress chemical reactions (as shown in Figure 3.10(a)) under appropriate trapping configurations. Below the Fermi temperature

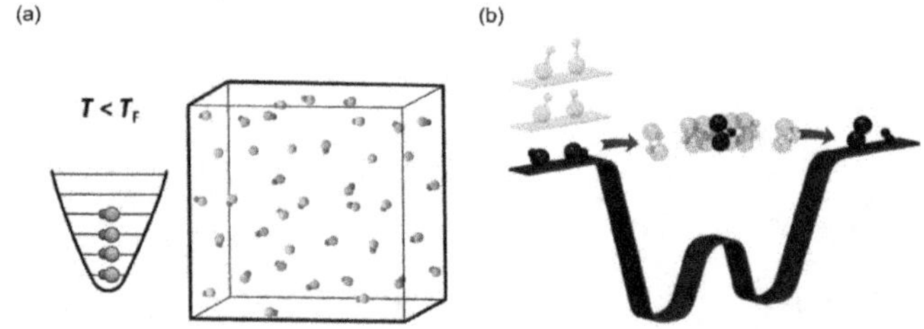

Figure 3.10 (a) An illustration of ultracold quantum gas of KRb molecules below the Fermi temperature, T_F. Because of Fermi–Dirac quantum statistics, the molecules can have a highly uniform density and be largely immune to destructive collisions, which results in low reaction rate. From Zelevinsky[2019]. Reprinted with permission from AAAS. (b) An illustration of the intermediate state $K_2Rb_2^*$ of the chemical reaction 2KRb $\rightarrow K_2Rb_2^* \rightarrow K_2 + Rb_2$. This intermediate state was observed for the first time using ultracold molecular methods. [Hu M.G.2019].

(which has a magnitude of tens of nK), this system can last for several seconds, providing more opportunities for convenient quantum-state engineering, precision measurements, and technological applications [De Marco2019].

Interestingly, when ultracold KRb molecules collide to form Rb_2 and K_2 molecules, the effects of the nuclear quantum state on the chemical reaction rate have also been observed. In particular, the reaction rate is more than 100 times faster when the nuclear spins in the molecules are disordered than when they are aligned all up or all down. This result can be attributed to the Pauli exclusion principle in an identical fermionic system, which can manifest as a repulsive barrier that may only be overcome by tunneling. Moreover, the ultracold temperature can force the KRb molecules to remain in the intermediate stage of the reaction for millions of times longer than usual, providing enough time to study chemical bond breaking and formation. This allows the capture and direct observation of the critical and elusive reaction intermediate state, $K_2Rb_2^*$, and its corresponding quantum state during the reaction via photoionization detection (Figure 3.10(b)), creating new ways to uncover full quantum effects on chemical reactions [Hu M.G.2019].

As mentioned in our earlier discussion of TST, potential barriers exist in most chemical reactions involving neutral atoms and molecules. However, at ultracold temperatures, the de Broglie wavelengths of the atoms and molecules are much larger than the interaction length. Both experimental results and theoretical calculations have confirmed that chemical reactions with potential barriers that cannot be overcome classically can still occur via quantum tunneling. Typical examples include F + H_2 reactions [Takayanagi1998] and Li + HF reactions [Wech2005]. Studies have shown that the reaction rates are significantly higher when the reactions are dominated by tunneling. This is true not only for reactions of light elements, such as hydrogen, but also for heavier elements, such as fluorine. It has also been observed that weak van der Waals interactions and external electric or magnetic modulation have little or no effect on the reaction process at high temperatures. However, they have a significant impact at temperatures close to absolute zero, which is an important experimental research frontier for tunable ultracold chemical reactions.

Plasma chemistry is another area where NQEs are important. For example, one of the central concepts of plasma chemistry is the Penning discharge, which is the ionization that occurs when atoms or molecules of one neutral species collide with those of another with a higher ionization potential. Such self-ionization only occurs during collisions, and nuclear quantum dynamics clearly plays a major role. These studies suggest that NQEs significantly influence the chemical reaction at ultralow temperatures or energies. In fact, when a sufficiently sophisticated experimental design is employed, the Penning discharge of metastable He atoms colliding with Ar and H_2 at different energies can be observed [Henson2012]. This experiment reveals an unusual peak in the Penning discharge reaction rate at 1 meV (11.5 K); whereas in the classical theory, the reaction rate decreases with temperature. The peak originates from scattering resonance as the particles tunnel through the angular momentum potential. In other words, when the collision energy matches the resonance state, particles tunnel through the barrier and are captured. Moreover, the increased probability of particle presence within the barrier eventually leads to a higher ionization rate.

3.2.3 Organic Chemical Reactions

Organic chemistry, or the chemistry of carbon compounds, is an important branch of chemistry. The transfer of hydrogen atoms between different organic molecules or functional groups can occur through positive ion (proton) transfer, atom transfer, or negative ion transfer. Because of the low mass of hydrogen, pronounced NQEs can be observed in these transfers, even at room temperature, manifesting as a large KIE. For organic reactions with H and D atoms, the rate-constant ratio k_H/k_D normally ranges from 6 to 10. In other words, a reaction involving C–H bonds is 6–10 times faster than a reaction containing C–D bonds. However, if quantum tunneling occurs during the reaction, the k_H/k_D value will be greater than 10. For example, in the reaction between nitropropane and 2,4,6-trimethylpyridine (shown in Figure 3.11), the α-H of nitropropane is deprotonated by the steric hindrance of pyridine, followed by iodine substitution. The KIE ratio of this reaction reaches 25 at room temperature (25°C), which clearly indicates that nuclear tunneling by hydrogen atoms occurs [Lewis1967].

Parallel reactions have complicated mechanisms and pathways that are common in complex organic systems. As discussed earlier, in classical chemical reactions driven by thermal fluctuations, the reactant molecules can only overcome the potential barrier when they gain sufficient energy. Because of the difference in the potential barrier heights along the different reaction pathways, the reactions can be either thermodynamically or kinetically controlled. For thermodynamically controlled reactions, the product selectivity depends on which product has a lower free energy. For kinetically controlled reactions, the product selectivity depends on which reaction pathway has a lower potential barrier. However, if quantum tunneling of nuclei is involved in the reaction, the effect of tunneling control on product selectivity must also be considered. In the case where quantum tunneling is prevalent, the product selectivity depends on which reaction pathway has a narrower barrier width.

In some special cases, the presence of quantum tunneling not only alters the product selectivity but can also result in products that are not obtainable through classical chemical reactions. For example, in the isomerization reaction of methylhydroxycarbene shown in

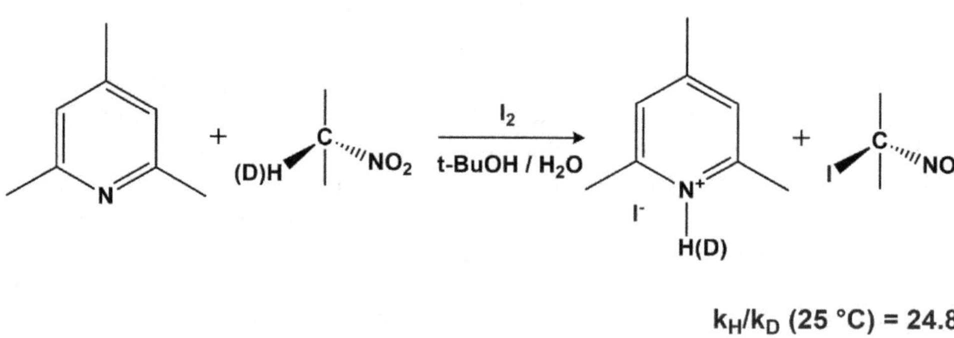

Figure 3.11 Iodination of nitropropane in the presence of 2,4,6-trimethylpyridine, showing isotopic effects

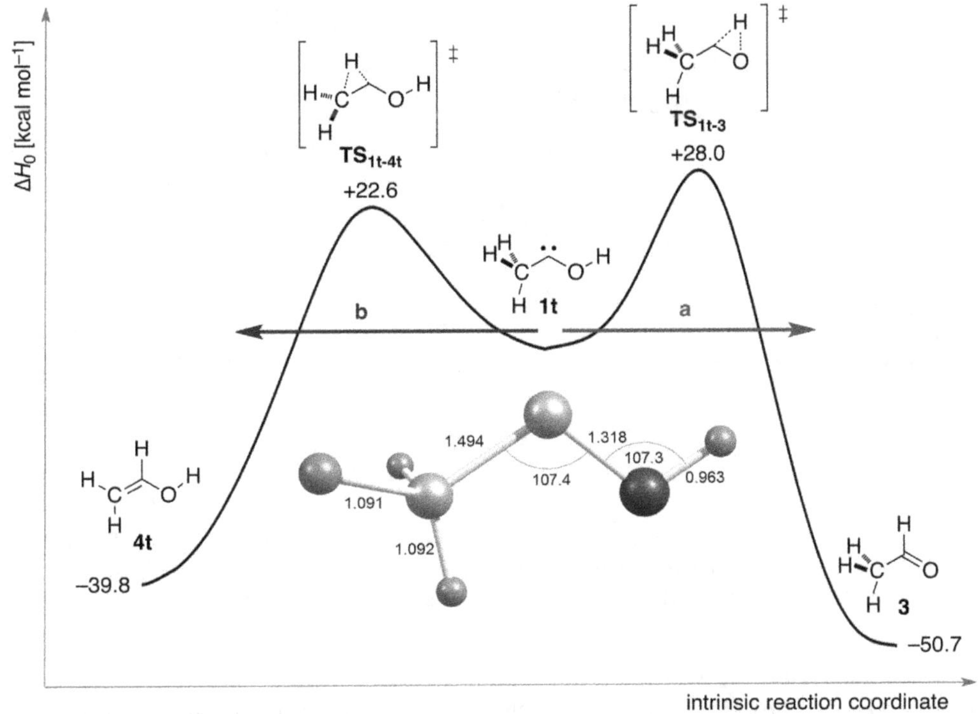

Figure 3.12 One-dimensional potential energy surface for the isomerization reaction of methylhydroxycarbene. Path "a" corresponds to the thermodynamically controlled reaction, and the reaction product is formaldehyde; path "b" corresponds to the kinetically controlled reaction, and the reaction product is an enol. From Schreiner[2011]. Reprinted with permission from AAAS.

Figure 3.12 [Schreiner2011], the possible reaction products are enols (see path "b") and formaldehyde (see path "a"). The former are kinetically controlled products, whereas the latter is a thermodynamically controlled product. However, since the formation of both products requires hydrogen transfer, which occurs in the methyl group and the hydroxyl group for the respective reactions, the reaction potential barriers for both are high. If the entire process is dominated by classical chemical reactions, these reactions are not expected to occur at low temperatures. However, numerous experimental results have shown that they can occur at temperatures even as low as 10 K, owing to the quantum tunneling effect of the hydrogen nuclei, and thermodynamically stable formaldehyde molecules can thus be produced. This is a typical result of a quantum tunneling-controlled reaction. Formaldehyde has a narrower potential barrier in its reaction pathway (see path "a" in Figure 3.12), which has a higher tunneling probability than the reaction pathway that produces an enol (see path "b" in Figure 3.12). If the H atom is replaced by a D atom (which is less likely to tunnel), no reaction is observed at 10 K, further supporting the claim that the reaction mechanism is tunneling controlled.

In organic chemistry, nuclear tunneling effects are most commonly observed in reactions that contain reactive intermediate states. Tunneling is not limited to hydrogen atoms.

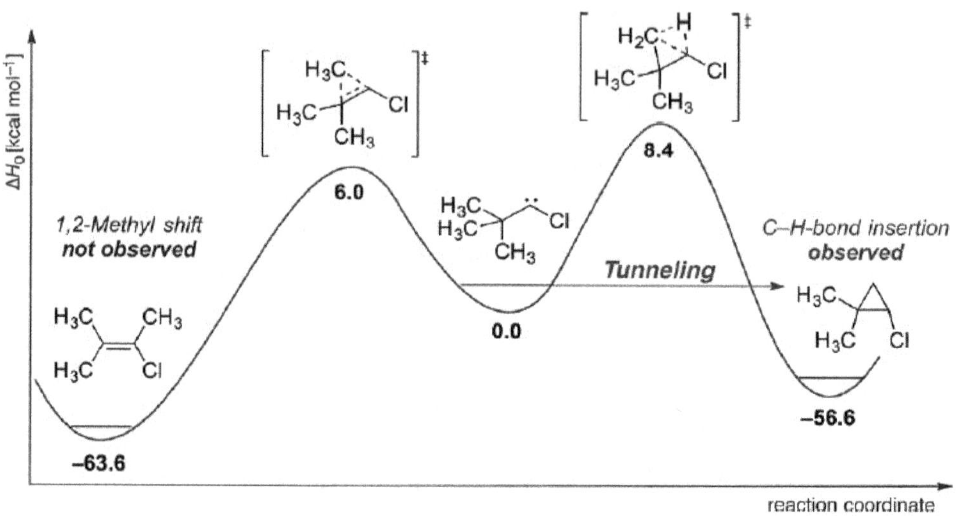

Figure 3.13 Potential energy barriers for the isomerization reaction of tert-butylchlorocarbene, showing quantum tunneling of the methyl group. Reprinted with permission from Schreiner[2017]. Copyright 2024 American Chemical Society.

The quantum tunneling of heavier atoms, such as carbon, and even of entire organic groups has been reported. For example, in the isomerization of tert-butylchlorocarbene [Zuev1994, Schreiner2017] shown in Figure 3.13, when the temperature is below 11 K the methyl group tunnels as a whole, generating a cyclopropane derivative that is forbidden for both thermodynamically and kinetically controlled reactions in the classical framework. In recent years, the KIEs of heavier elements (such as Li, B, C, N, O, F, S, Cl, and Br) have been investigated using quantum chemical calculations combined with NMR techniques to reveal the important role of zero-point energy and quantum tunneling in chemical reactions [Dale2021].

3.2.4 Electrochemical Reactions

The use of isotope effects in electrochemical reactions can be traced back to the early 1930s. Immediately after the discovery of deuterium, an isotope of hydrogen, by American scientist H. Urey in 1931, heavy water (D_2O) and semi-heavy water (HDO) were discovered [Brickwedde1982]. In natural water on Earth, the heavy water content is extremely low at ~0.0156%. Heavy water can be separated and prepared from water via a variety of methods, including electrolysis, distillation, chemical exchange, hydrogen-absorbing alloy adsorption separation, and laser separation [Brickwedde1982]. Electrolysis was the earliest method, using the H/D KIE.

The vibrational energy of a molecule is proportional to its vibrational frequency, which, in turn, is inversely proportional to the square root of the reduced mass. Because D_2O has a larger molecular mass than H_2O, it has a lower vibrational frequency. Theoretical calculations showed that the zero-point energy (or vibrational energy) of H_2O is 55.4 kJ mol^{-1}

at 0 K, which is greater than that of D_2O (40.3 kJ mol^{-1}). Because the potential energies of the two molecules are the same, the dissociation energy of D_2O is greater than that of H_2O. The bonding energy of the O–D bond (498.4 kJ mol^{-1}) is greater than that of the O–H bond (483.7 kJ mol^{-1}), indicating that D_2O has higher thermodynamic stability and is less likely to decompose [Svishchev1994]. During the electrolysis of water, because the decomposition speed of D_2O is relatively slow, a high proportion of H_2 will be generated at the cathode, thereby enriching the deuterium content in the water left behind. Thus, D_2O with sufficient purity can be obtained through multiple electrolysis processes in a cascade electrolytic cell. Based on this strategy, in 1933, G. N. Lewis succeeded in producing high-purity heavy water for the first time [Svishchev1994]. Subsequently, the methodology was further optimized and developed. After 1930, a large-scale water electrolysis facility was built in California that later became a U.S. government base for deuterium enrichment and production.

Electrolytes are indispensable in electrochemical energy storage devices such as batteries and supercapacitors. Compared with the utilization of organic electrolytes, which is common in commercial lithium-ion batteries, the use of water as an electrolyte medium has advantages in terms of cost and environmental safety. However, the narrow electrochemical stability window (1.23 V) of aqueous electrolytes is incompatible with many high-capacity electrode materials, which typically results in low energy density and poor cyclic lifespan [Xu J.J.2022]. In 2015, a breakthrough was made in the development of high-voltage aqueous Li-ion batteries [Suo L.M.2015]. A conceptually novel "water-in-salt" (WIS) electrolyte was invented that can significantly expand the electrochemical window of aqueous electrolytes by as much as 3 V. Since then, extensive research efforts have been made to further develop WIS and related aqueous electrolytes [Xu J.J.2022]. J. Chou et al. at the Institute of Chemistry, Chinese Academy of Sciences recently reported an innovative strategy, replacing H_2O with D_2O to attain a broader electrochemical window [Chou J.2022]. In their work, the researchers not only analyzed the differences in the zero-point energy and dissociation energy between D_2O and H_2O but also considered their differences in the H-bond network in the bulk solution (Figure 3.14). Detailed electrochemical measurements revealed that the D_2O-based electrolytes exhibited improved electrochemical windows with high anodic stability relative to high-voltage layered oxide cathode materials (Figure 3.15(a)). This is due to the stronger covalent bonds and hydrogen bonds of D_2O compared with H_2O. The results of molecular dynamics simulations reveal that, although the coordination environment around Li^+ ions is similar for both D_2O and H_2O solutions, the percentage of coordinated water molecules in the first solvation shell and the hydrogen-bond lifetime of the D_2O-based electrolyte are higher than those of the H_2O-based electrolyte because of the lower proportion of free water molecules and the stronger hydrogen bond (see Figure 3.15(b)). Consequently, the reaction kinetics for H_2 and O_2 evolution are significantly suppressed in the D_2O-based electrolyte, exhibiting a broadened electrochemical window. On this basis, high-voltage layered oxide cathode materials, including $LiCoO_2$ and $LiNi_{0.8}Co_{0.1}Mn_{0.1}O_2$, can operate well in a D_2O solution, displaying excellent cycling and reaction rate performance [Chou J.2022]. This work clearly demonstrates that an analysis of full quantum effects can inspire the design and development of high-performance aqueous batteries.

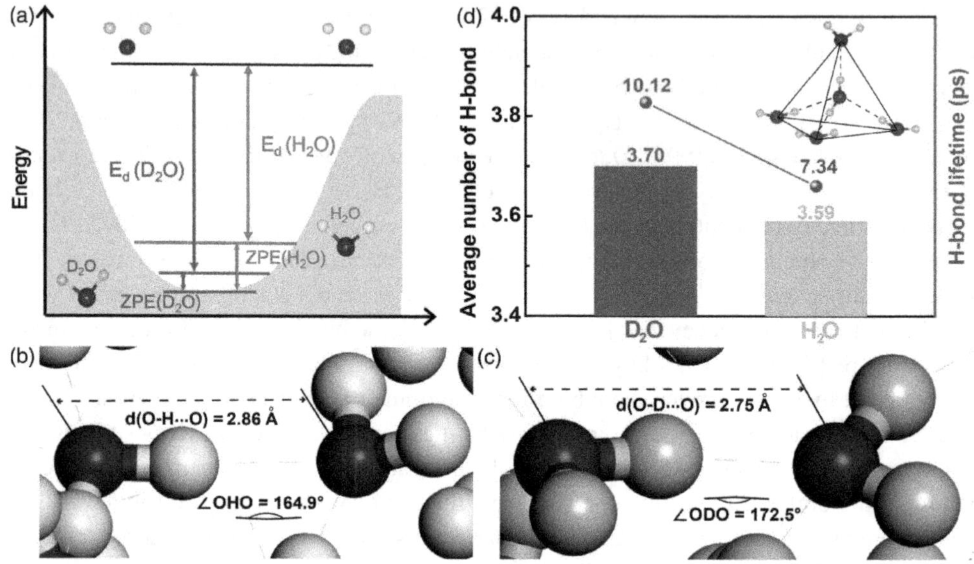

Figure 3.14 (a) Energy levels of D_2O and H_2O, showing the zero-point energy and dissociation energy E_d. (b) Schematic of intermolecular hydrogen bonds in H_2O. (c) Schematic of intermolecular hydrogen bonds in D_2O. (d) Calculated average numbers and lifetimes of hydrogen bonds in a tetrahedral configuration formed by D_2O and H_2O. From Chou J.[2022]. Reproduced with permission from John Wiley and Sons.

The impact of heavy water in electrochemical studies is not limited to energy storage devices. For example, electrochemical deuteration reactions with heavy water as a deuterium source have been reported recently [Jiao K.J.2020, Li P.E.2022]. Because of the distinct isotope difference of H/D, replacing H atoms in organic molecules with D can effectively modify the chemical properties of molecules and impart new functions, which has important potential for application in fields such as drug design and development. As the most readily available deuterium source, heavy water can be utilized for the *in-situ* generation of active deuterium species through electrolysis to initiate the deuteration reaction of organic compounds without the need for additional reducing reagents. Thus, this discovery provides a promising green synthetic path for preparing high-value deuterated compounds.

3.2.5 Origin of Life and Biochemical Reactions

There are numerous theories about the origin of life. The chemical theory proposes that hydrogen, methane, and water in the Earth's primordial atmosphere were the most fundamental chemicals involved in the emergence of life. It has been widely accepted that organic compound formation, a process essential to early life, is linked to the tunneling-dominated synthesis reactions of these substances in the very cold environment of outer space. Methane provided carbon, water provided the basic conditions for life, and hydrogen acted as a catalyst [D'Hendecourt1985]. Methane is conjectured to have formed when multiple

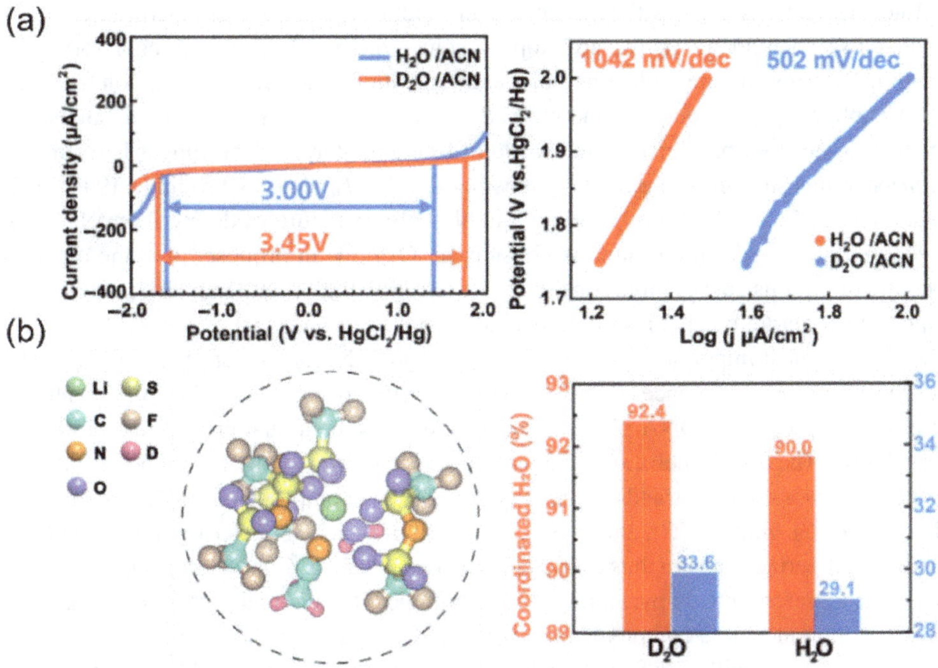

Figure 3.15 Hydrogen-isotope effects on electrochemical reactions. (a) Cyclic-voltammetry curves of high-concentration salt electrolytes in mixed H_2O/ACN and D_2O/ACN solvents. The Tafel curves of oxygen-emission reactions (OERs) are shown to the right of the cyclic-voltammetry curves. (b) Molecular dynamics simulations of the electrolyte solvation structure, with the elements Li, C, N, O, S, F, and D individually labeled. The percentage of coordinated water molecules in the first solvation shell and the lifetime of hydrogen bonds for D_2O and H_2O are plotted to the right of the molecular dynamics simulations. From Chou J.[2022]. Reproduced with permission from John Wiley and Sons.

hydrogen atoms present on the surface of interstellar ice combined with carbon atoms at extremely low temperatures. According to a mechanism proposed in 1985 [Yu Z.1985], 18 interstellar gas-phase reactions could have resulted in the formation of hydrogen molecules. One of these is a primitive reaction that does not involve molecular hydrogen ions. This is expressed as

$$H + H^{-} \rightarrow H_2 + e^{-}. \tag{3.5}$$

The remaining 17 reactions are secondary reactions of this one.

Theoretical calculations using the Heitler–London method show that a stable, mutually attractive potential exists in the H–H^{-} system during nuclear antisymmetric exchange (ψ-). It also crosses the interaction energy curve of hydrogen in such a way that the reaction in Eq. (3.5) can occur via quantum tunneling of hydrogen nuclei when the spacing between the two reactant atoms is sufficiently small. To support this claim, researchers have calculated the reaction rate constant and hydrogen-molecule density in the ionization front of stars, and the results are in good agreement with experimental observations.

Molecular biology studies the structure and function of biological macromolecules to elucidate live phenomena. Its research topics include the metabolism and regulatory functions of organisms and the molecular basis and regulation of genetic information transmission. The philosophy, principles, and methods of quantum mechanics are now extensively applied by molecular biologists in the world of living matter. E. Schrödinger is revered as a pioneer of this subject through his famous book, *What Is Life?* [Schrödinger1944]. J. D. Watson and F. H. C. Crick proposed the double-helix structure of deoxyribonucleic acid (DNA) in 1953. In 1963, the famous physicist P. O. Löwdin proposed that the quantum tunneling of protons causes mutations in DNA – another important landmark in the history of applying quantum mechanics to molecular biology.

DNA – the most important macromolecular system in biology – carries information that determines genetic inheritance, cell division, differentiation, growth, protein biosynthesis, and other life processes. DNA plays a crucial role in the continuation of life and biological evolution. The double-helix structure of DNA is stabilized via hydrogen bonds between nitrogenous bases on both strands of the structure. DNA is composed of four bases – adenine (A), cytosine (C), guanine (G), and thymine (T) – all of which are heterocyclic. A and G are derivatives of purines and are thus called purine bases, whereas C and T are related to pyrimidines and are referred to as pyrimidine bases. In typical double-helix DNA, each base pair consists of a purine and a complementary pyrimidine. A pairs with T to form two hydrogen bonds, while G pairs with C to form three hydrogen bonds (see Figure 3.16(a)). Both pairings are combinations of a double ring and a single ring, with small differences in diameter.

Abrupt mutations in DNA sequences are important in biological evolution. When studying DNA evolution, Löwdin proposed a new theory that attributed DNA mutations during replication to proton quantum tunneling due to full quantum effects [Löwdin1963]. According to this theory, a double potential well exists in every hydrogen bond system X–H……:Y (as shown in Figure 3.16(c)). There is a nonzero probability of proton transfer from donor X–H to acceptor Y due to quantum tunneling (see Figure 3.16(b)). When tunneling occurs, the keto-amine H-bond of the A–T base pair in DNA becomes an enol-imine H-bond, as depicted in Figure 3.17(a), and the A–T base pair transforms into an isomerized A^*-T^* base pair. Similarly, proton tunneling causes the G–C base pair to isomerize and transform into a G^*-C^* pair, as depicted in Figure 3.17(b).

After base isomerization, further changes in the specificity of the H-bonding pairs occur. The isomerized A^* and T^* bases will pair with the normal configurations of C and G, respectively. Therefore, when the DNA double strand unwinds and replicates, the A–T pair produces a G–C pair, and vice versa, instead of producing the same base pair.

Genetic mutations may occur if such errors occur in germ cells. Such mutations are the basis of biological evolution, and without them, there is no biological diversity. The probability of mutations in biological genes varies from 10^{-8} to 10^{-11}, which is essentially consistent with the early tunneling probability calculated using quantum mechanics. And recent work has found that the probability may be further improved (see discussion in Section 7.7).

If DNA base-pair changes occur in somatic cells, it usually causes the deterioration of cell function, contributing to the gradual aging of the entire organism. If the base-pair changes cause the mechanism that suppresses cell division to stop working, the cell will divide uncontrollably, which is a characteristic of cancer cells. (Evidently, there are many

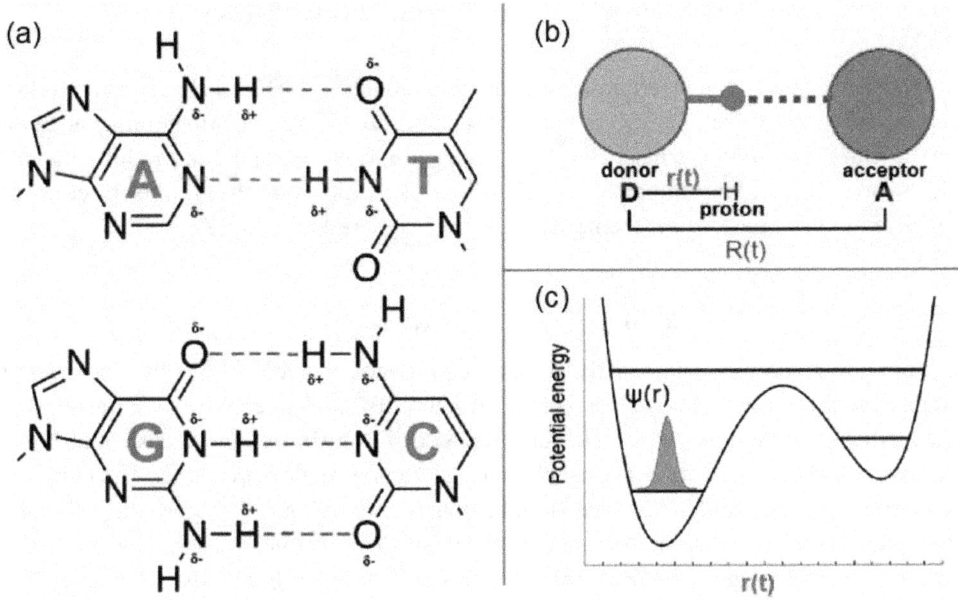

Figure 3.16 (a) H-bond pairing mode of bases in the primary structure of DNA. (b) and (c) Schematic of the double-well trap for H-bonding interactions. Reprinted figure with permission from Alexandrov[2012]. Copyright 2024 by the American Physical Society.

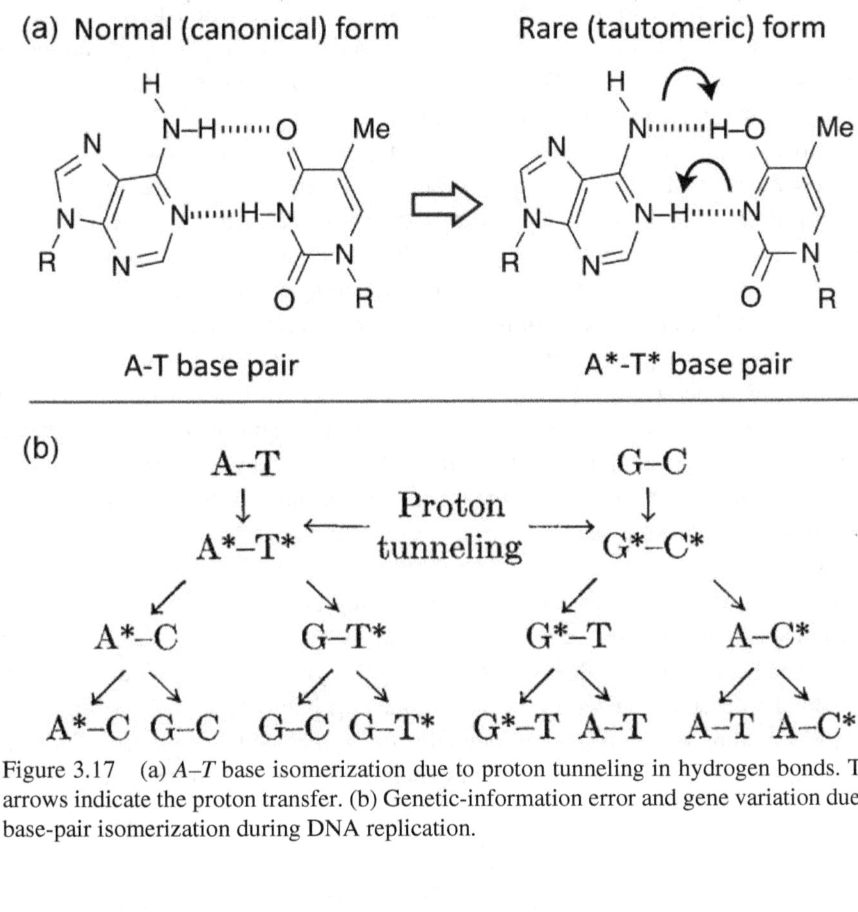

Figure 3.17 (a) *A–T* base isomerization due to proton tunneling in hydrogen bonds. The arrows indicate the proton transfer. (b) Genetic-information error and gene variation due to base-pair isomerization during DNA replication.

other possible causes of cancer; however, quantum tunneling in DNA is likely an important factor in spontaneous cellular carcinogenesis [Godbeer2015].) It is worth noting that all the genetic mutation processes discussed here occur in normal life and that the full quantum effects have far-reaching implications. Therefore, it is important that further research be conducted in this area. A new theoretical study will be presented in Section 7.7.

3.2.6 Enzyme-Catalyzed Reactions

Enzymes are macromolecular catalysts that are produced by biological cells. They have a specific catalytic role and a specific spatial structure. Biological enzymes are mostly proteins, although some are ribonucleic acid (RNA) or DNA. In living organisms, enzymes are involved in the catalysis of almost all material transformation processes. The hundreds of different reactions that take place simultaneously in a living cell require the assistance of various enzymes, which accelerate the otherwise slow chemical reactions, to complete them within biologically relevant times (microseconds or milliseconds). Enzymes have evolved in nature to ensure orderly metabolism, gene expression, and material transport in cells; however, they are also the most powerful tools for manipulating living organisms at a molecular level. Nevertheless, the physical and chemical mechanisms by which they achieve efficient catalysis remain unclear.

More than 4,000 types of enzyme catalysis are known. Based on the type of reaction, enzymes are categorized into six categories: oxidoreductases, transferases, hydrolases, lyases, isomerases, and ligases. Enzymes are composed of 20 amino acids, including essential and nonessential amino acids. Depending on their composition, enzymes can be divided into simple and complex enzymes. Simple enzymes are composed of proteins, whose basic building blocks are amino acids. The complex enzymes consist of proteins and cofactors. During enzyme catalysis, cofactors are responsible for electron and atom transfer, or for the transportational functional groups. Cofactors can be inorganic or organic. Inorganic cofactors are typically metal ions (particularly divalent metals). Organic cofactors, on the contrary, are organic compounds with smaller molecular masses than the proteins in the enzyme. For complex enzymes, both the enzyme protein and cofactor are indispensable, and neither can express enzyme activity alone.

The chemical structure of an enzyme is known as its primary structure. The spatial structures of enzyme proteins include secondary, tertiary, and quaternary structures that determine the catalytic functions and properties of the enzyme. The enzyme must bind to the substrate to function as a catalyst. First, the enzyme binds to the substrate to form the enzyme–substrate (E-S) complex. After catalysis, the complex decomposes into the product and enzyme.

In general, the binding of an enzyme to its substrate is highly specific to the type of enzyme and substrate. In 1894, E. Fischer proposed a lock-and-key theory to explain the binding mechanism. Fischer's theory states that an enzyme has a fixed conformation and can only bind to a substrate with a specific matching conformation. E–S binding has a lock-and-key binding relationship. Although this theory accounts for the specificity of E–S binding, it has some limitations and cannot explain issues such as transition-state stability during catalysis. In 1958, D. Koshland proposed the induced-fit theory of E–S binding. He argued that E–S binding is not a simple, rigid binding and that the conformation at the enzyme active site is flexible and modifiable. Before binding, the conformation of the active site is not necessarily suitable for substrate entry, but it can change to accommodate the substrate. Compared to the lock-and-key model, the induced-fit theory emphasizes the flexibility of the enzyme structure and the compatibility and adaptability of enzymes and substrates during E–S binding.

E–S binding is the first step in enzyme catalysis, which, like other types of chemical catalysis, is thought to accelerate reactions by reducing their activation energy. Two mechanisms are possible for induced-fit catalysis to occur: The uniform-binding mechanism, which increases the affinity of both the substrate and the transition state, and the differential-binding mechanism, which increases the affinity of the transition state alone. Most enzymes use differential binding, such that the enzyme weakly binds to the substrate at first and then changes its conformation to increase the transition-state binding affinity and stabilize the transition state, reducing the activation energy required to reach the transition state (see Figure 3.18).

The possibility of quantum tunneling in enzymatic reactions was suggested as early as the 1970s [Volkenshtein1972]. Since the mid 1980s, increasing evidence from

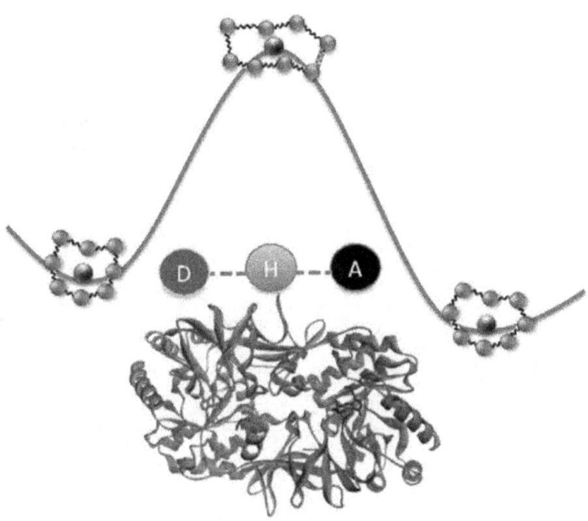

Figure 3.18 Schematic of NQEs in enzyme catalysis. On the potential energy surface, the left, middle, and right spheres surrounded by ring polymers (eight-bead ring polymers for each sphere) represent the reactants, transition state, and product, respectively. In the classical case, which does not consider zero-point vibration, the total energy of the reactant particles must be greater than the potential-barrier height to reach the transition state. In the quantum case, the reactant particles have a finite probability of penetrating the potential barrier due to the tunneling effect caused by the zero-point energy. The lower part of the figure depicts an example of alanine racemase catalyzing proton transfer, where D denotes the proton donor and A denotes the proton acceptor. Reprinted from Vardi-Kilshtain[2015]. Copyright 2024, with permission from Elsevier.

isotope-labeling experiments has suggested that significant KIEs can be observed in certain enzyme-catalyzed biochemical reactions. Moreover, it has been experimentally demonstrated that quantum tunneling of protons plays a critical role in enzyme catalysis and significantly increases the reaction rate (see Figure 3.18). In enzyme catalysis, the reaction-rate difference between H and D nuclei directly reflects the difference in their zero-point energies. J. Klinman's group at the University of California, Berkeley, conducted systematic comparisons of these rates in reactions such as the oxidation of benzyl alcohol to benzaldehyde catalyzed by alcohol dehydrogenase. The enzyme was able to create precise and compact active-site structures to facilitate quantum tunneling. For example, alcohol dehydrogenase was able to change its conformation to promote tunneling by bringing the hydrogen donor and acceptor sites closer together [Hong N.S.2018].

In-depth theoretical and experimental studies have found that the shape of the potential barrier significantly affects the tunneling process in enzyme catalysis. For instance, in the C–H bond breakage reactions of three organic amines (dopamine, benzylamine, and tryptamine) catalyzed by amine dehydrogenase, the reaction rates and KIE exhibited by different amines varied considerably. A law that is consistent with experimental observations can be derived theoretically only when the appropriate potential barrier shape is fully considered (see Figure 3.19) [Basran2001, Sutcliffe2002].

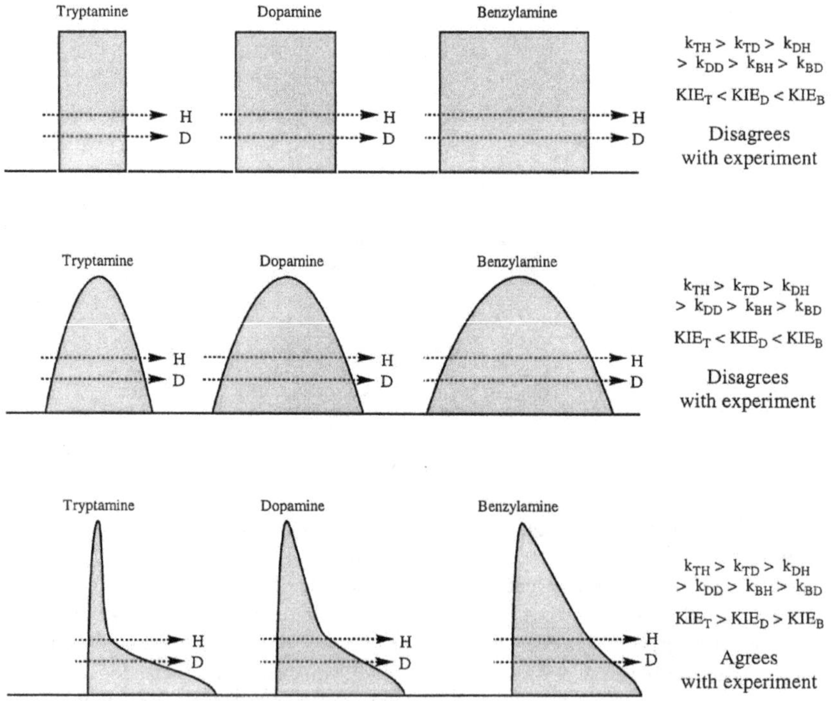

Figure 3.19 In the C–H breakage of organic amines (tryptamine, dopamine, and benzylamine) catalyzed by amine dehydrogenase, the shape of the potential barrier can affect the tunneling process. Figure created by Basran[2001] licensed under CC BY 4.0.

Furthermore, protons are not the only particles that undergo quantum tunneling during enzyme catalysis. Studies have shown that even oxygen nuclei in glucose oxidase can undergo quantum tunneling under certain physiological conditions [Klinman2018]. Evidently, it is important to examine role of full quantum effects play in fundamental processes and laws in the life sciences.

3.3 Nonadiabatic Effects

Nonadiabatic effects (NAEs) also play a unique and irreplaceable role in chemical reactions. Without nonadiabatic processes, we would not have such a colorful world as we do today. In fact, NAEs not only affect the electron transitions between different eigenstates through electron–nucleus coupling, altering the occupation numbers pertaining to the ground and excited states of the electrons, but also result in a dynamic nuclear transfer between splitting potential energy surfaces. Arguably, NAEs are strongly associated with the macroscopic evolution of the universe and the creation of life, as well as with microscopic nonequilibrium physicochemical processes. Until recently, the study of such processes in chemical reactions was mainly experimental, but considerable progress has been made theoretically.

3.3.1 Relaxation Dynamics

Neutral ethylene molecules (C_2H_4) and cations ($C_2H_4^+$) have been widely used as model systems for studying photodegradation and the action of radicals. For example, R. Locher et al. obtained information on the relaxation dynamics of ethylene cations using delay-dependent ion mass spectroscopy [Locher2014]. A pulse consisting of four harmonics in the 20–30 eV range was used to ionize the ethylene molecules and keep them in their ground state (as shown in Figure 3.20(a)). A tin plate was subsequently inserted in front of the UV beam to filter out the third and fourth harmonics, and the filtered pulse was then used to excite the ethylene cations to different excited states. Subsequently, a 1.6 eV probe pulse with a duration of 25 fs was applied. Information about the nonadiabatic dynamics of the system was obtained by varying the time delay of the pump and probe pulses.

Extreme UV pump light excited the ethylene cation system to the first, second, and third excited states. Based on the experimental results, ~95% of the electrons were confined to the ground state ($\tilde{X}^2B_{3u}$) or one of the first three excited states (A^2B_{3g}, B^2A_g, and C^2B_{2u}). Combining linear response time-dependent density functional theory (LR-TDDFT) with the PBE0 functional, Locher et al. analyzed the potential energy surfaces of the D0, D1, D2, and D3 states of the ethylene cation in the coordinate system of the C–H bond length and twist angle (as shown in Figure 3.20(d)). The dynamical evolution can be described by the relaxation of the nuclear wave packet on different potential energy surfaces. The final state determines the asymptotic fragment yield, which can be derived from the experimentally measured ion mass spectra.

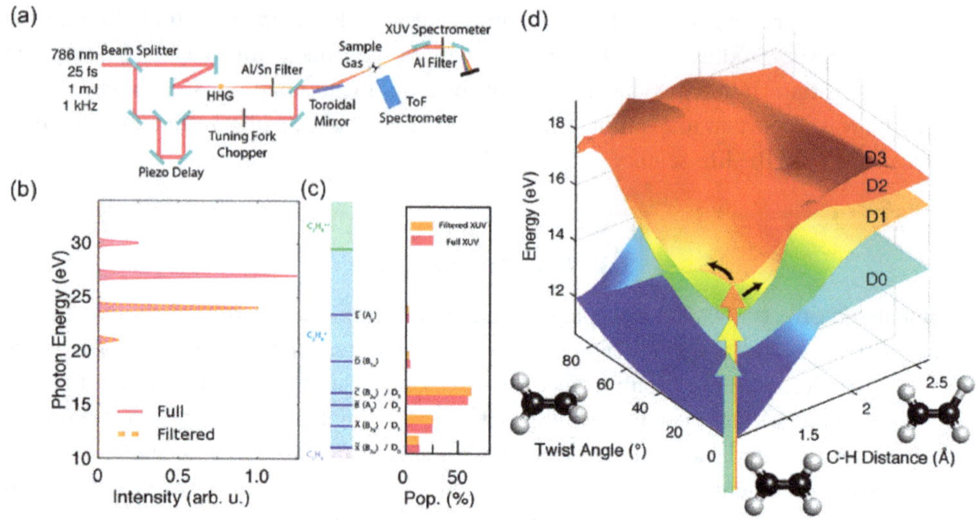

Figure 3.20 Relaxation-dynamics study of ethylene cations. (a) Schematic of the experimental setup. (b) Extreme-UV excitation spectrum. (c) Left: cationic states (blue) reachable at extreme-UV photon energies. Right: initial population of cationic states. (d) D0, D1, D2, and D3 potential energy surfaces along the C–H bond length and the twist angle. The reference zero-point energy in the experiment is the ground-state energy of the neutral molecule. Figure created by Bircher[2017] licensed under CC BY.

It was found that the dominant channels corresponded to ethylene cations ($C_2H_3^+$ and $C_2H_2^+$) formed by the loss of one or two hydrogen atoms. In contrast, C–C bond breaking and recombination (resulting in CH_3^+, CH_2^+, and CH^+) barely occurred. By varying the pump-probe time delay, the IR probe pulse changed the manner in which the nuclear wave packet relaxed on different potential energy surfaces, thus altering the fragmentation yield. The main products of the relaxation process, $C_2H_3^+$ and $C_2H_2^+$, were typically produced after ~25 fs. The additional production of CH_3^+ signifies the occurrence of ethylene-ethylidene ($CH_3CH=$) isomerization. The maximum rate of CH_3^+ production can be used to determine the upper bound of the characteristic isomerization time. Previously, this upper limit was estimated to be 50 ± 25 fs. Because their calibration technique was more accurate, Locher et al. were able to provide a more reliable result. They determined that the upper bound of the characteristic isomerization time was 30 ± 10 fs.

The observed temporal evolution of the population distribution of the four cationic states D0, D1, D2, and D3 (corresponding to states $\tilde{X}^2B_{3u}$, A^2B_{3g}, B^2A_g, and C^2B_{2u}, respectively) is shown in Figure 3.21. Evidently, the relaxation processes were mainly concentrated within the first 50 fs, with the average amount of D1 peaking at 15–20 fs before decaying to the ground state. The D1 time scale agreed with that of the IR-induced changes in the main products in the experiments (~15 fs). The IR pulse probed the presence of nuclear wave packets around the D1–D0 conical intersections, where the branching of different products was assumed to occur. This provided information on the timing of the relaxation process. It clearly showed the changes of the electronic distribution of the D0, D1, D2, and D3 states in photodegradation of ethylene molecules.

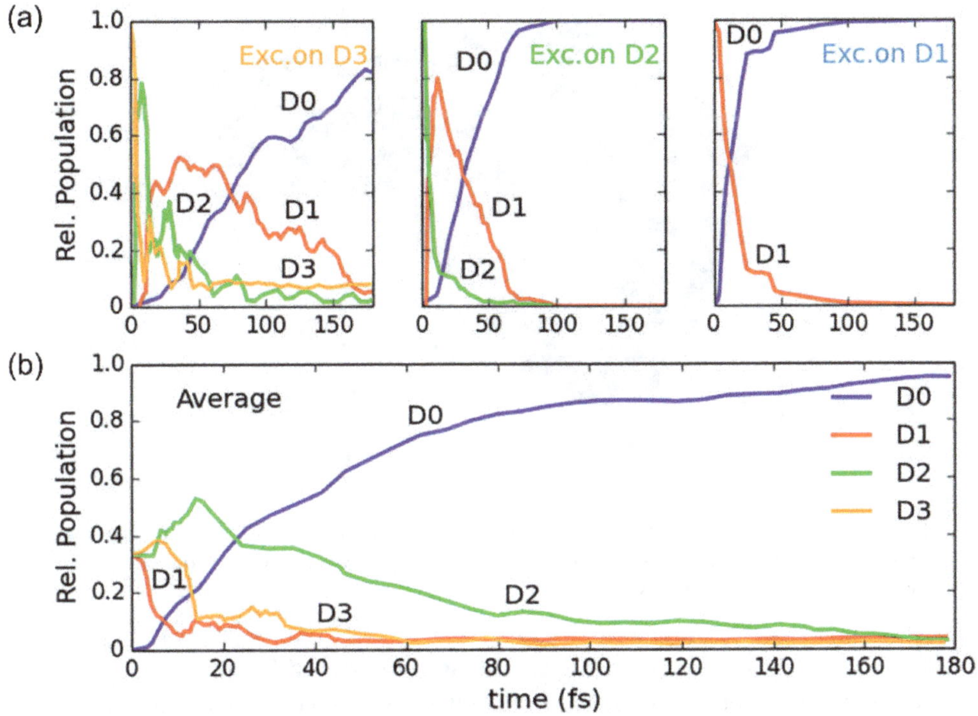

Figure 3.21 Yield of the four lowest states in the relaxation of ethylene cation as a function of time: (a) [left to right] corresponds to selective excitations at D3, D2, and D1, respectively; (b) represents the case of averaging over all initial excitations. Bircher[2017] licensed under CC BY.

3.3.2 Multi-Potential Energy Surface Evolution

Nonadiabatic processes often involve the quantum coupling of electrons and nuclei in motion. This means that the Born–Oppenheimer approximation, one of the fundamental assumptions on which traditional quantum chemistry is based, is no longer valid. As an example, we will discuss the nonadiabatic evolution among multiple potential energy surfaces by considering electron–nucleus quantum dynamic coupling in the chemical reaction of chlorine with hydrogen [Han K.L.2004].

The reaction of chlorine with hydrogen, a prototypical chemical reaction involving a nonadiabatic process, is of great importance in atmospheric and environmental chemistry. X. M. Yang's group at the Dalian Institute of Chemical Physics, Chinese Academy of Sciences, has made significant contributions to understanding chlorine–hydrocarbon interactions [Wang X.2008, Yang X.M.2009]. As shown in Figure 3.22, because of spin–orbit coupling, the ground state of the chlorine atom can be redivided into two new states, the ground state and excited state, which correspond to two reaction potential energy surfaces. If the Born–Oppenheimer approximation is adopted, the chemical reaction can proceed only along the direction of a single potential energy surface (as indicated by the curved arrow). However, X. M. Yang's group showed that the Born–Oppenheimer approximation no longer works at low collision energies (which are equivalent to low-temperature

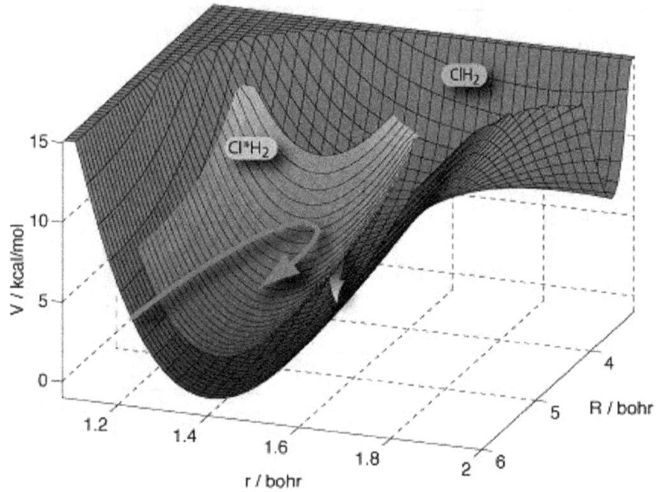

Figure 3.22 During the chemical reaction of chlorine with hydrogen, the processes induced by nonadiabatic effects only occur by crossing between two potential energy surfaces. Reprinted from Yang X. M.[2009].

conditions). For chlorine atoms, this reaction can only proceed from the potential energy surface of the spin–orbit excited state to that of the ground state through nonadiabatic coupling (indicated by the downward arrow in Figure 3.22).

As problems such as this cannot be studied within the framework of the Born–Oppenheimer approximation, other appropriate dynamical methods are required. The time-dependent wave packet method is a quantum-mechanical approach for dealing with nonadiabatic reaction processes. (It is worth noting that the conventional time-dependent approach can barely be applied to low-energy collisions of ultracold molecules.) The time-dependent wave packet method focuses on solving the time-dependent Schrödinger equation $i\hbar\frac{\partial\Psi}{\partial t} = \vec{H}\Psi$, and its starting point is the Born–Huang expansion (see Eq. (1.4)).

In the process of establishing a nonadiabatic reaction model for the ultrafast dynamics of photoexcited hydrogenated graphene, researchers have found that photoexcitation with full hydrogen coverage leads to rapid hydrogen desorption [Bang J.2013]. On the other hand, if the hydrogen coverage is low, hydrogen desorption does not occur, even under strong excitation. On-the-fly time-dependent density functional-dynamics simulations clarify that the excited H atoms are repelled by the surrounding H atoms at high coverages and thus are more likely to detach from the surface. At low coverages, however, they do not experience this repulsion and remain attached to the graphene [Bang J.2013].

3.3.3 Photochemical Reactions and Water Photodissociation

3.3.3.1 Photochemical Reactions

Photochemical processes involving small inorganic molecules in the gas phase, such as the photochemical reaction of H_2 molecules with halogen atoms, provide a typical example of nonadiabatic chemical reactions. Being the main focus of excited-state studies and an important topic for photochemical and photophysical research, these reactions are the subject of intensive research covering a broad range of physical chemistry and chemical physics, including cluster science, environmental chemistry, and atmospheric chemistry. In particular, photolysis is a central problem in physical chemistry. In fact, the photolytic process and the second half of the full collision process are identical. Therefore, understanding the nonadiabatic photolysis process provides insight into the mechanism of elementary collision reactions, and vice versa.

In photolysis, a bound-state molecule (whose state is described by a wavefunction that is zero at infinity) is broken down by absorbing one or more photons. In this process, the electromagnetic energy of light is transformed into the internal energy of the molecule, and when the transferred energy exceeds the binding energy of the weakest chemical bond within the molecule, the molecule dissociates. The reaction is as follows:

$$ABC + nh\nu \rightarrow (ABC)^* \rightarrow AB + C, \tag{3.6}$$

where AB and C are molecules or free radicals, and (ABC)* denotes the transition-state molecular configuration between the reactants and products at a higher energy.

Many photolytic processes involve two or more electronic states, and the adiabaticity/nonadiabaticity of the process depends on the interactions between these electronic energy levels. If the interaction is weak, then the reaction process is adiabatic. A nonadiabatic transition from one electronic state to another occurs if the interactions between the energy levels are strong or even the energy levels cross. In summary, nonadiabatic photolysis can be described as follows: When the interactions between these electronic energy levels are strong or even the energy levels cross, molecules in a given state can absorb one or more photons, resulting in a nonadiabatic electronic transition, which leads to further chemical bond breakage and dissociation of molecules.

3.3.3.2 Water Photodissociation

Theoretically, the photocatalytic dissociation of a single water molecule can be either adiabatic or nonadiabatic. In the adiabatic case, the water molecule remains on a single potential energy surface throughout the process, whereas in the nonadiabatic case, it evolves between multiple potential energy surfaces. In reality, the photodissociation of water molecules is usually strongly nonadiabatic.

The light absorption spectrum of a single water molecule is shown in Figure 3.23. When irradiated by UV light, water molecules absorb certain wavelengths of UV light and are excited from their electronic ground state ($\tilde{X}^1A_1$, often abbreviated as $\tilde{X}$) to any of the four lower-energy excited states. In increasing order of energy, these excited states are $\tilde{A}^1B_1$, $\tilde{B}^1A_1$, $\tilde{C}^1A_1$, and $\tilde{D}^1A_1$. During light absorption, an electron in the $1b_1$ or $3a_1$ orbital of a water molecule is excited to vacant states at higher energies. The first absorption peak of the water molecule is observed within a wavelength range of 150–200 nm. This peak corresponds to electron excitation from the ground state $\tilde{X}^1A_1$ to the first excited state, $\tilde{A}^1B_1$ ($\tilde{A}$), that is, from the $1b_1$ orbital to an antibonding orbital with a_1 symmetry. The second absorption peak is centered at ~128 nm, exhibiting a broad spectral distribution and many fine

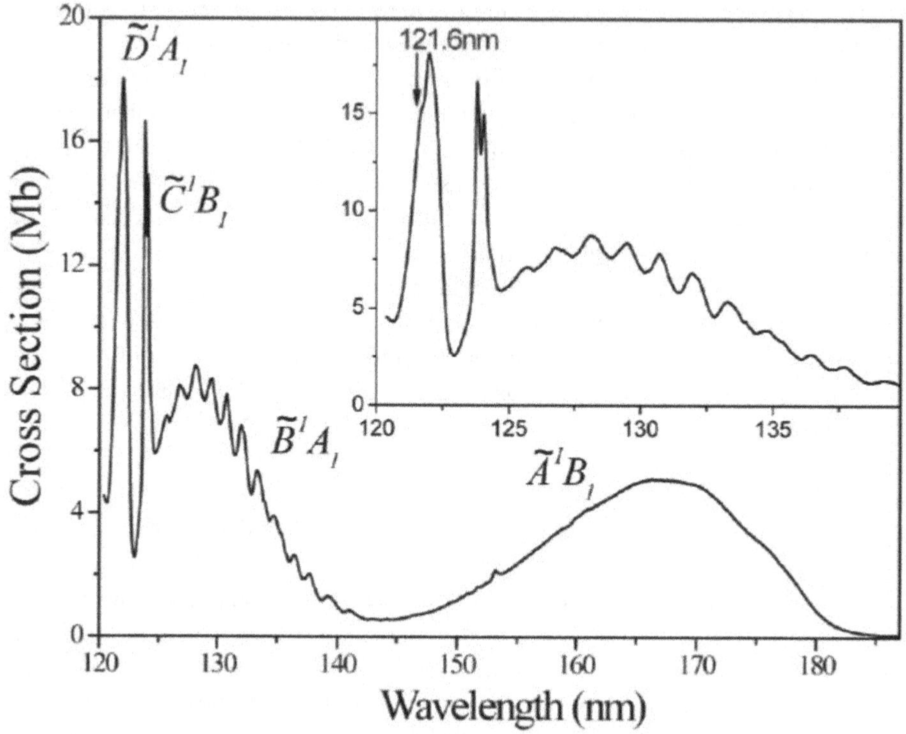

Figure 3.23 Absorption spectrum of H_2O molecule at room temperature (1 Mb = 1 × 10^{-18} cm^2). Light causes the ground state electrons of the water molecule to jump to four lower-energy excited states. Reprinted with permission from Yuan K.[2011]. Copyright 2024 American Chemical Society.

structures with a spacing of 810 cm^{-1}. This peak corresponds to the electron transition from the ground state $\tilde{X}^1A_1$ to the second excited state $\tilde{B}^1A_1$ ($\tilde{B}$). The fine spectral structure can be attributed to the unstable periodic motion of the water molecules, comprising both the shear mode of water molecule and the O–H stretching mode. The third absorption peak is located at ∼124 nm, corresponding to electronic excitation from the ground state $\tilde{X}^1A_1$ to $\tilde{C}^1A_1$ ($\tilde{C}$). Its excited-state wavefunction is characterized by a vertical transition to the 3p Rydberg state in the vertical Franck–Condon region. The fourth absorption peak is distributed around 122 nm and corresponds to the electronic excitation of the ground state $\tilde{X}^1A_1$ to $\tilde{D}^1A_1$ ($\tilde{D}$).

X. M. Yang's group studied the photodissociation dynamics of individual water molecules using a wavelength-tunable vacuum UV-light source [Yuan K.2011]. They measured the translational- and rotational-energy distributions of water-dissociation products using an H-atom Rydberg-tagging time-of-flight spectrometer. Once irradiated by a 157.6 nm laser, the majority of H_2O molecules in the ground state were excited to the state $\tilde{A}^1B_1$. Moreover, the water dissociated directly on a single potential energy surface, producing vibrationally excited electronic ground-state OH($X^2\Pi$) fragments and H-atoms. Irradiation with a 121.6

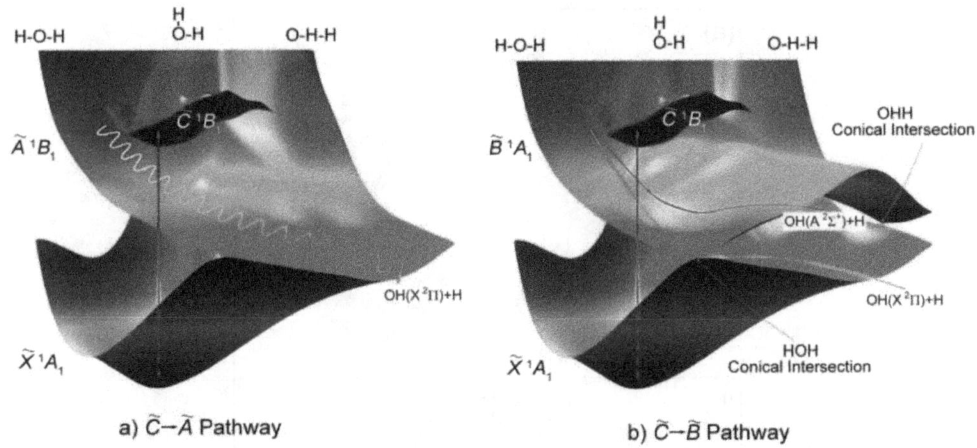

Figure 3.24 Potential energy surface distribution corresponding to the electronic ground state ($\tilde{X}1A1$) and the excited states (a) $\tilde{A}^1B_1$ and $\tilde{C}^1B_1$ and (b) $\tilde{B}^1A_1$ and $\tilde{C}^1B_1$, of the water molecule. The figure also shows the photodissociation pathways of (a) $\tilde{C}^1B_1 \rightarrow \tilde{A}^1B_1(\tilde{C} \rightarrow \tilde{A})$ and (b) $\tilde{C}^1B_1 \rightarrow \tilde{B}^1A_1(\tilde{C} \rightarrow \tilde{B})$. The conical intersections of potential energy surfaces corresponding to the excited-state $\tilde{B}^1A_1$ and the electronic ground-state $\tilde{X}^1A_1$ are also shown. Reprinted with permission from Yuan K.[2011]. Copyright 2024 American Chemical Society.

nm source and a wavelength-tunable UV laser (128–133 nm) excited the water molecules to the second excited state, $\tilde{B}^1A_1$. In this state, H_2O molecules are ionized through multiple channels. The most important of these is the channel where the H_2O molecule dissociates, via two conical intersections, between the excited state $\tilde{B}^1A_1$ and the ground state $\tilde{X}^1A_1$, producing $OH(X^2\Pi)$ residues that are not only in the ground state but also in the rovibrationally excited state. One of the conical intersections arises from the crossing between the ground state $OH(X^2\Pi)$ and the excited state $OH(A^2\Sigma^+)$ as the H atom linearly approaches the OH group in the H–O–H collinear conformation. The potential energy surfaces with this conical intersection become two discrete surfaces with an energy gap (i.e., the pathway is an avoided crossing) when the molecule in the asymmetric configuration is slightly bent. The other conical intersection originates from the second excited-state potential energy surface of the collinear O–H–H configuration. (Interestingly, the dissociation reaction pathways around these two conical intersections show significant interference effects, similar to the double-slit effect.) The population of the generated excited state $OH(A^2\Sigma^+)$ was found to be relatively small, suggesting that this is a minor channel. The potential energy surface geometry of the ground and excited states of the H_2O molecule, as well as the conical intersection distribution of the potential energy surfaces corresponding to the excited state $\tilde{B}^1A_1$ and the electronic ground state $\tilde{X}^1A_1$, are shown in Figure 3.24.

In addition, the authors measured the distribution of photodissociation products after nonresonant excitation of the H_2O molecules to the second excited state, $\tilde{B}^1A_1$. Figure 3.25 shows the translational energy distributions of the measured photodissociation products using five different laser wavelengths to excite electrons to the state $\tilde{B}^1A_1$ in the

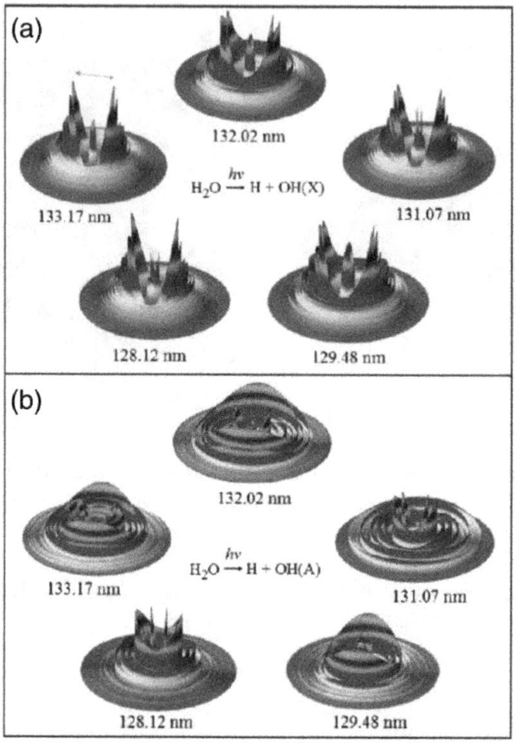

Figure 3.25 Angle-dependent translational energy (or OH-stretching vibrational energy) distributions of the second excited state $\tilde{B}^1A_1$ in H_2O photodissociation under nonresonant excitation at different wavelengths: (a) $H_2O\,(\tilde{B}^1A_1) \rightarrow H + OH\,(X^2\Pi)$ reaction pathway; (b) $H_2O\,(\tilde{B}^1A_1) \rightarrow H + OH\,(A^2\Sigma^+)$ reaction pathway. Reprinted with permission from Yuan K.[2011]. Copyright 2024 American Chemical Society.

H_2O molecules. The dissociation products include the ground-state residue $OH(X^2\Pi)$ (Figure 3.25(a)) and the excited-state residue $OH(A^2\Sigma^+)$ (Figure 3.25(b)). The horizontal arrow represents the electric field polarization direction of the laser. The shell distribution of the translational energy reveals different O–H stretching vibrationally excited states of the corresponding OH group, and the angular change in the distribution reflects the rotation of the group. The angular distributions of ground-state $OH(X^2\Pi)$ vibrational energies are dramatically different from those of the excited state $OH(A^2\Sigma^+)$. Under unstable resonant excitation at five different wavelengths, the angular distributions of the ground-state $OH(X^2\Pi)$ products were relatively similar. However, the vibrational energy and angular distributions of the excited state $OH(A^2\Sigma^+)$ are very different at different excitation wavelengths (H_2O dissociation via photoexcitation to the five different unstable resonance states on the $\tilde{B}$ state). This indicates that unstable resonant excitation has different (but extremely large) effects on the reaction pathways of different reaction products (such as the electronic ground state $OH(X^2\Pi)$ and excited state $OH(A^2\Sigma^+)$).

Photodissociation of the third excited state $\tilde{C}^1A_1$ is a typical rotation-dependent predissociation process. The third excited state exhibits the characteristics of a Rydberg state of atomic orbitals, and its lifetime is of the order of a few picoseconds. The $\tilde{C}^1A_1$

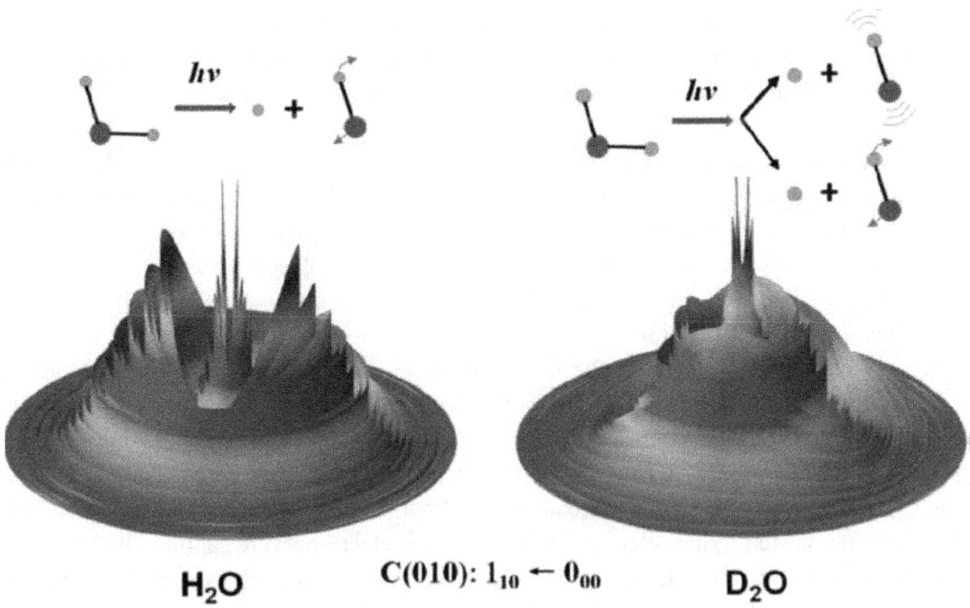

Figure 3.26 Vibrational-state distribution of the OH group from H_2O and D_2O photodissociation in the third excited state. Reprinted with permission from Chang Y.[2019]. Copyright 2024 American Chemical Society.

state of water molecules has two nonadiabatic dissociation pathways. In the first pathway, the third excited state can relax to the $\tilde{B}^1A_1$ excited state via electronic Coriolis coupling, resulting in an OH group with a high-energy rotational state and a low-energy vibrational state. Conversely, the third excited state can also decay directly to the $\tilde{A}^1B_1$ state through a homogeneous nonadiabatic process, leading to an OH group product with a high-energy vibrational excited state and a low-energy rotational state. The two photodissociation pathways and the corresponding potential energy-surface distributions are shown in Figure 3.24.

X. Yang's group further investigated isotope effects in the excited state's $\tilde{C}^1A_1$ photodissociation pathway [Chang Y.2019]. They measured the vibrational state distribution of OH groups during the photodissociation reactions of H_2O and D_2O (see Figure 3.26). After H_2O dissociates from the third excited state, $\tilde{C}^1A_1$, the OH group shows a prominent vibrational state distribution but a small number of rotational state distribution. However, the quantum distribution of the vibrational energy of OD from D_2O dissociation is significantly different from that of OH. In addition, linewidth measurements showed that the third excited state of H_2O had a lifetime of 330 fs, while the corresponding lifetime of the D_2O excited state was 3.0–4.2 ps, differing from that of H_2O by a factor of 10. These striking isotopic effects may be attributable to accidental degeneracy (and therefore strong resonance) in the H_2O molecule between the shear vibrational state of the third electronic excited state and the vibrational ground state of the fourth excited state. In the D_2O molecule, on the other hand, these two energy levels are separated by 270 cm^{-1} and are consequently out of resonance.

Water molecule photodissociation from the fourth excited state, $\tilde{D}^1A_1$, to the second excited state, $\tilde{B}^1A_1$, occurs mainly through rapid electronic relaxation and predissociation, followed by the dissociation reaction. During this process, there is almost no rotation of molecular groups.

Thus, experimental studies have revealed the nonadiabatic processes by which water molecules reach the four lowest-energy excited states and dissociate under UV irradiation. The likelihood of a water molecule decomposing into a single O atom and an H_2 molecule is relatively low. Typically, an individual gas-phase water molecule undergoes photodissociation only when exposed to extreme-UV irradiation, which significantly limits the scope of applications for this reaction.

However, in the presence of other materials in the environment, photodissociation of water molecules can be induced by interactions with solid surfaces (even under weak visible light). In general, this is considered a photodissociation reaction or photocatalytic water splitting by surface catalysis. In 1972, Fujishima and Honda first discovered photocatalytic water splitting on an n-TiO_2 electrode under UV light [Fujishima1972]. The TiO_2 electrode absorbed photons to produce electron–hole pairs. The electrons were captured by hydrogen ions, forming hydrogen gas, while the holes caused hydroxide ions to produce oxygen. Because TiO_2 has a large energy gap and does not absorb visible light, attempts have been made to enhance its light absorption through doping. Attempts have also been made to use oxide semiconductors with smaller energy gaps, such as Fe_2O_3 or PbO, as photocatalysts. Naturally, the photodissociation of water on the catalyst surface results in strong NAEs, as indicated by these experimental studies.

In addition, numerous experiments have shown that precious-metal nanoparticles, which produce strong plasmon resonances, can facilitate photocatalytic water splitting process. Their large absorption and scattering cross sections over a wide spectral range enable the effective use of solar energy, making them common practical photocatalysts. Radiation can directly excite localized surface plasmons, which in turn excite hot electrons via plasmon decay to drive chemical decomposition and synthesis. These processes are typical nonadiabatic processes. Plasmons can initiate the charge transfer and enhance the electric field. However, it is unclear which of these two effects is responsible for the high catalytic activity. Moreover, the dynamics of water dissociation are still poorly understood. These issues hinder further improvements in the efficiency of water photolysis.

Recently, S. Meng's group performed time-dependent density functional quantum dynamics simulations to systematically investigate the microscopic mechanism of water splitting on gold nanoparticles with diameters of ~2 nm, exposed to 1fs laser pulse [Yan L.2016, 2018]. Under the applied laser field, the water molecules on the gold nanoparticles rotated and split into hydrogen atoms and hydroxyl groups (Figure 3.27(a)). In the time-dependent evolution of the electron distribution near the water molecules (Figure 3.27(b)), the rapid injection of electrons from the metal to the antibonding state of the water molecules caused water destabilization and splitting. In addition, L. Yan et al. theoretically demonstrated for the first time that gold nanoclusters in a liquid-water environment could split water through rapid plasmon excitation. The nonadiabatic process of water splitting was realized on a 100fs time scale. This study discovered that, contrary to the classical

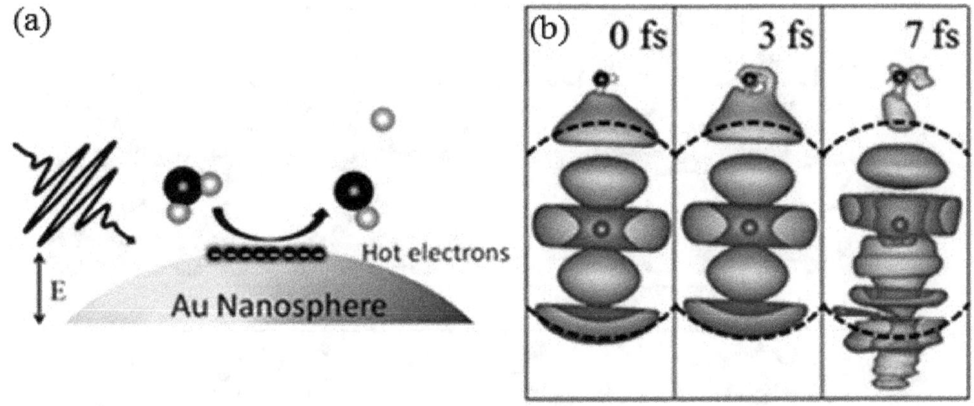

Figure 3.27 (a) Schematic of plasmon-induced water splitting on a gold nanosphere under an applied laser pulse. (b) Time-dependent evolution of the charge density of the system at the Fermi level. The dashed lines in (b) indicate the nanosphere's surface, and the gray dots indicate its center. Reprinted with permission from Yan L.[2016]. Copyright 2024 American Chemical Society.

assumption, the specific mechanism of metal-to-water charge transfer plays only a minor role in a liquid environment. The electric-field enhancement induced by the plasmons on the gold clusters is the main reason for water dissociation. They also observed the process by which H atoms from different water molecules collide with each other to produce H_2 molecules. Their simulations predicted a quantum efficiency of 0.06% for hydrogen production, which is in good agreement with the experimental results (0.05%). The results of this study suggest that the reaction process may have practical applications.

Furthermore, Yan et al. found that the water-splitting reaction rate depended linearly on the laser intensity, which suggests that the reaction rate is linearly correlated with the number of transferred hot electrons. Surprisingly, the wavelength dependence of the reaction rate on different-sized Au nanoparticles significantly deviated from the corresponding optical absorption spectra, suggesting that water splitting is not only related to light absorption but also exhibits quantum selectivity in the excitation mode. The results indicate that the odd modes are more effective for charge transfer than the even modes. By varying the nanoparticle diameters from 1.6 to 2.1 nm and repeating the simulations above, they found that the catalytic activity was the highest when the diameter was 1.9 nm. However, the reason for this size effect remains unclear. The efficiency can also be improved by altering the shape and structure of the nanoparticles. For example, by assembling Au nanorods of different sizes, panchromatic absorption can be realized with a total energy-conversion efficiency of 0.1%. Relevant studies have shown that in nonadiabatic processes, energy matching between hot electrons and adsorbates is crucial for achieving efficient water photolysis. These results provide a complete quantum-mechanical explanation for the nonadiabatic processes of plasmon-induced water splitting and hydrogen production.

Graphitic carbon nitride (g-C_3N_4) is a semiconductor photocatalyst that is active when irradiated by visible light. This example has recently attracted considerable attention

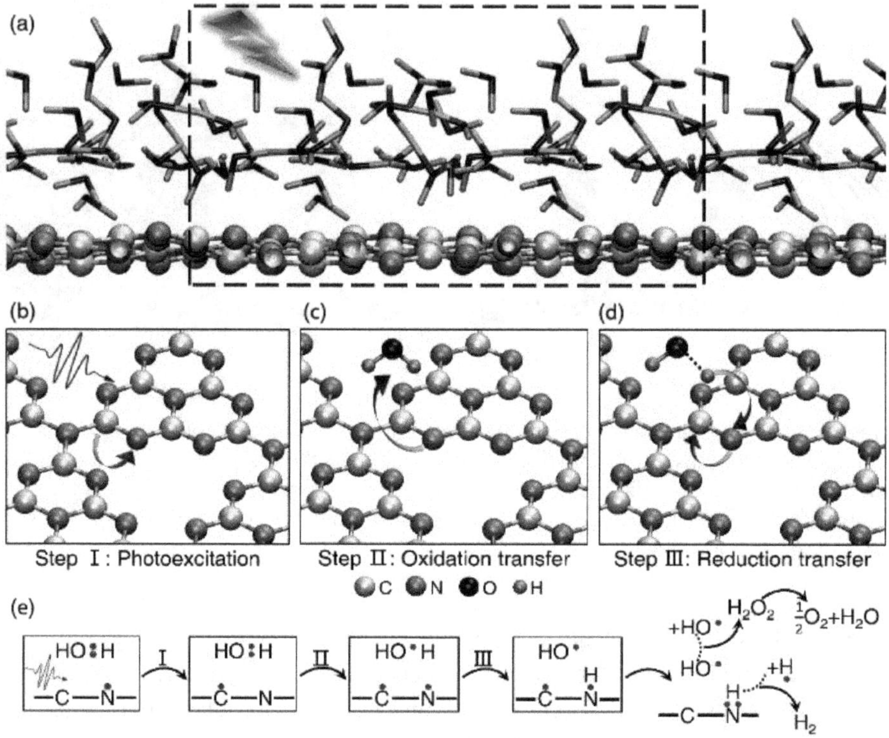

Figure 3.28 Three-step reaction mechanism of photocatalytic water splitting on g-C_3N_4 and the subsequent production of hydrogen and oxygen. Reprinted with permission from You P. W.[2021a]. Copyright 2024 American Chemical Society.

because of its excellent chemical stability, suitable energy bandgap, and nontoxicity. Its performance can be further improved by techniques such as embedding metallic atoms and carbon nanodots to obtain a higher hydrogen and oxygen release efficiency. For example, nanocomposites consisting of carbon nanodots and g-C_3N_4 can significantly improve the performance of photocatalytic water splitting under visible light via a two-step process with more efficient hydrogen and oxygen evolution. Its efficiency is close to the minimum required for commercial applications. Despite these remarkable advances, the underlying microscopic mechanisms remain unknown.

In 2021, S. Meng's group, in collaboration with E. G. Wang, developed a nonadiabatic molecular-dynamics approach to theoretically study the dynamical process of photocatalytic water splitting on semiconductors from first principles. S. Meng et al. also proposed a three-step reaction mechanism for g-C_3N_4 photocatalytic water splitting: photoexcitation, oxidation transfer, and reduction transfer (see Figure 3.28) [You P.W.2021a]. The study found that, after the formation of electrons and holes by photoexcitation, hole transfer from the valence band of g-C_3N_4 to water weakens the OH-bond. This is later accompanied by hole transfer in the opposite direction and proton motion, eventually leading to water splitting. The hydrogen atoms produced by the decomposition reaction are adsorbed on the nitrogen atoms of g-C_3N_4, whereas the remaining OH groups continue to react, forming hydrogen peroxide. The adsorbed hydrogen can recombine to form hydrogen gas through a Tafel-like reaction, but only if a high reaction potential can be overcome.

This work is the first time that real-time nonadiabatic dynamics of water photodissociation have been demonstrated on a practical semiconductor surface at the microscopic scale. Such a hole-dominated nonadiabatic mechanism of chemical bond breaking and recombination should result in an entirely new approach toward designing water photodissociation catalysts.

4

Basic Theory

Electronic Structure

In Chapters 2 and 3, we introduced the physical and chemical phenomena linked to full quantum effects in condensed matter systems and delivered an overview of cases beyond the purview of the Born–Oppenheimer approximation. In Chapters 4–6, we will delve into the theoretical and experimental methods presently employed to investigate full quantum effects, aiming to facilitate their effective utilization in future research.

As discussed in Chapter 1, full quantum effects have been gaining increasing attention in recent years. Several novel phenomena related to full quantum effects have been investigated experimentally and theoretically in condensed matter systems [Bove2009, Meng X.Z.2015, Guo J.2016a, Vuong2018, Borodin2022, Kundu2022]. The exploration of full quantum effects traces its origins back to J. Franck's experimental discovery of the selection rule for molecular light absorption in 1926 [Franck1926]. The advancement of full-quantum theoretical methods, coupled with increased computational power and progress in experimental techniques, has substantially propelled this research field over the years [Marx1996, Tuckerman1996, Marini2008, Zhang Q.F.2008, Ceriotti2009b, 2012, Li X.Z.2010, 2011, Giustino2017, Krüger2022]. Recently, the development of advanced technologies, such as the ultrahigh spatial-resolution imaging techniques and ultrahigh energy-resolution spectroscopy, as well as ultrafast time-resolved dynamics measurements have increased the precision of experimental results, especially under extreme conditions provided by external fields [Bove2009, Guo J.2014, 2016a, Meng X.Z.2015, Vuong2018, Borodin2022, Tian Y.2022]. These advances in the observation of nuclear quantum and nonadiabatic effects have, in turn, improved the theoretical treatment [Marini2008, Giustino2017]. Similar to the discovery and development process of other important fields of physics, there is a process of mutual reinforcement and constant optimization between theory and experiment. Establishing a clear theoretical framework is crucial to comprehending all these theoretical and experimental works.

While the Born–Huang expansion was proposed in the early 1950s, numerical methods had only recently been introduced as a research tool in physics, and first-principles calculation of electronic structure had not yet emerged. The condensed matter theorists typically used analytical methods to solve relatively simple model Hamiltonian. For this reason, although the theory of electron–phonon coupling based on second quantization had been partially developed, the computational study of full quantum effects in real systems based on the Born–Huang expansion remained far from the agenda.

It was not until the 1950s that N. Metropolis, E. Fermi, A. Rahman, and others began actively promoting the use of computers to implement numerical methods [Metropolis1953, Fermi1955, Alder1959, Rahman1964]. Some of the earliest problems treated in this way involved the interactions between atomic nuclei in condensed matter systems, at the time mainly described by empirical force-field models [Lennard-Jones1924, Buckingham1938]. Computational simulations were applied to increasingly complex chemical environments featuring bond breaking and recombination; empirical force fields expressible in simple analytical form could not describe the real physical state of these systems, and researchers instead used the Schrödinger equation to calculate the effect of the nuclear configuration on the electronic structure. Thus, the first-principles method of calculating electronic structure began, at first with a narrow understanding of the Born–Oppenheimer approximation (i.e., the adiabatic ball-and-stick model of electronic states, which works only for specific classical nuclear configurations). Even when the problem is significantly simplified, the quantum-mechanical description of an interacting many-electron system can be quite complex; hence, electronic structure calculation has undergone a long process of development. There are three main categories of electronic-structure calculation methods: the wavefunction method (see Section 4.3), the density functional theory (see Section 4.4), and the QMC method (see Section 4.5). J. A. Pople and W. Kohn shared the Nobel Prize in Chemistry in 1998 because of the successful development of the wavefunction method and the density functional theory [Hohenberg1964, Kohn1965, Hehre1972].

The development of these first-principles methods forms the foundation for the study of the full quantum effect beyond the Born–Oppenheimer approximation, which will be presented in Sections 4.3–4.5. In other words, full quantum theory is based on the accurate electronic structure calculation methods.

Before proceeding further, we remind the reader that the wavefunction method, the density functional theory, and the QMC calculations focus only on the ground state of electrons. A separate section (Section 4.6) is devoted to how these methods can be developed to study the excited states of electron. In this section, a method for calculating the excited states of electron will be specifically recommended: the many-body Green's function method based on the *GW* approximation. The introduction of these methods, which describe the properties of real condensed matter systems along with an accurate portrayal of the electronic structure, lays the foundation for our subsequent discussion on the full quantum theory.

4.1 Overview of Full Quantum Theory

In the following, we will use the term "full quantum theory" to denote a collection of theoretical and computational methods that go beyond the adiabatic ball-and-stick model of the electronic states. These methods consider all the quantum properties of condensed matter systems, including both nuclear and electronic degrees of freedom, as well as their quantum coupling. In Chapter 1, we repeatedly emphasized that the theoretical foundation for these methods is the correction of the narrow understanding of the Born–Oppenheimer approximation by the Born–Huang expansion. We will provide an overview of the full quantum theory in this section; in Section 4.2, we will discuss its derivation in detail, beginning with the Born–Huang expansion.

From the historical perspective, it is undeniable that the recent progress in the study of full quantum theory has been largely dependent on the calculational approach that emerged in the 1980s and the description of electronic structure in an adiabatic first-principles manner for nuclear configurations in the ball-and-stick model. In other words, the "full quantum" computational approach is based on the first-principles calculation for the properties of interacting electronic systems. With this consideration, after introducing the Born–Huang expansion, we will systematically discuss some electronic structure calculations using the ball-and-stick model of specific static nuclear configurations. Such a treatment can be simply regarded as a "partial" quantization of the real systems, targeting only the electronic quantum states. This will lay the groundwork for subsequent discussions, as all methods that consider nuclear quantum states and nucleus–electron quantum coupling are based on electronic structure calculations with static nuclear configurations.

With this background, full quantum effects can, in principle, be studied directly. However, as mentioned earlier, only simple model systems and gas-phase molecular reactions involving a small number of atoms are amenable to this approach; a rigorous full quantum treatment of large molecules and condensed matter systems is impractical at this stage [Zhang D.H.1994, Xiao C.2011]. When conducting research on a real system, it is necessary to simplify the Born–Huang expansion further. Depending on what aspects of the physical problem are retained (or temporarily ignored), the study of the full quantum effects can be roughly divided into the study of nuclear quantum effects and the study of nonadiabatic effects.

The first approach focuses on nuclear quantum effects, that is, on the difference introduced by describing the nucleus as a quantum particle rather than a classical particle. Strictly speaking, this difference can include the contribution of nonadiabatic effects, in which the electron–nucleus coupling is included, the potential energy surface is often splitted and the different surfaces are possibly crossed. In practical studies, nonetheless, when nuclear quantum effects are discussed, the electrons are assumed to always be in the ground state. In other words, electron transitions between different eigenstates as a result of electron–nucleus quantum coupling are neglected.

In the same way, in a rigorous treatment of the Born–Oppenheimer approximation at finite temperature, nuclear quantum effects are considered rigorously without taking electron transitions into account, and the nucleus on the Born–Oppenheimer potential energy surface in the ground state is described by the quantum wavefunction. By contrast, in a narrow understanding of the Born–Oppenheimer approximation, the adiabatic ball-and-stick model of the electronic states is used, and the nucleus on the ground-state Born–Oppenheimer potential energy surface is treated as a classical particle. There are nuclear quantum effects on both the statistical properties of a system at finite temperatures [Marx1996, Marini2008, Zhang Q.F.2008, Ceriotti2009b, 2012, Li X.Z.2010, 2011] and the dynamical properties [Miller1974, Tromp1986, Voth1989, Tuckerman1996, Shao J.S.1999, Shi Q.2003, Liu J.2006]. We will describe these methods in detail in Chapter 5.

The second approach focuses on nonadiabatic effects. At the theoretical level, it considers the quantum coupling during nucleus evolution between different Born–Oppenheimer potential energy surfaces and electronic transition between different electronic energy levels [Tully1990, Craig C.F.2005, Meng S.2008a, b]. Evidently, the nonadiabatic effects being studied here also contain contributions from nuclear quantum effects, similar to how the study of nuclear quantum effects includes some contributions from nonadiabatic effects. Nevertheless, given the current limits on computational capacity and the methodological complexity of the problem, the nucleus is often treated as a classical particle when studying nonadiabatic effects. In this case, these nuclei according to Newton's law evolve between potential energy surfaces. These will be discussed further in Chapter 5.

It should be noted that discussing nuclear quantum effects and nonadiabatic effects is similar to examining the difference between the narrow understanding of the Born–Oppenheimer approximation (the adiabatic ball-and-stick model of electronic states) and the rigorous Born–Huang expansion from two different perspectives. The dynamical processes of a quantum system coupled with all degrees of freedom can be classified as fast or slow. The underlying physical problem is how these types of process are related. The basic idea of the Born–Oppenheimer approximation is that the variables of a slow process can be treated as parameters when dealing with fast processes. After the Schrödinger equation for the fast processes is solved, it provides a potential for the slow processes so that the Schrödinger equation for the slow processes can also be solved. This is mathematically represented by the separation of variables of the overall wavefunction containing the nuclear and electron degrees of freedom, that is, by writing the wavefunction as a product of the electronic wavefunction (with the nuclear coordinates as parameters) and the nuclear wavefunction. The core physical idea is the adiabatic approximation, which ignores all nonadiabatic effects. By simplifying the Born–Oppenheimer approximation further using the ball-and-stick model of adiabatic electronic states, all nuclear quantum effects are also neglected.

Parallel to this simplification method is another route based on the Born–Huang expansion, which preserves the nonadiabatic transition. However, because nuclear motion is a slow process with respect to electron motion, it can still be treated classically by Newton's equations; the electrons can transition between different eigenstates as the nucleus moves. This is essentially a classical approximation of the slow process, and nuclear quantum effects are neglected. Moreover, if one further neglects the electron transitions between different eigenstates, the nonadiabatic effects can also be ignored. The result is the classical treatment of nuclear motion under the adiabatic approximation, the most commonly used method of current molecular simulation. In other words, depending on the specific focus of our investigation, we can decide whether to concentrate on nuclear quantum effects or nonadiabatic effects.

All these methods require an accurate description of the electronic structure in the interacting electron system. In particular, in a recent study of the condensed hydrogen phase transition at ultralow temperatures and ultrahigh pressures, Monacelli et al. highlighted that the atomic structures at various phases differ in enthalpy by less than 1 meV per atom. This sensitivity makes the electronic ground state highly responsive to the Born–Oppenheimer approximation, such as the choice of exchange-correlation functional in density functional theory calculations. Therefore, precise calculations, such as the QMC method, are required to accurately describe the potential energy surface of the electronic ground state [Monacelli2023]. Therefore, Sections 4.3–4.6 are dedicated to the discussion of electronic-structure calculation methods with the aim of laying a strong theoretical foundation. An accurate simulation of the full quantum effects for a real condensed matter system is represented schematically in Figure 4.1. Both nuclear quantum effects and nonadiabatic effects are crucial aspects of the simulation. The adiabatic approximation for the degrees of freedom of fast motions (the electron part) can be used to study nuclear quantum effects, and the classical approximation for the degrees of freedom of slow motions (the nucleus part) can be used to study nonadiabatic effects. When both approximations are based on accurate electronic structures, we obtain the most precise description of the real system at the quantum-mechanical level. In other words, rigorous full quantum simulation is performed along the [111] direction of the following 3D coordinate system. In fact, this concept is the fundamental principle of this book when establishing and discussing the full quantum theory. Currently, a few trade-offs are necessary in practical computations, depending on the research focus.

The differences among the Born–Huang expansion, the study of nuclear quantum effects (which is the rigorous Born–Oppenheimer approximation when nonadiabatic effects are ignored), the study of nonadiabatic effects, and the narrow understanding of the Born–Oppenheimer approximation are the central concepts of this book. We will present these

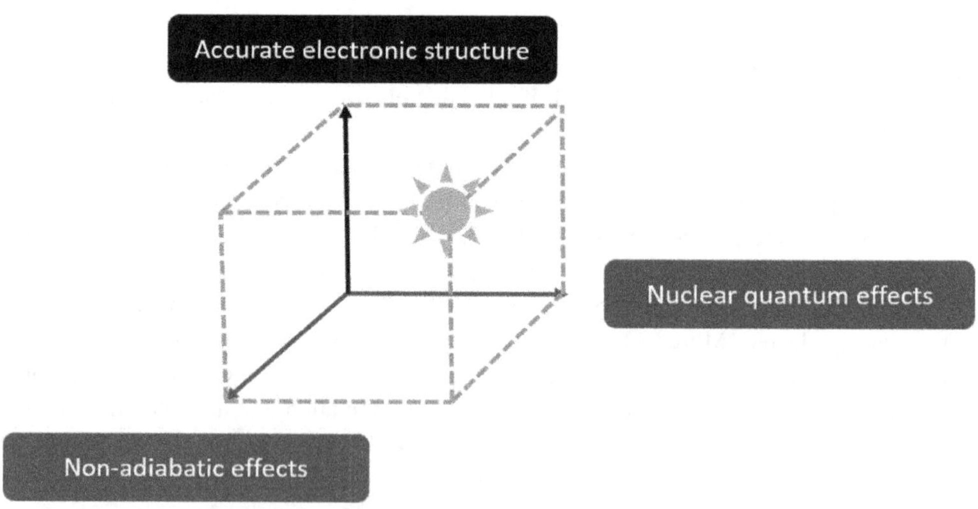

Figure 4.1 Simulating a real condensed matter system accurately at the quantum-mechanical level requires knowledge of three factors: nuclear quantum effects, nonadiabatic effects, and electronic structure. The sun in the figure represents a rigorous full quantum simulation, for which knowledge of all three factors is maximal.

concepts yet again, this time by analogy. Begin by imagining a coin. The narrow understanding of the Born–Oppenheimer approximation (the adiabatic ball-and-stick model of electronic states, neglecting both nuclear quantum effects and nonadiabatic effects) can provide us with the information that the coin is a metal disk of a certain size. Based on our experience, we can attempt to estimate the value of the coin from this, but the estimation is undoubtedly crude. The nuclear quantum effects represent the pattern or writing on the "heads" side of the coin, and the nonadiabatic effects to that on the "tails" side. In fact, the information on each side is related to that on the other in a consistent way; if the heads-side pattern indicates a one-dollar coin, then we do not expect the tails-side pattern to be that of a 50-cent coin. Looking at either the "heads" side or the "tails" side alone would not give us the complete picture. We can only produce a complete description of this coin if we combine the information from both sides. The overall details are what distinguishes the Born–Huang expansion from the narrow understanding of the Born–Oppenheimer approximation.

Just as it is natural to decide whether we look at heads first or tails first, the use of the expansion tends to begin by focusing on either nuclear quantum effects or nonadiabatic effects. To continue the coin analogy, the coupling of the two sides is naturally included in the framework of the Born–Huang expansion. From the perspective of a rigorous solution, one should not neglect nonadiabatic effects when studying nuclear quantum problems nor should one discard nuclear quantum effects when studying nonadiabatic problems. However, from the standpoint of practical calculation, the two must be discussed separately in order to promote relevant studies in real material systems, especially in condensed matter systems. At present, algorithms that unify both nuclear quantum effects and nonadiabatic effects under certain approximations are under development [Shushkov2012].

In Chapter 5, we will introduce computational methods for simulation of nuclear quantum effects and nonadiabatic effects, attempting to cover most approaches that are currently used. In terms of the physical picture, there are many similarities to the second-quantization approaches to electron–phonon coupling, which was proposed in the 1930s and 1940s and developed in the 1970s. However, there are also two major differences.

First, the second-quantization representation requires well-defined coherent states (i.e., the eigenstates of the quasiparticle annihilation operators) as the basis functions for the expansion of the electronic and nuclear vibrational states in coupled systems [Landau1933, Frohlich1950, Fan H.Y.1951, Bardeen1955, 1957, McMillan1968, Dynes1972, Allen1976, 1981]. Therefore, although these methods based on second quantization have found application in early studies of some model systems, in practical systems, lattice periodicity is not guaranteed, nonharmonic effects are significant, and tight-binding or near-free electron-gas models cannot accurately describe the electron coherent states. One must return to the original Born–Huang expansion for a theoretical description to achieve reliable conclusions. Moreover, despite the great success of model-based second-quantization theory in solid-state physics, it is often difficult to interpret the quantitative experimental data because multiple degrees of freedom are coupled and the microscopic details in the system are poorly understood. (For example, quantum paraelectricity, quantum spin liquids, quantum critical behavior, and quantum phase transition may involve coherent superposition and high entanglement of multiple spin configurations). A many-body theoretical calculation method based on the full-quantum Born–Huang expansion shows potential to overcome these limitations and is a promising research direction.

Second, there is a subtle difference between the specific processes described by these two languages. The language based on the Born–Huang expansion is widely used in theoretical chemistry and is self-consistent; the theory of electron–phonon coupling based on second quantization representation is extensively used in condensed matter physics and is also self-consistent. Electron–phonon coupling phenomena can be either adiabatic or nonadiabatic, depending on whether the electrons are on the ground-state or the excited-state potential energy surface when the phonon spectrum is calculated, and on whether the electron–phonon coupling induces electron transitions between different energy bands. Moreover, the real-time variation of phonon vibrations caused by electron transitions and the evolution of nuclei between potential energy surfaces induced by quantum coupling is also critical. In most current computational studies of electron–phonon coupling in condensed matter systems, factors such as the phonon vibration have not been explored thoroughly. In other words, the method used to study electron–phonon coupling in theoretical physics usually has no clear counterpart in theoretical chemistry and vice versa; the knowledge structures of the two research communities are quite distinct. It will take a long time to determine the fine details of these problems to achieve coordinated development.

In Chapter 5, we will provide a systematic review of the electron–phonon coupling theory that was developed in the 1970s (see Section 5.2). Some recent advances of this theory, especially its integration with first-principles electronic-structure calculation methods, will also be presented. Finally, we will return to the Born–Huang expansion. Starting from there, preliminary attempts to combine nuclear quantum effects with nonadiabatic effects in theoretical chemistry will be discussed. The content of this and Chapter 5 will be relatively familiar to readers with backgrounds in condensed matter physics and theoretical chemistry, respectively. However, we would like to encourage both to step out of their comfort zones and try to understand material with which they are not familiar. In the last section of Chapter 5, as a conclusion to the introduction of the full quantum theory, we will share a brief outlook on the development of full-quantum-effect simulation methods by taking into account both nuclear quantum effects and nonadiabatic effects in condensed matter systems.

The purpose of Chapters 4 and 5 is to help the reader grasp the crux of the theoretical problems involved in describing the physical and chemical properties of large molecular or condensed matter systems. The key is to go beyond the adiabatic ball-and-stick model of the electronic states. Mastering these concepts and methods is essential for the in-depth study of full quantum effects in condensed matter systems.

4.2 Born–Huang Expansion

To make this chapter self-contained, we rewrite the specific expression for the Born–Huang expansion given in the equation (Whatever it was in Chapter 1) as Eq. (4.1). Recall that the main idea is to expand the wavefunction of a system as a sum over the products of the electron and nuclear wavefunctions [Born1954]:

$$\Psi^{j}(\vec{r},\vec{R},t)=\sum_{n=1}^{\mathrm{el}}\chi_{n}^{j}(\vec{R},t)\,\Phi_{n}(\vec{r},\vec{R}) \tag{4.1}$$

Note that the electronic eigenstate $\Phi_n(\vec{r},\vec{R})$ in this equation is summed over the quantum number n. In the Born–Oppenheimer approximation, there is no such sum, and the overall wavefunction of the system can be written as a simple product of the electronic and nuclear wavefunctions, with the former parametrized by the nuclear configuration. Because it does not rely on the assumption that variables are separable, the expansion in Eq. (4.1) is more rigorous as long as the completeness of the electronic eigenstates is preserved. Moreover, the nuclear wavefunction $\chi_n^j(\vec{R},t)$ evolves with time, and its evolution is synchronized with the quantum coupling of the electron wavefunction. When the overall wavefunction is expanded in this way, all quantum-mechanical information about the electronic states, the nuclear states, and the coupling between them is naturally included.

In addition to the original formulation that M. Born and K. Huang published in their classic textbook in 1954 [Born1954], as we have mentioned previously, the Born–Huang expansion can be understood from a different perspective, assuming that the electronic state of the system can be described by a quantum state with a known wavefunction. For example, in a simple metal system, the electronic state can often be approximated as a free electron gas whose eigenstates are planewaves (or at least as a nearly free electron gas whose eigenstates are simple combinations of planewaves). The eigenstates constitute a complete basis of the electron wavefunction in Hilbert space. In the language of many-body quantum mechanics, the electron eigenstates are also eigenstates of the annihilation operator, that is, coherent states [Negele1998]. Notably, nuclear motion is also quantized. When the condition for simple harmonic motion is met, the vibrational states correspond to phonons in a periodic crystal. If the electronic states are diagonalized with respect to the inter-electron interactions and the nuclear vibrational states are diagonalized with respect to the corresponding Born–Oppenheimer potential energy surfaces, then as a result of electron–phonon coupling, we must further diagonalize with respect to the basis space corresponding to the product of the electronic and nuclear quantum states when describing the overall wavefunction. This is standard in traditional solid-state physics. In the language of nonrelativistic field theory (i.e., of many-body quantum mechanics), similar problems are described using a basis defined by second quantization and its associated creation and annihilation operators. After further diagonalization of the basis to include the electron–phonon interaction Hamiltonian, the eigenstates of the system become a linear superposition of the product of nuclear and electron eigenstates [Fan1951, Bardeen1955, Bardeen1957]. This linear superposition is physically equivalent to the Born–Huang expansion.

From the advent of quantum mechanics in the 1920s to the proposal of first-principles electronic-structure calculation in the 1980s, the language of nonrelativistic field theory was used to discuss elementary-excitation phenomena in electron–phonon coupling (e.g., superconductivity) [Landau1933, Frohlich1950, Fan H.Y.1951, Bardeen1955, 1957, McMillan1968, Dynes1972, Allen1976, 1981]. In metals, the electrons were treated as a free or nearly free electronic gas, and the eigenstates of nuclear motion were often described in terms of simple harmonic oscillations. For semi-metals, semiconductors, and insulators, electron states were usually described by the tight-binding model. The historical reason for this was the limited availability of the computing power required for solving condensed matter theoretical problems and second quantization provided the most straightforward approach. It was easy and effective when the electron and nuclear wavefunction eigenstates were both diagonalized, and it was amenable to analytic or simple numerical methods. However, a direct consequence of using this method was that theoretical physicists rarely returned to Eq. (4.1), the original wavefunction form of the Born–Huang expansion, to study the problem.

This may be a result of disciplinary practice, as one of the central goals of physics is to explore the composition and interaction laws of materials. In the pursuit of these objectives, the language of field theory has unparalleled advantages; for many twentieth-century physicists, it was the most natural way to describe the problem. Consequently, there were few attempts to describe electronic quantum states, nuclear quantum states, and the electron–nucleus quantum coupling in a first-principles way. Further, the Born–Huang expansion was not widely used as a research tool in the study of condensed matter.

By contrast, the Born–Huang expansion was used early on by theoretical chemists, especially for studying the dynamics of gas-phase reactions. (Indeed, chemists are often surprised to learn that physicists are only now discussing these problems in terms of the Born–Huang expansion.) There are two reasons for this which cannot be ignored, and they are mainly due to the difference in core problems (focal points of interest) between physics and chemistry.

First, the central problem of nuclear configuration in chemistry is that the system structure undergoes drastic splitting and restructuring during chemical reactions, manifested in the breaking and recombination of chemical bonds. As a result, the nucleus is rarely in the vicinity of the equilibrium position. The nuclear state itself therefore cannot be studied by expanding it in terms of simple analytic functions (e.g., the eigenstates of phonon vibrations near a perfect lattice configuration or the vibrational states of impurities close to their relaxation configurations). For the problems of maximum interest to chemists, it is inappropriate to use the language of field theory with second quantization. However, the central problems of condensed matter physics (at least until the 1980s) have involved electrons in solids interacting with the lattice (atoms) near the equilibrium states. Operations based on second quantization are suitable for this situation. They require analytic basis functions; they are both simple and capable of describing the nuclear state. Consequently, the language of field theory was very popular in early studies of solid-state physics problems.

Second, in chemical research, functionality at the molecular level is another topic of interest. For instance, the number of nuclei involved in kinetic problems of gas-phase reactions is very small; there are generally fewer than three or four atoms. In this case, describing the nuclear quantum state via wavefunctions is not a big issue in terms of computational cost. When dealing with small systems (consisting of a few atoms and electrons), the use of second-quantization theory based on model systems is not necessary when describing chemical reactions, and returning to the original Born–Huang expansion of the wavefunctions is more appropriate. This is why chemists are not interested in building "models" (such as physicists' tight-binding model of semiconductors) for small molecular systems. Instead, most effort is devoted to performing specific calculations for each different small system. As a result, the Born–Huang expansion was quickly applied in theoretical chemistry. Later, it was also frequently mentioned in the study of other problems (e.g., nonadiabatic problems) as a custom of theoretical chemistry research.

This is somewhat analogous to the situation in electronic structure theory, in which condensed matter physicists tend to use the language of second quantization and focus on model Hamiltonian systems, while theoretical chemists are accustomed to many-body wavefunction expansions based on Slater determinants and density functional theory. In physics, the focus is on model theories and the "big picture," whereas in chemistry, it is usually on the details of a specific system. Thus, theoretical chemists always emphasize the accuracy of electronic structure and nuclear-motion calculations for the specific system under study. It is essential to recognize and appreciate this difference as we learn the calculation methods of condensed matter physics and theoretical chemistry, so we can integrate them with each other. We will see that this difference is rooted in the fundamental problems faced by physics and chemistry, or in other words, these two subjects focus on different aspects.

Finally, we offer a more general perspective. An analogy can be drawn with the early stages of the establishment of quantum mechanics, where comparable divergences in thought may be discerned when exploring the relationship between the two formulations of quantum mechanics. Matrix mechanics allows us to analyze the physical meaning of mechanical operators. In the interaction regime, its transition to the language of field theory is relatively straightforward. Wave mechanics, on the other hand, does not depend on the model system when solving for mechanical expectation values and handles arbitrary quantum systems rather consistently, thus sharing similarities with the Born–Huang expansion.

The actual problem faced by these two formulations in the description of a many-body system containing multiple atoms and electrons is the problem of diagonalizing a Hamiltonian that contains all degrees of freedom (electronic and nuclear) in the complete many-body Hilbert space. In traditional condensed matter physics, it is customary to first choose a set of eigenstates as the complete basis, and then discuss the problem in the chosen basis using the language of second quantization. Conversely, when studying gas-phase reaction kinetics in chemistry, it is customary to solve the Schrödinger equation directly using the Born–Huang expansion, and simply diagonalize it using a complete set of basis vectors. These methodological differences are, to a great extent, due to differences in theoretical languages developed over a long period of time. This is an objective fact, and the different educations of researchers cannot be ignored, which bring about different habits of thought, impairing communications between different fields that have much to learn from one another.

In a condensed matter system, if the observed physical quantities are only relevant to particles and to elementary excitations that are stable on the observed time and energy scales, it would seem more straightforward to use the Born–Huang expansion directly. Based on such consideration, this book intends to address some problems in condensed matter physics, especially those related to nuclear quantum effects and nonadiabatic effects, from the Born–Huang perspective. We aim to resolve some theoretical doubts, summarize experimental findings, and attempt to solve the problems in a way that is different from the second-quantization representation.

The physical meaning of the Born–Huang equation is well-understood. Electrons and nuclei are known to exhibit quantum properties, and they are certainly coupled to each other. Therefore, if the quantum-mechanical framework is correct, calculations based on it should be in perfect agreement with experimental measurements. However, the complexity of the computational algorithms and the limitations of computing power compel the use of approximations. As mentioned in Chapter 1, these approximations were already known even before the Born–Huang expansion was proposed; for example, the Born–Oppenheimer or adiabatic approximation was proposed in 1927. Born and Oppenheimer noted that the mass of the nucleus is much greater than that of the electron, and the energy scale difference between the electronic eigenstates is relatively large. Therefore, as the nucleus moves, electrons do not undergo nonadiabatic transitions between different eigenstates at the time scale we are interested in (see left panel of Figure 1.35). This allows the separation of variables on the overall wavefunction and rewrite it as the product of the nuclear wavefunction and the electronic wavefunction. In the electronic wavefunction, the nuclear configuration appears as a parameter. This approach and the related adiabatic ball-and-stick model of electronic states are commonly used even today. In recent years, however, simulation has advanced; as theoretical descriptions reproduce experimental observation more closely and measurement accuracy is pushed to the limit [Meng X.Z.2015, Guo J.2016a, Vuong2018, Krüger2022], the ball-and-stick model seems less satisfactory [Wang E.G.2016, 2022]. Consequently, the study of the nuclear quantum effects and nonadiabatic effects included in the Born–Huang expansion has become a frontier research area in material science. The manipulation of these effects opens new avenues for the control of the quantum properties of real condensed matter.

The concepts of nuclear quantum effect and nonadiabatic effect were explained earlier (in Section 4.1), but the Born–Huang formula allows us to provide an explanation that is both mathematically simpler and more accurate. In Eq. (4.1), the nucleus is treated as a quantum object whose wavefunction $\chi_n^j(\vec{R},\ t)$ can include contributions from adiabatic and nonadiabatic terms. However, to simplify the problem, the summation over the electronic states n in Eq. (4.1) is often neglected, and only nuclear quantum effects on certain Born–Oppenheimer potential energy surfaces are considered. In other words, $\chi_n^j(\vec{R},\ t)$ is often replaced by $\chi^j(\vec{R},\ t)$. This means that most of the current studies on nuclear quantum effects do not strictly account for the nonadiabatic effects, ignoring the electron transition between potential energy surfaces.

Nonadiabatic effects, however, arise from the summation over n. Including it means including the nonadiabatic effects; excluding it (as do most current studies of nuclear quantum effects) means excluding them. Note that when understood in this way, the wavefunction $\chi_n^j(\vec{R},\ t)$ is not reduced to a classical trajectory $\vec{R}_n(t)$ of nuclear motion. Therefore, the calculation of nonadiabatic effects using Eq. (4.1) naturally contains nuclear quantum effects. In actual practice, however, their contribution is often ignored to simplify the problem; the quantum nuclear wavefunction $\chi_n^j(\vec{R},\ t)$ is reduced to the classical trajectory $\vec{R}_n(t)$, and the term $\sum_{n=1}^{\text{el}}\chi_n^j(\vec{R},t)\Phi_n(\vec{r},\vec{R})$ is replaced by $\sum_{n=1}^{\text{el}}\vec{R}_n(t)\Phi_n(\vec{r},\vec{R})$. The nuclear motion then satisfies the Newtonian equation instead of the Schrödinger equation. However, even under this approximation, Eq. (4.1) contains the summation of electronic states n. In other words, even if the nucleus is treated classically, it is still possible to have adiabatic and nonadiabatic contributions when starting from the Born–Huang expansion. Based on this observation, we can study the nonadiabatic effects by comparing $\sum_{n=1}^{\text{el}}\vec{R}_n(t)\Phi_n(\vec{r},\vec{R})$ and $\vec{R}(t)\Phi(\vec{r},\vec{R})$. Except in the case of a few extremely simple gas-phase reactions, studying the nuclear quantum effects while neglecting nonadiabatic effects or vice versa makes the calculation less rigorous (even though, in contemporary condensed matter physics, this lack of rigor is generally not considered important). Combining the two types of effects under certain approximations has become a new topic of research. In the following discussion, we will briefly explore this topic to the extent permissible by the length and capacity of this book.

Finally, let us emphasize the importance of understanding that, as condensed matter physics evolved, the philosophy behind the Born–Huang expansion was gradually accepted. When the Born–Huang expansion was proposed, first-principles electronic structure calculation methods (and other related methods for calculating physical properties) were still in their infancy, and the full quantum approach was yet to become the primary theoretical tool of material-science research. Many specific problems were still studied using simple models. For example, the ball-and-stick model was used to describe the nuclear configuration, the nearly free electron or tight-binding model was used for the electronic states, and the phonon concept was introduced to describe lattice vibration.

In the 1980s, theoretical foundations were established for exploring reliable electronic structure in condensed matter systems using first-principles methods. The Born–Oppenheimer molecular dynamics method also developed rapidly at the same time. Because of its success, there was a tendency to understand electronic structure by some first-principles method and describe the nucleus based on the narrow understanding of the Born–Oppenheimer approximation (i.e., the adiabatic ball-and-stick model of the electronic states). More specifically, for a particular nuclear configuration, the electronic structure was calculated self-consistently; then, the forces acting on the nucleus in this configuration were described by the Hellmann–Feynman theorem. In practice, the nuclear motion was either described completely in terms of classical mechanics (i.e., Born–Oppenheimer molecular dynamics and related methods) or perturbation theory was used when considering the quantum vibrations of the nucleus. In all these treatments, the electrons always remained in the ground state.

Such approaches have succeeded in explaining many experimental phenomena. However, there are serious underlying issues behind this success. It is important to understand that this description significantly deviates from the original representation of the Born–Huang expansion. Consequently, many incorrect comparisons are made between theoretical calculations and experimental observations. Moreover, researchers often pay little attention to (or show little understanding of) the approximations being made. This raises the usually unanswered question of whether the agreement between computational results and experimental results might come from offsetting computational errors at the computational level, over-interpretation of the experimental results, or the limitations of the experimental technique itself.

As an example, let us briefly consider how the vertical excitation energy between electronic states in a small system such as the ScO molecule is calculated. The nuclear states of the system in the electronic ground state and certain excited states are depicted in Figure 4.2.

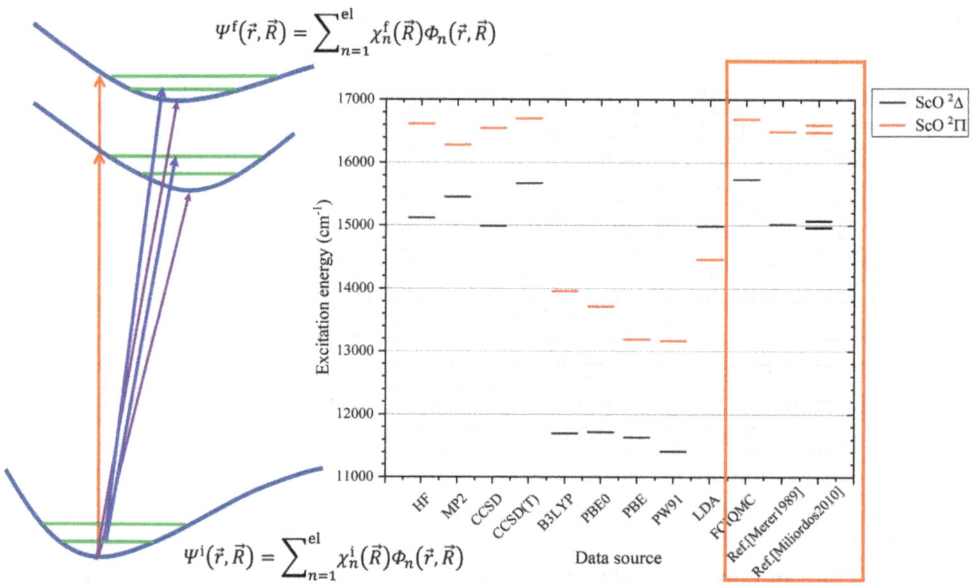

Figure 4.2 Photon transition in an ScO molecule. The figure on the right compares the theoretical and experimental vertical-transition excitation energies for the optimal nuclear configuration on the Born–Oppenheimer potential energy surface of the electronic ground state. The calculations were performed using different approaches on the electronic-structure level [data from Jiang T.H.2021]. The vertical transition corresponds to the process highlighted with the red arrow in the left panel. The purple arrows indicate the energy differences between minima of the Born–Oppenheimer potential energy surfaces for the electron excited state and ground state. Both of these are based on the adiabatic ball-and-stick model. Experimentally, the initial state of the photon transition contains the degrees of freedom for the nuclear wavefunction, and the same is true for the final state. The Franck–Condon principle provided a relatively simple picture of this phenomenon, that is, the real-space overlap of the nuclear wavefunctions in the initial and final states should be maximized. However, to establish a more rigorous theoretical description of this process, one has to return to the Born–Huang expansion. The initial and final states must be expressed separately so that they can be combined with the light-absorption transition-matrix elements of the real experimental process. Evidently, even this rigorous first-principles treatment may not yield the precise results observed in the actual system. However, many practical calculations produce results that are not in perfect agreement with the experiment; this is expected and is exactly what we need to focus on and improve by considering full quantum effects in future research.

This is a typical many-body system consisting of multiple electrons and nuclei, with their quantum states coupled to each other. The right side of the figure shows the excitation energies of the electronic states given by the most stable nuclear configuration on the Born–Oppenheimer potential energy surface of the electronic ground state. In the framework of density functional theory, the calculation results agree poorly with the experiment if the usual exchange-correlation functional approximation is used (see Section 4.4). This discrepancy is evident in the clear disparity between the theoretical calculations employing DFT methods with various approximations (ranging from B3LYP to LDA) and the experimental values shown in the figure. The traditional quantum-chemistry methods described in Section 4.3 (results from HF to CCSD(T)) yield slightly better results than the density functional theory calculations. However, the QMC results (FCIQMC, see Section 4.5) show even greater improvement. The reason for this is simple: The mean-field approximation is not effective in capturing electron correlation effects. The traditional quantum chemistry method, which is based on the concept of electron configuration, produces relatively better results. The results provided by the QMC method, on the other hand, are further improved because its wavefunctions are complete in terms of linear combinations of all possible electron configurations. As can be seen in the red box in Figure 4.2, the FCIQMC calculation results are close to the experimental results, both in terms of absolute values and the relative positions of the energy levels.

In fact, we should not expect perfect agreement between theory and experiment at this level, because the initial state before the photon transition and the final state after the transition need to be described quantum mechanically based on the Born–Huang expansion after considering nuclear degrees of freedom, instead of using nuclear configuration. The Born–Huang expansion is undoubtedly the most powerful tool that allows us to improve theoretical computation systematically. All one has to do is describe the complex many-body inter-electron interactions as accurately as possible on the electronic-structure level for a particular nuclear configuration. Then, using the Born–Huang expansion, their coupling to the nuclear quantum states can be considered rigorously. The final calculation can be compared with experimental values both in terms of the electronic and the nuclear quantum states.

While making such rigorous simulations may be difficult, the difficulty does not excuse mistakes at the conceptual level. We believe that, with the rapid development of theory and the substantial improvement of experimental techniques, the Born–Huang expansion will become the dominant paradigm in condensed matter physics in the coming years, especially for descriptions of the physical and chemical properties related to nuclear quantum effects and nonadiabatic effects. Moreover, refining first-principles electronic structure theory to include the electron and nuclear quantum states should be a major objective in the development of theoretical computational methods (Figure 4.1).

4.3 Wavefunction Method

The wavefunction method is one of the earliest and most straightforward quantum-mechanical methods for determining the electronic structure of materials. Its theoretical basis is the variational principle. For an interacting electronic system, let F_0 denotes the vacuum state, F_1 the complete Hilbert space of electronic wavefunctions when the number of electrons is 1, F_2 the complete Hilbert space of electronic wavefunctions when the number of electrons is 2, …, and F_N the complete Hilbert space of electronic wavefunctions when the number of electrons is *N*. Then, the complete Fock space formed by these interacting electrons is $F_0 \oplus F_1 \oplus F_2 \oplus \cdots \oplus F_N$ [Negele1998]. The span of the eigenstates Ψ_i of an *N*-electron system is complete in F_N. Thus, a trial *N*-electron wavefunction $|\psi\rangle$ can be written as a linear combination of these eigenstates

$$|\psi\rangle = \sum_i c_i \Psi_i. \tag{4.2}$$

When the system is described by this trial wavefunction, the expected value of the system energy is

$$\langle \psi | \hat{H} | \psi \rangle = \left\langle \sum_i c_i \Psi_i \middle| \hat{H} \middle| \sum_j c_j \Psi_j \right\rangle = \sum_j c_j E_j \left\langle \sum_i c_i \Psi_i | \Psi_j \right\rangle = \sum_j |c_j|^2 E_j. \tag{4.3}$$

If $|\psi\rangle$ is normalized, this becomes

$$\begin{aligned} \langle \psi | \hat{H} | \psi \rangle &= |c_0|^2 E_0 + \sum_{j=1} |c_j|^2 E_j = \left(1 - \sum_{j=1} |c_j|^2\right) E_0 + \sum_{j=1} |c_j|^2 E_j \\ &= E_0 + \sum_{j=1} |c_j|^2 (E_j - E_0). \end{aligned} \tag{4.4}$$

The expectation value can only take the minimum value E_0 if $c_j = 0$ holds for all $j \neq 0$. This means that, if we consider the variation of the trial wavefunction $|\psi\rangle$, the energy expectation of the system is only minimized when $|\psi\rangle$ is the ground-state wavefunction.

After the ground-state wavefunction is found, the trial wavefunction is subtracted from its projection on it, and the entire process is repeated, allowing us to determine the first excited-state wavefunction. In principle, we can solve for all eigenstates using variational methods; however, in practical calculations, we are mostly concerned with the electronic ground state. This principle also provides a reliable theoretical foundation for all variational-based wavefunction methods.

4.3.1 Hartree Method

R. B. Lindsay proposed a self-consistent-field method when using the Bohr model to solve the electronic structure of alkali metal atoms as early as 1924. He solved it using a variational method similar to the method mentioned previously [Lindsay1924]. Instead of using the wavefunction (since the Schrödinger equation had not yet been proposed), Lindsay utilized the electron orbitals of the Bohr model. After the introduction of the Schrödinger equation in 1926, D. R. Hartree immediately realized that Lindsay's technique could be used to solve the many-electron Schrödinger equation [Hartree1928]. In 1928, he proposed his famous method, treating the many-electron wavefunction as a product of single-electron wavefunctions. This is expressed as

$$\Psi_{1\cdots\mathrm{N}} = \psi_1 \times \psi_2 \times \cdots \times \psi_\mathrm{N}, \tag{4.5}$$

and the Hamiltonian of the interacting electron system as

$$\begin{aligned}\hat{H}_{\mathrm{e}} &= -\frac{1}{2}\sum_{i=1}^{\mathrm{N}}\nabla_i^2 + \sum_{\mathrm{i,j}}\mathrm{v}_{\mathrm{en}}(\vec{r}_i - \vec{R}_j) + \frac{1}{2}\sum_{i\neq i'}^{\mathrm{N}}\mathrm{v}(\vec{r}_i - \vec{r}_{i'}) \\ &= \sum_{i=1}^{\mathrm{N}}\hat{h}_i + \frac{1}{2}\sum_{i\neq i'}^{\mathrm{N}}\mathrm{v}(\vec{r}_i - \vec{r}_{i'})\,.\end{aligned} \tag{4.6}$$

Atomic units are used in this equation, where the electron mass $\mathrm{m}_0 = 1$ and the reduced Planck constant $\hbar = 1$. Assuming the single-electron wavefunction $\psi_{\mathrm{i}}(\vec{\mathrm{r}})$ is normalized, the requirement of the variational principle for the total energy is equivalent to the requirement that the L functional defined in Eq. (4.7) is variational.

$$\begin{aligned}L = &\sum_{i=1}^{\mathrm{N}}\int \mathrm{d}\vec{r}\psi_i^*(\vec{r})\hat{h}_i\psi_i(\vec{r}) + \frac{1}{2}\sum_{i\neq i'}^{\mathrm{N}}\iint \mathrm{d}\vec{r}\mathrm{d}\vec{r}'\psi_i^*(\vec{r})\psi_{i'}^*(\vec{r}')\mathrm{v}(\vec{r}-\vec{r}')\psi_i(\vec{r})\psi_{i'}(\vec{r}') \\ &-\sum_{i=1}^{\mathrm{N}}\epsilon_i\left(\int \mathrm{d}\vec{r}\psi_i^*(\vec{r})\psi_i(\vec{r}) - 1\right).\end{aligned} \tag{4.7}$$

For the variation of $\Psi_{1\ldots\mathrm{N}}$, if ψ_j from ψ_1 to ψ_{N} is changed to $\psi_j + \delta\psi_j$, then L will become $L + \delta L$. The variational principle requires that $\delta L / \delta\psi_j = 0$.

Considering this, we first need to determine $L + \delta L$; then we will solve for δL and $\delta L / \delta\psi_{\mathrm{j}}$. To find $L + \delta L$, we insert the transformed Hartree many-electron wavefunction into Eq. (4.7) to obtain the following equation:

$$\begin{aligned}L + \delta L = &\sum_{i\neq j}^{\mathrm{N}}\int \mathrm{d}\vec{r}\psi_i^*(\vec{r})\,\hat{h}_i\psi_i(\vec{r}) + \int \mathrm{d}\vec{r}\left(\psi_j^*(\vec{r}) + \delta\psi_j^*(\vec{r})\right)\hat{h}_j\left(\psi_j(\vec{r}) + \delta\psi_j(\vec{r})\right) \\ &+\frac{1}{2}\sum_{i\neq i'}^{\mathrm{N}}\iint \mathrm{d}\vec{r}\mathrm{d}\vec{r}'\psi_i^*(\vec{r})\psi_{i'}^*(\vec{r}')\nu(\vec{r}-\vec{r}')\psi_i(\vec{r})\psi_{i'}(\vec{r}') \\ &+\sum_{i\neq j}^{\mathrm{N}}\iint \mathrm{d}\vec{r}\mathrm{d}\vec{r}'\psi_i^*(\vec{r})\left(\psi_j^*(\vec{r}') + \delta\psi_j^*(\vec{r}')\right)\nu(\vec{r}-\vec{r}')\psi_i(\vec{r})\left(\psi_j(\vec{r}') + \delta\psi_j(\vec{r}')\right) \\ &-\sum_{i\neq j}^{\mathrm{N}}\epsilon_i\left(\int \mathrm{d}\vec{r}\psi_i^*(\vec{r})\psi_i(\vec{r}) - 1\right) \\ &-\epsilon_j\left(\int \mathrm{d}\vec{r}\left(\psi_j^*(\vec{r}) + \delta\psi_j^*(\vec{r})\right)\left(\psi_j(\vec{r}) + \delta\psi_j(\vec{r})\right) - 1\right).\end{aligned} \tag{4.8}$$

Taking the difference between Eqs (4.7) and (4.8) to the first order of $\delta\psi_j$, we have

$$\begin{aligned}\delta L = &\int d\vec{r}\delta\psi_j^*(\vec{r})\hat{h}_j\psi_j(\vec{r}) + \sum_{i\neq j}^{N}\iint d\vec{r}d\vec{r}'\psi_i^*(\vec{r})\delta\psi_j^*(\vec{r}')\nu(\vec{r}-\vec{r}')\psi_i(\vec{r})\psi_j(\vec{r}') \\ &-\epsilon_j\int d\vec{r}\delta\psi_j^*(\vec{r})\psi_j(\vec{r}) + c.c.\end{aligned} \tag{4.9}$$

In this way, because the variational principle requires $\delta L / \delta\psi_j = 0$, we can naturally obtain the equation $\psi_j(\vec{r})$ must satisfy

$$\hat{h}_j\psi_j(\vec{r}) + \sum_{i\neq j}^{\mathrm{N}}\int \mathrm{d}\vec{r}'\psi_i^*(\vec{r}')\nu(\vec{r}-\vec{r}')\psi_i(\vec{r}')\psi_j(\vec{r}) = \epsilon_j\psi_j(\vec{r})\,. \tag{4.10}$$

Each single-electron orbital in Eq. (4.5) must obey this equation to obtain a self-consistent solution.

In actual self-consistent calculation, all electrons are used together to produce a colloidal field. This field is then subtracted from the self-interaction of the electrons. Thus, Eq. (4.10) can be rewritten as

$$\hat{h}_j\psi_j(\vec{r})+\sum_{i=1}^{N}\int \mathrm{d}\vec{r}'\left|\psi_i(\vec{r}')\right|^2 v(\vec{r}-\vec{r}')\psi_j(\vec{r})- \int d\vec{r}'\left|\psi_j(\vec{r}')\right|^2 v(\vec{r}-\vec{r}')\psi_j(\vec{r})=\epsilon_j\psi_j(\vec{r}), \tag{4.11}$$

where the last term on the left side of the equation represents the self-interaction correction (SIC) of the electrons in state ψ_j. This term will be used many times in subsequent discussion of electronic-structure calculation methods.

4.3.2 Hartree–Fock Method

A drawback of the Hartree method is that it does not consider the exchange antisymmetry of the spatial part of the wavefunction, which must be satisfied by the electron as a fully homogeneous fermion. To include this property, V. A. Fock generalized Hartree's method shortly after its proposal using the mathematical concept of the Slater determinant [Slater1929, 1930b, Fock1930]. (Slater himself also noted in his 1930 paper that the Hartree method could be generalized in this manner.) The major difference between the Hartree–Fock and the Hartree method is that, instead of representing it using Eq. (4.5), the Hartree–Fock wavefunction of a many-electron system is expressed as

$$\Psi_{1\cdots N}=\frac{1}{\sqrt{\mathrm{N}!}}\begin{vmatrix}
\psi_1(\vec{r}_1) & \psi_1(\vec{r}_2) & \cdots & \psi_1(\vec{r}_i) & \cdots & \psi_1(\vec{r}_j) & \cdots & \psi_1(\vec{r}_N)\\
\psi_2(\vec{r}_1) & \psi_2(\vec{r}_2) & \cdots & \psi_2(\vec{r}_i) & \cdots & \psi_2(\vec{r}_j) & \cdots & \psi_2(\vec{r}_N)\\
\vdots & \vdots & \ddots & \vdots & \ddots & \vdots & \ddots & \vdots\\
\psi_i(\vec{r}_1) & \psi_i(\vec{r}_2) & \cdots & \psi_i(\vec{r}_i) & \cdots & \psi_i(\vec{r}_j) & \cdots & \psi_i(\vec{r}_N)\\
\vdots & \vdots & \ddots & \vdots & \ddots & \vdots & \ddots & \vdots\\
\psi_j(\vec{r}_1) & \psi_j(\vec{r}_2) & \cdots & \psi_j(\vec{r}_i) & \cdots & \psi_j(\vec{r}_j) & \cdots & \psi_j(\vec{r}_N)\\
\vdots & \vdots & \ddots & \vdots & \ddots & \vdots & \ddots & \vdots\\
\psi_N(\vec{r}_1) & \psi_N(\vec{r}_2) & \cdots & \psi_N(\vec{r}_i) & \cdots & \psi_N(\vec{r}_j) & \cdots & \psi_N(\vec{r}_N)
\end{vmatrix}. \tag{4.12}$$

By generalizing Eq. (4.7) and inserting in this many-body wavefunction, the Laplace operator can be applied based on the orthogonality of the single-body wavefunction. As a result, a Lagrangian can be obtained for use in the variational method. Applying the variational method, the single-electron orbital ψ_j becomes $\psi'_j = \psi_j + \delta\psi_j$, and the many-body wavefunction becomes

$$\frac{1}{\sqrt{\mathrm{N}!}}\begin{vmatrix}
\psi_1(\vec{r}_1) & \psi_1(\vec{r}_2) & \cdots & \psi_1(\vec{r}_i) & \cdots & \psi_1(\vec{r}_j) & \cdots & \psi_1(\vec{r}_N)\\
\psi_2(\vec{r}_1) & \psi_2(\vec{r}_2) & \cdots & \psi_2(\vec{r}_i) & \cdots & \psi_2(\vec{r}_j) & \cdots & \psi_2(\vec{r}_N)\\
\vdots & \vdots & \ddots & \vdots & \ddots & \vdots & \ddots & \vdots\\
\psi_i(\vec{r}_1) & \psi_i(\vec{r}_2) & \cdots & \psi_i(\vec{r}_i) & \cdots & \psi_i(\vec{r}_j) & \cdots & \psi_i(\vec{r}_N)\\
\vdots & \vdots & \ddots & \vdots & \ddots & \vdots & \ddots & \vdots\\
\psi'_j(\vec{r}_1) & \psi'_j(\vec{r}_2) & \cdots & \psi'_j(\vec{r}_i) & \cdots & \psi'_j(\vec{r}_j) & \cdots & \psi'_j(\vec{r}_N)\\
\vdots & \vdots & \ddots & \vdots & \ddots & \vdots & \ddots & \vdots\\
\psi_N(\vec{r}_1) & \psi_N(\vec{r}_2) & \cdots & \psi_N(\vec{r}_i) & \cdots & \psi_N(\vec{r}_j) & \cdots & \psi_N(\vec{r}_N)
\end{vmatrix}; \tag{4.13}$$

the Lagrangian L also changes accordingly. As in the Hartree method, $\delta L / \delta \psi_j = 0$, and the single-electron orbital must therefore satisfy the following equation:

$$\hat{h}_j \psi_j(\vec{r}) + \sum_{i=1}^{N} \int d\vec{r}' \left|\psi_i(\vec{r}')\right|^2 v(\vec{r} - \vec{r}')\psi_j(\vec{r}) \\ - \sum_{i=1}^{N} \int d\vec{r}' \psi_i^*(\vec{r}')\psi_j(\vec{r}')v(r_i - r_j)\,\psi_i(\vec{r}) = \epsilon_j \psi_j(\vec{r}) . \tag{4.14}$$

Note that the last term on the left side of Eq. (4.11) (the SIC term) has been replaced by a nonlocal exchange term originated in the antisymmetric nature of the Slater determinant under row and column exchange. An immediate consequence of this is that the Hartree–Fock method, although more accurate than the Hartree method, has a higher computational cost. The computational power required for the Hartree method is proportional to the cube of the number of electrons N, whereas that for the Hartree–Fock method is proportional to the fourth power. For this reason, practical calculations using the Hartree–Fock method were limited to small-molecule systems for a long time.

4.3.3 Post-Hartree–Fock Methods

Although inter-electron exchange interactions are included in the Hartree–Fock method, electron correlations are not. As a typical consequence, the total energy of the two-electron system in a He atom obtained by the Hartree–Fock method is 1.13 eV higher than the value calculated by the most accurate method for electronic structure calculations (the configuration-interaction method, described later). This discrepancy is much greater than the chemical accuracy (1 kcal/mol, ~43.4 meV). The main reason for it is that the Slater determinant can not fully describe the many-body wavefunction of a many-electron system with Coulomb interactions. In the case of the ground state of a many-electron system, for example, the configuration corresponding to a single Slater determinant is correlated with other configurations, which are Slater determinants produced by rearranging the occupation of electrons. Therefore, a single Slater determinant cannot accurately describe the ground-state wavefunction of a many-electron system. The real many-body wavefunction can be a linear combination of the Slater determinants corresponding to all possible configurations.

For this reason, the ultimate goal of the wavefunction theory has become to accurately characterizing the electron correlation effect based on the Hartree–Fock method. Such modified methods are known as post-Hartree–Fock or quantum chemistry methods; notable examples are the configuration-interaction methods, perturbative methods (e.g., the Moller–Plesset [MP] method), the coupled-cluster methods, and multi-reference methods [Moller1934, Purvis1982, Kendall1992]. One of the advantages of post-Hartree–Fock methods is that higher-order terms can be introduced systematically to approach the exact solution. Therefore, for the systems that can be handled by them, these higher-order methods are often used as the benchmark for electronic structure calculations. For example, the CCSD(T) method in coupled-cluster theory is widely recognized as the "gold standard" for non-strongly correlated system of quantum chemistry calculation because of its relatively high accuracy [Purvis1982]. Unfortunately, the computational cost of higher-order post-Hartree–Fock methods either increases exponentially with the size of the system or is proportional to its higher order polynomial. For example, in the MP2 method (one of the simplest), the computational cost is proportional to the fifth power of electron number *N*. This makes such methods particularly unsuitable for condensed matter systems [Szado1996]. Breakthroughs in the application of the post-Hartree–Fock method in condensed matter systems only appear recently, when new algorithms with design ideas that completely differ from the traditional quantization methods began to emerge [Booth2009]. Nevertheless, the accuracy of post-Hartree–Fock methods makes it likely that wavefunction theory will continue to play an important role in condensed matter theory.

4.4 Density Functional Theory

Density functional theory (DFT) is currently the most commonly used first-principles approach in condensed matter physics, theoretical chemistry, and computational materials science. It also serves as the electronic structure method upon which the majority of calculations involving full quantum effects depend. DFT was formally introduced in the 1960s, although it is descended from the Thomas–Fermi model [Fermi1927, Thomas1927], proposed in the late 1920s. J. C. Slater made a simple improvement to the theory in the 1950s; then W. Kohn, P. Hohenberg, and L. J. Sham rigorously refined it in the mid 1960s [Hohenberg1964, Kohn1965]. DFT starts with the idea that one can describe the electronic ground state of a system using the density of electrons instead of the many-body electron wavefunction. Thus, when describing a quantum system of many electrons, the problem of having to solve for the many-body electron wavefunction, a 3*N*-dimensional complex function, can be transformed into the problem of solving for the density of a three-dimensional real function, based on the Hohenberg–Kohn theorems.

Because of this simplification, in the case of the local or quasi-local exchange-interaction-functional approximation, the computational cost of DFT is proportional to the cube of *N*, comparable to that of the Hartree method (the simplest wavefunction method); the Hartree–Fock method, by contrast, has a computational cost proportional to the fourth power of *N* and does not account for the correlation interactions between electrons. In addition, Hohenberg and Kohn showed that the ground-state problem in interacting-electron systems can be solved by calculating the variation in density [Hohenber1964]. One year after the publication of these two key theorems, Kohn and Sham proposed another useful theoretical framework, namely the Kohn–Sham scheme [Kohn1965]. In the 1980s, the remaining barriers to a rigorous DFT (problems connected with finite-temperature theory, *N*-representability, *v*-representability, etc.) were addressed systematically by N. D. Mermin, M. Levy, E. H. Lieb, W. Kohn, and others [Mermin1965; Levy1982; Kohn1983, Lieb1983].

From a practical point of view, in early 1980s, the QMC algorithm, which is discussed in Section 4.5, provided reliable numerical results for the correlation interaction energy of homogenous electron-gas as a function of the electron density. The results provided the data for parameterizing the local density approximation (LDA), which is the simplest and the main functional used in early DFT calculations [Ceperley1980, Vosko1980,

Perdew1981]. LDA) [Ceperley1980, Vosko1980, Perdew1981]. As a result, DFT began to be widely used in practical solid-state calculations. With the development of the generalized gradient approximation (GGA) and hybrid functionals in 1980s and 1990s, DFT proved successful in the theoretical description of chemical reactions as well. These developments have cemented the irreplaceable central position of DFT in electronic-structure calculations. In the following subsection, we will briefly review some of the key developmental stages in DFT from a historical perspective. For a more specialized and detailed discussion, readers can refer to the monographs and review literature in this area [Parr1989, Koch2001, Cohen2012].

4.4.1 Thomas–Fermi Model and Extension

We begin with the Thomas–Fermi model. The first thing to note is that it is a model, not a theory. This very early model was developed soon after the Schrödinger equation and had tremendous success in describing single-electron systems such as the hydrogen atom (i.e., a nucleus plus an electron). Apart from spin (which was not yet understood as providing a degree of freedom), all experimental results on single-electron systems known in the 1920s could be explained by the analytical solution of Schrödinger's equation. Given this historical context, the next step was to apply Schrödinger's equation to monatomic systems with more than one electron. When the number of electrons is relatively large, a strictly analytical solution is no longer possible, and an empirical approach becomes necessary. This leads to the Thomas–Fermi model proposed by Thomas and Fermi in 1927 [Fermi1927, Thomas1927]. In an atom, the total energy of the many-electron system can be simply divided into three components: the kinetic energy T of the interacting electrons, the inter-electron colloidal interaction U^H, and the potential energy U^{ext} of the electron in the nuclear field. Assuming that all three can be written as functionals of the electron density, the energy takes the following form[1]:

$$E\left[n(\vec{r})\right]=T\left[n(\vec{r})\right]+U^{\text{ext}}\left[n(\vec{r})\right]+U^{H}\left[n(\vec{r})\right]. \tag{4.15}$$

It is clear that the last two terms have analytic expressions, but the first one does not and must therefore be approximated. In the original Thomas–Fermi model, this term is replaced by the kinetic energy of a noninteracting identical-electron system with the same density. Thus, Eq. (4.15) can be expressed as

$$E\left[n(\vec{r})\right]=\frac{3}{10}\left(3\pi^2\right)^{2/3}\int d\vec{r}\left[n(\vec{r})\right]^{5/3}+\int \mathrm{d}\vec{r}\, V_{\text{ext}}(\vec{r})n(\vec{r})+\frac{1}{2}\iint d\vec{r}d\vec{r}'\frac{n(\vec{r})n(\vec{r}')}{\left|\vec{r}-\vec{r}'\right|}. \tag{4.16}$$

In 1930, to characterize the effect of the inter-electron exchange antisymmetry on the total energy of the system, P. Dirac introduced another exchange term for the identical electron system at a specific density (the last term in Eq. (4.17)). The total energy of the system was then rewritten as

[1] At the time, it was assumed that the total energy is a functional of the electron density. However, this had not strictly been proven.

$$E[n(\vec{r})] = \frac{3}{10}(3\pi^2)^{2/3}\int d\vec{r}\,[n(\vec{r})]^{5/3} + \int d\vec{r}\,V_{\text{ext}}(\vec{r})n(\vec{r}) + \frac{1}{2}\iint d\vec{r}d\vec{r}'\frac{n(\vec{r})n(\vec{r}')}{|\vec{r}-\vec{r}'|} - \frac{3}{4}\left(\frac{3}{\pi}\right)^{\frac{1}{3}}\int d\vec{r}\,[n(\vec{r})]^{\frac{4}{3}}. \tag{4.17}$$

This equation can be solved self-consistently by variation. When modern computers were not yet available, this model was often used to solve for electron density distributions. Starting from Eq. (4.17), if the ionization energy μ of a particular many-electron atomic system is given, then the variational relation

$$\delta\left\{E[n(\vec{r})] - \mu\left[\int d\vec{r}\,n(\vec{r}) - N\right]\right\} = 0 \tag{4.18}$$

gives

$$\frac{1}{2}(3\pi^2)^{2/3}[n(\vec{r})]^{2/3} + V_{\text{ext}}(\vec{r}) + \int d\vec{r}'\frac{n(\vec{r}')}{|\vec{r}-\vec{r}'|} - \left[\frac{3}{\pi}n(\vec{r})\right]^{\frac{1}{3}} - \mu = 0\,. \tag{4.19}$$

If $V_{\text{ext}}(\vec{r})$ is known, then μ is constant, and the electron-density distribution can be easily obtained. The results can be further simplified when the electron-exchange term introduced by Dirac is excluded from consideration. These models contributed significantly to the understanding of many-electron atoms and small molecules in the twentieth century.

The Hartree–Fock wavefunction method matured almost concurrently with the Thomas–Fermi model. However, the fourth-power dependence of the computation cost on the number of electrons severely hindered its application in molecular and condensed matter systems. To simplify the calculations in the Hartree–Fock method, in 1951, Slater derived a density-functional form of it using a quantity called exchange hole density [Slater1951]. The results of this calculation were subsequently optimized using an adjustable parameter α.

$$E_{X_\alpha}[n(\vec{r})] = -\frac{9}{8}\left(\frac{3}{\pi}\right)^{\frac{1}{3}}\alpha\int d\vec{r}\,[n(\vec{r})]^{\frac{4}{3}}. \tag{4.20}$$

This method became known as the Hartree–Fock–Slater or X_α method. The main difference between X_α and the Thomas–Fermi–Dirac model is that the derivation of the latter is based on the concept of homogeneous electron gas, while the former is based on the exchange hole density. The coefficients in front of the exchange term are not identical in both methods, but their density dependencies are known. These methods are now regarded as early attempts at DFT calculation.

4.4.2 Density Functional Theory and the Exchange-Interaction Term

Modern density functional theory was proposed by W. Kohn, P. Hohenberg, and L. J. Sham in 1964–1965 [Hohenberg1964, Kohn1965]. The core of the theory consists of the two Hohenberg–Kohn theorems and the Kohn–Sham scheme. We will start with the theorems, at first ignoring degeneracy to simplify the discussion.

The first Hohenberg–Kohn theorem states that, in an interacting many-electron system, there exist one-to-one mappings between the ground-state many-body wavefunction $\Psi(\vec{r}_1,\vec{r}_2,\cdots,\vec{r}_N)$, the external potential $V_{\text{ext}}(\vec{r})$ (the potential difference is not a constant), and the ground-state electron density distribution function $n(\vec{r})$ (as illustrated in Figure 4.3). This can be proven by contradiction, placing the external potential $V_{\text{ext}}(\vec{r})$ in the middle to connect the other two quantities. First, let us examine the mapping between the external potential $V_{\text{ext}}(\vec{r})$ and the ground-state many-body wavefunction $\Psi(\vec{r}_1,\vec{r}_2,\cdots,\vec{r}_N)$. Suppose that the ground-state many-body wavefunction corresponds to two external potentials differing by more than a constant c. This means that $V_{\text{ext}-1}(\vec{r})-V_{\text{ext}-2}(\vec{r})\neq c$, although $\Psi[V_{\text{ext}-1}]=\Psi[V_{\text{ext}-2}]$. We then have

$$\begin{aligned}\hat{H}[V_{\text{ext}-1}]\Psi[V_{\text{ext}-1}]&=E[V_{\text{ext}-1}]\Psi[V_{\text{ext}-1}],\\ \hat{H}[V_{\text{ext}-2}]\Psi[V_{\text{ext}-2}]&=E[V_{\text{ext}-2}]\Psi[V_{\text{ext}-2}].\end{aligned} \tag{4.21}$$

This yields

$$\{\hat{H}[V_{\text{ext}-1}]-\hat{H}[V_{\text{ext}-2}]\}\Psi[V_{\text{ext}-1}]=\{E[V_{\text{ext}-1}]-E[V_{\text{ext}-2}]\}\Psi[V_{\text{ext}-1}], \tag{4.22}$$

where the difference in $\hat{\text{H}}$ is the external potential difference. As the kinetic energy and the potential energy have the same form in both Hamiltonians, we obtain

$$\{V_{\text{ext}-1}-V_{\text{ext}-2}\}\Psi[V_{\text{ext}-1}]=\{E[V_{\text{ext}-1}]-E[V_{\text{ext}-2}]\}\Psi[V_{\text{ext}-1}]. \tag{4.23}$$

The potential energy on the left is a multiplicative operator, which can be simply understood as directly multiplying non-differentiable operator. The factor on the right is proportional to the difference c between the two energies. Therefore, we have

$$V_{\text{ext}-1}-V_{\text{ext}-2}=c\,, \tag{4.24}$$

which contradicts the original assumption. It follows that the mapping on the left side of Figure 4.3 must be one-to-one.

Now, we must prove the uniqueness of the mapping on the right side of Figure 4.3 between the external potential $V_{\text{ext}}(\vec{r})$ and the electron-density distribution function $n(\vec{r})$. Since the mapping on the left side is already proven to be one-to-one, a given external potential gives rise to a unique ground-state many-body wavefunction. The many-body wavefunction also corresponds to a unique electron-density distribution, then it is impossible for one external potential to correspond to two electron-density distributions. Then, it suffices to show that one electron-density distribution corresponds to one external potential. In other words, no two external potentials $V_{\text{ext}-1}(\vec{r})$ and $V_{\text{ext}-2}(\vec{r})$ (unless differing only by a constant c) can give rise to the same $n(\vec{r})$.

This can also be proven by contradiction. Assume that there exist two external potentials giving rise to the same electron-density distribution. In this case, we have $V_{\text{ext}-1}(\vec{r})$ and its Hamiltonian $\hat{H}[V_{ext-1}]$ corresponding to the ground-state many-body wavefunction $\Psi[V_{\text{ext}-1}]$ and the ground-state energy $E_0[V_{\text{ext}-1}]$; there are also $V_{\text{ext}-2}(\vec{r})$ and its Hamiltonian $\hat{H}[V_{ext-2}]$ corresponding to the ground-state many-body wavefunction $\Psi[V_{\text{ext}-2}]$ and the ground-state energy $E_0[V_{\text{ext}-2}]$. According to the variational principle,

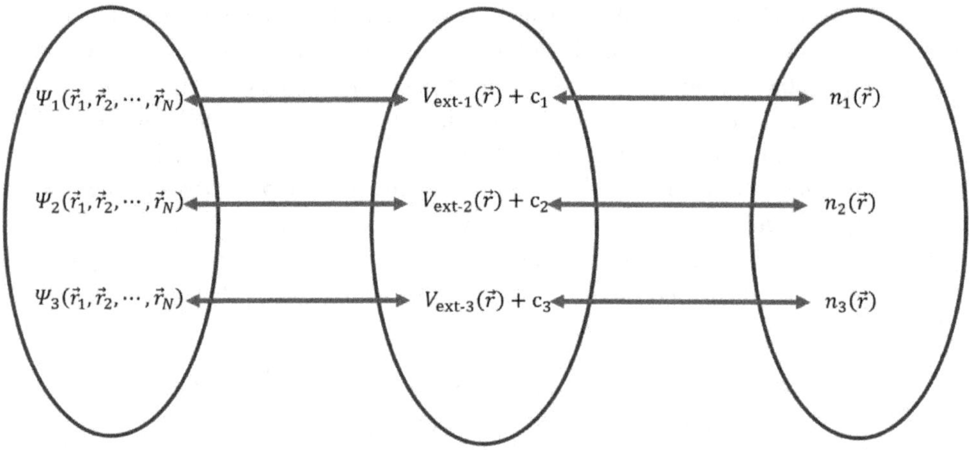

Figure 4.3 Schematic of the first Hohenberg–Kohn Theorem. In interacting electron systems, there exists one-to-one mappings between the ground-state many-body wavefunction $\Psi(\vec{r}_1,\vec{r}_2,\cdots,\vec{r}_N)$, external potential $V_{\text{ext}}(\vec{r})$ (the potential difference is not a constant), and ground-state electron-density distribution function $n(\vec{r})$.

$$E_0\left[V_{\text{ext}-1}\right]=\left\langle\Psi\left[V_{\text{ext}-1}\right]\middle|\hat{H}\left[V_{\text{ext}-1}\right]\middle|\Psi\left[V_{\text{ext}-1}\right]\right\rangle<\left\langle\Psi\left[V_{\text{ext}-2}\right]\middle|\hat{H}\left[V_{\text{ext}-1}\right]\middle|\Psi\left[V_{\text{ext}-2}\right]\right\rangle. \quad (4.25)$$

and

$$\begin{aligned}
&\left\langle\Psi\left[V_{\text{ext}-2}\right]\middle|\hat{H}\left[V_{\text{ext}-1}\right]\middle|\Psi\left[V_{\text{ext}-2}\right]\right\rangle\\
&=\left\langle\Psi\left[V_{\text{ext}-2}\right]\middle|\hat{H}\left[V_{\text{ext}-2}\right]+\hat{H}\left[V_{\text{ext}-1}\right]-\hat{H}\left[V_{\text{ext}-2}\right]\middle|\Psi\left[V_{\text{ext}-2}\right]\right\rangle\\
&=\left\langle\Psi\left[V_{\text{ext}-2}\right]\middle|\hat{H}\left[V_{\text{ext}-2}\right]\middle|\Psi\left[V_{\text{ext}-2}\right]\right\rangle\\
&+\left\langle\Psi\left[V_{\text{ext}-2}\right]\middle|\hat{H}\left[V_{\text{ext}-1}\right]-\hat{H}\left[V_{\text{ext}-2}\right]\middle|\Psi\left[V_{\text{ext}-2}\right]\right\rangle\\
&=E_0\left[V_{\text{ext}-2}\right]+\left\langle\Psi\left[V_{\text{ext}-2}\right]\middle|V_{\text{ext}-1}(\vec{r})-V_{\text{ext}-2}(\vec{r})\middle|\Psi\left[V_{\text{ext}-2}\right]\right\rangle=\\
&=E_0\left[V_{\text{ext}-2}\right]+\int d\vec{r}\left[V_{\text{ext}-1}(\vec{r})-V_{\text{ext}-2}(\vec{r})\right]n(\vec{r}).
\end{aligned} \quad (4.26)$$

Therefore,

$$E_0\left[V_{\text{ext}-1}\right]<E_0\left[V_{\text{ext}-2}\right]+\int d\vec{r}\left[V_{\text{ext}-1}(\vec{r})-V_{\text{ext}-2}(\vec{r})\right]n(\vec{r}). \quad (4.27)$$

In the same way,

$$E_0\left[V_{\text{ext}-2}\right]=\left\langle\Psi\left[V_{\text{ext}-2}\right]\middle|\hat{H}\left[V_{\text{ext}-2}\right]\middle|\Psi\left[V_{\text{ext}-2}\right]\right\rangle<\left\langle\Psi\left[V_{\text{ext}-1}\right]\middle|\hat{H}\left[V_{\text{ext}-2}\right]\middle|\Psi\left[V_{\text{ext}-1}\right]\right\rangle \quad (4.28)$$

and

$$\begin{aligned}
&\left\langle\Psi\left[V_{\text{ext}-1}\right]\middle|\hat{H}\left[V_{\text{ext}-2}\right]\middle|\Psi\left[V_{\text{ext}-1}\right]\right\rangle\\
&=\left\langle\Psi\left[V_{\text{ext}-1}\right]\middle|\hat{H}\left[V_{\text{ext}-1}\right]+\hat{H}\left[V_{\text{ext}-2}\right]-\hat{H}\left[V_{\text{ext}-1}\right]\middle|\Psi\left[V_{\text{ext}-1}\right]\right\rangle\\
&=\left\langle\Psi\left[V_{\text{ext}-1}\right]\middle|\hat{H}\left[V_{\text{ext}-1}\right]\middle|\Psi\left[V_{\text{ext}-1}\right]\right\rangle\\
&+\left\langle\Psi\left[V_{\text{ext}-1}\right]\middle|\hat{H}\left[V_{\text{ext}-2}\right]-\hat{H}\left[V_{\text{ext}-1}\right]\middle|\Psi\left[V_{\text{ext}-1}\right]\right\rangle\\
&=E_0\left[V_{\text{ext}-1}\right]+\left\langle\Psi\left[V_{\text{ext}-1}\right]\middle|V_{\text{ext}-2}(\vec{r})-V_{\text{ext}-1}(\vec{r})\middle|\Psi\left[V_{\text{ext}-1}\right]\right\rangle\\
&=E_0\left[V_{\text{ext}-1}\right]+\int d\vec{r}\left[V_{\text{ext}-2}(\vec{r})-V_{\text{ext}-1}(\vec{r})\right]n(\vec{r}),
\end{aligned} \quad (4.29)$$

which yield

$$E_0\left[V_{\text{ext}-2}\right] < E_0\left[V_{\text{ext}-1}\right] + \int d\vec{r}\left[V_{\text{ext}-2}(\vec{r}) - V_{\text{ext}-1}(\vec{r})\right]n(\vec{r}). \tag{4.30}$$

Adding Eqs (4.27) and (4.30), we obtain

$$E_0\left[V_{\text{ext}-1}\right] + E_0\left[V_{\text{ext}-2}\right] < E_0\left[V_{\text{ext}-1}\right] + E_0\left[V_{\text{ext}-2}\right], \tag{4.31}$$

which is clearly impossible. Therefore, the assumption that there are two external potentials with a nonconstant difference that can give rise to the same $n(\vec{r})$ must be false, and there is a one-to-one mapping between the external potential and the electron-density distribution.

Thus, the first Hohenberg–Kohn theorem is proven; the one-to-one correspondence of the three quantities is summarized in Figure 4.3.

The second Hohenberg–Kohn theorem states that the total ground-state energy can be calculated by taking the variation of the density distribution, with the minimum value corresponding to the total ground state energy and the ground state electron density distribution. More specifically, if a one-to-one mapping between the density and the wavefunction can be established, then any functional of the latter is naturally a functional of the former. In this case, the total energy can be written as

$$E\left[n(\vec{r})\right] = \int n(\vec{r})V_{\text{ext}}(\vec{r})d\vec{r} + F_{\text{HK}}\left[n(\vec{r})\right]. \tag{4.32}$$

The first term represents the energy of the electrons in the external potential, $V_{\text{ext}}(\vec{r})$. F_{HK} is the Hohenberg–Kohn functional, which describes the electron kinetic energy and the electron–electron interaction. After taking the variation of the total energy functional with respect to $n(\vec{r})$, the $n(\vec{r})$ that corresponds to the minimum E_0 is the ground state electron-density of this interacting-electron system. Three conditions must be satisfied: $n(\vec{r}) \geq 0$, $\int n(\vec{r})d\vec{r} = N$, and $n(\vec{r})$ can be generated by some $V_{\text{ext}}(\vec{r})$.

The last condition raises what is often referred to as the v-representability problem. In practice, satisfying such a requirement is difficult because for every $n(\vec{r})$ being generated, there is no guarantee that one can find a $V_{\text{ext}}(\vec{r})$ that corresponds to it. Until now, this has been a much-debated problem, especially when considering the question of whether an external potential can always be found for every density distribution function in interacting electron systems [Kohn1983, Chayes1985].

In practice, a method called Levy–Lieb constrained search is often used to convert the v-representability problem into the N-representability problem [Levy1982, Lieb1983]. When searching for the density, one does not simply look for it but adopts a stepwise scheme instead. The wavefunction is first classified according to the density by using the variational method on the ground-state energy with respect to the wavefunction. Each density distribution function can accommodate multiple many-body wavefunctions, and the many-body wavefunction with the lowest energy is selected to produce the total energy. Then, the variational principle is applied to the density. In this case, the variational space is simply an N-body wavefunction space, but the energy itself can still be expressed as a functional of the density distribution function such that

$$E_0 = \min_{\Psi}\left\langle \Psi \middle| \hat{H} \middle| \Psi \right\rangle = \min_{n(\vec{r})}\left\{ \min_{\Psi \to n(\vec{r})} \left\langle \Psi \middle| \hat{H} \middle| \Psi \right\rangle \right\}, \tag{4.33}$$

where

$$\min_{\Psi \to n(\vec{r})} \langle \Psi | \hat{H} | \Psi \rangle = \int n(\vec{r}) V_{\text{ext}}(\vec{r}) d\vec{r} + \min_{\Psi \to n(\vec{r})} \langle \Psi | \hat{T} + \hat{E}_{\text{ee}} | \Psi \rangle. \tag{4.34}$$

We define

$$F\left[n(\vec{r})\right] = \min_{\Psi \to n(\vec{r})} \langle \Psi | \hat{T} + \hat{E}_{\text{ee}} | \Psi \rangle, \tag{4.35}$$

so that

$$E_0 = \min_{n(\vec{r})} \left\{ \int n(\vec{r}) V_{\text{ext}}(\vec{r}) d\vec{r} + F\left[n(\vec{r})\right] \right\} \tag{4.36}$$

The $F[n(\vec{r})]$ here, similar to the $F_{\text{HK}}[n(\vec{r})]$ in the original Hohenberg–Kohn method, depends only on the ground-state density distribution.

The constrained-search method not only provides a solution to the v-representability problem mentioned earlier but, more importantly, links the density distribution function and the many-body wavefunction during the search; this allows the removal of the degeneracy assumption in the Hohenberg–Kohn theorem (because the selection of the many-body wavefunction with the smallest total energy is already included in Eq. (4.35)). Meanwhile, the Kohn-Sham scheme in DFT is more logically coherent than its early version. For this reason, recent studies of DFT have increasingly employed the constrained-search framework.

The next key to understanding DFT is the Kohn–Sham scheme. Here, we are dealing with the variation of the electron-density distribution function in the interacting electron systems (Eq. (4.36)); the density distribution function $n(\vec{r})$ is derived from some many-body wavefunction. Its functional $F[n(\vec{r})]$ can be divided into three terms:

$$F\left[n(\vec{r})\right] = T\left[n(\vec{r})\right] + J\left[n(\vec{r})\right] + E_{\text{ncl}}\left[n(\vec{r})\right], \tag{4.37}$$

where $T[n(\vec{r})]$ is the kinetic energy term, $J[n(\vec{r})]$ is the colloidal interaction between electrons (a classical potential energy term), and $E_{\text{ncl}}\left[n(\vec{r})\right]$ is the potential energy term in addition to the colloidal term.

If we draw inspiration from the Hartree–Fock method and introduce a series of interaction-free fictitious particles whose total density equals the density of the interacting electron system, then $T[n(\vec{r})]$ can be further divided into two parts: the kinetic energy $T_{\text{s}}[n(\vec{r})]$ of the interaction-free system, and a correction term $T_{\text{C}}[n(\vec{r})]$. It is expressed as

$$T\left[n(\vec{r})\right] = T_{\text{s}}\left[n(\vec{r})\right] + T_{\text{C}}\left[n(\vec{r})\right]. \tag{4.38}$$

The total $F[n(\vec{r})]$ can then be written as

$$F\left[n(\vec{r})\right] = T_{\text{s}}\left[n(\vec{r})\right] + J\left[n(\vec{r})\right] + T_{\text{C}}\left[n(\vec{r})\right] + E_{\text{ncl}}\left[n(\vec{r})\right]. \tag{4.39}$$

The first two terms on the right have analytic expressions. The first term can be written as

$$T_{\text{s}}\left[n(\vec{r})\right] = -\frac{1}{2} \sum_{i=1}^{N} \langle \varphi_i | \nabla^2 | \varphi_i \rangle, \tag{4.40}$$

where φ_i is a single-electron orbital of the interaction-free fictitious particles, which have the same density distribution function as the interacting system. The second term is defined as

$$J\left[n(\vec{r})\right]=\frac{1}{2}\iint \mathrm{d}\vec{r}\mathrm{d}\vec{r}'\frac{n(\vec{r})n(\vec{r}')}{\left|\vec{r}-\vec{r}'\right|}, \tag{4.41}$$

which describes the colloidal potential energy. The third and fourth terms in Eq. (4.39) ($T_C[n(\vec{r})]$ and $E_{ncl}[n(\vec{r})]$) are the kinetic energy and the potential energy corrections, respectively, due to electron interaction. Their sum is expressed as

$$E_{xc}\left[n(\vec{r})\right]=T_C\left[n(\vec{r})\right]+E_{ncl}\left[n(\vec{r})\right] \tag{4.42}$$

which corresponds to the correction of the total energy due to electron interaction. Therefore, the total $F[n(\vec{r})]$ can be expressed as

$$F\left[n(\vec{r})\right]=T_s\left[n(\vec{r})\right]+J\left[n(\vec{r})\right]+E_{xc}\left[n(\vec{r})\right], \tag{4.43}$$

where the unknown term $E_{xc}[n(\vec{r})]$ must be computed based on approximations.

Finally, the total energy must be optimized by varying the density, and this is mathematically equivalent to the definition of

$$E_0=\min_{n(\vec{r})}\left\{\int n(\vec{r})V_{ext}(\vec{r})d\vec{r}+T_s\left[n(\vec{r})\right]+J\left[n(\vec{r})\right]+E_{xc}\left[n(\vec{r})\right]\right\}. \tag{4.44}$$

Then, the following set of equations can be generated using the orbitals of noninteracting fictitious particles and a specific $E_{xc}[n(\vec{r})]$ functional:

$$\begin{aligned}&\left(-\frac{1}{2}\nabla^2+\int\frac{n(\vec{r}_2)}{\left|\vec{r}-\vec{r}_2\right|}d\vec{r}_2+V_{xc}(\vec{r})+V_{ext}(\vec{r})\right)\varphi_i(\vec{r})=\epsilon_i\varphi_i(\vec{r}),\\&V_{xc}(\vec{r})=\frac{\delta E_{xc}\left[n(\vec{r})\right]}{\delta n(\vec{r})},\\&n(\vec{r})=\sum_{i=1}^{N}\left|\varphi_i(\vec{r})\right|^2,\end{aligned} \tag{4.45}$$

where $V_{xc}(\vec{r})$ is the so-called exchange-correlation potential. After these equations are solved self-consistently, the total energy of the system is given by

$$E_0=\sum_{i=1}^{N}\epsilon_i-\frac{1}{2}\iint\frac{n(\vec{r}_1)n(\vec{r}_2)}{\left|\vec{r}_1-\vec{r}_2\right|}\mathrm{d}\vec{r}_1\mathrm{d}\vec{r}_2+E_{xc}\left[n(\vec{r})\right]-\int n(\vec{r})V_{xc}(\vec{r})d\vec{r}. \tag{4.46}$$

This constitutes the main body of density functional theory. As mentioned earlier, with the parameterization of the LDA exchange-correlation functional provided by the quantum Monte Carlo algorithm [Ceperley1980, Vosko1980, Perdew1981], DFT began to find a wide practical application in the early 1980s. Later, as the GGA and hybrid functional method advanced, DFT was increasingly used in the study of chemical reactions. This has established its unrivalled position in electronic-structure calculations for condensed matter physics, theoretical chemistry, and material simulation.

Figure 4.4 Jacob's ladder. The lowest rung, the local density approximation (LDA), only considers the local electron density. The generalized gradient approximation (GGA) accounts for the gradient of the density, and the meta-GGA further considers the kinetic-energy density. However, the hybrid functional method calculates the exchange-interaction energy using the orbitals of occupied states as the starting point [Seidl1996]. On the fifth rung, the orbital information of unoccupied states is also considered. From the bottom to the top, the form of the exchange-correlation functionals becomes progressively more complex, and the calculation results, in principle, become more accurate. From Zhang I. Y. [2021]. Reproduced with permission from John Wiley and Sons.

Nevertheless, although DFT is an exact method, the approximation of the exchange-interaction term may introduce unpredictable errors when applied to specific materials. Hundreds of different approximations have been proposed from the early 1980s to the present, and even if only the most commonly used ones are discussed, there are still dozens of them. In this book, we will not specifically cover these approximations. Instead, we will simply emphasize that DFT calculations are not black-box calculations without tunable parameters, and selecting the appropriate exchange-interaction terms based on the characteristics of the system is of the utmost importance.

The DFT framework has been depicted as Jacob's ladder (Figure 4.4), gradually approaching absolute accuracy [Poncé2014]. Although its path toward this goal is not as clear as that of the wavefunction theory, and there are still many areas that need improvement, we believe that DFT will remain an important foundation of full-quantum-effect calculations for many years to come.

4.5 Quantum Monte Carlo Method

The third widely used method for calculating the electronic structure of real materials is the QMC method. Its historical development differs slightly from that of the first two methods, which are derived from analytical theory. The QMC method is the result of progressive attempts made during the development process of modern computers.

The QMC method, as its name implies, is quantum mechanics plus the Monte Carlo (MC) algorithm. "Quantum mechanics" here refers to the fundamental principles used to describe microscopic physics in the nonrelativistic limit, specifically the solution of Schrödinger's equation. "Monte Carlo" refers to the numerical sampling method, which will be discussed below.

Two categories of problems in condensed matter research can be addressed by QMC. In the first category of problems, a model Hamiltonian is solved numerically; the focus is often on the effect of many-body interactions on the microscopic mechanisms behind physical properties, especially correlation effects. Problems in the second category involve solving the Schrödinger equation directly for real materials within the framework of first-principles electronic structure calculations. This part of the study certainly includes the correlation effects between electrons; however, as they are naturally included in the solution, they are rarely discussed extensively as a research focus. These studies focus on the exact calculation of the total or excitation energy of a real electron system, with the goal of accurately describing the electronic structure.

Based on this interpretation, if we were to redefine the QMC algorithm, we can say that it solves the fundamental equation of quantum mechanics using the MC algorithm. This equation can correspond to either the model Hamiltonian (commonly known as the QMC algorithm in traditional condensed matter theory) or the real Hamiltonian system (commonly known as the QMC algorithm in first-principles material calculations). This chapter focuses on obtaining the exact electronic structure of a real system and providing the basis for the full quantum theory; therefore, we mainly discuss the second case.

After understanding the meaning of QMC algorithm and mastering the basic principles of quantum mechanics, a natural question is: What is the MC algorithm? Although the history of this numerical sampling method can be traced back to the study of problems such as Buffon's needle, MC simulations were first implemented on computers in the late 1940s by E. Fermi, S. Ulam, and J. von Neumann aimed at studying neutron diffusion. In quantum mechanics, where most physical quantities are expressed as expectation values or other integrals, and MC methods are known to be an efficient approach to perform numerical integration of high-dimensional mathematical functions. As shown in the Eq. (4.7), the integral of function $g(x)$ can be transformed into an integral of the product of $f(x)$ and $P(x)$ over all values of variable x, where $P(x)$ is the distribution function and $f(x) = g(x)/P(x)$. This is approximately equivalent to summing the function values $f(x_i)$ and averaging them, where the discrete variables x_i satisfy the $P(x)$ distribution. When expressed as formula, it is presented as

$$\int dx g(x) = \int dx f(x) P(x) \approx \frac{1}{N} \sum_i f(x_i). \tag{4.47}$$

The rigor and convergence of MC are guaranteed by the law of large numbers and the central limit theorem. Because in this book we are primarily interested in the physical problems, we will not dwell on the mathematical basis of MC. Instead, we will focus on the relevant principles of QMC and the recent research progress using numerical methods based on Eq. (4.47). Readers interested in the underlying mathematics are encouraged to consult more specialized publications about probability and statistics. The MC method today has a broad range of applications in mathematics, physics, chemistry, engineering, artificial intelligence, and other fields.

Solving a quantum-physics problem by the QMC algorithm usually means finding the expectation value of some physical observable. Mathematically, this involves sampling the distribution that the variable x_i satisfies. The uniqueness in QMC calculations is reflected in the fact that the distribution of variables is often described by the wavefunction of the system.

QMC method is essentially the stochastic sampling of wavefunctions in the material. In contrast to the stochastic method, we can call other electronic structure methods such as post Hartree Fock and density functional theory are deterministic method. Deterministic methods usually start with defining a basis set, often composed of linear combination of atomic orbitals (LCAOs), Gaussians orbitals, or planewaves. The final computed wavefunction can be expressed as a linear combination of elements from this basis set. The linear expansion coefficients in front of the basis functions are found by diagonalizing the secular determinant, and the corresponding wavefunctions or Kohn–Sham orbitals are calculated. In the QMC method, on the other hand, the coefficients are obtained by expectation-value calculation via statistical sampling; the final expectation values of all physical quantities and their statistical errors can then be obtained by a formula similar to Eq. (4.47). With enough sampling points, the statistical error can be reduced so that it falls within a specified accuracy range. Thus, the QMC method is essentially statistical sampling, while the wavefunction and DFT methods solve the secular determinant self-consistently and deterministically.

It is also worth mentioning that QMC methods can break through the limit of a finite atomic-orbital basis set found in traditional wavefunction methods. QMC's numerical error becomes negligible if the real-space lattice-point sampling is sufficient. In fact, different QMC methods may adopt some special approximations that introduce statistical and systematic errors into the electronic structure calculation. As the new methods advance, the errors introduced by these approximations gradually become clear to researchers, hence can be effectively controlled, bringing us one step closer to the rigorous solution of the many-body Schrödinger equation of small-scale systems.

There are many different QMC varieties, including variational Monte Carlo (VMC), Green's-function Monte Carlo (GFMC), diffusion Monte Carlo (DMC), path-integral Monte Carlo (PIMC), auxiliary-field quantum MonteCarlo, and full-configuration-interaction quantum Monte Carlo (FCIQMC). As computational methods are closely linked to the development of computer science and technology, it is possible for algorithms that were previously considered inefficient to become promising choices when a new computational architecture emerges. Because QMC is well-adapted to the architecture of modern supercomputers, and particularly to large-scale parallel computing, it is playing an increasingly important role in computational physics. At present, it is no exaggeration to say that QMC, wavefunction methods, and DFT methods constitute the dominant triad of electronic structure calculation methods. For exact calculations involving condensed matter systems, the QMC method has unparalleled advantages in accuracy and is currently considered the gold standard for the electronic structure calculations in this domain.

The computational accuracy and efficiency of QMC method are mainly limited by its high computational cost. In particular, when the forces acting on the atomic nuclei in the system must be calculated, the QMC method can be extremely time-consuming. Nevertheless, the basic principles of quantum mechanics and statistical physics show that the statistical and dynamical properties of atomic nuclei will depend sensitively on the accuracy of the Born–Oppenheimer potential energy surface. Therefore, as we apply these methods to the real systems, such as high-pressure metallic-hydrogen, conventional DFT may produce an extremely inaccurate Born–Oppenheimer potential energy surface in some sensitive region, and QMC must be used instead to get reliable results [Morales2010, Gorelov2020].

This section addresses the electronic structure calculation of real materials that can be combined with the full quantum methods. Three QMC methods will be presented: VMC, DMC, and FCIQMC. Interested readers can refer to specialized books in computational physics to learn about other relevant QMC methods.

4.5.1 Variational MC Method

Similar to the wavefunction methods (traditional quantum chemical methods), VMC also uses variational principles to optimize many-body wavefunctions. However, it uses stochastic sampling-based variation, whereas the wavefunction methods introduce an analytic formula based on the variational principle and then solve it self-consistently, with each step involving the diagonalization of a secular determinant.

According to the variational principle, for any chosen trial wavefunction, the energy given by this trial wavefunction with the Hamiltonian of the system must be the upper limit of the ground state. Hence, the first problem is how to calculate the energy of a system for any given trial wavefunction. In practice, this is done by introducing a certain number of walkers and allowing them to random walk in the complete space according to the Markov process (typically using the Metropolis algorithm), one can then obtain the walker distribution corresponding to the absolute square of the trial wavefunction. The obtained trajectory of this particle distribution can then be used to calculate the expectation value of the system's total energy (Eq. (4.48)) and the statistical error (Eq. (4.49)) for this trial wavefunction. The energy expectation value is calculated by summing and averaging the local energies given by each group of walkers. The trial wavefunction can be continuously modified to reduce system's energy expectation value, thus approaching the exact solution of the ground-state wavefunction.

It is worth noting that, as the wavefunction of the system approaches the ground state, the variance of the energy expectation value calculated by MC sampling also approaches zero. Therefore, when optimizing the trial wavefunction, zero variance can also be used as the optimization goal.

W. L. McMillan first applied the VMC method in 1965, optimizing the many-body wavefunction of liquid helium [McMillan1965]. In this attempt, he used only three variational parameters to optimize the wavefunction. In subsequent studies, some additional variational parameters were added depending on the system being studied, but the number of parameters remained small until the early 1990s. Later, as numerical algorithms continued to improve, optimizing a large number of parameters to obtain a better variational wavefunction became possible. The reweighting method, the steepest-descent method, the stochastic reconfiguration method, and the linear method are representative techniques used to provide a good trial wavefunction; optimizing the trial wavefunction is also a necessary starting point for other QMC methods with greater accuracy, such as DMC [Wang E.G.1995].

We now describe how the wavefunction can be optimized with VMC, using the reweighting method as an example. Given a trial wavefunction Ψ_α that depends on a set of parameters α, with some complete basis set $|x\rangle$ such that $\Psi_\alpha(x) = \langle x|\Psi_\alpha\rangle$, the energy of the system and its error can be given by the following two equations:

$$E_\alpha = \frac{\langle \Psi_\alpha|\hat{H}|\Psi_\alpha\rangle}{\langle \Psi_\alpha|\Psi_\alpha\rangle} \approx \frac{1}{N}\sum_{i=1}^{N} e_{L,\alpha}(x_i) \tag{4.48}$$

and

$$\sigma^2_{\Psi_\alpha} = \frac{\langle \Psi_\alpha|(\hat{H}-E_\alpha)^2|\Psi_\alpha\rangle}{\langle \Psi_\alpha|\Psi_\alpha\rangle} \approx \frac{1}{N}\sum_{i=1}^{N} \left[e_{L,\alpha}(x_i) - E_\alpha\right]^2, \tag{4.49}$$

where the "local energy" is defined as

$$e_{L,\alpha}(x_i) = \frac{\langle x_i|\hat{H}|\Psi_\alpha\rangle}{\langle x_i|\Psi_\alpha\rangle}. \tag{4.50}$$

We can use configuration $\{x_i\}$ to label the electronic configuration obtained by sampling the system.

Reweighting means defining a ratio

$$R_{\alpha+\delta\alpha_k}(x) = \left|\frac{\Psi_{\alpha+\delta\alpha_k}(x)}{\Psi_\alpha(x)}\right|^2. \tag{4.51}$$

The numerator of this ratio is the wavefunction after introducing some small changes to the trial wavefunction, and the denominator is the trial wavefunction without the changes. The new energy and the corresponding error after the change can be written respectively as

$$E_{\alpha+\delta\alpha} = \frac{\sum_{i=1}^{N} e_{L,\alpha+\delta\alpha}(x_i) R_{\alpha+\delta\alpha}(x_i)}{\sum_{i=1}^{N} R_{\alpha+\delta\alpha}(x_i)} \tag{4.52}$$

and

$$\sigma^2_{\Psi_{\alpha+\delta\alpha}} = \frac{\sum_{i=1}^{N}\left[e_{L,\alpha+\delta\alpha}(x_i) - E_{\alpha+\delta\alpha}\right]^2 R_{\alpha+\delta\alpha}(x_i)}{\sum_{i=1}^{N} R_{\alpha+\delta\alpha}(x_i)}. \tag{4.53}$$

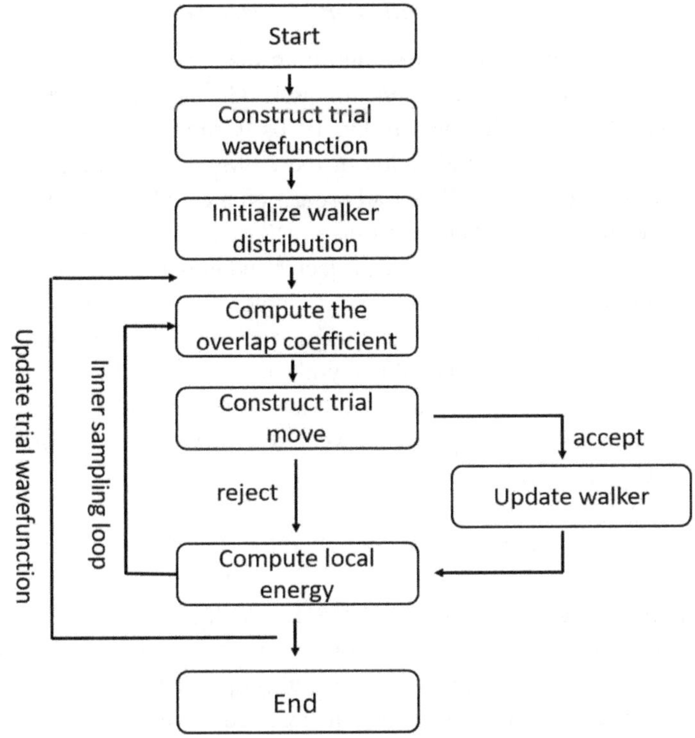

Figure 4.5 Flowchart of the variational Monte Carlo method.

By changing $\delta\alpha$ to minimize these, the trial wavefunction can be optimized.

A flowchart of the key parts of a basic VMC algorithm is shown in Figure 4.5.

It is worth noting that the result of optimizing the trial wavefunction using a variational method is highly dependent on the ansatz of the trial wavefunction. The Hartree–Fock wavefunction is the simplest ansatz. The main limitation is that it cannot account for correlation between electrons. To address electron correlation, a variety of wavefunction choices have been adopted, including the Jastrow [Jastrow1955], Gutzwiller [Gutzwiller1963], Haldane–Shastry [Haldane1988, Shastry1988], Laughlin [Laughlin1983], and Bardeen–Cooper–Schrieffer wavefunctions [Bardeen1957] and the resonating valence bond wavefunction developed by Philip W. Anderson et al. [Anderson1987a, b, Baskaran1988].

In Section 4.3, we introduced the self-consistent solution of the Hartree–Fock method. One major difference between the Hartree–Fock wavefunction and the complex trial wavefunction mentioned here is that it is very difficult to solve the latter self-consistently. The solution in the VMC case does not rely on the diagonalization of the secular determinant. This is the reason that it is better to use QMC methods for wavefunction optimization.

However, because of the limitations of trial wavefunction ansatz, optimization by the VMC method is unlikely to produce a rigorous ground-state wavefunction. As the need for computational accuracy gradually increased, methods that can approach the actual ground state more effectively were discovered. Among the best of these methods were DMC and FCIQMC, which are described below respectively.

4.5.2 Diffusion MC Method

The DMC method is a projection-based algorithm that uses imaginary-time evolution to obtain the system's ground-state wavefunction. R. C. Grimm, R. G. Storer, J. B. Anderson, and others pioneered its development in the 1970s [Grimm1971, Anderson1975, 1976]. DMC can be regarded as a generalization of the GFMC method in the continuum limit. GFMC, DMC, PIMC, and FCIQMC are all projection methods. In other words, they use a projector to extract the ground-state wavefunction Ψ_{ground}. The difference among them is in how the wavefunction is extracted by the projector and how it expands.

Take projection in the DMC method as an example, which corresponds to an imaginary time evolution process. The principle of imaginary time evolution is applicable to both one-dimensional and multi-dimensional many-electron systems.

As discussed in subsection 4.5.1, the ground state and all excited states of a quantum system form a complete basis in Hilbert space. Therefore, the exponential of the Hamiltonian operator can be expanded as follows:

$$\exp(-\tau \hat{H}) = \sum_i |\Psi_i\rangle \exp(-\tau E_i)\langle\Psi_i|. \tag{4.54}$$

Here, $\tau = it/\hbar$ denotes an imaginary time step. Every time this exponential operator acts on the trial wavefunction, the ground state component of the wavefunction increases exponentially. As τ approaches infinity (i.e., after the projector has been applied to the trial wavefunction a sufficient number of times), the resulted wavefunction is infinitely close to the ground state wavefunction. This is the general principle of a projector.

Note that a projector does not need to have an exponential form, and there are examples of projectors taking other forms in various MC algorithms. The efficiency and stability of different projectors vary, and it is necessary to find the right balance in practice. This is why there are numerous algorithms (with different advantages and disadvantages) that can be applied to various problems.

Random walkers that sample the Hilbert space of the system's electronic states must be defined for the DMC method. The projector defined in Eq. (4.54) corresponds to walker evolution in imaginary time. To build the MC sampling that corresponds to this imaginary-time evolution, we write the Green's function propagator corresponding to the projector shown above as follows:

$$G(x' \leftarrow x, \tau) \approx (2\pi\tau)^{-\frac{3N}{2}} \exp\left[-\frac{(x-x')^2}{2\tau}\right] \times P, \tag{4.55}$$

where

$$P = \exp\left[-\tau\left[V(x)+V(x')\right]/2\right]. \tag{4.56}$$

In Eq. (4.55), the exponential before P is a typical diffusion term. If only the diffusion term is considered, the walkers can perform a completely random walk in the full space. The corresponding walk probability depends on the distance between the two positions before and after the walk and on the step size defined during the walk. P can be considered a renormalization factor that describes the birth and death of the walker. In the specific evolution

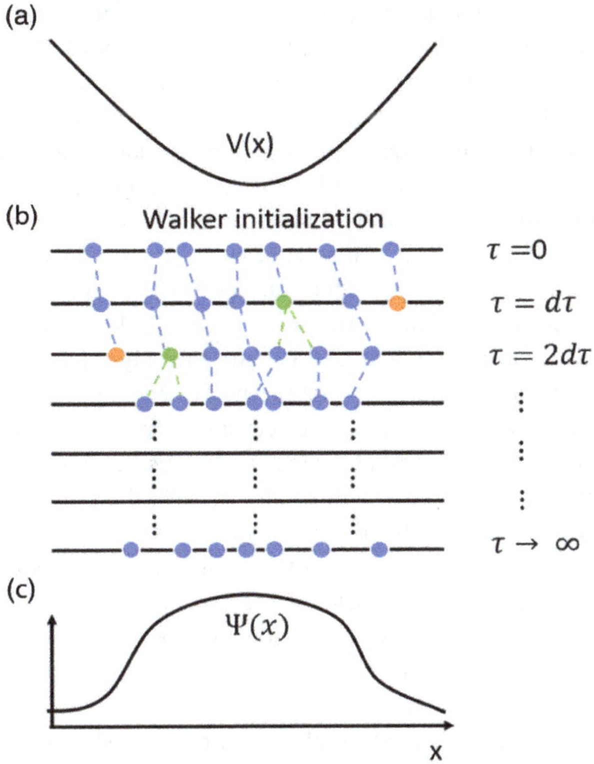

Figure 4.6 Evolution of a diffusion Monte Carlo simulation. Panel (a) shows a hypothetical one-dimensional potential, and Panel (b) shows the diffusive evolution of the walkers from the top down, starting from zero imaginary time. The evolution process can involve processes such as the birth (green) and the death (orange) of walkers. Panel (c) shows the distribution of the ground state wavefunction.

process, we can choose to continue the evolution with probability P whenever P is less than 1, setting the probability of walker death to $1-P$. When P is greater than 1, we generate a new walker at the same location with probability $P-1$ in addition to the diffusion evolution.

Figure 4.6 depicts a simple DMC simulation for a one-dimensional potential well. The system eventually reaches an equilibrium state, which is shown by the blue walkers in the bottom row of Figure 4.6. This final distribution is reflective of the ground-state wavefunction distribution.

Finally, let us briefly discuss an important limitation of the DMC, that is, the "sign problem." The wavefunction can be positive and negative in reality, but mathematically the walkers should only be sampled according to non-negative probability distributions. This is in fact a universal limitation of all projection methods. In second-quantization based methods, the matrix elements of the Hamiltonian operator in the projector can be positive or negative, and a negative matrix element can assign negative weights to the newly evolved walkers. A common consequence is that the statistical error will explode due to the "sign problem".

To address the sign problem in DMC, the fixed-node approximation is often introduced. Wavefunctions with positive and negative coefficients are separated in space, so that the walker evolution cannot cross the zero point of the wavefunction. This approximation has been incredibly successful and has extended the applicability of DMC methods significantly. One of the most notable applications is to the exact solution of a homogeneous electron gas: The results from this calculation were later used as reference data to fit the first LDA exchange-interaction functional in practice, directly contributing to the development of DFT [Ceperley1980, Vosko1980, Perdew1981]. Even today, the fixed-node approximation is still commonly used in DMC. In condensed matter systems, the wavefunction node from DFT is generally used as the node of the trial wavefunction for the fixed-node approximation, which limits the applicability of the DMC method to strongly correlated electron systems. Recently, J. Chen's research group at Peking University introduced a neural-network-based wavefunction ansatz that considerably reduced the error caused by the fixed-node approximation [Ren W.L.2022]. Their research paves the way for the DMC method to be applied to strongly correlated electron system in the future.

Furthermore, the neural-network approach overcomes certain limitations of the traditional wavefunction ansatz. In the traditional approach, the ground state wavefunction optimized by VMC only represents the ground state of the ansatz, and its corresponding nodal surfaces are only the nodal surfaces of the ansatz. Although DMC methods can, theoretically, go beyond the limitations due to the basis set, they are limited in other ways, and the resultant ground state is still strictly dependent on the wavefunction ansatz. With the advent of the neural network, the above-mentioned limitation can be overcome if the network is large enough to fully express the wavefunction of the system mathematically. The neural-network wavefunction method, combined with DMC, shows promise of becoming a systematic approach to obtaining the exact solution and providing accurate measurement of the Born–Oppenheimer potential energy surface in full quantum effect calculations.

4.5.3 Example: Exciton in Semiconductor Quantum Wells

Before recommending a different method for optimizing the trial wavefunction, we would like to combine the discussions from subsections 4.5.1 and 4.5.2 and provide the reader with a practical example of using variational and diffusion MC methods to solve a physical problem, that is, bound states of excitons in a semiconductor quantum well. We hope that this example will help the reader understand and master the QMC method.

Quantum wells were introduced by L. Esaki and R. Tsu in 1970 [Esaki1970]. Since then, these heterostructures have received extensive attention for their potential applications in the electronic and optical devices. In terms of both exciton binding energy and light-absorption modulation, their flexible electronic structure offers great advantages over the bulk materials from which they are made [Miller1985]. Initially, it was the type I semiconductor quantum wells that attracted the most interest. For instance, a $GaAs/Ga_{1-x}Al_xAs$ (type I) quantum-well system has a much larger exciton binding energy than that of its bulk materials. However, the rapid recombination of electrons and holes destroys the coherence of the excitonic state in this system because the electrons and holes are confined in the same regions of space. For this reason, research interest has shifted to the type II semiconductor quantum wells, which exhibit an indirect bandgap in real space (Figure 4.7). In this type of quantum well, electrons and holes are spatially separated into different layers. If the spatial separation between the two quantum wells is smaller than or close to the effective Bohr radius of the exciton, the Coulomb interaction between them is not significantly reduced. Excitons have been found to condense spontaneously to a collective ground state of the system when the exciton binding energy E_B is larger than the absolute value of the energy gap $|E_G|$. These exciton insulator phases may appear in the semiconductor–semimetal transition region [Datta1985].

G. Bastard et al. reported variational results for E_B in GaSb/InAs/GaSb heterojunctions with infinite potential barriers [Bastard1982]. They found that the value of the exciton binding energy decreases to one-fourth of the two-dimensional limit value. This result was subsequently improved by several groups that studied the GaSb/AlSb/InAs quantum well with a finite electron (hole) barrier height and in the presence of a magnetic field [Zhu X.1990, Xia X.1992]. In this heterostructure, the AlSb layer serves as an infinite energy barrier for both electron and hole wavefunctions; otherwise, virtual transitions between the bands would result in an increased dielectric constant, thus lowering the exciton binding energy. However, these improved variational results were estimated to still have an exciton binding-energy error of the order of 5%. This error was sensitive to the phase transition in the system; hence calculations with greater accuracy were required. In 1995, E. G. Wang and his colleagues accurately characterized the particle–particle correlation terms due to the Coulomb interactions and quantum confinement, and reported the first accurate result for the ground-state energy of excitons using the variational and diffusion quantum MC methods [Wang E.G.1995].

Figure 4.7 depicts the excitons formed by the electron–hole pair due to optical excitation in an $In_{1-x}Ga_xAs/GaSb_{1-y}As_y$ type II quantum-well system. For the electron, the zero point

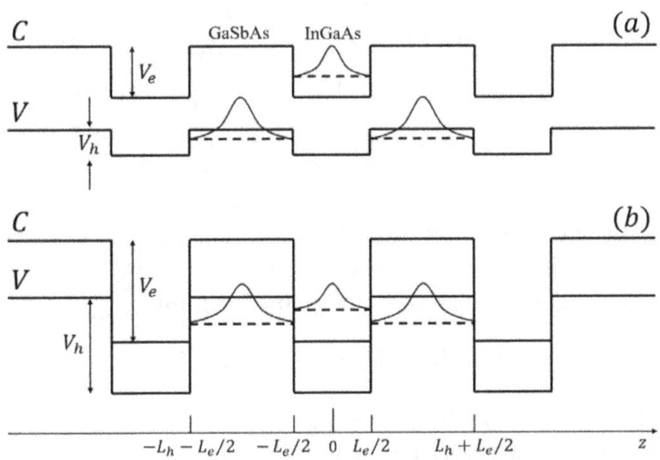

Figure 4.7 $In_{1-x}Ga_xAs/GaSb_{1-y}As_y$ type II quantum-well structure. L_e (L_h) denotes the width of the well (barrier) and V_e (V_h) is the effective potential of the electron (hole). (a) $\Delta E > 0$ case. (b) $\Delta E < 0$ case. Here, ΔE is the energy difference between the lower conduction band of $In_{1-x}Ga_xAs$ and the top valence band of $GaSb_{1-y}As_y$. The electron and hole wavefunctions are also shown. Reprinted from Wang E. G.[1995], with the permission of AIP Publishing.

of the energy is located at the bottom of the quantum well; for the hole, it is at the top. Therefore, the Hamiltonian of the exciton can be expressed as

$$H(\mathbf{r}_e,\mathbf{r}_h) = H(\mathbf{r}_e) + H(\mathbf{r}_h) + V(\mathbf{r}_{eh}) + U_e(z_e) + U_h(z_h), \tag{4.57}$$

where $\mathbf{r}_e$ and $\mathbf{r}_h$ are the positions of the electron and hole, respectively. The corresponding kinetic-energy term is $H(\mathbf{r}_i) = -(\hbar^2/2m_i)\nabla^2_{r_i}$, where $i=\{e,h\}$, and m_e, m_h are the corresponding effective masses. The Coulomb energy is $V(\mathbf{r}_{eh}) = -e^2/4\pi\epsilon_0\epsilon\,|\mathbf{r}_e - \mathbf{r}_h|$, where $\epsilon = \sqrt{\epsilon_1\epsilon_2}$ is the geometric mean of the dielectric constants of the two materials. $U_e(z_e) = V_e\Theta(|z_e| - nd - L_e/2)$ and $U_h(z_h) = V_h\Theta(|z_h - d/2| - nd - L_h/2)$ are the quantum-well potentials of the electron and hole, respectively. Here, $\Theta(\mathrm{x})$ is the Heaviside step function ($\Theta(\mathrm{x}) = 1,\ \mathrm{x} > 0;\ \Theta(\mathrm{x}) = 0,\ \mathrm{x} \le 0$). V_e and V_h are the respective heights of the potential well in the conduction band and valence band (as illustrated in Figure 4.7).

E. G. Wang et al. adopted a DMC method similar to that of P. J. Reynolds. They used the imaginary-time Schrödinger equation, with $\tau = it/\hbar$ denoting the imaginary time. Starting from a trial wavefunction $\Phi(\mathbf{R})$, they constructed a probability distribution function $f(\mathbf{R},\tau) = \Phi(\mathbf{R})\Psi(\mathbf{R},\tau)$. In this way, the imaginary-time Schrödinger equation for $\Psi(\mathbf{R},\tau)$ could be rewritten as a diffusion equation for f:

$$\frac{\partial f}{\partial \tau} = D\nabla^2 f - \nabla\cdot\left[fF(R)\right] + \left[E_0 - E_L(\mathbf{R})\right]f. \tag{4.58}$$

where $E_L(\mathbf{R}) = \Phi^{-1}(\mathbf{R})H\Phi(\mathbf{R})$ is the local energy, and $F(R) = D\nabla\ln|\Phi(\mathbf{R})|^2$ is the drift-velocity term imposed on the diffusion process. The drift term now guides the walk preferentially into the region where the wavefunction is large. The hermiticity of H can be used to obtain a τ-dependent expectation value of the system energy:

$$\mathrm{E}(\tau) = \frac{\langle\Phi(\mathbf{R})|H|\Psi(\mathbf{R},\tau)\rangle}{\langle\Phi(\mathbf{R})|\Psi(\mathbf{R},\tau)\rangle} = \frac{\int d\mathbf{R}f(\mathbf{R},\tau)E_L(\mathbf{R})}{\int d\mathbf{R}f(\mathbf{R},\tau)}. \tag{4.59}$$

This energy evolves with the imaginary time τ. As τ approaches infinity, the expectation value of the energy approaches the ground state energy of the system. The statistical error of an MC calculation generally depends on the number of simulation steps in the system; therefore, choosing the appropriate time-step interval can reduce systematic error such that it is within the statistical error of the MC calculation.

Generally, a better trial wavefunction accelerates convergence to the ground state. In this work, the trial wavefunction assumes the following form:

$$\Phi(\mathbf{r}_\mathrm{e},\mathbf{r}_\mathrm{h}) = f(\mathbf{r}_\mathrm{eh})g(\mathbf{r}_\mathrm{eh})\prod_{i=e,h}Q(z_i), \tag{4.60}$$

where $f(r) = \exp[-ar/(1+br)]$ and $g(r) = \exp[-a'r^2/(1+b'r)]$. Here, $Q(z_i)$ denotes the stepwise quantum confinement function:

$$Q(z_i) = \begin{cases} \cos\left[k_i\left(z_i - nd\dfrac{z_i}{|z_i|}\right)\right] & z_i \text{ in well} \\[2ex] \exp\left[-k_i'\left|z_i - nd\dfrac{z_i}{|z_i|}\right|\right] & z_i \text{ in barrier} \end{cases}. \tag{4.61}$$

where the parameters k_i and $k_i'(i = \{e,h\})$ can be obtained by solving the following characteristic equation:

$$\cos(qd) = \cos(k_i L_e)\cosh\left(k_i' \; L_h\right) - \frac{\xi - \xi^{-1}}{2}\sin\left(k_i L_e\right)\sinh\left(k_i' \; L_h\right), \tag{4.62}$$

where $\xi = k_i m_W^* / k_i' m_B^*$, and m_W^* and m_B^* are the effective masses at the well and the barrier, respectively. A commonly used boundary condition is that Q and $Q' / m_{W(B)}^*$ are continuous at the interface. The parameter a in $f(r)$ and $g(r)$ can be obtained by the cusp condition, and the other parameters b, a', b' can be obtained by VMC optimization. Finally, the optimized trial wavefunction $\Phi(\mathbf{r}_e, \mathbf{r}_h)$ is used in the DMC calculation.

E. G. Wang et al. explored the dependence of the exciton binding energy on different quantum-well widths $L(= L_e = L_h)$ and compositions (x, y). The material parameters were selected based on the experimental data of Sai-Halasz et al. [Sai-Halasz1977, 1978]. The resultant E_B values corresponding to different compositions at L = 50 Å are shown in Figure 4.8. Increases of x and y in the $In_{1-x}Ga_xAs/GaSb_{1-y}As_y$ type II quantum well lower the potential barrier and result in an enlarged overlap of the electron and hole wavefunctions. At the same time, we will see the increases of the binding energy.

Because of size quantization effects in the quantum well, the conduction-band edge of $In_{1-x}Ga_xAs$ is lower than the valence-band edge of $GaSb_{1-y}As_y$ at (x, y) ≤ 0.25 as the $In_{1-x}Ga_xAs/GaSb_{1-y}As_y$ system undergoes the transition from semiconductor to semimetal. If the quantum-well width is less than or comparable to the effective Bohr radius of the exciton, the reduction in exciton binding energy caused by space separation is not significant. At this point, with an appropriate choice of material parameters, it is possible to make the absolute value of the energy gap $|E_G|$ less than the exciton binding energy E_B, allowing the observation of various phase transitions of the system among the semiconductor state,

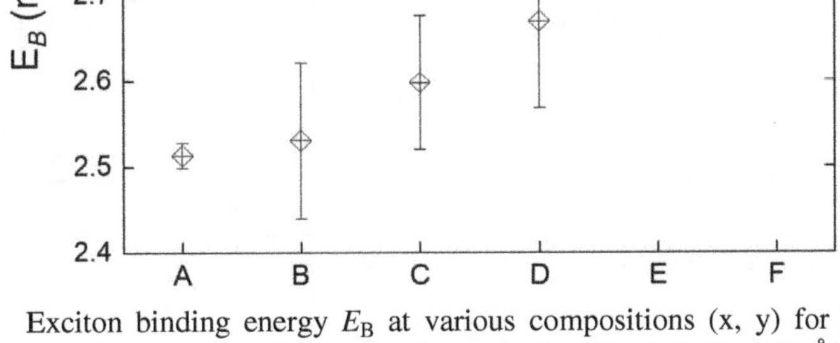

Figure 4.8 Exciton binding energy E_B at various compositions (x, y) for $In_{1-x}Ga_xAs/GaSb_{1-y}As_y$ type II quantum-well systems for a fixed well-width L = 50 Å). Here A = (0.00,0.00), B = (0.16,0.15), C = (0.25,0.23), D = (0.40,0.35), E = (0.55,0.45), and F = (0.62,0.64). Reprinted from Wang E. G.[1995], with the permission of AIP Publishing.

the excitonic insulator state with Bose condensate, and the semimetal state. In 1995, Cheng, J.P. et al.[1995] experimentally observed the far-IR magnetic spectra of InAs/$Al_xGa_{1-x}Sb$ quantum wells, providing the first evidence for the existence of a stable excitonic ground state in a spatially separated electron–hole system.

4.5.4 Full-Configuration-Interaction Quantum MC Method

The full-configuration interaction quantum MC method (FCIQMC) is a more recently developed method for solving electron correlation effects with greater precision [Booth2009]. FCIQMC, similar to DMC, approximates the ground state by projection. However, it differs from DMC in how the wavefunction is expanded, the form of the projector, the corresponding MC rules, and the way that the sign problem is handled.

The object on which the projector acts in DMC is the wavefunction in real space. It means the distribution of walkers in the DMC method reflects the wavefunction distribution in real space. In FCIQMC, on the other hand, the walkers are sampled in configuration space, as shown in Figure 4.9. Each configuration is a Slater determinant $|D_i\rangle$ consisting of N single-electron orbitals (both occupied and unoccupied). The set of all possible determinants constitutes the configuration space of the basis set; the determinants $|D_i\rangle$ themselves form a complete basis in terms of which the wavefunction of the system can also be expanded

$$|\Psi\rangle = \sum_i C_i |D_i\rangle . \tag{4.63}$$

Because the overall wavefunction is expanded in configuration–interaction space, C_i is known as the CI coefficient. This form of expansion is used in the configuration–interaction approach to quantum chemistry, similar to the post-Hartree–Fock method.

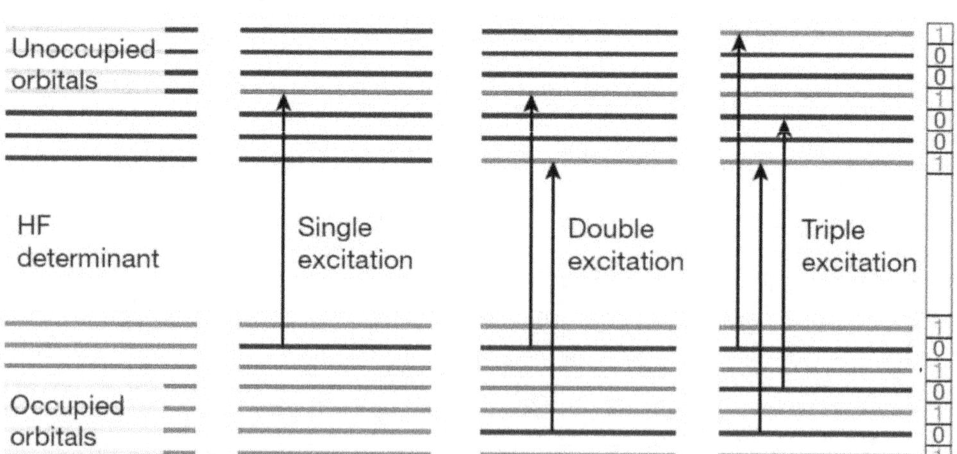

Figure 4.9 Many-body electron wavefunction can be written as a linear combination of determinants that are composed of single-electron Hartree–Fock orbitals (occupied, 1, and unoccupied, 0). The walkers of the FCIQMC method are sampled in the Hilbert space spanned by these determinants. From Booth[2013]. Reproduced with permission from SNCSC.

The projector of the FCIQMC method is different from that of the DMC method; it is simply $\hat{P} = 1 - \tau \hat{H}$. The ground state wavefunction can be obtained by applying this projector an infinite number of times. This is expressed as

$$\left|\Psi_{\text{ground}}\right\rangle = \lim_{n\to\infty} \hat{P}^n \left|\Psi_{\text{initial}}\right\rangle. \tag{4.64}$$

Because FCIQMC has a different projector, G. Booth et al. designed a new MC sampling method based on Eq. (4.64) (also see Figure 4.10). The walker distribution after reaching equilibrium is described by the CI coefficients in Eq. (4.63). In Figure 4.10, the hollow dots represent determinants without walker occupation, the solid dots represent determinants occupied by walkers, and the different shades of gray correspond to the population of walkers. According to Eq. (4.63), the walker distribution can uniquely determine the many-body wavefunction.

The MC evolution of the walkers involves three main processes: spawning, death/cloning, and pairwise annihilation. Generation is the birth of a new walker as the walker walks from an occupied determinant to a new unoccupied one (indicated by the gray arrow in the figure). A walker can also die (disappear) or be cloned with a certain probability.

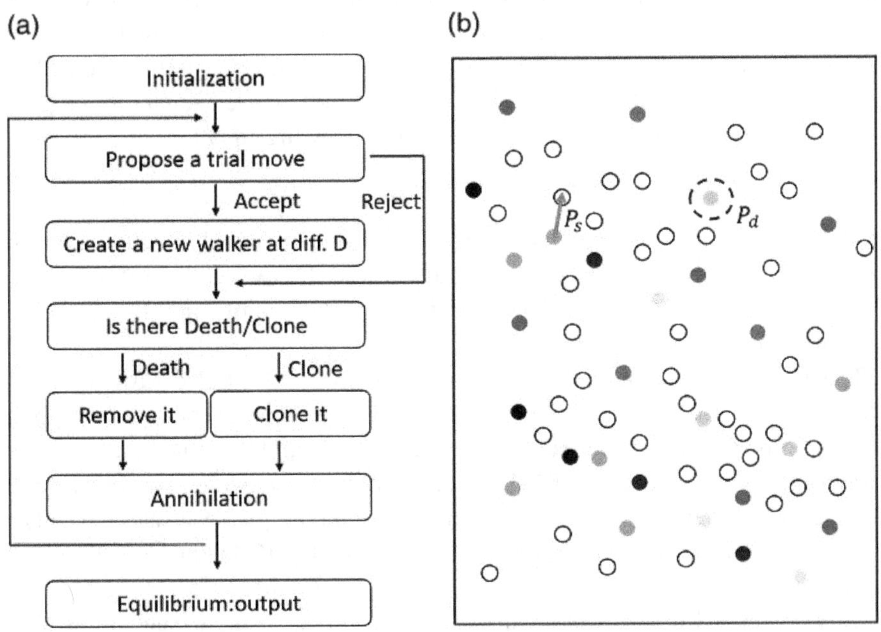

Figure 4.10 Panel (a) shows the flowchart of the FCIQMC process. Each dot in Panel (b) represents a determinant in the Hilbert configuration space, which contains a large number of unoccupied determinants (hollow dots). The filled dots represent finitely occupied determinants in space, and the shades represent the occupation number, which corresponds to the different configuration–interaction coefficients in Eq. (4.63). The Monte Carlo algorithm samples the occupation number of each determinant. The gray arrow shows an example of spawning, and the dashed circle shows an example of death/cloning.

Pairwise annihilation is the cancelation of walkers with opposite signs on a certain determinant; this may be caused either by the simultaneous spawning of walkers with positive and negative signs there, or by a spawned walker having the opposite sign from an existing walker on the determinant. Eventually, walkers on each determinant will only have the same sign.

In practical calculations, the probabilities of both spawning and death/cloning are determined by a matrix satisfying the following equation:

$$K_{i,j} \equiv \langle D_i | \hat{K} | D_j \rangle = \langle D_i | \hat{H} | D_j \rangle - E_{\mathrm{HF}} \delta_{ij}. \tag{4.65}$$

The value of $\hat{K}$ at the lowest energy is the correlation energy of the system's ground state, that is, the total energy minus the Hartree–Fock energy. The probability of generation from the ith to the jth cell in the MC process is given by

$$P_{\mathrm{s}}(j|i) = \frac{\delta\tau |K_{i,j}|}{P_{\mathrm{gen}}(j \mid i)}, \tag{4.66}$$

where $\delta\tau$ is the step size. If this probability is greater than 1, multiple new walkers will be generated. (The denominator $P_{\mathrm{gen}}(j \mid i)$ is the nonzero normalized probability of generating all possible excitations; its exact form is complicated and will not be discussed here.)

The sign of the newly spawned walker is determined by that of the matrix element $K_{i,j}$. When $K_{i,j}$ is negative, the sign of the walker spawned at the jth determinant is the same as that at the ith determinant; when $K_{i,j}$ is positive, the signs are opposite. The spawning proceeds from ith to jth cell; this process corresponds to the particle excitation in Figure 4.9.

The probability of death/cloning is given by the following equation.

$$P_d(i) = \delta\tau(K_{i,i} - S), \tag{4.67}$$

where S is a newly defined drift energy. If $P_d(i)$ is greater than 0, the ith walker dies; if $P_d(i)$ is less than 0, it is cloned with probability $|P_d(i)|$. Therefore, the drift energy can be used to control the total number of walkers in the system, allowing a faster approach to equilibrium.

FCIQMC as described above is theoretically rigorous, but it also suffers from a sign problem (albeit not the same one afflicting DMC) that prevents the solution from approaching the real ground-state wavefunction. To overcome this, D. Cleland et al. proposed the Initiator approximation [Cleland2010], which restricts the spawning process described in Eq. (4.66) so that it only occurs if the occupation number of the walker exceeds a certain threshold. In this way, the sign problem of the FCIQMC method can be effectively controlled.

The current FCIQMC method is capable of performing accurate calculations for some simple molecules and condensed matter systems. Recent work suggests that it shows promise for larger systems as well. For systems containing strongly correlated electrons, FCIQMC has undeniable accuracy advantages over other methods. In addition, the research group of J. Chen and E. G. Wang at Peking University, utilizing the quantum embedding theory, has extended the application of FCIQMC from the study of chemical molecular systems to nitrogen vacancy (NV) centers in diamond [Chen Y.L.2023]. They have also proposed a set of universal algorithms for applying the FCIQMC method to the calculation of forces acting on atomic nuclei, obtaining the potential energy surface of water molecules at the FCIQMC level for the first time [Jiang T.H.2022]. Such advances are crucial for the study of full quantum effects, as the electronic structures of many materials cannot be accurately calculated by mean-field methods such as DFT.

Due to its guaranteed computational accuracy, FCIQMC has become the benchmark method for accurately measuring Born–Oppenheimer potential energy surfaces and is internationally recognized as a new standard for determining the accuracy of other electronic-structure calculation methods. However, it should be noted that, unlike the DMC method introduced in subsection 4.5.3, the rigorous solution approximated by the FCIQMC method is limited to being the exact solution under a given basis set (i.e., the exact solution under a discrete Hilbert space, in the language of second quantization). In this regard, FCIQMC resembles another important branch of theoretical physics: The solution of lattice models. Strictly speaking, it is not an exact solution for an actual (continuous) condensed matter system, and the errors introduced by the basis set should be considered in practical calculations. However, as a measure of the accuracy of other electronic-structure methods, the results produced by FCIQMC are valuable as long as both methods use the same basis set. Basis-set error is an important topic in its own and is covered in many textbooks of first-principles calculation methods. Here, we only want to emphasize that it has a stronger effect on the total energy of the system than on the relative energy; the latter is more important for calculating a Born–Oppenheimer potential energy surface. If we select a sufficiently large basis set, our understanding of the characteristics of the physical properties based on the Born–Oppenheimer potential energy surface remain unaffected.

4.6 Excited States and *GW* Method

Not only the ground state but also the excited states of electron are needed in the calculation of full quantum effects. In Chapter 5, we will introduce the nonadiabatic effects, which contains information about the electronic excited states and includes the problem of transition between multiple electron states. For now, however, we will focus on the calculation of electronic excited states without considering the quantum motion of the nucleus. The combination of excited state calculations and the simulation of nuclear quantum effects is also crucial when considering full quantum effects on the excitation spectrum, the band structure of materials, and other properties, wherein electronic structure calculations with accurate treatment of excited states are essential.

The three methods of calculating the ground-state electronic structure presented above can be adapted to find the excited states as well. In this section, we will first provide a brief introduction to the extension of these ground-state methods; then, we will focus on a particular excited-state many-body Green's function method known as the *GW* approximation (also known as the *GW* method).

4.6.1 Electronic Structure of Excited State

We first consider the wavefunction approach. It is relatively straightforward to find excited states using the multi-configurational self-consistent field (MRSCF) method. Each configuration represents a different ground state and excited state; hence, excitation can be examined directly by considering multiple configurations simultaneously. A commonly used method for this is the state-averaged-multi-configurational-self-consistent field (SA-MRSCF) method. Similarly, we can also use the multi-state perturbation method along with a multi-configurational wavefunction to calculate the excited states, and the commonly used combination is called multi-state complete active space second-order perturbation theory (MS-CASPT2). Equation-of-motion coupled cluster (EOS-CC) theory is another wavefunction-based approach to the calculation of excited states that adds excitation operators to coupled cluster (CC) wavefunctions. At present, these methods all have unique advantages in their respective fields of expertise.

Within the framework of DFT, time-dependent density functional theory (TDDFT) can be directly extended to the calculation of excited states. Detailed discussion regarding this method will be presented in Chapter 5, where we discuss nonadiabatic effects.

It is also possible to calculate excited states using the QMC wavefunction approach. After obtaining the ground state wavefunction, we can separate the ground-state component from other components via orthogonalization. By projection-based MC evolution in the remaining space, the first excited state can be obtained. Higher excited states can be calculated in a similar way. Furthermore, we can conduct the MC evolution of the many-electron states along with the orthogonalization so that electron excitation can be considered in a single

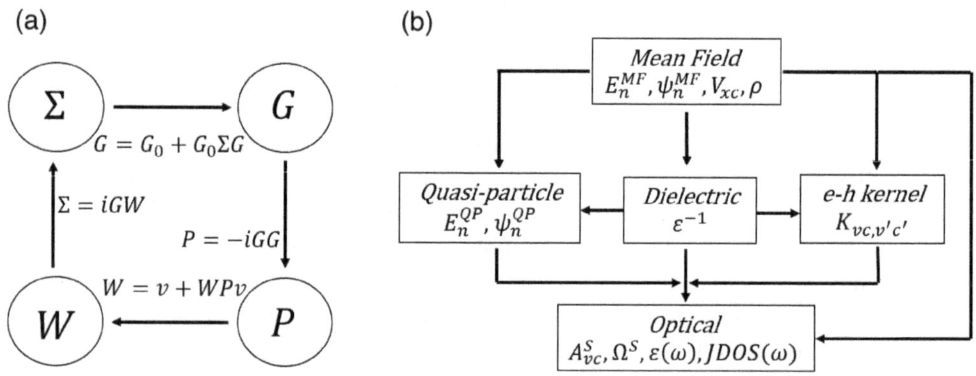

Figure 4.11 Theoretical framework and computational procedure for the *GW* approximation and Bethe–Salpeter equation (BSE) method. Panel (a) shows the self-consistent solution scheme for the system of Hedin equations under the *GW* approximation, where Σ, *G*, *P*, *W* are the self-energy, Green's function, polarizability, screened Coulomb potential, respectively. The simplified formulas next to the arrows show the relationships between these terms. The implementation of the *GW* and BSE calculations is shown in Panel (b). Starting from the mean-field calculation, we can obtain the dielectric function, the quasiparticle energy, and the wavefunction in the *GW* or G_0W_0 approximation, as well as the electron–hole interaction kernel. Finally, the spectral properties are calculated using the BSE.

calculation. It is also possible to realize such calculation in projection-based MC methods such as DMC and FCIQMC.

4.6.2 *GW* Method

We now discuss the many-body Green's function approach based on the *GW* approximation. First proposed by L. Hedin in 1965 [Hedin1965], this method has been successfully applied to real materials since the 1980s by S. G. Louie and others [Hybertsen1986]. It is essentially a perturbative expansion of the electron self-energy in terms of the Green's function (*G*) and the screened Coulomb interaction (*W*) that allows us to treat the quasiparticle–electron excitation properties of many systems accurately. The response of the electron system to light can also be obtained by solving the Bethe-Salpeter equation (BSE) in the *GW* approximation. The *GW* method has been applied to conventional semiconductor materials, insulators, metals, nanomaterials, polymers, and other systems. Particularly for the calculation of quasiparticle energy gaps, energy-band dispersion, and optical response, it is the most accurate electronic-structure algorithm available.

In practice, *GW* and BSE methods generally start from the ground-state solution of DFT theory. We assume that a set of single-particle Kohn–Sham orbitals ψ_n^{KS} has been obtained, along with the corresponding orbital energies E_n^{KS}. In the *GW* approximation, we calculate each quasiparticle excitation energy E_n^{QP} and wavefunction ψ_n^{QP} by solving the self-consistent equation

$$\left[-\frac{1}{2}\nabla^2 + V_{\mathrm{ion}} + V_{\mathrm{H}} + \Sigma\left(E_n^{\mathrm{QP}}\right)\right]\psi_n^{\mathrm{QP}} = E_n^{\mathrm{QP}}\psi_n^{\mathrm{QP}} , \tag{4.68}$$

where $\Sigma(E_n^{\mathrm{QP}})$ is the self-energy under the *GW* approximation. The initial guesses for E_n^{QP} and ψ_n^{QP} are given by E_n^{KS} and ψ_n^{KS}. The left panel of Figure 4.11 shows a simple relation diagram of the self-consistent solution under the *GW* approximation. As would be expected, the accuracy of this approach based on a many-body perturbative expansion depends on the reasonableness of the initial wavefunction and the convergence of the self-consistent solution. In many-body perturbation theory, as the expansion series diverges, it is impossible to approximate the real solution by truncating higher-order terms, as is done in the wavefunction theory.

To avoid the high computational cost of obtaining a self-consistent solution for real materials, Hybertsen and Louie adopted the single-step *GW* approximation (also known as G_0W_0). In G_0W_0, the quasiparticle equations are no longer solved self-consistently. Instead, their energies are corrected by first-order perturbation theory. Comparing the Kohn–Sham equation with the quasiparticle equation, we see that the difference arises from the self-energy $\Sigma(E_n^{\mathrm{QP}})$ and the exchange correlation energy V_{xc} in the Kohn–Sham equation. Hence, we can define a perturbative self-energy term $\Delta\Sigma = \Sigma\left(E_n^{\mathrm{QP}}\right) - V_{\mathrm{xc}}$. According to first-order perturbation theory, the quasiparticle energy can be expressed as

$$E_n^{\mathrm{QP}} = E_n^{\mathrm{KS}} + Z_n\left\langle\psi_n^{\mathrm{KS}}\middle|\Sigma\left(E_n^{\mathrm{QP}}\right) - V_{\mathrm{xc}}\middle|\psi_n^{\mathrm{KS}}\right\rangle, \tag{4.69}$$

where $Z_n = \frac{1}{1-Re\langle\psi_n^{KS}|\Sigma'(E_n^{KS})|\psi_n^{KS}\rangle}$ is a renormalization factor and Σ' is the partial derivative of the self-energy with respect to frequency. Under the G_0W_0 approximation, the self-energy can be computed from the reduced Hedin equations.

$$\mathrm{G}_0\left(x,x';\omega\right) = \sum_n \frac{\psi_n^{\mathrm{KS}}(x)\psi_n^{\mathrm{KS}*}(x')}{\omega-\epsilon_n}, \tag{4.70}$$

$$\mathrm{W}_0\left(\boldsymbol{r},\boldsymbol{r}';\omega\right) = \int dx''\varepsilon^{-1}(x,x'';\omega)v(\boldsymbol{r}''-\boldsymbol{r}'). \tag{4.71}$$

Here, ε^{-1} is the reciprocal of the dielectric function, which can be obtained from the Green's function G_0 and the bare Coulomb potential v.

We can calculate the spectrum more accurately by including the electron–hole excitation when solving the BSE:

$$\left(\mathrm{E}_c^{QP} - E_v^{QP}\right)A_{vc}^S + \sum_{v'c'}\left\langle vc\left|K^{eh}\right|v'c'\right\rangle = \Omega^S A_{vc}^S, \tag{4.72}$$

where A_{vc}^S is the wavefunction coefficient of the excitonic state S in the quasi-particle state representation, $\Psi(\boldsymbol{r_e},\boldsymbol{r_h}) = \sum_{c,v} A_{vc}^S\psi_c(\boldsymbol{r_e})\psi_v^*(\boldsymbol{r_h})$. Ω^S is the corresponding exciton energy, and K^{eh} is the electron–hole interaction kernel. After the exciton wavefunction and energy are determined, physical parameters related to spectral properties, such as the dielectric function, can be calculated. For example, the imaginary part of the frequency-dependent dielectric function is expressed as

$$\epsilon_2(\omega) = \frac{16\pi^2 e^2}{\omega^2}\sum_S\left|\vec{e}\cdot\sum_{vc}A_{vc}^S\left\langle v\right|\hat{O}_v\left|c\right\rangle\right|^2 \delta(\omega-\Omega^S), \tag{4.73}$$

where $\vec{e}$ is the unit vector in the polarization direction and $\hat{O}_v$ is the velocity operator along that direction.

The right panel of Figure 4.11 shows the computation process in the *GW* and BSE methods. In actual calculations, because the *GW* method does not consider the nearest-neighbor vertex correction, solving the equation self-consistently may result in error accumulation, and the final result is not necessarily better than the G_0W_0 approximation.

The description above is only the standard part of the *GW* method. In the last three decades, different approaches beyond *GW* approximation have been actively developed to solve the challenges resulting from its standard method and to further improve the accuracy of *GW* calculations. For example, one popular approach is the vertex correction method, which includes higher-order diagrams of the many-body expansion in the self-energy of the *GW* routine. In a different treatment, namely the cumulant expansion, higher-order terms are included in the Green's function. In addition, combining the *GW* method with other approaches to treat strong correlation effects has been demonstrated as a promising attempt. For example, post-Hartree-Fock wavefunction theory can provide better starting point for the *GW* calculations of excited states; dynamical mean field theory can also be used collaboratively with *GW* to improve the description of electron correlation effects. For more substantial reviews of the developments beyond *GW* method, we are happy to recommend several recent articles by some experts in this active field [Marom2016, Ghosh2018, Kent2018, Nilsson2018, Golze2019].

4.7 Electronic Structure Software

In the previous sections, we presented different methods of electronic-structure calculation. Most of these methods are now available to researchers in either open source or commercial software packages. The introduction and application of these computational tools has made first-principles calculation of electronic structure becomes the main part of the modern research in physics, chemistry, and materials science. Moreover, as we repeatedly emphasize in this book, electronic structure is the basis for full-quantum calculations. Therefore, in this section, we will briefly introduce and summarize some current electronic-structure calculation procedures used widely and discuss the challenges of these methods.

Let us begin with a brief historical overview. The earliest officially released program package for all the electronic structure calculation methods presented in this book was the Gaussian software package. This commercial software was initially released in 1970 by John Pople and his research group at Carnegie Mellon University. The release of the Gaussian program signified that electronic structure theory had officially entered the "computer age," marking the dawn of quantum computing in chemistry. Because of this achievement, John Pople shared the 1998 Nobel Prize in Chemistry with Walter Kohn, who proposed density functional theory, as mentioned earlier.

In the 1980s, the exchange correlation functional of DFT was gradually parameterized, and packages for DFT and time-dependent DFT were soon developed. *GW* methods were either integrated into existing DFT packages or developed as stand-alone programs. In particular, after 1990, as better density functional theory such as the plane-wave basis set and pseudopotential method were developed, DFT programs targeting solid material simulation were developed by several research teams. The emergence of these programs has considerably extended the scope of DFT applications; the DFT-based first-principles calculations thrived and became one of the main computing tools in condensed matter physics and materials physics research since then. MC methods are also well suited for computation, and QMC calculations were performed as early as 1965. However, in terms of general program development, QMC fell behind wavefunction theory and DFT, because the latter two methods were more efficient at the single-computer level. In the twenty-first century, as parallel computing on supercomputers became more widely used in scientific research, the advantages of QMC became more apparent, and software packages based on MC methods became increasingly popular.

The concepts of basis set, pseudopotential, and boundary condition play a vital role in both historic and contemporary electronic structure calculation programs. We will briefly review them before examining specific methods individually.

The basis-set concept has a long history in theoretical quantum chemistry. In 1928, when the wavefunction approach was in its infancy, Hartree carried out the first self-consistent wavefunction calculation manually [Hartree1928]. Wavefunctions were written in the form of analytic functions that could be manipulated directly. However, computers are only good with numbers, not analytic expressions. To perform calculations similar to Hartree's on a computer, we must first expand the wavefunction Ψ linearly in a set of basis functions χ_i.

$$\Psi = \sum_{i}^{M} c_i \chi_i \tag{4.74}$$

The expansion coefficients c_i can then be manipulated and solved using the various methods introduced in the previous sections; the calculation process involves only linear algebra and thus can easily be implemented on a computer. This idea was first proposed separately by C. C. J. Roothaan and G. G. Hall in 1951 [Hall1951, Roothaan1951]. By expanding the wavefunction with respect to a given set of basis functions, they rewrote the Hartree–Fock equations in the form of what became known as the Roothaan or Roothann–Hall equations, transforming the Hartree–Fock calculation into a linear algebraic eigenvalue problem.

The form proposed by Roothaan and Hall was general; therefore, a specific basis function is required to develop the Roothaan–Hall equations into a practical numerical method. In fact, even before the advent of numerical computation, choosing the basis function had been an important factor in the analytical treatment of wavefunction calculations. Because the wavefunction of a many-particle system was unknown, we had to assume a relatively reasonable form of the wavefunction that was easy to manipulate even for analytical calculations. As early as 1930, Slater suggested using the orbitals of hydrogen-like atoms as basis functions for electron wavefunctions in atoms. These orbitals are collectively known as Slater-type orbital (STO) [Slater1930b]. In 1950, S. F. Boys suggested using Gaussian-type orbitals (GTOs) instead; this further accelerates the integration step in the calculation [Boys1950]. The GTO was extensively adopted in computer programs for wavefunction theory, such as the Gaussian software developed by Pople.

The number of basis functions (M in Eq. (4.74)) directly determines the accuracy and efficiency of the calculation. If it is too small, then the correct wavefunction may not be adequately expressed; if it is too large, the matrix dimension in the linear-algebra operations (and therefore the computation time) will increase. The central issue in basis-set selection is determining the best balance between efficiency and accuracy.

For example, a Slater-type basis function is a hydrogen-like atomic orbital, which is expressed as

$$\chi_{n,l,m}^{SF}(r,\theta,\phi) = N_{n,l,m,\xi} Y_{l,m}(\theta,\phi) r^{n-1} e^{-\xi r}. \tag{4.75}$$

Here, r,θ,ϕ indicate the relative position of the electron with respect to some nucleus center in spherical coordinates. $N_{n,l,m,\xi}$ is the normalization factor, and $Y_{l,m}(\theta,\phi)$ is the spherical harmonic function. n,l,m represent the principal quantum number, the angular quantum number, and the magnetic quantum number of the hydrogen-like atomic orbitals, respectively. The parameter ξ controls the rate of exponential decay in the radial direction. These basis functions can reasonably describe the electron orbitals outside the nucleus and are the best way to minimize the number of basis functions. However, considering that a large number of two-center or four-center integrals need to be evaluated for the wavefunctions according to the wavefunction theory, choosing functions based on the Slater type can be inconvenient. As a result, Gaussian-based basis functions are more frequently adopted. It is expressed as follows:

$$\chi_{a,b,c}^{GF}(r,\theta,\phi) = N'_{a,b,c,\alpha} \mathbf{x}^{a} \mathbf{y}^{b} \mathbf{z}^{c} e^{-\alpha r^2}, \tag{4.76}$$

where a, b, c, α are the basis-function parameters. Employing Gaussian-type functions ensures quick execution of a large number of integrals.

To exploit both the computational advantages of Gaussian-type functions and the physical significance of hydrogen-like atoms, it is common practice today to assign the basis functions to hydrogen-like atomic orbitals first, then write each orbital as the best-fit combination of Gaussian-type functions on it. The basis set formed in this manner is often referred to as the atomic-orbital basis set.

The wavefunction of an entire polyatomic molecular or condensed matter system can be expressed as a LCAO. In wavefunction theory, where molecules are mostly studied, the LCAO method is the most commonly used, and it relies mainly on the more efficient GTOs. The left panel of Figure 4.12 shows a schematic of an atomic-orbital basis set describing a specific material.

Instead of localized atomic-orbital basis functions such as GTOs, the planewave basis set is often employed in the study of solids and other periodic systems. The planewave approach takes inspiration from Bloch's theorem in solid-state physics. According to this theorem, the electron wavefunction in a periodic potential can be written in the form of an amplitude-modulated planewave. That is

$$\phi_{\vec{k}}(\vec{r}) = \exp(i\vec{k}\cdot\vec{r})u_{\vec{k}}(\vec{r}). \tag{4.77}$$

Here, $\vec{k}$ is the electron wave vector, $\exp(i\vec{k}\cdot\vec{r})$ is the planewave corresponding to this wave vector, and $u_{\vec{k}}(\vec{r})$ is a periodic function of the lattice vector $\vec{R}_l$ satisfying $u_{\vec{k}}(\vec{r}) = u_{\vec{k}}(\vec{r}+\vec{R}_l)$. Because of its periodicity, $u_{\vec{k}}(\vec{r})$ can be linearly expanded further using a set of planewaves as basis functions, as shown in the following equation:

$$u_{\vec{k}}(\vec{r}) = \sum_{G} c_{\vec{k},\vec{G}} \mathrm{e}^{\mathrm{i}\vec{G}\cdot\vec{r}}. \tag{4.78}$$

We can then expand the Bloch wavefunction in terms of planewaves

$$\phi_{\vec{k}}(\vec{r}) = \sum_{\vec{G}} c_{\vec{k},\vec{G}} e^{i(\vec{k}+\vec{G})\cdot\vec{r}}, \tag{4.79}$$

where $\vec{G} = m_1\vec{b}_1 + m_2\vec{b}_2 + m_3\vec{b}_3$ is the lattice vector obtained by weighting the reciprocal lattice vectors $\vec{b}_1, \vec{b}_2, \vec{b}_3$ by scalar integers $m_1,\ m_2,\ m_3$.

In theory, the number of basis functions in the planewave basis set can be infinitely large. In practice, however, it is conventional to set an energy cutoff E_{cut} such that only planewaves below this energy cutoff are considered. In other words, only $\vec{G}$ that satisfies $\frac{\hbar^2}{2m}\vec{G}^2 < E_{cut}$ are taken into account. Higher-energy planewaves often correspond to electron wavefunctions that are more localized in real space. Therefore, when the properties of the local electrons in the system under consideration become more important, the plane-wave cutoff must be increased accordingly. The middle panel in Figure 4.12 shows a schematic of a plane-wave basis set for a specific material. The planewave method is universal and can accelerate the calculation via the fast Fourier transform. Its successful emergence was a key step in the large-scale application of DFT in computational condensed matter and material research. ABINIT, CASTEP, VASP, and QUANTUM ESPRESSO are some of the most influential computational programs based on the planewave method.

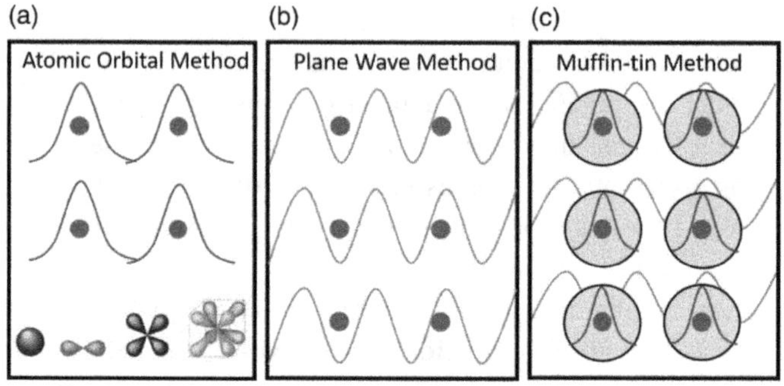

Figure 4.12 Schematic of the atomic-orbital method, the plane-wave method, and the muffin-tin approximation. The gray dots represent atomic positions, and the single-peak curves around nuclei correspond to the atomic orbitals (Panel (a)). The bottom of Panel (a) shows the shapes of typical s, p, d, and f orbitals. The light-color curves in Panels (b) and (c) represent plane-wave orbitals. The discs represent neighborhoods of the nucleus, with local orbitals inside the region and plane waves outside the region (Panel (c)).

The plane-wave and the Gaussian-atomic-orbital basis sets are two of the most common basis sets and, in some ways, represent the two extreme cases. In practical calculations, especially for DFT calculations of solid materials, there are also methods that combine the two ideas to optimize performance. A representative example is the linearized augmented plane wave (LAPW) method proposed by O. K. Anderson [Anderson O. K.1975]. The LAPW method is derived from the muffin-tin approximation proposed by J.C. Slater in 1937. This is an appropriate analogy as muffin tin is a bakeware with circular grooves used for baking Western muffins (see the right panel of Figure 4.12). The original augmented plane wave (APW) method is a direct result of the muffin-tin approximation. In the APW method, a cutoff radius is set around each nucleus. Within the cutoff, the wavefunction is represented by local atomic orbitals; outside it, the wavefunction is expanded by planewaves. Thus, the wavefunction can be expressed as

$$\phi_{\vec{k},\vec{G}}(\vec{r}) = \begin{cases} \dfrac{1}{\Omega} e^{i(\vec{G}+\vec{k})\cdot\vec{r}}, & r \in interstitial \\ \sum\limits_{lm} A_{l,m}(\vec{k}+\vec{G}) u_l(r^\alpha, \epsilon_l) Y_{l,m}(\vec{r}^\alpha), & r \in sphere \end{cases} \tag{4.80}$$

Within the framework of DFT, $u_l(r^\alpha, \epsilon_l)$ is the radial function derived from the Kohn–Sham equation for the radial part of the spherical region around the nucleus, and the $Y_{l,m}(\vec{r}^\alpha)$ are spherical harmonics. The problem with the APW method is that it turns the self-consistent field equations into a set of nonlinear equations in the subsequent solution, increasing the computational difficulty. To address this issue, Andersen proposed LAPW, a method that guarantees linearization. It is expressed as

$$\phi_{\vec{k},\vec{G}}(\vec{r}) = \begin{cases} \dfrac{1}{\Omega} e^{i(\vec{G}+\vec{k})\cdot\vec{r}}, & r \in interstitial \\ \sum\limits_{lm} \left[A_{l,m}(\vec{k}+\vec{G}) u_l(r^\alpha, \epsilon_l) + B_{l,m}(\vec{k}+\vec{G}) \dot{u}_l(r^\alpha, \epsilon_l) \right] Y_{l,m}(\vec{r}^\alpha), & r \in sphere \end{cases} \tag{4.81}$$

The derivative of $u_l(r^\alpha,\epsilon_l)$ with respect to the energy is included in LAPW ($\dot{u}_l(r^\alpha,\epsilon_l) = \partial u_l(r^\alpha,\epsilon_l) / \partial\epsilon \big|_{\epsilon=\epsilon_l}$) [Anderson O. K. 1975]. In both APW and LAPW, some localized atomic orbitals can be added (generally only within the cutoff radius) to improve the accuracy and efficiency. Frequently used schemes are LAPW + LO [Singh1991] and APW + lo [Sjöstedt2000], and both were proposed by S. H. Wei [Wei S.H.1985a, 1985b]. We will not discuss the details of these methods and interested readers can refer to the relevant literature and the latest presentations on LAPW computational procedures for in-depth information.

Furthermore, in the DFT framework, not only the basis-set representation of the wavefunction but also the expressions for the potential and density experienced by the electrons are required. In the original APW method, the potential within the radius cutoff was difficult to handle. However, after linearization is introduced to the calculation, the full potential inside and outside the cutoff can be easily considered in the LAPW framework. As a result, full-potential LAPW is often used directly in modern LAPW programs. The refinement made possible by this framework has significantly contributed to the development of DFT and of LAPW-based *GW* methods [Li X.Z.2014]. For simplicity, we will refer to them collectively as LAPW methods. In addition to Slater and Anderson, significant contributors to the development of LAPW methods have been made by many other researchers, including D.D. Koelling, A. Freeman, H. Krakauer, M. Weinert, S. H. Wei [Wei S.H.1985a, b], and D.J. Singh; their works have played crucial roles in advancing the first-principles study of solid materials. Currently, the Wien2k program package developed by P. Blaha et al. is one of the good representatives of the LAPW methods.

The basis-set elements discussed above have analytic forms with clear physical meaning. The numerical atomic orbital (NAO) basis set aims to optimize the basis set from an alternative direction. It is a nucleus-centered basis similar to the GAO basis set but no longer following the chemistry of hydrogen-like atomic orbitals in its design and optimization. Instead, it gains more freedom numerically; the ability to balance accuracy and efficiency is the main target of optimization. The success of NAO is due to the advancement of many underlying numerical algorithms. Therefore, although the NAO and traditional Gaussian-orbital basis sets share many similarities, we will nonetheless list them separately in Table 4.1 to highlight the difference.

Choosing a basis set is in fact discretizing the wavefunction. A straightforward approach would involve directly discretizing it on the lattice points in real space. In 1989, A.D. Becke suggested performing DFT calculations on real-space lattice points [Becke1989]. Since then, efficient discretization methods for real-space lattice points have been developed, as have other methods that can overcome the limitations of basis sets. However, unlike atomic-orbital basis-set and plane-wave methods, these discretization approaches have not become mainstream in DFT and wavefunction theory calculations, where real-space lattice points are only used to represent the electron-density distribution, and the wavefunctions are generally expanded in the form of traditional basis sets. The real-space lattice-point scheme is more frequently applied in methods such as QMC and time-dependent DFT.

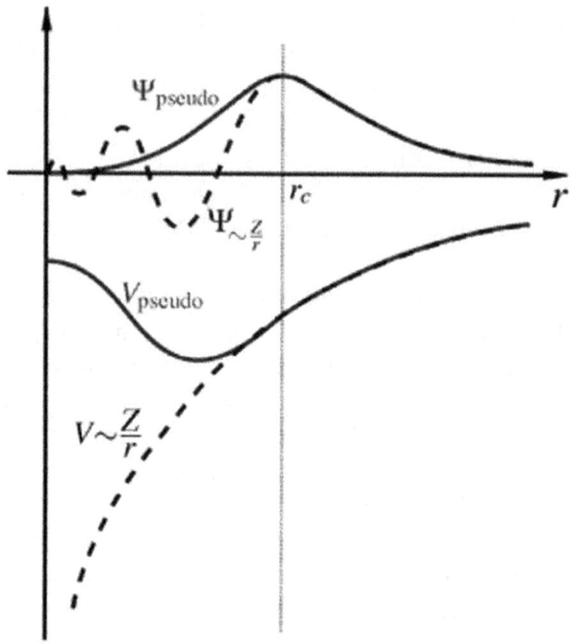

Figure 4.13 Basic idea of the pseudopotential approach. The dashed curves represent the all-electron potential (V) and the all-electron wavefunction (Ψ). The solid curves represent the pseudopotential (V_{pseudo}) and the pseudo-wavefunction (Ψ_{pseudo}). In the physically important region outside the cutoff radius r_c, the realistic and pseudo curves are identical. Z is the nuclear charge and r is the distance between the electron and nucleus.

Another commonly used approximation is the pseudopotential. Its main principle is to consider only certain chemically active valence electrons in the electronic structure calculation, while the electrons of the inner shell are replaced by an effective potential (Figure 4.13). By contrast, a calculation that considers all the electrons in the atoms of the material (but does not include treatment of many-electron correlations) is called an all-electron calculation. The external nuclear potential experienced by the electrons in the all-electron calculation is the Coulomb interaction, which is proportional to the number of charges within the nucleus and inversely proportional to the electron–nucleus distance. It should be noted that the scope of our discussion here is still limited to the single-electron wavefunctions of the DFT framework.

Under the Coulomb interaction, the inner layer electron wavefunction exhibits a shell-layer structure in the radial direction, and the amplitude of the wavefunction fluctuates vertically along that direction. In real materials, the inner-shell electrons are not directly involved in chemical processes such as bond formation, and they generally have much less influence on the material properties compared to the outer-shell electrons. The basic idea of the pseudopotential approach is to include a cutoff radius such that the potential outside the cutoff is consistent with the real Coulomb potential, while that inside it is a pseudopotential that has no real physical meaning. The only requirement for the pseudopotential is that the wavefunction outside the cutoff and the all-electron wavefunction be identical. After introducing the pseudopotential, only the active electrons outside the radius cutoff need to be considered in the electronic-structure calculation. For most materials, the pseudopotential approach can reduce the number of electrons to be simulated by approximately an order of magnitude.

The pseudopotential approach is especially popular for DFT calculations for solid materials, and it is an integral component in the development of relevant computation software. The norm-conserving pseudopotential proposed by D.R. Hamann, M. Schlüter, and C. Chiang in 1979 [Hamann1979] and the ultrasoft pseudopotential proposed by D. Vanderbilt in 1990 [Vanderbilt1990] are among the best-known schemes. The norm-conserving pseudopotential requires that the pseudo-wavefunction within the cutoff radius satisfy a certain charge-number conservation condition; in other words, the integral of the square of the wavefunction within the cutoff is the same as that of the all-electron wavefunction. The ultrasoft pseudopotential removes this requirement to ensure that the wavefunction within the cutoff is as smooth as possible; a smoother pseudo-wavefunction reduces the number of elements needed in the PW basis set and consequently speeds up the calculation.

The projector augmented wave method proposed by P. E. Blöchl borrows ideas from LAPW to further improve the accuracy of plane-wave ultrasoft pseudopotential calculations [Bloch11994]. A projector is constructed using the pseudo-wavefunction; then a linear transformation operator is established using the projector, the pseudo-wavefunction, and the all-electron wavefunction. The all-electron wavefunction can be obtained by applying this linear transformation operator to the pseudo-wavefunction. Since the introduction of the PAW method, the computational accuracy of the plane-wave ultrasoft pseudopotential method has been substantially improved. VASP, one of the most widely used DFT software packages, is a representative example of the PAW method.

Some quantum-chemical software based on wavefunction theory also contains an equivalent of the pseudopotential approach: the effective-core potential (ECP). The most widely used ECP libraries are those developed by the Stuttgart/Cologne team in Germany and the Los Alamos National Laboratory (LANL) in the United States. These are available in a standard file format along with their corresponding atomic-orbital basis sets, and are either directly compatible with most quantum chemistry programs, or are compatible after a simple file conversion. ECP is also a common technique for dealing with relativistic effects in wavefunction theory. As QMC methods have advanced and been applied to quantum chemistry, the demand for a more accurate effective-core potential has grown. In recent years, several research groups have proposed ECP methods that can better handle electron correlations, including BFD-ECP [Burkatzki2007], TN-ECP [Trail2017], and ccECP [Annaberdiyev2018] methods.

Table 4.1 lists some of the most commonly used programs for electronic structure calculation. Given the number of programs that have been developed and the rapid growth of the field, omissions and errors are unavoidable; the information presented here is for the readers' convenience only. We do not provide a description of the actual methods used in every program package, only those that are relevant to this chapter. As most computing programs support pseudopotentials, they are not described separately in the table. However, it should be noted that periodic DFT packages focusing on materials calculations

Table 4.1 *Introduction to commonly used electronic structure calculation procedures. Methods: wavefunction theory (WFT), density functional theory (DFT), time dependent density functional theory (TDDFT), quantum Monte Carlo (QMC), and* GW. Basis sets*: plane wave (PW), linearized augmented plane wave (LAPW), projector-augmented-wave (PAW), Gaussian atomic orbital (GTO), numerical atomic orbital (NAO), and real-space lattice points (Grid).* Boundary conditions*: three-dimensional periodic (3D), arbitrary-dimensional periodic (Any), no periodic (No), and periodic (Yes).* License*: Free (open source or similar permissive free license), Aca. (free for academic use), and Com. (paid license). The language tab lists programming languages used for development.*

Name	Method	Basis set	Boundary condition	Language	License
ABACUS	DFT	NAO	3D	C++	Free
ABINIT	DFT,TDDFT	PW	3D	Fortran	Free
ATOMLY	DFT	N.A.	3D	Other	Aca.
BerkeleyGW	GW	PW	3D	Fortran	Free
CALYPSO/ ATLAS	DFT	N.A.	3D	Fortran	Aca.
CASINO	QMC	Grid	Yes	Fortran	Aca.
CASTEP	DFT,TDDFT	PW	3D	Fortran	Aca.,Com.
CP2K	DFT,WFT,GW	GTO+PW	Any	Fortran	Free
CPMD	DFT	PW	3D	Fortran	Aca.
CRYSTAL	WFT,DFT	GTO	Any	Fortran	Aca.,Com.
Dalton	WFT,DFT	GTO	No	Fortran	Free
DFTK	DFT	PW	Any	Julia	Free
ELK	DFT,TDDFT	LAPW	Yes	Fortran	Free
FHI-aims	DFT,GW	NAO	Any	Fortran	Aca.,Com.
GAMESS	WFT,DFT	GTO	No	Fortran	Aca.,Com.
Gaussian	WFT,DFT	GTO	Any	Fortran	Com.
GPAW	DFT,TDDFT,GW	PW,NAO	Any	Python,C	Free
MOLCAS	WFT,DFT	GTO	No	Fortran,C++	Aca.,Com.
MOLPRO	WFT,DFT	GTO	No	Fortran	Com.
NWChem	WFT,DFT	GTO,PW	Yes	Fortran,C	Free
Octopus	DFT,TDDFT,GW	Grid	Any	Fortran,C	Free
ONETEP	DFT	PW	3D	Fortran	Aca.,Com.
OpenMX	DFT	NAO	Any	C	Free
ORCA	WFT,DFT	GTO	No	C++	Aca.,Com.
PSI	WFT,DFT	GTO	No	C,C++,Python	Free
PWmat	DFT,TDDFT	PW	3D	Fortran	Com.
PyQMC	QMC	Grid	Yes	Python	Free
PySCF	WFT,DFT	GTO	Yes	Python	Free

Table 4.1 (*cont.*)

Name	Method	Basis set	Boundary condition	Language	License
Qbox	DFT	PW	3D	C++	Free
Q-Chem	WFT,DFT	GTO	No	Fortran,C++	Aca.,Com.
Quantum ESPRESSO	DFT,TDDFT	PW	3D	Fortran	Free
QMCPACK	QMC	Grid	Yes	C++	Free
QWalk	QMC	Grid	Yes	Fortran	Free
SALMON	TDDFT	Grid	Yes	Fortran	Free
SIESTA	DFT,TDDFT	NAO	Yes	Fortran	Free
TDAP	TDDFT	NAO	Yes	Fortran	Aca.
TURBOMOLE	WFT,DFT	GTO	Yes	Fortran	Com.
VASP	DFT,GW	PW	3D	Fortran	Com.
WIEN2K	DFT,GW	LAPW	3D	Fortran,C	Com.
Yambo	TDDFT,GW	PW	3D	Fortran	Free

generally support the norm-conserving pseudopotential and the ultrasoft pseudopotential, whereas quantum-chemistry packages based on wavefunction theory generally support ECP. There are, however, some exceptions. The Wien2k package, for example, focuses on all-electron calculations. It should be emphasized that these computational programs are categorized in the table based on the previous methodological discussions in this section. In real computation, the contents are often mixed. The user must try to understand the nature of each method in detail, something that requires constant work and practice. After the introduction of the full-quantum theoretical approach in Chapter 5, how full-quantum computational methods can be developed from existing electronic structure computational programs will be discussed more clearly.

We have devoted this chapter to methods of calculating the ground and excited states of electronic structure, as well as to the various computational packages commonly used for implementing them. As emphasized at the beginning of the chapter, methods to simulate full quantum effects require accurate results from electronic structure calculations. Therefore, continuous improvement of methods for calculating electronic structure is essential for the study of full quantum effects in condensed matter systems. This information will be repeatedly emphasized throughout the book.

5

Theory for Full Quantum Effects

A key goal of condensed matter physics is to describe the quantum nature of electrons and nuclei as well as their many-body interaction of a real system in an accurate manner. The Born–Oppenheimer approximation, proposed in 1927, along with its associated ball-and-stick model, provides the means to address this challenge adiabatically. In recent years, with the experimental observation approaching the limit and theoretical simulation targeting the real system, the study of full quantum effects of condensed matter beyond the narrow understanding of Born–Oppenheimer approximation has been gradually put on the agenda.

In Chapter 4, we introduced various first-principles electronic-structure calculation methods and the software packages that implement them. This laid a solid foundation for the development of full quantum theory. In this book, our focus is not on further developing such traditional first-principles methods or discussing the challenges they are facing now, but on going beyond the narrow understanding of the Born–Oppenheimer approximation. Therefore, we will assume that these first-principles approaches are acceptable and focus on the full quantum theory and its associated numerical methods.

There are currently three main approaches to computing full quantum effects: Adding new algorithms to existing first-principles electronic-structure calculation packages; developing a higher-level package that handles full quantum effects using the Born–Oppenheimer potential energy surface given by an existing program; and developing a package that can handle both the calculation of electronic structure and full quantum effects. At present, the theoretical calculation of full quantum effects is far from being mature, and each of these three approaches has its advantages and disadvantages. As a result, these approaches will continue to coexist and evolve together in the study of the full quantum effects. Regardless of the approach taken, understanding the existing methods for electronic-structure calculations will definitely facilitate the future development of full quantum theory and full quantum computational methods.

As mentioned earlier, a rigorous theoretical framework for evaluating full quantum effects is provided by the Born–Huang expansion (Eq. (4.1)). However, in the case of real materials, it is still impractical to use this expansion directly. Therefore, further simplification is necessary. This can be carried out in two ways. First, we can assume that the electronic structure corresponds to a single Born–Oppenheimer potential energy surface (in most cases, that of the ground state), and focus on the thermodynamic and dynamical properties of the nucleus as a quantum particle, that is, on nuclear quantum effects. Alternatively, we can assume that the nuclear motion is described by Newtonian mechanics and focus on the quantum coupling of the evolution of the nuclei between different potential energy surfaces and electronic transitions between different eigenstates, that is, on nonadiabatic dynamics.

Section 5.1 will be devoted to methods of simulating nuclear quantum effects. In Section 5.2, we will show, as an example, how the first-principles calculation of electron–phonon coupling is realized in conventional condensed matter theory and briefly discuss its connection with the Born–Huang expansion. In this section, we will temporarily ignore the nonadiabatic processes of nuclei and electrons and focus on the electron–phonon coupling associated with nuclear quantum effects. Discussion of the real-time coupled electron–nucleus evolution using nonadiabatic simulation methods is saved for Section 5.3, where we return to the framework of the Born–Huang expansion. Finally, Section 5.4 will introduce some recent theoretical attempts to combine nonadiabatic effects with nuclear quantum effects and illustrate with an application example.

5.1 Simulation of Nuclear Quantum Effects

Thus far, our discussions on the electronic eigenstates have been limited to a specific, classical nuclear configuration. Treating this configuration as a variable, we can use these solutions to obtain the Born–Oppenheimer potential energy surfaces (which can include both the ground and excited states, although most published works are restricted to the former) under the ball-and-stick model. However, describing the nucleus within the framework of classical statistical physics [Nose1984a, b, 1986, Hoover1985, 1986, Martyna1992] results in an imperfect understanding of many-body systems. The nuclear motion depends on classical mechanics and cannot lead to quantum-mechanical electron transitions between different eigenstates.

This problem arose constantly from the late 1920s when quantum mechanics was established until the mid 1980s, but no suitable solution was found. The computational power required to solve the electronic structure self-consistently for each step in the evolution of a specific nuclear configuration exceeded what was available at the time. Then, in 1985, the Car–Parrinello molecular-dynamics method was introduced [Car1985]. This method, which relies on the degrees of freedom provided by a series of plane waves with a small fictitious mass of electrons, allows the time evolution of the coupled electron–nucleus system to be described in terms of a classical extended Lagrangian. In this way, the electronic structure can remain quasi-adiabatically close to the Born–Oppenheimer potential energy surface. The goal is to convert the quantum-mechanical problem into a classical mechanical problem, with the evolution process depending on the extended Lagrangian. A self-consistent electronic-structure calculation is required regularly every 10–100 timesteps to ensure that this evolution remains near the Born–Oppenheimer potential energy surface. In this way, the electronic structure can evolve with the nuclear motion almost adiabatically. The Car–Parrinello method is, therefore, able to describe phenomena and reaction processes, such as chemical bond breaking, that cannot be explained by conventional force-field methods.

From the late 1980s to the early 2000s, as the power of computer clusters increased and self-consistent electronic-structure algorithms were developed, the adiabatic Born–Oppenheimer molecular-dynamics method gradually replaced the Car–Parrinello method. Both approaches were further combined with other molecular dynamics methods, such as enhanced sampling and thermodynamic integration [Roux1995, Bartels1998, Laio2002, Gao Y.Q.2008], enabling adiabatic solution of the first-principles molecular simulation problem under the ball-and-stick model of electronic states. In spite of these developments, the theory of molecular dynamics still did not extend beyond the adiabatic approximation.

The most straightforward approach to the study of nuclear quantum effects is based on the Born–Huang expansion, which first neglects the electron transitions between different eigenstates and then describes the nuclear and electronic states simultaneously using the wavefunction method. In practice, however, this approach is not applicable to most condensed matter problems (except for a few small molecule reactions of gas systems [Zhang D.H.1994, Xiao C.2011]) because of its computational cost. For macromolecular or condensed matter systems, researchers must still resort to the path-integral formulation of quantum mechanics [Chandler1981, Berne1986, Pollock1987, Ceperley1995, Tuckerman1996, Marx1996, Zhang Q.F.2008, Li X.Z.2014]. This formulation utilizes the Green's function (propagator) concept discussed in many quantum-mechanics textbooks. By exploiting the similarity between the propagator and the density matrix and representing the latter in terms of path integrals, this treatment gives rise to the most commonly used numerical methods for nuclear quantum effects: the first-principles PIMD and PIMC methods. It also offers a rigorous theoretical framework for investigating nuclear quantum effects at the statistical level. Although there are still a number of technical difficulties in studying the dynamical (as opposed to statistical) properties of nuclear quantum effects in polyatomic systems, there is no doubt that this is a very promising (but challenging) area of research.

This section is divided into three subsections. The first subsection briefly describes the basic concepts of path integral, the second subsection discusses the simulation of nuclear quantum effects at the statistical level, and the final subsection focuses on the current status of simulation methods for dynamical properties. This book also briefly introduces several PIMD programs and software packages based on first-principle calculations in Appendix A. It focuses on the methods we developed, so as to facilitate readers to learn and use. Specifically, the first-principles path integral molecular dynamics method we developed earlier to deal with full quantum effects improves the efficient parallel simulation function of k-space sampling in the electronic structure calculation of periodic lattice systems in CPMD program. As a result, it can be applied to the calculation and simulation of condensed matter systems. The latest code adds the PIMD function to the VASP software, and the developed software code is fully compatible with VASP in terms of realizing the self-consistent calculation of the first principle electronic structure, evaluating the system energy and atomic force, etc.

5.1.1 Introduction to the Path-Integral Method

Quantum theory represents a radically counterintuitive way of thinking about the world. To develop a deeper understanding of quantum mechanics, Feynman proposed the path integral method, providing us with a fresh perspective [Feynman1949, 1953a, b, c, 1965]. The path-integral picture offers an intuitive illustration of the difference between the quantum and the classical world and reveals how the nature of quantum mechanics can be related to the laws of classical physics, such as $F = ma$.

In the classical world, the motion of an object is described deterministically. For example, if I hold a sandbag in my hand and throw it, then according to Newton's equations, the sandbag will follow a specific orbit of motion. When the sandbag reaches a certain point on its trajectory, you will be hit if you are standing there at that time; otherwise, you are completely safe. Whether an event occurs is determined by causality, and the probability of the event is either zero or one.

In the world of quantum mechanics, however, everything changes. Throwing a sandbag is analogous to the creation of a particle. After its creation, the particle can be annihilated at any point in space at any later time. Naturally, there must still be a causal relationship between the creation event a and annihilation event b. This relationship is described by a propagator K. In Schrödinger's formulation of quantum mechanics, the propagator has a concise analytic expression:

$$K(x_\text{b},t_\text{b};x_\text{a},t_\text{a}) = \sum_j \phi_j(x_\text{b})\phi_j^*(x_\text{a})e^{-\left(\frac{i}{\hbar}\right)E_j(t_\text{b}-t_\text{a})}. \tag{5.1}$$

Here, j is summed over all eigenstates ϕ_j. When event a occurs, the chance of event b occurring is equal to the square of the modulus of this propagator. The probabilistic nature of events is perhaps one of the most striking features of quantum mechanics. Quantum particles no longer move from one point to another along a defined path; in fact, we have to consider all possible paths and sum up the probabilities in the quantum world (because even if the same experiment is repeated multiple times, the position of the quantum particle being detected may be different each time). This method of summing over all possible paths is known as the Feynman path-integral method. It bridges the gap between classical and quantum physics, showing us how a classical sandbag following a well-defined path can be connected to the idea of total path summation in quantum mechanics.

However, we want to obtain the propagator without having to calculate all the eigenstates. This is where the concept of path integral comes into play. In the path-integral formulation, we can derive the propagator without the wavefunction. Instead, we exhaustively consider all paths between two events. Each path will contribute to the propagation, and the magnitude of its contribution is determined by its Euclidean action, which is the integral of the Lagrangian on the corresponding path. Therefore, the propagator becomes

$$K(x_\text{b},t_\text{b};x_\text{a},t_\text{a}) = \lim_{\varepsilon\to 0}\frac{1}{A}\iint\dots\int e^{\left(\frac{i}{\hbar}\right)S[x_\text{b},t_\text{b};x_\text{a},t_\text{a}]}\frac{\mathrm{d}x_1}{A}\frac{\mathrm{d}x_2}{A}\dots\frac{\mathrm{d}x_{P-1}}{A}, \tag{5.2}$$

where

$$S[x_b,t_b;x_a,t_a]=\int_{t_a}^{t_b}L(\dot{x},x,t)\,dt$$
$$=\left[\frac{m}{2}\left(\frac{x_1-x_a}{t_1-t_a}\right)^2-\frac{1}{2}\left(V(x_1)+V(x_a)\right)\right](t_1-t_a)$$
$$+\sum_{i=2}^{P-1}\left[\frac{m}{2}\left(\frac{x_i-x_{i-1}}{t_i-t_{i-1}}\right)^2-\frac{1}{2}\left(V(x_i)+V(x_{i-1})\right)\right](t_i-t_{i-1})$$
$$+\left[\frac{m}{2}\left(\frac{x_b-x_{P-1}}{t_b-t_{P-1}}\right)^2-\frac{1}{2}\left(V(x_b)+V(x_{P-1})\right)\right](t_b-t_{P-1}). \qquad (5.3)$$

In Eq. (5.2), $A=\left[2\pi i\hbar(t_b-t_a)/(Pm)\right]^{1/2}$ is a normalization factor. In Eqs (5.2) and (5.3), the path is divided into P segments. This technique is generally referred to as Trotterization.

Eq. (5.2) may not be intuitive, but we can attempt to understand it with the help of the one-dimensional system shown in Figure 5.1. Events a and b are marked as fixed points in the diagram. We divide t_b-t_a into P segments. One point x is chosen in each segment; connecting these points gives us a path. The action along this path is determined by Eq. (5.3), while Eq. (5.2) samples the coordinates of $t_1 \ldots t_{P-1}$ on the x -axis so that all

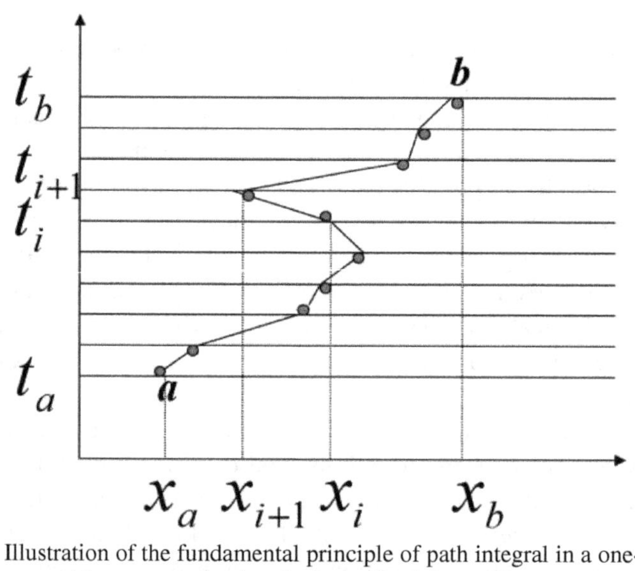

Figure 5.1 Illustration of the fundamental principle of path integral in a one-dimensional system, where a and b represent the already determined times. The propagator between the two events is determined by the sum of the contributions from all paths from a to b. (One such path is shown in the figure.) By dividing the time difference between the two events into segments and moving the position across the entire x-axis from left to right, all paths between the two time points can be sampled. Based on this, a propagator between a and b can be constructed.

paths are taken into account. In this way, we can obtain the propagator between events *a* and *b* numerically using a path integral. With sufficient sampling, the results should be identical to those obtained using the wavefunction method in Eq. (5.1).

5.1.2 Nuclear Quantum Effects on Statistical Properties

Equation 5.1 is equivalent to Eqs (5.2)–(5.3). The propagator of the quantum system is generally considered irrelevant when we want to simulate the statistical properties (i.e., when the system dynamics are not of interest). In this case, the physical quantity of interest is the density matrix of the quantum system at finite temperature. In the Schrödinger representation of quantum mechanics, the analytical expression for the density matrix is given by

$$\rho(x_b, x_a; \beta) = \sum_j \phi_j(x_b)\phi_j^*(x_a)e^{-\beta E_j}. \tag{5.4}$$

Here, $\beta = 1/(k_B T)$, where k_B is the Boltzmann constant, and T is the temperature. As in the case of the propagator, knowledge of the system wavefunctions is required to determine the density matrix. For large molecules or condensed matter systems, this is clearly impractical. Therefore, we need to take a different way to avoid wavefunction calculation.

The only difference between Eqs (5.4) and (5.1) is the exponent of the exponential function. For the propagator, the exponent term is an imaginary number depending on the time difference. For the density matrix, on the other hand, the exponential term is a real number depending on β. Therefore, we can treat $1/(k_B T)$ as an imaginary-time interval

$$-i\hbar\beta = t_b - t_a. \tag{5.5}$$

Using the Suzuki-Trotter expansion to truncate the imaginary time into P segments, the density matrix can be expressed in terms of an action analogous to that in Eq. (5.3) as follows:

$$\begin{aligned} S[x_b, x_a; -i\hbar\beta] = &-\left[\frac{mP^2}{2\beta^2\hbar^2}(x_1 - x_a)^2 + \frac{1}{2}\big(V(x_1) + V(x_a)\big)\right]\frac{-i\hbar\beta}{P} \\ &-\sum_{i=2}^{P-1}\left[\frac{mP^2}{2\beta^2\hbar^2}(x_i - x_{i-1})^2 + \frac{1}{2}\big(V(x_i) + V(x_{i-1})\big)\right]\frac{-i\hbar\beta}{P} \\ &-\left[\frac{mP^2}{2\beta^2\hbar^2}(x_b - x_{P-1})^2 + \frac{1}{2}\big(V(x_b) + V(x_{P-1})\big)\right]\frac{-i\hbar\beta}{P}. \end{aligned} \tag{5.6}$$

This is equivalent to

$$\begin{aligned} \frac{i}{\hbar}S[x_b, x_a; -i\hbar\beta] = &-\left[\frac{mP}{2\beta\hbar^2}(x_1 - x_a)^2 + \frac{\beta}{2P}\big(V(x_1) + V(x_a)\big)\right] \\ &-\sum_{i=2}^{P-1}\left[\frac{mP}{2\beta\hbar^2}(x_i - x_{i-1})^2 + \frac{\beta}{2P}\big(V(x_i) + V(x_{i-1})\big)\right] \\ &-\left[\frac{mP}{2\beta\hbar^2}(x_b - x_{P-1})^2 + \frac{\beta}{2P}\big(V(x_b) + V(x_{P-1})\big)\right]. \end{aligned} \tag{5.7}$$

If x_a is re-labeled as x_0 and x_b as x_P, Eq. (5.7) can be further reduced to

$$\frac{i}{\hbar}S[x_b,x_a;-i\hbar\beta]=-\beta\sum_{i=1}^{P}\left[\frac{mP}{2\beta^2\hbar^2}(x_i-x_{i-1})^2+\frac{1}{2P}\left(V(x_i)+V(x_{i-1})\right)\right]. \quad (5.8)$$

Then, similar to Eq. (5.2), the density matrix is equivalent to

$$\rho(x_b,x_a;\beta) = \lim_{\varepsilon\to 0}\frac{1}{A}\iint\ldots\int e^{-\beta\sum_{i=1}^{P}\left[\frac{mP}{2\beta^2\hbar^2}(x_i-x_{i-1})^2+\frac{1}{2P}(V(x_i)+V(x_{i-1}))\right]}\frac{dx_1}{A}\ldots\ldots\frac{dx_{P-1}}{A}, \quad (5.9)$$

where

$$A=\left[\frac{2\pi\beta\hbar^2}{Pm}\right]^{1/2}. \quad (5.10)$$

Here, the density matrix can be expressed as a path integral.

Equation 5.9 may appear to be complex, but it is highly suitable for molecular dynamics and Monte Carlo sampling. This is because we can associate the density matrix of a quantum system to a classical ring polymer with potential energy satisfying

$$V_{eff}=\sum_{i=1}^{P}\left[\frac{mP}{2\beta^2\hbar^2}(x_i-x_{i-1})^2+\frac{1}{2P}\left(V(x_i)+V(x_{i-1})\right)\right]. \quad (5.11)$$

The physical meaning of this expression is simple: For a polyatomic system where the inter-atomic interactions are determined by $V(x_i)$, we can generate several images (or beads) along its path. Neighboring images of the same atom are connected by springs with spring constant $\sqrt{mP}/(\beta\hbar)$, as shown in Figure 5.2. When the nuclear mass m is large and the temperature is high (i.e., β is low), the spring exerts a strong force; this drags all the images corresponding to the same atom to the same point in real space. Therefore, sampling the polymer and sampling the many-atom system will yield the same result. In

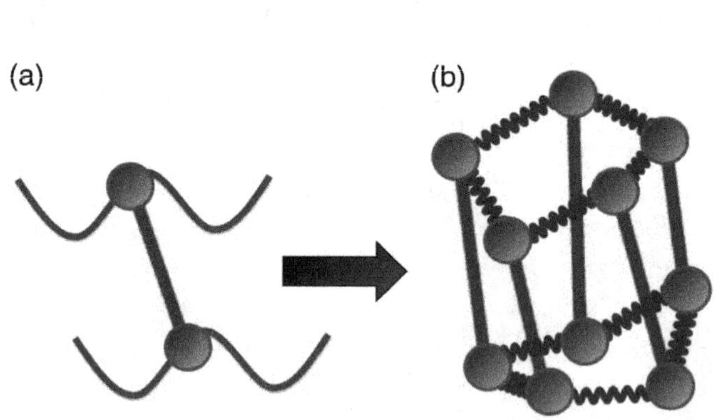

Figure 5.2 Schematic of path-integral statistical calculation using a hydrogen molecule (H_2) as an example. In the original wavefunction method, nuclear wavefunctions must be calculated for each atom, as depicted in (a) (gray curves). However, the wavefunction calculation can be avoided by constructing a ring polymer and letting its interaction potential satisfy Eq. (5.11).

other words, the nucleus can be regarded as a classical particle, and the system reaches its classical limit. However, when m is small and the temperature is low, the results obtained by sampling the ring polymer using molecular dynamics or Monte Carlo methods will be very different from the results obtained by sampling the classical many-atom system. This difference represents the nuclear quantum effect on the statistical properties of this system.

This formulation was proposed by R. P. Feynman in the late 1940s and early 1950s [Feynman1949, 1953a, b, c]. In the classic textbook *Quantum Mechanics and Path Integrals* that Feynman and A. R. Hibbs co-authored in the mid 1960s, the method was summarized in a very systematic way [Feynman1965]. However, researchers only began combining the path-integral approach with molecular simulations in the 1970–1980s, when molecular dynamics and Monte Carlo sampling methods became more widely used in physics and chemistry.

More specifically, in the early 1970s, W. H. Miller proposed the instanton method using the path-integral concept to describe nuclear tunneling in molecular reactions [Miller1974]. During the 1980s, D. Chandler, P. G. Wolynes, B. Berne, and others combined the imaginary-time path-integral method at finite temperature with molecular dynamics [Chandler1981, Berne1986] to simulate nuclear quantum effects on the statistical properties of real systems. D. M. Ceperley and E. L. Pollock combined numerical path integral with Monte Carlo methods and performed atomic-level simulations of superfluidity in liquid ^{4}He at low temperatures [Pollock1987, Ceperley1995]. Their work has provided a solid foundation for later studies of nuclear quantum effects based on molecular simulation.

It is worth noting that theoretical simulations in that era relied heavily on the force-field description of the interaction potential between atomic nuclei. One problem with this is that, when the change in electronic structure caused by the change of nuclear configuration is drastic (i.e. when chemical bond breaking occurs), the description of the forces acting on the nucleus and the total static energy of the polyatomic system in the configuration can be highly inaccurate. Therefore, the application of PIMD and PIMC simulations at the statistical level remained very limited.

In the mid 1990s, M. E. Tuckerman, D. Marx, M. Parrinello, M. L. Klein, and others combined the PIMD with the first-principles Car–Parrinello molecular-dynamics method [Tuckerman1996, Marx1996]. This progress was accompanied by equally important advances in sampling techniques. The vibration modes of the fictitious ring-polymer Hamiltonian in the path-integral representation have frequencies much higher than the eigenfrequencies of the system; hence, it is often necessary to reduce the sampling step size to ensure accurate simulation. For example, when simulating the quantum properties of protons in water at room temperature, 32 nodes of the ring polymer are generally required for the statistics of the nuclear quantum properties to converge. In this case, the step size of the molecular-dynamics simulation must be sufficiently small to characterize the vibrations between the different images. Using the normal-mode or staging transformations, one can separate the physical vibrational degrees of freedom from the high-frequency vibrational modes of the imaginary-time ring-polymer phase space [Pollock1984, Tuckerman1996]. With the transformed coordinates, the decoupled high-frequency component can be independently connected to different heat baths to sample every state. In this way, a path-integral molecular-dynamics simulation can use the same step size as a classical molecular dynamics simulation. In collaboration with G. Wahnstrom and S. W. Gao, E. G. Wang's group et al. further optimized the full quantum Car–Parrinello molecular dynamics software in 2005. They refined the parallel k-space sampling simulation for electronic-structure calculations in periodic lattice systems, thus enabling full quantum simulations in condensed matter.

In the twenty-first century, the Car–Parrinello method was gradually replaced by the Born–Oppenheimer method, and researchers began to sample systems directly using Born–Oppenheimer molecular dynamics or Monte Carlo sampling [Zhang Q.F.2008, Li X.Z.2010,2011, Ceriotti2011]. Meanwhile, techniques for calculating electronic structure evolved from being based solely on DFT to using a variety of calculation schemes, including the QMC method [Morales2010]. Also, at the beginning of this century, a ring-polymer contraction scheme was proposed by I. R. Craig and D. E. Manolopoulos [Craig I.R.2004, 2005a, b]. The main principle of this method is to find a ring-polymer representation with fewer nodes ($P' < P$). The time-consuming first-principles potential energy surface calculation only occurs on the P' ring polymer, while the P ring-polymer potential energy surface can be obtained by transforming the P' ring polymer to the normal-mode representation. This transformation can be performed in a computationally tractable reference system.

In 2009, M. Ceriotti et al. introduced colored-noise thermostat into PIMD methods [Ceriotti2009a, b]. By automatically designing a frequency-dependent effective temperature for the external heat bath, this method employs fewer ring polymers. Recently, these techniques have been further improved by combining them with enhanced sampling methods, such as the two-phase method and thermodynamic integration. Typical works employing this approach include those of A. von Lilienfeld's group at University of Toroto [Pérez2011] and E.G. Wang's group at Peking University [Chen J.2013, Feng Y.X.2015]. These studies advanced first-principles PIMD and PIMC methods, laying a rigorous theoretical foundation for studying nuclear quantum effects of condensed matter at the statistical level.

5.1.3 Nuclear Quantum Effects on Dynamic Properties

Nuclear quantum effects influence not only the statistical but also the dynamical properties of condensed matter systems, playing a crucial role in many physical, chemical, material, and biological problems. For example, key steps in some biological reactions are thought to be associated with electron and proton coupling transfer [Marcus1993, Huynh2007]. Similarly, intra- and intermolecular energy conversions in photophysical and photochemical processes play an indispensable role in new-energy materials [Berne1998]. There is no doubt that highly correlated electron and nuclear degrees of freedom at the quantum-mechanical level will become increasingly important in the material, life, energy, and environmental sciences in the future. The theoretical foundation for studying all these problems is quantum dynamics, which will be introduced in this section.

Electron dynamics will be discussed in Section 5.3, where we will cover nonadiabatic molecular simulations. Here we will focus only on dynamical problems related to nuclear quantum effects of atoms. Relevant numerical methods are divided into three categories: (1) semi-classical methods; (2) centroid molecular dynamics (CMD) and ring-polymer molecular dynamics (RPMD) methods; and (3) methods that obtain kinetic information from time-independent techniques. The literature frequently refers to the "hybrid quantum–classical" method, which treats some degrees of freedom quantum mechanically and others classically. Although this does not specifically imply that the electron degree of freedom is treated quantum mechanically and the nuclear degree of freedom is treated classically, such a treatment is often the case in practice. The nonadiabatic dynamics methods introduced in Section 5.3, such as surface hopping method, the mean-field method, and Ehrenfest dynamics method, all belong to this category. For this reason, the discussion of the hybrid quantum–classical dynamics will be saved for Section 5.3. This is perhaps one of the differences between this book and many others, which often treat hybrid dynamics as a subcase of quantum dynamics. The term "quantum dynamics" in this book, however, specifically refers to those methods that describe atomic nuclei quantum mechanically. Dynamic methods that study the quantum behavior of electrons are introduced in the nonadiabatic section.

5.1.3.1 Semiclassical Methods

In semiclassical methods, high order terms $\hbar$ are ignored in calculations. One of the earliest semiclassical theories was the WKB approximation – the three letters are the initials of G. Wentzel, H. A. Kramers, and L. Brillouin [Wentzel1926, Kramers1926, Brillouin1926]. Another was the van Vleck propagator method [van Vleck1928]. Both were proposed in the 1920s, early in the history of quantum mechanics. Developing a profound understanding of these methods not only helps solve specific problems on the research frontiers of contemporary physical science but also improves our knowledge of the fundamental concepts in the transition from classical physics to quantum mechanics.

Later, in the early 1970s, Professor W. H. Miller of the University of California, Berkeley, studied gas reactions in the stationary-phase approximation (SPA). He developed the classical S-matrix theory and the instanton method by neglecting the higher-order terms of $\hbar$ when solving the path-integral propagator [Miller1970, 1974]. The basic principle of the semiclassical initial-value representation (SC-IVR), which developed rapidly later, was also included in his 1970 article on semiclassical scattering theory [Miller1970, 2001,Herman1984, 1997]. IVR is one of the semiclassical methods with the promise of finding broad applications in condensed matter systems.

To discuss all these approaches comprehensively is clearly beyond the scope of this book. To provide the reader a clearer understanding of the key points, we will use IVR as an example and discuss its basic concepts in detail. In subsection 5.1.3.2, we will briefly introduce the principles of centroid and RPMD methods. Semiclassical ideas are also present in the instanton method [Miller1974]. However, because of its use of imaginary time and its connection with transition-state theory, the instanton method will instead be used as an example of how to derive dynamical information from time-independent methods.

When discussing dynamics, we are naturally concerned with time-dependent properties. We begin by recalling the quantum correlation function introduced in Chapter 1:

$$\mathrm{C_{AB}(t)} = \frac{1}{Z}\mathrm{Tr}\left[\hat{A}^{\beta} e^{i\hat{H}\ t/\hbar}\ \hat{B} e^{-i\hat{H}t/\hbar}\right], \tag{5.12}$$

where $\hat{H}$ is the Hamiltonian of the system, $\hat{A}$ and $\hat{B}$ are the operators of the physical observables, $Z = \mathrm{Tr}[e^{-\beta\hat{H}}]$ is the partition function, and $\hat{A}^{\beta} = e^{-\beta\hat{H}}\hat{A}$. For different dynamical properties, different quantum correlation functions are needed. For instance, the chemical reaction rate is related to the flow correlation function, and the IR spectrum is related to the dipole-moment correlation function. Correlation-function theory was developed and systematized by J. G. Kirkwood, M. S. Green, H. B. Callen, R. F. Greene, R. Kubo, R. Zwanzig, and others in the 1950s [Kirkwood1946, Green1952, 1954, Callen1951, 1952a, b, Kubo1957a, b, Zwanzig1965] and is now discussed in most statistical-physics textbooks. However, it remains challenging to calculate the quantum correlation function of a real system and use it to describe observables in experiments.

Equation 5.12 contains two terms that contribute to nuclear quantum effects: the Boltzmann distribution operator $e^{-\beta\hat{H}}$, which is determined by quantum statistical effects, as mentioned in subsection 5.1.2, and the time-evolution operator $e^{-i\hat{H}t/\hbar}$, which contains quantum-dynamical information.

IVR divides the time-dependent nuclear wavefunction into amplitude and phase components [Bohm1952]:

$$\phi(q_1,q_2,\cdots,q_D,t) = A(q_1,q_2,\cdots,q_D,t)\, e^{iR(q_1,q_2,\cdots,q_D,t)/\hbar}, \tag{5.13}$$

where D is the system dimensionality, and each component is evaluated separately. The time-dependent Schrödinger equation satisfied by the wavefunction can be expressed as:

$$\left(i\hbar\frac{\partial}{\partial t} + \frac{\hbar^2}{2m}\sum_{i=1}^{D}\frac{\partial^2}{\partial q_i^2} - V(q_1,q_2,\cdots,q_D)\right)\phi(q_1,q_2,\cdots,q_D,t) = 0. \tag{5.14}$$

Inserting Eq. (5.13) into (5.14) yields a complex equation, the real and imaginary parts of which are separately zero. In the case when $A(q_1,q_2,\cdots,q_D,t) \neq 0$, we have

$$\begin{aligned}&\frac{\partial R(q_1,q_2,\cdots,q_D,t)}{\partial t} + \frac{1}{2m}\sum_{i=1}^{D}\left(\frac{\partial R(q_1,q_2,\cdots,q_D,t)}{\partial q_i}\right)^2 + V(q_1,q_2,\cdots,q_D)\\ &-\frac{\hbar^2}{2m}\frac{1}{A(q_1,q_2,\cdots,q_D,t)}\sum_{i=1}^{D}\frac{\partial^2 A(q_1,q_2,\cdots,q_D,t)}{\partial q_i^2} = 0\end{aligned} \tag{5.15}$$

and

$$\begin{aligned}&\frac{\partial A(q_1,q_2,\cdots,q_D,t)}{\partial t} + \frac{1}{m}\sum_{i=1}^{D}\frac{\partial A(q_1,q_2,\cdots,q_D,t)}{\partial q_i}\frac{\partial R(q_1,q_2,\cdots,q_D,t)}{\partial q_i}\\ &+\frac{1}{2m}A(q_1,q_2,\cdots,q_D,t)\sum_{i=1}^{D}\frac{\partial^2 R(q_1,q_2,\cdots,q_D,t)}{\partial q_i^2} = 0.\end{aligned} \tag{5.16}$$

In this way, the nonlinear differential equation of the complex wavefunction is transformed into two nonlinear differential equations for the phase and amplitude component.

Under the semiclassical approximation, quadratic terms in $\hbar$ approach zero. Therefore, the last term on the left side of Eq. (5.15) (the coupling between the phase and amplitude functions) can be neglected. In this case, we will be dealing with the differential equation of an independent real function.

This transformation from a complex to a real function and the decoupling of the phase and amplitude functions considerably simplifies the calculation. Nevertheless, from a quantum-mechanical perspective, it is necessary to explain the physical meaning of $\hbar$ approaching zero: This physical idea is consistent with the WKB method mentioned in many textbooks when describing the 1D quantum tunneling problem. The underlying assumption here is that the amplitude $A(q_1,q_2,\cdots,q_D,t)$ of the wavefunction varies much less drastically in space than the phase component $R(q_1,q_2,\cdots,q_D,t)/\hbar$.

After the coupling term is neglected, Eq. (5.15) becomes

$$\frac{\partial R(q_1,q_2,\cdots,q_D,t)}{\partial t}+\hat{H}\left(q_1,\cdots,q_D,\frac{\partial R}{\partial q_1},\cdots,\frac{\partial R}{\partial q_D}\right)=0, \tag{5.17}$$

where

$$\hat{H}\left(q_1,\cdots,q_D,\frac{\partial R}{\partial q_1},\cdots,\frac{\partial R}{\partial q_D}\right)=\frac{1}{2m}\sum_{i=1}^{D}\left(\frac{\partial R(q_1,q_2,\cdots,q_D,t)}{\partial q_i}\right)^2+V(q_1,q_2,\cdots,q_D). \tag{5.18}$$

This has the typical form of a Hamilton–Jacobi equation.

By elementary calculus,

$$\mathrm{d}R(q_1,q_2,\cdots,q_D,t)=\frac{\partial R(q_1,q_2,\cdots,q_D,t)}{\partial t}\mathrm{d}t+\sum_{i=1}^{D}\frac{\partial R(q_1,q_2,\cdots,q_D,t)}{\partial q_i}\mathrm{d}q_i. \tag{5.19}$$

Therefore, after dividing both sides of Eq. (5.19) by dt and inserting the result into Eq. (5.17), we have

$$\frac{\mathrm{d}R(q_1,q_2,\cdots,q_D,t)}{\mathrm{d}t}=-\hat{H}\left(q_1,\cdots,q_D,\frac{\partial R}{\partial q_1},\cdots,\frac{\partial R}{\partial q_D}\right)+\sum_{i=1}^{D}\frac{\partial R(q_1,q_2,\cdots,q_D,t)}{\partial q_i}\frac{\mathrm{d}q_i}{\mathrm{d}t}. \tag{5.20}$$

Then, we introduce a fictitious momentum, such that

$$p_i=\frac{\partial R(q_1,q_2,\cdots,q_D,t)}{\partial q_i}, \tag{5.21}$$

the time derivative of which satisfies

$$\begin{aligned}\frac{\mathrm{d}p_i}{\mathrm{d}t}&=\frac{\mathrm{d}\left[\dfrac{\partial R(q_1,q_2,\cdots,q_D,t)}{\partial q_i}\right]}{dt}\\&=\frac{\partial^2 R(q_1,q_2,\cdots,q_D,t)}{\partial q_i\partial t}+\sum_{j=1}^{D}\frac{\partial^2 R(q_1,q_2,\cdots,q_D,t)}{\partial q_i\partial q_j}\frac{\mathrm{d}q_j}{\mathrm{d}t}.\end{aligned} \tag{5.22}$$

Using Eq. (5.17), we have

$$\frac{\mathrm{d}p_i}{\mathrm{d}t} = -\left[\frac{\partial \hat{H}(q_1,\cdots,q_D,p_1,\cdots,p_D)}{\partial q_i} + \sum_{j=1}^{D} \frac{\partial \hat{H}(q_1,\cdots,q_D,p_1,\cdots,p_D)}{\partial p_j}\frac{\partial p_j}{\partial q_i}\right] + \sum_{j=1}^{D}\frac{\partial p_i}{\partial q_j}\frac{\mathrm{d}q_j}{\mathrm{d}t}. \tag{5.23}$$

This is equivalent to

$$\frac{\mathrm{d}p_i}{\mathrm{d}t} + \frac{\partial \hat{H}(q_1,\cdots,q_D,p_1,\cdots,p_D)}{\partial q_i} = \sum_{j=1}^{D}\left[\frac{\mathrm{d}q_j}{\mathrm{d}t} - \frac{\partial \hat{H}(q_1,\cdots,q_D,p_1,\cdots,p_D)}{\partial p_j}\right]\frac{\partial p_j}{\partial q_i}. \tag{5.24}$$

If the right-hand side of this equation is set to zero, we have two mutually coupled dynamical evolution equations,

$$\frac{\mathrm{d}p_i}{\mathrm{d}t} = -\frac{\partial \hat{H}(q_1,\cdots,q_D,p_1,\cdots,p_D)}{\partial q_i} \tag{5.25}$$

and

$$\frac{\mathrm{d}q_j}{\mathrm{d}t} = \frac{\partial \hat{H}(q_1,\cdots,q_D,p_1,\cdots,p_D)}{\partial p_j}. \tag{5.26}$$

These equations can be solved for p_i, q_i, and $\hat{H}(q_1,\cdots,q_D,p_1,\cdots,p_D)$, after which Eq. (5.20) can be solved for $R(q_1,q_2,\cdots,q_D,t)$ in the form of a path integral:

$$\begin{aligned} R(q_1,q_2,\cdots,q_D,t) &= R\left(q_1',q_2',\cdots,q_D',0\right) \\ &+ \int_0^t \left[\sum_{i=1}^{D} p_i(\tau)\dot{q}_i(\tau) - \hat{H}\ \left(q_1(\tau),\cdots,q_D(\tau),p_1(\tau),\cdots,p_D(\tau)\right)\right] d\tau. \end{aligned} \tag{5.27}$$

Here, $p_i(\tau)$, $q_i(\tau)$ satisfy

$$\begin{aligned} q_i(0) &= q_i', \\ p_i(0) &= \frac{\partial R(q_1,q_2,\cdots,q_D,t)}{\partial q_i}\Big|_{t=0}, \\ q_i(t) &= q_i, \\ p_i(t) &= p_i, \end{aligned} \tag{5.28}$$

as well as the classical paths in Eqs (5.25) and (5.26). Although Eq. (5.27) gives the overall phase evolution, finding the $\{q_i'\}$ that satisfy all given conditions for a particular $\{q_i\}$ is not easy.

Now we look at the amplitude component $A(q_1,q_2,\cdots,q_D,t)$ of the wavefunction. In principle, if the phase component is determined, the amplitude can be found using Eq. (5.16), but the process is rather complicated. Since $A(q_1,q_2,\cdots,q_D,t)$ is a non-negative real function, we can transform it without any information loss into a density function ρ such that

$$\rho(q_1,q_2,\cdots,q_D,t) = A^2(q_1,q_2,\cdots,q_D,t). \tag{5.29}$$

This density distribution function satisfies a much simpler evolution equation. Therefore, in practical calculations, it is customary to use the density distribution function to describe the real part of the wavefunction evolution. Multiplying both sides of Eq. (5.16) by $2A(q_1,q_2,\cdots,q_D,t)$, we have

$$\frac{\partial \rho(q_1,\cdots,q_D,t)}{\partial t}+\sum_{i=1}^{D}\frac{\partial}{\partial q_i}\left[\rho(q_1,\cdots,q_D,t)v_i\right]=0. \tag{5.30}$$

This is the continuity equation of a fluid with the virtual velocity field during orbital evolution

$$v_i=\frac{p_i}{m}=\frac{1}{m}\frac{\partial}{\partial q_i}R(q_1,\cdots,q_D,t). \tag{5.31}$$

Once we have determined $(q_1,\cdots,q_D,p_1,\cdots,p_D)$ using Eqs (5.25) and (5.26), we can obtain particle-number conservation as

$$\rho\left(q_1(t),\cdots,q_D(t),t\right)=\left|\det\left[\frac{\partial\left(q_1(0),\cdots,q_D(0)\right)}{\partial\left(q_1(t),\cdots,q_D(t)\right)}\right]\right|\rho\left(q_1(0),\cdots,q_D(0),0\right). \tag{5.32}$$

This gives us the amplitude component of the wavefunction. The phase component is given by Eq. (5.27). Combining these, we find the wavefunction as

$$\begin{aligned}\psi_{\mathrm{SC}}(\boldsymbol{q},t)=A(\boldsymbol{q},t)e^{iR(\boldsymbol{q},t)/\hbar}&=\sqrt{\det\left[\frac{\partial \boldsymbol{q}(0)}{\partial \boldsymbol{q}(t)}\right]}A\left(\boldsymbol{q}(0),0\right)e^{i\left(R(\boldsymbol{q}(0),0)+R(\boldsymbol{q}(t),\boldsymbol{q}(0),t)\right)/\hbar}\\&=\sqrt{\det\left[\frac{\partial \boldsymbol{q}(0)}{\partial \boldsymbol{q}(t)}\right]}e^{iR(\boldsymbol{q}(t),\boldsymbol{q}(0),t)/\hbar}\psi\left(\boldsymbol{q}(0),0\right).\end{aligned} \tag{5.33}$$

Here, $\boldsymbol{q}\equiv(q_1,q_2,\cdots,q_D)$. The phase evolution satisfies,

$$R\left(\boldsymbol{q}(t),\boldsymbol{q}(0),t\right)=\int_0^t\left[\boldsymbol{p}(\tau)\cdot\dot{\boldsymbol{q}}(\tau)-H\left(\boldsymbol{q}(\tau),\boldsymbol{p}(\tau)\right)\right]\mathrm{d}\tau, \tag{5.34}$$

and each component of $\boldsymbol{q}(t)$ satisfies Eq. (5.28).

Over a short period of time, Eq. (5.33) is fully capable of predicting the correct wavefunction. However, over a longer period of time, "folding" can occur. This means that there is no one-to-one correspondence between $\boldsymbol{q}(t)$ and $\boldsymbol{q}(0)$, and multiple initial coordinates can be propagated to the same point in space after time t. A point for which this is true is known as a focal point. In this case, the phase R becomes a multi-valued function. The Jacobi determinant $\det\left[\frac{\partial \boldsymbol{q}(0)}{\partial \boldsymbol{q}(t)}\right]$ will diverge and may become negative, and the amplitude A (after taking zero values) is undetermined. Because this is a difficult problem, more advanced mathematical tools and subtle techniques are required to properly solve this [Littlejohn1992]. These details will be skipped here. Fortunately, the final result is not too far from intuition. Calculating the wavefunction requires summing over all possible paths:

$$\psi_{\mathrm{SC}}(\boldsymbol{q},t)=\sum_j\sqrt{\left|\det\left[\frac{\partial \boldsymbol{q}(0)}{\partial \boldsymbol{q}(t)}\right]_j\right|}e^{iR(\boldsymbol{q},\boldsymbol{q}_j,t)/\hbar-i\pi m_j(\boldsymbol{q},\boldsymbol{q}_j,t)/2}\psi(\boldsymbol{q}_j,0), \tag{5.35}$$

where $\boldsymbol{q}_j$ is the starting point of the jth path that ends at $\boldsymbol{q}$, and m_j denotes the number of times that the Jacobi determinant changes sign on that path.

Now that we have the time evolution of the wavefunction, it is possible to find the semi-classical expression for the propagator, which is the result of the wavefunction time evolution with an initial value $\delta(\boldsymbol{q}')$. Numerically, only the evolution time needs to be changed, from $0 \to t$ to $\delta t \to t+\delta t$. To achieve this, we need to first consider the propagator over a very short time δt:

$$K(\boldsymbol{q},\boldsymbol{q}',\delta t) = A(\boldsymbol{q},\boldsymbol{q}',\delta t)e^{iR(\boldsymbol{q},\boldsymbol{q}',\delta t)/\hbar}. \tag{5.36}$$

Because the time is very short, we can assume that $\boldsymbol{p}$ does not change. For classical particles, however, $\dot{\boldsymbol{q}} = \boldsymbol{p}/m$. Then,

$$R(\boldsymbol{q},\boldsymbol{q}',\delta t) \approx [\boldsymbol{p}\cdot\dot{\boldsymbol{q}} - H(\boldsymbol{q},\boldsymbol{p})]\delta t = \frac{m\dot{\boldsymbol{q}}^2}{2}\delta t - V(\boldsymbol{q})\delta t \approx \frac{m(\boldsymbol{q}-\boldsymbol{q}')^2}{2\delta t}. \tag{5.37}$$

This is the phase of a Gaussian function with imaginary variance. To ensure that the phase converges at $\delta t \to 0$, we require

$$A = \left(\frac{m}{2\pi i\hbar\delta t}\right)^{D/2}. \tag{5.38}$$

Here, A is a constant because the phase component is already written as a δ -like function. The $\boldsymbol{q}'$ is the only value of interest. This gives us the short-time propagator

$$K(\boldsymbol{q},\boldsymbol{q}',\delta t) = \left(\frac{m}{2\pi i\hbar\delta t}\right)^{D/2} e^{\frac{i}{\hbar}\left[\frac{m(\boldsymbol{q}-\boldsymbol{q}')^2}{2\delta t}\right]}, \tag{5.39}$$

which serves as the propagator of free particles.

Combining the above equation with Eq. (5.35), we have the long-time propagator

$$\begin{aligned} &K(\boldsymbol{q}'',\boldsymbol{q}',t+\delta t) \\ &= \sum_j \left(\frac{m}{2\pi i\hbar\delta t}\right)^{D/2} \sqrt{\left|\det\left[\frac{\partial \boldsymbol{q}}{\partial \boldsymbol{q}''}\right]_j\right|} e^{\frac{i}{\hbar}\left(\frac{m(\boldsymbol{q}_j-\boldsymbol{q}')^2}{2\delta t}+R(\boldsymbol{q}'',\boldsymbol{q}_j,t)\right)-i\pi m_j(\boldsymbol{q}'',\boldsymbol{q}_j,t)/2}. \end{aligned} \tag{5.40}$$

On the path from $\boldsymbol{q}'$ to $\boldsymbol{q}$, we have $\boldsymbol{p} = \frac{m(\boldsymbol{q}-\boldsymbol{q}')}{\delta t} = \frac{\partial R(\boldsymbol{q},\boldsymbol{q}',\delta t)}{\partial \boldsymbol{q}}$. Since this path is connected to the path from $\boldsymbol{q}$ to $\boldsymbol{q}''$, we have

$$\begin{gathered} \frac{\partial \boldsymbol{p}}{\partial \boldsymbol{q}} = \frac{m}{\delta t}, \\ \det\left[\frac{\partial \boldsymbol{p}}{\partial \boldsymbol{q}}\right] = \left(\frac{m}{\delta t}\right)^D. \end{gathered} \tag{5.41}$$

Substituting these into Eq. (5.40), we obtain

$$
\begin{aligned}
&K(\boldsymbol{q}'',\boldsymbol{q}',t+\delta t)\\
&=\sum_j\left(\frac{1}{2\pi i\hbar}\right)^{D/2}\sqrt{\left|\det\left[\frac{\partial \boldsymbol{p}}{\partial \boldsymbol{q}}\right]_j\right|}\sqrt{\left|\det\left[\frac{\partial q}{\partial \boldsymbol{q}''}\right]_j\right|}e^{\frac{i}{\hbar}R_j(\boldsymbol{q}'',\boldsymbol{q}',t)-i\pi m_j(\boldsymbol{q}'',\boldsymbol{q}',t)/2}\\
&=\sum_j\left(\frac{1}{2\pi i\hbar}\right)^{D/2}\sqrt{\left|\det\left[\frac{\partial \boldsymbol{p}}{\partial \boldsymbol{q}''}\right]_j\right|}e^{\frac{i}{\hbar}R_j(\boldsymbol{q}'',\boldsymbol{q}',t)-i\pi m_j(\boldsymbol{q}'',\boldsymbol{q}',t)/2},
\end{aligned}
\tag{5.42}
$$

where j sums up all paths that are connected to $\boldsymbol{q}'$ and $\boldsymbol{q}''$, and $\boldsymbol{p}$ denotes the initial momentum.

With the propagator, we can write the time evolution operator under the semiclassical approximation as

$$
\begin{aligned}
&e^{-i\hat{H}t/\hbar}\\
&=\int d\boldsymbol{q}_1\int d\boldsymbol{q}_2\,|\boldsymbol{q}_2\rangle\sum_j\left(\frac{1}{2\pi i\hbar}\right)^{D/2}\sqrt{\left|\det\left[\frac{\partial \boldsymbol{p}_1}{\partial \boldsymbol{q}_2}\right]_j\right|}e^{\frac{i}{\hbar}R_j(\boldsymbol{q}_2,\boldsymbol{q}_1,t)-i\pi m_j(\boldsymbol{q}_2,\boldsymbol{q}_1,t)/2}\langle\boldsymbol{q}_1|.
\end{aligned}
\tag{5.43}
$$

The above equation requires integrating over initial and final coordinates and summing all classical paths connecting these coordinates. Performing the summation is usually quite challenging, especially when the integration and summation are iterated over all possible paths. Therefore, we rewrite the equation in the initial-value representation [Miller2001]. This representation exploits the following property:

$$
\int d\boldsymbol{q}_2\sum_j\cdot=\int d\boldsymbol{p}_1\left|\det\left[\frac{\partial \boldsymbol{q}_2}{\partial \boldsymbol{p}_1}\right]\right|. \tag{5.44}
$$

In so doing, Eq. (5.43) can be rewritten as

$$
e^{-i\hat{H}t/\hbar}=\int d\boldsymbol{q}_1\int d\boldsymbol{p}_1\,|\boldsymbol{q}_2\rangle\left(\frac{1}{2\pi i\hbar}\right)^{D/2}\sqrt{\left|\det\left[\frac{\partial \boldsymbol{q}_2}{\partial \boldsymbol{p}_1}\right]\right|}e^{\frac{i}{\hbar}R(\boldsymbol{q}_1,\boldsymbol{p}_1,t)-i\pi m(\boldsymbol{q}_1,\boldsymbol{p}_1,t)/2}\langle\boldsymbol{q}_1|, \tag{5.45}
$$

where R and m are the results obtained by calculating the classical path from Eqs (5.15) and (5.16) using $\boldsymbol{p}_1$ and $\boldsymbol{p}_1$ as initial conditions. They are then integrated according to Eqs (5.33) and (5.36). As the only variables left in the integral are the initial values, the rest of the terms can be expressed as a function of them; this is why this method is called the "initial-value representation."

We have now successfully obtained the time evolution operator in the SC-IVR. The reader can test whether this procedure produces the correct results for free particles and harmonic oscillators, and check whether it yields approximately correct results in the general case.

To apply the IVR to physical quantities of interest, we can simply rewrite all time-evolution operators in the initial-value representation, as shown in Eq. (5.45), and obtain a phase space integral. By simulating and integrating the classical evolution, an approximate value of the physical quantity can be found. As an example, consider the problem of energy-spectrum calculation. The energy spectrum $I(E) \equiv \langle \chi | \delta(E-\hat{H}) | \chi \rangle$ for a given state $|\chi\rangle$ can be regarded as the proportion of different energy eigenstates in the state $|\chi\rangle$. For an energy eigenstate, the energy spectrum is a δ function shifted horizontally. In order to calculate $I(E)$, we must first write it as a time-evolution operator that takes the following form:

$$I(E) = \frac{1}{\pi\hbar} Re \int_0^{\infty} \mathrm{d}t e^{iEt/\hbar} \langle \chi | e^{-i\hat{H}t/\hbar} | \chi \rangle. \tag{5.46}$$

Substituting Eq. (5.45) into this expression yields

$$\begin{aligned} I(E) &= \frac{1}{\pi\hbar} \mathrm{Re} \int_0^{\infty} \mathrm{d}t e^{iEt/\hbar} \int \mathrm{d}\boldsymbol{q}_1 \int \mathrm{d}\boldsymbol{p}_1 \langle \chi \mid \boldsymbol{q}_2 \rangle \left(\frac{1}{2\pi i\hbar} \right)^{D/2} \\ &\sqrt{\left| \det \left[\frac{\partial \boldsymbol{q}_2}{\partial \boldsymbol{p}_1} \right] \right|} e^{\frac{i}{\hbar} R(\boldsymbol{q}_1, \boldsymbol{p}_1, t) - i\pi m(\boldsymbol{q}_1, \boldsymbol{p}_1, t)/2} \langle \boldsymbol{q}_1 \mid \chi \rangle \\ &= \frac{1}{\pi\hbar} \mathrm{Re} \int_0^{\infty} \mathrm{d}t e^{iEt/\hbar} \int \mathrm{d}\boldsymbol{q}_1 \int \mathrm{d}\boldsymbol{p}_1 \varphi_\chi^*(\boldsymbol{q}_2) \varphi_\chi(\boldsymbol{q}_1) \left(\frac{1}{2\pi i\hbar} \right)^{D/2} \\ &\sqrt{\left| \det \left[\frac{\partial \boldsymbol{q}_2}{\partial \boldsymbol{p}_1} \right] \right|} e^{\frac{i}{\hbar} R(\boldsymbol{q}_1, \boldsymbol{p}_1, t) - i\pi m(\boldsymbol{q}_1, \boldsymbol{p}_1, t)/2}. \end{aligned} \tag{5.47}$$

For a point $(\boldsymbol{q}_1, \boldsymbol{p}_1)$ in the phase space, the classical path is obtained using Eqs (5.25) and (5.26). Integrating the path according to Eq. (5.34) yields R, and calculating the Jacobi determinant yields the integrand. Simultaneously, the energy spectrum can be calculated by integrating $(\boldsymbol{q}_1, \boldsymbol{p}_1)$ in the phase space, followed by the Fourier transform.

It should be noted that the calculation procedure described above is only illustrative. In practical application, there would be more details to consider, such as the calculation of the Jacobi determinant. Moreover, the coordinate representation is not our only choice: We can also use the momentum representation or the coherent-state representation (which is, in fact, the most useful). Further explanation of this topic would require too much space; the interested reader should refer to the literature [Miller2001].

5.1.3.2 Centroid Molecular Dynamics and Ring-Polymer Molecular Dynamics

One advantage of the initial-value representation is that it provides a semi-quantitative description of quantum tunneling and quantum coherence effects, especially the latter. However, the extent to which this method can be applied to quantum dynamics simulations of large molecules and condensed matter systems is highly limited because the number of sampled orbitals grows exponentially with the system size. Moreover, the poor long-time numerical behavior of the stability matrix elements and the "sign problem" caused by the superposition of the classical orbitals further restrict the application of SC-IVR.

To solve these problems, H.B. Wang and Miller introduced a practical simplification in 1998 based on the van Vleck semiclassical propagator, linearizing the classical orbital difference of the time evolution operator $e^{-i\hat{H}t/\hbar}$ and its conjugate $e^{i\hat{H}t/\hbar}$ in Eq. (5.46) [Wang H.B.1998]. This simplified approach is called the linearized semiclassical initial-value representation (LSC-IVR). Because of its linearity, LSC-IVR cannot describe quantum coherence effects. Moreover, it requires the use of the Fourier transform, which is difficult to perform on large molecular systems. To address the second issue, Q. Shi and Geva proposed using the local harmonic oscillator approximation for molecular systems with insignificant imaginary frequencies [Shi Q.2003]. In 2009, J. Liu and Miller introduced the local Gaussian approximation to handle the case where the imaginary frequencies are non-negligible [Liu J.2009]. Their work ensured that LSC-IVR is applicable to ordinary molecular and condensed matter systems. The forward–backward semiclassical dynamics (FBSD) method, which is similar to LSC-IVR, was developed by Makri and J. S. Shao [Shao J.S.1999]. However, none of these approaches can accurately describe quantum coherence.

In addition to these methods, two other recent advances have been made in the dynamics simulation of condensed matter systems based on path integral methods. In 1993, J. S. Cao and G. A. Voth proposed the CMD method [Cao J.S.1993]. The concept of centroid (i.e., the mean value of all particle positions on the imaginary-time path of the imaginary-time path integral) was proposed by Feynman; J.S. Cao and Voth applied it to the study of dynamics [Feynman1965, Cao J.S.1993]. After the imaginary-time path integral has been used to find the centroid potential energy surface, the classical correlation function based on the potential energy surface can be related to the Kubo correlation function of quantum statistical mechanics. This discovery provides a theoretical foundation for applying CMD to the study of quantum dynamics [Cao J.S.1993, Voth1996].

The second method is the RPMD method proposed by D. E. Manolopoulos et al. This is also based on the Kubo quantum correlation function. The difference is that in statistical sampling, the bead mass can be assigned arbitrarily, but in RPMD, the actual mass of each atom is used. Moreover, the dynamical processes of each bead are tracked rather than those of the centroid and are used to calculate various correlation functions [Craig I.R.2004, 2005a, b, Habershon2013].

Both CMD and RPMD have been relatively successful, although they cannot perfectly describe nuclear quantum effects on chemical reaction rate corrections [Craig I.R.2005a, b]. Of particular note is the RPMD transition-state theory proposed by S. Althorpe et al. [2011], which provides a relatively comprehensive explanation of why RPMD is effective in describing quantum reaction rates [Hele2013]. However, neither CMD nor RPMD shows satisfactory performance in the theoretical simulation of molecular vibrational spectra [Witt2009]. The former encounters curvature problems at low temperatures. The concept of centroid fails at low temperatures in a ring potential, resulting in a fictitious redshift in the centroid dynamics when describing the associated vibrational frequencies. In the RPMD simulation, when the vibration between the beads matches the real vibrational spectrum of the molecules, the molecular vibrational spectrum inevitably affects the simulation result. These issues have severely limited the application of CMD and RPMD in the vibrational spectroscopy of real materials.

5.1.3.3 Obtaining Kinetic Information from Time-Independent Methods

One of the most important physical quantities which can be classified as kinetic information is the magnitude of the chemical reaction rate, which determines how the reaction proceeds and what the reaction product is. Therefore, developing an accurate theoretical description of the reaction rate has always been a research focus in theoretical chemistry. Because of its simple and efficient representation of the potential barrier, the well-known classical transition-state theory has long dominated reaction kinetics. In this theory, the reactant in statistical equilibrium must acquire enough energy to cross the potential barrier and reach the product state. Therefore, statistical mechanics can be used to describe the reaction rate directly. As the zero-time limit ($t \to 0$ limit) of the classical flow correlation function, it provides us with a typical example of how a time-independent method can handle kinetic information. To obtain the reaction rate in transition-state theory, local statistical information (or even static information in the coarse approximation) about the potential barrier height on the free-energy surface or the curvature of the potential energy surface at the transition-state point is needed. The success of classical transition-state theory has facilitated the development of quantum transition-state theory.

Nuclear quantum effects, dominated by quantum tunneling of light element atoms in such kinetic processes, have become increasingly important. As a result, chemical reaction theory that can accurately predict nuclear quantum effects in transition states is attracting more research interest in the field of theoretical chemistry.

The earliest examples of this type of method date to the 1970s. Using the Born–Oppenheimer approximation, W. H. Miller et al. derived an exact expression for the reaction rate that strictly considered nuclear quantum effects from quantum scattering theory. The resulting flux-side correlation function [Miller1975, 1983] became the starting point and theoretical foundation for numerous subsequent studies of quantum rate theory. In the SPA (also known as the saddle-point approximation), the zero-time limit is equivalent to the semiclassical limit, which yields the following rate expression [Miller1975]:

$$k(\beta)Q_{\mathrm{r}}(\beta) = (2\pi\hbar)^{-1/2}\left|\frac{\mathrm{d}^2\tilde{A}(\beta)}{\mathrm{d}\beta^2}\right|^{1/2} \times \prod_{i=1}^{f-1}\frac{1}{2\sinh\left(\dfrac{u_i}{2}\right)} e^{-\tilde{A}(\beta)/\hbar}, \tag{5.48}$$

where β is the inverse temperature $1/(k_{\mathrm{B}}T)$, Q_{r} is the partition function of the reaction state, $\tilde{A}(\beta)$ denotes the classical action of the instanton periodic orbit, and u_i is the local stability parameter of the instanton orbit. An instanton orbit is a periodic orbit that closes in imaginary-time space or on the inverse potential barrier after time $\beta\hbar$. In fact, it is the path of least action in the potential barrier region between the reactant and product state. This will be explained later. As can be seen from Eq. (5.48), quantum transition-state theory is time-independent, and the rate expression contains only the instanton action and the local information. Compared to classical transition-state theory (which only requires the energy difference between the transition state and the reactant state), quantum transition-state theory requires more static information about the potential barrier, including its shape.

In the early days, when available computational power was limited, finding the instanton orbit in imaginary-time space or on the inverse potential barrier was quite difficult except for some simple model systems. Later, the quantum instanton (QI) method, a type of quantum transition state-theory, was developed by Miller's group [Miller2003], and a path-integral sampling method was subsequently proposed [Zhao Y.2004]. This method adopts fewer approximations and avoids the problem of finding the instanton periodic orbit. As a result, it has higher accuracy in predicting the tunneling rate and can be easily extended to larger systems. However, fewer approximations often mean more computation; there are still few applications of QI compared to those of the instanton method. We will not elaborate on QI methods here but will focus on the saddle-point approximation, which is the core of all these methods and contains information about the nuclear quantum effects. For greater physical clarity, we introduce it along with the imaginary free energy (Im *F*) assumption.

It is worth pointing out that, compared to the rigorous theory derived from the flux-side correlation function, the Im *F* assumption does not seem reliable at first glance. This is because it depends on various assumptions and approximations and is an assumption by itself. Fortunately, it has been shown that this method produces exactly the same results as Eq. (5.49) [Althorpe2011]. Formally, it is also similar to the one-dimensional WKB method; it is based on a quite simple physical picture and has had many successful applications.

Instantons and Im *F* assumption are both ideas from nuclear physics, introduced into chemical reaction-rate calculations in the 1970s at almost the same time as quantum transition-state theory [Coleman1977]. When the resonant state in a reaction potential well crosses the potential barrier and becomes the product, its tunneling rate is the thermal average of the decay probability. For this reason, the reaction-state energy must be considered as a complex number, assuming that the reaction-state potential well is steep enough to form a resonant state.

In the following derivation, we will briefly explain the underlying principle of the Im *F* assumption, with reference to the literature [Richardson2009]. In a fully bound state (e.g., in a deep potential well), there exists a discrete set of bound state energy levels E_n^0 and a steady-state partition function $Q_r(\beta)$. However, because of tunneling, the system is not fully bound in practice, and states can tunnel through the potential barrier to the other side. The energy levels can be expressed in terms of complex numbers as

$$E_n = E_n^0 - i\hbar\Gamma_n / 2 \ . \tag{5.49}$$

Consequently,

$$\begin{aligned} |\Psi_n(t)|^2 &= \left|\Psi_n(0)e^{-\frac{iE_n t}{\hbar}}\right|^2 \\ &= |\Psi_n(0)|^2 e^{-\Gamma_n t}. \end{aligned} \tag{5.50}$$

The partition function can also be expressed in complex form as

$$\begin{aligned} Q(\beta) &= \sum_n e^{-\beta E_n} \\ &= \sum_n \left[\cos\left(\beta\hbar\Gamma_n / 2\right) + i\sin\left(\beta\hbar\Gamma_n / 2\right)\right] e^{-\beta E_n^0}, \end{aligned} \tag{5.51}$$

corresponding to the complex free energy

$$F = -\frac{1}{\beta}\ln Q(\beta) \ . \tag{5.52}$$

The tunneling rate is the Boltzmann average of the decay rates of all states, which is equivalent to

$$\begin{aligned} k(\beta) &= \frac{1}{Q_{\mathrm{r}}(\beta)} \sum_n \Gamma_n e^{-\beta E_n^0} \\ &\approx -\frac{2}{\beta\hbar Q_{\mathrm{r}}(\beta)} \sum_n \sin\frac{\beta\hbar\Gamma_n}{2} e^{-\beta E_n^0} \\ &= \frac{2}{\beta\hbar} \frac{\operatorname{Im} Q(\beta)}{\operatorname{Re} Q(\beta)} = -\frac{2}{\hbar} \operatorname{Im} F , \end{aligned} \tag{5.53}$$

assuming $\beta\hbar\Gamma_n / 2 \ll 1$. This approximation is reasonable within a certain range. For the lower energy levels, the potential barrier to be crossed is higher, so the decay rate Γ_n is lower. Thus, for the general case, $\operatorname{Im} Q(\beta) \ll \operatorname{Re} Q(\beta)$. For higher energy levels, the errors introduced by this approximation are negligible owing to the small Boltzmann factor $e^{-\beta E_n^0}$.

Equation 5.53 is the origin of the name "imaginary free-energy assumption." The partition function of a weakly bound system is complex. It is easy for us to understand the real part, which is simply the steady-state partition function $Q_{\mathrm{r}}(\beta)$ of the reaction state. The imaginary part, on the other hand, describes the decay process through the potential barrier. Later, when we carry out the saddle-point approximation of this quantum partition function under analytic continuation, we will see that the contribution from the unstable mode that corresponds to the reaction-path direction is imaginary.

The path integral is a natural choice for representing such a quantum partition function. In the coordinate representation (bearing in mind that all concepts in this subsection, including state, energy level, and operator, only apply to nuclear systems under the Born–Oppenheimer approximation), we have

$$Q(\beta) = \operatorname{Tr}\left[e^{-\beta\hat{H}}\right] = \int_{-\infty}^{+\infty} \mathrm{d}x \left\langle x \,|\, e^{-\beta\hat{H}} \,|\, x \right\rangle. \tag{5.54}$$

As discussed in subsection 5.1.2, in the actual numerical calculations, $\beta\hbar$ is divided into N copies, with $N-1$ completeness relations $\int |x\rangle\langle x| = 1$. By establishing a classical isomorphism using Trotterization, an end-to-end ring-polymer is obtained. In this way, the quantum partition function can be written as a discrete path integral as follows:

$$Q(\beta) \cong \left(\frac{m}{2\pi\beta_N\hbar^2}\right)^{\frac{N}{2}} \int \mathrm{d}\vec{x}\, e^{-\beta_N U_N(\vec{x})}, \tag{5.55}$$

where $\beta_N = \beta / N$; $\vec{x} = (x_1,\dots,x_N)$; x_i denotes the coordinate of the ith bead in the ring polymer; and $x_0 \equiv x_N$. The left and right sides of this equation become equal as N approaches infinity. The potential energy of the ring polymer is expressed as

$$U_N(\vec{x}) = \sum_{i=1}^{N} \left[\frac{m}{2\beta_N^2\hbar^2}(x_i - x_{i-1})^2 + V(x_i)\right]. \tag{5.56}$$

This indicates that any two adjacent beads are connected by a spring with the same spring constant. The cost of establishing a classical isomorphism, as shown in the discrete path integral above, is that a quantum system that originally had f degrees of freedom is now represented by a ring polymer with $N \times f$ degrees of freedom, and the integration of the partition function must be carried out in an $N \times f$ -dimensional space.

There is no practical way to evaluate the integral of this classical partition function directly. Nevertheless, using the saddle-point approximation (which corresponds to the method of steepest descent in applied mathematics), one can approximate the integral even in high dimensions.

In the one-dimensional case, the integral takes the form

$$\int_{-\infty}^{\infty} \mathrm{d}x\, e^{-s(x)/\hbar} \approx \int_{-\infty}^{\infty} \mathrm{d}x\, e^{-\left[S(\tilde{x})+\frac{1}{2}S''(\tilde{x})(x-\tilde{x})^2\right]/\hbar} = \sqrt{\frac{2\pi\hbar}{S''(\tilde{x})}}\, e^{-\frac{S(\tilde{x})}{\hbar}}, \tag{5.57}$$

where $\tilde{x}$ is the position corresponding to the minimum of the function $S(x)$, with $S''(\tilde{x}) = 0$, and $S''(\tilde{x}) > 0$. The approximate result becomes more accurate as $\hbar \to 0$. The integral value is determined by the $\tilde{x}$ position and the value of the second derivative of $S(x)$ at that point. Extending this observation to higher dimensions suggests expanding $S(x)$ to second order around the position of its extremum $\tilde{x}$ and evaluating the Gaussian integral. Ignoring the higher-order terms of $\hbar$, we see that this is, in fact, the simple harmonic approximation. It is also the reason why the instanton is called the semiclassical limit.

Generalizing this approach requires finding the minimum in the configuration space of the high-dimensional ring polymer with potential energy given by Eq. (5.56). For the reactant partition function, it is clear that the minimum corresponds to the case where all beads coincide at a single point of the reactant state. Applying the steepest-descent approximation is equivalent to determining a steady-state partition function under the simple harmonic approximation. $\mathrm{Im}\, Q(\beta)$, however, describes the information in the region of the potential barrier. We wanted to determine the first-order saddle point on the $N \times f$ -dimensional potential energy surface of the ring polymer. The transition state in the classical theory is a first-order saddle point on its potential energy surface; we are now seeking the first-order saddle point on the ring-polymer potential energy surface. There are many existing saddle-point-searching algorithms applicable to higher dimensional spaces. In fact, this saddle point is an instanton in the form of a ring polymer. Therefore, we can regard the saddle point as a quantum transition state in the ring polymer.

We can also look at the saddle point in Eq. (5.57) from a different perspective. The discrete form of the integral path action is $\beta\hbar U_N(\vec{x})$; its saddle point is the integral path with the least action in the real space with f degrees of freedom. Letting $\delta U_N(\vec{x}) = 0$, we have

$$V'(x_i) = m\frac{x_{i+1} - 2x_i + x_{i-1}}{\beta_N^2 \hbar^2}, \tag{5.58}$$

which can be viewed as the finite-difference form of the classical equations of motion in imaginary-time space, with $\beta_N \hbar$ being the imaginary-time step size, and the subscript i denoting the ith step. To satisfy the saddle-point condition, it must return to the starting point after taking N steps. When N approaches infinity, the imaginary-time step size approaches

0, and the trajectory becomes continuous. This classical trajectory of periodic motion in the imaginary-time space is the early instanton orbit. We can also write the Euclidean action of the continuous integral path $x(\tau)$ (where τ denotes the imaginary time) as

$$S[x(\tau)] = \int_0^{\beta\hbar} \left[\frac{1}{2} m \left(\frac{\mathrm{d}x}{\mathrm{d}\tau} \right)^2 + V(x(\tau)) \right] \mathrm{d}\tau. \tag{5.59}$$

This is a common form of action that can also be found in other tunneling-rate theories.

Now, let us consider the ring-polymer instanton before continuing with the derivation after the method of steepest descent. When all beads overlap at the highest point of the potential barrier $x_i = x^{\ddagger}$ (i.e., the transition state), the kth normal-mode frequency of the ring polymer is

$$\omega_k = \sqrt{\frac{4}{\beta_N^2 \hbar^2} \sin^2 \frac{|k|\pi}{N} - \omega_b^2}\ , \tag{5.60}$$

where $i\omega_b = \sqrt{V''(x^{\ddagger})/m}$ is the imaginary frequency of the transition state. Having a first-order saddle point means that there is only one imaginary frequency in ω_k. As N approaches infinity, the imaginary frequency $\omega_0 = i\omega_b$. Letting $\omega_{\pm 1} = 0$, we have

$$\beta_c = \frac{1}{k_B T_c} = \frac{2\pi}{\hbar \omega_b}. \tag{5.61}$$

This yields a transition temperature T_c that distinguishes the quantum tunneling process from the classical reaction process. When the temperature is higher than T_c, the spring is so stiff that all the beads are pulled to a single point in the transition state, and we are dealing with a classical transition-state scenario. When the temperature is below T_c, the configuration can no longer be described as a first-order saddle point on the ring-polymer potential energy surface. However, we can imagine a very soft spring that pulls the beads at both ends towards the bottom of both potential wells owing to the action of the potential energy. The spring stretches until it reaches equilibrium with the potential energy, and it hangs on the potential barrier like a necklace, forming an unstable equilibrium state.

Based on the optimized instanton configuration, one can examine the rate expression. The steepest-descent approximation takes the second derivative of the ring-polymer potential energy $U_N(\bar{x})$ (5.56) at the position of the instanton. For multivariable functions, this is a Hessian matrix; diagonalizing it yields $N \times f$ eigenfrequencies. Combined with the similar transformation of the generalized coordinates, the eigenfrequencies become $N \times f$ independent one-dimensional Gaussian integrals under the simple harmonic approximation. Since the saddle point is first order, there exists one and only one imaginary frequency, which corresponds to the vibrational mode in the reaction-path direction. Analytic continuation of the frequency gives an imaginary number, the numerical value of the imaginary part of the partition function. In addition, there exists a zero-frequency corresponding to the vibrational mode of the bead rotation. Directly integrating it yields a contribution $N\sqrt{B_N}$, where $B_N = \sum_{i=1}^{N} (\tilde{x}_{i+1} - \tilde{x}_i)^2$. The final rate expression is as follows [Andersson2009, Richardson2009]:

$$k_{\text{inst}}(\beta)Q_{\text{r}}(\beta) = \frac{1}{\beta_N \hbar}\sqrt{\frac{mB_N}{2\pi\beta_N\hbar^2}}\prod\nolimits_k^{\prime}\left|\frac{1}{\beta_N\hbar\eta_k}\right| e^{-\beta_N U_N(\tilde{x})}, \tag{5.62}$$

where η_k is the nonzero eigenfrequency of the Hessian matrix at the instanton position.

The advantage of the Im F instanton method is its relatively low computational cost. Only the optimization of the saddle point and the calculation of its Hessian matrix are required to capture the central part of the nuclear quantum effects. Combining the Im F method with the ring-polymer method solves the problem of finding the instanton orbit, identifies the path of the least action, and describes the quantum corner-cutting effects that the conventional WKB method fails to characterize. Nonetheless, as a semiclassical limit, the Im F instanton method relies heavily on the simple harmonic approximation, which can lead to significant errors for nonharmonic potentials or multiple reaction paths. In recent years, many improvements have been made to address this issue. For example, the relationship between different levels of the saddle-point approximation and the ring-polymer molecular-dynamic rate theory has been explored [Richardson2009], and a thermalized microcanonical instanton approach was proposed [Richardson2016a]. In retrospect, the ring-polymer instanton discussed here is, in fact, the result of applying the saddle-point approximation to the partition function at the potential barrier. The development of the ring-polymer instanton method opens up possibilities for applying instantons to other problems, such as tunneling splitting [Richardson2011]. Although quantitatively flawed, the ring-polymer instanton method has proven to be extremely helpful qualitatively [Richardson2016b].

5.2 First-Principles Electron–Phonon Coupling

In solid-state physics, the electron–phonon coupling theory based on second quantization was proposed much earlier than the first-principles numerical methods. It should be noted that most early theories of electron–phonon coupling were based on models. As research progressed, it became clear that simple model approximations (such as the atomic force-field model) could not adequately address the state of a condensed matter system during complex physical or chemical processes, especially in cases where the lattice periodicity of the system was not guaranteed or where nonharmonic effects were significant. This led to resorting to the first principles of quantum mechanics, so that the effect of the nuclear configuration on the electronic structure could be calculated and the physical nature of the electron–phonon coupling could be revealed. Therefore, from the perspective of full quantization, how to combine first principles with electron–phonon coupling theory has been a topic of considerable interest in recent years.

Many experimental observations in condensed matter spectroscopy correspond to frequency-dependent statistical properties. Therefore, in this section, we will focus on how the thermal and quantum fluctuations of the nucleus affect electronic properties. Here we will not consider electronic transitions caused by the nonadiabatic evolution of the nucleus. The description of the real-time electron–nucleus coupling will be the subject of the next section. Although the nonadiabatic description of photochemical or photophysical processes in finite-sized systems has shown promising results, in general, the application of nonadiabatic simulation methods in condensed matter systems is hindered by the need to treat the nucleus as a wavefunction when studying systems in the thermodynamic limit.

The following discussion will describe the first-principles electronic structure, starting from the DFT introduced in Section 4.4. In that section, it was assumed that the nucleus is fixed, and the effects of quantum and thermal fluctuations are excluded from consideration. If the fluctuations are included but are relatively small, we can adopt the simple harmonic approximation and describe the lattice dynamics in terms of phonons. Then, using perturbation theory, the corrections to the electronic properties due to the fluctuation-driven change in the electron effective potential can be derived. However, if the fluctuations are relatively large, the simple harmonic approximation will fail, and we must use a non-perturbative approach to characterize electron–phonon coupling effects. When studying realistic problems, the effects of lattice vibrations on the electronic structure cannot be neglected. This was clear in some of the examples in Chapter 2, where the full quantum effects on the electrical and optical properties of indirect bandgap semiconductor materials were discussed.

5.2.1 Perturbative Treatment of Electron–Phonon Coupling

For periodic solids, the electronic effective Hamiltonian of the Kohn–Sham system can be written as

$$\hat{H}_{\mathrm{e}} = \sum_i \hat{h}_i^{\mathrm{KS}}(\vec{r}_i;\vec{R}), \tag{5.63}$$

where

$$\hat{h}_i^{\mathrm{KS}}(\vec{r}_i;\vec{R}) = -\frac{1}{2}\nabla_{\vec{r}_i}{}^2 + v^{\mathrm{KS}}(\vec{r}_i) = -\frac{1}{2}\nabla_{\vec{r}_i}{}^2 + v_H(\vec{r}_i) + v_{xc}(\vec{r}_i) - \sum_{I\kappa}\frac{Z_\kappa}{\left|\vec{r}_i - \vec{R}_{I\kappa}\right|}. \tag{5.64}$$

The Σ_i in Eq. (5.63) denotes summation over all electrons within the system, $\vec{R}$ represents a set of atomic coordinates, $\vec{R}_{I\kappa}$ is the κth coordinate within the Ith unit cell, and Z_κ is the nuclear charge. Solving the Schrödinger equation for this Hamiltonian gives the total energy $E_e(\vec{R})$ of the electron subsystem. Moreover, the Hamiltonian with respect to the nucleus is written as

$$\hat{H}_{\mathrm{ion}} = -\sum_{I\kappa}\frac{1}{2M_\kappa}\nabla_{I\kappa}^2 + E_e(\vec{R}) + \frac{1}{2}\sum_{IJ\kappa\kappa'}\frac{Z_\kappa Z_{\kappa'}}{\left|\vec{R}_{J\kappa'} - \vec{R}_{I\kappa}\right|}. \tag{5.65}$$

In this equation, the last two terms denote the potential energy of the nucleus, and together they are known as the Born–Oppenheimer potential energy $U^{\mathrm{BO}}(\vec{R})$. In the case where the nuclear vibration with respect to the equilibrium position is relatively small, we can use the simple harmonic approximation and take $U^{\mathrm{BO}}(\vec{R}^0)$ as the zero-energy point. $\hat{H}_{\mathrm{ion}}$ can be expanded to second order in terms of the atomic displacement, so that

$$\hat{H}_{\mathrm{ion}}^{\mathrm{HA}} = -\sum_{I\kappa}\frac{1}{2M_\kappa}\nabla_{I\kappa}^2 + \frac{1}{2}\sum_{I\kappa\alpha}\sum_{J\kappa'\beta}\frac{\partial^2 U^{BO}}{\partial R_{I\kappa\alpha}\partial R_{J\kappa'\beta}}\Big|_{\vec{R}=\vec{R}^0} u_{I\kappa\alpha}u_{J\kappa'\beta}. \tag{5.66}$$

Here, $\vec{R}^0$ is the equilibrium position whose first-order expansion is zero; HA denotes the harmonic approximation; α and β are the Cartesian components; and $\vec{u}_{I\kappa} = \vec{R}_{I\kappa} - \vec{R}^0_{I\kappa}$ is the atomic displacement. Applying the canonical transformation [Ashcroft 1976] to Eq. (5.66), we have the following second-quantization form:

$$\hat{H}^{\text{HA}}_{\text{ion}} = \sum_{\vec{q}\nu}\left(\hat{b}^\dagger_{\vec{q}\nu}\hat{b}_{\vec{q}\nu} + \frac{1}{2}\right)\omega_{\vec{q}\nu}\,. \tag{5.67}$$

The $\hat{b}_{\vec{q}\nu}$ ($\hat{b}^\dagger_{\vec{q}\nu}$) here is the phonon annihilation (creation) operator, and $\omega_{\vec{q}\nu}$ is the νth-normal-mode vibration frequency of the wave vector $\vec{q}$.

Similarly, if we consider the atomic vibration and expand $\hat{H}_{\text{e}}$ to the second order with respect to the atomic displacement, we have

$$\hat{H}_{\text{e}} = \hat{H}^0_{\text{e}} + \hat{H}_1 + \hat{H}_2 = \hat{H}^0_e + \sum_i \hat{h}^{(1)}_i + \sum_i \hat{h}^{(2)}_i\,, \tag{5.68}$$

$$\hat{h}^{(\vec{R}=\vec{R}^0)} = \sum_{I\kappa\alpha} u_{I\kappa\alpha}\frac{\partial v^{\text{KS}}}{\partial R_{I\kappa\alpha}}|_{\vec{R}=\vec{R}^0}\,, \tag{5.69}$$

$$\hat{h}^{(2)} = \sum_{I\kappa\alpha, J\kappa'\beta} u_{I\kappa\alpha}u_{J\kappa'\beta}\frac{\partial^2 v^{\text{KS}}}{\partial R_{I\kappa\alpha}\partial R_{J\kappa'\beta}}|_{\vec{R}=\vec{R}^0}\,. \tag{5.70}$$

Rewriting the atomic displacement operator in second-quantization form [Ashcroft1976],

$$u_{I\kappa\alpha} = \sum_{\vec{q}\nu}\left(\frac{\hbar}{2NM_\kappa\omega_{\vec{q}\nu}}\right)^{\frac{1}{2}} e^{i\vec{q}\cdot\vec{R}_I}\xi_{\kappa\alpha}(\vec{q}\nu)\left(\hat{b}_{\vec{q}\nu} + \hat{b}^\dagger_{-\vec{q}\nu}\right), \tag{5.71}$$

where $\vec{R}_I$ denotes the position vector of the Ith unit cell, N denotes the number of single cells, and $\xi_{\kappa\alpha}(\vec{q}\nu)$ is the corresponding eigenvector. In this way, Eq. (5.69) becomes

$$\hat{h}^{(1)} = \sum_{I\kappa\alpha}\sum_{\vec{q}\nu}\left(\frac{\hbar}{2NM_\kappa\omega_{\vec{q}\nu}}\right)^{\frac{1}{2}} e^{i\vec{q}\cdot\vec{R}_I}\xi_{\kappa\alpha}(\vec{q}\nu)\left(\hat{b}_{\vec{q}\nu} + \hat{b}^\dagger_{-\vec{q}\nu}\right)\frac{\partial v^{\text{KS}}}{\partial R_{I\kappa\alpha}}|_{\vec{R}=\vec{R}^0}\,. \tag{5.72}$$

To derive the complete Hamiltonian of electron–phonon coupling in the second-quantization representation, we need to introduce the electron creation and annihilation operators. Choosing the Bloch function as the basis, $\hat{H}^0_{\text{e}}$ can be diagonalized:

$$\hat{H}^0_{\text{e}} = \sum_{n\vec{k}}\left(\varepsilon_{n\vec{k}} - \mu\right)\hat{a}^\dagger_{n\vec{k}}\hat{a}_{n\vec{k}}\,. \tag{5.73}$$

Here, $\varepsilon_{n\vec{k}}$ is the eigenvalue with the energy band index n and the electron quasi-momentum $\vec{k}$; $\hat{a}^\dagger_{n\vec{k}}$ ($\hat{a}_{n\vec{k}}$) is the electron creation (annihilation) operator; and μ is the chemical potential. The second-quantization representation of $\hat{H}_1$ in Eq. (5.68) is

$$\begin{aligned}\hat{H}_1 &= \sum_{nm\vec{k}\vec{k}'}\left\langle n\vec{k}\,|\,\hat{h}^{(1)}\,|\,m\vec{k}'\right\rangle\hat{a}^\dagger_{n\vec{k}}\hat{a}_{m\vec{k}'} \\ &= \sum_{nm\vec{k}\vec{k}'}\sum_{I\kappa\alpha}\sum_{\vec{q}\nu}\left(\frac{\hbar}{2NM_\kappa\omega_{\vec{q}\nu}}\right)^{\frac{1}{2}} e^{i\vec{q}\cdot\vec{R}_I}\xi_{\kappa\alpha}(\vec{q}\nu)\left\langle n\vec{k}\left|\frac{\partial v^{\text{KS}}}{\partial R_{I\kappa\alpha}}|_{\vec{R}=\vec{R}^0}\right|m\vec{k}'\right\rangle(\hat{b}_{\vec{q}\nu} \\ &\quad + \hat{b}^\dagger_{-\vec{q}\nu})\,\hat{a}^\dagger_{n\vec{k}}\hat{a}_{m\vec{k}'}\end{aligned} \tag{5.74}$$

where $\left|n\vec{k}\right\rangle = N^{-1/2} e^{i\vec{k}\cdot\vec{r}} u_{n\vec{k}}(\vec{r})$ is the Bloch function and $u_{n\vec{k}}(\vec{r})$ is the periodic component of the lattice. In the above equation, we note that

$$\sum_I e^{-i\vec{q}\cdot\vec{r}} \frac{\partial v^{KS}(\vec{r})}{\partial R_{I\kappa\alpha}} |_{\vec{R}=\vec{R}^0} = \sum_I e^{-i\vec{q}\cdot(\vec{r}-\vec{R}_I)} \frac{\partial v^{\mathrm{KS}}(\vec{r}-\vec{R}_I)}{\partial R_{0\kappa\alpha}} |_{\vec{R}=\vec{R}^0} \tag{5.75}$$

exhibits lattice periodicity, and the matrix element can be written as

$$\begin{aligned}
&\left\langle n\vec{k} \left| e^{i\vec{q}\cdot\vec{r}} \sum\nolimits_{I\kappa\alpha} e^{-i\vec{q}\cdot(\vec{r}-\vec{R}_I)} \frac{\partial v^{\mathrm{KS}}(\vec{r}-\vec{R}_I)}{\partial R_{0\kappa\alpha}} |_{\vec{R}=\vec{R}^0} \right| m\vec{k}' \right\rangle \\
&= \frac{1}{N_I} \int d^3\vec{r} e^{-i(\vec{k}-\vec{k}'-\vec{q})\cdot\vec{r}} \sum_{I\kappa\alpha} e^{-i\vec{q}\cdot(\vec{r}-\vec{R}_I)} \frac{\partial v^{\mathrm{KS}}(\vec{r}-\vec{R}_I)}{\partial R_{0\kappa\alpha}} |_{\vec{R}=\vec{R}^0} u_{n\vec{k}}^*(\vec{r}) u_{m\vec{k}'}(\vec{r}) \\
&= \frac{1}{N_I} \sum_J e^{-i(\vec{k}-\vec{k}'-\vec{q})\cdot\vec{R}_J} \int_\Omega d^3\vec{r} e^{-i(\vec{k}-\vec{k}'-\vec{q})\cdot(\vec{r}-\vec{R}_J)} \sum_{I\kappa\alpha} e^{-i\vec{q}\cdot(\vec{r}-\vec{R}_J-\vec{R}_I)} \\
&\quad \frac{\partial v^{\mathrm{KS}}(\vec{r}-\vec{R}_J-\vec{R}_I)}{\partial R_{0\kappa\alpha}} |_{\vec{R}=\vec{R}^0} u_{n\vec{k}}^*(\vec{r}) u_{m\vec{k}'}(\vec{r}) \\
&= \delta_{\vec{k}=\vec{k}'+\vec{q}} \int_\Omega d^3\vec{r} \sum_{I\kappa\alpha} e^{-i\vec{q}\cdot(\vec{r}-\vec{R}_I)} \frac{\partial v^{\mathrm{KS}}(\vec{r}-\vec{R}_I)}{\partial R_{0\kappa\alpha}} |_{\vec{R}=\vec{R}^0} u_{n\vec{k}}^*(\vec{r}) u_{m\vec{k}-\vec{q}}(\vec{r}) \\
&= \left\langle u_{n\vec{k}} \left| \sum\nolimits_{Ik\alpha} e^{-i\vec{q}\cdot(\vec{r}-\vec{R}_I)} \frac{\partial v^{\mathrm{KS}}(\vec{r}-\vec{R}_I)}{\partial R_{0\kappa\alpha}} |_{\vec{R}=\vec{R}^0} \right| u_{m\vec{k}-\vec{q}} \right\rangle_\Omega \delta_{\vec{k}=\vec{k}'+\vec{q}} \quad ,
\end{aligned} \tag{5.76}$$

where Ω indicates integration within the unit cell. Therefore, Eq. (5.74) can be expressed as

$$\hat{H}_1 = N_I^{-\frac{1}{2}} \sum_{nm\vec{k}} \sum_{\vec{q}v} g_{nmv}(\vec{k},\vec{q}) \hat{a}^\dagger_{n\vec{k}} \hat{a}_{m\vec{k}-\vec{q}} (\hat{b}_{\vec{q}v} + \hat{b}^\dagger_{-\vec{q}v}). \tag{5.77}$$

Note that we introduced the electron–phonon-coupling matrix element

$$g_{nmv}(\vec{k},\vec{q}) = \left\langle u_{n\vec{k}} \left| \nabla_{\vec{q}v} v^{\mathrm{KS}} \right| u_{m\vec{k}-\vec{q}} \right\rangle_\Omega, \tag{5.78}$$

where

$$\nabla_{\vec{q}v} v^{KS} = \sum\nolimits_{I\kappa} \left(\frac{\hbar}{2M_\kappa \omega_{\vec{q}v}} \right)^{\frac{1}{2}} e^{-i\vec{q}\cdot(\vec{r}-\vec{R}_I)} \xi_\kappa(\vec{q}v) \cdot \nabla_{\vec{R}_{0\kappa}} v^{KS}(\vec{r}-\vec{R}_I). \tag{5.79}$$

Similarly, the second-quantization representation of the second-order perturbative Hamiltonian in Eq. (5.68) takes the following form [Giustino2017]:

$$\hat{H}_2 = \frac{1}{N} \sum_{nm\vec{k}} \sum_{\vec{q}\vec{q}'vv'} g^{(2)}_{nmvv'}(\vec{k},\vec{q},\vec{q}') \hat{a}^\dagger_{n\vec{k}} \hat{a}_{m\vec{k}-\vec{q}-\vec{q}'} (\hat{b}_{\vec{q}v} + \hat{b}^\dagger_{-\vec{q}v})(\hat{b}_{\vec{q}'v'} + \hat{b}^\dagger_{-\vec{q}'v'}), \tag{5.80}$$

where

$$g^{(2)}_{nmvv'}(\vec{k},\vec{q},\vec{q}') = \frac{1}{2} \left\langle u_{n\vec{k}} \left| \nabla_{\vec{q}v} \nabla_{\vec{q}'v'} v^{\mathrm{KS}} \right| u_{m\vec{k}-\vec{q}-\vec{q}'} \right\rangle_{unit}. \tag{5.81}$$

Now, the total Hamiltonian of the system can be expressed as

$$\hat{H} = \hat{H}_0 + \hat{H}_1 + \hat{H}_2, \tag{5.82}$$

where

$$\hat{H}_0 = \hat{H}_{\mathrm{e}}^0 + \hat{H}_{\mathrm{ion}}^{\mathrm{HA}}. \tag{5.83}$$

In Eq. (5.82), $\hat{H}_0$ can be regarded as the unperturbed Hamiltonian, and $\hat{H}_1 + \hat{H}_2$ as the perturbation term. Thus, we can use many-body perturbation theory to obtain the electron–phonon coupling contribution to the Green's function and the self-energy of the electrons. Because the electron–phonon coupling occurs in a finite temperature interval, we can use the Matsubara Green's function for the electrons [Mahan1998], which is defined as

$$\begin{aligned} \mathcal{G}_{n\vec{k}}(\tau) &= -\frac{\mathrm{Tr}\left[e^{-\beta\hat{H}}\, T_\tau \hat{a}_{n\vec{k}}(\tau)\hat{a}^\dagger_{n\vec{k}}(0)\right]}{\mathrm{Tr}e^{-\beta\hat{H}}} \\ &= -\left\langle T_\tau \hat{a}_{n\vec{k}}(\tau)\hat{a}^\dagger_{n\vec{k}}(0)\right\rangle . \end{aligned} \tag{5.84}$$

Here, Tr denotes the trace of $\hat{H}$ with a complete set of eigenstates, $\beta = 1/k_B T$, and $\tau = it \in [-\beta, \beta]$ is the imaginary time. The creation and annihilation operators $\hat{a}^\dagger_{n\vec{k}}(\tau)$ and $\hat{a}_{n\vec{k}}(\tau)$ are defined in the interaction representation as

$$\hat{a}^\dagger_{n\vec{k}}(\tau) = e^{\tau\hat{H}}\hat{a}^\dagger_{n\vec{k}}e^{-\tau\hat{H}}, \tag{5.85}$$

$$\hat{a}_{n\vec{k}}(\tau) = e^{\tau\hat{H}}\hat{a}_{n\vec{k}}e^{-\tau\hat{H}}. \tag{5.86}$$

Introducing the matrix S [Mahan1998], we obtain

$$S(\tau_1,\tau_2) = T_\tau e^{-\int_{\tau_1}^{\tau_2} d\tau_1 \hat{\hat{H}}'(\tau)}, \tag{5.87}$$

where $\hat{H}' = \hat{H}_1 + \hat{H}_2$ and $\hat{\hat{H}}'(\tau) = e^{\tau\hat{H}_0}\hat{H}'e^{-\tau\hat{H}_0}$. Then, the Green's function can be rewritten as

$$\mathcal{G}_{n\vec{k}}(\tau) = -\frac{\mathrm{Tr}_0\left[e^{-\beta\hat{H}_0}\, T_\tau S(\beta,0)\, \hat{\hat{a}}_{n\vec{k}}(\tau)\hat{\hat{a}}^\dagger_{n\vec{k}}(0)\right]}{\mathrm{Tr}_0[e^{-\beta\hat{H}_0}S(\beta,0)]}, \tag{5.88}$$

where

$$\hat{\hat{a}}^\dagger_{n\vec{k}}(\tau) = e^{\tau\hat{H}_0}\hat{a}^\dagger_{n\vec{k}}e^{-\tau\hat{H}_0}, \tag{5.89}$$

$$\hat{\hat{a}}_{n\vec{k}}(\tau) = e^{\tau\hat{H}_0}\hat{a}_{n\vec{k}}e^{-\tau\hat{H}_0}. \tag{5.90}$$

The advantage of using Eqs (5.88)–(5.90) instead of Eqs (5.84)–(5.86) is that now we have a trace of $\hat{H}_0$ with a complete set of eigenstates. As a result, $\hat{H}_0$ can be easily solved at the expense of an additional S-matrix.

For the numerator of Eq. (5.88), computing the series expansion of the S-matrix yields

$$\begin{aligned} &\mathcal{G}_{n\vec{k}}(\tau) \\ &= -\sum_{n=0}^{\infty}(-1)^n \int_0^\beta d\tau_1 \int_0^\beta d\tau_2 \cdots \int_0^\beta d\tau_n \mathrm{Tr}_0\left[e^{-\beta\hat{H}_0}\, T_\tau \hat{\hat{H}}'(\tau_1)\cdots\hat{\hat{H}}'(\tau_n)\hat{\hat{a}}_{n\vec{k}}(\tau)\hat{\hat{a}}^\dagger_{n\vec{k}}(0)\right]_{\mathrm{connect}} \end{aligned} \tag{5.91}$$

Here, the subscript "connect" means that only connected, unequal Feynman diagrams are included in the expansion, and the unconnected diagrams are simply canceled out by the denominator. Using Wick's theorem to contract the annihilation operator, the Green's

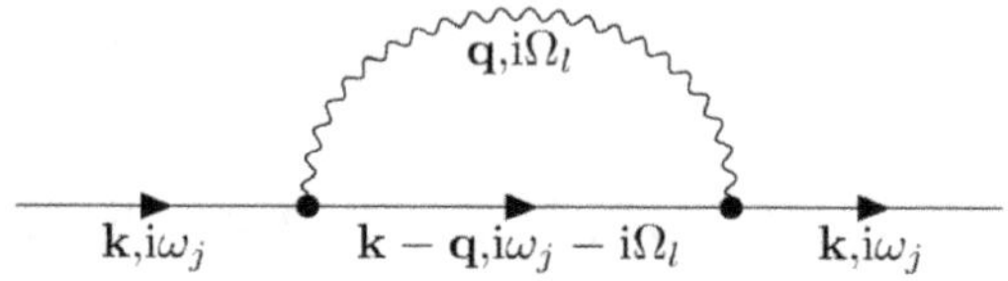

Figure 5.3 Second-order Green's functions corresponding to $\hat{H}_1$. The solid line represents the Green's function of the free electron, and the wavy line represents the Green's function of the free phonon. $\omega_j = (2j+1)\pi / \beta$ is the Matsubara frequency of the fermion, and $\Omega_l = 2l\pi / \beta$ is the Matsubara frequency of the boson. The dot • denotes the electron–phonon coupling matrix element g.

function can be obtained. Specifically, the first-order expansion of the perturbative Hamiltonian $\hat{H}_1$ will produce two pairs of creation and annihilation operators for electrons and unpaired phonon operators, and this term is zero. The second-order expansion of $\hat{H}_1$ produces three pairs of electron operators and paired phonon operators, which is a nonzero term. The Feynman diagram in frequency-momentum space is illustrated in Figure 5.3.

Based on this Feynman diagram, we can obtain the electron self-energy

$$\Sigma_{n\vec{k}}^{\mathrm{FM}}(i\omega_j) = -\frac{1}{\beta N}\sum_l \sum_{m\vec{q}\nu} \left|g_{nm\nu}(\vec{k},\vec{q})\right|^2 \mathcal{D}_{\vec{q}\nu}^0(i\Omega_l)\mathcal{G}_{m\vec{k}-\vec{q}}^0(i\omega_j - i\Omega_l), \tag{5.92}$$

where

$$\mathcal{D}_{\vec{q}\nu}^0(i\Omega_l) = -\frac{2\Omega_l}{\Omega_l^2 + \omega_{\vec{q}\nu}^2}, \tag{5.93}$$

$$\mathcal{G}_{m\vec{k}}^0(i\omega_j) = \frac{1}{i\omega_j - \varepsilon_{m\vec{k}} + \mu} \tag{5.94}$$

denote the Green's functions of the free electron and the free phonon, respectively. "FM" denotes Fan–Migdal. This term was used because (1) for semiconductors and insulators, as we will see later, the electron self-energy is consistent with the self-energy obtained by H. Y. Fan in 1951 using second-order time-dependent perturbation theory under a certain approximation [Fan H.Y.1951]; (2) for metals, this self-energy corresponds to the lowest-order electron–phonon coupling self-energy in Migdal's theory [Migdal1958].

The first-order expansion of $\hat{H}_2$ produces two electron propagators and two phonon propagators. This term is nonzero and can be expressed in the frequency-momentum space using the Feynman diagram in Figure 5.4.

Here, the electron self-energy is expressed as

$$\Sigma_{n\vec{k}}^{\mathrm{DW}}(i\omega_j) = -\frac{1}{\beta N}\sum_l \sum_{\vec{q}\nu} g_{nn\nu\nu}^{(2)}(\vec{k},\vec{q},-\vec{q})\mathcal{D}_{\vec{q}\nu}^0(i\Omega_l). \tag{5.95}$$

DW denotes Debye–Waller. In 1955, E. Antončík used the pseudopotential form factor modified by the Debye–Waller factor to calculate the temperature-dependent energy band [Antončík1955]. Equation 5.95 corresponds to the first-order term of the Antončík's method.

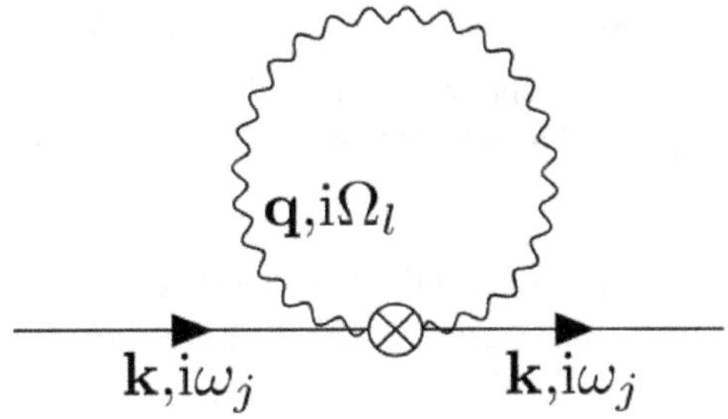

Figure 5.4 First-order Green's function corresponding to $\hat{H}_2$. ⊗ denotes the electron–phonon-coupling matrix element $g^{(2)}$.

Using the summation rule for the Matsubara frequency [Mahan1998], we have

$$-\frac{1}{\beta}\sum_{l}\mathcal{D}^{0}_{\vec{q}\nu}\left(\mathrm{i}\Omega_l\right)\mathcal{G}^{0}_{m\vec{k}-\vec{q}}(i\omega_j-i\Omega_l) = \left[\frac{N_{\vec{q}\nu}+f(\varepsilon_{m\vec{k}-\vec{q}}-\mu)}{i\omega_j-\varepsilon_{m\vec{k}-\vec{q}}+\mu+\omega_{\vec{q}\nu}}+\frac{N_{\vec{q}\nu}+1-f(\varepsilon_{m\vec{k}-\vec{q}}-\mu)}{i\omega_j-\varepsilon_{m\vec{k}-\vec{q}}+\mu-\omega_{\vec{q}\nu}}\right] \tag{5.96}$$

and

$$N_{\vec{q}\nu} = -\frac{1}{2}+\frac{1}{\beta}\sum_{l}\frac{1}{i\Omega_l-\omega_{\vec{q}\nu}}. \tag{5.97}$$

Here, $N_{\vec{q}\nu}$ and $f(\varepsilon_{m\vec{k}-\vec{q}}-\mu)$ are the boson occupation number and the fermion occupation number, respectively. The boson frequencies in both Eqs (5.92) and (5.95) can be summed up to obtain

$$\Sigma^{\mathrm{FM}}_{n\vec{k}}(i\omega_j) = \frac{1}{N}\sum_{m\vec{q}\nu}\left|g_{nm\nu}(\vec{k},\vec{q})\right|^2\left[\frac{N_{\vec{q}\nu}+f(\varepsilon_{m\vec{k}-\vec{q}}-\mu)}{i\omega_j-\varepsilon_{m\vec{k}-\vec{q}}+\mu+\omega_{\vec{q}\nu}}+\frac{N_{\vec{q}\nu}+1-f(\varepsilon_{m\vec{k}-\vec{q}})}{i\omega_j-\varepsilon_{m\vec{k}-\vec{q}}+\mu-\omega_{\vec{q}\nu}}\right], \tag{5.98}$$

$$\Sigma^{\mathrm{DW}}_{n\vec{k}} = \frac{1}{N}\sum_{\vec{q}\nu}g^{(2)}_{nn\nu\nu}(\vec{k},\vec{q},-\vec{q})(2N_{\vec{q}\nu}+1). \tag{5.99}$$

As calculating the second derivatives of the effective potential is extremely difficult, translational invariance and the rigid-ion approximation are used in practice, and second-order electron–phonon coupling matrix elements are represented by first-order electron–phonon coupling matrix elements [Cannuccia2012]. In this way, the DW self-energy becomes

$$\Sigma^{\mathrm{DW}}_{n\vec{k}} = \frac{1}{N}\sum_{m\vec{q}\nu}{}' \frac{\tilde{g}^{(2)}_{mn\nu}(\vec{k},\vec{q})}{\varepsilon_{n\vec{k}}-\varepsilon_{m\vec{k}}}(2N_{\vec{q}\nu}+1), \tag{5.100}$$

where the prime on the summation symbol means that m runs over all bands whose index is not equal to n. The term $\tilde{g}^{(2)}$ is the second-order electron–phonon coupling matrix element, which is a real number.

Thus far, we have considered the lowest-order nonzero Green's function corresponding to $\hat{H}_1$ and $\hat{H}_2$, while the Green's function expansion in Eq. (5.91) is of infinite order.

Figure 5.5 Feynman-diagram representation of the Dyson equation. The double line indicates the modified Green's function, and the single line indicates the Green's function with no interaction.

Higher-order Green's functions frequently correspond to multiple Feynman diagrams, which are quite difficult to calculate. To consider the higher-order correction to the Green's function, we can combine the two self-energy diagrams mentioned above. In other words, the infinite-order summation of these two self-energy diagrams yields the Dyson equation. Its Feynman diagrams are shown in Figure 5.5.

This is similar to the well-known *GW* approximation used to calculate the electron–electron interaction, as discussed in Chapter 4. Here, the Green's function for the interaction is written as

$$\mathcal{G}_{n\vec{k}}(i\omega_j)=\frac{1}{i\omega_j-\varepsilon_{n\vec{k}}+\mu-\Sigma_{n\vec{k}}^{\mathrm{FM}}(i\omega_j)-\Sigma_{n\vec{k}}^{\mathrm{DW}}}. \tag{5.101}$$

Note that experimentally measured frequencies are always real. Therefore, it is customary to extend the Matsubara Green's function analytically to the complex plane (so the frequency points remain close to the real axis). By substituting $i\omega\to\omega+i\delta\,\mathrm{sgn}(\omega)$ in Eq. (5.101), we have

$$\mathcal{G}_{n\vec{k}}(\omega)=\frac{1}{\omega-\varepsilon_{n\vec{k}}+\mu+i\delta\,\mathrm{sgn}(\omega)-\Sigma_{n\vec{k}}^{\mathrm{FM}}(\omega)-\Sigma_{n\vec{k}}^{\mathrm{DW}}}, \tag{5.102}$$

$$\begin{aligned}\Sigma_{n\vec{k}}^{FM}(\omega)=\frac{1}{N}\sum_{m\vec{q}v}\left|g_{nmv}(\vec{k},\vec{q})\right|^2\Bigg[&\frac{N_{\vec{q}v}+f(\varepsilon_{m\vec{k}-\vec{q}}-\mu)}{\omega-\varepsilon_{m\vec{k}-\vec{q}}+\mu+i\delta\,\mathrm{sgn}(\omega)+\omega_{\vec{q}v}}\\&+\frac{N_{\vec{q}v}+1-f(\varepsilon_{m\vec{k}-\vec{q}})}{\omega-\varepsilon_{m\vec{k}-\vec{q}}+\mu+i\delta\,\mathrm{sgn}(\omega)-\omega_{\vec{q}v}}\Bigg],\end{aligned} \tag{5.103}$$

where "sgn" denotes the sign function. From Eq. (5.102), a spectral function that can be directly compared with the experiment is obtained:

$$\begin{aligned}A_{n\vec{k}}(\omega)&=-\frac{1}{\pi}\mathrm{Im}\,\mathcal{G}_{n\vec{k}}(\omega)\\&=-\frac{1}{\pi}\frac{\mathrm{Im}\,\Sigma_{n\vec{k}}^{\mathrm{FM}}(\omega)}{\left[\omega-\varepsilon_{n\vec{k}}+\mu-\mathrm{Re}\,\Sigma_{n\vec{k}}^{\mathrm{FM}}(\omega)-\Sigma_{n\vec{k}}^{\mathrm{DW}}\right]^2+\left|\mathrm{Im}\,\Sigma_{n\vec{k}}^{\mathrm{FM}}(\omega)\right|^2}.\end{aligned} \tag{5.104}$$

Assuming $\mathcal{G}_{n\vec{k}}$ has a singularity at $E_{n\vec{k}}+i\Gamma_{n\vec{k}}-\mu$, the correction for the single-particle energy can be calculated from Eq. (5.102) as

$$E_{n\vec{k}}=\varepsilon_{n\vec{k}}+\mathrm{Re}\,\Sigma_{n\vec{k}}^{\mathrm{FM}}(E_{n\vec{k}}+i\Gamma_{n\vec{k}}-\mu)+\Sigma_{n\vec{k}}^{\mathrm{DW}}, \tag{5.105}$$

and

$$\Gamma_{n\vec{k}}=\mathrm{Im}\,\Sigma_{n\vec{k}}^{\mathrm{FM}}(E_{n\vec{k}}+i\Gamma_{n\vec{k}}-\mu). \tag{5.106}$$

Because the FM self-energy depends on the frequency, and the right-hand side of the equation depends on the quantity to be solved, a self-consistent iterative solution is necessary. In the case where the quasi-particle approximation is valid, the Taylor expansion can be applied to avoid the self-consistency issue:

$$E_{n\vec{k}} \approx \varepsilon_{n\vec{k}} + \mathrm{Re}\,\Sigma_{n\vec{k}}^{\mathrm{FM}}(\varepsilon_{n\vec{k}} - \mu) + \Sigma_{n\vec{k}}^{\mathrm{DW}} + \frac{\partial\, \mathrm{Re}\,\Sigma_{n\vec{k}}^{FM}(\omega)}{\partial \omega}\Big|_{\omega=\varepsilon_{n\vec{k}}-\mu} (E_{n\vec{k}} - \varepsilon_{n\vec{k}}). \quad (5.107)$$

The renormalization factor is defined as

$$Z_{n\vec{k}} = \frac{1}{1 - \dfrac{\partial\, \mathrm{Re}\,\Sigma_{n\vec{k}}^{FM}(\omega)}{\partial \omega}\Big|_{\omega=\varepsilon_{n\vec{k}}-\mu}}. \quad (5.108)$$

Then, we have

$$E_{n\vec{k}} \approx \varepsilon_{n\vec{k}} + Z_{n\vec{k}}\left[\mathrm{Re}\,\Sigma_{n\vec{k}}^{\mathrm{FM}}(\varepsilon_{n\vec{k}} - \mu) + \Sigma_{n\vec{k}}^{\mathrm{DW}}\right] \quad (5.109)$$

and

$$\Gamma_{n\vec{k}} \approx \mathrm{Im}\,\Sigma_{n\vec{k}}^{\mathrm{FM}}(\varepsilon_{n\vec{k}} - \mu). \quad (5.110)$$

From these two equations, it can be seen that the DW self-energy only contributes to the single-particle energy shift, while the FM self-energy not only shifts the single-particle energy but also provides the single particle with a finite lifetime. Let the renormalization factor in Eq. (5.109) be 1. In the literature, this approximation is known as the on-the-mass-shell approximation, and it holds when the quasiparticle approximation holds. Under this approximation, the form of the FM self-energy is consistent with the first-order perturbative Hamiltonian correction to the semiconductor electron energy according to H. Y. Fan's calculation using second-order time-dependent perturbation theory [Fan H.Y.1951].

In the history of semiconductor research, the FM and DW self-energies were long considered to be the same thing, leading to the misconception that it is sufficient to only include one self-energy in the calculation. In 1976, P. B. Allen et al. realized that both self-energies must be included in practical calculations, and that they are of equal magnitude [Allen1976]. In addition to the "on the mass shell" approximation, the adiabatic approximation must be made: The phonon frequency is negligible compared with the difference of the electron energies. To put it differently, the adiabatic approximation assumes that the nuclear motion is much slower than the electron motion. Assuming $\varepsilon_{n\vec{k}} - \varepsilon_{m\vec{k}-\vec{q}} \pm \omega_{\vec{q}\nu} \approx \varepsilon_{n\vec{k}} - \varepsilon_{m\vec{k}-\vec{q}}$ leads to the historically popular result regarding the temperature-dependent energy band based on Allen–Heine–Cardona theory [Allen1976, 1981]:

$$E_{n\vec{k}} = \varepsilon_{n\vec{k}} + \frac{1}{N}\sum_{m\vec{q}\nu}\left[\frac{\left|g_{nm\nu}(\vec{k},\vec{q})\right|^2}{\varepsilon_{n\vec{k}} - \varepsilon_{m\vec{k}-\vec{q}}} - \frac{1}{2}\sum_{m\vec{q}\nu}{}' \frac{\tilde{g}_{mn\nu}^{(2)}(\vec{k},\vec{q})}{\varepsilon_{n\vec{k}} - \varepsilon_{m\vec{k}}}\right](2N_{\vec{q}\nu} + 1). \quad (5.111)$$

Again, one summation has a prime, meaning that m runs over all bands whose index is not equal to n.

As can be seen from Eq. (5.111), the electron–phonon coupling renormalizes the energy spectrum of the electron. In fact, it also corrects the energy spectrum of the phonon. Under the Born–Oppenheimer approximation, the electronic ground state evolves adiabatically and can accumulate a geometric (Berry) phase characterized by the molecular Berry curvature. The molecular Berry curvature acts as a nonlocal effective magnetic field in the equation of nuclear motion, allowing the nuclear motion in real space to be affected by distant nuclei. As a result, the degeneracy of the optical branch at the Brillouin zone center is lifted in the phonon spectrum [Saparov2022].

In addition, electron–phonon coupling has a significant impact on the electron transport properties of metals. As the DW self-energy has no frequency dependence and no imaginary part, only the FM self-energy is considered in the study of transport properties. To describe the electron–phonon coupling properties of the system, the Eliashberg equation is defined [Allen1975] as

$$\alpha^2 F(\omega)=\frac{1}{N_{\uparrow}(\mu)}\sum_{mn\vec{k}}\sum_{\vec{q}\nu}\left|g_{nm\nu}(\vec{k},\vec{q})\right|^2\delta(\varepsilon_{n\vec{k}}-\mu)\delta(\varepsilon_{m\vec{k}-\vec{q}}-\mu)\delta(\hbar\omega-\hbar\omega_{\vec{q}\nu}), \quad (5.112)$$

where $N_{\uparrow}(\mu)$ is the density of electron states for a single spin on the Fermi surface. This equation is divided by the frequency and then integrated over it to obtain a dimensionless parameter λ that reflects the electron–phonon coupling strength. This is expressed as

$$\lambda=2\int_0^{\infty}\frac{\alpha^2 F(\omega)}{\omega}d\omega. \quad (5.113)$$

Using $\alpha^2 F(\omega)$ and λ, logarithmically averaged phonon frequencies can be defined as

$$\omega_{\log}=\exp\left[\frac{2}{\lambda}\int_0^{\infty}\log\omega\frac{\alpha^2 F(\omega)}{\omega}d\omega\right]. \quad (5.114)$$

From Eqs (5.113) and (5.114), we obtain the McMillan–Allen–Dynes formula for calculating the superconducting transition temperature T_c [McMillan1968, Allen1975]:

$$k_B T_c=\frac{\hbar\omega_{\log}}{1.2}\exp\left[-\frac{1.04(1+\lambda)}{\lambda-\mu^*(1+0.62\lambda)}\right], \quad (5.115)$$

where k_B is the Boltzmann constant and μ^* is a parameter that measures the Coulomb interaction. For the transport properties of the normal state, a spectral-function formula similar to Eq. (5.112) can be defined [Allen1978]:

$$\alpha_{tr}^2 F(\omega)=\frac{1}{N_{\uparrow}(\mu)}\sum_{mn\vec{k}}\sum_{\vec{q}\nu}\left|g_{nm\nu}(\vec{k},\vec{q})\right|^2 w_{n\vec{k},m\vec{k}-\vec{q}}\delta(\varepsilon_{n\vec{k}}-\mu)\delta(\varepsilon_{m\vec{k}-\vec{q}}-\mu)\delta(\hbar\omega-\hbar\omega_{\vec{q}\nu}). \quad (5.116)$$

where

$$w_{n\vec{k},m\vec{k}-\vec{q}}\equiv 1-\frac{\vec{v}_{n\vec{k}}\cdot\vec{v}_{m\vec{k}-\vec{q}}}{|\vec{v}_{n\vec{k}}|^2} \quad (5.117)$$

is a geometric factor depending on the velocity of two electrons ($\vec{v}_{n\vec{k}}$ and $\vec{v}_{m\vec{k}-\vec{q}}$) and the angle between them; $w_{n\vec{k},m\vec{k}-\vec{q}}$ determines the directional dependence of the electron scattering. The definition of the transport parameter of electron–phonon coupling is analogous to Eq. (5.113) [Allen1978]:

$$\lambda_{\mathrm{tr}} = 2\int_0^\infty \frac{\alpha_{\mathrm{tr}}^2 F(\omega)}{\omega} d\omega. \tag{5.118}$$

Then, $\alpha_{\mathrm{tr}}^2 F(\omega)$ and the lowest-order variational solution of the Boltzmann transport equation can be used to obtain the electrical resistivity [Allen1978, Grimvall1981]

$$\rho(T) = \frac{6\pi V}{e^2 k_B T N_\uparrow(\mu)\left\langle |\vec{v}|^2 (\mu)\right\rangle} \int_0^\infty \frac{\hbar\omega\alpha_{\mathrm{tr}}^2 F(\omega)}{\left(e^{\frac{\hbar\omega}{k_B T}} - 1\right)\left(1 - e^{-\frac{\hbar\omega}{k_B T}}\right)} d\omega, \tag{5.119}$$

where $\langle\cdots\rangle$ indicates that the electron velocity is averaged over the Fermi surface. At high temperatures (above the Debye temperature), the resistivity becomes [Allen1978]

$$\rho(T) = \frac{3\pi V \lambda_{\mathrm{tr}} k_B T}{\hbar e^2 N_\uparrow(\mu)\left\langle |\vec{v}|^2 (\mu)\right\rangle}. \tag{5.120}$$

In an actual electron–phonon coupling perturbation-theory calculation from first principles, the central quantity to be determined is the electron–phonon coupling matrix element. The most widely used method is density-functional perturbation theory [Baroni2001], which is based on the idea is that the nuclear vibration induces a perturbative potential that can change the ground-state electron density (and thus the ground-state electron wavefunction). The wavefunction change satisfies a certain equation that can be solved self-consistently. By tracing this solution back, the electron density and the effective potential change can be obtained. This method is suitable for lattice systems with small unit cells, and it allows us to obtain the electron–phonon coupling matrix elements at any point in the Brillouin zone. As electron–phonon coupling calculations often require obtaining dense matrix elements of electron–phonon coupling in the Brillouin zone, various interpolation techniques have been developed, such as Wannier function interpolation [Giustino2007]. For large systems with more complex structures, it is difficult to ensure convergence with density-functional perturbation methods.

Another method of calculating electron–phonon coupling matrix elements is the frozen-phonon supercell method [Dacorogna1985, Lam1986]. It is based on the simple idea of generating small atomic displacements according to a given phonon pattern and then calculating the corresponding change in the effective potential. This method has no convergence problem, and the computational steps are simple. However, for long-wavelength phonon modes, large supercells need to be constructed, which drastically increases the computational cost.

5.2.2 Non-Perturbative Treatment of Electron–Phonon Coupling

In subsection 5.1.2, both nuclear vibrations around the equilibrium position and the electron–phonon interaction in Hamiltonian were treated as small perturbations. In the following subsections, we will introduce non-perturbative methods for dealing with them. In particular, three representative methods will be discussed: the finite-difference method, the supercell-averaging method, and the effective-action theory. In these three methods, higher-order terms are included in the electron–phonon interaction. Thus, they are non-perturbative. Some algorithms still sample nuclear configurations around the equilibrium position, which we will specify in the subsequent discussion.

5.2.2.1 Finite-Difference Method

The finite-difference method is based on the simple harmonic approximation. The second-order Taylor expansion of the electron eigenvalues with respect to atomic displacement is given by [Capaz2005]

$$E_{n\vec{k}} = \varepsilon_{n\vec{k}} + \sum_{I\kappa\alpha} u_{I\kappa\alpha} \frac{\partial \varepsilon_{n\vec{k}}}{\partial R_{I\kappa\alpha}}|_{\vec{R}=\vec{R}^0} + \sum_{I\kappa\alpha,J\kappa'\beta} u_{I\kappa\alpha} u_{J\kappa'\beta} \frac{\partial^2 \varepsilon_{n\vec{k}}}{\partial R_{I\kappa\alpha}\partial R_{J\kappa'\beta}}|_{\vec{R}=\vec{R}^0} . \tag{5.121}$$

After thermal averaging of the atomic displacement, the first-order term sums up to zero, and only the second-order term remains. Using Eq. (5.71) [Allen1981], we have

$$E_{n\vec{k}} = \varepsilon_{n\vec{k}} + \sum_{\vec{q}\nu} \frac{\partial \varepsilon_{n\vec{k}}}{\partial N_{\vec{q}\nu}} \left(N_{\vec{q}\nu} + \frac{1}{2} \right), \tag{5.122}$$

$$\frac{\partial \varepsilon_{n\vec{k}}}{\partial N_{\vec{q}\nu}} \equiv \frac{\hbar}{2N\omega_{\vec{q}\nu}} \sum_{I\kappa\alpha,J\kappa'\beta} \frac{1}{\sqrt{M_\kappa M_{\kappa'}}} e^{i\vec{q}\cdot(\vec{R}_I-\vec{R}_J)} \xi_{\kappa\alpha}(\vec{q}\nu)\xi^*_{\kappa'\beta}(\vec{q}\nu) \frac{\partial^2 \varepsilon_{n\vec{k}}}{\partial R_{I\kappa\alpha}\partial R_{J\kappa'\beta}}|_{\vec{R}=\vec{R}^0} . \tag{5.123}$$

Note that while the atomic displacement is treated perturbatively, the electron–phonon coupling is not. In the actual calculation of Eq. (5.123), matching supercells are constructed for a given phonon pattern to produce a collective set of atomic displacements. Then, the second derivative of the electron eigenvalues with respect to the displacement is calculated using the second-order finite-difference formula. Similar to the electron–phonon coupling matrix element calculation using the frozen-phonon method, for long-wavelength phonons, a large supercell needs to be constructed. However, a non-diagonal supercell method has been recently proposed to generate atomic displacements, reducing the computational cost [Lloyd-Williams2015].

5.2.2.2 Supercell-Averaging Method

This method, which focuses on the statistical averaging of the desired physical quantity, is an adiabatic approximation method that does not consider the quantum dynamical nature of the nuclear motion. Through nuclear-configuration-sampling techniques, such as the simple harmonic approximation, molecular dynamics (MD), path-integral molecular dynamics (PIMD), and path integral Monte Carlo (PIMC) methods, different configurations are obtained for the calculation of relevant properties.

For the electron energy corrected by the nuclear motion, the temperature-dependent band gap can be expressed as

$$E_g(T) = \int \left|\chi(\vec{R})\right|^2 E_g(\vec{R}) d^3\vec{R}. \tag{5.124}$$

Here, $\chi(\vec{R})$ denotes the nuclear wavefunction. Under the simple harmonic approximation, since the phonon wavefunction can be obtained analytically, the above equation can then be transformed into a high-dimensional integral using canonical coordinates [Patrick2014]:

$$E_g(T) = \prod_{\nu} \int dx_\nu \frac{1}{\sqrt{2\pi \left\langle x_\nu^2 \right\rangle_T}} e^{-\frac{x_\nu^2}{2\left\langle x_\nu^2 \right\rangle_T}} E_g(\vec{R}), \tag{5.125}$$

$$\left\langle x_\nu^2 \right\rangle_T = (2N_\nu + 1) l_\nu^2, \tag{5.126}$$

where l_ν is the vibrational amplitude of the normal mode and N_ν is the bosonic occupation number. According to Eq. (5.71), $\vec{R}$ depends on the canonical coordinates x_ν. In Eq. (5.125), the exponent is a Gaussian function; therefore, the sampling efficiency is substantially better than with an MD-based method. Moreover, the computational cost does not increase at zero temperature. This method is more advantageous than the PIMD or PIMC sampling methods, which require many beads at low temperatures. Its disadvantage is that anharmonic effects are not taken into account.

The procedure for calculating the temperature-dependent spectrum is identical to that of Eqs (5.124)–(5.126), but the theoretical foundation is different because the spectrum involves transitions from ground state to excited states. According to Fermi's golden rule, the probability of an optically induced transition from initial state Ψ_i to final state Ψ_f is

$$W_{fi}(\omega) = \frac{2\pi}{\hbar} \left| \left\langle \Psi_f \right| \hat{M} \left| \Psi_i \right\rangle \right|^2 \delta(E_f - E_i - \hbar\omega). \tag{5.127}$$

Here, "initial" and "final" refer to the total quantum state of the electron–nucleus system and $\hat{M}$ denotes the electron-photon perturbation Hamiltonian. In the adiabatic approximation, the total wavefunction of the system can be rewritten as the product of the electron component and the nucleus component, such that

$$\Psi_{ku}(\vec{r}, \vec{R}) \approx \varphi_k(\vec{r}, \{\vec{R}\}) \chi_{ku}(\vec{R}), \tag{5.128}$$

where φ_k is the kth electron eigenstate, and χ_{ku} denotes the uth nuclear eigenstate on the kth Born–Oppenheimer potential energy surface. Substituting Eq. (5.128) into (5.127), we have

$$W_{ba}(\omega) = \frac{2\pi}{\hbar} \frac{1}{Z_a} \sum_{uv} e^{-\beta E_{au}} \left| \left\langle \chi_{bv} \right| M_{ba} \left| \chi_{au} \right\rangle \right|^2 \delta(E_{bv} - E_{au} - \hbar\omega). \tag{5.129}$$

Here, a and b denote the electron states, v and u denote the nuclear states, $Z_a = \sum_u e^{-\beta E_{au}}$ is the partition function of the nuclear evolution on the ath Born–Oppenheimer potential energy surface, and $M_{ba} = \left\langle \varphi_b \middle| \hat{M} \middle| \varphi_a \right\rangle$ is the electric-dipole-moment matrix element.

Solving the above equation can be difficult for extended systems for two reasons: (1) for crystals, integrating over each vibrational state is computationally intensive owing to the large number of vibrational modes; (2) obtaining the nuclear eigenstates on the electronic excited potential energy surface is challenging. To simplify the problem, M. Lax adopted the semiclassical approximation [Lax1952]. In particular, the number of vibrational energy levels is assumed large enough to be considered continuous. As a result, the difference between two discontinuous eigenenergy levels can be replaced by the classical potential energy difference, so that $E_{bv} - E_{au} \approx U_b^{\text{BO}}(\vec{R}) - U_a^{\text{BO}}(\vec{R})$. With this approximation, χ_{bv} in Eq. (5.129) can be summed up, and we obtain

$$W_{ba}^{\text{SC}}(\omega) = \frac{1}{Z_a} \operatorname{Tr} \hat{\rho}_a P_{ba}, \tag{5.130}$$

$$P_{ba}(\omega;\vec{R}) = \frac{2\pi}{\hbar} \frac{1}{Z_a} \left|M_{ba}(\vec{R})\right|^2 \delta\left(U_b^{\text{BO}}(\vec{R}) - U_a^{\text{BO}}(\vec{R}) - \hbar\omega\right), \tag{5.131}$$

where $\rho_a = e^{-\beta \hat{H}_a}$ is the density operator, and Tr denotes the trace of the nuclear eigenstates on the ath Born–Oppenheimer potential energy surface. Note that the imaginary part of the dielectric function can be obtained from P_{ba}, such that $\operatorname{Im} \varepsilon(\omega;\vec{R}) \propto \sum_b \left(\frac{1}{\omega}\right) P_{0a}(\omega;\vec{R})$. The real part can be determined by applying the Kramers–Krönig transformation to the imaginary part. Therefore, the temperature-dependent dielectric function is expressed as

$$\varepsilon(\omega,T) = \frac{1}{Z_0} \operatorname{Tr} \hat{\rho}_0 \mathrm{e}(\omega;\vec{R}). \tag{5.132}$$

Here, the subscript 0 denotes the electronic ground state. Therefore, the issue we are encountering here is the same as the one in Eq. (5.124), and sampling of the nuclear wavefunction at the electronic ground state is required.

Using the PIMD or PIMC methods, Eq. (5.132) can be written as [Della Sala2004]

$$\begin{aligned} \varepsilon(\omega,T) = \lim_{P\to\infty} \frac{1}{Z_P} \prod_{j=1}^{N} \left(\frac{M_j P}{2\beta\pi\hbar^2}\right)^{\frac{P}{2}} \int \mathrm{d}\vec{R}_1 \int \mathrm{d}\vec{R}_2 \cdots \int \mathrm{d}\vec{R}_P \\ \times \frac{1}{P} e^{-\beta \sum_{i=1}^{P} \left[\sum_{J=1}^{N} \frac{M_j}{2} \omega_P^2 x \left(\vec{R}_i^j - \vec{R}_{i-1}^j\right)^2 + \frac{1}{P} U_0^{BO}\left(\vec{R}_i^1,\cdots,\vec{R}_i^N\right)\right]} \sum_{i=1}^{P} \varepsilon(\omega;\{\vec{R}_i\}). \end{aligned} \tag{5.133}$$

Here, P is the number of beads, $\vec{R}_i$ denotes the nuclear configuration of the ith bead, and $\omega_P = \sqrt{P}/(\beta\hbar)$ determines the spring-interaction strength between neighboring beads. The partition function Z_P of the ring polymer system satisfies

$$\begin{aligned} Z_P = \prod_{j=1}^{N} \left(\frac{M_j P}{2\beta\pi\hbar^2}\right)^{\frac{P}{2}} \int \mathrm{d}\vec{R}_1 \int \mathrm{d}\vec{R}_2 \cdots \int \mathrm{d}\vec{R}_P \\ \times e^{-\beta \sum_{i=1}^{P} \left[\sum_{j=1}^{N} \frac{M_j}{2} \omega_P^2 \left(\vec{R}_i^j - \vec{R}_{i-1}^j\right)^2 + \frac{1}{P} U_0^{\text{BO}}\left(\vec{R}_i^1,\cdots,\vec{R}_i^N\right)\right]}. \end{aligned} \tag{5.134}$$

If we apply the simple harmonic approximation, Eq. (5.132) can be written in the same way as Eq. (5.125) [Zacharias2015]:

$$\varepsilon(\omega,T) = \prod_{v} \int dx_v \frac{1}{\sqrt{2\pi\left\langle x_v^2\right\rangle_T}} e^{-\frac{x_v^2}{2\left\langle x_v^2\right\rangle_T}} \varepsilon(\omega,\{\vec{R}\}). \tag{5.135}$$

M. Zacharias et al. successfully described phonon-assisted optical absorption and bandgap renormalization in indirect-gap semiconductors using the above equation

[Zacharias2015, 2016]. A detailed discussion of specific examples can be found in Section 2.1.

Electron–phonon scattering is an important source of electrical resistance. For liquid metals, resistance is often calculated using the Kubo–Greenwood formula [Kubo1957a,b, Greenwood1958]. This procedure is similar to that for finding the temperature-dependent dielectric function, in that it involves sampling and averaging over the configuration space. The difference is that it obtains the frequency-dependent photoelectric conductivity first and then extrapolates the zero-frequency limit using the Drude conductivity formula. For a finite number of supercells at a given point $\vec{\mathrm{k}}$ in the Brillouin zone, the distribution of electron eigenvalues is discrete. Therefore, this method exhibits prominent size effects [Pozzo2011].

In summary, the supercell-averaging method employs relatively simple calculation methods and can be applied to various electronic-structure calculations without treating the electron–phonon coupling perturbatively. However, when large supercells are required, the computational cost can be large. Furthermore, supercell averaging can only provide an ensemble average of the physical quantity, not the electron–phonon coupling matrix element. Its relationship with the perturbative electron–phonon coupling method introduced earlier is also unclear [Giustino2017].

5.2.2.3 Effective-Action Theory

Although the two methods mentioned above are currently more popular, a non-perturbative electron–phonon coupling approach based on effective-action theory has recently been developed by J. R. Shi's group at Peking University. By combining their approach with the PIMD simulation developed by X. Z. Li's group [Liu H.Y.2020], they have successfully predicted the superconducting-transition temperature of high-pressure liquid metallic hydrogen. This approach does not rely on the harmonic approximation and therefore does not involve the perturbative expansion of electron–phonon coupling. It can determine the effective electron–electron interaction and electron–phonon coupling parameter λ within conventional perturbative electron–phonon coupling theory, and its relationship with conventional perturbation theory is clearer. The main result of this theory is that the effective electron–electron interaction caused by electron–phonon coupling can be related to the T-matrix of the shielding ion potential due to fluctuation. In other words, the following Bethe-Salpeter equation can be established:

$$W_{11'} = \Gamma_{11'} + \frac{1}{\hbar^2 \beta} \sum_2 W_{12} \left|\bar{\mathcal{G}}_2\right|^2 \Gamma_{21'}. \tag{5.136}$$

Here, the numerical subscripts indicate matrix elements, such that $1 \to (\omega_n, \vec{k})$. The chosen basis is the time-dependent plane wave $\varphi_{\omega_n \vec{k}}(\vec{r}\tau) = 1/\sqrt{\hbar\beta V}\exp(-i\omega_n\tau + i\vec{k}\cdot\vec{r})$, where τ is the imaginary time, and ω_n is the fermionic Matsubara frequency. W denotes the effective electron–electron interaction, $\bar{\mathcal{G}}$ is the average electron Green's function, and $\Gamma_{11'}$ denotes the scattering amplitude of the electron pair from the initial state $1 = (\omega_{n_1}, \vec{k}_1)$ and $\bar{1} = (-\omega_{n_1}, -\vec{k}_1)$ to the final state $1'$ and $\bar{1}'$. $\Gamma_{11'}$ can be obtained from the T-matrix of the shielding ion potential as

$$\Gamma_{11'} = -\beta \left\langle \left| \mathcal{T}_{11'}[\vec{R}(\tau)] \right|^2 \right\rangle_{\mathrm{C}}. \tag{5.137}$$

Here, $\vec{R}(\tau)$ denotes the nuclear trajectory on the imaginary-time axis, the subscript C indicates the classical ensemble averaging determined by the PIMD, and $\mathcal{T}_{11'}[\vec{R}(\tau)]$ denotes the T-matrix of electron scattering by the fluctuating ion potential from 1 to 1′. $\mathcal{T}$ satisfies the following Lippmann–Schwinger equation:

$$\hat{\mathcal{T}}[\vec{R}(\tau)] = \hat{\mathcal{V}}[\vec{R}(\tau)] + \frac{1}{\hbar}\hat{\mathcal{V}}[\vec{R}(\tau)]\hat{\bar{\mathcal{G}}}\hat{\mathcal{T}}[\vec{R}(\tau)], \tag{5.138}$$

where $\hat{\mathcal{V}}[\vec{R}(\tau)] = \hat{\mathcal{V}}_{ei}[\vec{R}(\tau)] - \hat{\bar{\Sigma}}$ is the scattering potential and $\hat{\mathcal{V}}_{ei}[\vec{R}(\tau)]$ is the imaginary-time-dependent shielding ion potential. $\hat{\bar{\Sigma}}$ is the self-energy corresponding to $\hat{\bar{\mathcal{G}}}$. The frequency-dependent electron–phonon-coupling parameter λ can be obtained from W. Then, the superconducting transition temperature is predicted by solving the Eliashberg equation. Using this equation, J. R. Shi et al. found that the transition temperature of high-pressure liquid metallic hydrogen is higher than room temperature. For $\hat{\mathcal{V}}_{ei}[\vec{R}(\tau)]$, they used a model potential with the linear shielding approximation; this is a reasonable choice for hydrogen, which has no inner electrons. However, further investigation is needed to extend this approach to other elements.

5.3 Simulation of Nonadiabatic Effects

In Sections 5.1 and 5.2, we discussed the current progress of the study of nuclear quantum effects on condensed matter at the statistical and dynamics levels. We also introduced the first-principles study of electron–phonon interaction. Strictly speaking, this topic should include nonadiabatic effects, in particular, the effects of electronic transitions on the nuclear quantum states. As a part of the full quantum theory, nonadiabatic approaches to the real-time evolution of electron–nucleus interaction will be discussed in the present section. In fact, nuclear quantum effects and nonadiabatic effects are inseparable in the study of practical physical and chemical problems, and several recent studies that consider them in a unified way will be introduced in Section 5.4.

According to full quantum theory, nonadiabatic effects describe the cases in which the fast component of the motion is not necessarily in the same state, for example, the ground state of a many-body system. The fast component may experience a change due to the change in the slow component, causing nonadiabatic electron transitions [Tully1998, Hack2000, Cederbaum2008]. Evidently, the changes in the motion of the fast component also affect the evolution of the slow component, and there is a quantum coupling between them. In the description of real processes, it is impossible to completely separate the fast component evolution (usually electron motion) from the slow component evolution (usually nuclear motion). In the quantum evolution of a many-body system, all electronic and nuclear degrees of freedom must be quantized, and the coupling between nuclear and electron motion must be taken into account. The quantum coupling between fast and slow components is collectively known as the nonadiabatic coupling [Hu C.P.2007]. In the Born–Huang expansion, the coupling terms are obtained by summing over all electronic configurations, and the nuclear component evolves with time. The evolution of the nuclear wavefunction is related not only to the concurrent electron eigenstate but also to other electron eigenstates.

In principle, the most appropriate way to describe nonadiabatic effects in a polyatomic system is to use a wavefunction that includes both electronic and nuclear degrees of freedom. The evolution of this common many-body wavefunction reflects the mutual influence of electrons and nucleus. In actual calculations, a wavefunction containing both the electronic and nuclear degrees of freedom can be written as a linear superposition of multiple electron eigenstates according to the Born–Huang expansion. Further, the coefficient in front of each electron eigenstate is the nuclear wavefunction component that describes the corresponding nuclear state.

We have already thoroughly discussed this approach in Section 4.2. Its main limitation is that it has a high computational cost, and it requires determining the electron eigen-energies and eigen-wavefunctions (the latter is necessary for calculating the nonadiabatic coupling terms) for all nuclear configurations. As in the usual electron cases, the computational cost grows exponentially with the number of nuclei if the nuclear quantum state is described by the wavefunction. Evidently, one can also discard the wavefunction approach and describe nuclear quantum states numerically using semiclassical or path-integral methods. However, while these methods are successful in computing statistical properties, many practical difficulties arise when the dynamical properties of the system are calculated numerically, as we have discussed in subsection 5.1.3. Therefore, the practical application of computational methods that simultaneously quantize electronic and nuclear motion is currently very limited, especially for condensed matter systems with a large number of particles [Beck2000].

Nevertheless, we can always attempt to solve the real problem by adopting a suitable approximation. As in the Born–Oppenheimer approximation, we can treat the slow component (nuclear degrees of freedom) classically. In this way, nonadiabatic effects can be described approximately throughout the evolution of the system. This is known as the method of nonadiabatic simulation, and it is currently the most commonly used method for treating nonadiabatic effects in the literature.

It is therefore reasonable to argue that the essence of understanding the electron–nucleus nonadiabatic coupling based on classical nuclear motion lies in the idea that both the electrons and the nuclei need to be described in a unified, quantum-mechanical way. This idea is the central theme of this book. When we consider slow-component evolution adiabatically but retain its quantum properties, our research is focused on the nuclear quantum effects. When we consider the evolution of the slow component classically but retain its nonadiabatic properties, our research focus has shifted to the nonadiabatic effects. From the perspective of the Born–Huang expansion, both are improvements over the adiabatic ball-and-stick model of the electronic structure, only with different emphases.

In this section, we will introduce nonadiabatic molecular dynamics simulation methods based on classical nuclear trajectories (the customary methods in most research work). Although in the previous discussion, we often referred to the evolution of the electron and nucleus systems as a dynamic process, it is worth noting that electron–nucleus nonadiabatic effects must also be included when considering the thermal and statistical properties of the entire system near the equilibrium state. Similar to the nuclear quantum effects, nonadiabatic effects manifest themselves at both the dynamic and the thermodynamic levels. As discussed in subsection 2.2.1, nonadiabatic effects can also affect the frequency of atomic vibrations, but we will not emphasize this in detail in our subsequent discussion.

Studies on nonadiabatic effects in finite-sized isolated systems, including the free H_2 molecule, HF molecule, and benzene ring, have been carried out by D. H. Zhang [Zhang D.H.1994]; W. H. Fang and J. S. Shao [Shao J.S.1999]; J. Liu [Liu J.2006]; and other research teams [Shi Q.2003]. S. Meng's group and others have studied nonadiabatic effects in larger nanosystems and periodic solid systems (including nanotubes, 2D materials, surfaces, interfaces, and bulk materials), using model calculations and first-principles simulations [Meng S.2008a, Lian C.2018, Zheng Q.J.2019]. These methods can be broadly divided into three categories: trajectory-based surface hopping, Ehrenfest dynamics, and other hybrid quantum–classical dynamics methods. In subsections 5.3.1–5.3.3, we will discuss each of these in more detail.

5.3.1 Trajectory-Based Surface-Hopping Method

Surface hopping is a semiclassical nonadiabatic method introduced by J. C. Tully et al. [Tully1990]. Nonadiabatic effects in this method are described using the electronic transitions between the ground state and the different excited states, as well as the evolutions of the nuclei among the different potential energy surfaces. The essential difference between this method and those within the Born–Oppenheimer approximation is that, in the latter, the nucleus remains on the ground-state surface and follows Newtonian mechanics. In the trajectory-based surface-hopping method, by contrast, the nucleus can evolve and hop on different potential energy surfaces. However, within each potential energy surfaces, classical trajectories are often used in practice. As a result, quantum tunneling effects and the zero-point motion of the nucleus are often neglected. The physics underlying the trajectory-based surface-hopping method is easy to visualize, and the implementation of the method is simple; however, multiple potential energy surfaces must be calculated in advance, including that of the excited states. Furthermore, if the electron excitation process is too complex and cannot be approximated by a few excited-state potential energy surfaces, the method will no longer be applicable [Barbatti2011].

In essence, the trajectory-based surface-hopping method is a hybrid quantum-classical approach to the nonadiabatic evolution of physical systems. It utilizes the classical evolution equations (Newton's equations) to describe the motion of the nuclear degrees of freedom. The nucleus is only allowed on one specific potential energy surface at a time, but it can hop to another potential energy surface at any time and continue to evolve. The probability of hopping depends on the distribution of the fully quantized wave packet on a particular classical trajectory during the evolution. Thus, the wave-packet splitting can be described by the statistical distribution of independently evolving classical trajectories (Figure 5.6).

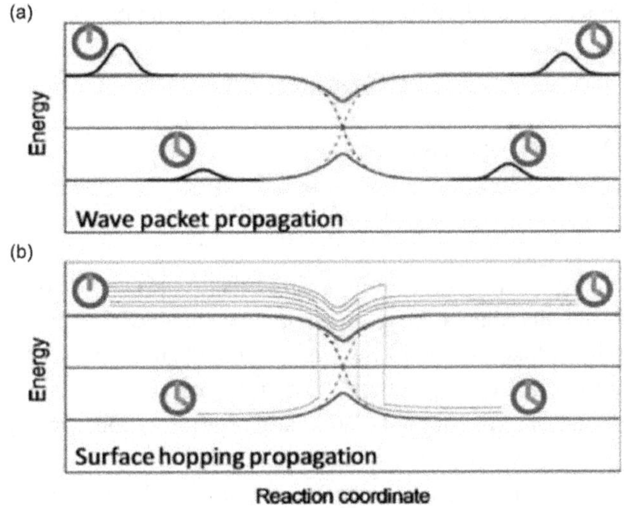

Figure 5.6 (a) Schematic of the dynamic evolution of a wave packet. The solid and dashed curves represent potential energy surfaces in the adiabatic and diabatic basis-vector representations, respectively. The clock near the wave packet provides the current time. (b) In the trajectory-based surface-hopping method, the evolution of the wave packet is represented by a series of randomly generated classical trajectories that satisfy certain distribution conditions. From Barbatti, M. [2011]. Reproduced with permission from John Wiley and Sons.

Assuming that the nuclear motion follows a classical path $\boldsymbol{R}^{\mathrm{c}}(t)$, the electron wavefunction that evolves with time can be written as a superposition of electronic eigenfunctions Φ_j that are determined by the classical nuclear configuration $\boldsymbol{R}^{\mathrm{c}}$. This is expressed as

$$\varphi(\boldsymbol{r},\boldsymbol{R}^{\mathrm{c}},t)=\sum_j c_j(t)\Phi_j(\boldsymbol{r};\boldsymbol{R}^{\mathrm{c}}(t)), \tag{5.139}$$

where the Φ_j satisfy

$$\langle\Phi_k \mid \Phi_l\rangle_{\boldsymbol{r}} \equiv \int \Phi_k^*(\boldsymbol{r};\boldsymbol{R}^{\mathrm{c}}(t))\,\Phi_l(\boldsymbol{r};\boldsymbol{R}^{\mathrm{c}}(t))\mathrm{d}\boldsymbol{r}=\delta_{kl}. \tag{5.140}$$

Substituting the approximate wavefunction in Eq. (5.139) into the time-dependent Schrödinger equation of the electron, we have

$$\left(i\hbar\frac{\partial}{\partial t}-\hat{H}_e\right)\varphi(\boldsymbol{r},\boldsymbol{R}^{\mathrm{c}},t)=0\ , \tag{5.141}$$

which can be translated into the following form:

$$i\hbar\frac{dc_k}{dt}+\sum_j\left(-H_{kj}^{\mathrm{c}}+i\hbar\boldsymbol{F}_{kj}^{\mathrm{c}}\cdot\boldsymbol{v}^{\mathrm{c}}\right)c_j=0\ . \tag{5.142}$$

This is the evolution equation for the coefficients c_j, which is also called the semiclassical time-dependent Schrödinger equation. Here, the Hamiltonian matrix element can be written as:

$$H_{kj}^{\mathrm{c}} \equiv \left\langle\Phi_k\middle|\hat{H}_{\mathrm{e}}\middle|\Phi_j\right\rangle_{\boldsymbol{r}}, \tag{5.143}$$

where $\boldsymbol{F}_{kj}^{\mathrm{c}}$ represents the strength of the nonadiabatic coupling between the electronic states k and j, such that

$$\boldsymbol{F}_{kj}^{c} \equiv \left\langle \Phi_k \left| \nabla_{\boldsymbol{R}} \right| \Phi_j \right\rangle_{\boldsymbol{r}}. \tag{5.144}$$

Using Eqs (5.141)–(5.144), we have:

$$\boldsymbol{F}_{kj}^{\mathrm{c}} \cdot \boldsymbol{v}^{\mathrm{c}} = \left\langle \Phi_k \left| \frac{\partial}{\partial t} \right| \Phi_j \right\rangle_{\boldsymbol{r}}, \tag{5.145}$$

where $\boldsymbol{v}^{c}$ is the velocity vector of the atom. Using Eqs (5.145) and (5.141), the evolution of wavefunction coefficients, that is, Eq. (5.142), can be easily derived.

In Eq. (5.139), if we take the electron wavefunction Φ_j to be the eigenstate corresponding to the classical nuclear configuration $\boldsymbol{R}^{\mathrm{c}}$, then Φ_j becomes the adiabatic basis vector Φ_j^{a}. In this representation, known as the adiabatic representation, the Hamiltonian matrix is diagonal: $H_{kj}^{\mathrm{c}} = V_k \delta_{kj}$, where each nonzero matrix element V_k corresponds to the kth potential energy surface. If we choose another set of orthonormal basis vectors Φ_j^{d}, such that $H_{kj}^{\mathrm{c}} = W_{kj}$ and $\boldsymbol{F}_{kj}^{\mathrm{c}} = 0$, then Φ_j^{d} is the diabatic basis vector. The corresponding representation is called the diabatic representation. Both representations can be used to solve the coefficient-evolution equation, that is, Eq. (5.142). The adiabatic representation is more commonly used in practical applications because of its clarity and simplicity. There are advantages in using diabatic representation, as it tends to stabilize the process of dynamic evolution. However, it is important to note that the diabatic basis vector is not unique and must be skillfully constructed. In practical simulations, either of the two representations may be used, depending on which is more convenient.

In the trajectory-based surface-hopping method, the nucleus moves classically, that is, the nuclear motion on the potential energy surface is driven by Newton's equation, and the nucleus moves along the lth potential energy surface. This is expressed as

$$M \frac{\mathrm{d}^2}{\mathrm{d}t^2} \boldsymbol{R}^{\mathrm{c}} = \boldsymbol{f}^{\mathrm{c}} = -\nabla_{\boldsymbol{R}} H_{ll}^{\mathrm{c}}. \tag{5.146}$$

To determine the trajectory of the electron, the transition probability between states must be calculated. In fact, using different methods to estimate the probability of the electron transitioning from one state l to another state k is equivalent to using different trajectory-based surface-hopping methods. The simplest approach is to specify that electron transition from one state to another occurs once the energy difference between the two electron states is less than a given value. However, this is slightly inaccurate, and it would be more reasonable to determine the transition probability between different electron states based on the evolution equation of the coefficients in front of the electron orbital basis. This can be performed analytically for the Landau–Zener model [Wittig2005], though some assumptions and simplifications are included in the calculation. For example, this method requires that the two electronic energy levels must approach each other linearly at a fixed rate.

The most commonly used method for determining the transition probability is the fewest-switch surface-hopping method proposed by J. C. Tully in 1990 [Tully1990]. The number of electron transitions is minimized in this method, given that the probability of electron transitions follows a certain distribution. Therefore, the probability of transitioning from electron state l to k is expressed as

$$P_{l\to k} = \frac{\text{Increase in the } k \text{ state occupation number due to the } l \text{ to } k \text{ transition}}{l \text{ state occupation number}}. \quad (5.147)$$

Note that the transition probability is no longer proportional to the occupation number itself, but to the change in the occupation number. Consequently, any mutually offsetting l-to-k and k-to-l transition contributions are excluded from consideration, which is what is meant by the term "fewest-switch." The occupation number of a given electron state is determined by the diagonal elements of its corresponding density matrix

$$\rho_{lk}(t) = c_l c_k^*. \quad (5.148)$$

Let us consider the evolution of the density matrix ρ_{kj}. Substituting it into Eq. (5.142), we have

$$i\hbar\dot{\rho}_{kj} = \dot{c}_k c_j^* + c_k \dot{c}_j^* = \sum_l \left\{\rho_{lj}(H_{kl}^{\mathrm{c}} - i\hbar \boldsymbol{F}_{kl}^{\mathrm{c}} \cdot \boldsymbol{v}^{\mathrm{c}}) - \rho_{kl}(H_{lj}^{\mathrm{c}} - i\hbar \boldsymbol{F}_{lj}^{\mathrm{c}} \cdot \boldsymbol{v}^{\mathrm{c}})\right\}, \quad (5.149)$$

where the relations $\boldsymbol{F}_{kl}^{\mathrm{c}*} = -\boldsymbol{F}_{lk}^{\mathrm{c}}$, and $\boldsymbol{F}_{ll}^{\mathrm{c}} = 0$ have been used.

In particular, we obtain the evolution equation for the diagonal matrix elements of the density matrix:

$$i\hbar\dot{\rho}_{kk} = \sum_{l\neq k} b_{lk}, \quad (5.150)$$

where

$$b_{lk} = \frac{2}{\hbar}\mathrm{Im}(\rho_{lk} H_{lk}^{\mathrm{c}}) - 2\,\mathrm{Re}(\rho_{lk} \boldsymbol{F}_{lk}^{\mathrm{c}} \cdot \boldsymbol{v}^{\mathrm{c}}). \quad (5.151)$$

Now, we can prove that Eq. (5.147) is satisfied if the transition probability from electron state l to k is written as

$$P_{l\to k} = \max\left\{0, \frac{\Delta t}{\rho_{ll}} b_{lk}\right\} = \max\left\{0, \frac{2\Delta t}{\rho_{ll}}\left(\frac{1}{\hbar}\mathrm{Im}(\rho_{lk})H_{lk}^{\mathrm{c}} - \mathrm{Re}(\rho_{lk})\boldsymbol{F}_{lk}^{\mathrm{c}} \cdot \boldsymbol{v}^{\mathrm{c}}\right)\right\}. \quad (5.152)$$

In practice, to ensure that the transition from electron state l to k occurs, two conditions must be met:

(1) A random number r_t between [0,1] is generated, which satisfies:

$$\sum_{n=1}^{k-1} P_{l\to n}(t) \le r_t \le \sum_{n=1}^{k} P_{l\to n}(t); \quad (5.153)$$

(2) The energy corresponding to the electron state l and k obeys

$$V_k\left(\boldsymbol{R}^c(t)\right) - V_l\left(\boldsymbol{R}^c(t)\right) \le \frac{M(\boldsymbol{F}_{kl}^c \cdot \boldsymbol{v}^c)^2}{2(\boldsymbol{F}_{kl}^c)^2}. \quad (5.154)$$

The second relation ensures that the total energy of the system does not increase after the electron transition. To satisfy it, the kinetic energy of each nucleus must be adjusted linearly after the electron transition, based on the magnitude and direction of $\boldsymbol{F}_{lk}^{\mathrm{c}}$, so that the total kinetic energy of the nuclear system should decrease by $V_k\left(\boldsymbol{R}^c(t)\right) - V_l\left(\boldsymbol{R}^c(t)\right)$.

This guarantees total energy conservation in the entire system before and after the electron transition. In fact, it is equivalent to exerting an additional force $\boldsymbol{F}_{lk}^c$ on the nucleus during the electron transition. To simplify the process, it is also possible to adjust the magnitude of nuclear velocity $\boldsymbol{v}^c$ in its natural direction so that the total energy is conserved before and after the transition. As a result, complicated calculation of $\boldsymbol{F}_{lk}^c$ is no longer required, and the condition in Eq. (5.154) is changed to

$$V_k\left(\boldsymbol{R}^c(t)\right)-V_l\left(\boldsymbol{R}^c(t)\right)\leq E_{kin}(\boldsymbol{v}^c). \tag{5.155}$$

In practice, the density matrix of the electron states is represented by Eq. (5.149) and depends only on the evolution of the coefficient c_k. All coefficients c_k are driven by the classical nuclear trajectories at state l, causing the system to maintain an excessively strong and false coherence between different electron states. A common way to solve this problem is to adopt the mean-field approximation and introduce decoherence corrections to the trajectory-based surface-hopping method. For example, after solving Eq. (5.142), the magnitude of the coefficient c_k can be manually adjusted so that

$$c_k' = c_k \exp\left(-\frac{\Delta t}{\tau_{kl}}\right)\forall k\neq l, \tag{5.156}$$

$$c_l' = c_l\left[\frac{1-\sum_{k\neq l}\left|c_k'\right|^2}{\left|c_l\right|^2}\right]^{1/2}, \tag{5.157}$$

$$\tau_{kl}\equiv\frac{\hbar}{\left|V_{kk}-V_{ll}\right|}\left(1+\frac{\alpha}{E_{kin}}\right), \tag{5.158}$$

where l is the current electron state, E_{kin} is the total kinetic energy of the nucleus, and α is a manually adjusted empirical parameter with a magnitude of ~3 eV. Introducing the decoherence correction makes the statistical average of the multiple simulated trajectories closer to the real evolution of the electron state.

It is worth noting that the nuclear zero-point motion and quantum tunneling effects are often neglected in the trajectory-based surface-hopping method because of the use of classical nuclear trajectories. The coherence between the electron states is also misrepresented. The coherence between the electron states is too strong before the transition, and the randomness and independence of the trajectories after the transition can completely remove the electron coherence. As a result, electron state coherence only approximately obeys the assumed statistics. In addition, Tully et al. have also shown that the occupation of various electronic states approximately obeys the quantum Boltzmann distribution and the detailed-balance condition. It is not known whether the various decoherence corrections have any effect on this approximate detailed-balance distribution.

For statistical significance, a large number of trajectories must be considered. In actual simulations, hundreds to thousands of trajectories are often selected for a simulation time of a few picoseconds. However, whether or not these parameters are sufficient to produce an accurate result is still an open question. Obtaining sufficiently convergent statistics can be extremely computationally intensive, especially when combined with first-principles electronic-structure methods without model potentials. Currently, there are only a few algorithms that truly combine trajectory-based surface-hopping methods with first-principles calculations. Popular approaches in the literature are either based on model Hamiltonians or use electronic ground-state potential energy surfaces for nuclear configuration evolution and then track the corresponding nonadiabatic changes of the electronic states on the given nuclear trajectory [Tully1998, Hack2000, Zheng Q.J.2019].

5.3.2 Ehrenfest Dynamics Methods

The Ehrenfest dynamics method is a class of methods named after the Austrian physicist P. Ehrenfest. In these methods, nuclear evolution is classical but the electron motion is obtained by solving the time-dependent Schrödinger equation.

The Ehrenfest dynamics method is based on the Ehrenfest theorem, which describes the evolution of quantum systems and can be derived from the Schrödinger equation [Ehrenfest1932]. Assuming that the wavefunction. Satisfies the Schrödinger equation

$$i\hbar\frac{\partial}{\partial t}\varphi(x,t)=\hat{H}\varphi(x,t) \tag{5.159}$$

we have

$$-i\hbar\frac{\partial}{\partial t}\varphi^*(x,t)=\varphi^*(x,t)\hat{H}^*=\varphi^*(x,t)\hat{H}. \tag{5.160}$$

For any physical operator A, the evolution of the expectation value then satisfies

$$\begin{aligned}\frac{d}{dt}\langle A\rangle&=\frac{d}{dt}\int\varphi^*(x,t)\,A\varphi(x,t)dx^3\\&=\int\frac{\partial\varphi^*(x,t)}{\partial t}A\varphi(x,t)dx^3+\int\varphi^*(x,t)\frac{\partial A}{\partial t}\,\varphi(x,t)dx^3+\int\varphi^*(x,t)\,A\frac{\partial\varphi(x,t)}{\partial t}dx^3\\&=-\frac{1}{i\hbar}\int\varphi^*(x,t)\,\hat{H}A\varphi(x,t)dx^3+\left\langle\frac{\partial A}{\partial t}\right\rangle+\frac{1}{i\hbar}\int\varphi^*(x,t)\,A\hat{H}\varphi(x,t)dx^3\\&=\frac{1}{i\hbar}\langle[A,\hat{H}]\rangle+\left\langle\frac{\partial A}{\partial t}\right\rangle.\end{aligned} \tag{5.161}$$

Eq. (5.161) is Ehrenfest's theorem.

For an ordinary system, the Hamiltonian is written as

$$\hat{H}(x,p,t)=\frac{p^2}{2m}+V(x,t)\ . \tag{5.162}$$

Using Ehrenfest's theorem, we can easily obtain the evolution equation of the expectation value of the momentum for the particle:

$$\frac{\mathrm{d}}{\mathrm{dt}}\langle p\rangle=\frac{1}{i\hbar}\langle[p,\hat{H}]\rangle+\left\langle\frac{\partial p}{\partial t}\right\rangle=\frac{1}{i\hbar}\langle[p,V(x,t)]\rangle. \tag{5.163}$$

Inserting the operator definition of p into the commutator, we have,

$$\begin{aligned}\frac{\mathrm{d}}{\mathrm{dt}}\langle p\rangle &= -\int \varphi^*(x,t)\nabla\big(V(x,t)\varphi(x,t)\big)dx^3 + \int \varphi^*(x,t)V(x,t)\nabla\varphi(x,t)dx^3 \\ &= -\int \varphi^*(x,t)\big(\nabla V(x,t)\big)\varphi(x,t)dx^3 - \int \varphi^*(x,t)V(x,t)\nabla\varphi(x,t)dx^3 \\ &\quad + \int \varphi^*(x,t)V(x,t)\nabla\varphi(x,t)dx^3 = -\int \varphi^*(x,t)\big(\nabla V(x,t)\big)\varphi(x,t)dx^3 \\ &= -\langle \nabla V(x,t)\rangle.\end{aligned} \tag{5.164}$$

This equation is an immediate corollary of Ehrenfest's theorem. From it, we see that the time derivative of the expectation value of the particle momentum is directly determined by the expectation value of the gradient of the potential function. If the quantum system is treated as a classical system, we can use the expectation value of the particle momentum as the momentum of the fictitious classical particle. Then, Eq. (5.164) becomes the equation of motion that the classical particle must satisfy, and the corresponding dynamical process is called Ehrenfest dynamics.

Equation 5.164 is similar to Newton's equation, but they are not identical: The force in Newton's equation is the potential energy gradient of the mean particle position $\langle x\rangle$ (i.e., the position of the classical point particle)

$$\frac{d}{dt}\langle p\rangle = -\nabla V\big(\langle x\rangle,t\big), \tag{5.165}$$

which is not the same as Eq. (5.164).

Note that the right-hand side of Eq. (5.164) is the expectation value of the potential gradient with respect to a wavefunction that is evolving with time. For a condensed matter system consisting of nuclei and electrons, the wavefunction of interest is the overall electron–nucleus wavefunction. In practical applications, the nucleus is generally treated as a classical particle, that is, the nuclear wavefunction is approximated by a δ function. In this way, the left side of Eq. (5.164) is the momentum change of the atom, and the right side is the expectation value of the potential gradient of the atom as the electron wavefunction evolves with time. In other words, the right-hand side of the equation corresponds to the force function of the atom averaged over the electron wavefunction. The time-dependent electron wavefunction corresponds to a linear superposition of the electron eigenfunctions for a given nuclear configuration and time, and it includes contributions from the ground state, excited states, and nonadiabatic processes; consequently, the atomic force corresponding to this wavefunction is an average of forces on multiple potential energy surfaces. Consequently, this approach is also known as mean-field theory.

The Ehrenfest dynamics method is a direct attempt to overcome the limitations of the Born–Oppenheimer approximation [Tully1998, Hack2000, Lian C.2018]. The idea of the Ehrenfest method is quite old, but it was not until the 1980s, when time-dependent density functional theory (TDDFT) [Runge1984] was proposed and developed, that it began to find widespread practical applications.

TDDFT-based molecular dynamics is currently the most widely used Ehrenfest dynamics method. It is implemented as follows: Starting with the atomic structure optimized by DFT, a molecular dynamics simulation based on the Born–Oppenheimer approximation is carried out; then the electron-occupation state is changed so that some electrons occupy empty states, generating holes at the original energy levels to simulate the occupation of excited-state electrons. Subsequently, TDDFT is used to calculate the electronic wavefunction evolution of the entire system at all energy levels, and the atomic forces corresponding to the new wavefunction are used to obtain the atomic positions and velocities at the next time step.

According to the principles of quantum mechanics, the evolution of the electron and nucleus system satisfies the following time-dependent Schrödinger equation [Meng S.2008a, b]:

$$i\hbar \frac{\partial \Psi(\{r_j\},\{R_J\},t)}{\partial t} = \hat{H}_{\text{tot}}(\{r_j\},\{R_J\},t)\Psi(\{r_j\},\{R_J\},t), \tag{5.166}$$

where $\hbar$ is the Planck constant, Ψ is the wavefunction of the system, r_j is the coordinate of the *Jth* electron, and R_J is the coordinate of the *Jth* nucleus. In general, the system Hamiltonian contains the kinetic energy of the electron, the kinetic energy of the nucleus, the Coulomb repulsion between electrons, the Coulomb repulsion between nuclei, the Coulomb attraction between electrons and nuclei, and the contribution of the external potential field. Therefore, it can be written as

$$\begin{aligned}\hat{H}_{\text{tot}} = &-\sum_j \frac{\hbar^2}{2m}\nabla_j^2 - \sum_J \frac{\hbar^2}{2m}\nabla_J^2 + \frac{1}{2}\sum_{i\neq j}\frac{e^2}{|r_i - r_j|} + \frac{1}{2}\sum_{I\neq J}\frac{Z_I Z_J}{|R_I - R_J|} - \sum_{j,J}\frac{eZ_J}{|r_j - R_J|} \\ &+ U_{\text{ext}}(\{r_j\},\{R_J\},t).\end{aligned} \tag{5.167}$$

In practical calculations, the electron motion is described by TDDFT, while the nucleus is described classically. Thus, the nuclear density distribution is given by $\rho_J(R,t) = \delta\left(R - R_J(t)\right)$. The equations of motion for the electrons and nuclei are

$$i\hbar \frac{\partial \phi_j(r,t)}{\partial t} = \left[-\frac{\hbar^2}{2m}\nabla_r^2 + v_{\text{ext}}(r,t) + \int \frac{\rho(r',t)}{|r-r'|}dr' - \sum_i \frac{Z_J}{|r-R_J|} + v_{\text{xc}}\left[\rho\right](r,t)\right]\phi_j(r,t) \tag{5.168}$$

and

$$M_J \frac{d^2 R_J(t)}{dt^2} = -\nabla_{R_J}\left[V_{\text{ext}}^J(R_J,t) - \int \frac{Z_J\rho(r,t)}{|R_J - r|}dr + \sum_{I\neq J}\frac{Z_I Z_J}{|R_J - R_I|}\right], \tag{5.169}$$

where ϕ_j is the Kohn–Sham orbital of the electron under the single-particle approximation, ρ is the total electron density, and v_{xc} is the exchange-correlation functional of electrons. M_J and Z_J are the mass and charge of the *Jth* nucleus, respectively.

By repeating this process, we can obtain the real-time dynamics of electron and nucleus evolution within the system. Since the atomic coordinates and velocities at each time step can be calculated from the state of the system in the previous step, there is no need for a predefined potential energy surface.

The Ehrenfest dynamics method is widely used in practice because of its clear physical interpretation. However, it requires solving the time-dependent Schrödinger equation, which leads to higher computational cost. Recently, S. Meng et al. at the Institute of Physics, Chinese Academy of Sciences, developed a computational method based on TDDFT and the Ehrenfest theorem to describe the real-time evolution of the electron wavefunction coupled to atomic motions [Meng S.2008a, b, Lian C.2018]. This method can be used to conduct large-scale studies (with a spatial scale >500 atoms and a time scale >1,000 fs) of optical absorption and dynamic processes of excited-state electrons in semiconductors and nanomaterials. This constitutes significant progress in studying semiclassical nonadiabatic problems.

5.3.3 Other Quantum–Classical Dynamic Methods

Hybrid quantum–classical dynamics methods are also widely used. The main idea is to describe the total system as a quantum system coupled to a classical environment. The total Hamiltonian consists of three parts: the free Hamiltonian of the quantum system for the electrons, the potential energy surface of the classical system for the nuclei, and the coupled electron–nucleus Hamiltonian. The total Hamiltonian can evolve with time. The previously introduced trajectory-based surface-hopping method and the Ehrenfest dynamics method belong to this category. Although these are the two most popular methods, there are others in which the time evolution can take very different forms. Those include methods that utilize the time dependence of the density matrix, Liouville dynamics, and the time dependence found in linearization algorithms [Miller1970, Ben-Nun1998, Huo P.2012].

For example, the density matrix satisfies the following equation:

$$i\hbar\frac{\partial}{\partial t}\rho=\left[\hat{H},\rho\right]. \tag{5.170}$$

In some cases, it is possible to obtain information about the quantum evolution of the system from the density matrix and its semiclassical counterpart. Suppose that the system Hamiltonian is

$$\hat{H}=T(\boldsymbol{p})+V_0(\boldsymbol{x})+\sum_{n,m}|\phi_n\rangle V_{nm}(x)\langle\phi_m|, \tag{5.171}$$

where $\boldsymbol{p}$ and $\boldsymbol{x}$ are the momentum and position of a nucleus-like classical particle. This is analogous to a classical system of harmonic oscillators, such that

$$T(\boldsymbol{p})+V_0(\boldsymbol{x})=\sum_{j=1}^{N}\frac{\omega_j}{2}\left(p_j^2+x_j^2\right). \tag{5.172}$$

Now we assume that $|\phi_n\rangle$ is an eigenstate of the electron-like quantum system

$$|\phi_n\rangle\rightarrow|0_1,\ldots,1_n,\ldots,0_M\rangle. \tag{5.173}$$

It is possible to rewrite $|\phi_n\rangle$ in terms of particle creation and annihilation operators. This is expressed as

$$|\phi_n\rangle\langle\phi_m|\rightarrow a_n^{\dagger}a_m. \tag{5.174}$$

We can define

$$X_n=\frac{1}{\sqrt{2}}(a_n^{\dagger}+a_n), \tag{5.175}$$

$$P_n=\frac{i}{\sqrt{2}}(a_n^{\dagger}-a_n). \tag{5.176}$$

It can be easily shown that the Hamiltonian in Eq. (5.171) can be transformed into the following form [Huo P.2012]:

$$\hat{H} = h_0(\boldsymbol{x}, \boldsymbol{p}) + \frac{1}{2}\sum_{nm}(X_n X_m + P_n P_m) V_{nm}(\boldsymbol{x}), \tag{5.177}$$

where

$$h_0(\boldsymbol{x}, \boldsymbol{p}) = T(\boldsymbol{p}) + V_0(\boldsymbol{x}) - \frac{1}{2}\sum_{n} V_{nn}(\boldsymbol{x}). \tag{5.178}$$

In this way, the total hybrid quantum–classical system can be approximated by a fully classical system, and the system evolution can thus be solved using existing classical methods. This is the basic idea of linearization algorithms. Currently, owing to limited computational capacity, these methods are mainly used for theoretical calculations of model systems and have not been extensively applied to simulations of realistic condensed matter systems.

5.4 Joint Study of Nonadiabatic Effects and Nuclear Quantum Effects

In the previous sections, we repeatedly established the importance of combining nonadiabatic effects and nuclear quantum effects. Although joint study of these effects in real systems is still in its infancy, increasing research evidence shows that a full quantum theory combining them is necessary for the accurate understanding of experimental observations for natural physical and chemical processes.

The combined study of nonadiabatic and nuclear quantum effects should include both statistical thermodynamic and dynamic properties. Theoretically, because nonadiabatic effects are essentially quantum-mechanical in nature, quantum coherence can only be described by the rigorous wavefunction approach. Unfortunately, the computational power required for simulating a large system is prohibitive; hence, a variety of hybrid quantum–classical methods are currently used. These methods do not take full quantum effects sufficiently into account and therefore are far from being perfect. Furthermore, combining both effects in dynamical simulations is even more difficult.

There are currently two main research directions: The first one considers time-dependent dynamic properties such as electron occupancy and atomic/molecular motion, and the other calculates the rate of nonadiabatic reactions. Both essentially study the quantum correlation function given by Eq. (5.12). To agree with most of the currently available literature, we will describe some of the theoretical attempts using the above approaches. However, the reliability of these methods is still under debate. Limited by computing restraints and by the difficulty in constructing high-quality diabatic potential energy surfaces, they are presently only applicable to model systems.

5.4.1 Nonadiabatic Dynamic Properties

In order to obtain dynamic information that includes both nonadiabatic effects and nuclear quantum effects, a number of studies have recently been conducted from a practical perspective. For example, the RPMD method might be combined with the nonadiabatic simulation method discussed in Section 5.3.

One method begins with the nonadiabatic partition function

$$Z \propto \lim_{N\to\infty} \int d\boldsymbol{R} d\boldsymbol{P} e^{-\beta_N H_{rp}} \mathrm{Tr}_\mathrm{e} \prod_{\alpha=1}^{N} e^{-\beta_N \hat{V}(R_\alpha)}. \tag{5.179}$$

Here, H_{rp} denotes the Hamiltonian of the free ring polymer, which can be written as

$$H_{\mathrm{rp}} = \sum_{\alpha}^{N} \frac{P_\alpha^2}{2M} + \frac{M}{2\beta_N^2 \hbar^2} \left(R_\alpha - R_{\alpha-1}\right)^2. \tag{5.180}$$

If $\hat{V}$ represents the diabatic potential energy matrix and Tr_e denotes the trace of the electron degrees of freedom, we can assume that

$$e^{-\beta_N V_{\mathrm{eff}}(\vec{R})} = \mathrm{Tr}_\mathrm{e} \prod_{\alpha=1}^{N} e^{-\beta_N \hat{V}(R_\alpha)}; \tag{5.181}$$

the partition function can then be written as

$$Z \propto \lim_{N\to\infty} \int d\boldsymbol{R} d\boldsymbol{P} e^{-\beta_N H_N^{\mathrm{MF}}}, \tag{5.182}$$

where $H_N^{\mathrm{MF}} = H_{\mathrm{rp}} + V_{\mathrm{eff}}(\boldsymbol{R})$.

In the mean-field ring-polymer molecular dynamics (MF-RPMD) method [Hele2011], each bead is driven by an effective potential. The ring polymer is propagated by the equivalent Hamiltonian H_N^{MF}, and the corresponding quantum correlation function can be calculated. In essence, MF-RPMD averages a multi-energy Hamiltonian and turns it into a Hamiltonian that does not contain information about the electron states.

Another approach is the ring-polymer surface hopping (RPSH) method [Shushkov2012]. It assumes that all beads of the ring polymer are on the same adiabatic potential energy surface and uses the equation

$$i\dot{c}_i = \left[\frac{1}{N}\sum_{k=1}^{N} V_i(R_k)\right] c_i - i\sum_j \left[\frac{1}{N}\sum_{k=1}^{N} \dot{R}_k \cdot d_{ij}(R_k)\right] c_j \tag{5.183}$$

to propagate the electron state coefficients. Letting $\bar{R}$ denotes the centroid of the ring polymer, this becomes

$$i\dot{c}_i = V_i(\bar{R})c_i - i\sum_j \dot{\bar{R}} \cdot d_{ij}(\bar{R})c_j. \tag{5.184}$$

(Symbols not specifically defined have the same meanings as in subsection 5.3.1.) If the electron state coefficients satisfy conditions similar to those in subsection 5.3.1, a transition will occur. However, in RPSH, the velocity of each bead must be adjusted along the direction of the nonadiabatic coupling to ensure that the system conserves its total energy. Because nuclear quantum effects are incorporated according to Tully's models mentioned above, the RPSH method provides more accurate electron-occupancy information than the classical trajectory-based surface-hopping method in the low-energy regime [Shakib2017].

However, RPSH assumes that all beads are on the same adiabatic potential energy surface; therefore, it is not as rigorous as the MF-RPMD method in a statistical sense.

Combining RPSH with quantum statistics, T. F. Miller's group proposed a method called isomorphic Hamiltonian ring-polymer potential surface hopping (Iso-RPSH) [Tao X.2018]. In this method, a new isomorphic Hamiltonian $\hat{H}_N^{\text{iso}} = H_{\text{rp}} + \hat{V}^{\text{iso}}(\boldsymbol{R})$ is constructed, with $\hat{V}^{iso}(\boldsymbol{R})$ satisfying

$$\text{Tr}_{\text{e}} e^{-\beta_N \hat{V}^{\text{iso}}(R)} = \text{Tr}_{\text{e}} \prod_{\alpha=1}^{N} e^{-\beta_N \hat{V}(R_\alpha)} \equiv \mu(\boldsymbol{R}). \tag{5.185}$$

In this way, the quantum statistics are preserved. It is also possible to have the entire ring polymer on the eigenenergy surface $\hat{V}^{\text{iso}}(\boldsymbol{R})$ hop to a different surface, so that electron-occupation information can be obtained. Taking two electron states as an example, $\hat{V}^{\text{iso}}(\boldsymbol{R})$ can be written as

$$\hat{V}^{iso}(\boldsymbol{R}) = \begin{bmatrix} V_1^{\text{iso}}(\boldsymbol{R}) & K_{12}^{\text{iso}}(\boldsymbol{R}) \\ K_{12}^{\text{iso}}(\boldsymbol{R}) & V_2^{\text{iso}}(\boldsymbol{R}) \end{bmatrix}. \tag{5.186}$$

In the region without nonadiabatic coupling, $\hat{V}^{\text{iso}}(\boldsymbol{R})$ must have the same form as the potential in the usual RPMD method, such that

$$V_i^{iso}(\boldsymbol{R}) = \frac{1}{N}\sum_{\alpha=1}^{N} V_i(R_\alpha). \tag{5.187}$$

The expression for the non-diagonal element can be obtained based on Eq. (5.182):

$$\left(K_{12}^{iso}(\boldsymbol{R})\right)^2 = \text{a}\cosh^2\left[e^{\frac{\beta}{2}\left(V_1^{\text{iso}}(R)+V_2^{\text{iso}}(R)\right)}\mu(\boldsymbol{R})/2\right]/\beta^2 - \left(V_1^{\text{iso}}(\boldsymbol{R}) - V_2^{\text{iso}}(\boldsymbol{R})\right)^2/4. \tag{5.188}$$

It can be seen that both $V_{\text{eff}}(\boldsymbol{R})$ and $\hat{V}^{\text{iso}}(\boldsymbol{R})$ are complicated. In fact, for each step of MF-RPMD propagation, $3N-6$ matrix multiplications must be carried out. This substantially increases the computational cost of the dynamic propagation in both the MF-RPMD and the Iso-RPSH methods, hindering their practical application.

Another type of method integrates RPMD with the methods introduced in Eqs (5.173)–(5.178) in subsection 5.3.3 [Stock1997]; examples of this approach are nonadiabatic ring-polymer molecular dynamics (NRPMD) [Richardson2013], mapping-variable ring-polymer molecular dynamics (MV-RPMD) [Ananth2013], and coherent-state-mapping ring-polymer molecular dynamics (CS-RPMD) [Chowdhury2017]. Quantum statistics is rigorously preserved in all three methods, which allows them to describe the nonadiabatic dynamic information more accurately. For example, CS-RPMD discretizes the electronic degrees of freedom into N beads in the same way as the nuclear degrees of freedom. This is expressed as

$$Z \propto \lim_{N\to\infty} \int d\boldsymbol{R} d\boldsymbol{P} d\vec{\boldsymbol{p}} d\vec{\boldsymbol{q}} \Gamma e^{-\beta_N H_{\text{cs}}}, \tag{5.189}$$

where

$$\Gamma = \prod_{\alpha=1}^{N} \frac{1}{2}\left(\boldsymbol{q}_\alpha - i\boldsymbol{p}_\alpha\right)^T \left(\boldsymbol{q}_{\alpha+1} - i\boldsymbol{p}_{\alpha+1}\right) e^{-\frac{1}{2}\left(\boldsymbol{q}_\alpha^T \boldsymbol{q}_\alpha + \boldsymbol{p}_\alpha^T \boldsymbol{p}_\alpha\right)} \tag{5.190}$$

and

$$H_{\text{cs}} = H_{\text{rp}} + \sum_{nm} V_{nm}(\boldsymbol{R}_{\alpha})\left[\frac{1}{2}(\boldsymbol{q}_{\alpha m}\boldsymbol{q}_{\alpha n} + \boldsymbol{P}_{\alpha m}\boldsymbol{P}_{\alpha n}) - \delta_{nm}\right]. \tag{5.191}$$

In terms of dynamic evolution, CS-RPMD is similar to RPMD in that they both use H_{cs} to describe the position and momentum propagation of the nuclei and their corresponding electrons. The dynamic information of the nucleus and electron occupation in this process can be represented by $\boldsymbol{R}$, $\boldsymbol{P}$, $\boldsymbol{q}$, and $\boldsymbol{p}$.

5.4.2 Nonadiabatic Reaction Rates

In essence, the reaction rate is also a quantum correlation function and can be expressed as

$$k(T)Q_{\text{r}}(T) = \lim_{t\to\infty} C_{fs}(t). \tag{5.192}$$

The right-hand side of the equation can be rewritten as

$$C_{fs}(t) = \text{Tr}\left[e^{-\beta\hat{H}/2}\hat{F}\, e^{-\beta\hat{H}\ /2}e^{i\hat{H}t/\hbar}\,\hat{h}(\hat{x}-s)e^{-i\hat{H}t/\hbar}\right], \tag{5.193}$$

where $\hat{h}$ is the Heaviside function; $\hat{F} = \frac{i}{\hbar}\left[\hat{H},\hat{h}(\hat{x}-s)\right] = -\frac{i\hbar}{2m}\left\{\delta(\hat{x}-s)\frac{d}{dx} + \frac{d}{dx}\delta(\hat{x}-s)\right\}$ is the flux operator; and s is the dividing surface of the reaction. In order to obtain a nonadiabatic rate, we can use a similar derivation as done by adiabatic RPMD. Notably, $C_{fs}(t)$ is treated classically and $\hat{x}$ and $\hat{p}$ are considered commutative numbers rather than non-commuting operators. In this way, the propagation can be determined by the classical Hamiltonian canonical equation. By calculating the trace $\text{Tr}[\ldots\ldots]$, the equation is transformed into a phase-space integral problem. This is expressed as

$$C_{fs}^{\text{cl}}(t) = \frac{1}{2\pi}\int \text{d}p\text{d}x e^{-\beta H(p,x)}\frac{p}{m}\delta(x-s)h(x_t - s). \tag{5.194}$$

To include the quantum effects of the Boltzmann operator, Eq. (5.194) can be rewritten in terms of path integrals. This gives

$$C_{fs}^{N}(t) = \frac{1}{(2\pi)^N}\int d\boldsymbol{R}d\boldsymbol{P}e^{-\beta_N H_N^{\text{MF}}(\boldsymbol{R},\boldsymbol{P})}\frac{\bar{P}}{m}\delta(\bar{R}-s)h(\bar{R}_t - s), \tag{5.195}$$

where H_N^{MF} is consistent with Eq. (5.182). Eq. (5.195) yields the rate according to MF-RPMD [Hele2011]. For specific calculations, the following Bennett–Chandler decomposition [Bennett1977, Chandler1978] is usually adopted:

$$k_{\text{MF-RPMD}} = k_{\text{MF-QTST}} \cdot \kappa_{\text{MF-RPMD}}, \tag{5.196}$$

where the time-independent quantity

$$k_{\text{MF-QTST}} = \frac{1}{Q_{\text{r}}}\int \text{d}\boldsymbol{R}\text{d}\boldsymbol{P}e^{-\beta_N H_N^{MF}(\boldsymbol{R},\boldsymbol{P})}\frac{\bar{P}}{m}\delta(\bar{R}-s)h(\bar{P}) \tag{5.197}$$

is called the quantum transition rate, and the other term in Eq. (5.196)

$$\kappa_{\text{MF-RPMD}} = \frac{\lim_{t\to\infty} C_{fs}^{N}(t)}{\lim_{t\to 0+} C_{fs}^{N}(t)} \tag{5.198}$$

is the dynamic correction term. In this way, the calculation of the rate no longer requires an optimal choice of dividing surface, thus reducing the computational cost.

It is also possible to combine RPSH with Iso-RPSH to obtain state-resolved reaction rates [Shushkov2012, Tao X.2019]. The state-resolved dynamical-correction term can be calculated as

$$\kappa_{i\to f,\gamma}^{(iso)} = \frac{\lim\limits_{t\to\infty}\int \mathrm{d}\boldsymbol{R}\mathrm{d}\boldsymbol{P} e^{-\beta_N H_{N,\gamma}^{(\mathrm{iso})}(R,P)} \frac{\overline{P}}{m}\delta(\overline{R}-s)h(\overline{R}_t - s)\delta_{i\theta_{-t}}\delta_{f\theta_t}}{\int \mathrm{d}\boldsymbol{R}\mathrm{d}\boldsymbol{P} e^{-\beta_N H_{N,\gamma}^{(\mathrm{iso})}(R,P)} \frac{\overline{P}}{m}\delta(\overline{R}-s)h(\overline{P})}, \tag{5.199}$$

where $\delta_{i\theta_{-t}}$ indicates that the reactants are in the ith electron state before the reaction, $\delta_{f\theta_t}$ means that the products are in the fth electron state after the reaction, and γ denotes the electron state of the ring polymer on the dividing surface. The ring polymer is propagated according to RPSH or Iso-RPSH. By statistical averaging over a large number of trajectories, the state-resolved dynamical-correction term can be computed. Then, the relation

$$k_{i\to f}^{(\mathrm{iso})} = k_{\mathrm{MF-QTST}} \sum_{\gamma} \kappa_{i\to f,\gamma}^{(\mathrm{iso})} \tag{5.200}$$

can be used to determine the state-resolved reaction rate.

Strictly speaking, in the nonadiabatic case, the classical approximation used here is not appropriate. This is because the Hamiltonian at this point is a matrix, and the evolution of operator $\hat{x}$ within this Hamiltonian cannot be solved classically. In fact, in the nonadiabatic limit, this method can be problematic [Menzeleev2014, Lawrence2019a, b]. In order to obtain satisfactory nonadiabatic reaction rates, many solutions have been recently proposed by theoretical chemists [Richardson2015, Lawrence2018, 2020, Thapa2019]. However, the applicability of these solutions is limited to model calculations; their extension to real condensed matter systems will require further development.

5.4.3 Example: Proton Transfer in Water Dimers of H_2O–H_2O^+

In 2023, a team led by Prof. S. Meng at the Institute of Physics Chinese Academy of Sciences, made a significant advance in the computational methods for nonadiabatic full quantum dynamics simulations in condensed matter. The method simultaneously describes the quantum effects of atomic nuclei and nonadiabatic transitions of electrons, and has been successfully applied to the study of proton transfer in water dimers under excitation [Zhao R. J. 2023].

Today, there are two most popular approaches in the field of molecular dynamics (MD) simulations of materials: (i) The Born–Oppenheimer Molecular Dynamics (BOMD) approaches based on adiabatic approximations, including classical MD methods with empirical interatomic potentials, *ab initio* molecular dynamics with interatomic potentials calculated by first principles (Car-Parrinello MD included), and PIMD with adiabatic potential energy surfaces. In general, these methods do not explicitly calculate the evolution and transition of electronic degree of freedoms between different states – that is, electrons are forcefully "frozen" in their adiabatic ground state. (ii) Hybrid quantum-classical dynamics approaches which incorporate the quantum dynamic evolution of electrons but treat the nuclei as classical point particles. However, these two groups of methods have significant limitations, especially in the cases when the systems have strong coexisting nonadiabatic effects and nuclear quantum effects. More importantly, the high dimensionality of the problem and large number of degrees of freedom forbids the easy implementation of these methods to simulate real materials, which usually require a large increase of computation at the first-principles level in order to reach an accurate description of the physical process. It is thus desirable to develop an efficient algorithm for a full quantum description of condensed matters and their dynamics.

The S. Meng's group develops a new method for nonadiabatic full quantum dynamics simulations via theoretical derivation and comparison with exact quantum-mechanical results [Zhao R. J. 2023]. The new method, termed as Ring-Polymer Molecular Dynamics

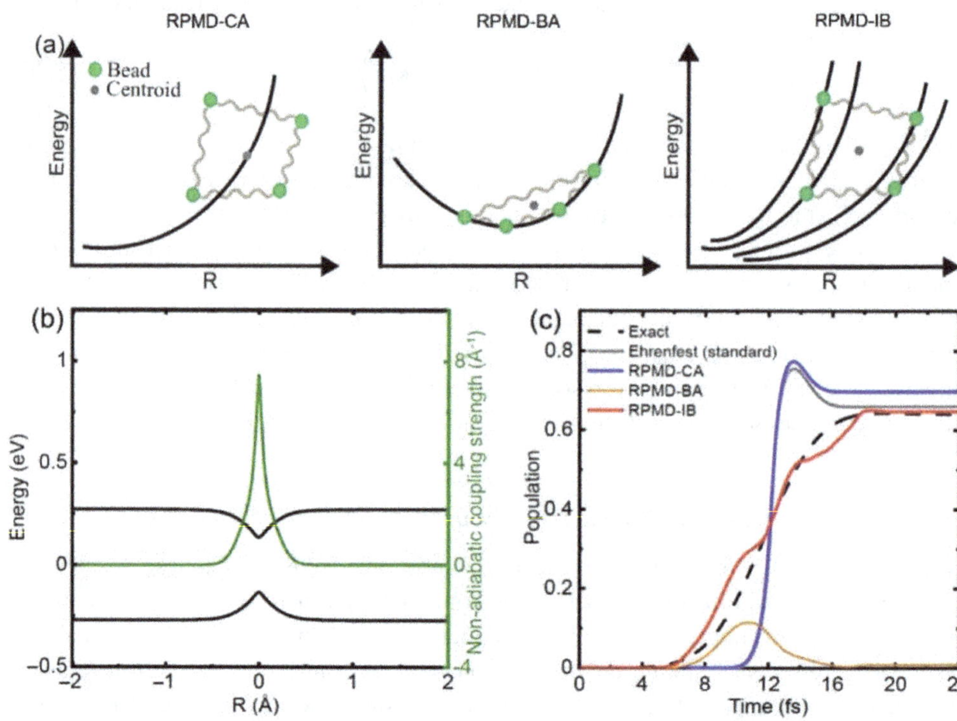

Figure 5.7 Comparison between nonadiabatic quantum dynamics simulations and the exact quantum solution. (a) Schematic diagrams for nonadiabatic path integral simulations with centroid approximation (RPMD-CA), bead approximation (RPMD-BA), and independent bead approximation (RPMD-IB). (b) Two adiabatic potential energy curves (black lines) and the nonadiabatic coupling strength (green line) as a function of position R for the avoided crossing model. (c) The real-time electronic population in the excited state obtained by different methods. The dashed line denotes the results from the exact quantum theory. The red line denotes the results from the new method, RPMD-IB. Reprinted figure with permission from Zhao R. J.[2023]. Copyright 2024 by the American Physical Society.

based on Independent Beads, can treat the quantum effects of electrons and nuclei on the same footing. The main idea is to use Ehrenfest theorem for each path in the path integral representation of nuclear quantum trajectories, therefore along each path of nuclear trajectories there is an electronic configuration independently evolving according to quantum mechanics (Figure 5.7). Comparing with the other nonadiabatic dynamic methods using the classical path and the averaged path, this approach ensures simultaneous description of nuclear quantum effects as well as the excitation of the electrons during evolution (i.e., nonadiabatic effects). Moreover, this approach has an advantage that it can be easily implemented in first-principles computer codes such as the Time-Dependent Ab-initio Package (TDAP) developed in S. Meng's group, ready for use in the quantum-mechanical simulations of molecules and materials comprising hundreds of atoms.

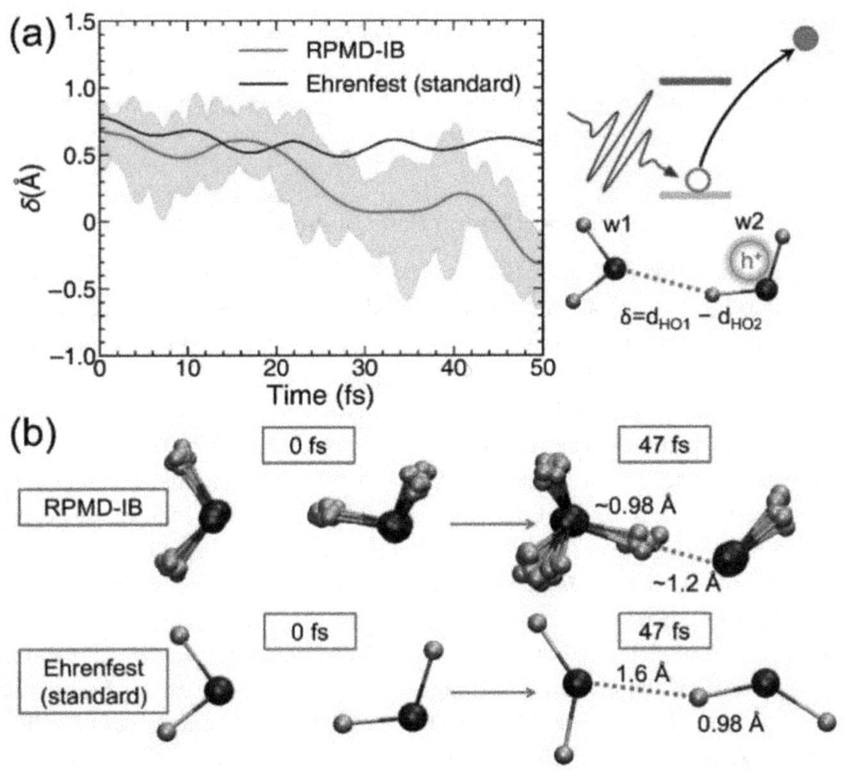

Figure 5.8 (a) The distance difference of the hydrogen atom (H) and its two nearest oxygen atoms (O1, O2) in the water dimer upon laser excitation. Shadows indicate the quantum fluctuation of all beads. Schematic setting is shown in the right panel. (b) The snapshots at 0 and 47 fs in the excited-state dynamics. In the RPMD-IB simulation, the configurations of all beads are presented. At 47 fs, the entire nuclear wavepacket of the split hydrogen atom approaches the oxygen atom (O1) of the first water molecule, indicating the formation of hydronium (H_3O^+) and an OH radical, while in standard Ehrenfest dynamics such a process is absent. Reprinted figure with permission from Zhao R. J.[2023]. Copyright 2024 by the American Physical Society.

The new approach has been adopted to simulate a series of processes induced by photoionization. These include the proton transfer in water dimers of H_2O–H_2O^+, splitting of a water molecule in water clusters and liquid water, charge transfer between water and titania surface. Additionally, it has been used to model the passage of high-energy hydrogen ions through graphene membranes, etc. The simulations revealed some unexpected new insights into these processes. For instance, it was found that within 50 fs after intense laser illumination of the water dimer, the ionized water transfers proton to its neighbor water molecule (Figure 5.8). However, a parallel simulation of standard Ehrenfest dynamics based on classical nuclei does not show the proton transfer in a short time upon photo illumination. This means only when the quantum motions of nuclei are correctly incorporated can nonadiabatic proton transfer occur. The proton transfer rate predicted by the new approach agrees well with the experimental observations, thus justifying the reliability of the new approach. In addition, the obtained results also agree well with those from the exact quantum mechanics theory. We expect that the new method will pave the way for a better understanding of the full quantum behaviors in condensed matters including the nuclear quantum effects and nonadiabatic effects and thus help to solve important puzzles such as unconventional superconductivity.

5.5 Perspectives on Full Quantum Theory

Following the ideas mentioned in Section 4.1, we have systematically discussed a large number of current theoretical simulation methods for full quantum effects in condensed matter. However, owing to the many-body quantum nature of condensed matter systems, rigorously solving real problems is currently unrealistic. In practice, suitable numerical approximations must be adopted. A key feature of condensed matter computational physics is to find the best balance between computer power and computational accuracy. Therefore, the primary task in each historical period of theoretical condensed matter research has been the optimization and development of computational methods that achieve the best possible accuracy with the computer power available at the time; the present case is no exception. Under existing computing conditions, most research will still have to focus on either nuclear quantum effects or nonadiabatic effects.

In the case of nuclear quantum effects, the existing methods are generally rigorous (the treatment of the electron exchange term is relatively mature) from a statistical point of view, but they cannot describe the exchange interactions between fermionic nuclei. This exchange problem is negligible in most practical systems. The main problem at present is how to incorporate nuclear-quantum-effect simulation tools into accurate electronic-structure calculation methods; this is especially true for systems of which the Born–Oppenheimer potential energy surface of the electronic ground state cannot be precisely described by conventional DFT. QMC and quantum chemistry electronic-structure calculation methods are viable choices for problems of this type. At the dynamic level, as we have mentioned in subsection 5.1.3, there is still considerable room for improvement in existing algorithms (including semiclassical methods, CMD, ring-polymer molecular dynamics, and instanton methods).

In the case of nonadiabatic effects, the existing methods are far from perfect. The trajectory-based surface-hopping method is practically useful but has no rigorous theoretical foundation. The Ehrenfest dynamics method violates the basic principles of quantum mechanics in the long-time limit. Other hybrid quantum–classical dynamical methods have limited applications. Therefore, developing effective simulation methods and establishing a clearer theoretical framework for the study of nonadiabatic processes remains an urgent issue.

When combining nuclear quantum effects with nonadiabatic effects, the first-principles approach of electron–phonon coupling and related studies in condensed matter systems is relatively mature, as we have established in Section 5.2. However, in that section, our focus was only on nuclear quantum effects. As noted in Section 5.4, studies that combine nuclear quantum effects with nonadiabatic effects are mostly from the field of theoretical chemistry; they are still in their infancy, especially when the complexity of the many-atom system exceeds the current computational capacity. Notably, for most of the methods mentioned here, completely different languages are used by condensed matter physicists and theoretical chemists when discussing full quantum effects. In condensed matter physics, the current understanding of electron–phonon coupling based on the Born–Huang expansion has many unresolved problems. Theoretical-chemistry methods, although more rigorous, encounter difficulty in solving practical problems in condensed matter systems. Unifying these two languages to study full quantum effects in real condensed matter systems will remain a challenging task for a long time to come.

6

Experiments for Full Quantum Effects

Experimentally studying full quantum effects is challenging. In any comprehensive study of the nature of full quantum effects, understanding the details of the experimental observations is essential. While physical and chemical properties related to full quantum effects are common in condensed matter, they are highly sensitive to the experimental conditions, particularly the local environment. Therefore, immense care is necessary in experimental design, sample preparation, detection techniques, and data analysis. Extreme conditions and intense external fields are also often necessary for full quantum research.

Nuclear quantum effects can be experimentally examined as zero-point corrections to the values of different observables caused by zero-point vibrations, nuclear quantum fluctuations, tunneling, and delocalization. Theoretical studies are often very useful for experimental observations in this research area. Many published theoretical articles provide detailed descriptions and analyses of experiments involving nuclear quantum effects [Milonni1991, Tuckerman1997, Benoit1998, Miyazaki2004, Marx2006, Zhang Q.F.2008, Ceriotti2016, Meisner2016, Guo J.2017, Markland2018, Fang W.2019]. Isotope substitution is one of the most effective approaches among all these experimental methods. The results of evaluating the statistical and dynamical properties of condensed matter systems before and after isotope substitution reflect, to some extent, the combined nuclear quantum effects on the actual physical observables.

The study of nonadiabatic effects typically involves observing the changes in physical quantities during nonequilibrium processes, such as the breaking and recombination of chemical bonds in photo-induced reactions, and exploring the relationship between reactants and products. Spectroscopy and photoemission spectroscopy have been the most prevalent and direct methods for studying excited-electron transitions. In recent years, the development of ultrafast laser technology has made it possible to capture the entire nonadiabatic quantum coupling process between electronic and nuclear motions. By combining this with other detection techniques, problems involving ultrafast charge transfer, optically induced phase transitions, and charge density waves can be investigated in detail. Isotope substitution is also a powerful tool for studying nonadiabatic effects. For example, it affects the reaction rates and products, especially in the presence of optical field. As in nuclear-quantum-effect research, theoretical support is essential in research on nonadiabatic effects. In recent years, the development of nonadiabatic quantum dynamics has been a key theoretical advancement.

In some cases, full quantum effects can be observed directly in experiments. For instance, nuclear quantum tunneling is observed when the reaction rate at low temperatures converges to a constant value. Nuclear quantum fluctuations can be detected when the nuclear momentum distribution deviates from classical statistics. Nonadiabatic processes such as Franck–Condon phenomena can also be directly measured in spectroscopic experiments. However, in certain experimental measurements, it is challenging to separate the full quantum effects from that predicted by the simulations based on the adiabatic ball-and-stick model of electronic states. Therefore, observing the changes in the physical properties of condensed matter after isotope substitution is crucial.

In the study of nuclear quantum effects, replacing hydrogen with deuterium can induce changes in the physical properties of a real hydrogen-containing system. Because hydrogen and deuterium have the same number of protons, the Born–Oppenheimer energy landscape of the system remains the same after isotope substitution. However, as the atomic mass of deuterium is approximately twice that of hydrogen, nuclear quantum effects experienced by deuterium at a given temperature are relatively weaker. This leads to less significant zero-point corrections caused by the quantum fluctuations and zero-point vibration of the nuclei and to less quantum tunneling and delocalization when deuterium is substituted for hydrogen. Even minor changes on full quantum effect parameters can alter the actual physical and chemical properties of the condensed matter system, making precise measurements and subtle explanations critical. The specific heats, melting points, and boiling points of un-deuterated and heavy water are easily measured parameters that exhibit such isotope dependence. The nuclear quantum effects behind these macroscopic parameter changes can only be understood at the level of full quantum mechanics in the microscopic world.

In addition, for problems closely related to the local environment, such as the hydrogen-bonding strength of water at the gas–liquid interface, the isotope effect changes the bond strength of an atom and its isotope at different temperatures. In general, the hydrogen-bonding strength of deuterium (D) is greater than that of normal hydrogen (H), but at temperatures above 493 K, the H bonding strength becomes stronger. To investigate this bonding strength change, appropriate experimental design and precise experimental measurements are necessary [Horita1994]. High-resolution real-space imaging techniques such as scanning tunneling microscopy [Kumagai2015, Guo J.2016a, Lin C.2019] have recently replaced nuclear magnetic resonance (NMR) and X-ray or neutron scattering [Ubbelohde1955] as the preferred tools for this purpose.

Isotope-substitution experiments also provide important clues in the study of nonadiabatic effects. Because of the different atomic masses of isotopes, the electron excitation or decay caused by nonadiabatic effects differs before and after substitution. The dynamical effects on atoms are also altered by the changes in electron–nucleus quantum coupling. The peak-position shift, peak-intensity change, and spectral broadening in NMR; the change in superconductivity caused by electron–phonon coupling; and the products of optically induced nonequilibrium reactions caused by electron–nucleus evolution are all directly related to NAEs in many-body quantum processes. By effectively utilizing external field conditions, we can design an experiment that either manifests or suppresses full quantum effects, as desired. By measuring small changes in physical quantities in the corresponding modulation processes and comparing them with theoretical predictions of the full quantum theory, we can extract important information about nonadiabatic effects directly.

This chapter presents commonly used experimental methods and external-field platforms for studying full quantum effects on the physical and chemical properties of condensed matter systems. The experimental methods include NMR, neutron scattering, X-ray scattering spectroscopy, photoemission spectroscopy, electron scattering, and scanning probe technology. The extreme conditions produced by the external-field platforms of big research facilities include extremely low temperatures, ultrahigh pressures, high magnetic fields, and ultrafast, ultraintense light fields, as well as synergetic combinations of these. Finally, the chapter concludes with a discussion on the outlook for future development of experimental techniques for the study of full quantum effects.

6.1 Technical Approach

Compared to the theoretical approach to the full quantum effect, systematic experimental techniques for studying the full quantum effect are still in the development stage. This section focuses on ways to overcome the measurement limits of existing experimental methods. By exploiting the advantages of different techniques, it is possible to observe manifestations of full quantum effects in many physical phenomena through existing experimental methods.

6.1.1 Nuclear Magnetic Resonance

Nuclear magnetic resonance (NMR) is an essential spectroscopic tool for studying molecular structures at the atomic scale. The energy levels of nuclei with non-zero spins undergo Zeeman splitting in the presence of a constant external magnetic field. As a result, the spins become quantized and align themselves along the field direction. Classically, the magnetic moments of the nuclei process around the external magnetic field are at Larmor frequency ω_0. When a radio frequency (RF) electromagnetic wave with the same angular frequency ω_0 is applied in a direction perpendicular to the field, the nuclear magnetic moment can absorb the radiative energy of the wave through resonance and transition from a lower to a higher energy state. This phenomenon is called NMR.

NMR can be observed using either a frequency sweep method or a field sweep method. The frequency sweep method detects the resonance signal by fixing the applied magnetic field strength and changing the RF frequency, whereas the field sweep method detects the resonance signal by fixing the frequency and gradually changing the magnitude of the magnetic field strength. Using either of these methods, the resonance frequency of the sample can be measured. Because atoms of different elements have different resonance frequencies, acquiring the resonance signal allows us to identify the elements within the sample. However, if nuclei of the same element exist in different chemical environments within the sample, their surrounding electron clouds will produce different magnetic shielding. As a result, the NMR spectral lines will be shifted; the magnitude of this "chemical shift" is denoted by δ.

By measuring the chemical shifts of specific nuclei, such as ^1H or ^{13}C, the corresponding chemical environment or chemical structure can be predicted (Figure 6.1(a)). For instance, L. Dubrovinsky's group at the University of Bayreuth used ^1H-NMR spectroscopy to

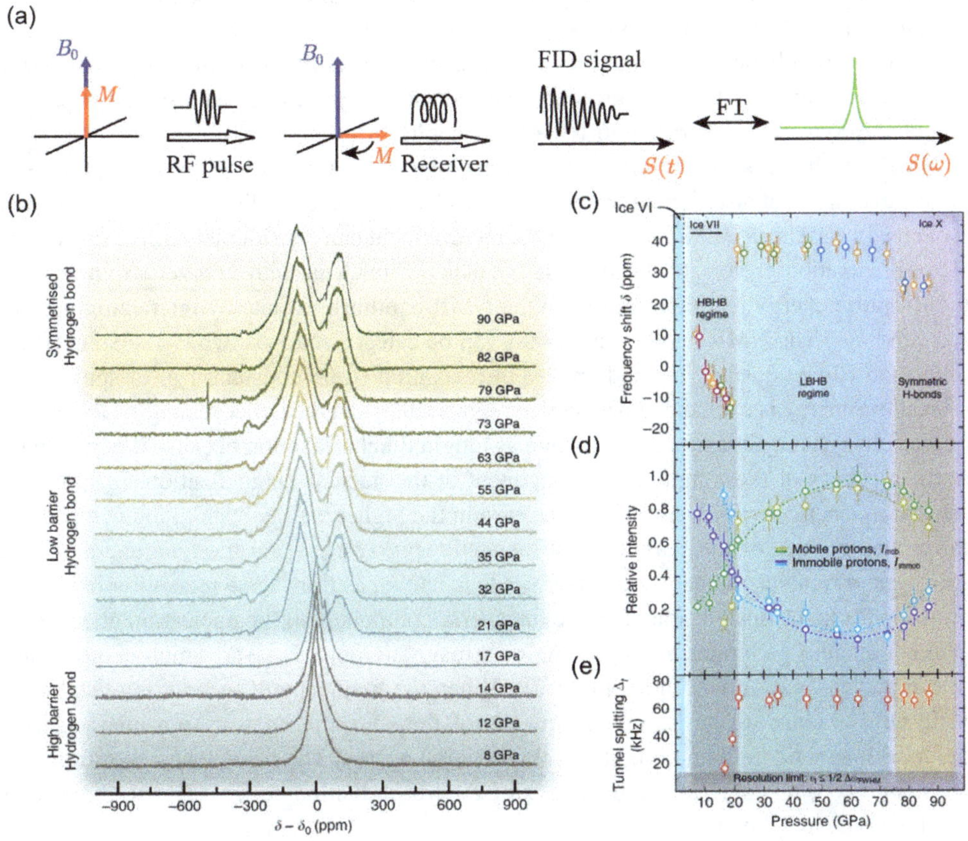

Figure 6.1 Experimental observations of nuclear quantum effects and symmetric hydrogen bonds in ice X. (a) Principle of NMR and detection of the free induction decay (FID) signal. A static magnetic field B_0 is applied to the sample, and the originally degenerate nuclear energy level splits into different energy level states; the nuclear magnetic moment (M) aligns with the magnetic field. Irradiating the sample with a radio-frequency field at the Larmor frequency flips the nuclear magnetic moment from the Z-axis to the X–Y plane. The FID signal is measured in the time domain using a coil located in the X–Y plane, and Fourier transformation (FT) is applied to obtain the NMR spectrum in the frequency domain. (b) Schematic of the chemical shift δ in the ^1H-NMR spectrum as a function of pressure for $|B_0| = 9.3$ T. (c) Schematic of the chemical shift of the ^1H-NMR spectral line, (d) Schematic of localized-proton and tunneling-proton densities, and (e) Tunnel splitting of ^1H-NMR spectral line due to proton tunneling, as a function of pressure. The lowest starting pressure for which nuclear quantum effects cause proton tunneling is ~20 GPa. Figure created by Meier[2018] licensed under CC BY 4.0.

study the phase transition of high-pressure ice (ice X). Their results clearly demonstrated hydrogen-bond symmetrization triggered by nuclear quantum effects. As the pressure increased from 10 GPa to 90 GPa, the ice gradually transitioned from phase VII to phase X. The NMR signal (Figure 6.1(b)) showed a unimodal shape with respect to δ when the pressure was less than or equal to 17 GPa but became bimodal when the pressure was greater than or equal to 21 GPa. In the bimodal region, the shape of the signal continued to change with increasing pressure. When the pressure was relatively low, the probability of proton tunneling was small, and no signal of tunnel splitting could be detected. However, high pressures enhanced the quantum tunneling, and the resultant splitting gave rise to the bimodal structure. As the pressure continued to increase, the hydrogen bonding became completely symmetrized (Figure 6.1(b–e)) [Meier2018].

NMR relaxation is a useful and highly effective tool for studying not only long-term statistical averages but also dynamical processes such as proton tunneling in condensed matter systems. In certain cases, the changes in the environment can considerably affect relaxation. Here, "relaxation" refers to the process by which a nucleus with non-zero spin absorbs the RF pulse energy and then returns to the NMR equilibrium state by interacting with its surroundings. The NMR relaxation process can be categorized into two types: spin–lattice relaxation (time constant T_1) and spin–spin relaxation (time constant T_2). In spin–lattice relaxation, energy is exchanged between the spin system and the lattice, changing the energy of the spin system. This process is known as longitudinal relaxation because it corresponds to the recovery of the longitudinal component of the nuclear magnetization. In spin–spin relaxation, energy is transferred from the spin in the higher energy state to the neighboring lower energy state of the same type of magnetic spin system, and the total energy of the spin system remains unchanged. This process is known as transverse relaxation because it corresponds to the disappearance of the transverse component of the nuclear magnetization.

Each nucleus generates a fluctuating local magnetic field, $B_{Loc}(t)$, which is modulated by the nuclear motion within the molecule. When the frequency of the varying local magnetic field is close to the Larmor frequency of the spin, the nuclear spin interacts with the local magnetic field and longitudinal relaxation occurs. The resonance-relaxation time constant is typically determined by analyzing the free induction decay (FID), the spin echo, and the gradient echo signal.

Nuclear quantum tunneling is another example of full quantum effect. The field cycling of NMR spectroscopy is often used to study proton tunneling in molecules in a hydrogen bonding network [Horsewill2002, 2006, Noble2009]. By measuring the T_1 of protons as a function of magnetic field, the proton transfer rate can be accurately determined. Based on this principle, in 1997, D. F. Brougham et al. at the University of Nottingham suggested that the dynamical process of proton transfer along the direction of hydrogen bonds in powdered dimeric benzoic acid is dominated by phonon-assisted quantum tunneling at low temperatures [Brougham1997]. Subsequently, concerted two-proton tunneling effects were observed in the hydrogen bonds in single crystals of dimeric benzoic acid [Jenkinson2003, Xue Q.2004]. Furthermore, the variations in T_1 with respect to the magnetic field and temperature were measured to detect concerted proton tunneling in calixarene molecules; the tunneling rate was determined to be $\sim 8 \times 10^7\ s^{-1}$ [Brougham1999].

In addition, R. Damle's group used hydrogen NMR spectroscopy to evaluate the relationship between T_1 and temperature in the range from 6.6 K to 389 K and observed quantum rotational tunneling of methyl groups in $[(CH_3)_4N]_2SeO_4$ and $(CH_3)_4NGeCl_3$ powder at low temperatures [Mallikarjunaiah2007, 2008]. As an experimental technique sensitive to the local environment, NMR has proven to be an effective tool in the study of nuclear quantum effects.

6.1.2 Neutron Scattering

Neutrons in condensed matter interact mainly with nuclei and are, therefore, an effective tool for studying nuclear quantum effects. Neutron scattering, which can be classified as elastic, quasi-elastic, or inelastic depending on the energy exchanged between neutrons and other matter during the scattering process, is an irreplaceable experimental tool for studying the structure and dynamics of condensed matter. Elastic scattering typically uses neutrons with wavelengths in the range of 0.1–10 Å, which closely matches the interatomic distances in condensed matter systems. Because neutrons have spins, in addition to the atomic structure, elastic neutron scattering can be used to obtain information about the arrangement of magnetic moments in a sample (see Figure 6.2(a)). In inelastic scattering, neutrons exchange energy with the sample through excitations such as phonons and magnons. The energy of the incident neutrons is usually between 1 meV and 1 eV, which is comparable to the energy of the excitations under study. Inelastic neutron scattering is, therefore, an ideal tool for studying dynamical processes in many excited states. Quasi-elastic scattering occurs when the energy exchange is relatively small and can be observed as a broadening of the neutron energy spectrum. It is widely used to study diffusion dynamics and related topics. As neutrons do not carry charges, they do not interact with the outer electron clouds of atoms. However, they are sensitive to light elements and isotopes (unlike X-rays, whose scattering amplitude decreases with decreasing atomic number). Thus, neutron scattering is more useful than X-ray scattering for accurately identifying light elements and distinguishing isotopes in materials.

Neutrons interact differently with nuclei depending on their spin states. For the hydrogen nucleus with a single proton, this spin-dependency is strong, resulting in a large incoherent neutron scattering cross section. This property is often exploited to study full quantum effects such as quantum fluctuation and quantum tunneling in hydrogen-containing materials. Theoretical calculations have shown that nuclear quantum effects are essential for hydrogen diffusion in metals [Dupios2017]. As early as 1981, inelastic neutron scattering spectra measured on $NbO_{0.013}H_{0.016}$ at 0.07 K and 5 K showed that as the temperature increases, the inelastic neutron spectrum for energy gain gradually increases, whereas that for energy loss decreases. This phenomenon is caused by energy transfers arising from protons tunneling between two interstitial crystalline sites with different energies. Thus, proton quantum tunneling is prominent in metallic niobium [Wipf1981]. The same technique was also used to study rotational tunneling in molecules, for example, in the observation of rotational tunneling among the three hydrogen atoms of the methyl group ($-CH_3$) [Prager1997]. In ice, concerted

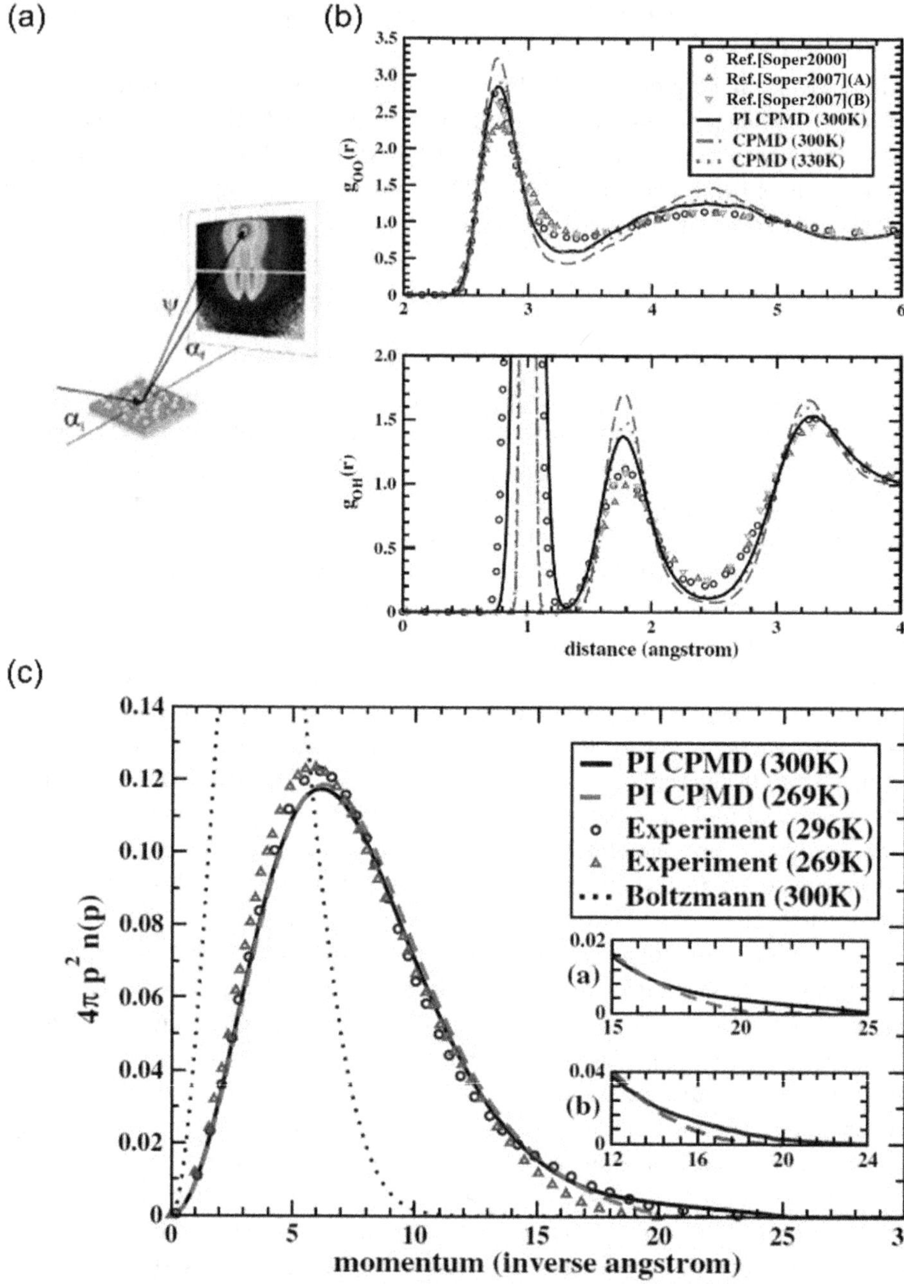

Figure 6.2 Nuclear quantum effects in liquid water. (a) Schematic of neutron scattering. After the neutron beam is scattered, its energy and momentum are changed by an amount related to the structural and dynamical properties of the sample. Analyzing the scattering data allows researchers to study the sample's structure and dynamics. (b) Radial distribution functions of O–O (top) and O–H (bottom) in liquid-water molecules. The markers denote data from neutron and X-ray experiments. The solid and dashed curves correspond to molecular dynamics simulations that include (PI CPMD) or exclude (CPMD) nuclear quantum effects. (c) The experimental proton radial distribution function in water compared to path-integral molecular dynamics simulations (full quantum effect calculations). Reprinted figure with permission from Morrone[2008]. Copyright 2024 by the American Physical Society.

proton tunneling along the O–O bond was also observed. In the hexagonal (I_h) and cubic (I_c) phases of ice [Huang X.2023], 4% and 2% of hydrogen atoms, respectively, were observed to be involved in concerted proton tunneling at 5 K. This tunneling is believed to occur in ordered hexagonal hydrogen–oxygen rings, which are present in small amounts in ice [Bove2009].

Neutron scattering is also used to study nuclear quantum effects in elements other than hydrogen. At low temperatures, silicon single crystals exhibit an abnormal negative coefficient of thermal expansion. This anomaly can be explained by combining inelastic neutron scattering data with theoretical calculations that account for nuclear quantum effects [Kim D.S.2018].

When high-energy (~1–100 eV) incident neutrons are used in high-momentum-transfer (~30 Å^{-1}) scattering experiments, interatomic interactions such as those mediated by phonons are negligible because of the large momentum transfer between the scattered neutrons and the particles under study and the extremely short scattering-interaction time. The scattering process can be viewed as incoherent and treated using the impulse approximation, in which the system's energy and momentum are conserved; the initial momentum of the particles under study can be derived by measuring the neutrons' energy and momentum before and after scattering. This scattering process is called deep inelastic neutron scattering [Hohenberg1966, Evans1993, Bafile1998]. It is widely used to study nuclear quantum effects [Reiter2002, Giuliani2012, Ceriotti2016, Markland2018, Bakó2021]. Deep inelastic neutron scattering spectroscopy allows researchers to study the momentum distribution and quantum motion of protons in a sample [Reiter2004]; it can also reveal the quantum motion of relatively heavier nuclei such as D or O [Andreani2005]. A comparison of the average kinetic energies of nuclei along different molecular axes obtained in deep inelastic neutron scattering studies of water and ice with the results of full quantum calculations reveals the momentum distribution to be anisotropic with a competing effect among the molecular axes. This provides direct experimental proof of competing nuclear quantum effects [Andreani2013, Senesi2013a,b].

With high incident neutron energies, scattering data can be collected over a very wide momentum-transfer domain (~0.05–100 Å^{-1}), facilitating the calculation of the RDF of nearest atomic neighbors. Figure 6.2(b) and (c) shows experimentally obtained RDFs for O–O, O–H, and H–H in water, as well as the momentum distribution of the hydrogen nuclei. Direct comparisons of these results with PIMD simulations have improved our understanding of the full quantum effect. Similar experimental techniques have been extensively applied to the study of nuclear quantum effects in bulk water and other complex aqueous systems [Morrone2008].

Deep inelastic neutron scattering measurements of the proton momentum distribution have revealed that, at low temperatures, the proton state of water confined in carbon nanotubes differs from that of ice I_h, ice IV, and high-density amorphous ice [Reiter2006, 2012, Kolesnikov2016]. The protons in the confined water are in a quantum state of coherent delocalization [Reiter2006], and their momentum distribution lacks the high-momentum tail that is characteristic of the stretching modes of covalently bonded water molecules (see Figure 6.3). In the beryl crystal ($Be_3Al_2(SiO_3)_6$), along the c-axis (which has a channel connecting cages of ~5 Å diameter that may contain alkali cations and water), integrated inelastic neutron scattering intensities of characteristic peaks mostly exhibit an inverse correlation with the temperature (see Figure 6.4). This is characteristic of systems exhibiting quantum tunneling, thus further confirming the existence of proton quantum tunneling in confined water. Projection of the proton momentum distribution of confined water onto the xy- and yz-planes additionally reveals the coherent delocalization of protons in the ground state [Kolesnikov2016, 2018].

This spectroscopic method is also applicable to the study of water in critical states, such as supercooled water. Using deep inelastic neutron scattering spectroscopy, C. Andreani's group at the University of Rome studied nuclear quantum effects on the hydrogen nuclear momentum distribution in supercooled water. They found that protons in supercooled

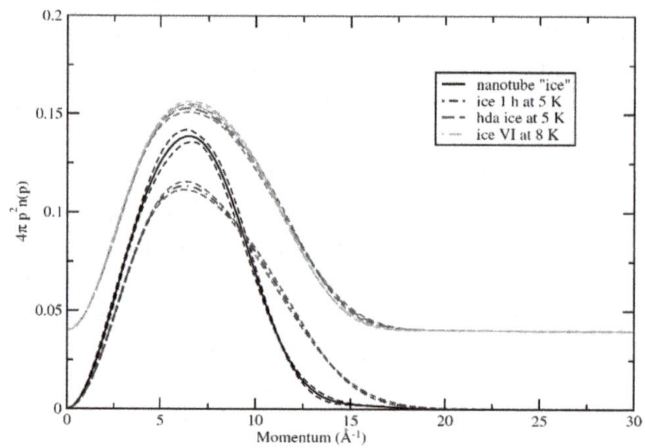

Figure 6.3 Neutron scattering spectral curve of water confined in carbon nanotubes (black curve) lacks the high-momentum tail seen after 15 Å^{-1} in various forms of ice. Reprinted figure with permission from Reiter[2006]. Copyright 2024 by the American Physical Society.

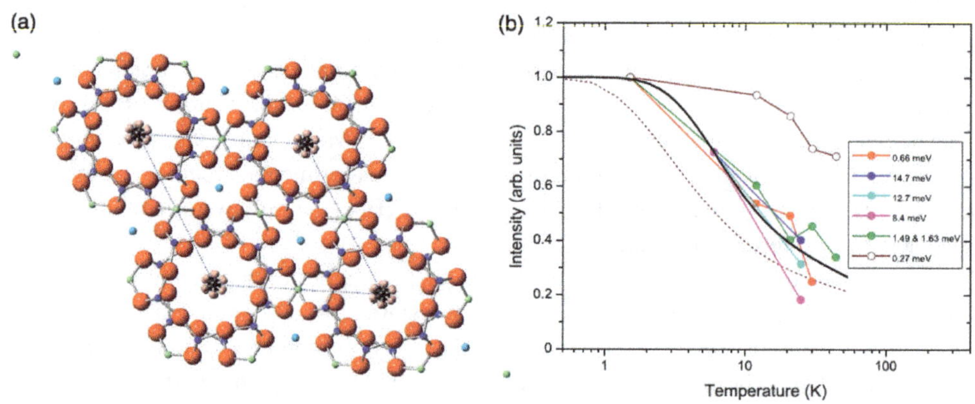

Figure 6.4 Beryl ($Be_3Al_2(SiO_3)_6$). (a): crystal structure along the c-axis; (b): temperature dependence of the quantum tunneling neutron scattering intensity. Reprinted figure with permission from Kolesnikov[2016]. Copyright 2024 by the American Physical Society.

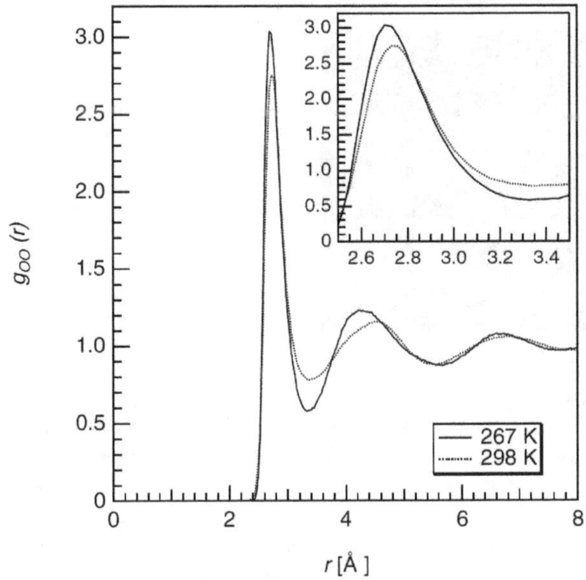

Figure 6.5 O–O radial distribution functions for supercooled (solid curve) and ambient (dashed curve) water. The inset shows an enlarged view of the first peak: the O–O distance in supercooled water is 2.7 Å, which is slightly less than the value of 2.76 Å measured in ambient water. Reprinted figure with permission from Pietropaolo[2008]. Copyright 2024 by the American Physical Society.

water have a larger mean kinetic energy (7 kJ/mol above the ambient value) and a smaller first neighbor O–O distance compared with liquid and solid water (Figure 6.5). They concluded that protons in supercooled water are located closer to the centers of the two oxygen atoms and therefore have a higher probability of quantum tunneling. In more extreme cases, the protons in supercooled water may exist in a fully delocalized state, that is, at the midpoint of the two oxygen atoms, exhibiting full symmetry which is directly related to NQEs [Pietropaolo2008, 2009].

6.1.3 X-ray Scattering

In 1895, German physicist W. C. Röntgen discovered X-rays while studying cathode rays and was awarded the first Nobel Prize in Physics (1901). An X-ray is a type of electromagnetic wave with a short wavelength ($\lambda = 10^{-3}$–10^{1} nm) and high energy. It interacts with matter and produces absorption and scattering effects; this has led to the development of various X-ray-based spectroscopic detection methods such as X-ray diffraction (XRD), X-ray absorption spectroscopy (XAS), and X-ray scattering spectroscopy (XRS). When X-rays enter a crystal, they are scattered by the lattice. As shown in Figure 6.6(a), the scattered waves from each atomic center interfere with each other, resulting in diffraction patterns that are reflective of the atomic distribution inside the crystal. Therefore, the XRD method is widely used to determine crystal structure.

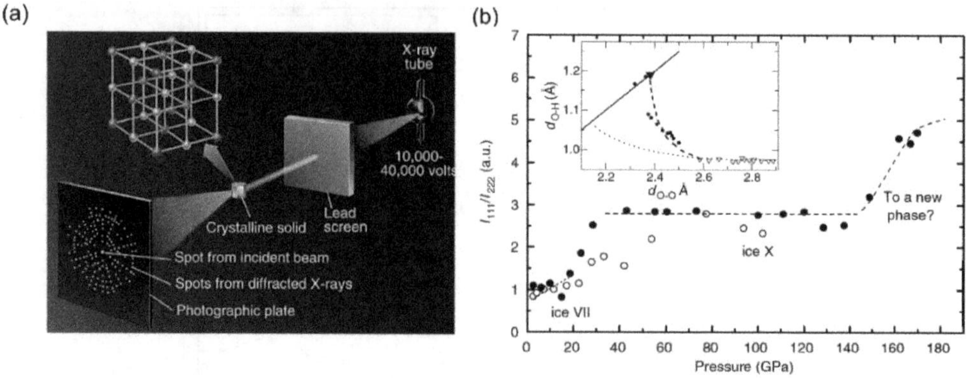

Figure 6.6 Discovery of ice with symmetric hydrogen bonding (ice X). (a) Schematic of X-ray diffraction. A beam of X-rays enters the crystal and is scattered by the periodic lattice. The X-rays scattered by different atoms interfere with each other, producing diffraction patterns. (b) Ratio of the (111) diffraction peak intensity to the (222) peak intensity (I_{111} / I_{222}) plotted against pressure. The dotted line represents the simple empirical model based on the extrapolation of the constant linear variation of d_{O-H}. The dashed line is a reasonable interpolation between experimental data. The symmetrization line, $d_{O-H} = d_{O-O}/2$, is indicated (solid line). When the pressure exceeds 60 GPa, the O–O spacing decreases drastically, leading to the formation of ice X with symmetric hydrogen bonds. From Loubeyre[1999]. Reproduced with permission from SNCSC.

As the solid state of water, ice has a very complex phase diagram. At present, there are as many as 20 types of experimentally recognized bulk-phase ice. XRD is commonly used to study the structure and phase transitions of ice under high pressures. As shown in Figure 6.6, the (111) diffraction peak indicates the position of hydrogen atoms and the ordering of their distribution. Measuring the variation in the ratio of the (111) peak intensity to the (222) peak intensity (I_{111} / I_{222}) with pressure reveals that I_{111} / I_{222} = 3 when the pressure is greater than 60 GPa. This indicates that the hydrogen nucleus is completely delocalized, that is, the hydrogen is in the middle of the two oxygen atoms, forming ice with symmetrized hydrogen bonds (ice X) (see Figure 6.6(b)). The XRD method provides direct experimental evidence for the discovery of ice X [Loubeyre1999].

In addition to diffraction, X-ray-based scattering techniques are crucial tools for understanding the microstructure of materials, especially liquids. Information about the short-range ordered structure of liquids can be obtained using the radial, coordination number, and angular distribution functions. XRS is based on the inelastic light-scattering induced by inner or valence-band electrons. Similar to X-ray absorption, the inner electrons are excited by X-rays to an unoccupied state, leading to energy loss in the scattered radiation, including Rayleigh scattering, Compton scattering, Raman scattering, and Brillouin scattering. Among these scattering processes, the Raman band reflects band edges of the K-absorption spectrum, giving the XRS spectrum a shape similar to that of the K-absorption (XAS) spectrum.

In 1969, A. H. Narten et al. at Oak Ridge National Laboratory determined the coordination number of water to be ~4.7 by using X-ray scattering; this is consistent with the common belief that liquid water has a localized tetrahedral structure. In other words, a water molecule is connected to four surrounding water molecules by hydrogen bonds [Narten1969]. In 2004, the tetrahedral model was challenged by A. Nilsson's research group at Stockholm University. Through XAS and XRS, they found that water molecules contain only two types of strong hydrogen bonds: donor and acceptor. They proposed a "loop model" for the first shell in the liquid water. However, R. J. Saykally's group at the University of California, Berkeley, analyzed the temperature dependence of X-ray absorption near-edge fine structure in liquid water and argued that, at room temperature, liquid water is still best understood in terms of tetrahedral structure. For supercooled water at low temperatures, a distorted hydrogen bonding network structure was observed [Smith2004]. Although these studies have provided ample experimental evidence concerning the structure of liquid water, there is still much debate over the subject, and the role played by nuclear quantum effects remains mostly unknown.

In addition, XRS and XAS are used extensively to study the isotope effects in liquid water and its temperature dependence. The difference in the spectral features of room-temperature liquid H_2O and D_2O has been compared to those of liquid H_2O at two different temperatures (2°C, 22°C). The spectral features of D_2O were found to be very similar to those of the low-temperature H_2O [Bergmann2007], verifying that at room-temperature the hydrogen-bond strength of liquid D_2O is greater than that of H_2O. Recently, Nilsson's group at Stockholm University used femtosecond X-ray pulses to probe micrometer-sized supercooled water droplets in a vacuum. They found that nuclear quantum effects influence the thermodynamic response and correlation length. At the temperature with the maximum correlation length (229 K for water (H_2O) and 233 K for heavy water (D_2O)), a significant isotope effect was further observed [Kim K.H.2017]. Y. Harada et al. from S. Shin's research group at the University of Tokyo used high-resolution resonance inelastic X-ray scattering at the O K-edge to probe the stretching vibrations of liquid water molecules. Through isotopic substitution, they explored the nuclear quantum effects on the hydrogen bond strength, substantiating that the hydrogen bond strength of D_2O is greater than that of H_2O [Harada2013].

Vibrational spectra recorded by inelastic X-ray scattering can also be used for direct detection of nonadiabatic effects in different systems. An important example is given by the high-precision phonon spectrum of heavily doped diamond measured by Caruso et al. (Figure 6.7) [Caruso2017]. Phonon dispersion plots (i.e., plots of the phonon energy as a function of momentum) can be obtained using first-principles calculations with the adiabatic approximation, but the theoretical results differ significantly from the experimental measurements. The agreement with the experiment can be significantly improved by going beyond the adiabatic approximation and including nonadiabatic electron–phonon and phonon-self-energy interactions. This result clearly indicates that significant nonadiabatic effects are present in heavily hole-doped diamond, causing the phonon vibration energy to be modified by renormalization. For a doping concentration of 1.4×10^{21} cm^{-3}, the energy change reaches ~15 meV, which is around 10% of the total phonon energy. Therefore, the nonadiabatic effect cannot be neglected in this case.

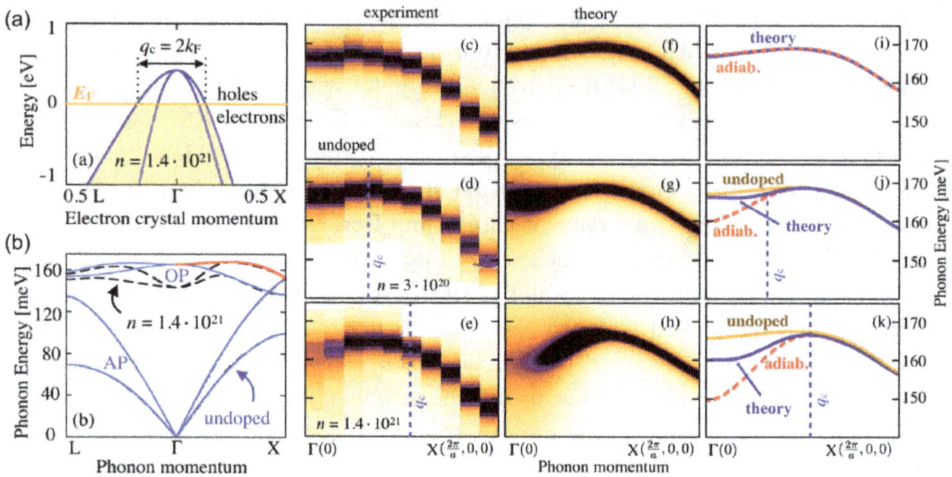

Figure 6.7 (a) Electron energy band and (b) the corresponding phonon spectrum of hole-doped diamond. (c)–(e) The phonon spectra of diamond at three doping concentrations measured by inelastic X-ray scattering. (f)–(h) The phonon spectra after considering non-adiabatic effects. (i)–(k) Phonon spectra calculated by including (blue solid curve) and excluding (red dashed curve) nonadiabatic effects. Reprinted figure with permission from Caruso[2017]. Copyright 2024 by the American Physical Society.

The ultrafast X-ray diffraction technique, which was developed in conjunction with ultrafast laser technology, allows high temporal-resolution detection of the momentum-space structure corresponding to the atomic structure in real space in condensed matter. Furthermore, as X-rays are sensitive to the electronic structure of the material, this technique can directly detect changes in electronic states on extremely short time scales. It can be used to analyze the evolution of atomic and electronic structures simultaneously and is capable of decoupling nuclear quantum from nonadiabatic effects to some extent. Currently, X-ray diffraction is one of the most advanced methods for studying full quantum effects in condensed matter systems. Figure 6.8 presents the nonadiabatic evolution of the electronic states during the excitation of the C_2H_4 molecule, measured using ultrafast X-ray diffraction. The molecule crosses the conical intersection between the excited-state and ground-state potential energy surfaces 7 fs after excitation. It transitions to the potential energy surface of ground electronic state nonadiabatically, altering the lattice-space configuration [Zinchenko2021].

However, the scattering power (cross section) of X-rays increases with the atomic number, making X-ray scattering methods generally less sensitive to light elements, especially hydrogen. Neutron scattering may provide a better option: Neutrons can interact directly and strongly with atomic nuclei, as protons have a very large incoherent neutron scattering cross section. Electron-scattering techniques are also very useful for the study of light elements, as the cross section for electron scattering is typically four to six orders of magnitude higher than that for X-ray scattering. Overall, combining

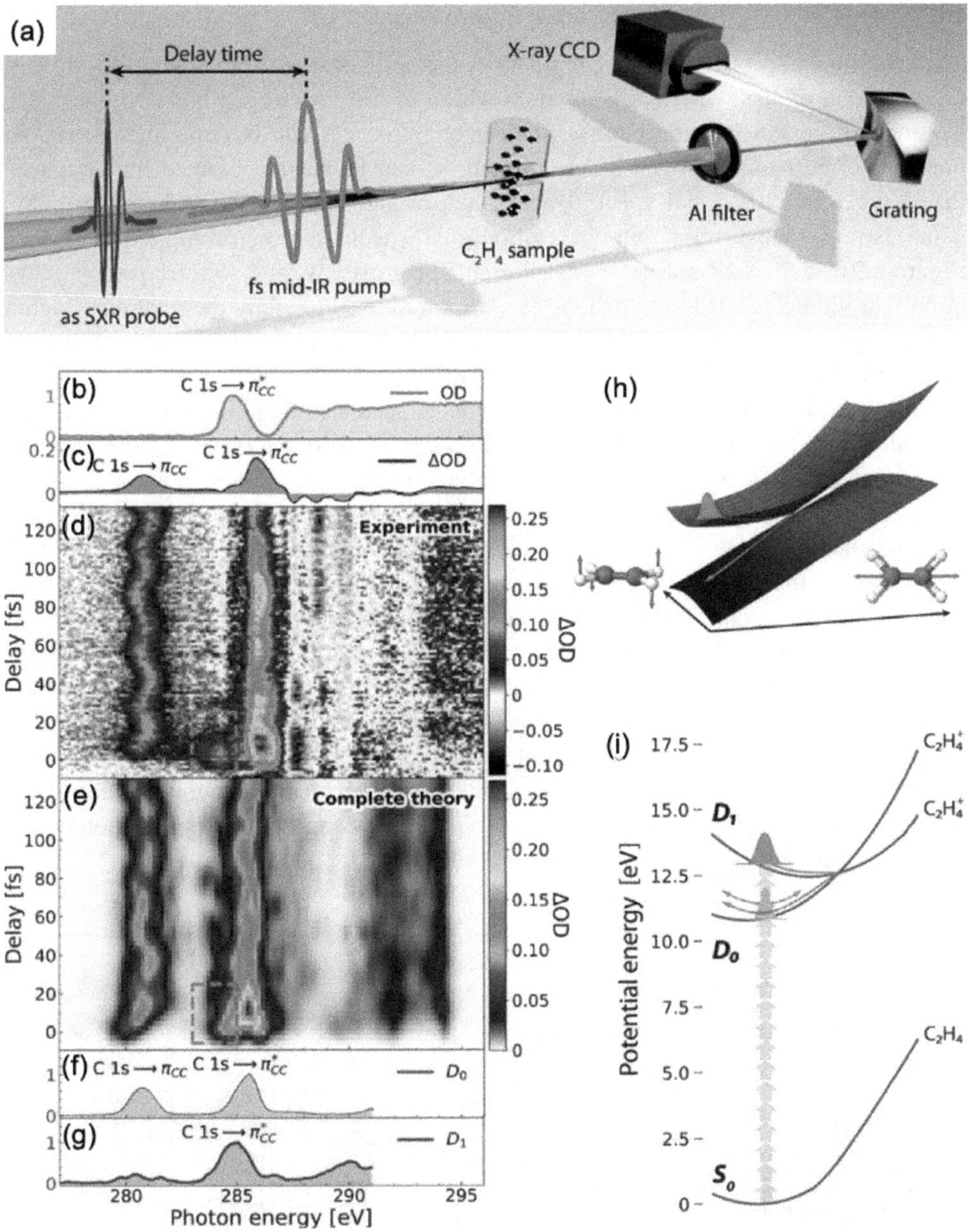

Figure 6.8 Ultrafast X-ray diffraction measurements of nonadiabatic processes in electrons of laser-excited C_2H_4 molecules. From Zinchenko[2021]. Reprinted with permission from AAAS.

X-ray scattering, neutron scattering, and electron scattering allows accurate characterization of the microstructure of light element systems and investigation of full quantum effects in dynamical processes.

6.1.4 Spectroscopy

The vibrational spectra of hydrogen bonds (especially of the X–H stretching and bending vibration modes, where X is an atom which forms a hydrogen bond with a hydrogen atom.) are sensitive probes for nuclear quantum effects. In experiments, nuclear quantum effects on hydrogen bond interactions can be easily obtained by comparing the changes of vibrational spectral peaks before and after hydrogen-bond formation using isotope substitution. Optical vibrational spectroscopy techniques (shown in Figure 6.9), such as IR spectroscopy, Raman spectroscopy, and sum frequency generation vibrational spectroscopy (SFG-VS), are often used to study the quantum fluctuations of hydrogen nuclei and their effects on the hydrogen bond strength [Hirsch1986, Aoki1996, Goncharov1996, 1999, Nagata2012].

IR spectroscopy is commonly used to study molecular vibrations. When a sample is irradiated by continuous broad-spectrum IR light, absorption occurs if the radiation energy is exactly equal to the energy difference between different states of the molecule. Consequently, the transmitted light intensity in the corresponding absorption region is diminished. Because the vibrational energy of the molecule is greater than the rotational energy, the frequency region corresponding to IR absorption differs for different energy-state excitations. The region of the absorption spectrum corresponding to molecular vibrations – the mid-IR – is the most widely studied region. In general, the term "infrared spectrum" commonly refers to the absorption spectrum in the mid-IR region. In addition to the condition that the IR frequency must match the frequency of the excited vibration, the dipole moment of the molecule must remain changeable during the vibration for IR absorption to occur. The alternating electric field generated by the changing dipole moment interacts with the electromagnetic radiation to produce absorption. Therefore, polar molecules are all IR active.

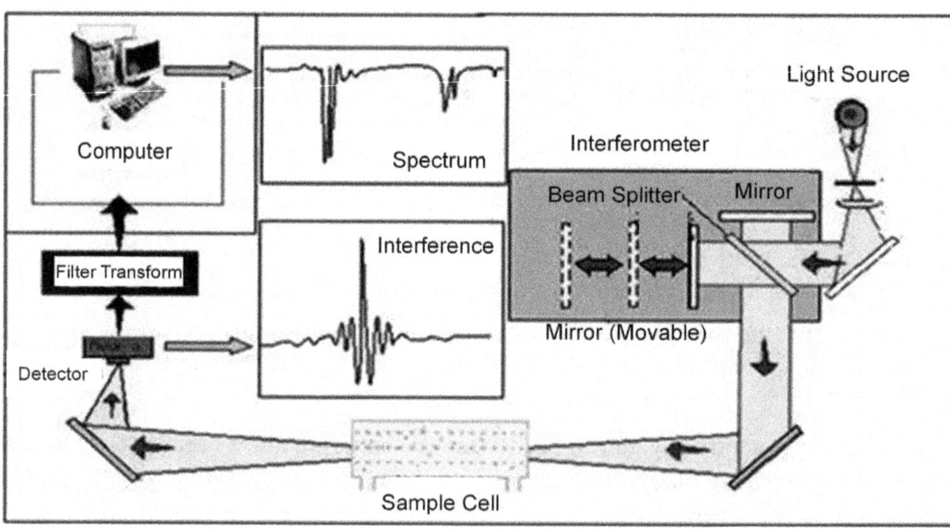

Figure 6.9 Schematic of IR and Raman spectroscopy.

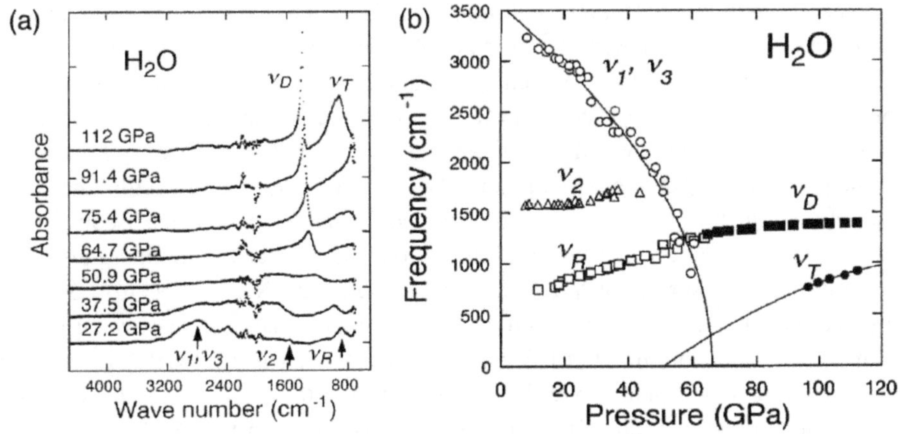

Figure 6.10 Infrared spectroscopy measurements of the structural phase transition in ice at high pressures. Panel (a) shows the measured vibrational spectra of ice at different pressures. Panel (b) shows the pressure dependence of the vibrational peak position. For H_2O, the O–H stretching vibrational peak disappears at 62 GPa, indicating that the ice structure has transitioned from phase VII to phase X, which exhibits hydrogen bond symmetrization. Because of nuclear quantum effects, the corresponding phase transition occurs at 70 GPa in D_2O molecules. Reprinted figure with permission from Aoki[1996]. Copyright 2024 by the American Physical Society.

Water is one of the common polar molecules. Using IR spectroscopy, it is possible to study the symmetrized hydrogen bond formation when ice VII transitions to ice X as the pressure is increased [Aoki1996, Goncharov1996]. At room temperature, if the pressure is greater than 2 GPa, ice is in the VII phase. The crystal has a bcc structure, and each water molecule forms a tetrahedral structure with its four neighbors. The potential energy surface where the protons move along the hydrogen bond direction can be described as a double potential well, with a finite height potential barrier in the middle between the wells. As the pressure gradually increases, the distance between the oxygen atoms in adjacent molecules is reduced, the height of the potential barrier decreases, and the probability of proton tunneling increases. When the pressure is large enough, the double potential well shrinks to a single potential well. In other words, ice VII transforms into ice X, a phase with symmetrized hydrogen bonds. The stretching vibration frequency of O–H in H_2O measured by IR spectroscopy decreases with enhanced pressure (Figure 6.10). When the pressure increases to 60 GPa, a phase transition occurs [Aoki1996]. (For D_2O, this phase-transition pressure increases to 70 GPa due to stronger hydrogen bonds and weaker nuclear quantum effects [Goncharov1996].) It is worth mentioning that J.M. Zhao's research group at the Institute of Physics, Chinese Academy of Sciences, has recently made some progress in the study of ultrafast spectroscopy under high pressure conditions; it is expected that the ultrafast dynamics of excited states in water under high pressure will be studied in the near future [Wu Y.L.2020, 2021].

Raman spectroscopy is complementary to IR spectroscopy in the study of low-frequency vibrational, rotational, and other modes of a sample. When a beam of monochromatic light enters a medium, molecular vibrations and elementary excitations (such as optical phonons in solids) interact with the photons. As a result, the frequency of the scattered light is different from that of the incident light, and the scattering is inelastic. By analyzing the magnitude of the photon energy transferred, information about the vibrational modes in the medium can be uncovered. The molecular vibrational mode is only Raman active when the polarizability changes along a fixed direction during the molecular motion.

The phase transitions of ice can be studied using Raman spectroscopy. Proton-disordered ice VII can change into proton-ordered ice VIII at low temperatures. Under sufficiently high pressures, ice VIII can change into ice X; Raman spectroscopy can be used to verify the formation of symmetrized hydrogen bonds (characteristic of ice X) [Goncharov1999, Yoshimura2006]. When the pressure reaches 60 GPa, the O–H symmetric stretching vibrational mode disappears, and the translational modes exhibit frequency and damping anomalies [Goncharov1999], as shown in Figure 6.11.

Brillouin scattering, similar to Raman scattering, is inelastic, but in this case, it is the acoustic phonons that interact with the photons. Because the energy of the acoustic phonons is lower than that of the optical phonons, the frequency shift in the Brillouin spectrum is smaller than that in the Raman spectrum. A. Polian and M. Grimsditch measured the Brillouin frequency shift of longitudinal phonons as a function of pressure. Anomalous behavior of the longitudinal sound velocity was observed at a pressure of ~44 GPa. Polian and Grimsditch also plotted the logarithm of the effective longitudinal elastic modulus against that of the density; at 44 GPa, the relation ceased to be linear, and the first derivative changed abruptly. Their observations signified the generation of ice X [Polian1984].

The spectroscopic methods discussed thus far lack interfacial selectivity. As a result, it is difficult to distinguish the surface signal from the bulk-phase signal. By contrast,

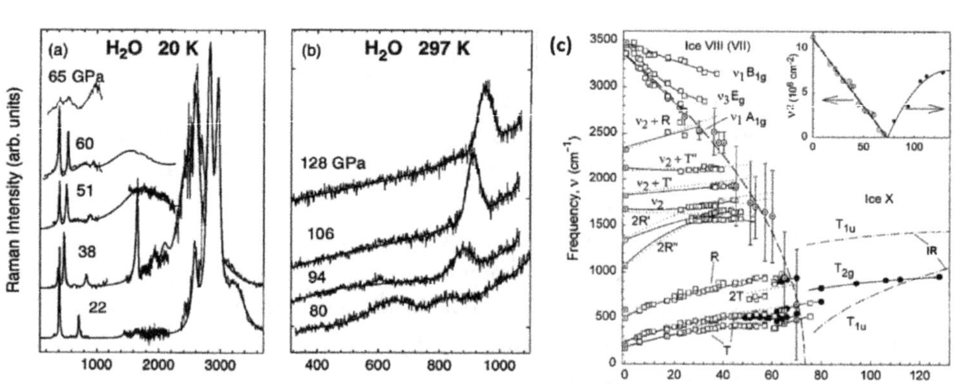

Figure 6.11 Raman spectral measurements of the phase transition in ice at high pressures. (a) Raman spectra of ice at different pressures at 20 K. (b) Raman spectra of ice at different pressures at 297 K. (c) Pressure dependence of Raman spectral peak position. The hollow and filled dots represent measurements at 20 K and room temperature, respectively. The inset shows the phase transition that occurs at ~60 GPa. Reprinted figure with permission from Goncharov[1999]. Copyright 2024 by the American Physical Society.

second-harmonic generation and SFG-VS possess interface selectivity and involve second-order nonlinear optical processes. Consequently, they have been extensively applied to studying structure, molecular orientation, and dynamical processes at interfaces. Y. R. Shen at the University of California, Berkeley, played a crucial role in the development of these methods [Shen Y.R.2006]. Experiments have shown that the interface-sensitive spectroscopic technique is contact-free and thus does not destroy the object under study during the measurement. This method is also local, being sensitive to only a few atomic layers at the interface. It can generate signals in symmetry-breaking regimes at the nanometer scale [Zhao H.2021].

In typical SFG-VS experiments, a beam of visible light at a fixed frequency (ω_{vis}) and a beam of coherent tunable IR light (ω_{IR}) are incident on the test sample. When ω_{IR} matches a molecular vibrational transition, energy resonance occurs, causing molecules at lower vibrational energy levels to be excited to higher energy levels. Molecules at higher vibrational energy levels are further excited to the imaginary state by visible light, which subsequently relax to the vibrational ground state. During this process, photons are emitted, and a sum-frequency signal is produced. The frequency of the signal equals to the sum of the frequencies of the two incident beams (see Figure 6.12(a)). Because the sum-frequency signal is blocked by centrosymmetric objects, SFG-VS is interface selective and is often used to study physical properties at interfaces.

The O–H stretching vibrations at the gas–liquid interface have recently been extensively studied using SFG-VS. These experiments can help us obtain information on the microstructure and dynamics of interfacial water. Based on experimental observations,

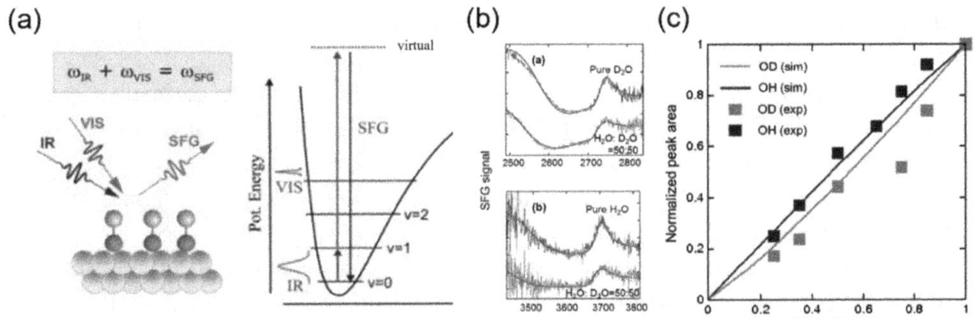

Figure 6.12 Nuclear quantum effects on the orientation of water molecules at the liquid–vapor interface. (a) Mechanism of sum frequency generation vibrational spectroscopy (SFG-VS). When two beams of light are incident on the material, the light resonates with a vibrational energy level of the molecules. The molecules in the current energy level are excited to a higher one, and those at the high vibrational energy level are further excited to the virtual state by visible light. Then, these unstable molecules relax to the vibrational ground state. The frequency of the light emitted in the relaxation process is equal to the sum of the frequencies of the incident photons. (b) SFG-VS spectral signal of water molecules at the water–vapor interface. (c) Free O–H and O–D bond peak areas as a function of isotope concentration. The free OD peak area is smaller than that of OH. Consequently, the number of O–H bonds oriented upward toward vapor is greater than that of OD. Reprinted figure with permission from Nagata[2012]. Copyright 2024 by the American Physical Society.

characterizing the movement of O–H stretching vibration peak positions and the line shape of the peaks allows us to acquire information on the hydrogen bond interactions, as well as the coupling strength between the O–H vibrational dipoles of water molecules [Shen Y.R.2006, Stiopkin2011, Schaefer2016]. In 2011, I. V. Stiopkin et al. found that the free O–D stretching vibrations on the liquid-water surface are only affected by intra-molecular vibrational dipole coupling [Stiopkin2011], resulting in narrower linewidths that can be directly distinguished from the bonded O–D stretching vibrational mode in the spectrum. In 2012, M. Bonn's group at the Max Planck Institute further investigated the relation between the peak areas of the free O–H and O–D stretching and the isotope concentration using SFG-VS. They found that nuclear quantum effects affect the orientation of surface water, that is, OH is oriented upward toward the vapor-phase molecules, while O–D is oriented downward toward the bulk-phase molecules (Figure 6.12(b–c)). This phenomenon is directly related to nuclear quantum effects on the hydrogen-bond strength, where the hydrogen bond strength of D near the surface is greater than that of H [Nagata2012]. In addition, spectroscopic techniques have been often used to study full quantum effects in chemical reactions, such that the reaction mechanism that is dominated by quantum tunneling can be studied via KIEs [Meisner2016].

As the transition frequencies of molecular vibrational energy levels are generally located in the IR and visible wavelength range, while those of molecular rotational energy levels are in the microwave or submillimeter range, rotational spectroscopy has better energy resolution than vibrational spectroscopy. Therefore, rotational spectroscopy is commonly used to study the fine structure and chemical composition of molecules. Common rotational spectrometers include absorption spectrometers (used for probing in the millimeter and submillimeter bands) and Fourier transform microwave spectrometers (often used to obtain rotational emission spectra in the microwave band). The FID signal detected in the Fourier transform spectrometer is similar to the RF signal in NMR experiments. The test molecules are distributed on the ground state of the rotational energy level by ultrasonic expansion. As the molecules return to the ground state after being excited by the microwave source, the FID signal is emitted. Fast Fourier transform of the FID signal yields the power spectrum in the frequency domain.

Rotational spectroscopy is also commonly used to study nuclear quantum effects. At ultralow temperatures, molecules exist in the ground-state energy levels, and the system exists in the corresponding ground-state configuration. However, quantum tunneling can change the molecular configuration. Even subtle rotation processes can result in rotational spectral splitting. The magnitude of this energy-level splitting is in the order of MHz and THz [Keutsch2003]. B. H. Pate's group measured the rotation spectrum of the water hexamer using a chirped-pulse broadband spectrometer. In collaboration with S.C. Althorpe's group at University of Cambridge, they discovered that the splitting of the rotation spectrum results from coordinated tunneling of water molecule pairs, and that the "geared" motion of molecules involves the concerted breaking of two hydrogen bonds; these results have been confirmed by quantum simulations including NQEs [Richardson2016b].

Vibrational spectra provide not only information about nuclear quantum effects but also direct evidence for nonadiabatic effects. A well-known example is the nonadiabatic

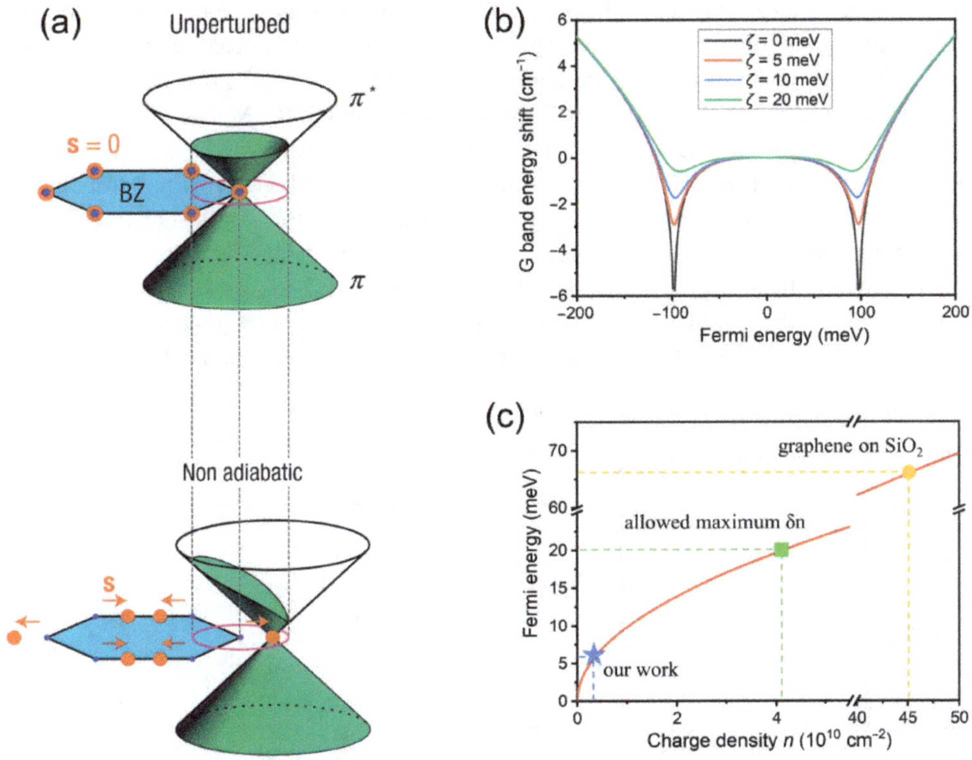

Figure 6.13 Raman spectroscopy shows that the adiabatic Born–Oppenheimer approximation no longer holds in graphene. (a) Upper panel: energy-band dispersion of a perfect graphene lattice; Lower panel: energy-band dispersion of graphene in the presence of E_{2g} phonon lattice vibrations under the nonadiabatic approximation. (b) Theoretical prediction of the logarithmic Kohn anomaly in phonons; the logarithm of phonon energy decreases as the Fermi energy approaches half of the phonon energy. (c) Using high-quality graphene devices encapsulated with boron nitride, Zhao Y. C. et al. compared the maximum charge fluctuations allowed by the logarithmic Kohn anomaly with the usual graphene charge fluctuations on silicon oxide. Reprinted figure with permission from Zhao Y. C.[2020]. Copyright 2024 by the American Physical Society.

effect in graphene (shown in Figure 6.13). In general, the electron–nucleus interaction can be described by the Born–Oppenheimer approximation, in which the system electrons are assumed to be in an instantaneous eigenstate at all times. However, when the phonon energy of the system approaches the electron Fermi energy, the electron–phonon interaction deviates from the adiabatic Born–Oppenheimer approximation. Nonadiabatic electron–phonon coupling can lead to many novel physical phenomena, such as high-temperature nonadiabatic superconducting pairing or electron correlation effects.

Monolayer graphene is theoretically predicted to be an ideal system to study nonadiabatic electron-phonon coupling in the two-dimensional limit because of its linear dispersion relation and strong electron–phonon coupling and because it exhibits the logarithmic Kohn anomaly (as the Fermi energy approaches half of the phonon energy, the logarithm of the phonon energy is reduced) [Lazzeri2006]. Although many attempts have been made to investigate this important problem, none have been successful, because the charge fluctuation is ~10^{11} cm^{-2}; to observe the logarithmic Kohn anomaly, the charge fluctuation is required to be less than 4.15×10^{10} cm^{-2} [Pisana2007, Yan J.2007, Malard2008, Das2008].

Recently, G. Y. Zhang's group at the Institute of Physics, Chinese Academy of Sciences, has prepared high-quality monolayer graphene devices with boron nitride encapsulation via the dry transfer method and has achieved the smallest charge fluctuations in the existing literature (i.e., less than 3.3×10^9 cm^{-2}). Using these devices, they uncovered significant E_{2g} phonon softening when the Fermi energy was close to half of the E_{2g} phonon energy; the Fermi-energy-dependent lifetime of the monolayer graphene increased by more than a factor of 2, as shown in Figure 6.14 [Zhao Y. C. 2020]. This

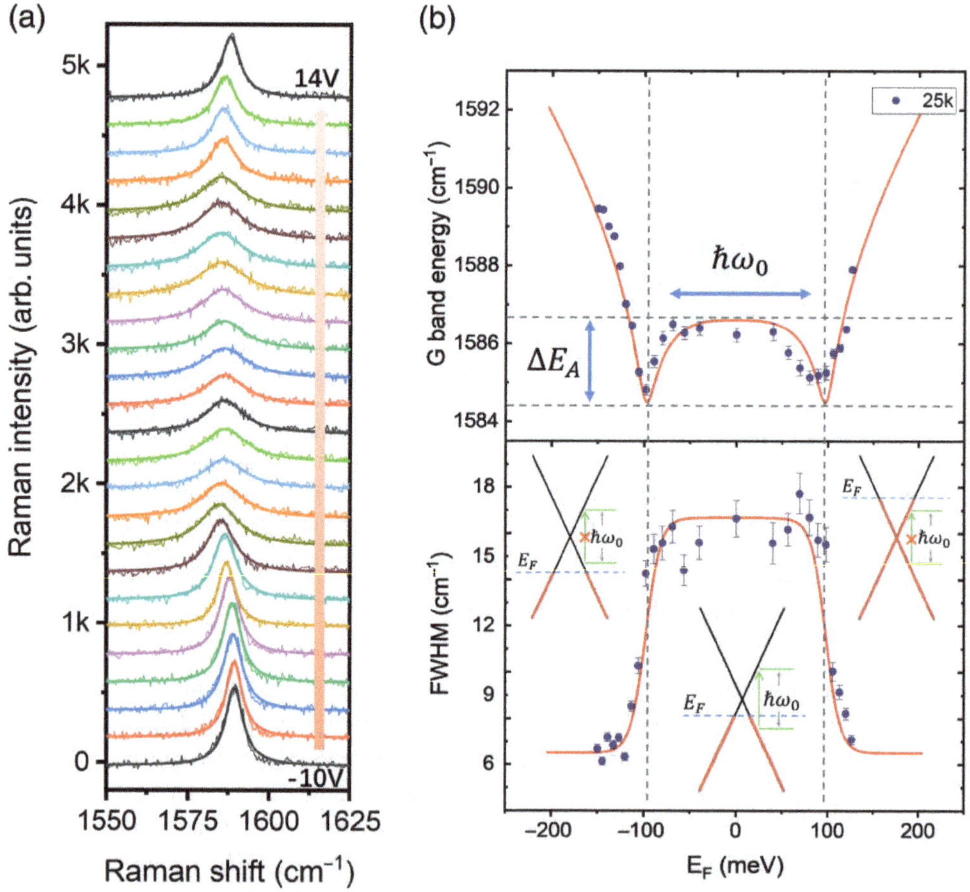

Figure 6.14 Phonon spectrum measurement in monolayer graphene. (a) Raman spectra of graphene under different back-gate voltages (electron Fermi energies). (b) Upper panel: evolution of phonon energy as a function of Fermi energy; lower panel: evolution of phonon full width half maximum (inverse of lifetime) as a function of Fermi energy. Reprinted figure with permission from Zhao Y. C.[2020]. Copyright 2024 by the American Physical Society.

experimental result confirms the theoretical predictions of nonadiabatic electron–phonon coupling and logarithmic Kohn anomaly. Furthermore, the experiment directly shows that the simple Born–Oppenheimer adiabatic approximation does not hold in a monolayer graphene. These observations can help us better understand the physical properties of graphene and its heterojunctions, such as the superconductivity of magic-angle graphene and the linear temperature resistance.

Other nonlinear optical effects (such as self-phase modulation in space) also reflect the broad-spectrum linear-energy-band electronic structure of graphene [Wu R.2011]. With ultrafast spectroscopic techniques, it is possible to realize broad-band photo-acoustic generation in graphene sheets and clearly uncover its physical mechanism [Tian Y.2015].

The methods discussed above are mainly used to study full quantum effects in the static state of light-element materials. Ultrafast laser pump-probe techniques, on the other hand, allow the study of full quantum effects in the excited state. These methods detect the transient intensity change after the probe light passes through the sample by scanning the delay between the arrival times of the pump and probe pulses, given that the arrival time of the probe and pump light pulses are different. In this way, full quantum effects in the excited-state energy levels and ultrafast dynamical processes of materials can be studied [Wu Q.2020].

Ultrafast time-resolved transient absorption spectroscopy is a common ultrafast laser pump-probe technique, and it is often used to detect probe-light absorbance by excited-state water molecules and to record the time-dependent particle population at each excited energy level; fitting the experimental result yields the potential energy function $V(r, R)$ of the O–H···O interaction between water molecules, where r denotes the O–H bond length and R denotes the O–O spacing. With this method, it is possible to uncover the non-harmonic characteristics of the potential energy and to probe the quantum delocalization of the H nucleus between O–O bonds in the vibrationally excited water molecule (see Figure 6.15) [Bakker2002]. This has crucial implications for understanding full quantum effects in proton transfer and water decomposition at room temperature.

6.1.5 Photoemission Spectroscopy

Photoemission spectroscopy is the most direct and powerful experimental tool for studying the microscopic electronic structure of novel quantum states and materials (including solid surfaces, high-temperature superconductors, topological quantum materials, and two-dimensional quantum materials). Many macroscopic physical properties of condensed matter are governed by the microscopic motion of electrons and nuclei; the electronic structures of materials must be studied if numerous novel physical phenomena are to be understood, controlled, and utilized. To obtain complete information about the electronic states in materials, knowledge of the following three fundamental parameters is generally required: energy, momentum, and spin. Photoemission spectroscopy is the only tool that can directly measure these parameters, making it indispensable in the experimental study of full quantum effects and helpful in developing related theories.

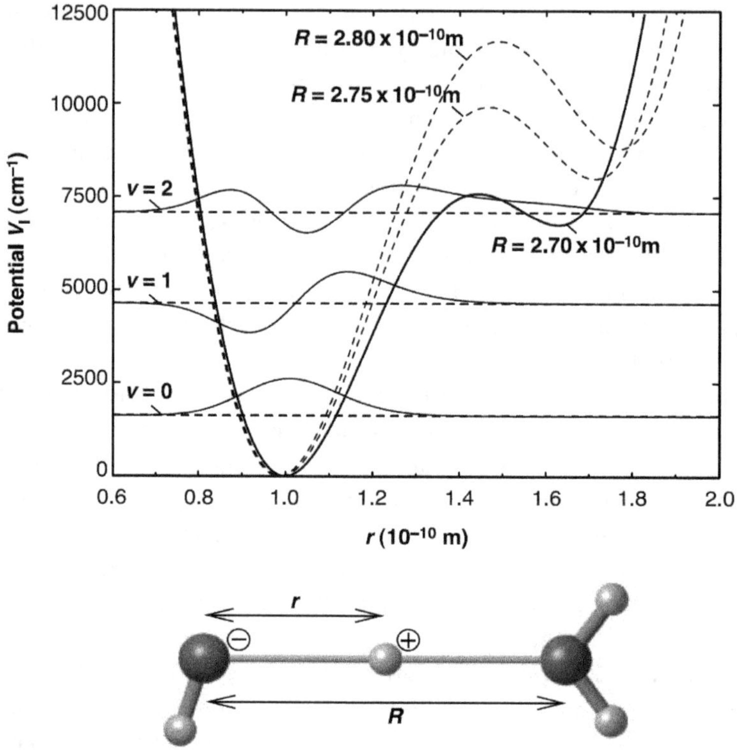

Figure 6.15 Using ultrafast pump-probe spectroscopy to detect quantum delocalization of protons in liquid water. Significant hydrogen-atom delocalization can be observed experimentally in the vibrational mode of the second excited state. $V(r, R)$ is the interaction potential energy. The quantum calculation is consistent with the experimentally measured pump-probe transient absorption spectrum of D_2O.

By measuring the same electron energy band position at different temperatures, one can directly identify the nuclear quantum effects on electronic structure renormalization if the temperature dependence of the electron band position deviates from linearity and reaches saturation at low temperatures. Combining photoemission spectroscopy with ultrafast laser techniques allows measurement of the time dependence of band positions and occupancy numbers. Further combining it with other methods (such as ultrafast X-ray or electron diffraction) allows for simultaneous measurement of the corresponding lattice changes, thus directly demonstrating nonadiabatic effects on the electron motion coupled with the lattice changes within the material during time evolution. For this reason, photoemission spectroscopy has become one of the most important experimental tools for the observation, characterization, and *in situ* regulation of novel quantum states and quantum materials.

To obtain momentum information, it is necessary to measure the angle-resolved photoemission distribution; this is achieved by using angle-resolved photoemission spectroscopy (ARPES). An ARPES measurement system is generally composed of four major components: light source, electron energy analyzer (see Figure 6.16), ultrahigh vacuum system,

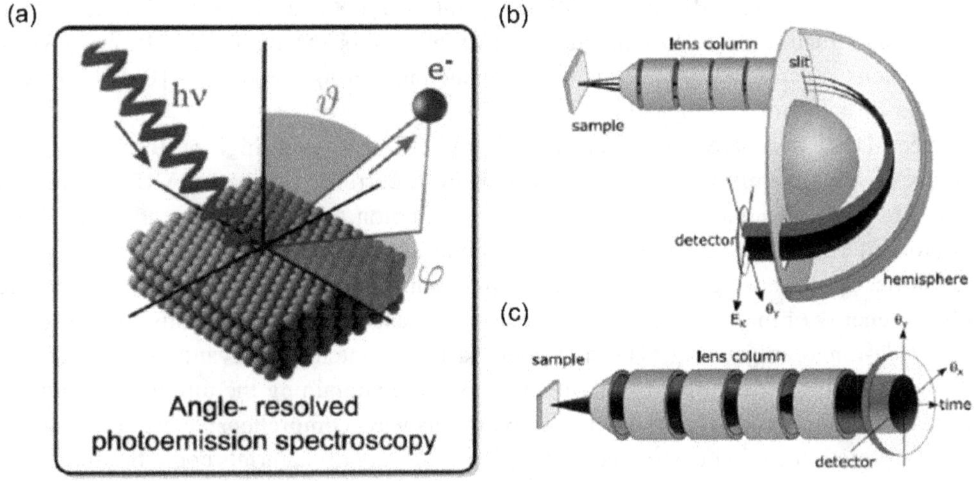

Figure 6.16 (a) Principle of ARPES, which is based on momentum and energy conservation during the photoelectric effect. Commonly used electron-energy analyzers are (b) hemispherical or (c) time-of-flight. Reprinted figure with permission from Sobota[2021]. Copyright 2024 by the American Physical Society.

sample holder, and temperature control system. In addition to these key components, a complete photoemission spectroscopy system also includes units for ultrahigh-vacuum sample transfer and preparation, ultrahigh vacuum sample measurement, and sample manipulation and cooling.

Over the past three decades, ARPES techniques have advanced significantly, leading to important breakthroughs in the study of high-temperature superconducting materials, new topological quantum materials, and two-dimensional materials. However, the performance requirements of photoemission spectroscopy have become increasingly demanding. Existing techniques utilize synchrotron radiation sources and should be further improved in terms of energy resolution, bulk-electron state detection, spin resolution, and experimental efficiency.

Recently, X.J. Zhou's group at the Institute of Physics, Chinese Academy of Sciences, successfully developed the first UV laser-based angle-resolved photoemission spectrometer with ultrahigh energy resolution [Liu G.D.2008]. Its superior performance and other unique advantages have made this spectrometer a leading experimental tool worldwide. Subsequently, they successfully developed a spin-resolved angle-resolved photoemission spectrometer using a deep-UV laser and a time-of-flight energy analyzer, as well as a large-momentum ultralow-temperature (<1 K) angle-resolved photoemission spectrometer [Zhou X.J.2018]. Achieving ultrahigh energy resolutions better than 1 meV in synchrotron photoemission spectroscopy has been a long-pursued dream that is difficult to realize. With the use of deep-UV lasers, such resolution can be achieved, allowing surface-insensitive photoemission spectroscopy measurements in the laboratory. Furthermore, the laser intensity can be increased so that it is hundreds to thousands of times stronger than the synchrotron radiation, thus considerably improving measurement efficiency and data quality.

Using deep-UV-laser ARPES, X.J. Zhou's group has obtained important results on the electronic structure of high-temperature superconductors [Meng J.Q.2009], many-body interactions [Zhang W.T.2008], the superconducting pairing mechanism [Bok2016], and the spin structure of topological insulators [Xie Z.J.2014].

Deep-UV lasers are cost-effective, highly efficient, and high-performance light sources for photoemission spectroscopy. However, current laser-based ARPES methods still have two obvious shortcomings. First, the photon energy is low, and second, the photon energy is single-sourced with a limited adjustment range, unlike synchrotron radiation. In the next-generation of laser photoemission spectroscopy technology, the photon energy of the laser needs to be further increased, with the possibility of performing wide-range adjustment. The new method must match the advantages of methods using synchrotron radiation as a light source while maintaining the ultrahigh resolution of laser-based photoemission spectroscopy. In this way, comprehensive electronic structure information could be obtained, and a greater variety of materials could be studied. In the development of next-generation laser photoemission spectroscopy, these extensions are inevitable.

Shining high-energy femtosecond lasers on rare gases can produce higher-order harmonics with energies ranging from tens to hundreds of eV; such lasers have been used increasingly in time-resolved ARPES. Advances such as the emergence of kilowatt lasers and new-generation high-resolution, highly sensitive electron-energy analyzers have made it possible to develop high-resolution extreme-ultraviolet-laser ARPES with wide-range photon energy adjustment. Compared to existing hemispherical energy analyzers, time-of-flight analyzers reduce the light-source intensity requirement by two orders of magnitude; their momentum-detection efficiency is more than a hundred times that of conventional hemispherical analyzers. This makes it possible to develop two-dimensional momentum-detecting photoemission spectroscopy with wide-range high-resolution photon-energy adjustment, which may revolutionize ARPES technology.

Photoemission electron microscopy (PEEM) is a technique that performs real-time imaging and measures the momentum of the photoelectron distribution emitted from the sample surface using an on-the-fly imaging system. Ultraviolet light, X-rays, near-IR femtosecond lasers, and extreme-ultraviolet attosecond lasers can all be used as excitation light sources for PEEM. Femtosecond-laser ultrafast PEEM has been applied in many fields, including surface physics, semiconductor physics, nanophotonics, and the investigation of photoelectric materials, magnetic materials, and surface catalysis. In particular, it has been applied to the study of interfacial charge transfer and ultrafast dynamic processes such as excitation, relaxation, transfer, and transport of carriers in the plasmon nanostructure and a variety of materials (two-dimensional, nonlinear, and semiconductor). Full quantum effects, especially nonadiabatic effects, play a crucial role in these processes. Therefore, PEEM is also one of the major techniques for studying full quantum effects.

Using time-resolved ARPES measurements, Hellmann et al. studied nonadiabatic dynamical processes in several typical charge density wave systems driven by ultrafast lasers [Hellmann2012]. They found that the energy band near the Fermi level changed significantly at 40–70 fs, corresponding to the closing of the energy gap and the vanishing of the charge density wave (as shown in Figure 6.17). This implies the existence of a nonadiabatic process in which the electron system responds directly to the external field at ultrafast time scales. Another example is the direct observation of dark-exciton formation using PEEM momentum imaging by Madeo et al. (shown in Figure 6.18). The ultrafast laser excites electrons in the K-valley, generating electrons in the conduction band and holes in the valence band. Within 300 fs, excited-state electrons in the conduction band of the K-valley are transferred to the Q-valley with the assistance of phonons, forming dark excitons [Madeo2020]. This is a typical nonadiabatic process.

Various spectroscopic techniques have found broad applications in the study of nuclear quantum effects and nonadiabatic effects and can even be applied to some complex systems. However, they share a common problem: Their spatial resolution is

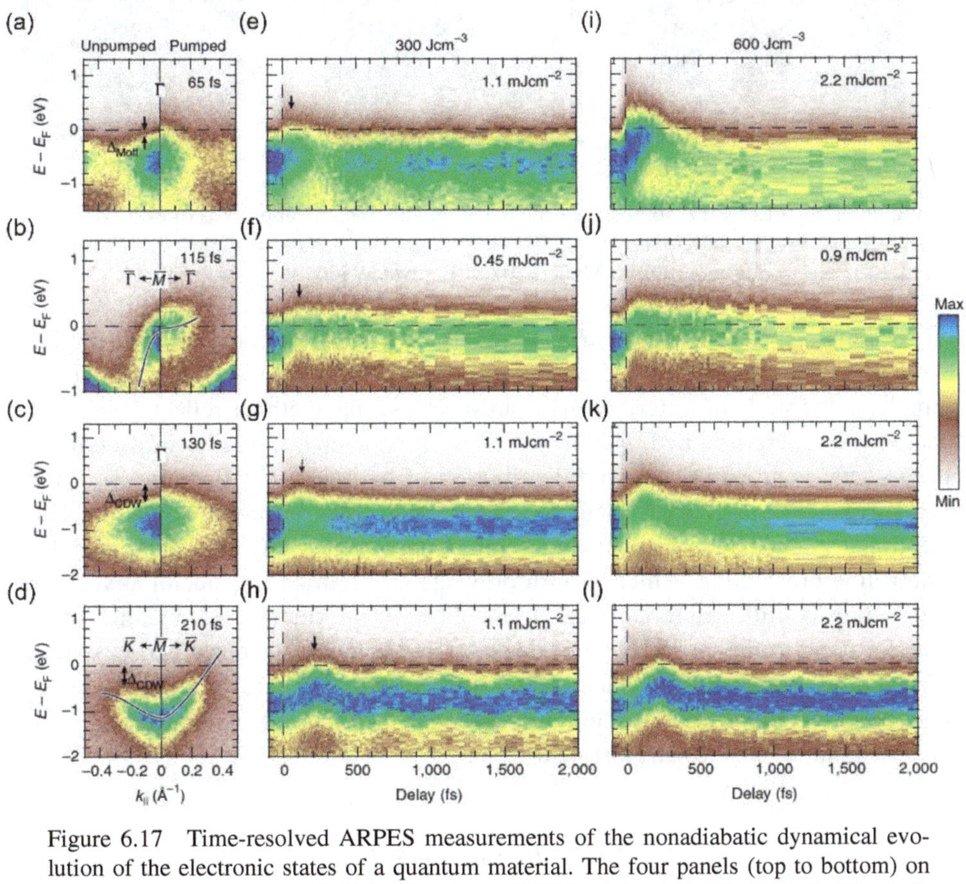

Figure 6.17 Time-resolved ARPES measurements of the nonadiabatic dynamical evolution of the electronic states of a quantum material. The four panels (top to bottom) on the left correspond to the 1T-TaS_2 (Γ-point), 1T-$TiSe_2$ (M-point), Rb:1T-TaS_2 (Γ-point), and 1T-TaS_2 (M-point), respectively. The time evolution of the electron energy band near the Fermi energy level and its intensity after ultrafast laser excitation at both intensities is illustrated in the figure. The change in electronic state occurs very quickly, indicating that it is a nonadiabatic electronic process. From Hellmann[2012]. Reproduced with permission from SNCSC.

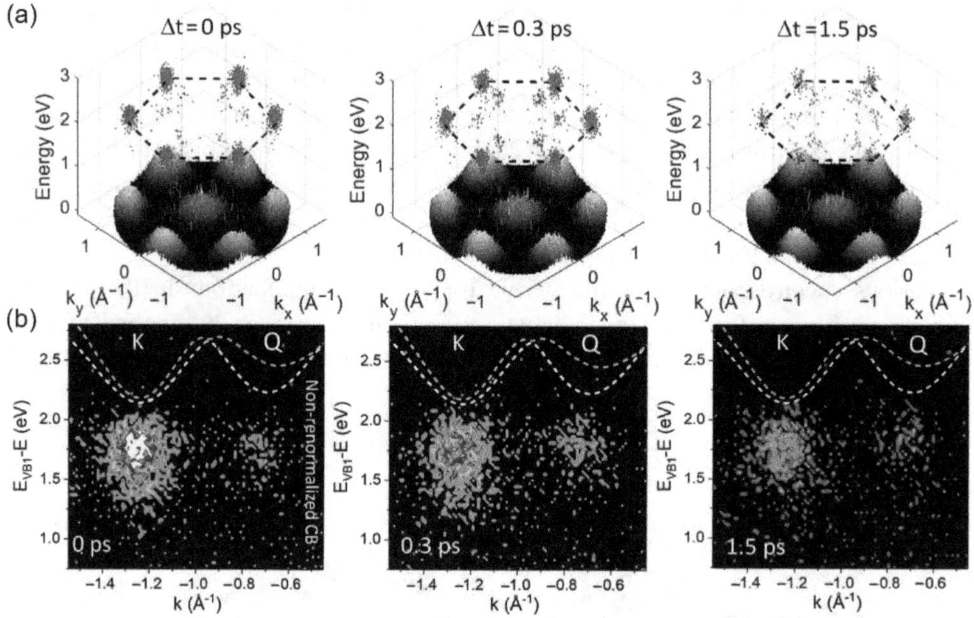

Figure 6.18 PEEM measurement of the nonadiabatic dynamical process of excited electron transfer from the K-valley to the Q-valley in WSe_2. A dark exciton state is formed after the electron transfer. From Madeo[2020]. Reprinted with permission from AAAS.

limited, ranging from hundreds of nanometers to tens of micrometers, and the information obtained is subjected to the superposition effect of many atoms/molecules. For example, in the study of water, results from spectroscopic methods reflect the averaging effect of the superposed hydrogen bonds. The intrinsic characteristics of each individual hydrogen bond and the spatial distribution of hydrogen-bond configurations cannot be revealed through these methods. The quantum state of the hydrogen nucleus is exceptionally sensitive to the local environment. This sensitivity can lead to significant spectral line broadening, which in turn prevents an accurate and quantitative characterization of the full quantum effects. In localized or interfacial systems, the influence of the local environment on the nuclear quantum state also becomes particularly pronounced. Therefore, it is necessary to probe the full quantum effects at the single-atom/single-bond level with high resolution to uncover the physical roots of phenomena in condensed matter physics and chemistry.

6.1.6 Electron Scattering

In subsections 6.1.2 and 6.1.5, we focused on experimental techniques exploiting the interaction of neutrons or photons with materials for studying the full quantum effects in condensed matter systems. Similarly, when a beam of electrons hits a material, the electrons interact with the material in a complex way, leading to energy and momentum changes in the scattered electrons (Figure 6.19(a)). The energy loss and momentum transfer can provide considerable information about the intrinsic properties of the sample material, including its electronic and phonon structures.

Electron scattering instruments can be categorized either by the energy level of the electron source or by the type of scattered electrons detected. Depending on the electron source and the detection mode, experiments differ in spatial, momentum, and energy resolution. Common experimental methods include transmission electron microscopy, scanning electron microscopy, low-energy electron diffraction, reflective high-energy electron diffraction, and MeV ultrafast electron diffraction. They provide a range of valuable information about such intrinsic properties of materials as crystal structure, element distribution, electronic structure, and lattice vibrations on various space, energy, and time scales.

In particular, the transmission electron microscope (TEM), as one of the most common tools in research, is known as the eye of materials science and nanotechnology. TEM has multiple operation modes, allowing for imaging, diffraction, and spectroscopic characterization of the sample. Nowadays, TEMs with aberration correctors routinely achieve spatial resolutions better than 0.05 nm. When equipped with a monochromator and cold-field-emission gun, the TEM can achieve an energy resolution better than 5 meV, and its momentum resolution can be adjusted over a wide range depending on the operating mode. Using ultrafast laser pumping, a time resolution better than 500 fs can be realized, and the performance can be enhanced by an additional order of magnitude if TEM is combined with other derivative technologies. Further combining TEM with in-situ probing methods enables investigation of the dynamical and nonequilibrium processes of atomic, electronic, and phonon structures in materials under various external fields. These unique capabilities of TEM make it a powerful platform for the study of full quantum effects.

Time-resolved ultrafast electron scattering techniques can detect the evolution of crystal structures and electronic states in nonadiabatic processes. In principle, ultrafast electron scattering is similar to ultrafast optics and ultrafast X-ray spectroscopy. However, the probe-beam source in the electron scattering method is an electron beam with ultrashort pulses, and its detection mode can be either electron diffraction [Barwick2008], electron imaging [Lorenz2014], or electron energy-loss spectroscopy [Carbone2009]. This makes it possible to study the structural phase transitions, morphological evolution, and electronic state evolution of materials on ultrafast time scales. After nearly two decades of development, these methods have become increasingly mature, and interested readers can consult the relevant literature [Zewail2009, Tian Y.2021] for further reading.

On the other hand, high electron energy loss spectroscopy (EELS), an inelastic electron-scattering technique, enables the simultaneous detection of lattice vibrations and local electronic structures. Therefore, although not many relevant experimental studies have been reported, EELS can, in principle, reveal nuclear quantum effects on electronic states and electron–phonon coupling at the atomic scale. In fact, EELS based on electron microscopy was invented decades ago, but its application was limited by the instrumental energy resolution. As a result, EELS could only be used to study core-electron transitions and the collective behavior of electrons; it could not be used to study lattice vibrations (phonons), which have much smaller energies. In recent years, the development of monochromator technology has significantly improved the energy resolution of electron microscopes [Krivanek2014], making it possible to detect lattice vibration with unprecedented spatial resolution. As lattice vibration is one of the main manifestations of nuclear quantum effects, the latest EELS based on scanning transmission electron microscopy (STEM-EELS) opens new possibilities for the study of full quantum effects.

To obtain information on the full quantum effects, one commonly used approach is to observe the effects of isotopic substitution on electronic properties. In particular, it is possible to distinguish between isotopes based on the phonon-spectrum shift due to the mass difference. Because phonons are described by their dispersion curves, high momentum resolution is required to study isotope effects on phonons, in addition to high spatial and energy resolution. Traditional isotope detection methods such as Raman [Giles2018] and IR [Vuong2018] spectroscopy with energy resolution on the order of 1 cm^{-1} can provide very accurate vibrational information but have limited spatial resolution. State-of-the-art tip-enhanced Raman spectroscopy (TERS) [Chen X.2019] and scattering-type scanning near-field optical microscopy (s-SNOM) [Mastel2018] have achieved nanoscale spatial resolution, but the small momentum transfer of photons prevents these methods from obtaining high-momentum vibrational behavior in the entire Brillouin zone. However, theoretical studies have found that the phonon energy difference between isotopes within the first Brillouin zone increases as the momentum increases [Hladky-Hennion2005]. Therefore, the phonon changes measured at the boundary of the Brillouin zone are usually much more informative about nuclear quantum effects than those measured at zero momentum. Currently, the energy resolution of STEM-EELS can reach below 10 meV and the momentum transfer offered by the high-energy electron beam is thousands of times greater than that of optical methods, making STEM-EELS one of the most suitable approaches for studying isotopes at the atomic scale.

For example, a research team at the Oak Ridge National Laboratory first utilized STEM-EELS to demonstrate the feasibility of identifying isotopes in an electron microscope. They measured the vibrational spectra of nano-sized liquid water encapsulating between boron nitride sheets and were able to distinguish the vibrational spectrum of H_2O from that of D_2O (Figure 6.19(b)) [Jokisaari2018]. This team further identified the bonding information of ^{13}C-, ^{12}C-, and O-labeled alanine [Hachtel2019]. A vibrational energy redshift of 4.8 meV was observed in the ^{13}C-labeled C–O asymmetric stretching mode relative to the ^{12}C-labeled case (Figure 6.19(c)). These studies based on dipole detection could achieve isotope identification with a spatial resolution ranging from hundreds to tens of nanometers. In fact, theoretical work predicts that EELS isotope detection may even reach atomic-scale resolution [Konecna2021]. Later, Senga et al. identified ^{12}C and ^{13}C isotopes with sub-nanometer spatial resolution in a single layer of carbon atoms [Senga2022].

The spatial resolution of STEM-EELS is mainly limited by the delocalization effects of the detection signal. As the size of the electron beam can be even less than that of a hydrogen atom, the excited region can be very small. The lattice vibration signal includes both a highly delocalized dipole scattering signal (such as phonon polaritons) and a localized impact (collisional) scattering signal [Allen2018]. Generally, the contribution of dipole scattering dominates the small momentum transfer. As the transfer momentum increases, the intensity of dipole scattering decreases rapidly while the proportion of impact scattering increases, exhibiting better local characteristics in the phonon spectrum. Based on this, researchers have proposed different solutions, such as using a circular frequency-selective diaphragm [Dwyer2017] or off-axis spectrum sampling [Hage2019, 2020] to improve the collection efficiency of large momentum transfer signals, reduce the collection of small momentum transfer signals, and thus further enhance the local characteristics of lattice vibration signals. Recently, these ideas have recently been realized in experiments. Several research teams in the US, UK, and China have achieved atomically resolved vibrational spectrum measurements in several material systems by using these strategies [Hage2019, Venkatraman2019, Hage2020, Shi R.C.2020, Qi R.S.2021a,b].

In 2022, P. Gao's group at Peking University designed multiple isotopically purified boron nitride nanostructures and demonstrated atomic-level characterization of isotopic interfaces by STEM-EELS [Li N.2022]. As shown in Figure 6.19(d), the elemental composition and atomic structure on both sides of the hexagonal boron nitride h-^{10}BN/h-^{11}BN isotopic-heterojunction interfaces were identical, differing only by one neutron mass in B (~4% of the h-BN single-cell mass). Using STEM-EELS, they verified that, for the out-of-plane phonon mode ZO, the energy of h-^{10}BN was ~2.5 meV higher than that of h-^{11}BN. The intensity of the ZO mode showed atomic-level periodic variations (~0.34 nm) as the ratio of impact scattering changed. Across the isotopic interface, the transition length of the ZO phonons between the center and the boundary of Brillouin zone was significantly different because of electron–phonon interactions. This study demonstrated for the first time that STEM-EELS can achieve atomic-level resolution not only in identifying isotopes but also in detecting electron–phonon interactions, providing new opportunities to probe nuclear quantum effects (especially isotope effects).

High spatial resolution is not the only advantage of STEM-EELS for studying isotopic nuclear quantum effects. It also has a wide energy-detection window, because of the very high energy of the electron microscope's excitation beam. The lower energy limit of the window depends on the microscope's energy resolution, while the upper limit can reach several thousand eV. Therefore, the high-energy-resolution STEM-EELS technique enables simultaneous local detection of lattice vibrations and electronic structure to analyze their relation, allowing effective exploration of full quantum effects in the electronic states and electron–phonon coupling. In addition, the electron scattering cross section is typically four to six orders of magnitude higher than the photon scattering cross section in nanoscale optics. Thus, the signal strength (signal-to-noise ratio) of EELS is much better than that of other optical detection methods when studying samples with only a few atoms. In light elements, the cross section for inelastic scattering is much greater than that for elastic scattering, resulting in a strong EELS signal. This is particularly useful because full quantum effects are usually more pronounced in light element systems. In this sense, EELS is an extremely suitable research tool for studying full quantum effects.

In addition, electrons can excite transition modes more than photons. They are also capable of exciting and detecting some dark optical modes [Li Y.2020]. Further, since the photon momentum is typically two orders of magnitude smaller than the momentum in the crystal's Brillouin zone, only the signal at the center of the zone can be detected by optical spectroscopy. This issue does not exist for electrons, enabling the detection of signals from the entire Brillouin zone. This advantage of EELS is also evident in the study of light–phonon coupling modes (i.e., phonon and plasmon polaritons) in few-layer polar materials [Li N.2021]. In fact, the largest energy difference of the vibrational signal (phonon bandgap) occurs at the Brillouin zone boundary. The phonon bandgap width reflects the magnitude of the mass difference between atoms and their neighbors (as shown in Figure 6.19(e)). Therefore, quantitatively comparing the phonon bandgap widths measured at the boundaries of the Brillouin zone should be more accurate than optical spectroscopy for determining isotope types and concentrations at the nanoscale.

EELS is not only capable of detecting scattered electrons during large momentum transfer but also of providing high momentum resolution. Although scanning tunneling spectroscopy can also reach atomic-level spatial resolution when measuring phonons, it only provides information about the state density of the vibration signal integrated over the entire momentum space and therefore cannot properly resolve the momentum. EELS, on the other hand, is capable of achieving adequate spatial and momentum

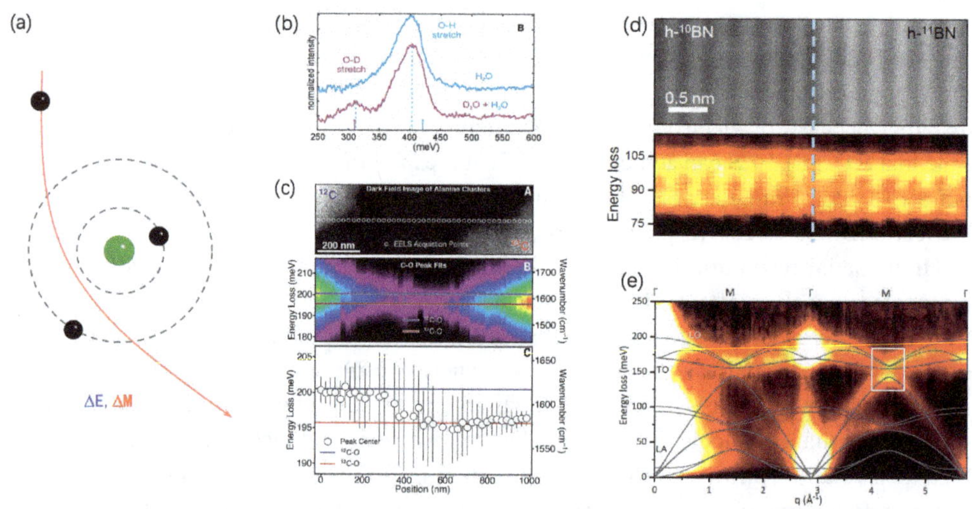

Figure 6.19 Potential application of STEM-EELS to the study of full quantum effects. (a) Energy and momentum changes of fast electrons during scattering in the electron microscope. (b) Isotope-sensitive EEL spectra of H_2O and D_2O. (c) Vibrational spectra of ^{12}C- and ^{13}C-labeled alanine. (d) Atomically resolved vibrational spectrum at the h-^{10}BN/h-^{11}BN interface. (e) Phonon-dispersion map of h-^{10}BN. The phonon bandgap difference between the optical branch and the acoustic branch of the different isotope samples is most pronounced at the M point, the boundary of the Brillouin zone. Figure 6.19(b) from Jokisaari[2018]. Reproduced with permission from John Wiley and Sons; Figure 6.19(c) from Hachtel[2019]. Reprinted with permission from AAAS; Figure 6.19(d) created by Li N.[2023] licensed under CC BY 4.0; Figure 6.19(e) Reprinted from Li N.[2022], with permission of the author.

resolution simultaneously. More specifically, the recently developed four-dimensional EELS [Qi R.S.2021a] can balance spatial and momentum resolution based on the needs of practical problems. Though the uncertainty principle prevents achieving the best possible spatial and momentum resolution simultaneously, this method allows a reasonably close approach to the diffraction limit. It overcomes the shortcoming of conventional spectroscopy – the inability to simultaneously achieve high momentum resolution and nanoscale spatial resolution – thus enabling local phonon dispersion measurements of individual nanostructures and interfaces [Qi R.S.2021a,b]. As the strength of electron–phonon coupling in materials strongly depends on the momentum, momentum resolution is crucial for studying full quantum effects. In short, the rapidly evolving STEM-EELS technique is expected to open new avenues in the study of atomic-scale full quantum effects.

6.1.7 Scanning Probe Techniques

In 1981, G. Binnig and H. Rohrer at IBM Zurich laboratory invented the scanning tunneling microscope (STM). They shared the 1986 Nobel Prize in Physics with E. Ruska, the inventor of the TEM. STM is an experimental technique developed based on electron quantum tunneling (Figure 6.20(a)). It uses an atomically sharp metal tip as the scanning probe; the substrate is usually made of metal. A piezoelectric tube is used to move the tip and scan in the x, y, and z directions. The metal tip gradually approaches the metal sample to a distance smaller than 1 nm, so the electron wavefunctions of tip and sample overlap. When a bias voltage is applied between them, the tip and sample function as electrodes, and electrons flow from one to the other through quantum tunneling. In most STM devices, the tip is grounded, so the bias voltage is applied to the sample. When a positive bias voltage is applied, electrons tunnel from the occupied state of the tip to the unoccupied state of the sample; when a negative bias voltage is applied, they tunnel from the occupied state of the sample to the unoccupied state of the tip. The tunneling current varies exponentially with the tip–sample distance; when this falls by 0.1 nm, the tunneling current increases by an order of magnitude, providing ultrahigh spatial resolution. Using STM, one can observe the arrangement of individual atoms on the material surface in real space, as well as the physical and chemical properties related to the electronic behavior of the surface. As a result, it is widely used for research in surface science and nanoscience.

Recently, STM has also been applied to the study of high-resolution structural imaging, dynamics, and nuclear quantum effects in light element systems at surfaces. In 2000, W. Ho's group at the University of California, Irvine, used STM to image and perform the first real-space vibrational spectroscopy measurements of hydrogen atoms (the smallest of all atoms in terms of mass and size) adsorbed on metal surfaces, successfully identifying individual hydrogen and deuterium atoms (Figure 6.20(b)). Subsequently, they used STM to trace the diffusion rate of a single hydrogen atom on the metal surface at different temperatures. Their results suggest that, when the temperature is below 60 K, the hydrogen-atom diffusion rate is independent of temperature. This experimental work directly confirms the quantum tunneling behavior of hydrogen atoms on metal surfaces at low temperatures (Figure 6.20(c)) [Lauhon2000].

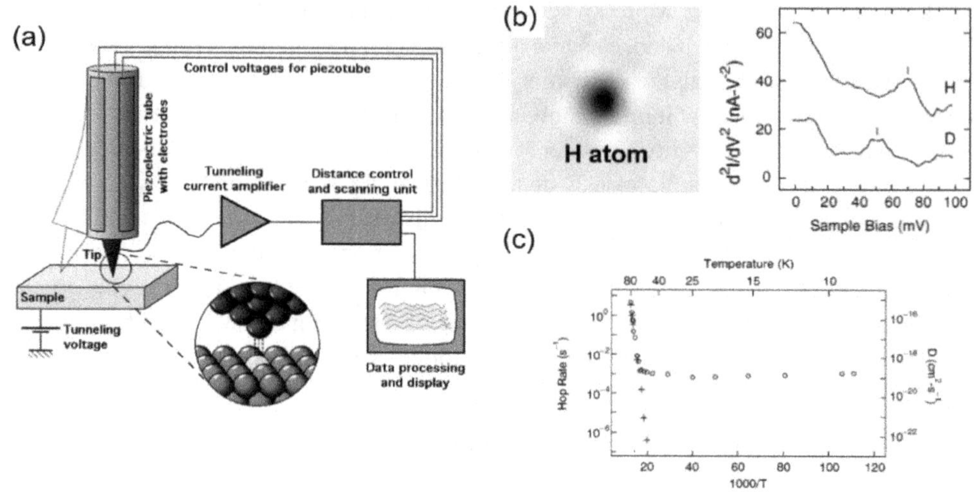

Figure 6.20 Quantum tunneling of a single hydrogen atom on a metal surface measured by STM. (a) Schematic of STM. (b) Left: high-resolution STM image of a single hydrogen atom on the Cu(001) surface. Right: For a single H(D) atom, the vibrational peak position of the STM-IETS spectrum represents the H(D)–Cu stretch mode; the difference between them enables isotope identification. (c) Temperature dependence of hydrogen-atom diffusion rate. The normal hydrogen atom's behavior changes from classical motion to quantum tunneling at 60 K, while the heavier deuterium atom always follows the classical Arrhenius law in the temperature range investigated. In other words, under this experimental condition, the deuterium atom does not exhibit quantum tunneling dominated by nuclear quantum effects. Reprinted figure with permission from Lauhon[2000]. Copyright 2024 by the American Physical Society.

In addition, STM has been used to study quantum tunneling effects during proton transfer in hydrogen-bonding systems. STM tips can manipulate proton transfer in such systems. This allows us to further identify structural changes before and after the proton transfer via STM imaging, and then track and record the kinetic process using tunneling current. Finally, isotope effects of proton transfer rate and the dependence on temperature, voltage, electric field, tunneling current, and other control factors can be measured to determine the proton transfer mechanism. In general, quantum tunneling shows a significant isotope effect and does not need to overcome the classical reaction barrier. Therefore, the proton transfer rate tends to be independent of the control parameters. H. Okuyama's team at Kyoto University has reported a series of important results while studying nuclear quantum effects in water on a Cu(110) surface. They found that the interchange of roles between hydrogen bond donors and acceptors in water molecular dimers is related to quantum tunneling of protons [Kumagai2008]. Moreover, they found that the dynamic switching of the O–H adsorption orientation on the Cu(110) surface also originates from the quantum tunneling of protons [Kumagai2009].

All the research work mentioned above focused on tunneling of individual protons. However, hydrogen atoms in a hydrogen-bonding network are not independent of each other, and they are typically strongly correlated. As a result, proton transfer in hydrogen-bonding systems can involve many-body quantum processes.

Using tip manipulation, T. Kumagai et al. prepared a one-dimensional $H_2O\text{-}(OH)_n$ (n = 2–4) chain configuration on a Cu(110) surface. By vibrational excitation, they induced sequential proton transfer in the chain water structure. Furthermore, they discovered that a single porphycene molecule on the Ag(110) surface spontaneously tautomerizes. In particular, when the temperature is below 10 K, the transition rate is temperature independent. However, the transition rate decreases by two orders of magnitude after deuterium substitution. Partial deuterium substitution experiments further revealed that the tautomerization process is dominated by double proton "stepwise" tunneling [Koch2017].

Using a sub-molecular-resolution imaging technique that they developed, Y. Jiang's group at Peking University directly identified hydrogen-bonding directionality in real space, and observed two water tetramers with different O–H orientations (chiralities) [Guo J.2014]. By tracking the proton transfer within a single tetramer cluster on the NaCl(001) surface in real time, they directly observed the dynamical process of proton tunneling within the water cluster. Through complete and partial isotope substitution, combined with full quantum calculations, they confirmed that this was a novel coherent quantum process: concerted tunneling of four protons [Meng X.Z.2015]. At cryogenic (liquid-helium) temperatures, STM has been used to trace in real space the quantum tunneling effect of heavy nuclei – including C [Lin C.2019], Cu [Repp2003], and Co [Stroscio2004] – as well as molecules such as CO [Heinrich2002].

Although much progress has been made in the study of nuclear quantum effects using STM, understanding in this area is still very limited. The main challenge is that STM is insensitive to nuclei and inner-shell electrons and can only be used to study the conductive substrates, a shortcoming that can be overcome by noncontact atomic force microscopy (NC-AFM).

In 1998, F. J. Giessibl from the University of Regensburg invented the qPlus-type force sensor, using a quartz tuning fork (Figure 6.21(a)) to replace the conventional silicon cantilever, achieving high spatial resolution [Giessibl2003]. The qPlus sensor has a high-quality factor and can operate stably at sub-Å amplitudes (<100 pm), allowing the tip to be placed very close to the sample surface. This can substantially improve the sensitivity to short-range forces between the tip and the atom (see Figure 6.21(b)), allowing direct atomic-level imaging of chemical bonds (as shown in Figure 6.21(c, d)) [Giessibl2003, Gross2009]. The research groups led by L. Gross at the IBM-Zurich Research Center and M. Crommie at the University of California, Berkeley, used qPlus NC-AFM to study the backbone structure of organic molecules and to obtain order imaging, as well as to identify reaction products in surface chemistry [de Oteyza2013, Gross2012]. X. H. Qiu's group at the National Center for Nanoscience has precisely analyzed the intermolecular hydrogen bond configuration in 8-hydroxyquinoline for the first time [Zhang J.2013]. Y. Sugimoto's group at the University of Tokyo applied qPlus-AFM to the study of surface water and identified the hydrogen bond configuration of one- and two-dimensional water clusters on the Cu(110) surface [Shiotari2017]. Recently, Y. Jiang's group at Peking University has developed non-invasive AFM, based

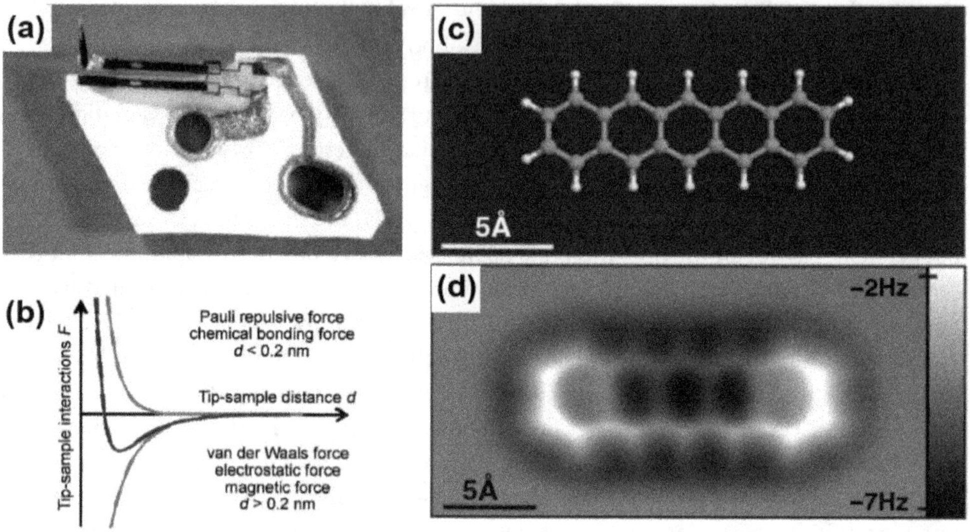

Figure 6.21 (a) qPlus-type force sensor; its design employs a quartz tuning fork. (b) The force profile between tip and sample (The black solid line is the sum of the two light grey solid lines). When the tip–sample distance is large, long-range interactions such as the van der Waals, electrostatic, and magnetic forces dominate; when it is small, short-range interactions such as Pauli repulsion and chemical bonding dominate. (c) Structural model of pentacene molecule; (d) AFM image of the pentacene molecule. Real-space imaging of the molecular backbone was achieved. Figure 6.21(a) reprinted figure with permission from Giessibl[2003]. Copyright 2024 by the American Physical Society. Figure 6.21(cd) from Gross[2009]. Reprinted with permission from AAAS.

on higher-order electrostatic forces, achieving increased force sensitivity to the level of femtonewton. By performing precise chemical modification to the tip and modulating the charge distribution at the tip apex, it is possible to detect higher-order electrostatic interactions between the multipolar tip and water molecules. This enables non-invasive imaging of surface water molecules and hydrated ionic systems, as well as direct identification of hydrogen nuclei in real space [Peng J.B.2018a,b]. The development of such innovative AFM techniques significantly advanced the study of nuclear quantum effects in weakly interacting light-element systems.

Inelastic electron tunneling spectroscopy (IETS) is a spectroscopic technique based on STM. It focuses on the identification of individual molecular vibrational modes by probing the coupling between the tunneling electrons and vibrations of a single molecule [Stipe1998, Ho2002]. The principle of IETS is illustrated in Figure 6.22. Similar to vibrational spectroscopy, it can be used to study nuclear quantum effects at the single-molecule level. The invention of this technique has eliminated the bottleneck of low spatial resolution in conventional vibrational spectroscopy. However, since the electron vibration coupling of molecules is generally weak, the inelastic-scattering cross section for the tunneling electrons tends to be small, and the vibration-induced conductance of the molecules only changes by a few percent [Baratoff1988]. To detect a 2% conductivity change, the stability of the tunnel junction needs to be better than 1 pm.

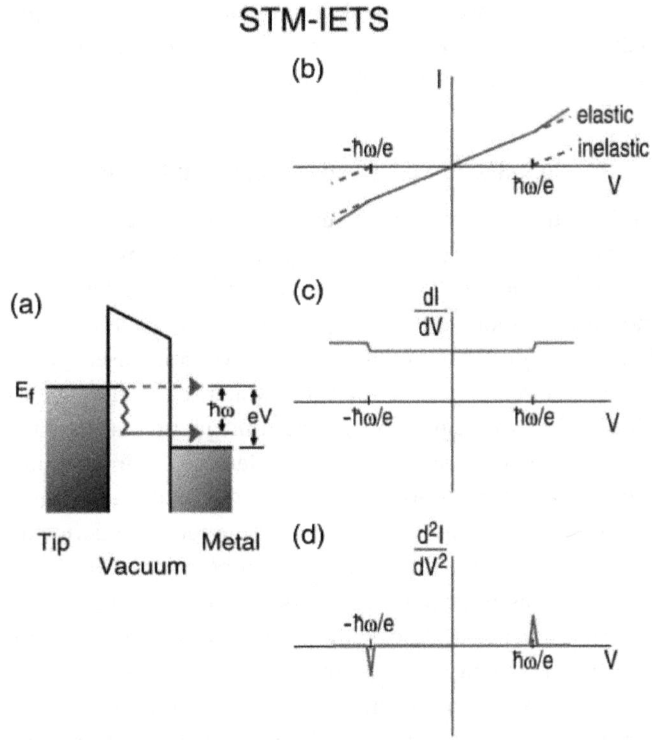

Figure 6.22 Principle of inelastic electron tunneling spectroscopy (IETS). As can be seen in (b), the vibration-induced tunneling-current change is difficult to measure from the *I–V* curve. However, it can be measured from the *dI/dV* curve in terms of conductance jumps, as can be seen from (c). In (d), the tunneling-current changes are more visible as a peak and a dip in the d^2I/dV^2 curve. Reprinted from Ho[2002], with the permission of AIP Publishing.

Therefore, IETS measurement requires a scanning probe with very high stability. To improve the signal-to-noise ratio of IETS, the tip can be used as the top gate. By tuning the coupling between the tip and the molecule, the position and broadening of molecular orbital can be adjusted, and the DOS near the Fermi energy level can be increased to enhance the IETS signal via the resonant inelastic tunneling effect [Persson1987, Guo J.2016a]. In this way, IETS gains sufficient sensitivity to detect small frequency shifts of vibrational modes caused by hydrogen-bonding interactions, and the technique can be used for quantitative characterization of nuclear quantum effects at the atomic scale.

Using TE-IETS, Y. Jiang's group at Peking University obtained the first high-resolution vibrational spectrum of a single water molecule; moreover, the strength of individual hydrogen bonds was also measured. By combining controlled isotope substitution with full quantum computational simulations, the contribution of the nuclear quantum effects on the hydrogen bond was found to be around 14%, which is as much as the thermal energy at room temperature [Guo J.2016a]. These studies suggest that the nuclear quantum effect of hydrogen is strong enough to have a decisive influence on the structure and properties of water. Further analysis has also revealed that the anharmonic nuclear quantum effects of hydrogen weaken weak hydrogen bonds and strengthen strong ones. This physical picture can be generalized to other hydrogen-bonded systems, clarifying the long-debated role of full quantum effects in hydrogen bonding.

6.2 Extreme External-Field Conditions

It is well known that physical parameters such as temperature, pressure, magnetic field, and optical field have significant effects on multi-atom systems, giving rise to new states of matter. Understanding the effects of external fields on these states is of paramount scientific and technological importance. In fact, the main way in which physics has benefited humanity has been through the regulation of states of condensed matter. For example, the regulation of the liquid and gaseous states of water led to the invention of the steam engine and mechanization, the regulation of electromagnetic fields led to the invention of the electric motor and electrification, and the regulation of the electron energy bands in solids led to the invention of the transistor, integrated circuits, and computerization.

From the earliest telescopes and microscopes to immense modern experimental facilities such as the Large Hadron Collider, instrumental platforms for realizing or observing extreme external conditions have played key roles in the major scientific discoveries of every era. As materials science advances, the continuous development of more advanced experimental environments using such platforms has become paradigmatic. Using them to study full quantum effects will significantly expand the scope of research in condensed matter physics. Effectively incorporating external field platforms and different measurement techniques to form a complex interweaving of experimental methods can open avenues for broader and more profound research and applications in physical science. In particular, it will allow us to search for new phenomena beyond the traditional Born–Oppenheimer approximation regime, explore new states of condensed matter, and discover new laws related to full quantum effects. Using material regulation to promote sustainable development of human society has been and always will be the ultimate goal of physics.

6.2.1 Extremely Low Temperature

Temperature is one of the most fundamental physical quantities. The lower the temperature, the easier it is to discover and observe different aspects of quantum-mechanical phenomena. Low-temperature physics led to the discoveries of superconductivity, superfluidity, the integer quantum Hall effect, and the fractional quantum Hall effect. In general, nuclear quantum effects are much weaker than electron quantum effects and are therefore best observed when thermal fluctuations are minimized.

Refrigerators are instruments used to achieve a low-temperature environment in physics research. The refrigeration performance is mainly reflected in the ability to approach the zero-temperature limit, an ability that depends not only on cooling methods but also on the environmental heat leakage. For a given cooling power, the optimal performance of the refrigerator in an extremely-low-temperature environment is limited by the source of maximum heat leakage. We live in a "high temperature" environment of 300 K. To

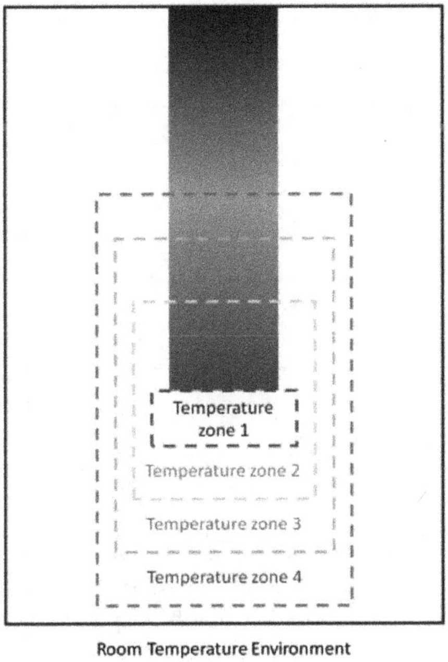

Figure 6.23 Basic principle of extremely-low-temperature equipment: Different cooling methods are used to build a temperature gradient from room temperature to extremely low temperature. A modern extremely-low-temperature refrigerator generally uses a combination of methods, including gas expansion, pulse tube cooling, Joule–Thomson expansion, liquid evaporation, dilution refrigeration, Pomeranchuk cooling, and adiabatic demagnetization. Each method has its own cooling power and can be used to reach a different characteristic temperature. The number of cooling methods involved increases as the final target temperature approaches the absolute-zero limit.

maintain a temperature gradient from room temperature to an extremely low temperature with minimal heat leakage, high-performance refrigerators employ a nested structure with multiple cooling mechanisms (Figure 6.23). Each cryogenic environment in the refrigerator is protected by another warmer cryogenic environment. Therefore, extremely-low-temperature refrigeration relies on a set of technologies applicable to a wide temperature range. The major milestones in the historical development of cryogenics technology and low-temperature physics have each involved the rapid expansion of the experimental parameter space, resulting in a flood of novel observations.

The expansion of the low-temperature parameter space can be traced back to Faraday. After his pioneering attempts at gas liquefaction, oxygen and nitrogen gas were liquefied by other researchers in 1883. This was the first time a temperature below 100 K was achieved. Although the practical goal of refrigeration was the preservation of fresh meat, the scientific goal was to verify whether permanent gases exist, that is, gases that never liquefy even at absolute zero. After the liquefication of H_2, helium was successfully liquefied in 1908; it remains liquid even at the zero-temperature limit. This example shows that, although permanent gases were never found, the permanent liquid state does exist.

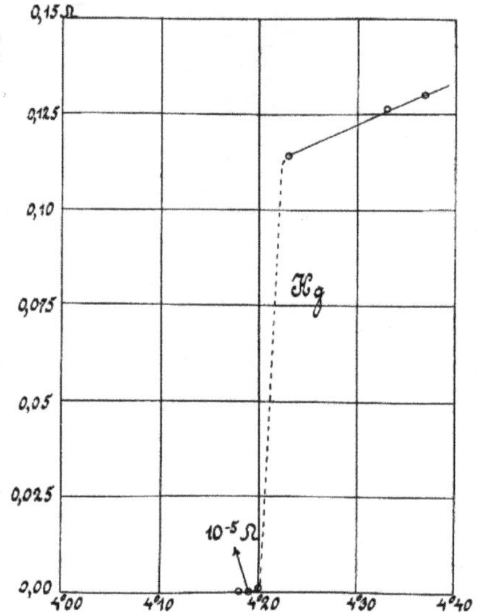

Figure 6.24 First experimental evidence of superconductivity, published in 1911 [Onnes1911]. This result was discussed within the University of Leiden, Netherlands, on November 25, 1911.

Most helium on Earth is ^{4}He, which is composed of two protons and two neutrons. In addition to ^{4}He, helium also has a stable isotope, ^{3}He, with two protons and one neutron. ^{3}He and ^{4}He are both ideal for physics research, but for different reasons. ^{4}He nuclei are experimentally accessible bosons with simple structure, while ^{3}He nuclei are the most easily studied fermions that are not electrons. Once ^{4}He had been liquefied, superconductivity was soon discovered in this newly available low-temperature environment (Figure 6.24).

Gaseous helium, which is colorless and odorless, is the simplest inert gas; liquid helium, however, has extremely complex properties. Since its temperature is close to absolute zero, liquid helium has played a key role in low-temperature physics, both as a research object and as a coolant. After the liquefaction of ^{4}He, the depressurized refrigeration of liquid ^{4}He has led to an important scientific breakthrough in human civilization. The lowest temperature found in nature is 2.73 K, which is the temperature of the cosmic background radiation. The field of low temperature is one of the few cases where a laboratory can easily break the natural limits of a physical parameter. Humans cannot break nature's limits on high temperature, maximum or minimum length, or greater or smaller mass, but they broke the natural limits on low temperature a century ago.

P. Kapitza discovered that ^{4}He becomes a superfluid as the temperature approaches 2 K [Kapitza1938] (see Figure 6.25). In addition to superconductivity and ^{4}He superfluidity, the integer quantum Hall effect and the fractional quantum Hall effect are also experimental discoveries unpredicted by theorists. These four phenomena remain important in physics today, especially in frontier research on topological properties and many-body problems. For example, in 2013, Q. K. Xue's group reported experimental

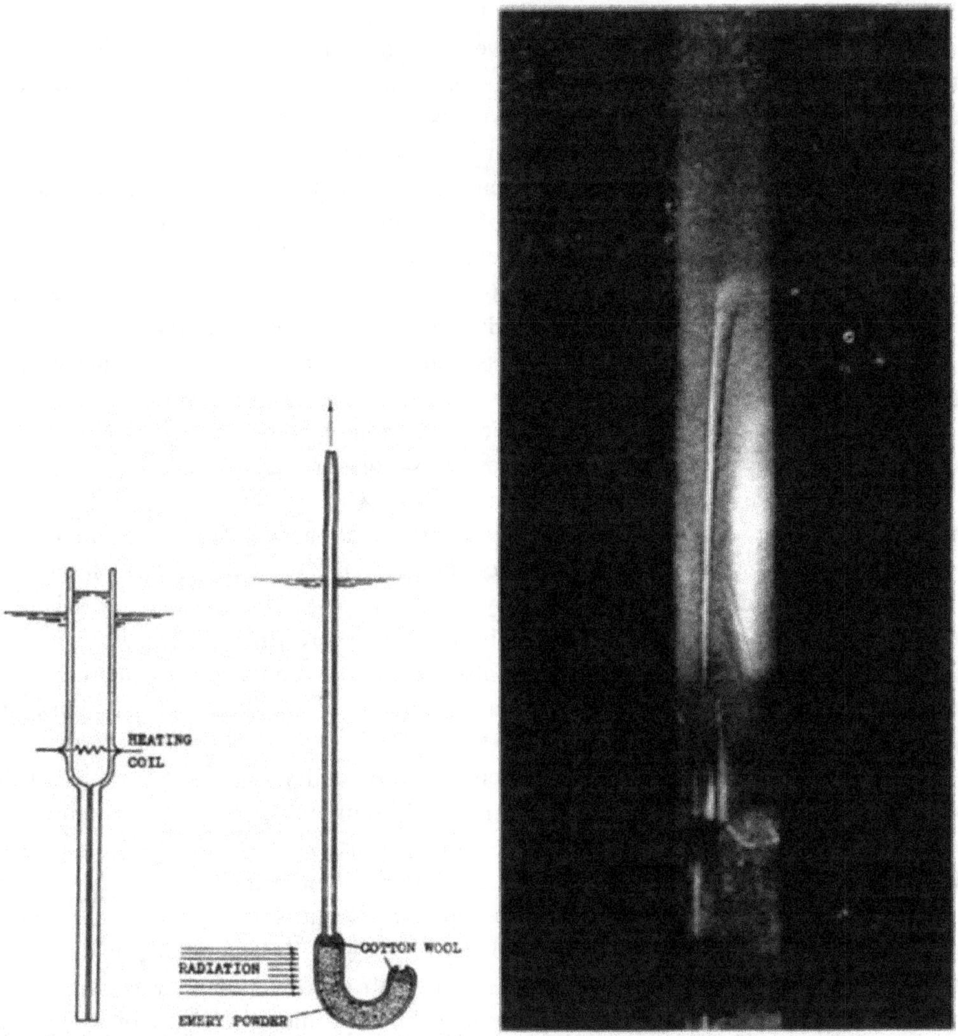

Figure 6.25 Fountain effect caused by superfluidity. After the experimentalist has established a temperature gradient across a porous material through which only a viscosity-free superfluid can pass, the fluid can produce a macroscopic movement that overcomes gravity. From Allen[1938]. Reproduced with permission from SNCSC.

observation of the quantum anomalous Hall effect in a magnetic topological insulator [Chang C.Z.2013]. Low-temperature technology has even become a foundation of the new International System of Units.

The cryogenic environment has also provided the conditions for verifying important theoretical predictions. The zero-sound phenomenon predicted by Landau in 1957 was confirmed by sound-speed experiments under newly attainable experimental conditions in 1966 [Abel1966]. After the BCS theory succeeded in explaining superconductivity, the corresponding superfluidity of ^{3}He was expected; with the advancement of cryogenic techniques, it was finally observed near 1 mK in 1972 [Osheroff1972]. However, the cooling method used in that experiment involves compression of the ^{3}He solid–liquid coexistence state, which only provides single-shot cooling. This makes it difficult to study materials other than ^{3}He. Therefore, this method (called Pomeranchuk cooling) is not widely used to produce ultralow-temperature environments.

The use of mK-level extremely-low-temperature environments in fundamental scientific research was rare until the advent of the dilution refrigerator in the 1960s and its subsequent commercialization over the next 50 years. Dilution refrigerators utilize the principle that a ^{3}He-^{4}He mixture undergoes phase separation in the zero-temperature limit. After phase separation, the ^{3}He-rich and ^{4}He-rich phases stratify under gravity. ^{3}He absorbs heat during dilution as it crosses the interface from the high concentration phase to the low one.

The liquefaction of ^{4}He in 1908 and the realization of dilution refrigeration in 1966 are two important milestones in the history of cryogenics. The former began the era of low-temperature physics at the quantum level, while the latter provided a stable ultralow-temperature environment. Dilution refrigeration is currently the dominant method for obtaining stable temperatures on the order of 10 mK, and a dilution refrigerator is the best commercially available cryogenic equipment for cutting-edge scientific research. In 1997, fractional charge was observed at 57 mK using a dilution refrigerator [de-Picciotto1997]. Dilution refrigerators also have many applications in quantum technology, such as superconducting quantum computation and topological quantum computation (see Figure 6.26).

Dilution refrigerators require the protection of pre-cooling environments. One of the most important pre-cooling environments is provided by liquid ^{4}He. As cryogenic technology advances, cryogen-free cooling technology has gradually replaced technology based on liquid ^{4}He. Refrigeration without consuming liquid helium has become a mainstream trend in instrumental development, because helium is a scarce natural resource extracted mainly as an operational by-product of a few oil or natural gas fields. While the consumption of liquid helium by conventional refrigerators is similar to gas for conventional vehicles, cryogen-free refrigeration technology allows refrigerators to operate while consuming only electricity. Today, dilution refrigerators based on cryogen-free technology are preferred by researchers. The widespread application of cryogen-free refrigeration instruments occurred mainly in this century and is a milestone almost as important as ^{4}He liquefaction and dilute refrigeration.

Temperature environments below 1 mK cannot be obtained commercially, as the construction and operation of the equipment depends on specialized expertise. There are only a few laboratories in the world that can achieve temperatures below 1 mK. Nuclear adiabatic demagnetization is currently the refrigeration method with the lowest ambient temperature. This method of refrigeration utilizes the fact that the entropy of nuclear spins can be changed by an external magnetic field. The final temperature of adiabatic demagnetization depends on the ratio of the precooling temperature to the initial external magnetic field. Precooling is generally provided by a dilution refrigerator. Nuclear adiabatic demagnetization refrigeration is the last mainstream refrigeration

Figure 6.26 Dilution refrigerator that provides an extremely-low-temperature environment for superconducting quantum computing at Google.

technology that is not yet cryogen-free. The eventual elimination of all helium consumption in refrigeration will represent another major technological landmark.

Cryogenic experiments require not only funding support but a high level of accumulated technical knowledge, and the training period for the required operating personnel is long. In 2011, the International Center for Quantum Materials at Peking University proposed building a nuclear adiabatic demagnetization refrigerator that does not consume liquid helium. After 10 years of efforts, X. Lin and R. R. Du successfully designed and built such a refrigerator and achieved an extremely low temperature (0.090 mK) with it (Figure 6.27). There are currently four cryogen-free nuclear adiabatic demagnetization refrigerators around the world, in contrast to the large number of refrigerators using other cooling methods. The refrigerator at Peking University holds the world record to date for obtaining the lowest temperature without helium consumption.

Establishing extremely-low-temperature conditions provides powerful platforms for probing the full quantum effects of condensed matter systems. This is because in normal circumstances, quantum fluctuations are so weak that they are drowned out by thermal fluctuations, making it impossible to observe full quantum effects. In particular, thermal fluctuations are largely suppressed in the low-temperature environment, and phenomena such as zero-point energy corrections and quantum tunneling induced by nuclear quantum

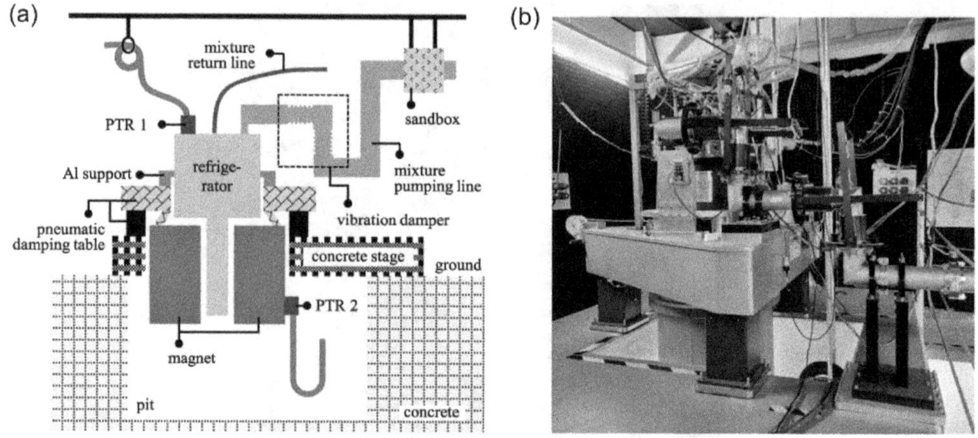

Figure 6.27 Cryogen-free nuclear adiabatic demagnetization refrigerator built at Peking University. It has achieved the lowest temperature of any cryogen-free refrigerator in the world to date. Reprinted from Yan J.[2021], with the permission of AIP Publishing.

effects predominate. The various transport properties associated with full quantum effects, as well as the new states of matter caused by ultrafast nonadiabatic phase transitions, are all worthy topics for further research.

6.2.2 Ultrahigh Pressure

Similar to temperature, pressure is a fundamental physical parameter that determines the states of matter. In the universe, pressures vary by over 50 orders of magnitude. The lowest measurable pressure is that between galaxies (~10^{-26} GPa); the highest computable pressure is that at the center of a neutron star (~10^{25} GPa) (see Figure 6.28). The pressure on Earth ranges from 10^{-4} GPa at the surface to ~360 GPa at the core. In the laboratory, novel full quantum states can be achieved by changing pressure. For example, the application of high pressure shortens the interatomic distances in condensed matter and reduces the atomic transfer potential for light elements, thus facilitating nuclear quantum tunneling and quantum delocalization. Consequently, it can enhance the electron–phonon coupling by increasing the electronic DOS near the Fermi level and modifying low-frequency phonons of non-hydrogen elements. This is a common phenomenon in the hydrogen-rich superconductors that have been studied extensively in recent years. In addition, as mentioned above, as the pressure increases, the nucleus can overcome the Coulomb repulsion through full quantum effects, triggering new states of matter such as superfluidity and supersolidity.

High pressure can significantly affect the physical and chemical properties of condensed matter. Since the publication of *The Physics of High Pressure* by Bridgman in 1931 [Bridgman1931], high-pressure science has been developing for nearly a century into an important subdiscipline that plays a crucial role in understanding the laws of nature and improving human welfare as well as industrial productivity. However,

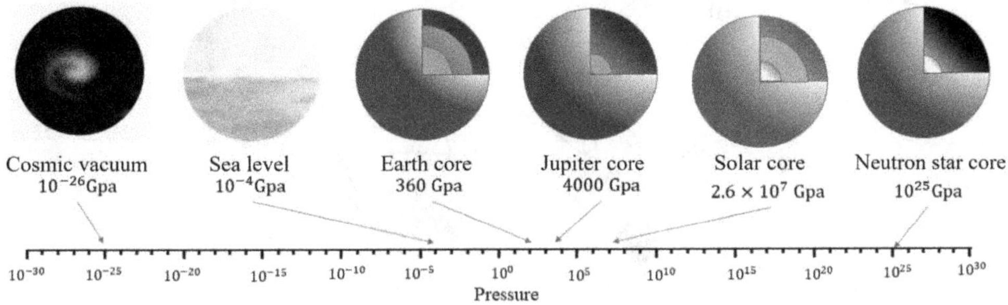

Figure 6.28 In the universe, pressure values span 50 orders of magnitude. The maximum static pressure that can be achieved in the laboratory is ~900 GPa.

the potential of high pressure as a scientific research tool for chemical syntheses and physical-property measurements has not been fully explored, largely because of the complexity and specialized nature of high-pressure techniques.

High pressures can be classified as static or dynamic according to how they are generated. Static high pressure is realized by mechanical compression and can be maintained over a long period of time. Dynamic high pressure, on the other hand, is generated by rapid, short-lived processes such as explosions, collisions, and intense irradiation. Here, we will focus on the progress achieved in recent decades using static high-pressure techniques.

The static pressure inside a diamond anvil cell (DAC) can now reach up to 900 GPa, which is more than 10 times higher than that generated by any other method [Ulmer1987]; laser heating can raise the temperature of the sample in a DAC to over 5,000 K at high pressures [Hemley1987]. Under such extreme conditions, the density, electronic structure, and free energy of elements in chemical reactions are dramatically altered. This can result in extraordinary changes in chemical equilibrium and material properties and thus lead to a variety of new compounds and novel quantum states. In addition, many solid materials synthesized under high-pressure and high-temperature conditions can be quenched to form thermodynamically metastable states that remain dynamically stable indefinitely under ambient conditions. Such a high-pressure synthesis method is of great practical value in obtaining useful metastable materials such as artificial diamond.

The development of high-pressure science has partly been driven by advances in laboratory techniques. Over the years, various static high-pressure devices based on different principles have been developed for applications in different pressure ranges; some of them have even been commercialized. For example, compact, self-clamped piston-cylinder devices that can generate pressures up to 3 GPa [Loveday2012] have been widely employed in conjunction with commercial refrigerators and superconducting magnets to measure physical properties such as resistance and magnetization at low temperatures and under strong magnetic fields [Thompson1984]. Large-volume devices typically rely on hydraulic presses to generate high pressures on the compressed samples. For example, the Bridgman high-pressure device can generate a maximum pressure of ~20 GPa by using two oppositely compressed conical anvils [Loveday2012]. Large-volume multi-anvil presses, typically

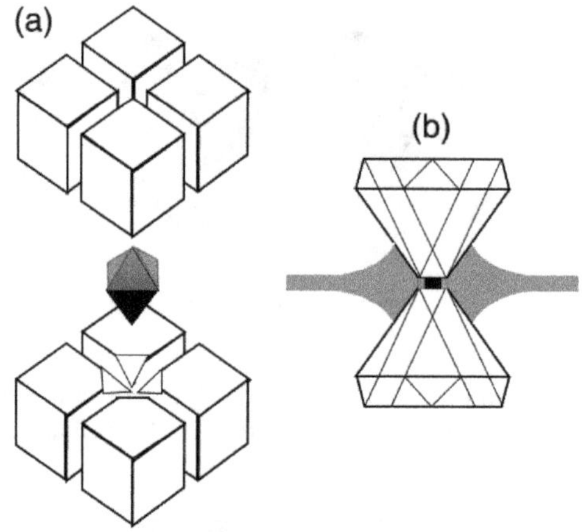

Figure 6.29 (a): The key components of a large-volume multi-anvil press are eight cubes with truncated corners compressing an octahedral pressure medium. The size of the sample is typically less than 1 cm. (b): The configuration in a diamond anvil cell. A small hole (black) with a diameter of 250–300 μm is drilled in the middle of a metal gasket (gray) and used as a sample chamber. The sample is compressed between the flat culets (small "tip" surfaces of the diamonds) when a load is applied to the large, flat "table" surfaces at the top and bottom.

with cubic or octahedral (left panel of Figure 6.29) geometries, are now in common laboratory use. A maximum pressure of 40 GPa has been achieved using a two-stage multi-anvil module modified by Walker et al. [Walker1990]. The advantages of a multi-anvil press over opposite-anvil cells such as DACs include a larger sample space (approximately a few cubic millimeters) and a better hydrostatic pressure environment (i.e., one able to produce isotropic pressure). The former is preferred for chemical synthesis experiments, while the latter allows for precise measurements and regulation of physical properties under high pressures [Cheng J.G.2014].

Currently, the DAC is the device that can generate the highest static pressure [Bassett2009, Mao H.K.2018]. It ordinarily consists of a pair of gem-quality diamonds (less than 1/2 carat), as in the Bridgman device. The sample chamber is a small hole drilled in the middle of a stainless-steel or nichrome gasket placed between the diamond anvils (see right panel of Figure 6.29). Not only can a DAC reach very high pressures, but it can also be easily cooled to low temperatures (<2 K). On the other hand, samples under high pressure in a DAC can be heated up to 5,000 K using IR-laser or external heating.

The DAC is especially suitable for material characterizations at high pressures because the diamond anvils are transparent to X-rays and to most light from the UV to the IR. Therefore, spectroscopic techniques, such as X-ray diffraction, Raman, IR, and UV/Vis reflection and absorption spectroscopy, can be integrated with DAC to characterize new structural phases at different pressures and temperatures and to determine the phase boundaries. DAC-based measurements of transport properties such as resistance have become routine in most laboratories. Recently, magnetization measurements using a DAC have also been conducted to study the Meissner effect of superconductors under high pressures. Although the DAC has a smaller sample space than a multi-anvil device, it is still widely used in chemical syntheses because it allows for easy measurement of crystal structure at high pressures [Walsh2018].

6.2.2.1 Material Synthesis under High Pressure

In chemical syntheses involving gaseous components, for example, the Haber–Bosch method used to synthesize ammonia in industry or the chemical reactions in autoclaves, the pressure rarely exceeds 10^{-2} GPa. By contrast, multi-anvil presses and DACs that utilize a solid pressure-transmitting medium can generate pressures from tens to hundreds of GPa. At such high pressures, many elements that do not react with each other at atmospheric pressure may combine to form new compounds. For instance, the alkali metal potassium can form a variety of compounds with transition-metal elements [Atou1996]; iron and bismuth can form the binary compound $FeBi_2$ at high pressures [Walsh2016] and even the inert-element xenon can form compounds with transition-metal elements [Stavrou2018]. The high-pressure environment thus introduces new possibilities for the development of inorganic chemistry, and research in this area is rapidly growing. Two material systems synthesized at high pressures that have significant implications for other subjects are briefly described below. These are the superhard materials and the ABO_3 oxides, which possess considerable practical value because they can be quenched and remain stable under ambient conditions.

Superhard materials have always been a popular research topic in high-pressure synthesis. Nevertheless, the only superhard materials that are fully commercialized are cubic diamonds and boron nitride synthesized at high pressures (6 GPa) and temperatures (1,200 K) using transition-metal catalysts [Badding1998]. Synthetic diamonds are generally used as abrasives. Cubic boron nitride, on the other hand, is used for processing ferrous metals, because iron reacts with carbon at high temperatures. Recently, it has been reported in the literature that nanotwinned cubic boron nitride and diamond synthesized at high pressure and temperature exhibit improved hardness and thermal stability that surpass those of the corresponding single-crystal forms [Tian Y.J.2013, Huang Q.2014]. In addition, the use of transition-metal flux agents at high temperature and pressure allows the growth of diamond single crystals with higher chemical and isotopic purity than natural diamond. Similarly, high-purity hexagonal boron nitride synthesized at high pressure has a lower oxygen content and fewer carbon impurities than a single crystal grown at atmospheric pressure, and it therefore shows better insulating properties.

Oxides with the chemical composition ABO_3 (where A and B can be alkali metals, alkaline-earth metals, transition metals, or subgroup elements) are important in material science, in solid state chemistry, and especially in geophysics: They include the silicate compounds that are among the main components of the Earth's lower mantle. ABO_3 oxides can adopt various structures (including the pyroxene, lithium niobate, and perovskite structures), depending on the pressure and temperature [Navrotsky1998]. In general, perovskites and perovskite-like structures with high density are stable under high pressure. One of the most significant breakthroughs in recent high-pressure studies is the discovery of the perovskite–post-perovskite phase transition in magnesium

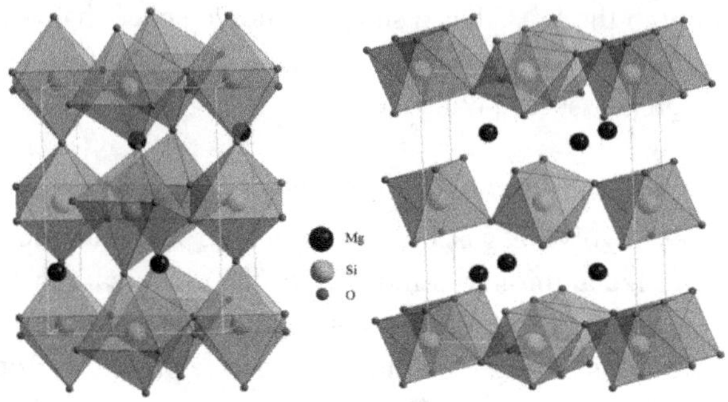

Figure 6.30 (a) Perovskite structure and (b) post-perovskite structure.

silicates at high temperatures and pressures that are close to the conditions in the Earth's lowermost mantle [Murakami2004, Oganov2004] (see Figure 6.30). This discovery successfully revealed the mineral structure in the depths of the Earth and explained the long-standing puzzles about the discontinuity of seismic-wave propagation in the Earth's lower mantle.

The ABO_3 oxides containing transition metals can exhibit many intriguing physical properties, including high-temperature superconductivity, colossal magnetoresistance, ferroelectricity, and multiferroic properties. These rich physical phenomena arise because the partially filled d orbitals in the transition metals have strong electron–electron/electron–lattice interactions, and the electronic and magnetic structures can be effectively tuned by varying the crystal structure and chemical bonds as well as the crystalline electric fields. As pressure can determine the crystal structure of ABO_3 oxides, high-pressure synthesis is vital for discovering new quantum materials of this class. It has also played an important role in the exploration of cuprate superconductors. Recent high-pressure studies have focused on ABO_3 oxides containing $4d/5d$ elements at the B-site, for which the presence of intrinsic strong spin-orbit coupling produces novel magnetic and ferroelectric properties different from those of their $3d$ counterparts and closely associated with full quantum effects [Cheng J.G.2011, Shi Y.G.2013].

6.2.2.2 Phase Change under High Pressure

High pressures can drastically reduce the interatomic distances in a substance and significantly change its physical properties. In condensed matter physics, researchers are mostly interested in pressure-driven phase transitions between the solid, liquid, and gas phases, and in pressure-driven lattice transitions between solid phases. Even in the case of simple elements, there are rich phenomena associated with pressure-induced phase transitions that are still incompletely understood. One example is the α–γ phase transition in cerium, which occurs at ~1 GPa. While the crystal structure remains unchanged, the lattice volume is reduced by 15% [Ramirez1971]. This phase transition has been studied for many years but its mechanism is still debated [de' Medici2005].

The electrical and magnetic properties of condensed matter can be significantly altered by pressure-induced electronic phase transitions, including the metal–insulator transition, superconducting transition, paramagnetic–ferromagnetic/antiferromagnetic transitions, and charge/spin density waves. The underlying mechanisms for these transitions may involve modifications of crystal and/or electronic structure due to the shrinkage of atomic spacing and changes in electron–electron or electron–lattice interactions. In the case of metal–insulator transitions, the pressure can turn semiconductors such as silicon and germanium into metals via Wilson phase transitions; it can also turn alkali metals such as lithium and sodium into semiconductors and insulators, respectively [Ma Y.M.2009, Matsuoka2009].

The modulation of electron–electron/electron–lattice interactions through pressure is of paramount importance in the study of quantum materials. Compression of condensed matter systems such as heavy-fermion materials can lead to significant electron correlation or quantum critical behaviors. Heavy-fermion systems are generally intermetallic compounds containing *f*-electron elements such as cerium, ytterbium, and uranium. The conduction band electrons of these elements have large effective masses due to the Kondo effect. The application of high pressure can significantly change the magnetic interaction between the *f*-electron magnetic moments as well as the Kondo effect between the *f* electrons and the conduction-band electrons. Even at a pressure of around 1 GPa, quantum critical phenomena can be induced in heavy fermion systems using a simple piston cylinder device [Ferriere2001]. Pressure regulation has thus become a significant and routine experimental tool in condensed matter study with novel quantum behaviors, in which full quantum effects play key roles.

Remarkable new superconductors continue to be discovered after a century of development. Room-temperature superconductivity has long been the dream of condensed matter physicists, and this dream can be fulfilled by conducting high-pressure experiments. Over the past few years, chemical syntheses and physical-property regulations of materials at high pressures have significantly contributed to superconductivity research. Figure 6.31 shows the transition temperature (T_c) of some typical superconducting materials as a function of the year in which it was first measured. In the cuprates with a transition temperature above 150 K and all the hydrogen-rich materials, superconductivity was observed at high pressures. In Chapter 7, the phase diagram of hydrogen under pressure and the progress on the synthesis of metallic hydrogen will be described. Because pressure strongly modifies the electronic structure of crystals, insulating elements such as sulfur and boron are transformed into superconductors at high pressures [Struzhkin1997, Eremets2001]; manganese- and chromium-based compounds with strong magnetic properties can also exhibit superconductivity under high pressures [Wu W.2014, Cheng J.G.2015].

Pressure regulation is of great importance for the study of unconventional superconductivity, which cannot be described satisfactorily by the BCS theory based on electron–phonon coupling. Typical unconventional superconductors include copper oxides, iron-based high-T_c superconductors, heavy-fermion superconductors, and organic superconductors. Currently, the highest superconducting transition temperature of a cuprate (164 K) was obtained by applying a pressure of 30 GPa to $HgBa_2Ca_2Cu_3O_8$ [Gao L.1994]. The effect of pressure is particularly prominent in iron-based superconductors. For example, superconductivity can be induced in 122-type parent compounds such as $BaFe_2As_2$ [Chu C.W.2009], and the T_c of 1111-type superconductors with the

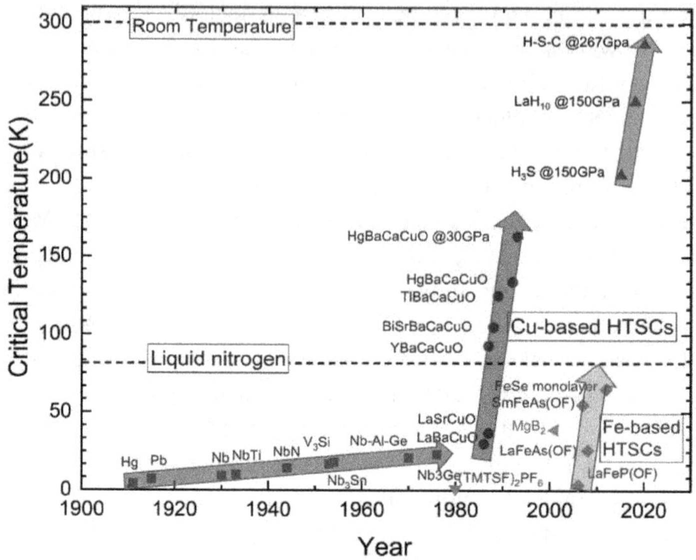

Figure 6.31 Representative superconducting materials, with transition temperatures and dates of discovery. (The transition temperature of H–S–C remains to be verified by further experiments.) Reprinted from iScience, 24, Chao Y.[2021], Copyright 2024, with permission from Elsevier.

composition $LaO_{1-x}F_xFeAs$ can be raised to 40 K by applying pressure [Takahashi2008]. The T_c of the type-II superconductor FeSe can increase with pressure at a rate of 9 K/GPa [Medvedev2009].

Smaller pressures may have a significant effect on organic and heavy-fermion systems. The first organic superconductor, $(TMTSF)_2PF_2$, was discovered at a pressure of 1.2 GPa [Jérome1980], and the T_c of fullerene Cs_3C_{60} was found to be as high as 40 K at 1.43 GPa [Palstra1995]. As mentioned above, pressure can significantly alter the magnetic interaction between the *f*-electron magnetic moments and also the Kondo effect between *f*-electrons and conduction-band electrons. In a heavy-fermion system, a small pressure can drive the quantum critical transition and induce unconventional superconductivity. Pressure-induced superconductivity has been found in both cerium- and uranium-based heavy-fermion materials [Jaccard1992].

At present, the most promising route to realizing room-temperature superconductivity involves hydrogen-rich compounds at ultrahigh pressures and uses the conventional BCS-type mechanism. Such compounds exhibit hydrogen symmetrization at high pressures, which is apparently caused by full quantum effects. An in-depth discussion of this topic will be presented in Chapter 7.

In experimental materials science, the object being studied is as important as the method. First, we need to discover or synthesize a material that can exhibit quantum properties and then use various characterization and regulation tools to reveal the quantum-mechanical laws behind its behavior. In general, an experimental method is only useful for achieving one of these two objectives. High-pressure research, on the other hand, can contribute to both as it is capable of synthesizing new materials and regulating their physical properties. Therefore, the external-field research platform provided by high-pressure technology is expected to play a more significant role in the discovery of new states of matter and novel materials with full quantum effects.

6.2.3 High Magnetic Field

The temperature and pressure techniques that were introduced in the previous sections can change the physical properties of materials by changing their thermodynamic parameters. Magnetism is another fundamental property of matter. By altering the electron state energies and magnetic moments through external magnetic fields, we can study the electromagnetic properties of atoms and molecules. Higher magnetic fields allow greater electronic state change within a material, which can lead to novel phenomena and provide more opportunities for scientific research and applications. Further, ultrahigh magnetic fields can be used to enhance the polarization of nuclear spins, thus improving the signal-to-noise ratio and spectral resolution of NMR signals. This can be used to probe the nuclear quantum effects. As a result, high magnetic fields are now indispensable for research in condensed matter physics, materials science, chemistry, life science, medicine, and other fields.

Magnetic fields above 20 T are generally classified as "high." Currently, high magnetic fields are generated by utilizing electric current to generate a magnetic field in the space around a conductor. The higher the current, the stronger the magnetic field. To avoid problems such as the thermal effect of the current and structural damage caused by electromagnetic force, a specific spatial and temporal distribution of the magnetic field must be achieved. This can be accomplished by creating different conducting-coil structures using different technical approaches.

High magnetic fields can be also categorized as steady or pulsed magnetic fields. The former has constant intensity and an arbitrarily long duration. The latter is generated in the coil space as the pulsed current flows through the winding coil; the pulse duration typically ranges from milliseconds to hundreds of milliseconds.

6.2.3.1 High Steady Magnetic Field Technology

Magnets that generate a high steady magnetic field can be divided into three types: water-cooled (or resistive) magnets, superconducting magnets, and hybrid magnets. In 1936, F. Bitter at MIT first proposed a water-cooled coil design (the Bitter coil) that was used to make resistive magnets. With a power supply of 10 MW, a magnetic field of 20 T was produced [Bitter1936]. Based on this result, the world's first national high-magnetic-field laboratory was built at MIT in 1960. Later, high magnetic field laboratories were set up in France, the Netherlands, Japan, China, and other countries.

Advances in high magnetic field technology have significantly contributed to the development of condensed matter physics. In 1980, K. von Klitzing observed the integer quantum Hall effect in a silicon–metal-oxide field-effect transistor using a 18 T magnetic field [von klitzing1980]. In 1982, D. C. Tsui observed the fractional quantum Hall effect in a 2D GaAs-AlGaAs electron gas using a 21.6 T magnetic field [Tsui1982]. Both the integer and fractional quantum Hall effects were observed under steady high magnetic fields. These researchers were awarded the Nobel Prize in Physics in 1985 and 1998, respectively. In 2005, the half-integer quantum Hall effect was observed in high-quality graphene using superconducting magnets in the laboratories of A. K. Geim and P. Kim [Novoselov2005, Zhang Y.B.2005]; this attracted widespread research interest in the field of graphene. S. K. Novoselov and A. K. Geim won the Nobel Prize in Physics in 2010.

In recent years, as the high-temperature superconducting materials and magnet technology continue to develop, fully superconducting magnets (made of either low-temperature or high-temperature superconducting materials) are now being used to generate high steady magnetic field conditions. In 2017, the National High Magnetic Field Laboratory in the United States successfully finished the development of a 32 T fully superconducting magnet. In 2019, the Institute of Electrical Engineering of the Chinese Academy of Sciences developed a fully superconducting magnet with a 32.35 T central magnetic field strength. To generate higher magnetic fields, M. F. Wood at Oxford University and D. B. Montgomery at MIT proposed the concept of hybrid magnets, in which superconducting coils are added to the periphery of a water-cooled magnet. This reduces electrical energy losses while ensuring that the entire magnet can generate higher magnetic fields. In the 1990s, the hybrid magnet developed by the National High Magnetic Field Laboratory produced a steady magnetic field of 45 T. In 2019, the National High Magnetic Field Laboratory tested a high-temperature superconducting coil using a water-cooled magnet and reported a central magnetic field of 45.5 T, the highest steady magnetic field produced to date.

a Water-Cooled Magnets

The magnetic field strength that a conventional electromagnet – a solenoid with a metallic core – can generate is limited by the saturated magnetic induction strength and temperature rise of the core material. Therefore, in case of a magnetic field above 3 T, a hollow solenoid without a core is employed instead. A large steady current passes through the solenoid to generate a strong magnetic field. Although this solves the saturation problem, most of the energy is still consumed inside the magnet, generating a high amount of Joule heating. To solve this problem, researchers have proposed a rapid cooling technique using a liquid medium; magnets that use this technique are called water-cooled magnets. Such a magnet consists of perforated circular conducting plates (Bitter disks) that are stacked and interleaved with insulating material to form a thick coil (as shown in Figure 6.32). Bitter magnets, widely used in high magnetic field laboratories because of their simple construction and robustness, have proven to be a reliable technology.

In the early 1980s, the Max Planck Institute (MPI) and the Centre National de la Recherche Scientifique (CNRS) developed the polyhelix magnet [Schneider-Muntau1981]. Its structure reduces the stress and provides an alternative magnet structure capable of generating high steady magnetic fields. To further reduce the cooling-coil stress, a new type of Bitter magnet sheet has been designed by the National High Magnetic Field Laboratory. The shape and location of the cooling holes were

Figure 6.32 Schematic of a Bitter magnet.

optimized. This has allowed gradual increases in the high steady magnetic field strength of the water-cooled magnet to 27, 30, and 33 T [Bird M.D.1996a,b, Bird A.D.2004]. In 2015, the Steady High Magnetic Field Science Center of the Chinese Academy of Sciences successfully increased the magnetic field strength of water-cooled magnets to 38.5 T [Gao B.J.2016]; in 2017, the National High Magnetic Field Laboratory increased the strength of water-cooled magnets to 41.4 T, which is the highest magnetic field strength of water-cooled magnets to date [Toth2018].

b Superconducting Magnets

In 1911, the Dutch physicist H. K. Onnes discovered that mercury becomes superconducting at 4.2 K. Since then, superconductors have attracted widespread attention owing to their zero resistance, ideal diamagnetism, and zero electric power consumption. Superconducting magnets (magnets made from coils of superconducting wire) generally consist of different nested superconducting materials. Compared to the water-cooled magnet, the superconducting magnet has advantages such as high stability, high uniformity, and low operating cost that allow it to find practical applications in engineering, biomedical, and other scientific instrumentation. However, the critical current and magnetic field of superconducting materials limit the ability of superconducting magnets to achieve high magnetic field strength. For a long time, the maximum magnetic field strength of superconducting magnets remained around 20 T.

At present, there are two types of low-temperature superconducting materials suitable for large-scale applications: NbTi and Nb_3Sn. Magnets with a working capacity below 10 T are generally made of NbTi, whereas 10–20 T magnets are made of NbTi and Nb_3Sn. In recent years, as the performance of the second-generation high-temperature superconducting material REBCO has continued to advance, it has become possible to combine low-temperature and high-temperature superconducting materials to create fully superconducting magnets. This technology is the result of research advances in superconducting materials, magnet insulation, coil technology, and quench protection. It is considered the most promising technical solution for achieving higher magnetic fields.

In 2016, the RIKEN Institute finished developing a fully superconducting magnet with a central magnetic field of 27.6 T [Yanagisawa2016]. In 2017, the High Field Laboratory for Superconducting Material in Tohoku University reached 24.57 T [Awaji2017]. In the

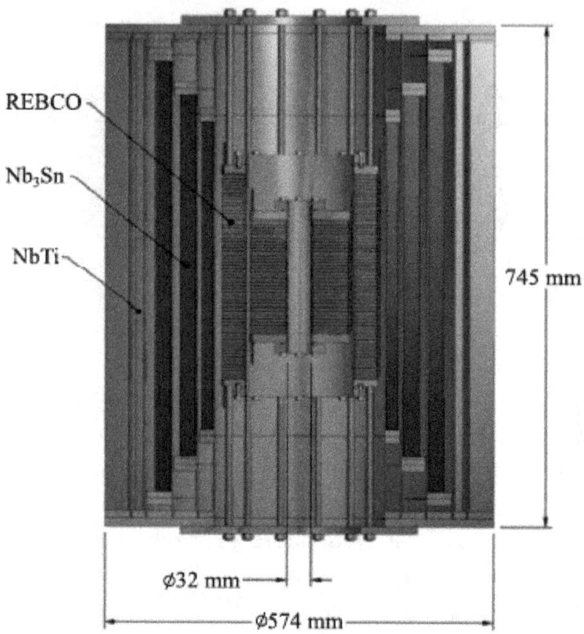

Figure 6.33 Structural diagram of the 32 T fully superconducting magnet at the National High Magnetic Field Laboratory, USA.

same year, the National High Magnetic Field Laboratory successfully completed a 32 T fully superconducting magnet (shown in Figure 6.33). In 2019, the Institute of Electrical Engineering of the Chinese Academy of Sciences successfully developed a 32.35 T superconducting magnet [Liu J.H.2020], which is the full superconducting magnet with the world's highest magnetic field.

Currently, the National High Magnetic Field Laboratory is planning to construct a 40 T fully superconducting magnet [Bai H.Y.2020] with a cold bore of 34 mm and a homogeneity of 500 ppm uniformity within a 1-cm-diameter spherical volume. Taking the advantages and disadvantages of each material into account, it was decided that four high-temperature superconducting materials (insulated REBCO, no-insulation REBCO, Bi2212, and Bi2223) will be used in parallel to develop the magnet.

c Hybrid Magnets

The Bitter magnet is a resistive magnet that requires a high-power supply system. It generates considerable heat during operation and is cooled by the water in the cooling channels. The large size and low current density of water-cooled magnets are major factors that limit their application. By contrast, superconducting magnets are more stable and commercially available. However, their performance is limited by the critical current and the critical magnetic field of superconducting materials. To generate a stronger magnetic field, the concept of hybrid magnet was introduced. In 2000, the hybrid magnet at the National High Magnetic Field Laboratory produced a

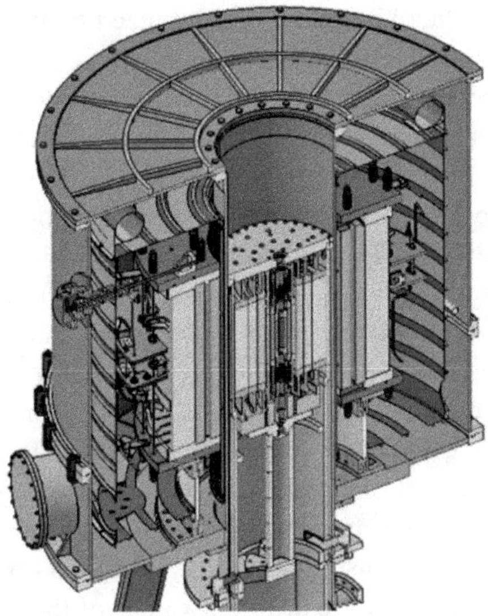

Figure 6.34 Structural diagram of the 40 T hybrid magnet at the High Magnetic Field Laboratory, Chinese Academy of Sciences.

steady high magnetic field of 45 T [Pugnat2014]. In 2016, hybrid magnets at the High Magnetic Field Laboratory, Chinese Academy of Sciences, produced 40 T magnetic field [Kuang2018] (Figure 6.34). In 2019, the National High Magnetic Field Laboratory tested a high-temperature superconducting coil using a water-cooled magnet and reported a central magnetic field of 45.5 T [Hahn2019].

The strong magnetic fields generated by such hybrid magnets offer unprecedented opportunities for precise physical measurements under extreme conditions. For example, the quantum Hall effect at room temperature was observed in graphene using a 45 T steady magnetic field at the Geim's laboratory [Novoselov2007]; Sasa Zaric et al. reported optical splitting in spectra of single-walled carbon nanotube under 45 T, confirming the existence of the Aharonov–Bohm effect of the electromagnetic vector potential for the first time [Zaric2004].

6.2.3.2 Pulsed High Magnetic Field Technology

Pulsed high magnetic field technology dates back to the beginning of the twentieth century. As early as 1924, the Soviet physicist P. Kapitza used a lead-acid battery to achieve a 50 T magnetic field in a 1-mm-diameter coil [Kapitza1924]. However, the limitations of the available measuring instruments were an obstacle to further progress. By the 1960s, techniques had improved, and scientists were able to measure the Haas–van Alphen and magnetoresistance effects under pulsed fields. As a result, pulsed high magnetic field laboratories began to be established in Europe, the United States, and Japan [Jones2000, The LNCMP-team2004]. At the end of the 1990s, scientific developments and increasing demand for higher magnetic fields initiated a new wave of international interest in implementing pulsed strong magnetic field technology. In particular, the discovery of high-temperature superconductivity in 1986, recognized by the Nobel Prize in Physics in 1987, heavily influenced the development of this field.

Recently, China has become a worldwide leader in pulsed strong magnetic field technology, owing to the unique technologies invented by L. Li's group at the Wuhan National High Magnetic Field Center. In collaboration with S. Jia's research group at Peking University, L. Li's group observed the non-trivial annihilation behavior of Weyl fermions in TaP [Zhang C.L.2017]. Meanwhile, by working with J. Wang's group at Peking University, they discovered the third type of logarithmic periodic quantum oscillation in $ZrTe_5$ [Wang H.C.2018]. Because the magnetic exchange interactions within these materials are significant, pulsed fields are effective in revealing a variety of quantum states that are closely related to full quantum effects, such as quantum phase transitions, spin liquids, Bose-Einstein condensation, and quantum magnetization platform.

The waveform characteristics of a pulsed magnetic field are determined by the needs of scientific experiments. The performance of a pulsed magnetic field is typically determined by the magnetic field strength, pulse duration, stability, and pulse-repetition frequency. Different research targets may require different values of these magnetic field characteristics. During the generation of pulsed high magnetic field, the pulsed power supply provides energy, the pulsed magnet generates the magnetic field as the current flows through it; both play an indispensable role in the development of pulsed high magnetic field technology. Therefore, upgrading either the pulsed power supply or the pulsed magnet can give rise to new scientific opportunities.

a Ultrahigh Pulse Magnetic Fields

Before ultrahigh magnetic field strengths are produced, the magnet and power supply must be configured so that the structural damage caused by the high magnetic stress can be managed. For a 50 T magnetic field, the magnetic stress is 1 GPa; for a 100 T magnetic field, the stress becomes 4 GPa. This value exceeds the material strength of any practical conductors that are available today. Early pulsed magnets were made of copper wire, which had a mechanical strength of ~300 Mpa; only a magnetic field less than 40 T could be produced. In 1986, S. Foner at MIT introduced high-strength, high-conductivity copper–niobium alloy wire into pulsed magnets for the first time, achieving a peak magnetic field of 68 T. Later, a copper–stainless steel composite wire capable of generating a magnetic field of 80 T was developed at the University of Oxford and the Laboratoire National des Champs Magnétiques Intenses in Toulouse. In addition, Russia made some breakthroughs in copper–niobium alloy, and researchers in Japan successfully developed wires made of copper–silver alloy.

Although increased wire strength improves the magnetic field, it is often accompanied by decreased conductivity. When the wire strength reaches 1 GPa, it is impossible to further increase the magnetic field strength without loss of conductivity. Therefore, we must resort to other methods to increase the magnetic field. In the 1990s, F. Herlach at Katholieke Universiteit Leuven proposed a layered reinforced pulsed magnet structure that could increase the magnetic field strength from 60 T to ~80 T. This technical solution is now used in all ultrahigh pulsed magnets around the world.

However, the layered reinforcement technique cannot address the magnetic stress problem caused by magnetic fields above 80 T. The only way to reduce the magnetic stress is to increase the volume of the pulsed magnet. However, increasing the volume of the pulsed magnet means increasing its inductance and energy. From the perspective of power supply, powering a large-volume pulsed magnet requires both high power and high energy, which is unrealistic for a pulsed power supply. Consequently, multi-stage pulsed high magnetic field systems with multiple coils and multiple power supplies have been proposed. The basic idea of the multi-stage pulsed high magnetic field system is to use two or more coils, each with an independent pulsed power supply. The outer coil is powered by a high-energy pulsed power supply to produce a long background magnetic field; the inner coil is powered by a low-energy, high-power supply. Since the two coils are powered separately, they can be designed and optimized independently to have low power requirements for producing ultrahigh pulsed magnetic fields.

In 2003, the "ARMS" project funded by the European Union built the world's first dual-coil ultrahigh magnetic field system [Jones2006]. The system was tested at the Laboratoire National des Champs Magnétiques Intenses in Toulouse, and a magnetic field of 76 T was achieved. In 2012, the National High Magnetic Field Laboratory produced a 100.7 T magnetic field using a four-coil structure, which currently holds the world's record for the highest pulsed magnetic field. The innermost coil consists of high-strength, high-conductivity copper–niobium alloy wire supplied by Russia. The innermost coil is powered by a capacitor and the outer three coils consist of Glidcop wire and are powered by a pulsed generator. Pulsed high magnetic field research facilities in Dresden (Germany), Toulouse (France), and Wuhan (China) have also adopted the structural design with two coils and two capacitor power supplies; the strongest magnetic fields achieved were 95.6 T, 90 T, and 94.8 T, respectively. Scientists have been relentlessly searching for ultrahigh magnetic fields.

b Long-Pulse Magnetic Fields

The development of long-pulse magnetic fields is mainly limited by the power-supply capacity. Currently, the commonly used power sources include capacitors and pulse generators.

Capacitor power supplies are typically operated above 20 kV, with high power and relatively low energy storage, making them suitable for generating ultrahigh short-pulsed magnetic fields. At present, the High Magnetic Field Laboratory in Dresden has the world's largest capacitor power supply, with a maximum voltage of 24 kV and a total energy storage of 80 MJ. The capacitors developed by the Wuhan National High Magnetic Field Center and the Laboratoire National des Champs Magnétiques Intenses in Toulouse can store energies of 28.7 MJ and 14 MJ, respectively.

A pulse-generator power supply can store hundreds of megajoules of energy, but its power is low. Using a rectifier system, the magnetic field waveform can be easily adjusted, making it possible to generate long-pulse magnetic fields, flat-top magnetic fields, or magnetic fields with other special waveforms. The largest pulse-generator power supply is located at the National High Magnetic Field Laboratory. It has a peak power output of 1,430 MVA and an energy output of 650 MJ. Using a three-coil magnet, it can produce waveforms with a peak magnetic field of 60 T, a full pulse width of 2,000 ms, and a flat-top time of 100 ms. The Wuhan National High Magnetic Field Center shares a 100 MJ/100 MVA pulse generator with the on-campus tokomak J-TEXT. When used as a pulsed magnetic field power source, the generator is used in conjunction with two rectification systems with a no-load voltage of 3 kV and a full-load current of 50 kA [Ding H.F.2012]. With this system, a magnetic field waveform with a peak strength of 50 T, a full pulse width of 1,000 ms, and a flat top time of 100 ms has been achieved [Xu Y.2014].

c Highly Stable Flat-Top Magnetic Fields

NMR and similar experiments where the magnetic field must remain extremely stable for an extended period of time have been customarily carried out under steady-magnetic-field conditions. However, producing very strong but completely steady magnetic fields is not feasible. As state-of-the-art experiments came to require ever higher magnetic field strengths, scientists began to improve the stability of pulsed magnetic fields for use in NMR studies.

With a capacitor power supply, it is difficult to achieve a stable magnetic field because the discharge current cannot be controlled. To resolve this issue, the High Magnetic Field Laboratory in Germany adopted an approximate approach. They used a 43 MJ capacitor and a large pulsed magnet that weighs over 2,000 kg to generate a magnetic field with a peak intensity of 55 T and a pulse duration of 1,500 ms [Weickert2012]. However, the large capacitor and magnet led to cooling times of more than 8 hours, resulting in low experimental efficiency. The Wuhan National High Magnetic Field Center proposed a new method to achieve a highly stable flat-top magnetic field using a capacitor power supply [Jiang F.2014]. The system consisted of two capacitor banks and a coupled pulse transformer. A record-breaking flat-top magnetic field of 64 T was achieved, providing unique experimental conditions for NMR and specific-heat studies.

d High Pulse-Repetition-Frequency Magnetic Fields

In X-ray scattering, X-ray diffraction, and neutron scattering experiments, large amounts of data are collected. Therefore, the cooling time between pulses in the magnet needs to be as short as possible. Further, the challenges faced during high repetition frequency power supply need to be solved to improve the efficiency of the experiment. The cooling time of pulsed magnets generally ranges from several minutes to several hours, which cannot meet the requirements of such experiments. Researchers at the University of Oxford installed a copper disk at the end of the magnet coil, connecting the coil to an external 77 K liquid nitrogen environment to improve the heat transfer efficiency from the magnet to the surrounding. The Laboratoire National des Champs Magnétiques Intenses in Toulouse and the Wuhan National High Magnetic Field Center utilized the fact that the inner windings of the magnet are separated under the action of electromagnetic force and designed axial liquid-nitrogen cooling channels inside the coils. In this way, the cooling medium was introduced directly into the interior of the magnet, considerably improving the cooling speed. The Wuhan National High Magnetic Field Center proposed a multi-module sequential power supply technology to achieve a maximum 45 T magnetic field at a pulse repetition rate of 60 Hz.

Additionally, there is another class of pulsed strong magnetic fields, which are commonly referred to as destructive pulsed magnetic fields or megagauss-magnetic fields. As we have seen, neither steady nor conventional pulsed magnetic fields can reach a field strength higher than 100 T because of the magnetic stress. In the destructive pulsed magnetic field method, the coil is loaded with millions of amperes of current in microseconds, generating a strong magnetic field before the magnet is destroyed. Current technical approaches include the single-turn coil method, the electromagnetic flux-compression method, and the explosive flux-compression method. These three methods have strict laboratory requirements and only a few laboratories are capable of carrying out research work in this area. The single-turn coil method is studied by the Institute for Solid State Physics in the University of Tokyo, the National High Magnetic Field Laboratory, and the Laboratoire National des Champs Magnétiques Intenses in Toulouse, with a magnetic field strength of 150–300 T. Research on electromagnetic flux compression is mainly conducted by the Institute for Solid State Physics at the University of Tokyo, which has reached magnetic field strengths of 600–1,000 T. Owing to their high experimental difficulty, only a few experiments have been conducted using the single-turn coil and electromagnetic flux-compression methods; current research in this area is mainly focused on the quantum phase transition of low-dimensional resistive magnets [Miyata2011, Zhou X.G.2020]. Explosive magnetic flux compression generates the strongest magnetic fields, with a strength of 2,700 T. This method was first implemented by scientists from the Soviet Union. As its technical difficulties and experimental conditions are even more demanding, it is now scarcely studied.

In recent years, various laboratories around the world have focused on the development of steady and pulsed strong magnetic fields. These efforts have opened a new avenue in condensed matter physics research. As a result, broader external-field research platforms are being established for regulating full quantum effects and discovering new states of matter.

6.2.4 Ultrafast and UltraIntense Lasers

Among all external field conditions, manipulation using optical fields has distinctive characteristics. Today's ultrafast techniques of optical observation and manipulation fall into three sub-categories: ultrafast laser methods; ultrafast spectroscopy and dynamics employing ultrashort pulses; and strong-field plasma physics employing ultraintense pulses. The ultrahigh speed and ultraintense optical field that characterize the latter two methods are different from the other extreme conditions discussed previously. Each of them represents an independent physical dimension and offers an additional option for creating extreme conditions.

Nonadiabatic electron transition process and nuclear quantum state evolution generally occur at femtosecond and picosecond intervals and must be probed at ultrafast timescales. Additionally, nuclear quantum effects in equilibrium and nonequilibrium states may differ markedly. Ultrafast laser spectroscopy plays a unique role in these two aspects. Under the strong field generated by a laser, not only is the reaction barrier height decreased, but also the nuclear quantum motion is accelerated. In this case, full quantum effects become more significant. Thus, the lowered quantum potential barrier will promote quantum tunneling and delocalization of nuclei, and the accelerated atomic motion will lead to breaking and recombination of chemical bonds.

Ultrafast spectroscopy with ultrahigh time resolution has now been integrated with numerous other experimental methods, leading to the development of many new techniques. Similar to various other external field techniques, ultrafast spectroscopy represents a comprehensive interweaving and cross-integration of technologies. Presently, scientific breakthroughs have enabled the achievement of most technological milestones. Its significance extends beyond the progress of the technology itself; it also facilitates broader scientific research. These advances, in turn, impose additional demands on the development of future technologies.

There are two main development routes for ultrafast spectroscopy. One combines the ultrafast spectroscopy with other experimental techniques to extend its application. In this area, time-resolved THz ultrafast spectroscopy, time-resolved ARPES, time-resolved transmission electron microscopy, time-resolved scanning tunneling microscopy, time-resolved high-pressure ultrafast spectroscopy, time-resolved X-ray diffraction, time-resolved imaging techniques, time-resolved Raman spectroscopy, etc., have been developed based on traditional ultrafast spectroscopy such as ultrafast pump-probe spectroscopy and time-resolved emission spectroscopy. These experimental techniques are commonly used in the study of high-temperature superconductors, strongly correlated systems, topological quantum materials, magnetic materials, two-dimensional materials, multiferroic materials, nanostructures, organic molecules, energy and environmental materials, and others. Together, they have improved our understanding of material structure, optical property control, and photoelectric conversion processes; they have even contributed to the emergence of a new research field, the study of excited states [Tian Y.C.2016, Wu Q.2020]. Many important scientific questions have arisen in this new field, including the evolution path of hidden quantum states in momentum space and transient correlation effects; ultrafast spectroscopy and excited state science complement and contribute to each other in the effort to address them.

Although the aforementioned efforts have not solved all the problems of full quantum effects in condensed matter physics, such as the mechanism of high-temperature unconventional superconductivity, correlation effect of non-Landau images, confirmation of the quantum spin liquid, emergence phenomena, quantum phase transition, and the correlation effect of topological quantum states, they have already provided a clearer research direction.

This is because, first, it has opened an avenue for the study of full momentum space. Conventional research has focused on the vicinity of the Fermi surface and the region of momentum space below it, but now the region above the Fermi surface can also be explored, filling in the unknown "one-third" of momentum space. In this way, the full picture of a physical process can be revealed, including all quantum properties.

Second, it advances the study of bosons. Most studies of electricity involve the detection of fermionic electrons and holes. Occasionally, the role of bosons is determined indirectly when the study of interactions is also focused on the detection of fermionic single-particle states. However, directly detecting bosons is now possible, although such direct detection is not always complete. For example, magnetic oscillators are less well understood than phonons [Zhao J.M.2004, 2006, Kirilyuk2010], and orbital waves are even less studied. Condensed matter is composed of fermions and bosons, and studying both of these will help us better solve many puzzles.

Third, it promotes the study of coherent states. Ultrafast spectroscopy technology can transfer laser coherence to solids. With coherence modulation, the coherence can be transferred to both the lattice and the electron, which allows us to observe many novel quantum states that cannot be achieved by other techniques, such as transient superconductivity-like phenomena [Fausti2011], transient ferroelectric polarization [Nova2019], laser-induced electronic coherence [Wu Y.L.2015], and laser-induced phase transitions [Baldini2020].

Fourth, it enables the study of nonlinear interactions. The study of nonlinearity in optics is extensive and it can affect interactions between elemental excitations in condensed matter physics.

Fifth, it also facilitates the study of interactions with multiple degrees of freedom: In the past, an experimental tool could only be used to study one degree of freedom, but ultrafast spectroscopy can detect many and understand their interactions. Finally, it has promoted the study of ultrafast processes. Expanding older time-integrated detection methods to reveal the details of physical processes as they unfold in real time enables the study of chemical transition states that are closely related to full quantum effects.

The other route for the technological development of ultrafast spectroscopy is the improvement of time resolution. A wide variety of physical processes in condensed matter physics arise from the interaction of atoms and electrons in four degrees of freedom: charge, lattice, spin, and orbit. The strength and spacing of the interaction determine whether the time scale of most physical processes is femtosecond (fs), picosecond (ps), or nanosecond (ns). At present, the physical processes corresponding to each time scale are roughly understood, and each of them can be reliably separated and confirmed by ultrafast spectroscopy. In general, the interaction times and characteristic lifetimes of heavier particles are longer. For instance, electron–phonon scattering has a shorter time scale than phonon–phonon scattering, and the spin relaxation time scale with phonon coupling is much longer than the spin relaxation lifetime without phonon coupling. Since electron is the lightest particle, the time scale of electron–electron scattering is often the shortest, with a time scale of a few femtoseconds or even subfemtoseconds. For the detection of these processes, traditional femtosecond ultrafast spectroscopy is insufficient.

The recently developed attosecond laser has opened additional research possibilities. Since the ultrafast laser pulses are shorter than the physical processes described above, it enables research at the sub-femtosecond level [Ferray1988, Hentschel2001, Paul2001]. P. Agostini, F. Krausz, and A. L'Huillier won the Nobel Prize in Physics in 2023. However, a light source alone is not sufficient; further development of attosecond spectroscopy is needed to probe physical phenomena and processes on the attosecond scale. Considerable progress has been made in this direction. Conversely, the integration of ultrafast spectroscopy with other experimental methods and various external fields has produced significant contributions to attosecond research.

Figure 6.35 Photograph of the Extreme Light Infrastructure Attosecond Light Source (ELI-ALPS) in Europe.

As consecutive technological milestones are reached, this newly emerged and thriving field is expected to develop even more rapidly in the near future [Hao W.J.2021]. Figure 6.35 shows the Extreme Light Infrastructure Attosecond Light Source (ELI-ALPS) in Europe. The first large attosecond facility in China (the second in the world) is currently being developed, coordinated by Z. Y. Wei at the Institute of Physics and W. Zhao at Xi'an Institute of Optics and Mechanics, Chinese Academy of Sciences; its construction has taken place at the Songshan Lake Materials Laboratory in Guangdong.

It is worth mentioning that the attosecond ultrafast laser has also facilitated the development of time-resolved deep-ultraviolet, vacuum-ultraviolet, soft X-ray, hard X-ray, and γ-ray spectroscopy and imaging. As higher-order harmonic frequency multiplication of photons in the THz band is frequently employed to generate attosecond pulses, the attosecond laser has also indirectly promoted the development of THz spectroscopy. In another decade or two, ultrafast spectroscopy will probably be advanced enough to cover the entire spectral band, bringing physics research to a new level. In 2016, the Murnane's group used the atto-ARPES technique to investigate the effect of Ni(111) energy-band structure on the photoelectron relaxation time (shown in Figure 6.36) [Tao Z.S.2016].

Ultraintense laser-pulse and strong-field plasma physics is a subject that has developed in parallel with ultrafast spectroscopy and dynamics. Driven by the electric and magnetic fields of a transient intense light field, outer or even inner electrons within atoms, molecules, or solids can overcome the work function and be ejected. The strong electric-field component also pulls on the potential energy surface, making it easier for the electron to detach and leading to occasional multi-photon-induced transitions. At the same time, because the ultraintense laser pulse itself is a periodic light field, its electric field draws electrons back to the sample, "bombarding" it. During this process, positive and negative charges are separated and form a plasma. (Note that this is different from the plasmon formed by unescaped electrons in solids under a light field.)

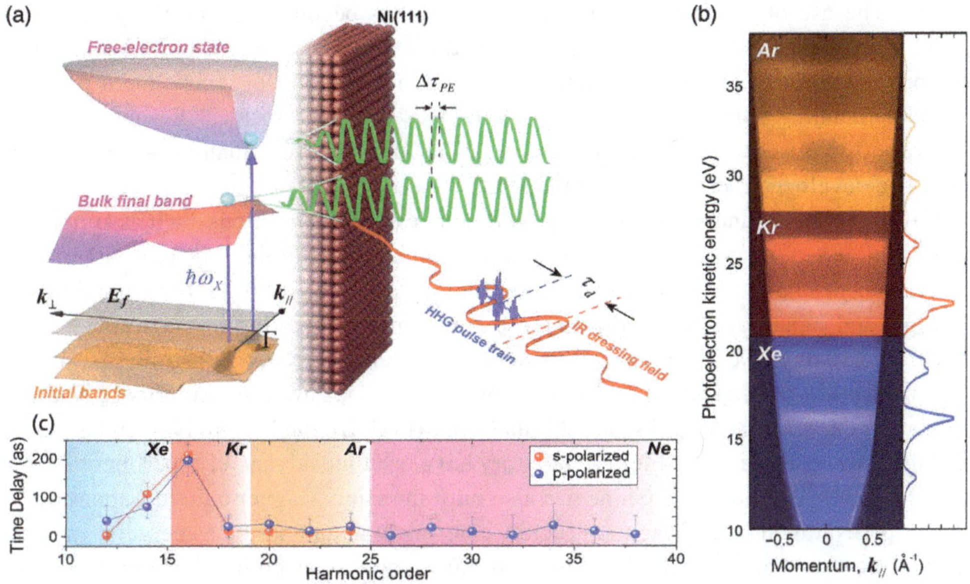

Figure 6.36 Time-resolved ARPES measurements of the effect of Ni(111) energy band structure on photoelectron relaxation. (a) Schematic of the experimental principle; (b) static ARPES excited by s-polarized higher-order harmonics; (c) optical excitation delay for different polarizations. From Tao Z. S.[2016]. Reprinted with permission from AAAS.

As the transient intensity of the ultrafast laser pulse increases, the advantages of a strong light field for the study of plasma and other branches of physics become apparent. It can be used to investigate vacuum fluctuations, laser nuclear fusion, and new particle beam sources and light sources; it can even provide a novel research platform for modern cosmology by simulating astrophysical processes at desktop scale [Liao G.Q.2019].

The field of ultraintense physics is advancing into the macroscopic realm, paralleling the way ultrafast physics is being applied in the microscopic domain. There is potential for ultrafast ultraintense laser technologies to offer robust tools for detection and modulation in the future. Integrating these technologies with conventional detection methods is poised to further expand the frontiers of fundamental and applicational scientific research of full quantum effects.

6.2.5 Synergetic Extreme Conditions

In the previous sections, we focused on the development of platforms that create extreme conditions in one variable at a time. Generating multiple extreme physical conditions at once to exploit their synergy offers even more opportunities for novel research. In fact, many scientific breakthroughs were made in the past through the concurrent use of conditions such as extremely low temperatures, ultrahigh pressures, strong magnetic fields, and ultrafast or ultraintense electromagnetic fields. Such work has led to major discoveries in physics, materials, chemistry, and life sciences and to the awarding of Nobel

Prizes. The use of synergetic extreme conditions has become a particularly important trend in frontier condensed matter physics and material science research. As shown in the previous section, some synergetic extreme conditions not only provide a platform for implementing various experimental techniques but are also ideal for achieving the control of material properties required if high-performance full-quantum-effect devices are to be developed (see Chapter 11).

Given the importance of generating synergetic extreme conditions such as simultaneous ultralow temperature, ultrahigh pressure, strong magnetic field, or ultrafast ultraintense laser irradiation, many countries and regions in the world (e.g., the United States, Europe, and Japan) are engaged in intense competition, investing considerable human, material, and financial resources in this field. Some famous research institutions, such as the National High Magnetic Field Laboratory in Florida, Lawrence Livermore National Laboratory and Los Alamos National Laboratory (LANL), the Institut Neél in Grenoble, and the European Magnetic Field Laboratory have installed advanced experimental facilities for extreme conditions. China has also built the High Magnetic Field Laboratory in Hefei, the National High Magnetic Field Center in Wuhan, and the Synergetic Extreme Condition User Facility (SECUF); the last has campuses at both the Huairou Science City/Huairou Comprehensive National Science Center in Beijing and at Jilin University in Changchun. In addition, the Attosecond Light Source, which contains a series of light sources and various time-resolved measurement facilities, is being built at Songshan Lake, Guangdong.

SECUF is an advanced user facility that can generate extreme conditions such as ultralow temperature, ultrahigh pressure, strong magnetic field, or ultrafast ultraintense light field. It is used to expand the range of the parametric space for these conditions and to carry out cutting-edge research in condensed matter physics and material science through physical property measurements, quantum state manipulations, and ultrafast time-resolved physical and chemical studies. It will promote the discovery of new states of matter, new phenomena, and even new laws of nature. The use of synergetic extreme conditions is expected to help find new high-temperature superconductors, understand the mechanism of unconventional superconductivity, develop the core technologies of quantum computing, and achieve ultrafast tracking and control of nonequilibrium physical and chemical reactions dominated by full quantum effects.

SECUF at the Huairou Comprehensive National Science Center has the following objectives: For a single extreme condition, it aims to achieve a strong magnetic field of more than 26 T generated by entirely superconducting magnets, an ultralow temperature of less than 1 mK, an ultrahigh pressure of more than 300 GPa, and an ultrafast optical field with a duration of less than 100 as. For synergetic extreme conditions, it aims to achieve a B/T (magnetic field/temperature) value of 10,000 T/K, a B·P/T (magnetic field·pressure/temperature) value of 2,800 T-GPa/K, and a P·T (pressure·temperature) value of 60,000 GPa-K, providing a variety of synergetic extreme conditions for material preparation, physical property characterization, quantum manipulation, and ultrafast dynamics studies. Figure 6.37 shows the experimental station combining oxygen molecular-beam epitaxy and pulse laser deposition for thin film growth, together with *in situ* low-temperature

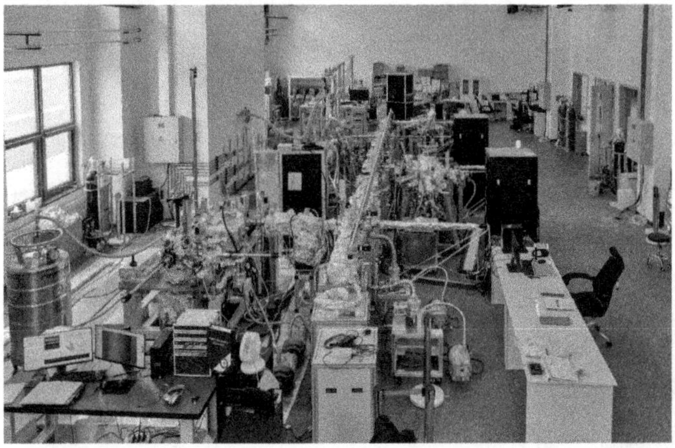

Figure 6.37 Photograph of the low-temperature scanning tunneling spectroscopy and angle-resolved photoemission spectroscopy experimental station at Huairou Comprehensive National Science Center.

scanning tunneling spectroscopy and ARPES for physical property characterization. In 2020, researchers at the Institute of Physics, Chinese Academy of Sciences and the Center for High Pressure Science & Technology Advanced Research in Beijing used the synergetic experimental condition apparatus to heat ammonia borane in a symmetric diamond anvil cell with a laser. The ammonia borane decomposed to produce hydrogen gas. By reacting this with a La flake, $LaH_{10+\delta}$ was synthesized at 165 GPa and 1,700 K. In this material, they observed hydrogen symmetrization induced by quantum tunneling. The superconducting transition at $T_c \approx$ 240–250 K was confirmed, providing one of the few internationally recognized pieces of experimental evidence for near-room-temperature superconductivity (Figure 6.38) [Hong F.2020].

Led by L. Lu at the Institute of Physics, Chinese Academy of Sciences, SECUF was designed to have four experimental platforms: the Extreme Condition Physical Property Measurement Platform; the Extreme Condition Quantum Manipulation Platform; the Ultrafast Dynamic and Imaging Platform at Huairou Science City; and the High-Temperature High-Pressure Bulk Material Research Platform at Jilin University. These platforms consisted of 22 experimental stations in total.

The Extreme Condition Physical Property Measurement Platform is primarily utilized for characterizing and studying the physical properties of materials, encompassing full quantum effects under ultralow temperatures, strong magnetic fields, and ultrahigh-pressure conditions. It furnishes experimental evidence crucial for addressing key physical challenges in condensed matter physics and material science research, such as comprehending the mechanism of high-temperature superconductivity and exploring room-temperature superconductors and new topological quantum materials. The platform consists of nine experimental stations: a diamond-anvil-cell station and a cubic-anvil station, both for ultralow temperature, ultrahigh pressure physical property measurements; an ultralow temperature high-magnetic-field quantum oscillation experimental station;

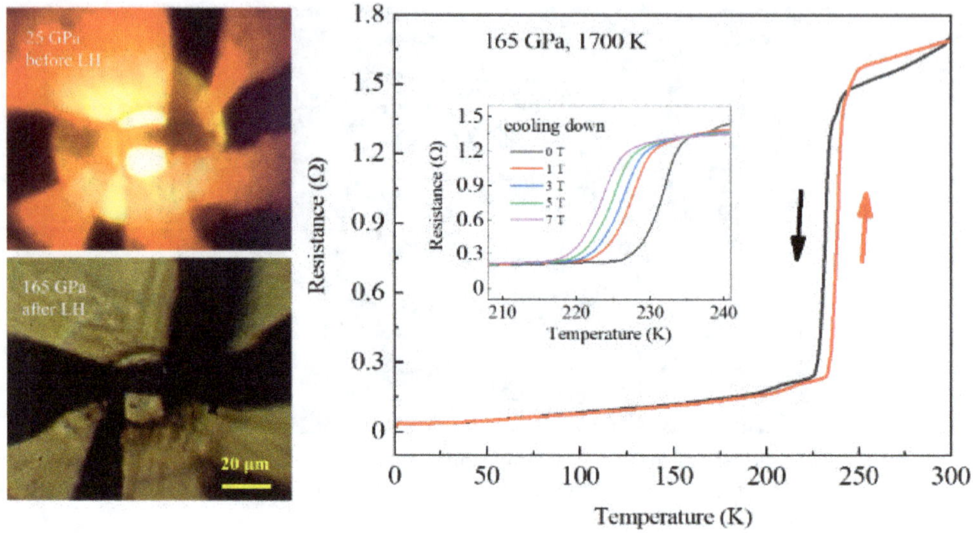

Figure 6.38 Near-room-temperature superconductivity observed in $LaH_{10+\delta}$ under high pressure, with a transition temperature $T_c \approx 240–250$ K. From Hong F.[2020].

a station for IR and terahertz spectroscopy under extreme conditions; a station for Raman-spectroscopy under extreme conditions; a high-magnetic-field NMR experimental station; an ultralow-temperature, high-magnetic-field scanning-tunneling-microscopy station; a low-temperature *in situ* scanning-tunneling and ARPES experimental station; and a high-pressure synergetic measurement station.

The Extreme Condition Quantum Manipulation Platform is primarily employed to manipulate quantum processes under various extreme conditions, such as ultralow temperatures and high magnetic fields. Its objective is to surmount the limitations of classical manipulation in superconducting quantum computing, topological quantum computing, nanoelectronics, and others, thereby establishing a new quantum form of manipulation and facilitating the development of innovative quantum devices. The platform comprises four experimental stations: a sub-millikelvin experimental station; an ultralow-temperature experimental station for superconducting quantum devices; an ultralow-temperature, high-magnetic-field quantum transport and manipulation experimental station; and a low-temperature, high-magnetic-field electron transport and spectroscopic measurement station.

The Ultrafast Dynamic and Imaging Platform is an ultrafast scientific research platform that incorporates a wide range of ultrafast radiation sources with final optics for the measurement of dynamical behavior in atoms, molecules, solids, and surfaces. It is used to study ultrafast physical and chemical processes such as chemical reactions, catalytic processes, photosynthesis, and photovoltaic reactions, which are closely related to nonadiabatic effects. This platform contains five experimental stations: a femtosecond-dynamics experimental station; an attosecond-dynamics experimental station; an ultrafast X-ray dynamics experimental station; an ultrafast electron microscopy station; and an ultrafast electron diffraction experimental station.

The High-Temperature High-Pressure Bulk Material Research Platform is mainly used for the synthesis and characterization of bulk materials under high temperature and pressure. The platform includes three experimental stations: an experimental station for extreme-condition measurements of solid environments at high temperature and high pressure; an experimental station for extreme-condition measurements of liquid environments at high temperature and high pressure; and an experimental station for nonequilibrium extreme-condition measurements at high pressure.

6.3 Perspectives on Full Quantum Experiments

Modern experimental techniques have revealed many full quantum effects on the physical and chemical properties of condensed matter. However, these techniques have their limitations. For example, scanning probe microscopy and transmission electron microscopy, despite their high spatial resolution, have several shortcomings. Compared with other spectroscopic methods, scanning probe microscopy has only a limited ability to identify chemical composition. It is also much less sensitive than IR and Raman spectroscopy for detecting characteristic molecular vibrations. In transmission electron microscopy, the high-energy electrons tend to destroy samples; hence, the application of this technique to the study of light-element and weakly interacting systems is extremely restricted. Limited by the circuit bandwidth, the time resolution of both is on the nanosecond or near-picosecond scale, whereas the quantum motion and quantum state evolution of electrons and nuclei typically occur at ultrafast time scales (at the femtosecond or even attosecond level). As a result, capturing these dynamic processes experimentally is quite challenging.

The study of condensed matter systems has revealed their increasingly diverse and complex physical and chemical properties. However, sub-molecular-scale imaging through scanning-probe techniques has primarily been conducted in ultrahigh-vacuum and low-temperature environments. It is still challenging to extend this methodology to atmospheric environments at room temperature, or even to solutions. Furthermore, experimental techniques face hurdles when applied to interfacial systems, especially those involving electrochemical solid–liquid interfaces. These systems are strongly affected by environment, morphology, the presence of local electrolytes, and atomic, molecular, and electronic structure. It is not feasible to investigate all microscopic details of an interface using a single experimental technique. To comprehensively understand solid–liquid interfaces and other intricate systems, analytical methods must be developed that integrate the most advantageous traits of several techniques, including X-ray diffraction, photoemission spectroscopy, sum frequency vibrational spectroscopy, and scanning probe microscopy (as depicted in Figure 6.39). By enhancing experimental methods, researchers can acquire more extensive and powerful tools for observing full quantum effects in interfaces at the microscopic scale.

TERS exploits scanning-probe technology's ultrahigh spatial resolution and Raman spectroscopy's ability to identify molecular chemical groups. In 2013, J. G. Hou and Z. C. Dong demonstrated single-molecule Raman imaging at the sub-nanometer level for the first time. The probe technique, capable of sensitive chemical recognition, was optimized to have a spatial resolution below one nanometer (~5 Å) [Zhang R.2013]. By improving

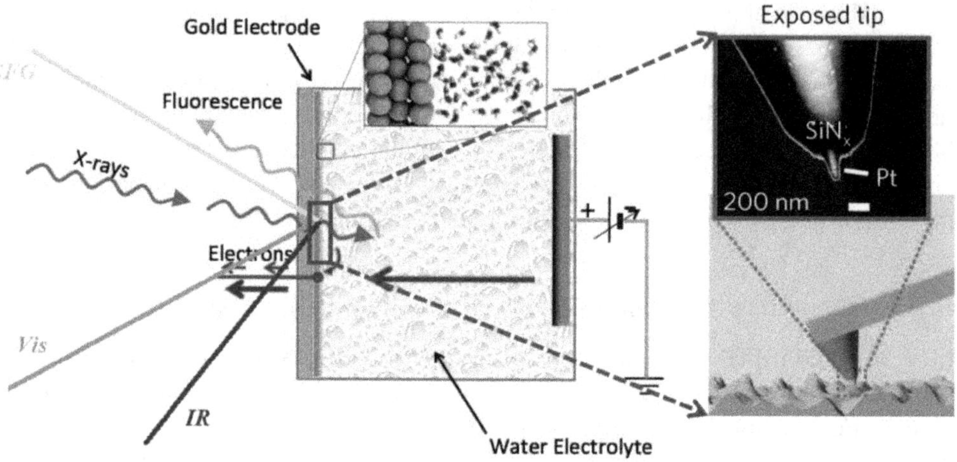

Figure 6.39 Schematic of an experimental study of an electrochemical interface.

their cryogenic (liquid helium) ultrahigh-vacuum TERS system and fine tuning the highly localized plasmon field at the tip, they eventually reached spatial resolutions on the level of a single chemical bond (~1 Å) and completely imaged various intrinsic molecular vibrational modes in real space [Zhang Y.2019].

K. H. Wu's group at the Institute of Physics, Chinese Academy of Sciences, systematically investigated the local vibrational spectra of two-dimensional materials using low-temperature STM–tip-enhanced Raman spectroscopic apparatus of their own design. A Raman spatial resolution of 0.5 nm and Raman signal enhancement of 10^9 were obtained in a monolayer silicene sheet [Sheng S.X.2017]. Raman spectroscopy measurement of a single quantum Si_{13} cluster was also carried out. Combining TERS with the high spatial resolution of the scanning probe technique allowed the Raman spectral frequency shifts caused by weak bond-length changes in localized surface structures to be detected under local stress [Sheng S.X.2018].

Z.Q. Tian's group at Xiamen University also employed surface-enhanced Raman spectroscopy to study more complex electrochemical reaction systems and electrochemical interfaces of water, successfully identifying reactive sites at the sub-nanometer scale [Zhong J.H.2017]. In addition, they detected the Raman signals of interfacial water on a single-crystal gold electrode surface and observed *in situ* transitions between two interfacial water configurations during the hydrogen evolution reaction. In other words, as the interfacial water shifted toward negative potentials, the molecules changed from a structurally "parallel" configuration to first a "one-H-down" and then a "two-H-down" configuration [Li C.Y.2019]. This experimental method will help uncover the mechanism behind interfacial reactions and the catalytic activity associated with the interfacial configuration, which is crucial to the study of electrochemical and energy-related interface problems. In the future, TERS based on the scanning probe method will also contribute to research on nuclear quantum and nonadiabatic effects in more complex light-element molecules and condensed matter systems.

Currently, the investigation of atomic-level full quantum effects mainly focuses on simple model systems in ultrahigh vacuum at low temperature. High-resolution characterization in atmospheric room-temperature environments poses a greater challenge. The nitrogen vacancy (NV) color center of diamond is a solid-state quantum probe at the atomic scale with stable quantum states prone to coherent manipulation [Mamin2013]. An NV probe can explore samples non-invasively; it is compatible with a wide range of systems, including biological systems, and is sensitive enough to detect the magnetic field generated by a single nuclear spin. By utilizing diamond color centers, J. F. Du's group at the University of Science and Technology of China developed quantum magnetic-sensing technology with high sensitivity and spatial resolution [Ding Z.2020]. If NV probes are combined with STM techniques, it will be possible to detect nuclear spin at the atomic/molecular level and carry out high-sensitivity and high-resolution experimental studies of full quantum effects in atmospheric and solution environments.

As previously mentioned, the time scale of atomic motion is on the order of picoseconds or femtoseconds, while that of electron motion is on the order of attoseconds. Most dynamical processes in material systems are due to electron motion. For example, the collective motion of electrons in nonadiabatic processes results in plasmons with nanoscale oscillation on a time scale of a few hundred attoseconds and a relaxation time ranging from a few to tens of femtoseconds. Therefore, experimental tools with ultrafast time resolution are urgently required for understanding the dynamical behavior of electrons.

Combining conventional detection methods with ultrafast laser technology, especially for the characterization of nonadiabatic processes in nonequilibrium states, is currently an important research direction in experimental full quantum physics. In general, electrons, lattices, orbitals, and spins are coupled together in the equilibrium state, and it is difficult to distinguish their effects on material properties or phase transitions. Using time-resolved methods to differentiate these effects has been a central goal of ultrafast science. However, the change in electronic state occurs at the femtosecond scale, and the change in the lattice, which is affected by phonon frequency, occurs at the scale of 100 femtoseconds. Currently available methods, such as ultrafast X-ray diffraction and ultrafast electron diffraction, have time resolutions of only ~300 femtoseconds. Therefore, distinguishing the contributions of electrons and lattices in the time domain and probing them separately still require significant effort. Regarding this, improving the spatial resolution of this method also remains a significant challenge.

Because of circuit bandwidth limitations, the time resolution of conventional scanning-probe methods is typically limited to nanoseconds or even near picoseconds. One of the more feasible approaches for improving it involves combining the pump-probe ultrafast laser technique (shown in Figure 6.40) with the scanning-probe technique [Terada2010]. This method can track the dynamical evolution of a single quantum state in real time, achieving atomic-level spatial and femtosecond-level temporal resolutions.

In addition to ultrahigh-resolution detection methods, researchers utilize different external field conditions in their studies. Various large-scale scientific facilities offer extreme-condition platforms that overcome the physical limitations of traditional detection methods and can reveal the full quantum effects hidden in the details of the

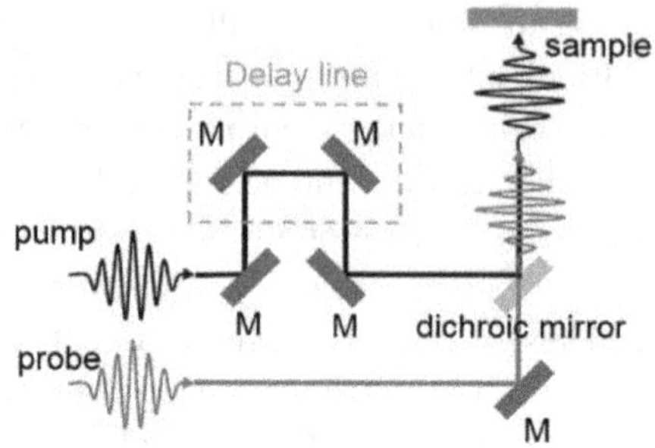

Figure 6.40 Circuit diagram of the pump-probe method using an ultrafast pulsed laser. It can be combined with a variety of other experimental tools to achieve ultrafast time-resolved physical detection.

observations. For example, next-generation attosecond light sources will allow the study of the quantum-mechanical laws associated with electron motion in real time, and will help in exploring the entire process of interactions and effects of nuclear and electronic quantum states. Combining an attosecond laser system with a TEM provides a new way to conduct in-depth study of full quantum dynamical processes in electronic, atomic, molecular, and functional systems, as well as a novel approach to precisely manipulating electron, spin, and lattice motions within condensed matter (Figure 6.41). By using the latest electron pulse compression method, an ultrashort electron beam with reduced timing jitter can be obtained. Combined with a periodically pulsed laser, it is possible to solve the key technical issues that limit the time resolution of ultrafast electron diffraction, and to improve the resolution to a record-breaking scale of tens of femtoseconds. The time resolution of ultrafast electron diffraction based on an attosecond laser beam is expected to reach this scale, which is an order of magnitude better than that of conventional ultrafast electron diffraction (with a time resolution ~300 femtoseconds). This method will therefore be applicable to a wider variety of research targets; the spatial detection range extends from nanometers to mesoscopic scale, and the dynamical time detection range can reach from tens of femtoseconds to nanoseconds. In this way, studies can be carried out for both equilibrium and nonequilibrium systems. With cutting-edge ultrafast transmission electron microscopy, it is possible to conduct a large number of microscopic ultrafast dynamics experiments related to full quantum phenomena such as zero-point correction of atomic vibrations and molecular rotations, nonadiabatic processes in chemical bond breaking and recombination, superconductivity related to electron–phonon coupling in solids, and optically induced quantum phase transitions. Many of these are frontier research problems for full quantum effects in condensed matter physics, chemistry, materials science, and other fields.

Combining the ultrafast ARPES and PEEM methods with an attosecond laser enables direct observation of electron transition and ionization, the dynamics of the Auger effect, electron tunneling in atoms, charge transport, valence electron motion, and other physical phenomena. This combined method can also be used to probe the time evolution of atomic/molecular photoionization, the formation and evolution of nanoscale plasmon

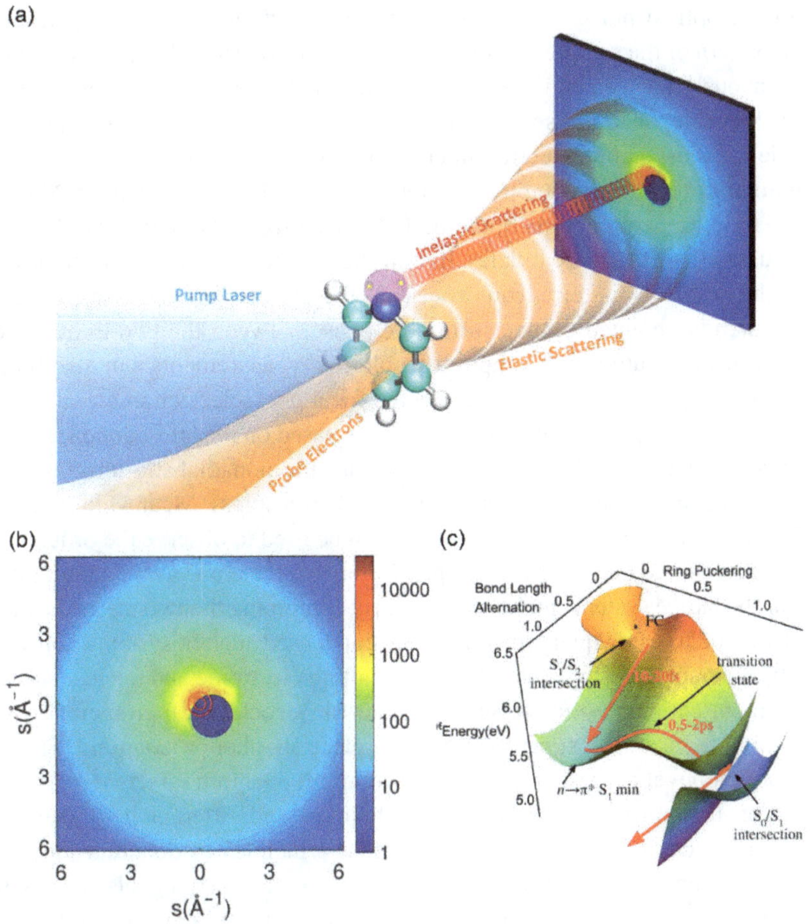

Figure 6.41 Experimental method based on ultrafast electron diffraction allows for direct tracking of atomic structure and electronic state changes during molecular reactions, thus revealing the entire dynamical processes of nuclear quantum effects and nonadiabatic effects in chemical processes. From Yang J.[2020]. Reprinted with permission from AAAS.

fields, intervalley electron scattering in low-dimensional materials, electron dynamics of high-temperature superconductors, chemical bond formation and breakage, and many-body quantum kinetics in catalytic reactions. In biology, it can be applied to the study of protein formation and decomposition, biochronometry, and photosynthesis. This new experimental method is even capable of manipulating electron transitions in nanoscale plasmons and low-dimensional semiconductors, which is crucial to the design of high-sensitive quantum devices, and addressing some of the core problems of full quantum condensed matter physics.

An attosecond ultrafast PEEM measurement system with high real-space resolution can adopt a dual-beam pump-probe scheme, where the attosecond probe light is focused with a toroidal lens to achieve imaging from the monochromator slit to the sample. The pump light is focused with a lens, and a delayed optical path is set up. Spatiotemporal overlap of the optical pulses can be achieved by adjusting this path and observing the coherence pattern of the sample. Further, as long as the attosecond and femtosecond lasers are simultaneously focused on the sample, spatial and temporal overlap of the two optical pulses is achievable, enabling spatially resolved and time-resolved measurements using a magnetic-lens-focused photoelectron microscope. In the reciprocal-space ultrafast PEEM system with high resolution, pump detection is achieved by introducing a Mach–Zendel interferometer into the optical path; the time delay between the pump and detection light can be adjusted with the optical delay line in the interferometer. The accuracy is determined by the adjustment step of one fourth of the optical carrier pulse period, and the adjustment step is obtained by measuring the phase between the two light pulses. High reciprocal-space-resolution and temporal-resolution measurements can also be achieved using an electron lens and aberration-corrected focal photoelectron microscope.

Ultrafast spectroscopy and related technologies based on the attosecond laser can probe the electronic structure of the sample and capture the nonadiabatic ultrafast dynamical processes of the electron system (including the electron spins). Meanwhile, ultrafast electron diffraction and ultrafast X-ray diffraction can be used to observe the ordered state and quantum motion in the lattice of the sample. In this way, the ultrafast quantum dynamical behavior of the lattice system can be observed. By combining the two methods, the dynamical processes of the full quantum effects in condensed matter systems can be studied from multiple perspectives by coupling or decoupling electronic and nuclear states. This is extremely important in the study of such complex physicochemical problems as molecular reactions, high-temperature superconductivity, electron–phonon coupling, charge order, and other atomic-level processes are dependent on full quantum effects.

In addition to improved measurement techniques and external field facilities, clear experimental protocols based on the fundamental principles of quantum mechanics and precise sample preparation are necessary to achieve these research goals. This will require advances in atomic-scale sample preparation techniques, such as molecular beam epitaxial growth and *in situ* characterization. The development of microfabrication techniques is also crucial for quantum physical property measurements and device research.

Humanity's insatiable curiosity about the mysteries in nature has led to the use of widely varying experimental techniques and external field conditions, and each method is being pushed toward its limit. Experimental physicists have become accustomed to carry out the designed research protocols using various perfect samples on extreme-condition platforms in these external-field environments (e.g., ultralow temperature, ultrahigh pressure, high magnetic fields, and ultrafast, ultraintense light fields) by skillfully utilizing these advanced single or integrated experimental techniques with extreme detection sensitivity. Different methods are being combined with various external field extreme conditions to form an effective interweaving of technological network. This allows integrated experimental scenarios to be better simulated by theorists. These experimental and theoretical studies will undoubtedly help us to fully reveal the role of full quantum effects in condensed matter, especially for light element systems. Further advances in basic research using the full quantum effect as a new parameter will allow us to develop and manipulate novel quantum materials and states and to design new full quantum devices.

After introducing the main theoretical and experimental approaches to full quantum effects in Chapters 4–6, we would like to return to the original purpose of this book, which can be broadly summarized in two aspects. First, we hope that this book will provide the reader with a well-rounded view and a clear foundation for further studies of complex condensed matter systems, particularly when the physical properties cannot be described by using the ball-and-stick model of adiabatic electronic states. In this way, readers can choose the correct theoretical and experimental tools at each step when solving a research problem. Second, we hope that this book will contribute to the development of this newly emerging field and steer discussions in the right direction. Most of the problems in this field can be described as "many-body" and "quantum" in nature, reflecting the core value of condensed matter physics; in P. W. Anderson's words, "More is different!" (Anderson1972). We aspire that, in the future, the readers of this book will remember it as an early comprehensive review that contributed to the establishment of the full quantum physics of condensed matter. Currently, we aim to spark interest in this research field and encourage readers to participate in the study of full quantum effects in condensed matter physics.

After studying the theoretical and experimental approaches to full quantum effects, we will next use several typical element (or comparatively light element) systems as examples. The following chapters will provide a detailed discussion on how to discover and understand new phenomena and properties induced by full quantum effects in some important condensed matter systems.

7

Typical Full-Quantum-Effect System

Hydrogen

As explained in the earlier part of this book, full quantum effects are more pronounced in condensed matter systems composed of light elements, especially those at the top of the periodic table. Therefore, we choose these elements as typical systems of the study of full quantum effect and give a systematic discussion for each element. When two or more light elements are present in the same compound, the situation may become even more challenging, as each element may exhibit a different degree of full quantum behavior. Such complications are encountered in studying the formation and diffusion of hydrogen on the surfaces of carbon materials, which will be discussed in subsection 9.2.2 (containing H and C elements) and the diffusion of water clusters on metal surfaces, which will be discussed in subsection 10.2.1 (containing H and O elements as well as metal substrate). In such situations, one light element typically manifests more dramatic full quantum effects than the others. We will therefore place our discussion of such cases in the section focusing on the dominant light element. For example, in this chapter, which focuses on full quantum effects in hydrogen element, we will introduce the hydrogen-bonding structure of water, the concerted quantum tunneling of protons in water cluster, superconductivity in hydrogen-rich compound, the problem of proton transport in inorganic and organic perovskites, and spontaneous mutation of genes in living organisms related to proton tunneling.

Hydrogen is the most abundant element in the universe. Hydrogen atoms are ~100 times more abundant than all other types of atoms summed; consequently, hydrogen accounts for more than 75% of the total weight of all elements in the universe, even though it has the smallest atomic size and lowest atomic mass. Similarly, dense hydrogen systems are the lightest materials in terms of molecular weight. Because of their low atomic weights, dense hydrogen systems (or pure dense hydrogen materials) and hydride compounds (in which the majority of atoms are hydrogen) are widely used for studying full quantum effects.

In physics research, the dense hydrogen system is a particularly representative material, although experimental studies of it remain highly controversial [Mao H.K.1994]. Theory predicts that condensed hydrogen under high pressures exhibits novel phase transitions and rich physical properties such as metallization, superconductivity, and superfluidity, all of which are closely related to full quantum effects [Monacelli2023]. In retrospect, the study of hydrogen essentially reflects the brief history of modern condensed matter physics: since the advent of quantum mechanics, people have gradually moved into the realm of full quantum research, starting from the simplest element in the periodic table. Our understanding of the problems in condensed matter physics is constantly improving as the research progresses.

Hydride compounds have also recently attracted considerable research interest. As the most abundant hydride compound on Earth, water plays an irreplaceable role in various human activities. Indeed, the human body is mostly composed of water. Water is typically used to study hydrogen bonding interactions because the quantum tunneling and quantum fluctuations of hydrogen nuclei in water reduce the influence of the classical potential barrier for proton transfer. This changes not only the interaction strength but also the symmetry and network configuration of hydrogen bonds, affecting macroscopic properties and causing water to exhibit anomalous characteristics. Studying full quantum effects in water not only helps uncover how the microscopic hydrogen-bonding structure leads to these anomalous properties but also opens new research directions for studying other condensed matter systems composed of hydrogen bonds (including soft matter and living systems). Recent studies on sulfur hydride, lanthanum hydride, and other hydrides have also led to the discovery of many new physical phenomena, such as high-pressure superconductivity near or possibly above room temperature. The superconducting transition temperatures and pressures of these materials show significant isotope dependence, which clearly indicates that quantum delocalization of hydrogen nuclei and change of electron–phonon quantum coupling affect the superconducting phase transition.

Although this chapter will focus mainly on full quantum effects in dense hydrogen and hydride compounds, it is known that the study of proton (H^+) transport and diffusion is of considerable importance in material science and life science. Therefore, we will also discuss proton transport in systems, where hydrogen does not dominate in content. In particular, we will cover some conditions where hydrogen exists as a minority medium, like proton transport through 2D materials and proton diffusion in bulk materials, which are both closely associated with full quantum effects.

7.1 Nuclear Quantum Effects in Dense Hydrogen

High-pressure dense hydrogen exhibits novel phase transitions and physical properties (such as metallization, superconductivity, and superfluidity) that are closely related to full quantum effects. In Chapter 2, we discussed the phase transition of molecular solid hydrogen. The hydrogen phase diagram is the "holy grail" of high-pressure condensed matter physics. Here, we do not intend to review the entire history of hydrogen research but rather to discuss the progress related to nuclear quantum effects that has taken place in the last three decades. Because of experimental limitations, some topics related to dense hydrogen remain controversial. Nevertheless, existing experimental studies have gradually extended the hydrogen phase diagram to pressures higher than 300 GPa [Mao H.K.1994], and theorists have made predictions far beyond this pressure range [Guo J.Q.2022]. The discussion presented in this section will focus on these recent findings.

One notable example of the phenomena connected with dense hydrogen is the insulator–metal phase transition predicted by E. Wigner and H. B. Huntington in 1935 [Wigner1935]. Structural transitions directly affect the physical properties of dense hydrogen. The boundaries and relative stability of different structures in the phase diagram are extremely sensitive to the free energy of each structure. Many of these competing structures differ in enthalpy by less than 1 meV per atom across a broad range of pressures [Monacelli2023]. Because the mass of a hydrogen atom is small, quantum fluctuations and quantum tunneling occur simultaneously, affecting both the phase stability and phase boundaries [Pickard2007, Drummond2015, Monacelli2023]. This quantum behavior is especially prominent when the temperature is extremely low. Thus, Ashcroft et al. at Cornell University predicted that hydrogen, similar to helium, will form a molecular hydrogen superfluid at low temperatures under nuclear quantum effects [Ashcroft2000, Babaev2004].

In addition, quantum interference phenomena are common in chemical reactions. The simplest example of this would be the atom–molecule collision reactions between hydrogen and deuterium. Quantum interference induced by nuclear quantum effects has been observed in several experimental studies on such reactions [Xie Y.2020].

Hydrogen is the only element whose isotopes are still given different names: D denotes deuterium (hydrogen-2 or ^{2}H) and T denotes tritium (hydrogen-3 or ^{3}H). Since "P" represents phosphorus, it is no longer used as a symbol for protium (hydrogen-1 or ^{1}H), and protium is simply written as H. There are also several unstable isotopes with very short half-lives (e.g., hydrogen-4), but they are not the focus of this book. Substituting one isotope for another (such as D for H) can lead to different physical phenomena, known as the isotope effect (IE). In many cases, isotopic substitution is a direct and effective way to probe nuclear quantum effects. For example, the isotope effect can be used to control chemical reaction rates, as we have discussed in Chapter 3.

Before describing nuclear quantum effects on the physical properties of dense hydrogen, let us briefly review the structure and phase diagram of hydrogen. At room temperature and atmospheric pressure, hydrogen is a molecular gas. As the pressure increases, the hydrogen molecules combine to form a molecular solid. When the pressure is further increased, the individual molecules gradually decompose to form atomic solid hydrogen [McMahon2011, Liu H.Y.2012, Labet2012a, b]. The existence of atomic solid hydrogen is a theoretical prediction and the pressure and temperature required for its formation have not been realized experimentally. As the temperature increases, the molecular solid hydrogen and the atomic solid hydrogen melt into molecular liquid hydrogen and atomic liquid hydrogen, respectively. Figure 7.1 shows the four phases of dense hydrogen: molecular solid, molecular liquid, atomic solid, and atomic liquid.

Although there is no direct experimental evidence, many theoretical calculations have shown that the atomic and electronic structural phase transitions are correlated to some extent [Babaev2004, Morales2010, McMahon2011]. For instance, it was demonstrated computationally that, when the molecular solid and the molecular-liquid hydrogen are completely or partly transformed into the atomic-solid and the atomic-liquid forms, the hydrogen becomes metallic. Therefore, it is crucial to study the nuclear quantum effects

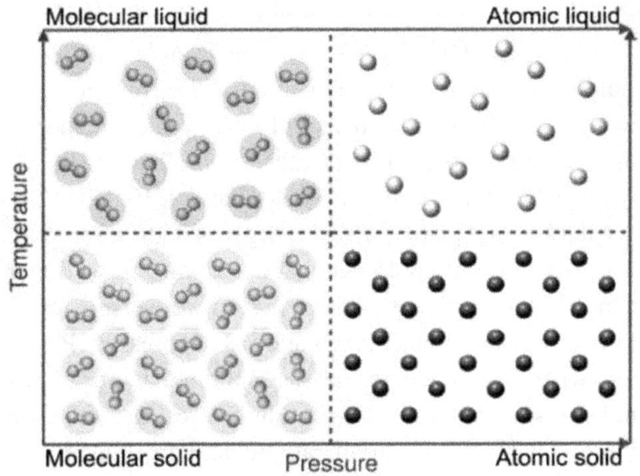

Figure 7.1 Four main regions of the condensed-hydrogen phase diagram: molecular solid, molecular liquid, atomic solid, and atomic liquid. As the pressure increases, molecular hydrogen becomes atomic hydrogen. As the temperature rises, the solid phases become liquid phases. Reprinted from Fang[2019], reprinted by permission of the publisher (Taylor & Francis Ltd, www.tandfonline.com).

on the phase diagram of dense hydrogen. In this section, we will organize our discussion in terms of experimental observations and computational simulations. We will also briefly review how the understanding of nuclear quantum effects in different regions of the condensed-hydrogen phase diagram has evolved over the past three decades.

7.1.1 High-Pressure Phase Diagram

Most experimental studies of high-pressure dense hydrogen employ diamond anvil cells. This technology advanced considerably in the 1980s and early 1990s. As described earlier, the pressures achievable in the laboratory can now reach 900 GPa [Dubrovinsky2022]. A variety of experimental characterization methods, including Raman spectroscopy, IR spectroscopy, XRD, and neutron scattering, can be integrated into the high-pressure technique. Over the past three decades, several groups have applied these methods in experimental studies of dense hydrogen, including I. F. Silvera's group at Harvard University [Silvera1981, Cui L.J.1994], H. K. Mao's group and R. J. Hemley's group at Carnegie Mellon University [Mao H.K.1994, Goncharov1995, Mazin1997], and E. Gregoryanz's group at the University of Edinburgh [Liu X.D.2017]. In these experiments, nuclear quantum effects were probed using H/D isotope substitution. The most direct manifestation of nuclear quantum effects is the change in the position of boundaries in the phase diagram, which can be directly observed by spectroscopic methods; in the molecular solid regime of phase diagram, the phase transitions between different structures are indicated by discontinuous changes of the hydrogen vibrational modes. Internationally, considerable effort has gone into the study of nuclear quantum effects in high-pressure condensed hydrogen.

7.1.1.1 Phases I, II, and III of Molecular Solid Hydrogen

When the pressure increases and hydrogen transforms from gas to solid, the solid hydrogen still exists in the molecular form, with a variety of possible arrangements. Four structures of molecular solid hydrogen – phases I, II, III, and IV – have been experimentally confirmed thus far [Mao H. K.1994, Eremets2011, Howie2012a, b, Zha C.S.2012]. Figure 7.2 shows the experimentally measured boundaries between phases I, II, and III, as well as the triple coexistence point of the I–II–III phases. Combining this with information from Figure 2.2, it is possible to infer that (for both hydrogen and deuterium) phase I is orientationally disordered where the molecules can freely rotate. Phase II, conversely, corresponds to the orientationally ordered case where molecular rotation is suppressed. Phase III can be simply described as the case where the molecules possess a limited degree of orientational order in a certain direction. H/D isotope-substitution results suggest that the changes in the boundaries between the three phases reflect significant nuclear quantum effects. However, these effects manifest themselves in completely different ways at different phase boundaries.

If we focus on the transition between phases I and II along the isotherms, we can observe that the transition pressure of molecular solid deuterium is much lower than that of hydrogen; the difference is tens of gigapascals. If we look at the same phase transition along the isobars, the transition temperature of molecular solid hydrogen is lower than that of deuterium by tens of kelvins. By contrast, the I–III and II–III phase boundaries are insensitive to the temperature change, and the phase transition pressure of dense deuterium is higher than that of dense hydrogen. The most direct reason for this is that the mass of the deuterium

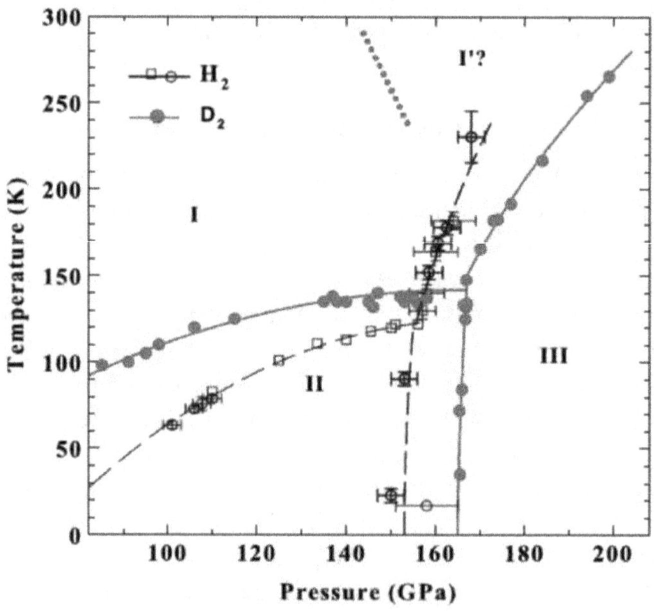

Figure 7.2 Experimentally measured phase diagrams of molecular solid hydrogen (hollow light gray) and deuterium (filled deep gray), showing phases I, II, and III. Reprinted from Goncharov[2011], with the permission of AIP Publishing.

atom is twice that of the hydrogen atom, leading to much weaker nuclear quantum effects. In general, the phase diagram suggests that nuclear quantum effects make phase II molecular solid hydrogen less stable than the corresponding phase of molecular solid deuterium, so that it occupies less space in the phase diagram than deuterium.

The nuclear quantum effects on the I–II phase transition are easy to understand intuitively because the molecules in phase I are rotating freely, whereas those in phase II have a fixed orientation. This strongly suggests that nuclear quantum effects make the molecular rotation easier in phase II. Nevertheless, it is difficult to understand all the transitions among phases I, II, and III using such a simple description. Phase III molecules are also oriented; hence, it is difficult to reveal the mechanism of nuclear quantum effects in phases II and III experimentally. In addition, the transition from phase I to phase III seems to exhibit anomalous features, for example, the phase boundary for deuterium is in the region with higher temperature and pressure. We will discuss the empirical aspects of these phenomena first; detailed simulation results of nuclear quantum effects on molecular rotation of dense hydrogen will be discussed in subsection 7.1.2.

Recently, H. K. Mao's group at the Center for High Pressure Science & Technology Advanced Research in Beijing investigated the high-pressure phase diagrams and structures of various mixtures of hydrogen isotopes. They found that mixing hydrogen and deuterium at different concentrations resulted in the formation of a series of molecular solid solutions at high pressures and low temperatures [Liu X.D.2020], as shown in Figure 7.3. Although the hydrogen–deuterium solid solutions undergo high-pressure structural phase transitions similar to those of pure hydrogen, their phase-transition pressures are significantly higher, revealing unique nuclear quantum effects in dense hydrogen.

7.1.1.2 Molecular Solid Hydrogen: Phase IV

In recent years, as higher pressures have been attained experimentally, a new molecular solid phase of hydrogen – phase IV – has been discovered, which forms at room temperature only when the pressure is high. Its structure consists of alternating layers of orientationally disordered hydrogen molecules and graphene-like hydrogen atoms (as shown in Figure 7.4) [Pickard2007, Eremets2011, Howie2012a]. By carefully comparing Raman spectra taken during the III–IV transition, R.T. Howie et al. predicted the presence of significant nuclear quantum tunneling in phase IV molecular solid hydrogen at room temperature.

Figure 7.5 illustrates the Raman peak positions and the full width at half maximum (FWHM) of the intramolecular vibrational mode ν_1 of hydrogen molecules. The discontinuous change of the Raman peaks in the left panel of Figure 7.5 represents the III–IV transition. The FWHM in the right panel of Figure 7.5 increases substantially after the phase transition. In an ordered molecular solid phase, hydrogen molecules can break and recombine. If every three hydrogen molecules constituting a six-membered ring in phase IV simultaneously break and recombine, the resulting crystal structure is completely equivalent to the original one; thus, there are two degenerate states. Their interconversion is significantly accelerated by nuclear quantum tunneling. In addition, the six-membered ring of the phase IV is more similar to the perfect atomic six-membered ring than other

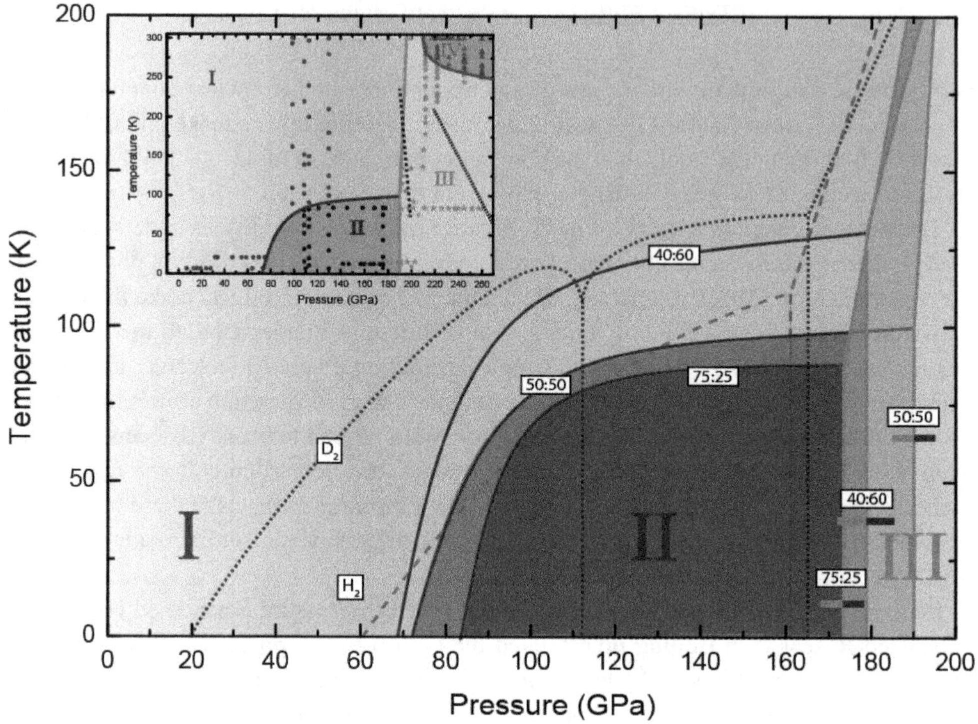

Figure 7.3 High-pressure phase diagrams of hydrogen–deuterium solid solutions with different concentrations. Reprinted from Liu X.D.[2020].

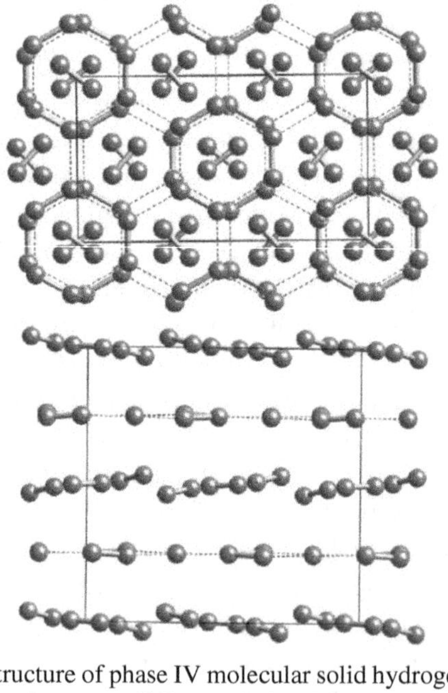

Figure 7.4 Proposed structure of phase IV molecular solid hydrogen. The upper and lower panels show the structure from two different angles. Phase IV is composed of alternating layers of hydrogen molecules and quasi-atoms.

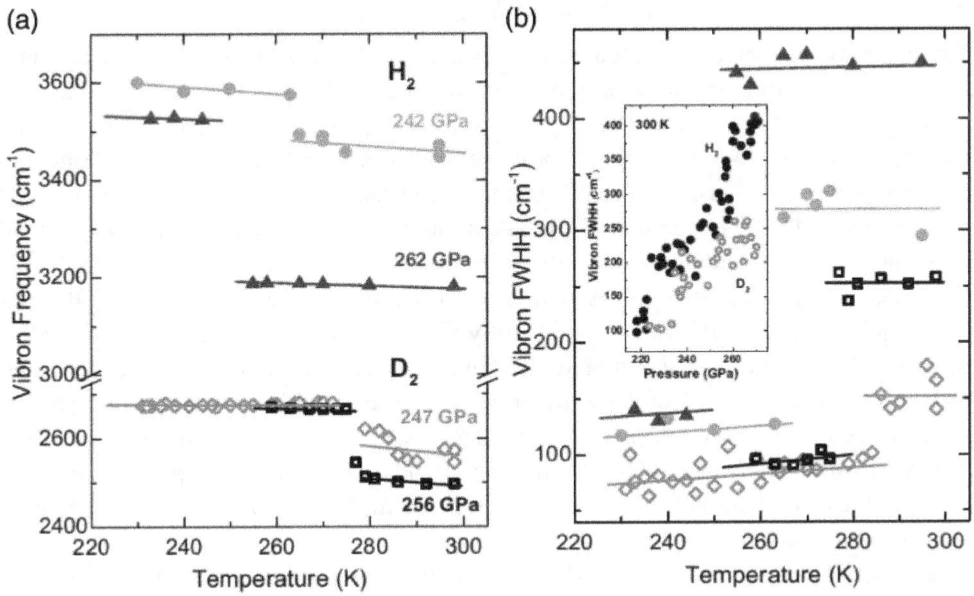

Figure 7.5 Raman spectral measurements of phase IV molecular solid hydrogen and deuterium. The peak position and FWHM of the ν_1 Raman vibrational mode are shown on (a) and (b), respectively. The inset on the right panel shows the pressure dependence of the FWHM at a temperature of 300 K. Reprinted figure with permission from Howie[2012b]. Copyright 2024 by the American Physical Society.

structures. Therefore, the potential barriers and displacements that need to be overcome are relatively small, which leads to more pronounced nuclear quantum tunneling.

Nuclear quantum effects are also reflected in the anomalous decrease of the III–IV transition temperature as the pressure increases; this indicates that the pressure reduces the distance between the hydrogen molecules in the six-membered rings, making their partial dissociation more likely to occur.

7.1.2 Phase Transition Simulation

The experimental studies described in subsection 7.1.1 have undoubtedly contributed to our understanding of the physical and chemical properties of dense hydrogen. However, because extremely high-pressure conditions are difficult to achieve, theoretical studies are far ahead of experiments. In many instances, new structures and phase transitions have been predicted well before they were experimentally verified [Monacelli2023]. The limited pressure and temperature ranges used in experimental studies over the last 30 years have hindered the full exploration of nuclear quantum effects on the hydrogen phase diagram; much larger phase diagram regimes have been explored by computational simulations. The most commonly used computational methods in this research area are DFT (Section 4.4) and first-principles PIMD (Section 5.1), although more accurate QMC methods are also under development.

7.1.2.1 Molecular Solid Hydrogen

The experimental research presented in subsection 7.1.1 has shown that nuclear quantum effects (in particular, nuclear quantum properties and the quantization of rotational modes) influence the rotation of hydrogen molecules in phases II and III of molecular solid hydrogen and therefore affect the stability of the molecular structure and the boundary between these phases. To explore this situation theoretically, S. Biermann et al. utilized first-principles PIMD to simulate the lattice structure of molecular solid hydrogen at 50 K, taking nuclear quantum effects into account. They found that, at a pressure of 350 GPa, the hydrogen nuclei were exceptionally delocalized because of quantum fluctuations, and the structure differed significantly from the classical case [Biermann1998a, b]. Later, H. Kitamura et al. simulated the nuclear quantum effects on molecular rotation in phases I, II, and III of dense hydrogen [Kitamura2000]. They found that phase-I molecules can rotate very freely, while the degree of this rotation in phases II and III is differently affected by nuclear quantum effects, leading to different suppression in rotating speed and orientation. They concluded that quantum localization effects can occur in molecular rotation.

To further clarify this issue, X.Z. Li et al. added a van der Waals force correction to obtain a more accurate electronic structure, and according to this, they further conducted PIMD simulation. The result confirmed that the fixed molecular orientation in phase II of the molecular solid hydrogen is lost because of nuclear quantum effects [Li X.Z.2013]. Classically, hydrogen molecules (red objects in Figure 7.6(a)) exhibit a fixed orientation that does not change significantly over the simulation time scale. When nuclear quantum effects were considered, the distribution of hydrogen molecules (yellow objects in Figure 7.6(e)) was isotropic in a certain plane, showing no specific orientation. A similar situation was observed for dense deuterium molecules. Moreover, as the temperature increased from 50 K to 150 K, an increased orientational disorder was observed (compare Figures 7.6(c) and (g)). The findings by X.Z.Li et al. contradict the results reported by H. Kitamura et al., indicating that nuclear quantum effects are very sensitive to the electronic structure of the system. In other words, nuclear quantum effects are sensitive to the Born–Oppenheimer potential energy surface obtained from the first-principles calculation. An accurate Born–Oppenheimer potential energy surface is a necessary condition for the theoretical discussion of full quantum effects. X.Z. Li et al. corroborated this fact: A small improvement to DFT results by introducing a van der Waals interaction correction can lead to completely opposite results for nuclear quantum effects. Thus, the theory can only predict the full quantum effects appropriately by simulations that approach the exact electronic structure. Only in this way can we develop a proper understanding of the experimental discoveries.

Advances in theory are not limited to DFT results. J. Chen et al. compared the nuclear quantum effects on phase IV calculated by more accurate QMC method (see Section 4.5) with those calculated by methods based on density functional theory [Chen J.2014b]. They found that the DFT methods tend to underestimate the dissociation barrier of phase-IV hydrogen molecules (as shown in Figure 7.7).

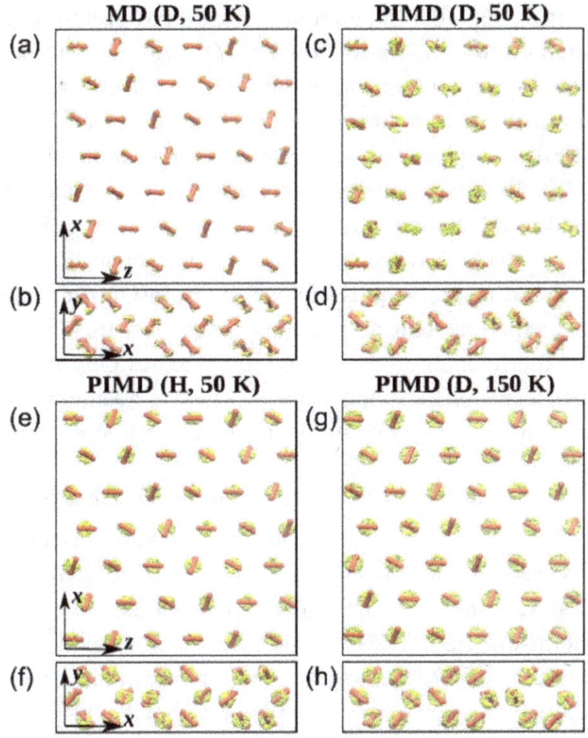

Figure 7.6 First-principles molecular dynamics (MD) and path integral molecular dynamics (PIMD) simulations of phase II molecular solid hydrogen after considering the nuclear quantum effects. The PIMD calculation is performed with van der Waals interaction corrections. Used with permission of AIP Publishing, from Li X. Z.[2013], Copyright 2024; permission conveyed through Copyright Clearance Center, Inc.

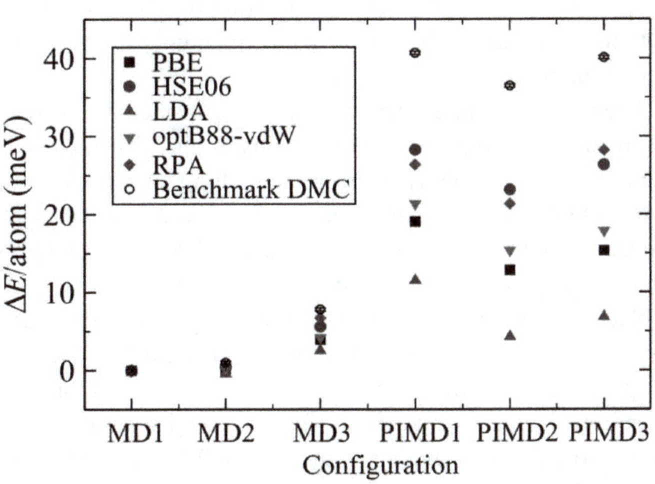

Figure 7.7 Total energy difference per atom for different phase IV configurations calculated by MD (classical nuclei) and PIMD (quantum nuclei) methods at a pressure of 250 GPa and a temperature of 300 K. Reprinted from Chen J.[2014b], with the permission of AIP Publishing.

Although some previous DFT calculations found rapid proton transfer and hydrogen molecule dissociation corresponding to "low barrier" without considering nuclear quantum effects, this result was completely accidental because the errors in the DFT electronic structure canceled out those introduced by omitting the nuclear quantum effects. This example not only further emphasizes the importance of accurate electronic structures for obtaining correct descriptions of nuclear quantum effects but also proves that only by taking nuclear quantum effects into consideration can we give a right description of the phase transition of condensed hydrogen. The influence of nuclear quantum effects on the electronic structure can significantly change the Born-Oppenheimer potential energy surface. Although calculations that do not consider nuclear quantum effects may sometimes appear to fit experimental observations, such studies only provide a classical picture, which is wrong because the importance of nuclear quantum effects is completely ignored. Reliable conclusions can only be established by adopting a full quantum picture.

7.1.2.2 Solid–Liquid Phase Transition of Dense Hydrogen

The aforementioned discussion focused on nuclear quantum effects in molecular solid hydrogen, and the research in this area was also the first to be performed. This is because the melting temperature of dense hydrogen is higher at relatively low pressures; in general, nuclear quantum effects are less pronounced in the relatively high temperature region. However, in 2004, S. A. Bonev et al. reported a rapid decrease in the melting temperature of dense hydrogen as the pressure exceeded 200 GPa and predicted the existence of low-temperature quantum-liquid hydrogen [Bonev2004]. Interestingly, the hypothesis of a low-temperature quantum-liquid and the theoretical prediction of superconductor/superfluid transitions proposed by Ashcroft et al. both pointed to liquid metallic hydrogen in the high-pressure and low-temperature region of the hydrogen phase diagram, which is still difficult to attain experimentally [Babaev2004]. In this situation, it becomes more important to understand the melting of dense hydrogen in this regime using computational simulations, because nuclear quantum effects become more significant when the melting temperature decreases.

In 2013, J. Chen et al. obtained the melting curve of metallic hydrogen at pressures above 500 GPa by first-principles PIMD simulations [Chen J.2013, 2014]. They found that nuclear quantum effects significantly reduced the melting temperature of dense hydrogen (Figure 2.3). At 500 GPa, the nuclear quantum effects reduced the melting temperature of dense hydrogen from 300 to 200 K. At pressures of 900 GPa and above, the melting temperature of hydrogen decreased to the low-temperature limit of 50 K in the simulations (displayed by black triangle in the inset of Figure 2.3). Nuclear quantum effects in such low-temperature regions are prominent. More importantly, nuclear quantum effects caused the melting temperature to decrease with increasing pressure, and this trend continued up to 1 TPa. By contrast, if nuclear quantum effects are neglected, the melting temperature curve almost levels off (with an average value of ~300 K) at 500 GPa and does not fall below 200 K across the entire pressure range.

The predicted properties of liquid hydrogen at low temperatures owing to nuclear quantum effects have attracted considerable research interest. In particular, liquid forms

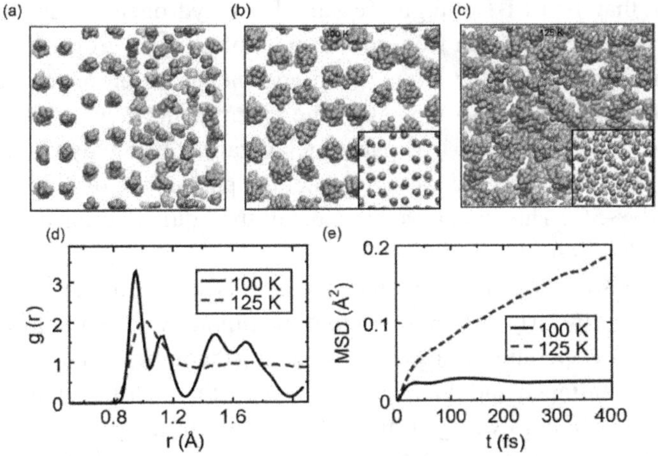

Figure 7.8 First-principles PIMD simulations of solid–liquid coexistence at 700 GPa. (a) The starting structure. (b) When T ≤ 100 K, the system remains in the solid state; (c) when T ≥ 120 K, the system enters the liquid state. (d) The radial distribution function: there are several sharp peaks on the T = 100 K curve (solid line), whereas the curve at T = 125 K (dashed line) is much flatter. This loss of structure with rising temperature indicates that a transition from solid to liquid occurs. (e) The mean-square displacement (MSD) is constant at 100 K, suggesting a solid, but rises linearly with time at 125 K, suggesting a liquid. Figure created by Chen J.[2013] licensed under CC BY 3.0.

of metallic and superconducting hydrogen have been predicted to exist, constituting a novel liquid state directly caused by nuclear quantum effects.

J. Chen et al., for example, investigated solid–liquid coexistence in the pressure range 500–800 GPa by using first-principles PIMD. Figure 7.8 illustrates the simulation results at 700 GPa. It is clear that, when T ≤ 100 K, the solid state persists in the system (as shown in Figure 7.8(b)); when T ≥ 120 K, the system transforms into a liquid state (as shown in Figure 7.8(c)). This characteristic can also be understood from the RDF shown in Figure 7.8(d): There are several sharp peaks on the curve at T = 100 K (black), whereas the one at T = 125 K is much flatter. This indicates that, as the temperature increases from 100 K to 125 K, the system becomes less structured, suggesting a transition from solid to liquid. The time-dependent mean square displacement (MSD) also supports this conclusion (see Figure 7.8(e)) [Chen J.2013].

7.1.2.3 Liquid–Liquid Phase Transition of Dense Hydrogen

Besides the important role in solid–solid and solid–liquid phase transitions, nuclear quantum effects in the liquid–liquid phase transition (LLPT) of dense hydrogen (i.e., the transition between molecular-liquid and atomic-liquid hydrogen) have also become a popular research topic in recent years. As mentioned in the first part of this chapter, it is generally believed that dense hydrogen can have two different liquid phases; the liquid–liquid transition is considered the key to understanding hydrogen metallization and the dissociation of hydrogen molecules into hydrogen atoms. Previous theoretical studies have suggested that the LLPT of high-pressure dense hydrogen is a dual phase transition process, involving the transition from insulating molecular hydrogen to metallic atomic hydrogen. This implies that the dissociation and metallization of the liquid molecular hydrogen occur simultaneously. However, the experiments conducted by Nellis, Weir, and their coworkers on liquid-hydrogen metallization using dynamic high-pressure methods have shown that, in the metallic state, the percentage of dissociated molecular hydrogen is only ~5%. This indicates that most of the liquid hydrogen is still in the form of molecules [Weir1996, 1998, Nellis1999].

One of the reasons why nuclear quantum effects in the liquid hydrogen were not extensively studied in early research was that the liquid hydrogen at low pressures only appears when the temperature is above 1,000 K. As theoretical studies on high-pressure liquid hydrogen continued and the negative slope of the melting curve was discovered, it was realized that liquid hydrogen might form at room temperature or lower. As can be seen from the high-pressure hydrogen phase diagram shown in Figure 2.3, after considering the nuclear quantum effects, the melting curve from 500 to 1,200 GPa given by PIMD simulations (PIMD curve in the inset) has a clear negative slope. The melting curve without nuclear quantum effects (MD curve in the inset) given by molecular dynamics simulations does not reflect this trend. Therefore, it can be inferred that there exists a low-temperature liquid induced by nuclear quantum effects in the region between the two curves. In this way, the liquid–liquid phase transition of dense hydrogen could be conveniently studied around 600 K.

Some studies have found that the nuclear quantum effects on hydrogen atoms are quite pronounced in this temperature range. Figure 7.9 compares the MD simulation with the PIMD simulation; both are based on the Born–Oppenheimer potential energy surface

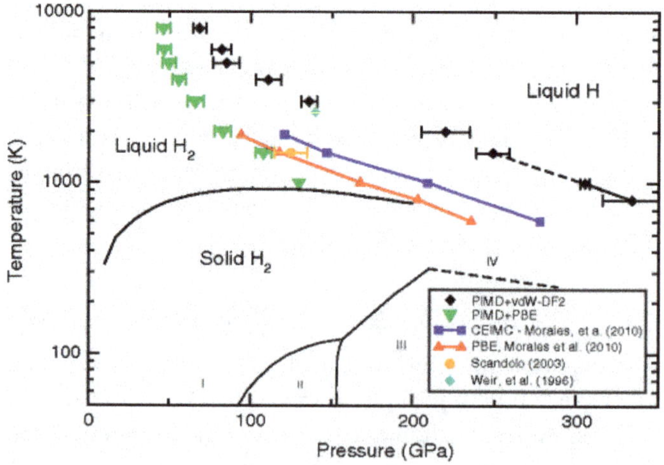

Figure 7.9 Phase diagrams of dense hydrogen given by different calculation methods, showing the liquid–liquid phase transition. Data labeled with PIMD are the results of first-principles PIMD simulation after considering nuclear quantum effects. Reprinted figure with permission from Morales[2013]. Copyright 2024 by the American Physical Society.

given by DFT. We see that nuclear quantum effects decrease the liquid–liquid transition pressure of molecular hydrogen by at least 50 GPa around 1,000 K. Moreover, this result contradicts the phenomenological understanding of nuclear quantum effects, making hydrogen molecules more susceptible to dissociation. This suggests that nuclear quantum effects cannot simply be regarded as an increase of thermal fluctuations, but rather as completely new quantum effects.

The different liquid–liquid phase transition curves shown in Figure 7.9 reflect not only whether nuclear quantum effects are considered but also the influence of different calculation methods. For example, the black and green curves represent the results obtained using two different density functionals. This example conveys an important message: The accurate description of the electronic structure (i.e., the description of the Born–Oppenheimer potential energy surface) is important for the study of liquid-hydrogen phase transitions and nuclear quantum effects. With the development of more accurate and efficient calculation methods such as QMC, it is now possible to discuss liquid–liquid phase transitions and the associated full quantum effects based on more accurate electronic structure. In the last few years, D. M. Ceperley's group at the University of Illinois at Urbana-Champaign, S. Sorella's group at Scuola Internazionale Superiore di Studi Avanzati (SISSA), and J. Chen's and E. G. Wang's groups at Peking University have conducted in-depth studies in this area [Morales2013, Mazzola2015, Guo J.Q.2022].

Recently, J. Q. Guo et al. explored the relationship between the dissociation and metallization of liquid hydrogen using PIMD calculation based on QMC method and DFT calculations [Guo J.Q.2022]. In Figure 7.10(a), the horizontal axis represents the molecular fraction of liquid hydrogen, and the vertical axis represents the fundamental gap of liquid hydrogen. The green dashed line in the figure represents the position where the molecular–atomic phase transition of liquid hydrogen occurs ($x = 0.5$), whereas the red dashed line represents the position where the band gap opens ($x = 0.8$). The plot shows that when $x = 0.8$, the fundamental gap reaches 0. This means that when ~20% of hydrogen molecules dissociate into hydrogen atoms, the liquid hydrogen enters the metallic phase. Although this appears to be consistent with the experimental results [Weir1996, 1998, Nellis1999], there is one important difference. In previous high-pressure experiments, the temperature was very high, and the presence of the molecular metallic phase in the high-temperature region was well explained by statistical theory [Li R.Z.2015]. However, in the latest study conducted by J.Q. Guo et al., the separation of the two liquid–liquid hydrogen phase transitions (i.e., molecular–atomic and insulating–metallic) occurs in the first-order phase transition region at lower temperatures (<1,000 K). The hydrogen-metallization phase transition in the low-temperature region is the main goal of research on superconducting hydrogen. The metallic molecular-liquid hydrogen discovered by J.Q. Guo et al. covers an interval with a pressure of tens of gigapascal and a temperature of hundreds of Kelvin (the MMH region in Figure 7.10(a)). Figure 7.10(b) shows the liquid–liquid phase transition in the temperature–pressure phase diagram, which corroborates the above statement. The pink region corresponds to molecular metallic liquid hydrogen (MMH).

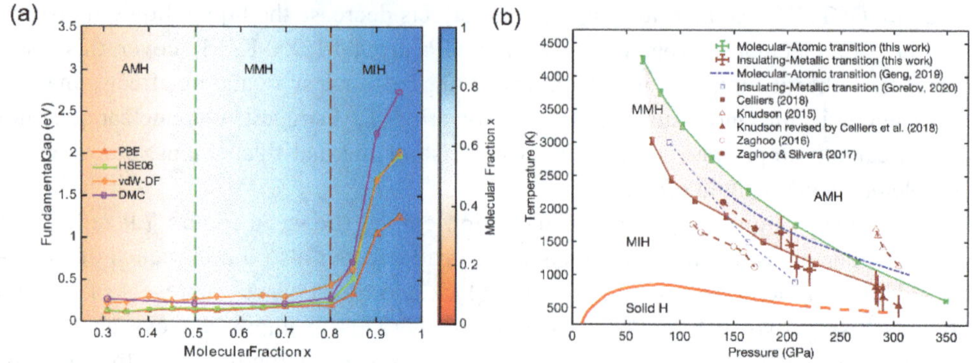

Figure 7.10 (a) Fundamental gap as a function of molecular fraction of liquid hydrogen. The solid lines in different colors are the results obtained using different calculation methods. (b) Temperature–pressure phase diagram of hydrogen, including computational data and data from other literature. AMH, MMH, and MIH mark the atomic metallic hydrogen, molecular metallic hydrogen, and molecular insulating hydrogen phases, respectively. Figure created by Guo J.Q.[2022] under CC BY 4.0.

This result not only contradicts the conventional understanding of the liquid–liquid phase transition of hydrogen but also implies that there are two types of nuclear quantum effects in the LLPT: one is responsible for the structural phase transition and the other affects metallization (i.e., the physical-property phase transition). The former mainly comprises nuclear quantum fluctuations and quantum tunneling, whereas the latter mainly comprises electron–phonon coupling, electron motion, and band gap renormalization caused by zero-point vibrations. Therefore, some of the anomalous aspects of nuclear quantum effects on the hydrogen LLPT are, in fact, due to the existence of two different and competing phase transitions. Based on this discussion, it is believed that the influence of NQEs on LLPT does not require the structural transition and the physical-property transition to occur simultaneously. Although more detailed theoretical study, additional computational research, and further experimental verification are required to confirm these findings, it can be expected that the full quantum effects on the two liquid-hydrogen phase transitions are significant, as have already been observed in solid hydrogen.

Monacelli et al. computed the phase diagrams of dense hydrogen and deuterium at low temperatures and high pressures using state-of-the-art methods to include both many-body electronic correlation and the quantum anharmonic motion of protons recently [Monacelli2023]. They found that the long-sought atomic metallic hydrogen phase – predicted to host room-temperature superconductivity – might form at pressures above 577 GPa. For the atomic metallic deuterium phase, the onset pressure was 63 GPa higher. They predicted that the molecular system would undergo a first-order phase transition from conductive phase III to metallic phase VI before fully entering the atomic metallic phase. The phase transition pressures are at 410 ± 20 GPa and 442 ± 16 GPa for dense hydrogen and deuterium, respectively.

Because of limitations in experimental technique, theoretical work is far ahead in the study of dense-hydrogen phase transitions. We believe that in this frontier field, new phenomena induced by full quantum effects will be discovered and experimentally verified, dramatically improving our understanding of the dense hydrogen phase. This will be a very exciting direction.

7.2 Nuclear Quantum Effects in Water

In the examples in Chapter 1, we considered some of how water, as a typical hydride compound, manifests full quantum effects. In Chapter 2, we also introduced some anomalous macroscopic thermodynamic phenomena associated with full quantum effects on liquid and solid water. A deeper understanding of these phenomena requires the development of not only more efficient full quantum simulations, but also sophisticated experimental imaging and spectroscopic methods that can provide key information at the single-atom or single-molecule level, such as data about the strength and orientation of hydrogen bonds, and individual hydrogen tunneling and delocalization. As we will see in subsections 7.2.1–7.2.5, recent years have witnessed substantial progress in this highly interdisciplinary research area, which is also one of our core contents.

Our emphasis will be on basic science, mainly related to the atomic or sub-molecular details of water from the perspective of full quantum effects, and on the physical mechanism behind the various novel properties of water based on the bonding structure and interaction. The results discussed will undoubtedly prove helpful for the more efficient use of hydrological resources; a detailed discussion of the applications of water in energy and environmental sciences will be presented in Chapters 9 and 10, respectively.

"What is the structure of water?" was one of the 125 scientific questions of the twenty-first century presented in the 125th anniversary series of *Science*. S. Meng and E. G. Wang conducted intensive research on basic scientific questions about water, as described in their book *Water: Basic Science* [Meng S.2023]. The structure and dynamics of water are complex, largely because of the existence of multiple competing interactions of various strengths, such as covalent bonds, hydrogen bonds, and van der Waals forces. The hydrogen bonding interaction, which has an intermediate strength, is important. In 2011, the International Union of Pure and Applied Chemistry (IUPAC) proposed a new definition of the hydrogen bond: an attractive interaction between a hydrogen atom from a molecule or a molecular fragment X–H in which X is more electronegative than H, and an atom or a group of atoms in the same or a different molecule, in which there is evidence of bond formation [Arunan2011]. However, our understanding of the nature of the hydrogen bond is far from complete.

It is commonly believed that the hydrogen bond is the result of the classical electrostatic interaction. However, it has recently become clear that the nuclear quantum effects of hydrogen also play a non-negligible role. Because of the small mass of the hydrogen nucleus, these full quantum effects are obvious even at room temperature. The quantum tunneling and quantum fluctuations of hydrogen change the confining effect of the classical barrier, thus enhancing or weakening the strength of the hydrogen bond interaction and altering the configuration of the hydrogen-bonding network. This effect influences not only the properties of water but also the general physical and chemical properties of other hydrogen-bonding systems. Therefore, it is not sufficient to understand hydrogen bonds from classical electrostatic interactions alone; the quantum nature of hydrogen nuclei must also be considered. Studying water science from the perspective of full quantum effects opens up new ways to understand the microscopic structure and the macroscopic anomalous properties of water and provides important clues for understanding other hydrogen-bonding systems as well.

7.2.1 Quantum Fluctuations of Hydrogen Bond

One of the most direct ways to probe nuclear quantum effects on hydrogen bonds is isotopic substitution. As shown in Table 7.1, replacing H with D or T in H_2O significantly changes many macroscopic properties and characteristics of water. For example, 12.7% more energy is required to decompose the gaseous dimer of D_2O than that of H_2O. [Rocher-Casterline2011a, b, Ch'ng2012]; the melting point of D_2O is 3.82 K higher than that of H_2O [Lide1999, Nilsson2010, Kim K.H.2017]; D_2O reaches maximum density at a temperature 7.21 K higher than that of H_2O [Kudish1972, Hill1982]; D_2O has a higher viscosity coefficient and a smaller diffusion coefficient than H_2O [Kudish1972, Holz2000, Price2000, Hardy2001, Kim K.H.2017]; D_2O has a higher specific heat capacity at 25°C than H_2O [Nakamura1995, Kim K.H.2017]. These macroscopic properties and characteristics of gaseous and liquid water suggest that the nuclear quantum effects associated with replacing H by D strengthen the hydrogen bonds between water molecules, compared with those in H_2O.

However, water has other macroscopic properties that conflict with this image. For instance, the triple point of H_2O is 3.25 K higher than that of D_2O [Nakamura1995, Nilsson2010, Kim K.H.2017], and the tension of D_2O on the gas–liquid surface is less than that of H_2O [Kim K.H.2017]. These results suggest that nuclear quantum effects can also strengthen the hydrogen bonds of water.

Currently, there is no unified physical picture of nuclear quantum effects on hydrogen-bonding interactions between water molecules. However, several review articles have been published, summarizing the current status of theoretical and experimental work on nuclear quantum effects in bulk-phase and interfacial water [Marx2006, Paesani2009, Marx2010, Ceriotti2016, Guo J.2017].

It is important to determine how much nuclear quantum effects contribute to hydrogen bonding. This is a central question in the study of nuclear quantum effects and will undoubtedly be crucial for unraveling the mystery of water. In the early 1950s, it was observed that the strength of intermolecular interactions in hydrogen-bonded systems would change when H was replaced by D. This is known as the Ubbelohde effect [Ubbelohde1955]. Using sum-frequency spectroscopy, M. Bonn's group at the Max Planck Institute for Polymer Research in Mainz found that the orientation of surface water molecule was changed by nuclear quantum effects and attributed this to the differences in hydrogen-bonding strength between D and H [Nagata2012]. Michaelides' group (then at University College London, now at the University of Cambridge) found that nuclear quantum effects tend to weaken the weak hydrogen bonds and strengthen the strong ones (as shown in Figure 7.11(a)) [Li X.Z.2011]. However, the question of exactly how nuclear quantum effects influence hydrogen-bonding strength is by no means a simple one.

Table 7.1 *Isotope effects in water*

Physical properties	H_2O	D_2O	T_2O	$H_2{}^{18}O$	References
O–H(D) bond energy (kJ/mol) (gaseous state)	458.9	466.4 (1.6%)			[Maksyutenko2006]
Dipole moment (*D*) (gaseous state)	1.855	1.855			[Dyke1973]
Symmetrical stretching vibration frequency (cm^{-1}) (gaseous state)	3657.1	2671.6	2237.2	2237.2	[Nilsson2010, Tennyson2014]
Bending vibration frequency (cm^{-1}) (gaseous state)	1594.7	1178.4	995.4	1588.3	
Asymmetric stretching vibration frequency (cm^{-1}) (gaseous state)	3755.9	2787.7	2366.6	3741.6	
Energy of water dimer decomposition (kJ/mol) (gaseous state) at 10 K	13.22	14.88 (12.7%)			[Rocher-Casterline2011a, b, Ch'ng2012]
Melting point, T_m (K) (1atm)	273.15	276.97 (1.40%)	277.64 (1.64%)	273.46 (0.11%)	[Nilsson2010, Kim K.H.2017]
Temperature of maximum density (TMD) (K)	277.13	284.34 (2.60%)	286.55 (3.40%)	277.36 (0.08%)	[Kudish1972, Hill1982]
Critical point (K)	647.10	643.85 (−0.50%)	641.66 (−0.84%)		[Nilsson2010, Kim K.H.2017]
Molar density (mol/L)	55.35	55.14 (−0.38%)	55.08 (−0.49%)	55.42 (0.13%)	[Nakamura1995, Kim K.H.2017]
Molar density$_{TMD}$ (mol/L)	55.52	55.22 (−0.53%)	55.17 (−0.63%)	55.59 (0.13%)	[Kell1977, Kim K.H.2017]
Vapor/liquid surface tension (N/m)	0.007198	0.07187 (−0.15%)			[Kim K.H.2017]
Specific heat capacity C_v [J/(K mol)]	74.54	84.42 (13.2%)			[Nakamura1995, Kim K.H.2017]
Translational diffusion (Å^2/ps)	0.230	0.177 (−23.0%)		0.219 (−6.1%)	[Holz2000, Price2000]
Rotational diffusion (rad^2/ps)	0.104	0.086 (−17%)			[Hardy2001]
Dielectric constant relaxation time (ps) (20°C)	9.55	12.3 (29%)			[Eisenberg1969]
Viscosity	0.8904	1.0966 (23.2%)		0.9381 (5.4%)	[Kudish1972, Kim K.H.2017]
pH/pD values	7.00	7.43	7.6		[Lide1999, Kim K.H.2017]

Note: The above parameters are measured at 25°C unless stated otherwise. 1 atm = 1.01325×10^5 Pa.

To quantitatively characterize nuclear quantum effects on the hydrogen bond strength, Y. Jiang's group at Peking University designed brilliantly a single-hydrogen-bonding model (see Figure 1.36(a)). Using their newly developed TE-IETS (see subsection 6.1.7 for details), they defeated the limitation of the signal-to-noise ratio with traditional technique, and determined the strength of the hydrogen bond formed between a water molecule and a NaCl substrate by measuring the redshift of the water molecule's O–H stretching mode. Then, they observed the nuclear quantum effects on the hydrogen bond strength at the single-bond level using precisely controlled H–D isotope substitution. The experimental results showed that the nuclear quantum component of the hydrogen bond can reach up to 14%, which is comparable to the contribution of kinetic energy of thermal motion at room temperature. At the same time, they found that the full quantum effect of hydrogen atom satisfies a unified law – it weakens the weak hydrogen bonds and strengthens the strong ones (as shown in Figure 7.11(c)) [Guo J.2016a], and this is consistent with the theoretical simulations by Michaelides' group.

To explore the physical basis for this phenomenon better, Li's group also at Peking University carried out *ab initio* PIMD simulations. In Figure 7.11(d), the black line represents the difference in the strength of the hydrogen bond O–H···O (E_H) and the strength of the hydrogen bond O–D···O (E_D) as a function of their average value ($E_{average} = (E_D + E_H)/2$). Their calculations are consistent with experiments. The fundamental reason is the nuclear quantum effects, due to the principle of uncertainty in quantum mechanics, which cause the hydrogen atoms in water molecules to exhibit significant nonharmonic zero-point vibrations. The stretching vibration shortens the hydrogen bond length (which means the bond strength is enhanced), while the bending vibration reduces the hydrogen bond angle (which means the bond strength is weakened). Therefore, the effects on the hydrogen bond of the two vibration modes are opposite, which together determine the nuclear quantum effects on the hydrogen bond strength as indicated in Figure 7.11(b) [Ceriotti2016]. Then, the delicate interplay between covalent bond stretching and H-bond bending determines the bonding energy change ($E_H - E_D$) upon isotope substitution. To further verify this reasoning, they defined a projection of the OH (OD) covalent bond length along the intermolecular axis of the H bonds (inset of Figure 7.11(d)), and studied the ratio between the averaged OH and OD projections as a function of the tip height both before and after the isotope substitution (shown by the red line in Figure 7.11(d)). The zero-point motion of O–H (O–D) stretching increases the projection, whereas that of H-bond bending decreases the projection, due to the anharmonic nature of the potentials. The ratio between the averaged OH and OD projections (red curve) correlates well with the bonding energy difference (black curve). This result presents a unique picture to understand nuclear quantum effects on hydrogen bond strength. Based on this comprehensive microscopic picture, they validated the proposed physical nature (Michaelides group) of nuclear quantum effects on the strength of hydrogen bonds in water [Li X.Z.2011]. The research teams led by Y. Jiang and X. Z. Li have revealed from both experimental and theoretical perspectives that full quantum effects weaken weak hydrogen bonds and strengthen strong ones, thereby changing the structure of the hydrogen bond network. This universal physical picture can be used for the understanding of all hydrogen bond systems. Recently, researchers at the

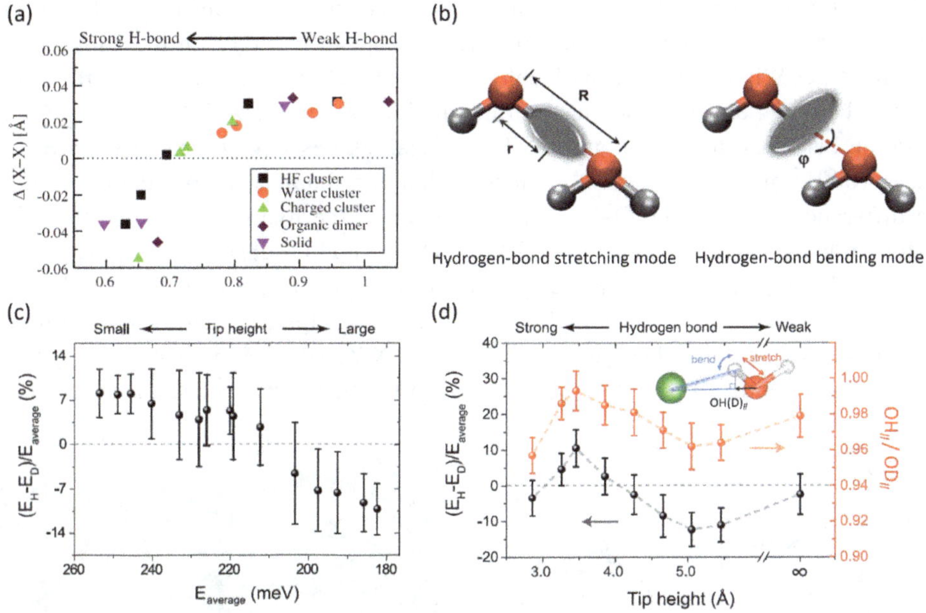

Figure 7.11 Results of theoretical and experimental studies of how nuclear quantum fluctuation affects the hydrogen bond strength. (a) Nuclear quantum effects on hydrogen bond strength in different hydrogen bonding systems, obtained by full quantum calculations. The y-axis is the differences between the shortest heavy-atom distances from PIMD and MD simulations denoted by Δ(X - X). X-axis is the ratio of the X-H stretching frequency in the H-bonded cluster to that in the free monomer. (b) O–H stretching of the hydrogen bond between the oxygen atoms of two water molecules: shared delocalized proton motion shortens the hydrogen bond length, strengthening the bond and slowing down the dynamic process. However, bending or twisting vibrations reduce the angle between the oxygen atom and the two nearest hydrogen neighbors, weakening the hydrogen bond. The overall nuclear quantum effects on the hydrogen bond strength depend on the outcome of the competition between these two vibrational modes. (c) Relative difference between hydrogen bond strengths of O–H···O (E_H) and O–D···O (E_D) as a function of their average ($E_{average} = (E_D + E_H) / 2$). The result indicates that nuclear quantum effects of hydrogen tend to weaken weak hydrogen bonds and strengthen strong hydrogen bonds. The errors in the figure are calculated by averaging experimental measurements on seven different HOD water molecules. (d) Relative energy difference of single hydrogen bond before and after isotope substitution as a function of their average, obtained using full quantum calculations (black dashed line, bottom). The red dashed line (top) represents the ratio between the lengths of OH and OD covalent bonds projected along the molecular axis, as a function of the tip height. These theoretical simulations further exemplify nuclear quantum effects on the hydrogen bond strength. Figure 7.11(a) reprinted from Li X.[2011]; Figure 7.11(b) reprinted with permission from Ceriotti[2016]. Copyright 2024 American Chemical Society; Figure 7.11(c) and (d) from Guo J [2016a]. Reprinted with permission from AAAS.

SLAC National Accelerator Laboratory, using ultrafast electron diffraction, observed the enhancement of hydrogen bonds in liquid water caused by nuclear quantum effects on femtosecond time scales, a process known as quantum tug [Yang J.2021].

For water on a solid surface, the interaction between the water molecules and the substrate causes the intermolecular distance to depend strongly on the substrate lattice structure. When the water molecule spacing (O–O spacing) decreases, the hydrogen bond strengthens, and the quantum fluctuation associated with the O–H stretching vibration becomes more pronounced. Theoretical studies have found that, on Pt(111) and Ru(0001) metal substrates, the nuclear quantum fluctuation considerably reduces the difference between the O–H covalent bond and the hydrogen bond in the mixed system of hydroxide ions and water ($H_3O_2^-$). On a Ni(111) substrate, the O–H covalent bond and hydrogen bond are indistinguishable. Each hydrogen atom is shared by two adjacent oxygen atoms, and the hydrogen nucleus is completely delocalized, forming a symmetric hydrogen bond (Figure 7.12(a–h)) [Li X.Z.2010]. Experimentally, symmetric hydrogen bonds (H–O···H···O–H) were also found in water clusters on a Cu(110)

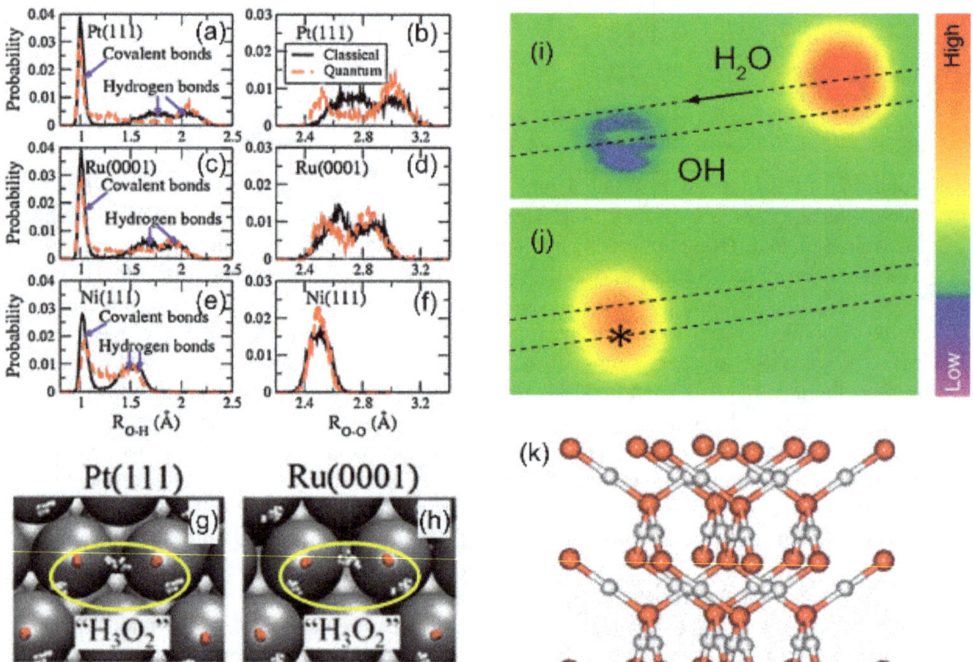

Figure 7.12 Formation of symmetric hydrogen bonds after delocalization of hydrogen nuclei. (a)–(h) Length distributions of ((a), (c), and (e)) O–H and ((b), (d), and (f)) O–O bonds in the mixed hydrogen-bonding network system of H_2O and OH on metal surfaces. The red (black) curves are the results of full quantum effect simulations (classic molecular dynamics simulations). Ni(111) has the smallest lattice constant. The hydrogen nucleus between the two oxygen atoms is completely delocalized, forming a symmetric hydrogen bond. (i) and (j) Symmetric hydrogen bond in the H_2O–OH system on the Cu(110) surface, HO–H–OH. (k) Ice X phase of water at high pressure, showing symmetric hydrogen bonds. Reprinted Figure 7.12(a)–(h) with permission from Li X. Z.[2010]. Copyright 2024 by the American Physical Society; Reprinted Figure 7.12(i) and (j) with permission from Kumagai[2010]. Copyright 2024 by the American Physical Society.

surface by H. Okuyama's group at Kyoto University (Figure 7.12(i, j)) [Kumagai2010]. Therefore, nuclear quantum effects not only alter the hydrogen bonding strength but also the hydrogen bonding network configuration in water.

7.2.2 Concerted Proton Tunneling

On the microscopic scale, the hydrogen-bonding configuration of water is often not static but undergoes a constant process of bond breaking and reformation, and information and energy are often passed back and forth through the hydrogen-bonded network via protons. Therefore, the study of hydrogen-bonding dynamics, and in particular of proton transfer in the hydrogen-bonding networks, is crucial for understanding the physical and chemical properties of water under different conditions. Because of the small mass of the hydrogen nucleus, significant quantum tunneling effects can be observed during proton transfer.

In 2000, quantum tunneling of a single H atom on the Cu(001) surface was observed via STM by Ho's group at the University of California, Irvine. The transition temperature from classical thermal diffusion to quantum tunneling diffusion was determined by isotopic substitution experiments at 60 K (cf. Figure 6.20) [Lauhon2000]. Proton tunneling was further discovered in the OH system by Okuyama's group at Kyoto University. Hydrogen bond donors and acceptors were observed to interchange their roles in water molecular dimers; this is associated with proton tunneling [Kumagai2008, 2009]. These phenomena are shown in Figure 7.13(a–e).

All the research results mentioned above involved the quantum tunneling of individual protons. However, the protons in the hydrogen-bonding network are not independent of each other. Indeed, they are usually strongly correlated. Therefore, proton transfer in hydrogen-bonding systems can involve many-body quantum behavior. In 2017, M. Koch et al. reported "stepwise" tunneling of multiple protons in the $H_2O\text{-}(OH)_n$ (n = 2–4) hydrogen-bonding chain structure on the Cu(110) surface and in the porphycene molecule on the Ag(110) surface [Koch2017]. Moreover, quantum tunnelling diffusion that could also be accomplished through a "concerted" behavior was predicted theoretically more than two decades ago [Gerritzen1984]. In subsequent NMR experiments [Brougham1999] and neutron scattering experiments [Prager1997], coordinated proton tunneling within a cyclic array of four hydrogen bonds in solid p-tert-butyl calixarene and rotational quantum tunneling between three hydrogen atoms in methyl (-CH3) were observed. Nevertheless, no definitive experimental evidence for concerted quantum tunneling has been found on the atomic scale in real space.

In 2015, Y. Jiang's group at Peking University combined sub-molecular-level STM imaging with real-time detection techniques to track proton transfer within a single water-tetramer cluster on a NaCl(001) surface at 5K and observed concerted proton tunneling within the water cluster *in situ*. Within the water tetramer, the hydrogen bonding network has both clockwise (lower left panel of Figure 1.38(a)) and counterclockwise (lower right panel of Figure 1.38(a)) chirality states. Both can appear randomly in the experiment, but it is possible to switch between the two chirality states controllably by varying the STM tip height, as shown in Figure 1.38. They further proved the concerted quantum tunneling of protons by isotope substitution and found that the two chirality states can be switched by controlling full quantum effects [Meng X.Z.2015].

Through first-principles PIMD calculations, Li's group at Peking University has uncovered the physical mechanism of the full quantum effects in this system [Feng Y.X.2018b]. According to the simulation results given in Figure 7.13(f), the transition potential barrier obtained by treating the protons in the water tetramer classically is greater than 0.7 eV (black and red lines in Figure 7.13(f)). When the potential barrier is this high, chirality switching is impossible at low temperatures. However, when the four protons are treated quantum mechanically, the corresponding tunneling barrier decreases dramatically. In particular, compared to the potential barrier of stepwise tunneling, the potential barrier of concerted tunneling is the lowest because of nuclear quantum effects: its height is less than 0.2 eV (green line in Figure 7.13(f)).

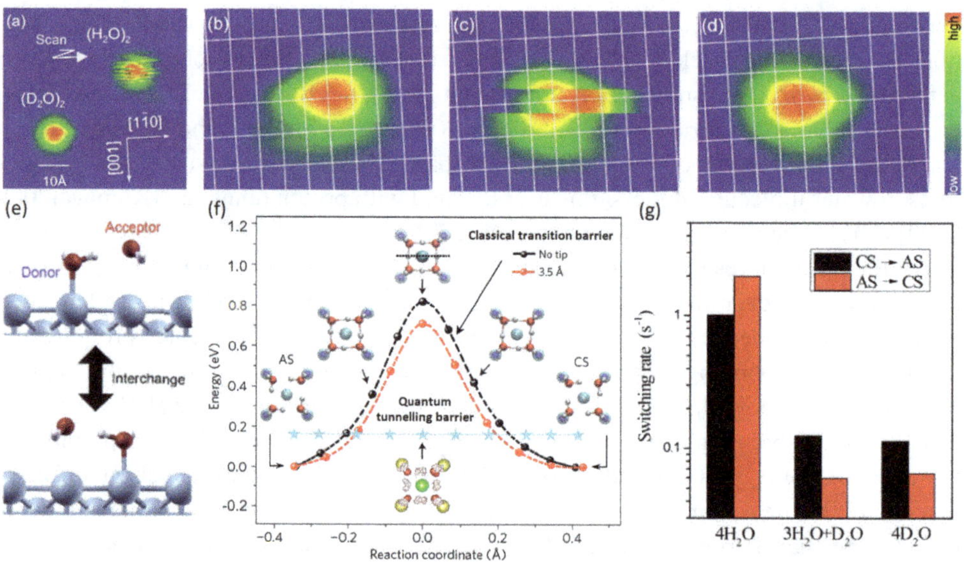

Figure 7.13 Quantum tunneling of protons. (a)–(e) Hydrogen-bond donor–acceptor exchange process in water dimer on Cu (110) surface, achieved through single-proton tunneling. (f) Potential energy barriers of water tetramer on NaCl (100) surface obtained using first-principles MD and PIMD methods. The protons in the water tetramer are treated both classically and quantum mechanically. The classical result with the STM tip-position set at infinity is shown by the black line, and that with the STM tip positioned at 3.5 Å above the surface by the red line. The height of the concerted-tunneling barrier after considering full quantum effects (green line) is lower than the stepwise-tunneling barrier height, which suggests that chirality switching can only be achieved through concerted tunneling of four protons. (g) The chirality-switching rate after partial or complete isotope substitution of protons in water-tetramer hydrogen bonds on the NaCl (100) surface. Reprinted Figure 7.13 (a)–(e) with permission from Kumagai[2008]. Copyright 2024 by the American Physical Society; Figure 7.13(f) from Meng X. Z.[2015]. Reproduced with permission from SNCSC; Figure 7.13(g) reprinted from Feng Y. X.[2018b], with the permission of AIP Publishing.

This result clearly demonstrates that concerted proton tunneling is a coherent quantum process and is more likely to occur than stepwise proton tunneling. They also investigated the effects of partial and complete isotope substitution. By replacing a single H atom on the hydrogen bonds of the four water molecules in the tetramer with a D atom, the potential barrier for concerted tunneling increases significantly, reducing the chirality-switching rate. This is in good agreement with the experimental results (as shown in Figure 7.13(g)).

Subsequently, using techniques such as rotational spectroscopy and dielectric relaxation measurements, S. C. Althorpe's group at the University of Cambridge and researchers from the Hefei Institute of Physical Science observed concerted proton tunneling in gas-phase water clusters and bulk-phase ice, respectively [Yen F.2015, Wu H.L.2016]. Their research further confirms the universality of concerted proton tunneling. Such fully quantized correlated nuclear effects are important for understanding the zero-point entropy and phase transitions of ice, as well as signaling and enzymatic catalysis in biological systems. For example, it has long been puzzling about the origin of residual entropy in ice near the absolute zero. It is expected that the protons in ice tend to be ordered when the temperature approaches the absolute zero, leading to zero entropy. The observed concerted proton tunneling may promote the rapid proton transfer in ice and induce the proton disordering even at the absolute zero, thus explaining the origin of the zero-point entropy in ice.

7.2.3 Bulk Water and Ice

Bulk ice is known as a crystal with the richest structures in nature. Currently, ~20 phases of bulk ice have been predicted theoretically; huge experimental efforts have been made to identify these ice structures, which has been a frontier topic in condensed matter physics, and 18 of them have been verified before. The ice commonly encountered in daily life has a hexagonal structure (I_h). Whether water can freeze to form pure cubic ice (I_c) has long been disputed. This is mainly due to the fact that the process of ice growth is often accompanied by defects. This problem has been exacerbated by the experimental difficulty of distinguishing true I_c from a mixture of cubic and hexagonal ices [Moore2011, Lupi2017]. However, using cryogenic transmission electron microscopy combined with low-dose imaging, X.D. Bai's group at the Institute of Physics, Chinese Academy of Sciences, recently demonstrated the separate crystallization of I_c and I_h from water vapor deposited at 102 K [Huang X.2023]. The surface tension and the nucleus–substrate interaction encourage preferential nucleation of I_c at small sizes; with increasing icing time, the proportion of I_h increases in the polymorphic ice. Moreover, X.D. Bai's group discovered various defect structures in I_c, which were supported by molecular dynamics simulations. The direct observation of ice formation at the molecular level in real space thus considerably extends our knowledge of water.

Nuclear quantum effects in bulk ice and water have attracted considerable attention recently. Approximately 20 years ago, D. Marx at Ruhr-Universität Bochum predicted that ice X would exhibit significant full quantum effects at high pressure. Some of the important manifestations are the quantum tunneling of hydrogen nuclei and symmetric hydrogen bonds caused by the zero-point motion of hydrogen nuclei. Later, Drechsel-Grau and Marx proposed that the concerted proton tunneling also exists in ice I_h [Drechsel-Grau2014, 2017]. Experimentally, spectroscopic and diffraction methods have been used by several research groups to prove the existence of symmetric hydrogen bonds in ice X at high pressures (Figure 7.12(k)), which are caused by the reduced spacing of adjacent oxygen atoms at high pressures and the complete delocalization of hydrogen nuclei caused by nuclear quantum effect. In addition, L. E. Bove et al. used neutron scattering techniques to reveal that the proton dynamical behavior in ice I_h may be a concerted quantum tunneling process [Bove2009]. Subsequently, researchers at the Hefei Institute of Physical Science, Chinese Academy of Sciences, made dielectric relaxation measurements and observed concerted proton tunneling in bulk ice [Yen F.2015]. T. Egami's team at Oak Ridge National Laboratory used inelastic X-ray scattering methods to resolve the motion of water molecules. They found that hydrogen bonds among water molecules behave not in random but in highly coordinated ways, and that quantum effects significantly affect their dynamical behavior [Iwashita2017].

The nuclear quantum effects are also significant for bulk water. They not only alter the structure and dynamical behavior of the hydrogen-bonding network via delocalization of hydrogen nuclei but also change the density and configuration of the water. As shown in Figure 7.14, when nuclear quantum effects are considered, the peaks of the RDFs are flattened, indicating a more disordered local structure of bulk water. It is worth noting that the first peak in the O–H RDF $g_{OH}(r)$ (shown as the black dashed curve in Figure 7.14 (low panel)) is significantly wider than that predicted classically. This peak is located near 1 Å, which corresponds to the O–H covalent bond in the water molecule, implying that the nuclear quantum effects in bulk water induce prominent delocalization of hydrogen nuclei [Ceriotti2012]. If nuclear quantum effects are not considered, the predicted first peak of g_{OH} (r) is unlikely to be broadened by only changing the exchange-correlation functional in DFT, such as by adding a van der Waals correction based on PBE or by using higher-precision functional meta-GGA. This result reinforces the point that nuclear quantum effects of hydrogen can also play a very important and unique role in bulk water.

The quantum fluctuations of hydrogen nuclei can also lead to a wider distribution of electronic energy levels in bulk water. For example, R. Car's group at Princeton University and X. Wu's group at Temple University found that the X-ray absorption spectrum of water becomes wider and smoother after considering the nuclear quantum effect of hydrogen atoms [Sun Z.R.2018]. Similar results were observed by P. Wernet's group in Germany when they compared the absorption spectra of water and heavy water [Schreck2016]. Z. R. Sun et al. found that the X-ray absorption spectrum of water generally shows three absorption edges: the pre-edge, main-edge, and post-edge (Figure 7.15). The pre-edge gives information about the short-range structure around the excited water molecules. The quantum fluctuations of the hydrogen nuclei distribute

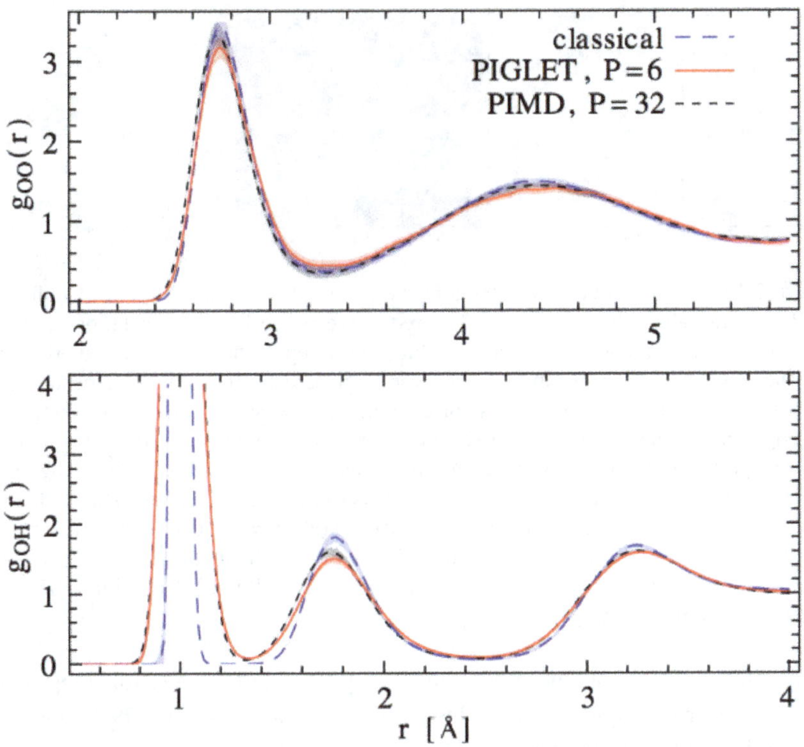

Figure 7.14 Nuclear quantum effects on the (high panel) O–O and (low panel) O–H radial distribution functions of bulk water. The blue dashed curves show the results of a simulation not considering nuclear quantum effects. The red and dashed black curves are the PIMD simulation results using different density-functional approximations that account for nuclear quantum effects. Reprinted figure with permission from Ceriotti[2012]. Copyright 2024 by the American Physical Society.

the excited states to wider energy levels, resulting in spectral broadening. Moreover, the more delocalized hydrogen results in excited electron states with lower energy levels. Consequently, the hydrogen-bonded network of bulk water becomes more disordered under the influence of quantum fluctuations, and the spectrum becomes smoother. These results were also obtained by comparing the absorption spectra of water and heavy water in Wernet's experiments [Schreck2016].

Nuclear quantum effects also have a non-negligible impact on the properties of supercooled water [Pietropaolo2008, Nilsson2010]. C. Andreani's group at the University of Rome studied nuclear quantum effects on the nuclear momentum distribution of hydrogen atoms in supercooled water using deep inelastic neutron scattering spectroscopy. They found that the nearest neighbor O–O distance is smaller in supercooled water than in liquid or solid water, and that the hydrogen nuclei are delocalized. As a result, the supercooled water exhibits an excess mean kinetic energy of 7 kJ/mol [Pietropaolo2008]. Using X-ray scattering, Nilsson's group at Stockholm University found that nuclear

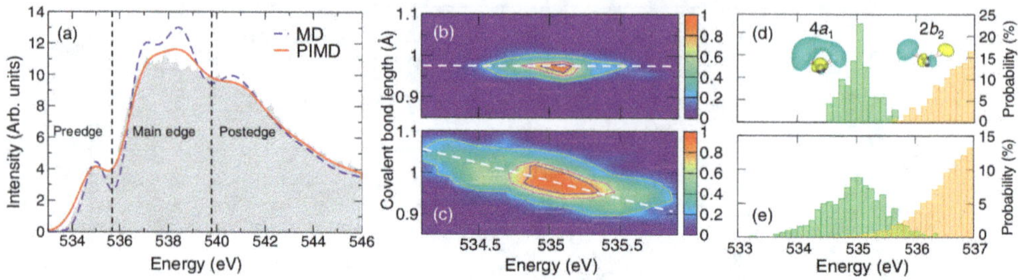

Figure 7.15 Nuclear quantum effects in the X-ray absorption spectrum of bulk water. (a) The X-ray absorption spectrum. The red curve is the PIMD calculation result after considering nuclear quantum effects. The blue curve represents the MD calculation result, and the shaded area corresponds to the experimental data. (b) and (c) respectively give the distribution of the energy levels for excited electron state with $4a_1$ symmetry in the pre-edge and of the corresponding O–H covalent bond lengths. (c) indicates that when nuclear quantum effects in hydrogen are included, the O–H covalent bond lengths have an inverse linear relationship to the energy levels of the excited electron states, that is the longer the bond length the lower the energy. (d) and (e) show the energy level distributions of different excited electron states, where (e) is the result of full quantum calculation. Reprinted Figure 7.15(a) with permission from Sun Z. R.[2018]. Copyright 2024 by the American Physical Society.

quantum effects affect the thermodynamic response and correlation length of supercooled water. For example, the correlation length is maximal at 229 K and 233 K for normal and heavy water, respectively [Kim K.H.2017].

Many anomalous properties of liquid water arise from the dynamic reorganization and relaxation processes of its hydrogen-bonding network as well as the redistribution of energy. Accurate description of ultrafast dynamical processes in water molecules is essential for understanding hydrogen bonds and many liquid-phase chemical reactions. These studies require *in situ* temporally and spatially resolved measurements. Currently, most studies on vibrational relaxation processes of liquid-water molecules are based on ultrafast spectroscopy experiments. For example, K. Ramasesha et al. at MIT used ultrafast two-dimensional broadband IR spectroscopy to confirm that intermolecular motion of water molecules occurs ~50 fs before vibrational relaxation, but the dynamics of the structural changes during this time period is still unknown [Ramasesha2013]. Recently, a research group at the National Accelerator Laboratory using liquid ultrafast electron scattering achieved atomic-level spectroscopic characterization of the hydrogen bond enhancement induced by nuclear quantum effects in liquid water [Yang J.2021]. They found that, when an excited water molecule starts to vibrate, each hydrogen atom pulls the oxygen atoms in the neighboring water molecules closer, then repels them through quantum effects (as shown in Figure 7.16). This study provides atomic-level spectroscopic observations of hydrogen bonding enhancement induced by nuclear quantum effects in liquid water, which is known as the quantum tug between water molecules. Previous experimental and theoretical work have also reported that nuclear quantum effects can change the hydrogen-bonding strength and configuration at the single bond level [Guo J.2016a, Li X.Z.2011].

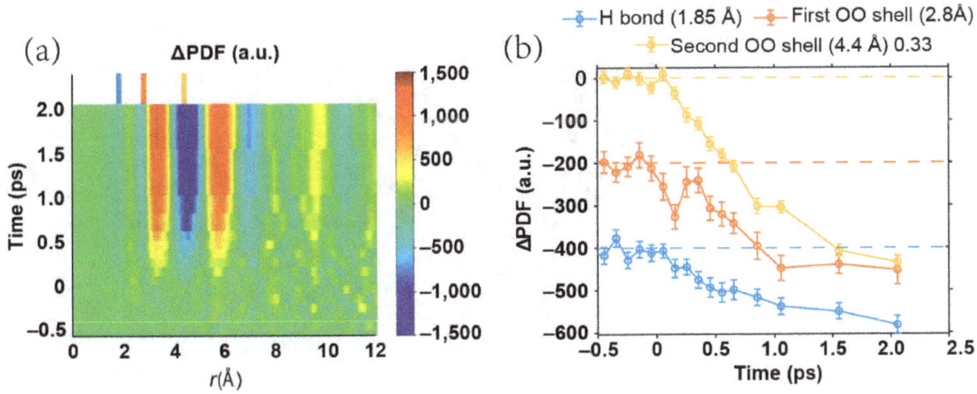

Figure 7.16 The "quantum tug" phenomenon in water. (a) Time-resolved difference pair distribution function (PDF) for different atomic pair distances, obtained after Fourier sine transformation. (b) Time-resolved difference PDF of hydrogen-bond length, first oxygen–oxygen shell distance, and second oxygen–oxygen shell distance. From Yang J. [2021]. Reproduced with permission from SNCSC.

7.2.4 Confined Water and Ice

Nuclear quantum effects tend to become more pronounced in water when it is spatially confined; this gives rise to many anomalous physical and chemical properties [Kapil2022]. For example, proton transfer is significantly accelerated in nanochannels of confined water [Roux1991a, b, Dellago2003], and the confined water in carbon nanotubes exhibits anomalous properties including low dielectric constants [Fumagalli2018] and a wider range of phase-transition temperatures [Agrawal2017]. At the microscopic level, G. F. Reiter et al. found that the zero-point energy of water confined in carbon nanotubes is very different from that of bulk water at room temperature [Reiter2012, 2013]. Because of the zero-point energy, the proton quantum state of water confined in carbon nanotubes at low temperatures differs from those of ice I_h, ice IV, and high-density amorphous ice. In particular, its spectral lines lack the high-momentum tail that is characteristic of the molecular covalent bond and stretching mode in bulk water, and its quantum states are coherently delocalized (see Figure 6.3) [Reiter2006]. The confined water molecules within beryl were studied by A. I. Kolesnikov et al. using deep inelastic neutron scattering techniques (Figure 7.17(a)) [Kolesnikov2016]. In the projection of the water proton momentum distribution onto the xy-plane (Figure 7.17(b)), they observed a second maximum in the radial direction. This corresponds to a coherent delocalized proton state because of quantum tunneling. It is a completely new state that has not been observed in water previously. These studies suggest that the nuclear quantum effects in confined water can be modulated and enhanced at the nanoscale.

Other interesting phenomena arise from the flow of water on the surface of another material or in a confined space [Shen R.2010]. A typical example is the study of the nanoscale friction generated by placing water on a surface made of graphite (a multilayer carbon allotrope) or graphene (a monolayer carbon allotrope). Water has also been placed

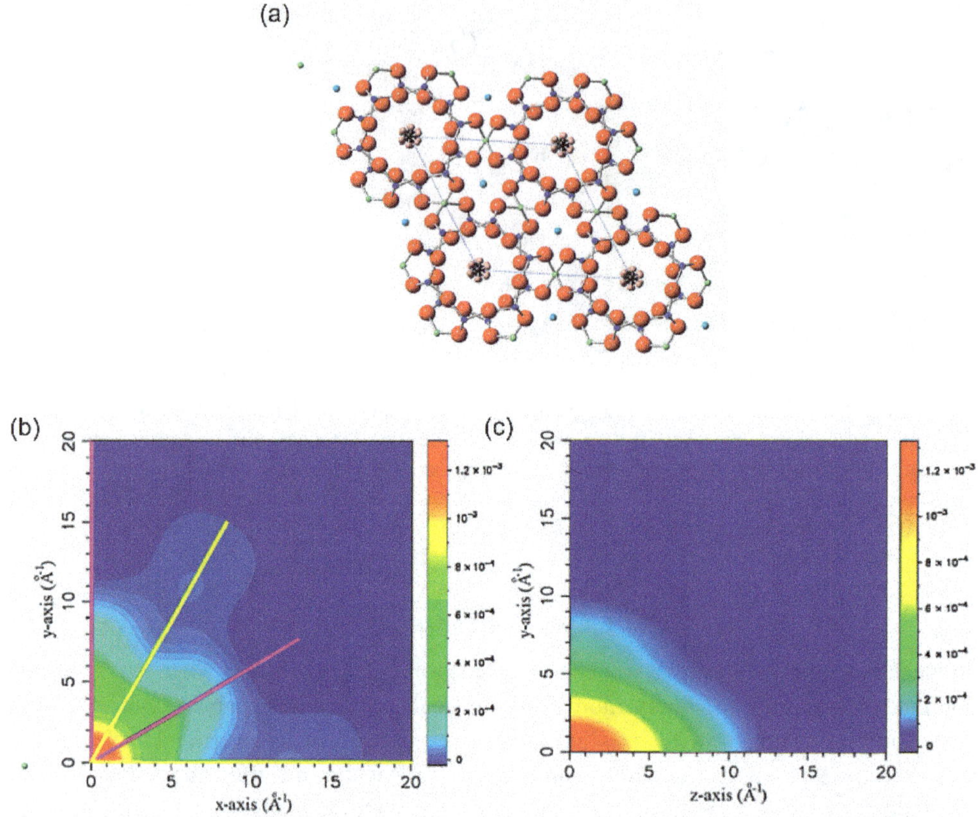

Figure 7.17 Nuclear quantum effects on confined water. (a) Structure of confined water in beryl. Blue spheres are Si atoms, red spheres are O atoms, cyan spheres are Al atoms, green spheres are Be atoms, and pink spheres are H_2O molecules. (b) and (c) Projections of water proton momentum distribution onto the (b) xy and (c) yz planes in confined water, measured by deep inelastic neutron scattering (DINS). The second maximum in the radial direction in (b) indicates that the hydrogen nuclei are coherently delocalized because of quantum tunneling. Reprinted figure with permission from Kolesnikov[2016]. Copyright 2024 by the American Physical Society.

in carbon nanotubes of different diameters to observe its motion and frictional behavior. Extensive experimental observations have found that the friction coefficient of water on graphite surfaces or within multiwalled carbon nanotubes larger than 30 nm in diameter is much larger than that on graphene surfaces or within smaller-diameter multiwalled carbon nanotubes [Maali2008, Secchi2016]. Theoretical work in this area is generally based on classical models, and the motion of water molecules on a smooth solid surface is studied in a potential energy surface obtained by the Born–Oppenheimer approximation. However, such models cannot explain all the questions raised by the above experiments.

Recently, N. Kavokine et al. found that this novel quantum friction is a result of full quantum effects, which cannot be accurately described by theories that treat the nucleus classically, such as the Born–Oppenheimer approximation [Kavokine2022]. Kavokine et al. also developed a new theory of nanoscale quantum friction beyond the classical fluctuation-dissipation theorem by considering a contribution due to the coupling of the charge fluctuation to the electronic excitation of the interfacial material. The new theory encompasses both quantum fluctuations and nonadiabatic effects to some extent, and is a typical manifestation of the dominance of full quantum effects. The theory is not only capable of calculating differences in the friction coefficients of water on different material surfaces and in confined spaces but also explains why water inside large-diameter carbon nanotubes behaves similarly to the frictional behavior of water on graphite. This research shows the equally important role of the full quantum effects in the development of hydrodynamic and nanomechanical devices.

7.2.5 Quantum Phase Transitions in One-Dimensional Water Chains

Quantum phase transitions are phase changes in a quantum system at a temperature infinitely close to zero. Unlike classical phase transitions, quantum phase transitions (QPTs) are not induced by temperature changes. Instead, they are affected by quantum fluctuation, one aspect of the full quantum effects. At the QPT point, the ground-state properties of the system change drastically. This phenomenon has been reported in heavy fermionic compounds, magnetic materials, and unconventional superconductors [Bianconi2001, Si Q.2010, Sachdev2012].

T. Serwatka et al. recently demonstrated the occurrence of a QPT in a one-dimensional water chain using a full quantum simulation approach based on a rigorous wavefunction representation [Serwatka2023]. Figure 7.18 depicts the phase diagram of a one-dimensional water structure. When the water–water distance (R) in this structure is large, the local dipoles of water molecules are in a disordered phase (shown on the right side of Figure 7.18). This implies that the molecular interactions are very weak and that the Hamiltonian is dominated by the rotational kinetic energy of individual molecules. The kinetic energy term of the system tends to delocalize the ground state wavefunction, leaving water molecule dipole in a disordered phase. In this disordered phase, the molecular dipoles do not adopt any specific orientation and can rotate almost freely. However, as the water–water distance R in the one-dimensional water structure decreases, the dominance of the kinetic energy declines, and the quantum fluctuation of the ground state increases. Consequently, the molecular dipole–dipole interaction increases and starts to gradually select a uniform orientation. The water–water distance R in a one-dimensional water chain acts as a nonthermal parameter that can effectively mediate the interaction strength. Eventually, at a critical distance R_c, the system crosses the QPT point, creating an ordered dipole phase of water molecules with parallel and aligned orientations (shown on the left side of Figure 7.18).

Finite size scaling analysis was used to investigate the Schmidt gap $\Delta\lambda$. The results reveal that $\Delta\lambda$ closely follows the scaling law and is therefore a suitable order parameter for describing the 1D water chain system. The continuous change in $\Delta\lambda$, as shown in Figure 7.18, indicates that the transition point is the critical point of QPT. For the critical exponent, Serwatka et al. obtained $\nu = 1.004 \pm 0.004$ and $\beta = 0.118 \pm 0.008$, which agrees

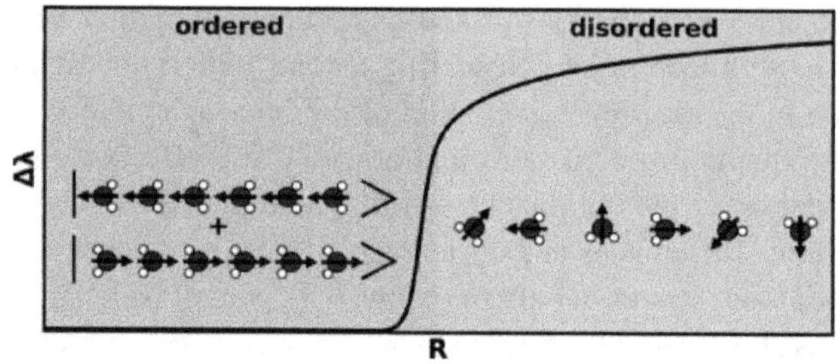

Figure 7.18 Phase diagram of the one-dimensional water system. The right and left sides correspond to the disordered and ordered phases, respectively. The solid line represents the Schmidt gap $\Delta\lambda$ as a function of the water–water distance R. $\Delta\lambda$ represents the order parameter describing the quantum phase transition. The characteristic of disordered phase is that when the molecular distance R in the one-dimensional water structure is large, the orientations of the water molecular dipoles are dispersed. In other words, the dipoles do not adopt any specific direction and are free to rotate. In contrast, as R decreases, the system enters the ordered phase and the dominance of the rotational kinetic energy of the water molecules decreases. The intermolecular dipole interaction increases and the dipoles begin to align to a uniform electric dipole moment (black arrow), leading to a twofold degenerate ground state of the system (equally weighted superposition of the left- and right-polarized states). Reprinted figure with permission from Serwatka[2023]. Copyright 2024 by the American Physical Society.

well with the (1+1)-dimensional Ising universality class of $\nu = 1.0$ and $\beta = 0.125$. These results indicate that the QPT in a one-dimensional linear water chain belongs to the universality class of the (1+1)-dimensional Ising model.

Serwatka et al. also extensively studied QPTs in a linear water molecular chain by analyzing the excitation spectrum, entanglement entropy, and dipole–dipole correlation function. The energy differences between the ground state and the first and second excited states are shown in Figure 7.19(a) and (b), respectively. After crossing the QPT point, the first full symmetry (FS) energy gap vanishes as the ground state becomes a twofold degenerate state owing to the inversion symmetry of the chain. For the second FS energy gap and all higher excited state energy gaps, an avoided crossing that becomes infinitely sharp in the infinite-chain limit is observed. In fact, at finite temperatures (the quantum critical temperature can reach up to 10 K for one-dimensional water chains), QPT can occur if the quantum fluctuation is greater than or close to the thermal fluctuation. In experiments, quasi-QPTs have been recently reported in one-dimensional water chains encapsulated in carbon nanotubes [Ma X.2017].

The long-range correlation effect near the QPT can be described by the von Neumann entanglement entropy $S_{\nu N} = -\mathrm{tr}(\rho_A \ln \rho_A)$ (Figure 7.19(c) and (d)). It was assumed that A and B denote the two halves of the chain and ρ_A denotes the reduced density matrix of region A. The von Neumann entanglement entropy represents the entanglement entropy between regions A and B; it increases as the entanglement between A and B increases (and becomes zero when there is no entanglement). As shown in Figure 7.20, $S_{\nu N}$ diverges at the QPT point, indicating that the system is in a maximally entangled state. In this state, there are quantum fluctuations on all length (and time) scales [Vojta2003, Sachev2011].

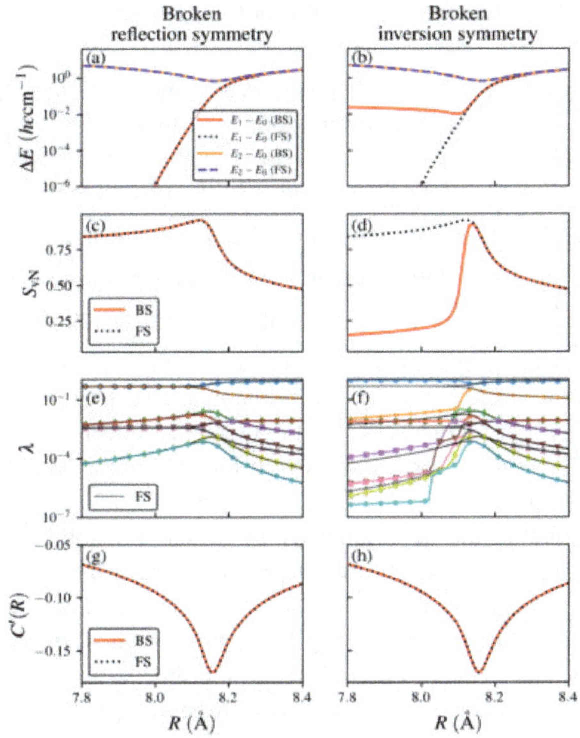

Figure 7.19 (a) and (b): Energy difference between the ground state and the first and second excited states, where FS and BS denote the water chains of full and broken symmetries, respectively. (c) and (d): the von Neumann entanglement entropy for FS and BS chains. (e) and (f): the first ten Schmidt values for FS and BS chains. (g) and (h): derivatives of the nearest-neighbor dipole–dipole correlation function $C(R) = (1/N-1)\sum_{i=1}^{N-1}\left\langle \mu_i^z \mu_{i+1}^z \right\rangle$. In all calculations, a small electric field with a strength of $\left|\vec{E}\cdot\vec{\mu}\right| = 10^{-7} E_h$ is applied to the two edge molecules. In the simulation, the system contains 300 water molecules. Reprinted figure with permission from Serwatka[2023]. Copyright 2024 by the American Physical Society.

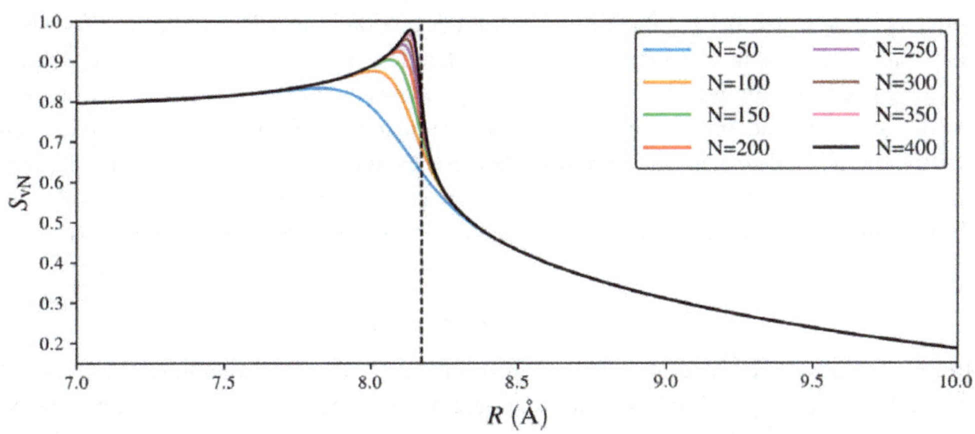

Figure 7.20 The von Neumann entropy as a function of the water–water distance R for one-dimensional water chains with different numbers of water molecules N. Reprinted figure with permission from Serwatka[2023]. Copyright 2024 by the American Physical Society.

Owing to the multiple properties of this phenomenon, the quantum properties of a system are only related to its symmetry and dimensionality, thus leading to the universality of different systems. This universality can be further explored by calculating the dipole correlation function $\langle \mu_i^z \mu_j^z \rangle$ between sites i and j. In the disordered phase, the correlations are short-ranged, and the correlation function decays exponentially $\left(\langle \mu_i^z \mu_j^z \rangle \sim e^{-\frac{|i-j|}{\xi}}\right)$ for large site spacing $|i-j|$. At the QPT, the correlation function follows a power law such that $\left(\langle \mu_i^z \mu_j^z \rangle \sim -1/|i-j|^{\eta}\right)$. The critical exponent of $\eta = 0.253 \pm 0.002$ was obtained by fitting.

To examine the effect of symmetry breaking, a very small electric field is applied to the two edge water molecules to break a specific symmetry. This can be accomplished in two ways: with an XY-polarized transverse field to break the reflection symmetry in the plane containing the chain axis or with a Z-polarized longitudinal field to break the inversion symmetry of the chain. In the first case, as shown in Figure 7.19, breaking the reflection symmetry has no effect on the QPT (indicated by the red line in Figure 7.19(a)). Breaking the inversion symmetry, by contrast, fundamentally changes the ordered quantum phase, whereas the disordered phase is unaffected (shown by the red line in Figure 7.19(b)). It lifts the degeneracy of the ground state, causing the equally weighted superposition of left- and right-polarized states (left side of Figure 7.18) to split into two states with opposite polarization. Because the degeneracy is lifted, most of the entanglement entropy is removed (Figure 7.19(d)). Considering the effect of inversion symmetry breaking, the degeneracy of the ground state for an ordered quantum phase is lifted, creating two many-body states with opposite polarizations. This implies that the one-dimensional water molecular chain may exhibit fascinating ferroelectric behavior, and these physical properties could provide an ideal platform for the study of quantum devices.

In addition, by examining the von Neumann entanglement entropy of one-dimensional water chains with different symmetries, a large plateau was observed after the QPT, owing to the twofold degeneracy of the ground state. Meanwhile, even though the central bond of the water chain is gradually weakened, its entanglement entropy in the ordered phase remains ln(2), which is the value of the twofold degenerate state. This indicates that when the bipartite chain is in the ordered phase, the entanglement between the two halves cannot be eliminated, even though the interaction strength between the two halves can be infinitely small. This phenomenon, which results from pure quantum effect, also suggests that even if the central bond is weakened considerably, for example, by removing a central water molecule or inserting fullerene cages, the two halves of the chain would still be entangled. This entanglement could find a practical application in quantum information processing.

7.3 Superconductivity of Hydrides

Research on superconducting phase transitions in hydride compounds has focused on high-pressure and doping-induced superconductivity. The high-pressure-induced superconductivity utilizes the pressure to reduce the interatomic distance, modify the interatomic interaction, enhance the electronic orbital hybridization between different molecules, and increase the density of electronic states near the Fermi surface. As a result, high pressure strengthens the electron–phonon coupling, causing materials that are non-superconducting at ambient pressure to exhibit superconducting properties. By contrast, doping-induced superconductivity involves altering the energy band structure of the original system by introducing electrons or holes. Introducing charge carriers produces new phonon modes, enhancing the electron–phonon coupling and enabling the doped material to display superconducting properties even at ambient pressures.

In 1968, Ashcroft at Cornell University predicted that metallic hydrogen would exhibit high-temperature superconductivity owing to its exceedingly high Debye temperature and strong electron–phonon interactions [Ashcroft1968]. He proposed that under pressures of ~500 GPa, dense hydrogen could transform into a superconductor, with a transition temperature close to room temperature. Although it has been reported that diamond anvil cells can achieve pressures close to 1 TPa, the highest static pressure attainable for hydrogen is still ~500 GPa [Loubeyre2020]. Consequently, despite significant efforts over the years, metallic hydrogen has never been realized in the laboratory. So far, inducing a stable superconducting transition in hydrogen through high pressure remains a technical challenge.

Many researchers have taken an alternative approach, focusing on hydride compounds instead. These compounds involve non-hydrogen elements that form specific crystal structures under relatively "low" high-pressure conditions. For example, sulfur forms a bcc structure in sulfur hydrides at 220 GPa. Calcium also adopts a bcc structure in clathrate calcium hydride at 150 GPa [Wang H.2012]. When hydrogen is compressed within the interstices of these crystals, it exhibits interatomic interactions similar to those in metallic hydrogen, even at relatively low pressure. Additionally, the non-hydrogen elements in the compound contribute low-frequency phonons, further enhancing the electron–phonon coupling and raising the superconducting transition temperature.

Early research in this area focused on ionic metal hydrides such as Th_4 H_{15}, PdH, $Pd_{0.55}Cu_{0.45}H_{0.7}$, and $NbH_{0.69}$. However, these compounds have low hydrogen contents, and their superconducting transition temperatures are all below 16 K. It was not until 2004, after Ashcroft proposed the application of chemical pre-compression, that a new wave of research on the superconductivity of hydrides was initiated [Ashcroft2004]. In 2015, M. Eremets's group at the Max Planck Institute of Chemistry, discovered that sulfur hydride (H_3S) could achieve superconductivity at 203 K and 220 GPa [Drozdov2015]. This breakthrough set a new record for superconducting transition temperatures. It also revealed a significant isotope effect: sulfide hydrides have a superconducting transition temperature of 195 K at 141 GPa, whereas sulfide deuterides exhibit 150 K at 155 GPa (see Figure 7.21(a), (b)). Subsequent studies revealed that the quantum fluctuation of the hydrogen nuclei facilitated the formation of symmetric hydrogen bond. This quantum delocalization phenomenon has a considerable impact on the superconducting phase transition diagram of the sulfide hydride system at high pressures [Errea2016].

Eremets's group later discovered another class of clathrate hydrides that are capable of achieving even higher superconducting transition temperatures. For instance, lanthanum hydride (LaH_{10}) exhibits a superconducting transition temperature of 250 K at 170 GPa and also demonstrates a significant isotope effect [Drozdov2019] (Figure 7.21(c), (d)). Researchers from George Washington University, found that the same lanthanum hydride can appear superconducting at 280 K (7°C) under a pressure of 202 GPa [Somayazulu2019]. Subsequently, scholars from countries such as Italy and Spain demonstrated that nuclear quantum effects of hydrogen not only play a decisive role in the stability of the lanthanum hydride superconducting phase but also enhance the electron–phonon coupling strength [Errea2020]. These clathrate hydride compounds possess a "host–guest" structure. The covalent bonds of hydrogen atoms form the central structural unit, a metal atom ("guest") is located at the center of the primary "host" hydrogen cage, and the host–guest interaction is ionic. Several clathrate hydride superconductors, including LaH_{10}, YH_9, and YH_6, have been discovered, exhibiting superconductivity temperatures of 260 K, 243 K, and 224 K, respectively. These discoveries have ignited the hope for the realization of room-temperature superconductors.

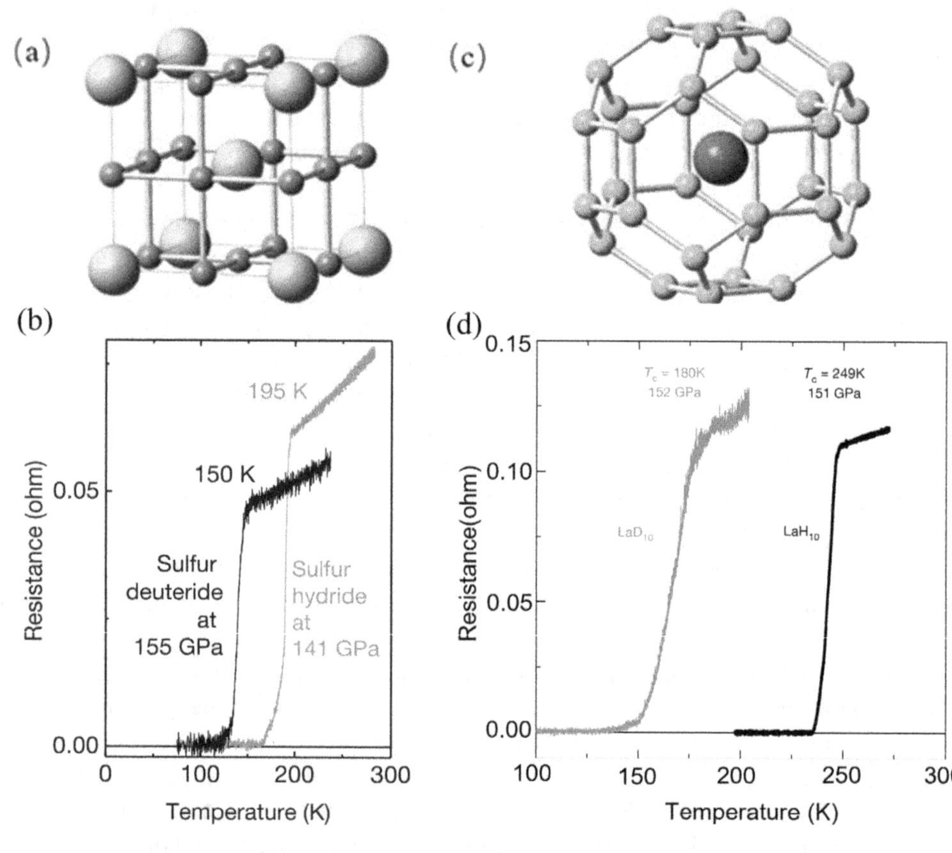

Figure 7.21 Superconductivity of sulfur hydride and lanthanum hydride: Crystal structure and isotope effect. (a) Crystal structure of sulfur hydride (H_3S), where the smaller spheres are H and the larger spheres are S. (b) Experimental evidence of isotope effect on the temperature dependence of resistance in sulfur hydride. (c) Crystal structure of clathrate lanthanum hydride (LaH_{10}), where the small spheres are H and the big sphere is La. (d) Experimental evidence of isotope effect on the temperature dependence of resistance in lanthanum hydride. Figure 7.21(b) from Drozdov[2015]. Reproduced with permission from SNCSC; Figure 7.21(d) from Drozdov[2019]. Reproduced with permission from SNCSC.

Despite significant progress in the quest for room-temperature superconductivity in hydride compounds, high pressure remains necessary for achieving superconducting phase transitions. The reduction of this pressure requirement is an urgent challenge. Scientists from the Institute of Physics of the Czech Academy of Sciences and the University of Marseille have recently found a potential solution. They discovered that in the hydrogen-bonding network of 1D benzoquinone-diimine molecules, the nuclear quantum effects of protons lead to the delocalization of π-electrons in the axial direction of the molecular chain. This delocalization enhances the cohesive energy between molecules and creates unique electronic states within the gap at the end of the chain [Cahlik2021], as illustrated in Figure 7.22. Because of the pronounced nuclear quantum effects in low-dimensional hydride compounds, the hydrogen bonds of benzoquinone-diimine molecules may approach covalency and become metallic at ambient pressure. Consequently, nuclear quantum effects have the potential to significantly reduce the required pressure for the superconducting phase transition in this hydride. This experimental finding provides a valuable clue for the further study of hydride superconductivity.

Many studies on hydride superconductivity have been mainly theoretical, owing to the experimental challenges associated with achieving high-pressure conditions. Numerous research groups worldwide have employed first-principles density functional theory to make predictions, providing valuable guidance for experimentalists. Related theoretical work covers a large number of hydrogen-rich compounds, and we summarize these results in Table 7.2.

One particular class of clathrate hydride compounds exhibits remarkably high superconducting transition temperatures. Y. M. Ma's group at Jilin University initially predicted this phenomenon in calcium hydride (CaH_6) [Wang H.2012]. Subsequently, they extended their predictions to several rare earth hydrides with clathrate structures similar to CaH_6, including LaH_{10}, YH_6, YH_9, and YH_{10}, all of which were identified as potential high-temperature superconductors [Liu H.2017, Peng F.2017]. These works found that at a pressure of 150 GPa, the hydrogen-atom spacing in calcium hydride is 1.24 Å; this is close to the spacing in solid hydrogen. Bader charge-transfer analysis reveals that each metallic calcium atom transfers 1.02 electrons to the hydrogen cage, further strengthening the bonds between different hydrogen atoms. Analysis of the phonon DOS indicates that hydrogen atom vibrations account for ~84% of the superconducting effect. These results indicate that CaH_6 is a new high-temperature superconductor.

Furthermore, the theoretically calculated electronic energy-band structure exhibits a high electronic DOS near the Fermi surface in calcium hydride, accompanied by strong electron–phonon coupling (Figure 7.23). These phenomena have attracted considerable interest from experimentalists because of the implication that calcium hydride holds promise as a potential high-temperature superconductor. Recently, the predicted high-temperature superconductivity of CaH_6 has been experimentally confirmed, with a measured superconducting transition temperature of 215 K [Li2022, Ma2022].

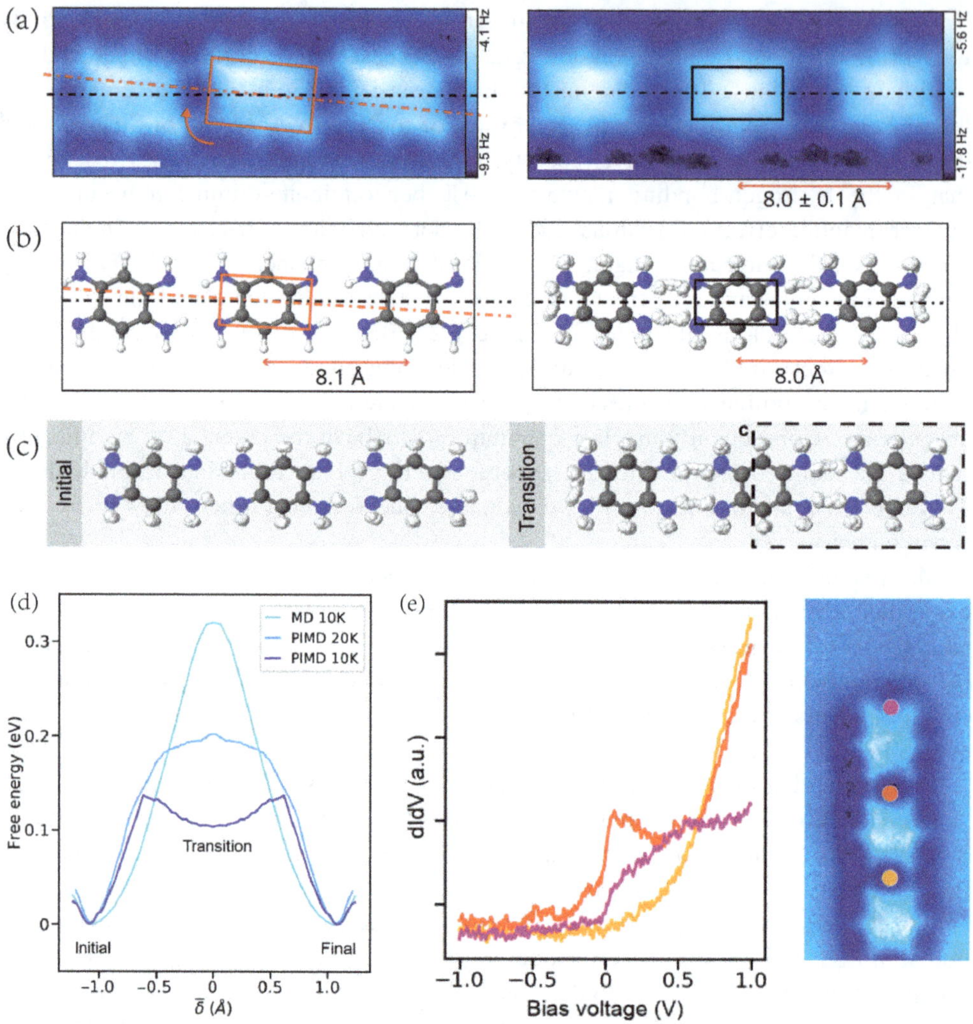

Figure 7.22 High-resolution AFM images of 1D benzoquinonediimine molecules with different hydrogen-bonding networks, their corresponding atomic structures, and scanning tunneling spectroscopy evidence for the presence of electronic states in the gap. (a) High-resolution AFM images of a 1D benzoquinonediimine molecule with canted hydrogen bonds (left) and symmetric hydrogen bonds (right). (b) Atomic structure model corresponding to (a); the plot on the right is a PIMD calculation of the transition state. (c) PIMD structure for the initial state (δ = −1 Å) and the transition state (δ = 0 Å). (d) Potential barrier curves for proton transfer calculated using classical MD (10 K) and quantum PIMD (10 K and 20 K), with the average value of all hydrogen bonds as the horizontal coordinate. (e) scanning tunneling spectroscopy results corresponding to the marked positions in the right panel, showing the presence of electronic states in the gap. Reprinted with permission from Cahlik[2021]. Copyright 2024 American Chemical Society.

Table 7.2 *Calculated superconducting transition temperatures of hydride compounds under high pressure.*

Hydride	Pressure (GPa)	Superconducting transition temperature (K)
CaH_6[Wang H.2012]	150	235
YH_6[Li Y.W.2015]	120	264
YH_{10}[Liu H.2017, Peng F.2017]	300	310
SiH_4[Eremets2008]	300	17.5
GeH_4[Gao G.2008]	220	64
SnH_4[Tse2007]	120	80
MH_2 (M = V or Nb) [Chen C.2014]	100	4 or 1.5
C_6H_6[Zhong G.H.2018]	195	20

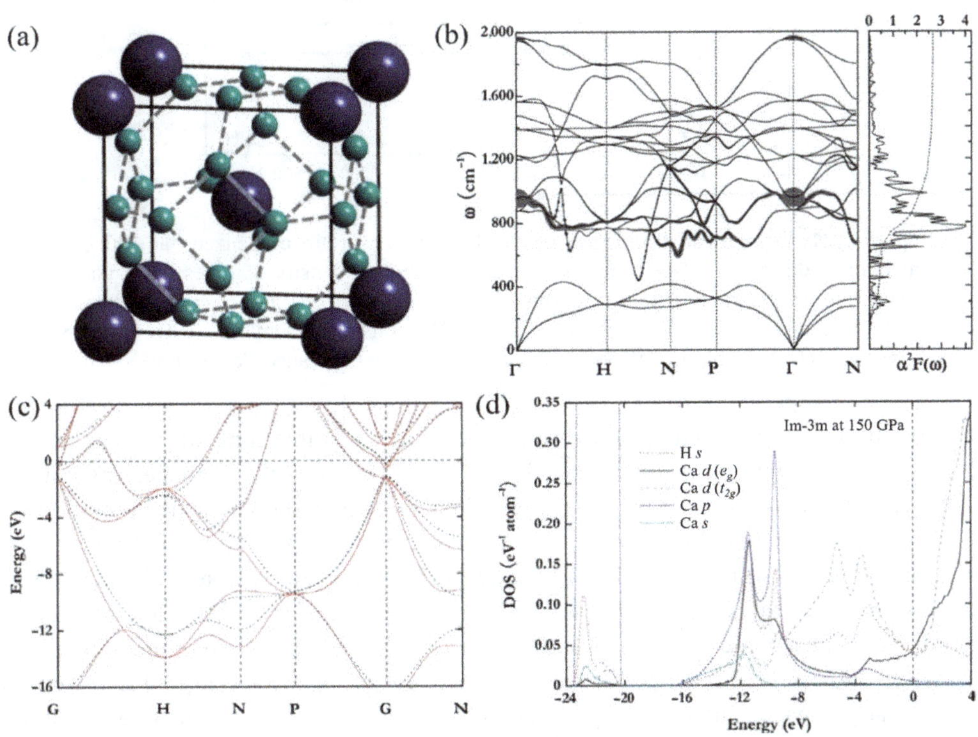

Figure 7.23 Superconductivity of calcium hydride (CaH_6). (a) Crystal structure, where the green spheres are H and the indigo spheres are Ca. (b) Phonon density of states. (c) Electronic energy-band structure. (d) Electronic density of states. Reprinted from Wang H.[2012].

In most hydride compounds, superconductivity arises from the coupling between electron and phonon modes, primarily driven by the vibrations of hydrogen atoms. However, studies have revealed that in aromatic hydrocarbon compounds, the phonon

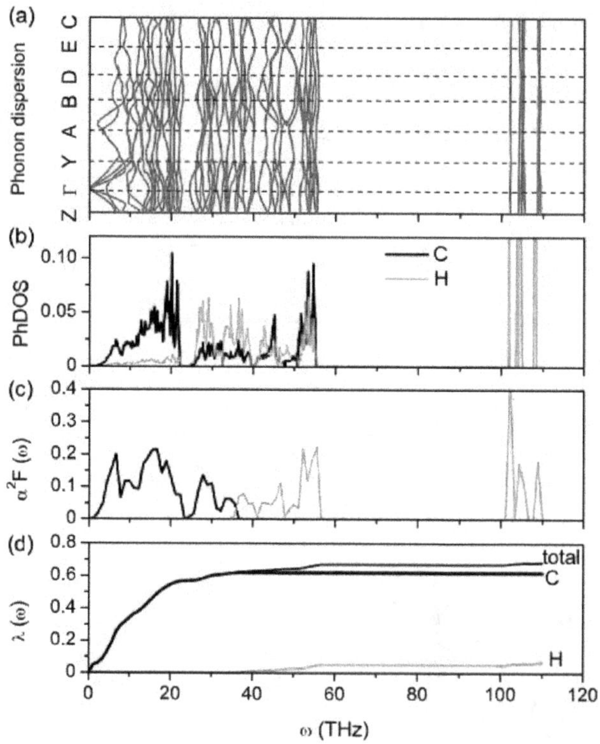

Figure 7.24 Numerical calculation of the $P2_{1/c}$ space group of benzene at 190 GPa. (a) Phonon dispersion of benzene. (b) Calculated phonon density of states for C and H. (c) Eliashberg spectral function $\alpha^2 F(\omega)$ for C and H. (d) Electron–phonon coupling integral $\lambda(\omega)$ for C and H. Used with permission of AIP Publishing, from Zhong G. H.[2018], Copyright 2024; permission conveyed through Copyright Clearance Center, Inc.

contribution to superconductivity originates from the vibrations of carbon atoms. Theoretical calculations conducted by G. H. Zhong's group at the Shenzhen Institute of Advanced Technology, Chinese Academy of Sciences, have shown that the contribution of hydrogen-atom vibrations to the superconductivity of C_6H_6 is minimal under high pressures (as shown in Figure 7.24) [Zhong G.H.2018]. Their theoretical calculations demonstrate that the superconductivity of C_6H_6, like that of most hydride compounds, stems from enhancement of interatomic bonds at high pressures. Further theoretical calculations indicate that increased numbers of benzene rings allow charge delocalization over a larger space; this gradually raises the high-pressure superconducting transition temperature of polycyclic aromatic hydrocarbons (PAHs) from 5–7 K to nearly 30 K as the number of benzene rings increases (Figure 7.25).

The delocalization of hydrogen nuclei in condensed matter systems has garnered significant attention. However, according to conventional BCS theory, achieving high-temperature superconductivity requires additional conditions, such as a modified electron density near the Fermi surface and enhanced electron–phonon coupling. As mentioned earlier, delocalization of hydrogen atoms has been observed in bulk ice at a high pressure

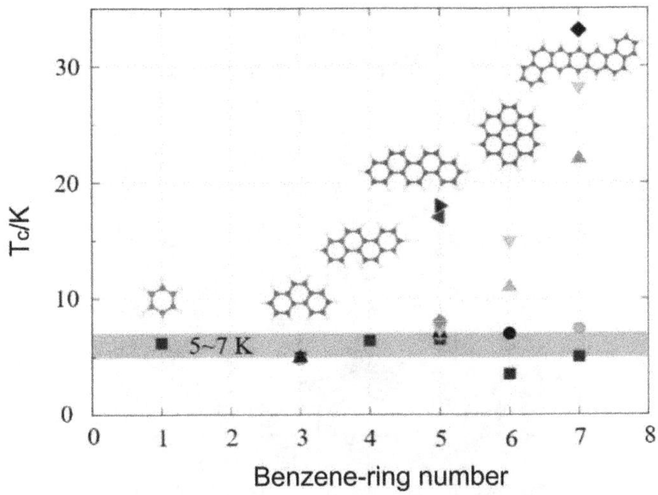

Figure 7.25 Superconducting transition temperature T_c of aromatic hydrocarbons as a function of benzene-ring number. Figure created by Zhong G. H.[2015] licensed under CC BY 4.0.

of ~60 GPa. Recently, Y. Jiang and J. Chen at Peking University observed hydrogen delocalization without high pressure in the hydrogen-bonding network of 2D ice composed of high-density hydrated protons. It is very exciting to see that one can control bonding style by nuclear quantum effects and make the O–H covalent bonds and hydrogen bonds indistinguishable in this symmetric hydrogen configuration, which results in an entirely new full quantum state (Figure 7.26) [Tian Y.2022]. This result suggests that in certain low-dimensional hydride materials, the delocalized hydrogen structure can be effectively stabilized through spatial confinement, including surface/interface conditions. The symmetrization of hydrogen bonds at low (even atmospheric) pressure may lead to metallization and potential superconductivity transitions. This phenomenon is expected to become more pronounced when low-dimensional materials are doped with hydrogen [Wang E.G.2022]. These predictions await further experimental verification. Nonetheless, because of full quantum effects, the distorted structure distribution of light-element (primarily hydrogen) atoms will generate new quantum states of matter that deserve further investigation and in-depth research.

In addition to applying high pressure, doping with alkali metals can also induce superconductivity in hydride compounds. High pressure modifies the lattice constant of a crystal, thereby influencing its physical properties. Doping, by contrast, is a chemical method that alters the energy-band structure of the material, leading to changes in its electronic properties and potential for superconductivity.

The superconductivity of alkali-metal-doped fullerenes was first discovered in 1991; since then, doping has been extensively utilized in superconductivity research on hydride compounds. Theoretical calculations have shown that doping hydrides with alkali metals can induce superconductivity. For example, potassium doping of C_6H_6

Figure 7.26 Schematic of the new atmospheric-pressure 2D ice surface with symmetrical hydrogen nuclei, as predicted by first-principles path-integral molecular dynamics simulations including full quantum effects.

achieves a stoichiometric ratio similar to that of $K_2C_6H_6$, yielding a material with a theoretical superconducting transition temperature of 6.2 K [Zhong G.H.2015]. The introduction of potassium provides additional electrons to the benzene ring, which occupy the low-energy orbitals and elevate the Fermi level so that the new Fermi surface intersects with a different energy band. This enhances the possibilities for superconductivity.

Furthermore, potassium doping reduces the spacing between benzene molecules, making it even smaller than the spacing formed under high pressures. This strengthens interactions between the benzene rings and broadens the energy levels. The calculated phonon DOS reveals that the vibrations of carbon and doped potassium atoms both contribute to the superconductivity (as shown in Figure 7.27).

Researchers at Okayama University discovered that picene ($C_{22}H_{14}$) doped with alkali metals exhibits a superconducting transition temperature ranging from 7 to 18 K, depending on the doping concentration [Mitsuhashi2010]. Superconductivity in PAHs such as alkali-metal-doped phenanthrene ($C_{14}H_{10}$) [Wang X.F.2011] and dibenzo pentacene ($C_{30}H_{18}$) has been successively demonstrated by research groups led by X. H. Chen at the University of Science and Technology of China and G. F. Chen at the Institute of Physics, Chinese Academy of Sciences [Xue M.2012].

Recently, a research group performed controlled electron doping of a water layer using alkali metals to induce metallicity in insulating water. This may open a path to high-temperature superconductivity if solid electron–phonon coupling can be introduced into the system [Mason2021].

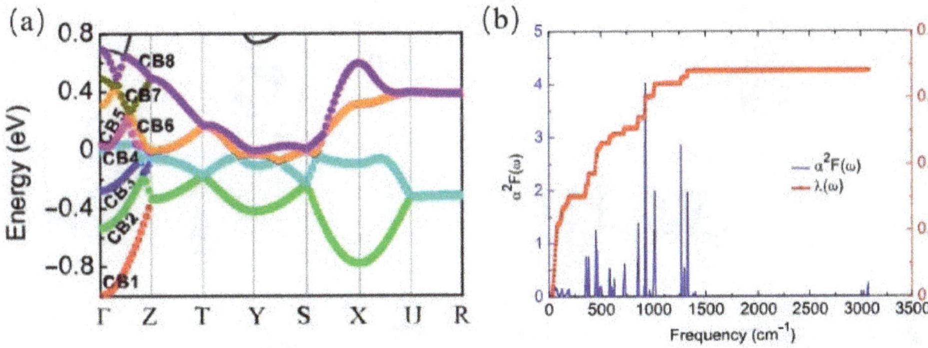

Figure 7.27 Energy band structure and electron–phonon coupling of $K_2C_6H_6$. (a) Energy band structure of $K_2C_6H_6$. (b) The Eliashberg spectral function α^2 F(ω) of $K_2C_6H_6$ and the electron–phonon coupling integral $\lambda(\omega)$ as functions of frequency ω. Figure created by Zhong G. H.[2015] licensed under CC BY 4.0.

7.4 Nonlinear Optical Property of Mineral Hydrides

Dihydrogen phosphates constitute a class of mineral crystals with excellent optoelectronic and nonlinear optical properties that can be used to prepare frequency converters for high power lasers. The general chemical formula for a dihydrogen phosphate is MY_2XO_4, where M = K^+, Rb^+, or NH_4^+, Y = H or D, and X = P. Dihydrogen phosphate crystals contain many hydrogen bonds and undergo a structural phase transition at low temperatures. Studies have shown that the low-temperature phase of crystals containing alkali metal cations is ferroelectric, whereas the low-temperature phase of crystals containing ammonium ions is antiferroelectric.

Ammonium dihydrogen phosphate (ADP) and potassium dihydrogen phosphate (KDP) are representative dihydrogen phosphates. Recently, it was demonstrated experimentally that ADP crystals are superior to KDP crystals in several aspects when they are made into fourth-order harmonic generators and are used in inertial confinement fusion devices [Ji S.H.2013]. Hydrogen bonds affected by nuclear quantum effects play a key role in determining the physical properties of these crystals. Therefore, the differences in the optical properties of ADP and KDP are most likely the result of different local hydrogen-bonding networks in the crystals.

Choudhury et al. studied the structures of ADP and KDP crystals using single-crystal neutron diffraction (see subsection 6.1.2 for experimental methods). They were particularly interested in the physical properties of the deuterated ADP and KDP crystals obtained after isotope substitution [Choudhury2018]. The primary result of deuterium substitution was altered hydrogen bond strength: it was found that the change in hydrogen bond length after deuterium substitution was approximately four times greater in ADP (0.082 Å) than in KDP (0.022 Å). This indicates that the isotope effect in ADP is significantly stronger than that in KDP.

Figure 7.28 shows the electron density distribution in the crystal. In KDP, the electrons of the K–O bonds show no signs of delocalization. In ADP, by contrast, electron

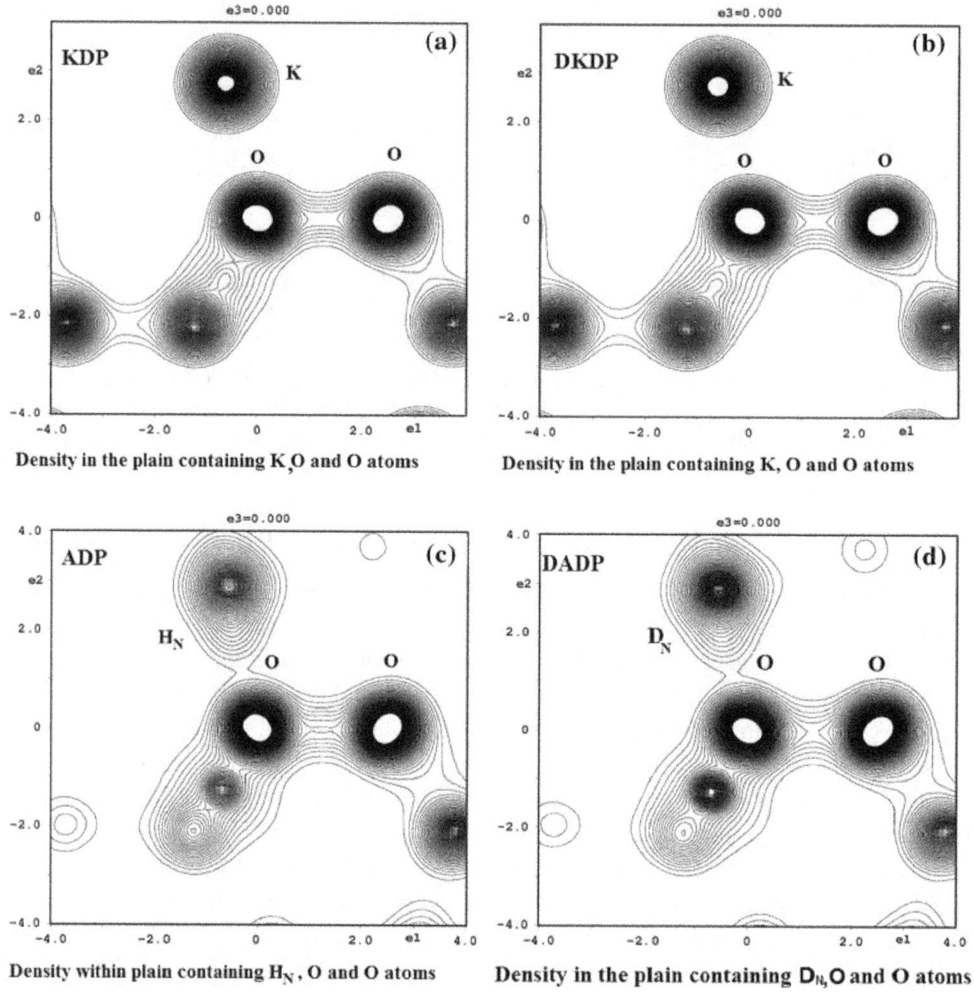

Figure 7.28 Charge-density contour plot formed by the superposition of electron density for each atom in (a) KDP; (b) deuterated KDP (DKDP); (c) ADP; and (d) deuterated ADP (DADP) crystals. From Choudhury[2018]. Reproduced with permission from SNCSC.

delocalization is present in the H_N–O bonds, that is, some covalent bonds are formed between the ammonium ions and the O atoms. This affects the hydrogen bonds, giving rise to anharmonic characteristics. Because the hydrogen bonds are weaker, they are more susceptible to the isotope effect. As the atomic mass of deuterium is larger than that of hydrogen, deuterium substitution reduces the hydrogen vibrational frequency. The calculated ratio of O–H and O–D bond vibrational frequencies, f(O–H)/f(O–D) is greater for KDP than for ADP, indicating that the O–H bonds in ADP exhibit stronger anharmonic effects and more pronounced full quantum effects. This is believed to be the main reason for the higher polarizability and larger nonlinear optical coefficient of ADP.

7.5 Proton Transport through Two-Dimensional Materials

Over the past 20 years, the rapid development of 2D materials has attracted significant interest from both the academic and industrial communities. In this section, we limit our discussion to the problem of proton transport in the 2D materials related to full quantum effects. For an introduction to 2D semiconductor devices, readers may refer to Section 11.6.

In 2004, the discovery of graphene changed the paradigm of research on low-dimensional quantum systems [Novoselov2004]. Monolayer graphene is a 2D material that is only one atom thick. Theoretical studies have shown that a monolayer of graphene with a perfect lattice structure is impermeable to all atoms and molecules in the atmospheric environment [Bunch2008]. For example, the theory has also predicted that it would take billions of years for hydrogen atoms to pass through a monolayer of graphene at room temperature. To drive a hydrogen atom or a proton through a graphene monolayer, the particle must have a sufficiently large kinetic energy [Stolyarova2009, Banhart2011]. However, in 2014, Geim's research team at the University of Manchester used mass spectroscopy measurement to demonstrate that monolayers of graphene and hexagonal boron nitride (h-BN) have unexpectedly high proton permeabilities at room temperature [Hu S.2014]. This result indicates that graphene and h-BN have promising potential for applications in hydrogen separation, fuel cells, and many other technologies involving hydrogen energy use. Subsequent theoretical work also found that graphyne membranes have higher proton penetration rate than graphene membranes [Xu J.Y.2019].

Furthermore, Geim et al. found that the proton transport is related to the electron density distribution of the monolayer film. The nitrogen atoms in the h-BN layer are more attractive to valence electrons than the boron atoms. Consequently, the electron density in the lattice is nonuniform, and protons can pass through the regions with low-density electrons. The symmetric structure of carbon atoms in graphene, by contrast, results in a denser and more uniform electron distribution that is more difficult for protons to penetrate. Therefore, the proton transport capacity of h-BN is significantly better than that of graphene at room temperature (Figure 7.29). Furthermore, Geim et al. found that monolayer molybdenum disulfide films have an even higher electron density than others, and thus worse proton conductivity (see Figure 7.29). In addition, theoretical studies suggest that the partial hydrogenation of graphene during proton transport, which makes the lattice slightly sparse, facilitates proton passage [Feng Y.X.2017, Bartolomei2019]. In all these studies, first-principles PIMD calculations suggest that nuclear quantum effects aid the penetration of protons through graphene membranes [Feng Y.X.2017].

Based on the above research, Geim et al. also found that monolayer graphene and h-BN can be used for the isotope separation of hydrogen ions [Lozada-Hidalgo2016]. They demonstrated that protons pass much more quickly through the membrane than deuterons (as shown in Figure 7.30); the separation factor is on the order of 10 at room temperature. The origin of the isotope selectivity is the lower zero-point energy of the deuterons, 60 meV below that of the protons. The zero-point oscillation of a particle with a mass of about 4,000 times that of an electron is purely caused by full quantum effects, and can be a physical source of room temperature transport. The difference in

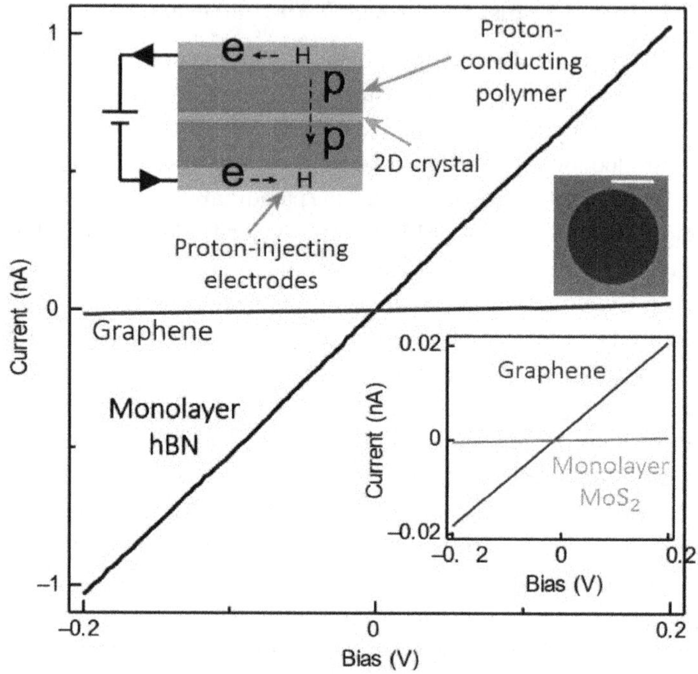

Figure 7.29 Current–voltage (I–V) characteristics for monolayer films of h-BN, graphene, and MoS_2. The upper inset shows the experimental setup. The middle inset (scale bar 1 μm) shows an electron micrograph of a graphene membrane before the deposition of the Nafion. From Hu S.[2014]. Reproduced with permission from SNCSC.

zero-point energies is equivalent to the difference in potential barriers during transport, which ultimately affects the rates at which protons and deuterons are transported through the barrier. This work thus provides an effective way to realize application of hydrogen isotope enrichment (See Section 11.1 for further discussion.)

S. Zhang et al. at Tianjin University collaborated with Geim's group and used electrochemical pumping to demonstrate experimentally that the separation factor of hydrogen isotopes can reach up to ~8 [Lozada–Hidalgo2017]. This nontoxic and environment-friendly method of isotope separation has low energy consumption and a small footprint, making it suitable for industrial applications. S. Zhang's group also found that illuminating a Pt-nanoparticle-decorated graphene membrane with low-intensity visible light strongly enhances proton transport [Lozada–Hidalgo2018]; this is known as the photo-proton effect. These findings suggest new ways to fabricate membranes for artificial leaves.

In a recent study, Geim et al. continued collaborating with S. Zhang's group and discovered that monolayer graphene and boron nitride films can be used to prepare perfect proton-conducting membranes [Mogg2019a]. In the permeation experiment, hydrochloric acid solutions with different concentrations were placed on both sides of the micron-sized monolayer membrane, and the relationship between the membrane potential and the concentration gradient of the hydrochloric acid solution was measured. Using the Nernst–Planck formula, the proton transport number was calculated to be 1. This implies that

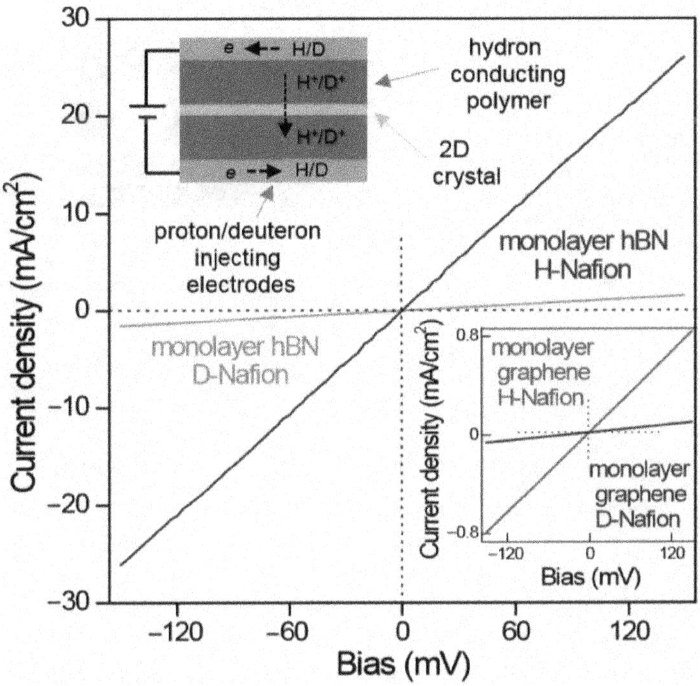

Figure 7.30 Current–voltage (I–V) characteristics for proton and deuteron transport through h-BN and graphene. From Lozada-Hidalgo[2016]. Reprinted with permission from AAAS.

the current through the membrane was due to proton transport alone. Other ions were blocked by the membrane, even chloride ions (which are the ion hydrates with the smallest diameter), confirming 100% proton selectivity in single-atom membranes. In addition, Geim et al. exchanged protons with potassium ions that exist naturally in mica membranes [Mogg2019b], creating ~5Å-wide tubular channels within the mica crystal (see Figure 7.31). Further DFT calculations corroborated the formation of this tubular structure. Because the interior of the channels contains only hydroxyl groups, which facilitate proton transport, ion-exchanged mica can also be a novel proton-conducting membrane.

To better understand proton transport through 2D materials, a series of computational studies have been performed by other research groups. The trans-membrane transport barriers for protons, firstly calculated by first-principles molecular dynamics simulations, are 1.25–1.40 eV for graphene and 0.7 eV for monolayer boron nitride membranes; these theoretical results are much larger than the experimentally measured values of 0.8 eV (graphene) and 0.3 eV (monolayer boron nitride membranes). Classical molecular dynamics simulations ignore nuclear quantum effects and therefore require correction. In fact, nuclear quantum effects are critical to the proton transfer. They effectively lower the potential barrier, making proton diffusion slightly more favorable, and then accelerate isotope separation. Figure 7.32 shows the results calculated from the Wigner–Landau–Lifshitz formula after considering nuclear quantum effects. These results clearly show that the proton

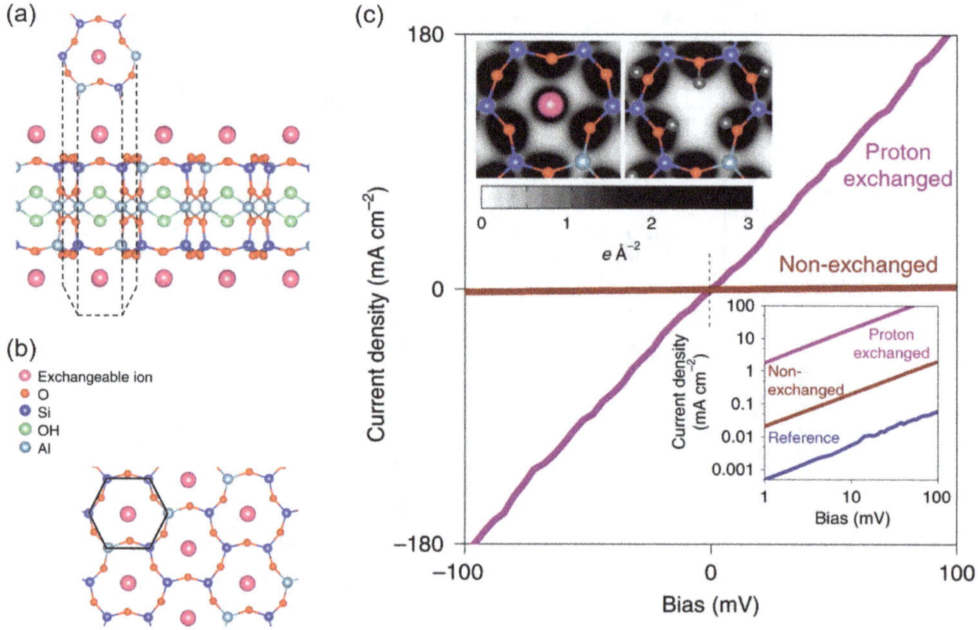

Figure 7.31 Ion-exchanged mica as a proton-conducting medium. (a) Cross-sectional schematic of a mica monolayer; the mica crystal consists of Si–O tetrahedra and Al–O octahedra, forming a 2D sheet. Exchangeable ions (magenta balls) can be adsorbed onto these aluminosilicate layers. (b) Plane-view schematic of the corresponding structure. Si, Al, and O atoms can pierce through the tubular channels (see 3D projection indicated as dashed lines in (a)). The hexagonal channels contain hydroxyl groups and exchangeable ions. (c) The current–voltage (I–V) characteristics before and after proton exchange, indicating the proton transport capacity of mica. From Mogg[2019a]. Reproduced with permission from SNCSC.

tunneling potential barrier is much lower than the classical barrier height [Poltavsky2014, 2018]. Again, this example proves that nuclear quantum effects cannot be neglected when studying proton transport in 2D material membranes.

Recently, a theoretical study conducted by Li's group at Peking University further showed that, if nuclear quantum effects are considered, partial hydrogenation of 2D graphene and h-BN membranes can effectively reduce the potential barrier for proton transport [Feng Y.X.2017]. Comparative calculations show that the sp^2 hybridized orbitals of pure graphene and h-BN induce strong chemisorption of protons. Consequently, protons are generally bound to the material surface, which is not conducive to proton transport in the plane of the 2D material. However, hydrogenation can transform the sp^2 hybridization of carbon atoms into sp^3 hybridization. This effectively weakens the strong proton chemisorption exerted by graphene and transforms it into relatively weak physisorption, thus significantly reducing the potential barrier for proton passing through. The same applies to 2D h-BN materials. In addition, partial hydrogenation causes lattice distortion and increases transport-pore size on the hexatomic rings of the graphene and h-BN planes. These results suggest that both 2D graphene and h-BN enhance proton transport after certain chemical modifications, as shown in Figure 7.33.

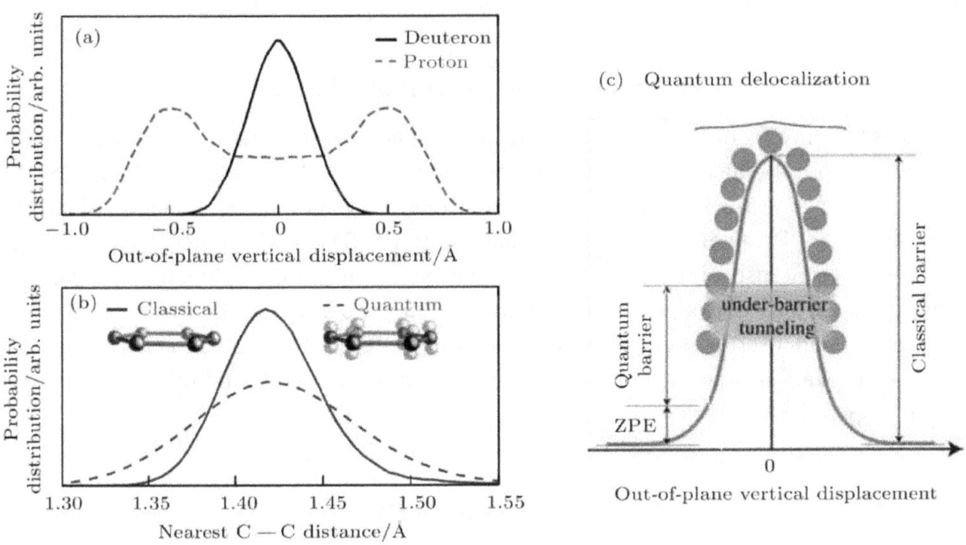

Figure 7.32 Schematic of the transition state during quantum tunneling through a 2D material. (a) Probability distribution of protons and deuterons in the direction perpendicular to the plane of the hexatomic ring. (b) Classical and quantum simulated probability distributions as a function of distance between two nearest atoms. (c) Quantum tunneling mechanism. Figure 7.32(a) reprinted from Poltavsky[2018], with the permission of AIP Publishing; Figure 7.32(b–c) created by Poltavsky[2014] licensed under CC BY 4.0.

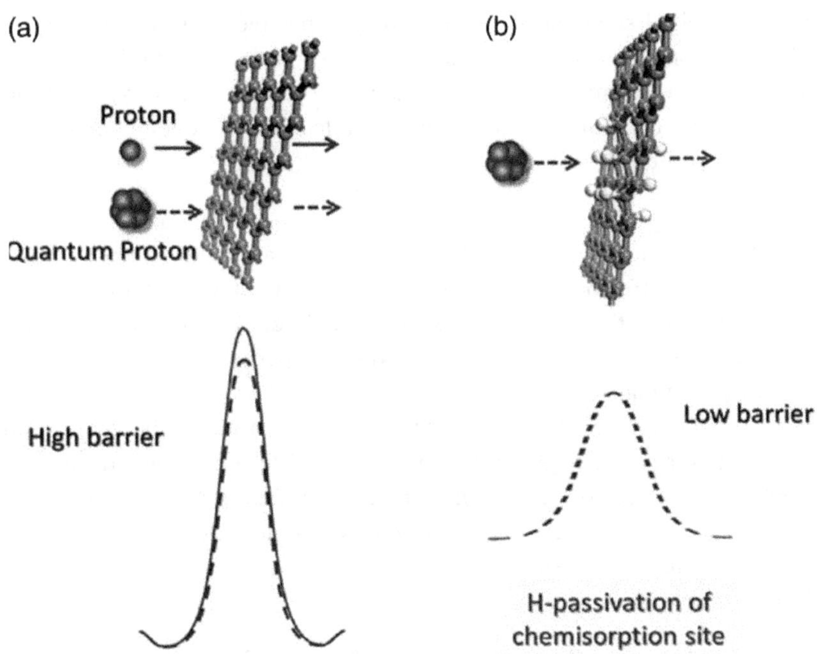

Figure 7.33 (a) Schematic representation of proton transport through ideal and (b) partially hydrogenated graphene, and the corresponding transport potential barriers in these two cases. The solid and dashed curves represent the proton transport potential barriers obtained by classical and quantum simulations, respectively. Reprinted with permission from Feng Y. X.[2017]. Copyright 2024 American Chemical Society.

7.6 Proton Transport in Bulk Materials

In fields ranging from physics science to life sciences, it is critical to study the proton migrations in bulk materials. To study the influence of nuclear quantum effects (NQEs) on the proton transport process in condensed matter systems, Q.F. Zhang et al. further optimized Car–Parrinello Molecular dynamics (CPMD) based on a full quantum simulation program, especially improving k-space sampling in the calculation of the electronic structure of periodic lattice systems, so that it can be applied to computational studies of condensed matter systems [Zhang Q.F.2008]. The description of the relevant specific derivations and the computational procedures can be found by interested readers in Appendix A.

7.6.1 Transition Rate

The fundamental migration mechanism governing long-range proton transport in periodic structures has been extensively studied in theory. In general, the hydrogen-atom transition rate k between two adjacent lattice sites can be simply described as:

$$k = \frac{k_B T}{h} \frac{Z^{\ddagger}}{Z}, \tag{7.1}$$

where k_B is Boltzmann's constant, h is Planck's constant, and Z and $Z^{\ddagger}$ denote the partition functions of the initial state and transition state, respectively. In the harmonic approximation, the vibrational characteristic phonon energies of the solid in the steady state and transition state can be expressed as $\{\hbar\omega_i\}_{i=1}^{N}$ and $\{\hbar\omega_i^{\ddagger}\}_{i=1}^{N-1}$, respectively. According to the Eq. (3.1), in the classical limit, Arrhenius reaction rate of corresponding migration reaction process can be rewritten as:

$$k = k_0^{cl} e^{-\Delta E / k_B T}, \tag{7.2}$$

where the activation energy ΔE is equal to the classical migration potential V_m, that is, to the potential energy difference between the stable state and the transition state. It is obvious that this classical method excludes the quantum effect of the hydrogen nuclei, and the discrepancy between the theoretical values and the experimental results arises from the classical treatment of nuclei.

In 1970, C. P. Flynn and A. M. Stoneham proved that the Arrhenius formula could still be used to describe the quantum tunneling rate of discontinuous transition between self-trapped ground states of protons in neighboring double potential wells if the surrounding phonon excitations can be described by classical physics [Flynn1970]. Then the approximate quantum transport rate is

$$k = \frac{1}{\hbar}\left(\frac{\pi}{4E_c k_B T}\right)^{\frac{1}{2}} J_o^{\,2} e^{-E_c / k_B T}, \tag{7.3}$$

where J_0 is the transition matrix. The activation energy E_c is not the conventional self-trapped ground-state potential (shown in Figure 7.34(a)) but the coincidence energy, i.e., the energy required to construct the coincidence configuration, in which hydrogen exists in the quantum delocalized state in the double potential well (as shown in Figure 7.34(b)). This method of calculating the transport rate considering quantum tunneling effects is regarded as one of the important full quantum calculational techniques. It requires solving for a number of parameters, such as the lepton matrix in the Flynn-Stoneham model. Generalized density function theory provides the best approach to obtaining them. An accurate description of the quantum delocalized states of atomic nuclei is the key to the solution of these problems.

7.6.2 Quantum Correction

In 2004, P. G. Sundell et al. used first-principles calculations to obtain the recombination energies and corresponding transition matrices required to construct coincidence configurations [Sundell2004]. They performed specific calculations for two systems: hydrogen in metal bulk materials (H/Nb, H/Ta) and hydrogen on a metal surface (H/Cu(001)). They started by considering a single hydrogen atom interacting with many metal atoms on their lattice sites of the host materials. The state of motion of the hydrogen nucleus in the main lattice can be described by a wave function $\psi_n(r;\{R\})$, where $\{R\}$ denotes the positions of the metal atoms and r denotes the position of the hydrogen atom. When the hydrogen atom is in the quantum state n, the expectation value of the force acting on the metal atom I can be written as:

$$\boldsymbol{F}_I = -\left\langle \psi_n(r;\{R\}) \middle| \nabla_I V(r;\{R\}) \middle| \psi_n(r;\{R\}) \right\rangle \tag{7.4}$$

where $V(r;\{R\})$ is the interaction potential. The simulation is performed in two steps. First, the hydrogen nucleus is treated as a classical point particle to determine the lattice configuration of the self-trapped state. The force in Eq. (7.4) can be simplified to the DFT framework's Hellmann–Feynman interaction in this way. In the coincidence configuration, the delocalized hydrogen nucleus can be represented by two point-like particles with equal mass, located at the minima of adjacent potentials with symmetrical displacement. In this case, the force $\boldsymbol{F}_I$ equals the weighted average of the Hellmann–Feynman interaction. In the second stage, the energies of hydrogen atoms at various positions within the lattice unit are determined by extensive DFT calculations, whereas the 3D equipotential energy surface corresponding to these energies is constructed for both self-trapped and coincidence configurations (as shown in Figure 7.34(c) and (d)), where Figure 7.34(a) and (b) are their wave functions, respectively. Finally, by solving the Schrödinger equation for the hydrogen atom motion, the ground-state and excited-state energies are obtained, and also the transition matrix of the coincidence configuration.

Table 7.3 lists the results of a proton transfer in bulk materials and on surfaces. V_{st} is the potential energy obtained by allowing the metal atom positions to relax around a hydrogen atom located at a stable interstitial site. V_c is the additional potential energy required to transform the self-trapped configuration into the lowest symmetric migration configuration. V_m is the migration barrier energy calculated by the classical method. As can be seen from the table, the potential energy term V_c for the coincidence configuration is much smaller than that of the other two cases, whereas the kinetic energy terms for the

Table 7.3 *Potential energies for self-trapped configuration* (V_{st}), *coincidence configuration* (V_c), *and classical migration* (V_m). K_c *and* K_m *are the kinetic energy contribution terms (zero-point energy).* $E_c \equiv V_c + K_c$ *and* $E_m \equiv V_m + K_m$ *are the total energies. Unit: meV.*

	Bulk				Surface	
Energy	H/Nb	D/Nb	H/Ta	D/Ta	H/Cu(001)	D/Cu(001)
V_{st}	189		185		26	
V_c	19		19		11	
V_m	148		180		126	
K_c	+4.6	+3.8	+9.4	+7.1	−2.6	−1.7
K_m	−22	−17	−15	−12	+53	+43
E_c	24	23	29	26	8.6	9.5
E_m	126	131	165	168	180	170

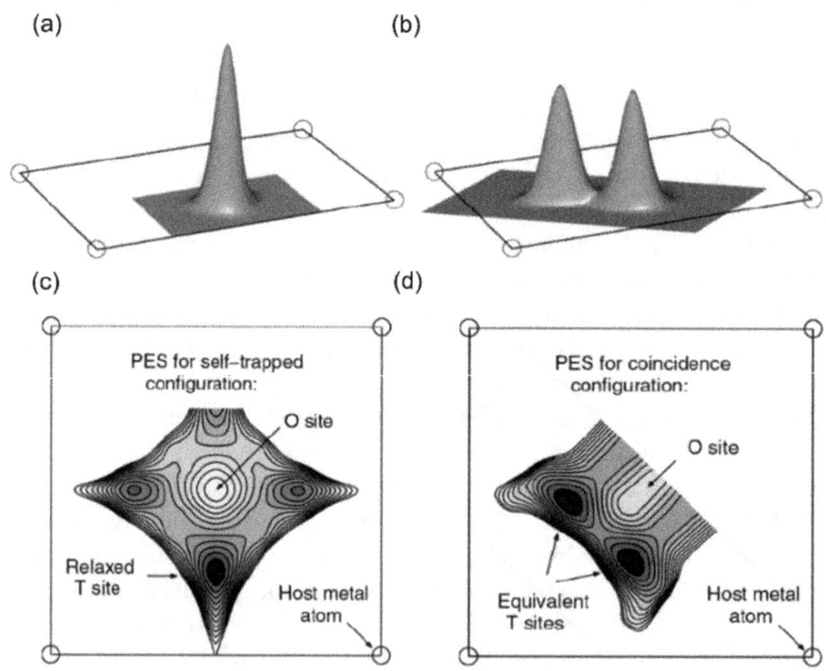

Figure 7.34 Proton transport subject to quantum tunneling. (Top): Wave functions in the (001) plane for H (a) self-trapped at T site and (b) in the coincidence configuration between two neighboring T sites in Ta metal. (Bottom): Potential energy surfaces of the (001) plane for H/Ta in (c) self-trapped and (d) coincidence configurations.

self-trapped configuration and the coincidence configuration are similar. For the two bulk materials corresponding to Nb and Ta, the Flynn–Stoneham theory is applicable in the temperature range 100 K < T < 200 K [Flynn1970]. The predicted coincidence energy E_c and the transition matrix J_0 agree well with the experimental results. The calculated results at the metal surface are also in agreement with the experimental values. Note that in this approach the hydrogen nucleus is described by a wave function $\psi_n(r;\{R\})$ to study the activation energy of proton transport. Based on this, the coincidence energy and the corresponding transition matrix required to construct the coincidence configuration are obtained using the first-principles calculation method. This method is a useful theoretical approach for investigating the full quantum process of hydrogen atom transport, as demonstrated by the excellent agreement between calculated and experimental results.

The simulation of quantum vibrational energies in stable and transition states is a key step in understanding the quantum effects in proton transport. According to the classical transition mechanism, in which only the potential barrier is considered, the pre-exponential factor in Eq. (7.2) can be written as

$$k_0^{cl} = \frac{1}{2\pi}\frac{\prod_{i=1}^{N}\omega_i}{\prod_{i=1}^{N-1}\omega_i^{\ddagger}}. \tag{7.5}$$

This is generally regarded as the characteristic attempt frequency. The vibrational energy levels become discrete and are broadened because of the slight reduction in mass caused by a proton defect. By replacing the classical partition function in the equation, the potential can be corrected for quantum-mechanical effects. The corrected activation energy is given by:

$$\Delta \mathrm{E} = V_m + \frac{1}{2}\sum_{i=1}^{N-1}\hbar\omega_i^{\ddagger} - \frac{1}{2}\sum_{i=1}^{N}\hbar\omega_i. \tag{7.6}$$

It is the sum of the classical migration potential and the zero-point vibrational energy (ZPE) correction. After taking the quantum-mechanical transitions into account, the quantized pre-exponential factor becomes:

$$k_0^{qm} = \frac{k_B T}{h}\frac{\prod_{i=1}^{N}\left(1-e^{-\hbar\omega_i/k_B T}\right)}{\prod_{i=1}^{N-1}\left(1-e^{-\hbar\omega_i^{\ddagger}/k_B T}\right)}. \tag{7.7}$$

In the high temperature limit, Eq. (7.7) reduces to the formula for the classical pre-exponential factor (Eq. (7.5)). By contrast, when the temperature is relatively low, the pre-exponential factor can be approximated as $\frac{k_B T}{h}$, which is independent of the transport-ion mass and the properties of the host lattice.

The incorporation of NQEs into activation energy calculations via ZPE correction enables the discovery of novel functional materials. Although recent theoretical studies of adsorbent and catalytic materials have paid more attention to quantum vibrational effects on transition states, the ZPE correction is often non-negligible when small molecules are involved in the calculation of activation energy [Peterson2010, Shi C.2014].

7.6.3 Perovskite-Type Oxides

Proton transport in perovskite oxides furnishes a typical example of nuclear quantum effects on hydrogen migrations in bulk materials. Using the theoretical methods derived above, the NQEs on the proton transport in this system can be calculated. As shown in Figure 7.35(a), long-range proton transport in perovskite oxide can be decomposed into two basic processes: the transfer (T) process between two neighboring oxygen atoms, and the reorientation (R) process around a single oxygen atom. As it is difficult to characterize these microscopic processes accurately in experiments, theoretical simulations are necessary for describing microscopic proton conduction. The stable and transition states, proton-transport path, and transport potential of $BaZrO_3$ especially need to be investigated in detail. The consensus of various early theoretical studies is that the transfer process is slower than the reorientation process in this type of oxides, making it the rate-controlling step that limits the long-range proton transport. This result can be easily understood in a classical model, as the transfer process causes hydrogen bond breaking and recombination, whereas the reorientation process does not; therefore, the transfer process requires more energy. In the experimental IR spectra, a large red-shift of O–H bond stretching mode has been observed, which indicates that significant breaking and recombination of hydrogen bonds were involved in proton transport. This experimental result suggests that the transfer process should be faster than the reorientation process, in contrast to the result of classical theory. Thus, there are still many unresolved differences between the experimental results and classical theoretical calculations without considering nuclear quantum effects.

By introducing a ZPE correction, P. G. Sundell et al. found that the potential barriers for the transfer and reorientation processes of protons in perovskite oxides are reduced by 0.12 eV and 0.04 eV, respectively, which is closer to the experimental values [Sundell2007]. Similar simulation methods have been applied to doped perovskite systems. These results suggest that nuclear quantum effects play a non-negligible role in the proton-transport properties of perovskite oxides.

In perovskite systems such as $BaZrO_3$, both quantum fluctuations and quantum tunneling are crucial to the proton transfer process. Therefore, it is particularly important to consider both types of NQEs on this system simultaneously using the first-principles PIMD method. Q.F. Zhang et al. employed fully quantum-mechanical molecular dynamics simulations to investigate the NQEs of hydrogen atoms on the proton transfer (T) and reorientation (R) processes in $BaZrO_3$ [Zhang Q.F.2008, 2010]. Because it is extremely difficult to describe these two fundamental processes using Cartesian coordinates, they introduced a different reaction coordinate for each process in the following studies. The difference between the distances from hydrogen nucleus H^+ to the oxygen atoms O1 and O2 (i.e., $\Delta\vec{R} = \left|\vec{R}_{O1,H}\right| - \left|\vec{R}_{O2,H}\right|$) was chosen as the reaction coordinate for the T process. The angle θ between the H–O bond and the O–Ba bond was chosen as the reaction coordinate for the R process. In this research, the first-principles centroid PIMD method was used. Since the energy barriers of both the T and R processes are not very low, protons in direct molecular dynamics simulations can only move near the stable site and cannot reach or even approach the transition state within the simulation time, which is on the order of picoseconds. Therefore, the energy barriers cannot be calculated using the probability distribution of the protons during their transport. To address this problem, Q.F. Zhang et al. combined the constrained molecular-dynamics method with the PIMD method by calculating the averaged constrained forces when the reaction coordinates in the reaction path have fixed values. Integrating the averaged constrained force yields the free energy difference

$$W(\xi_2) - W(\xi_1) = \int_{\xi_1}^{\xi_2} d\xi' \left\langle \frac{\partial H}{\partial \xi} \right\rangle_{\xi'}^{cond}, \tag{7.8}$$

where H is the Hamiltonian of the system motion, and $\left\langle \frac{\partial H}{\partial \xi} \right\rangle_{\xi'}^{cond}$ is the averaged constrained force within the constrained system. The reaction coordinate is fixed at a certain value and its evolution rate is always zero, that is, $\xi(\vec{R}) = \xi'$ and $\dot{\xi}(\vec{R}, \dot{\vec{R}}) = 0$. Because of the constraints, the Lagrange multiplier λ is introduced and the Lagrangian for the motion of the system can be extended to

$$L'(\{\vec{R}\},\{\dot{\vec{R}}\}) = L(\{\vec{R}\},\{\dot{\vec{R}}\}) - \lambda\sigma(\{\vec{R}\}), \tag{7.9}$$

where $\sigma(\vec{R}) = \xi(\vec{R}) - \xi'$. The value of λ can be found using the Rattle algorithm for solving the equations of motion in molecular dynamics simulations [Andersen1983]. The averaged constrained force is equal to

$$\frac{dW}{d\xi'} = \frac{\left\langle Z^{-1/2}\left[-\lambda + k_B T G\right]\right\rangle}{Z^{-1/2}}, \tag{7.10}$$

where

$$Z = \sum_I \frac{1}{M_I}\left(\frac{\partial \xi}{\partial \vec{R}_I}\right)^2, \tag{7.11}$$

$$G = \frac{1}{Z^2}\sum_{I,J} \frac{1}{M_I M_J}\frac{\partial \xi}{\partial \vec{R}_I}\cdot\frac{\partial^2 \xi}{\partial \vec{R}_I \partial \vec{R}_J}\cdot\frac{\partial \xi}{\partial \vec{R}_J}. \tag{7.12}$$

Using the above method, Q.F. Zhang et al. derived the potential barriers of the proton transfer and reorientation processes in both the classical molecular dynamics and centroid PIMD simulations. It is worth noting that in the PIMD simulations, the centroid of the proton needs to be fixed at a certain reaction coordinate, $\xi_c(\{\vec{R}^S\}) = \xi_c'$, whereas the noncentroid mode is free to move.

The free energy evolution curves for the T and R processes are shown in Figure 7.35(c). The four dashed curves represent the quantum potential barriers calculated using the fully quantum-mechanical method at temperatures of 100 K, 200 K, 300 K, and 600 K. It can be seen that when the temperature is high, the difference between the quantum and classical potentials is small. This indicates that the nuclear quantum effect is suppressed at high temperatures and that classical thermal motion dominates the proton-transport process.

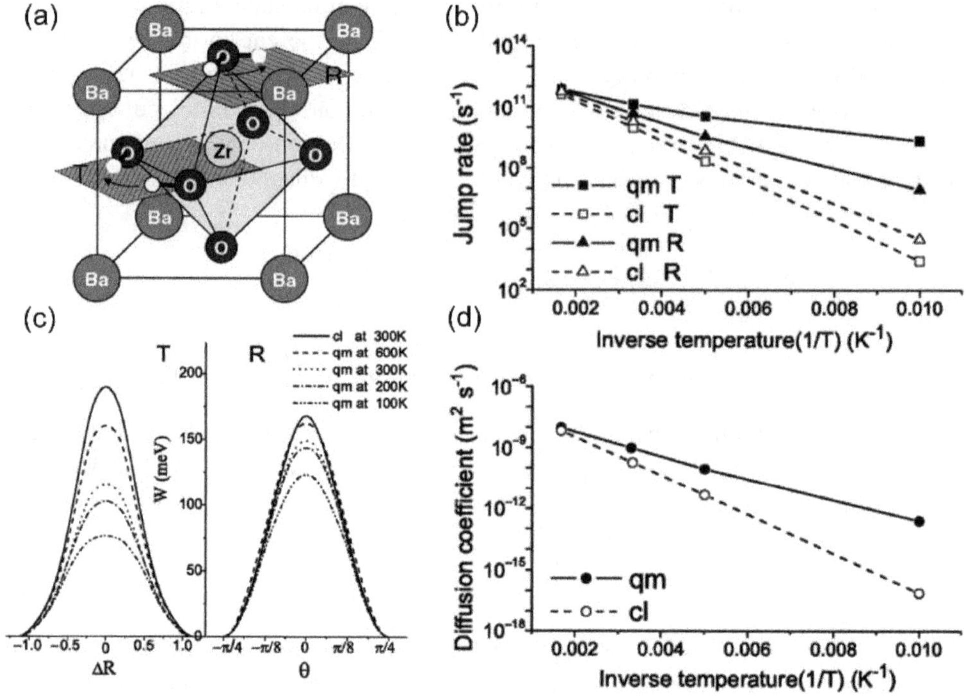

Figure 7.35 Proton transport in $BaZrO_3$. (a) Schematic showing the meaning of the transfer (T) and reorientation (R) processes. The small white balls represent the stable positions of the protons. (b) Classical (cl) and full quantum (qm) proton jump rates k_T (for the transfer process) and k_R (for the reorientation process) at different temperatures. (c) Classical and full quantum potentials for the T and R processes at temperatures of 100 K, 200 K, 300 K, and 600 K. (d) Classical and full quantum proton diffusion coefficients at different temperatures. Reprinted figure with permission from Zhang Q. F.[2008]. Copyright 2024 by the American Physical Society.

By contrast, at low temperatures, the quantum potential barrier is lower, suggesting that the quantum vibration and quantum tunneling effects are significant.

It is clear that the energy barrier exhibits a larger variation for the T process than for the R process when the quantum vibrations of the nucleus are considered. The potential barrier of the T process is also much smaller than that of the R process at low temperatures. When T = 100 K, the quantum potential barrier of the T process is 110 meV lower than that obtained by the classical dynamics method, whereas the quantum potential barrier of the R process is only 35 meV lower than the value obtained from the classical method. Therefore, NQEs play different roles in the T and R processes, which can be attributed to the different ZPE changes during the reaction. The ZPE difference between the stable state and the transition state of protons qualitatively reflects the quantum effect of vibrations on the energy barrier.

Although the transfer and the reorientation are the fundamental processes of proton migrations in $BaZrO_3$, they are profoundly different in nature. The transfer of protons is the process of breaking one O–H bond and forming another O–H bond. In this process, the stretching of the O–H bonds that are close to the transition state is softened, which significantly reduces the zero-point vibrational energy of the transition state. By contrast, during the R process, there is no bond breakage or recombination. As a result, the quantum vibrational effects cancel out during the rotation process of H atom, and the change in the energy barrier is relatively small.

From the above discussion, it is clear that quantum vibration effects are mainly reflected in changes in interatomic chemical bonds. An earlier study of quantum vibration effects by M. Parrinello and others illustrates this further [Tuckerman2002]. Using theoretically calculated potential barriers, both the classical proton transition rates k_T^{cl}, k_R^{cl} and the quantum transition rates k_T^{qm}, k_R^{qm} were obtained at 100, 200, 300, and 600 K. The results are shown in Figure 7.35(b). Because the transition rate depends mainly on the potential barrier, the potential barrier of the T process is larger than that of the R process under the classical approximation; therefore, $k_T^{cl} < k_R^{cl}$. Nuclear quantum effects cause the T process potential barrier to decrease significantly, whereas the decrease in the potential barrier of the R process is smaller. Therefore, when the temperature is not very high, the potential barrier of the T process becomes smaller than that of the R process, and $k_T^{qm} > k_R^{qm}$; the R process becomes the slower process. We see that the quantum vibration effect of the nucleus can completely reverse the relative speed of the two proton transport processes.

Using the transition rates of the two processes, the long-range proton-transport rate D in $BaZrO_3$ can be written as

$$D = \left(\frac{a^2}{6}\right) k_T k_R / \left(k_T + k_R\right). \tag{7.13}$$

The classical and quantum long-range diffusion rates at different temperatures are plotted in Figure 7.35(d). To realize long-range proton transport, both the T and R processes must occur. Therefore, both k_T and k_R determine long-range transport. When $k_T \ll k_R$ or $k_T \gg k_R$, Eq. (7.13) yields $D \approx \left(a^2/6\right) k_T$ or $D \approx \left(a^2/6\right) k_R$, respectively. Thus, the magnitude of D depends on which process is slower. The rate of the long-range proton transport is therefore determined by the T process in the classical approximation and by the R process in the full quantum simulation including nuclear quantum effects. This study also illustrates how the results yielded by the numerical simulation including full quantum effects are in perfect agreement with the experimental results of far-IR spectroscopy.

A similar method can be used for the quantitative simulation of quantum tunneling. The centroid of the reaction coordinate, $\xi_C = \frac{1}{P}\sum_{S=1}^{P} \xi^{(S)}$, is fixed at the transition state of the T and R processes ($\xi_c = \xi^{\#}$), whereas the coordinates of each pathway segment are free to evolve. Calculating the distribution $P(\xi)$ of the quantum path allows us to probe the proton quantum effects and study proton transport properties near the transition state. The distribution of the quantum paths in the reaction coordinates near the transition state at 100, 200, 300, and 600 K is shown in Figure 7.36(a). It becomes narrower as the temperature increases and eventually converges to the classical limit ($P(\xi) = \delta(\xi - \xi^{*})$). With decreasing temperature, the delocalization of the quantum paths becomes more prominent, and the distribution becomes closer to the stable state. This indicates that quantum

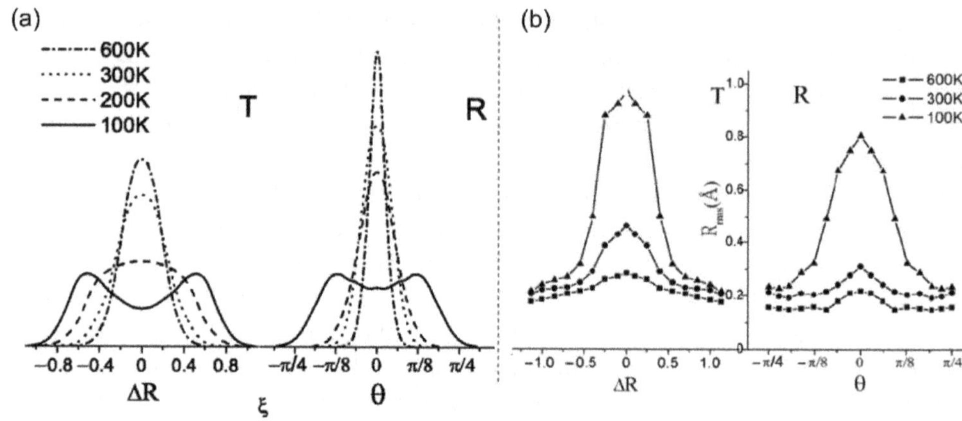

Figure 7.36 (a) The distribution of the quantum paths after fixing the centroid in the transition states of the T and R processes at 100 K, 200 K, 300 K, and 600 K. (b) The imaginary-time rms correlation functions $R_{rms}\left(\frac{\hbar\beta}{2}\right)$ along the transfer and reorientation pathways in reaction coordinates at 100 K, 300 K, and 600 K. Figure 7.36(a) reprinted with permission from Zhang Q. F.[2008]; Copyright 2024 by the American Physical Society; Figure 7.36(b) reprinted from Zhang Q. F.[2010], with permission of the author.

tunneling dominates the proton transport process at low temperatures where the thermal motion of protons is suppressed. At 100 K, the maximum probability distribution of "quantum nuclei" is not located at the top of the transition state, but on its sides. To examine the quantum tunneling effect during proton motion in detail, the rms displacement correlation function for protons during the T and R processes at 100 K, 300 K, and 600 K was calculated using the following equation:

$$R_{\text{rms}}(\tau) = \left\langle \left(v(0) - v(\tau) \right)^2 \right\rangle^2 . \tag{7.14}$$

For the T process, $v = \Delta R$; for the R process, $v = \left|\vec{R}_{OH}\right|\theta$. Figure 7.36(b) shows R_{rms} values for $\tau = \hbar\beta/2$ at different temperatures. $R_{rms}\left(\frac{\hbar\beta}{2}\right)$ can quantitatively characterize the dispersion of quantum nuclei, which increases when quantum tunneling becomes more pronounced. According to the figure, lower temperatures correspond to higher R_{rms} values. As protons approach the transition state, the quantum distribution of nuclei becomes more discrete. At 100 K, when $\Delta R \approx \pm 0.3$Å and $\theta \approx \pm 13°$, the R_{rms} values increase drastically, indicating that the quantum tunneling is considerably enhanced. Therefore, these reaction coordinates can be considered as the boundary between the T and R processes where quantum tunneling occurs. Comparison shows that the quantum tunneling of the T process is more prominent than that of the R process. At 300 K, the quantum tunneling of the T process is still very significant, and its boundary is consistent with that observed at 100 K [Zhang Q.F.2010].

Proton migrations in organic–inorganic hybrid perovskite materials (e.g., $CH_3NH_3PbI_3$) have also attracted extensive experimental and theoretical research interest [Egger2015,

Chen Y.F.2017]. Hybrid perovskites are the most popular solar photovoltaic materials with excellent photovoltaic efficiency. However, their poor optical and thermal stability limit their industrial application. An important constraint affecting the operating stability of photovoltaic materials is ion migration (including proton migration).

Recently, combined experimental and theoretical research has shown that the nuclear quantum effects in $CH_3NH_3PbI_3$ are gradually enhanced as the temperature decreases, and that proton migrations gradually change from a classical thermal process to a quantum tunneling process [Feng Y.X.2018a]. Experimentally, the dynamical behavior of proton diffusion can be characterized by measuring the relaxation time of a constant current. Theoretically, NQEs on the proton transport can be found by combining first-principles MD and PIMD calculations with a transition-state search algorithm. Comparison between the experimental and theoretical results (see Figure 7.37) reveals that, above 270 K, the logarithm of the transition rates k_{ion} for proton has an inverse linear relationship to the reciprocal of the temperature, suggesting that the classical hopping dominates. When 140 K < T < 285 K, the logarithm of the transition rates still decreases linearly as (1/T) increases, but the activation energy of proton migration decreases from 0.4 eV to 0.1 eV. In this temperature range, the classical hopping of protons still dominates. When 75 K < T < 140 K, the linear dependence of log k_{ion} on (1/T) gradually vanishes, signifying

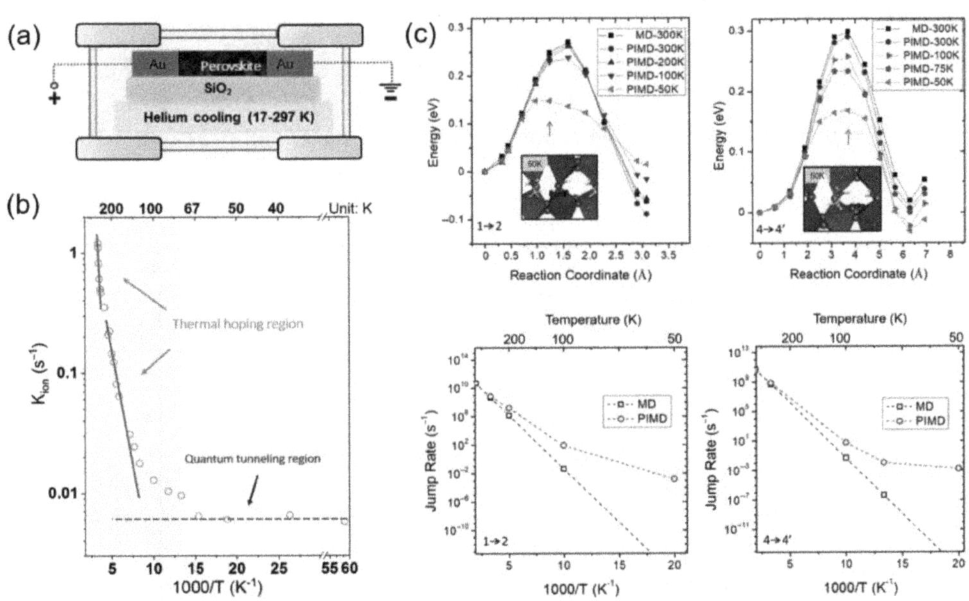

Figure 7.37 (a) Plane structure of Au/MAPbI$_3$/Au on silica substrate. The measurement setup is placed in a vacuum system consisting of a cryogenic stage and electrode wires. (b) The ion transition rate k_{ion} as a function of temperature T (ln k_{ion} vs. 1000/T). (c) The free energy profile and the proton jump rates at different temperatures. The free energies are calculated by both first-principles molecular dynamics (MD) and path integral molecular dynamics (PIMD). Reprinted with permission from Feng Y. X.[2018a]. Copyright 2024 American Chemical Society.

the transition from classical hopping to quantum tunneling. When T < 80 K, the transition rates no longer depend on the temperature. The nuclear quantum effects of protons become more pronounced, and the system enters a stage of deep quantum tunneling.

7.7 Spontaneous Mutations of DNA

As discussed in subsection 3.2.5 on the biochemical reactions, DNA base pairs are affected by quantum tunneling. In a recent study of the genetic stability of DNA, L. Slocombe et al., by considering proton tunneling caused by nuclear quantum effects, found that the probability of genetic mutations is much higher than previously understood [Slocombe2022].

According to the Wigner–Moyal–Caldeira–Leggett (WM-CL) equation [Wigner1932, Caldeira1983], the Wigner function $W(q,p,t)$ satisfies

$$\frac{\partial W}{\partial t} = -\frac{p\partial W}{m\partial q} + \frac{\partial V \partial W}{\partial q \partial p} - \frac{\hbar^2 \partial^3 V \partial^3 W}{24 \partial q^3 \partial p^3} + \mathcal{O}\left(\hbar^4\right) + \gamma \frac{\partial p W}{\partial p} + \gamma m k_B T \frac{\partial^2 W}{\partial p^2}, \tag{7.15}$$

where the first four terms on the right correspond to the Schrödinger dynamics, and the last two correspond to dissipation and decoherence. The initial state of the nonequilibrium distribution in the reaction well can be expressed as

$$W(q,p,t=0) = \frac{1}{\mathcal{N}}\left(1-\hat{h}(q)\right)\exp\left(-\frac{\mathcal{H}}{E_\omega}\right), \tag{7.16}$$

where $\mathcal{H} = p^2/(2m) + V(q)$, $\mathcal{N}$ is a normalization constant, and $\hat{h}(q)$ is the Heaviside step function.

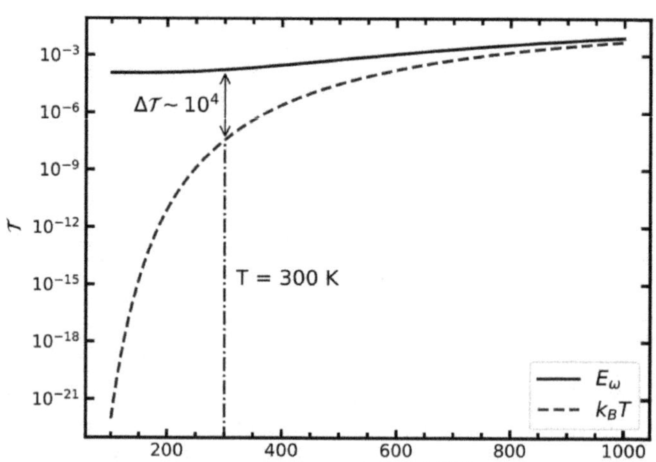

Figure 7.38 Tautomeric occupation probability T predicted classically (dashed curve) and with correction due to full quantum effects (solid curve). Figure created by Slocombe[2022] licensed under CC BY 4.0.

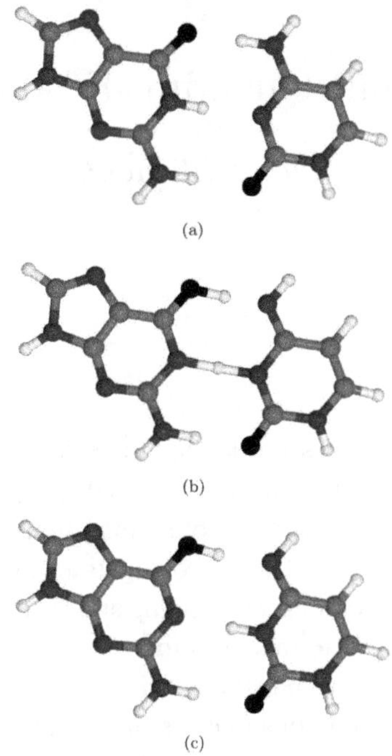

Figure 7.39 Transfer of a proton within a guanine–cytosine (G-C) base pair. Figure created by Slocombe[2022] licensed under CC BY 4.0.

By numerically solving Eq. (7.15), Slocombe et al. showed that hydrogen-bond exchanges between DNA duplexes occur much more frequently than expected, with a probability of ~1.73×10^{-4} (see Figure 7.38). Previously, it has been widely believed that nuclear quantum effects are negligible in the warm, humid, and complex environment inside biological cells and that proton tunneling does not occur in living organisms. Nevertheless, this study shows that it is still possible for the quantum behavior of protons to occur in the local cellular environment through quantum fluctuations, which allow them to overcome the potential barrier from one side of the duplex to the other (as shown in Figure 7.39). This means that when DNA is split into single strands – a process that unfolds rapidly – some protons may be trapped on the wrong side. If this occurs in the first step of intracellular replication before the two strands are broken, the error will be passed down, leading to DNA mismatches in which the bases on one strand do not complement those on the other. This could result in spontaneous mutations.

In condensed matter study, nuclear quantum effects are generally only considered at low temperatures. However, Slocombe et al. showed that in some special cases, such as life activities, full quantum effects cannot be neglected, even near room temperature. Thus, their work has expanded the scope of full quantum effect research and opened new pathways in the development of soft condensed matter physics.

8

Typical Full-Quantum-Effect Systems

Other Elements

In Chapter 7, we extensively discussed full quantum effects in hydrogen – the lightest element in the periodic table – and in hydride compounds composed of hydrogen atoms. We also discussed the full quantum effects on proton migrations in various materials, of which hydrogen atoms exist as a minority. These investigations naturally lead to the question of whether condensed matter containing other light and heavier elements can also exhibit full quantum effects. The answer to this question is yes. Using helium (He), lithium (Li), carbon (C), boron (B), oxygen (O), and silicon (Si) as examples, this chapter explores their novel physical and chemical properties that are dominated by full quantum effects. This includes studying condensed matter formed by relatively light elements (e.g., helium, lithium, carbon, and boron) or relatively heavy elements (e.g., oxygen and silicon), as well as examining compounds formed by them.

8.1 Helium (He)

Helium is the second lightest element. Therefore, it is natural to expect that full quantum effects will manifest themselves through unique physical phenomena in condensed systems composed of helium. We have already introduced helium in Chapter 2 as an example illustrating the properties of superfluids (quantum liquids) and supersolids (quantum solids) in our discussion on full quantum effects in physical problems.

Liquid helium can only become solid when the pressure exceeds 25 GPa and the temperature is extremely low. Remarkably, at atmospheric pressure, it remains in a liquid state even if its temperature approaches absolute zero. This is why helium is called a permanent liquid. This phenomenon cannot be understood within the framework of classical physics. No other element exhibits this behavior, making the study of helium a highly active and ongoing research field in physics. Since the successful liquefaction of helium in 1908, it has remained a prominent area of investigation.

Helium has two stable isotopes, ^{4}He (helium 4) and ^{3}He (helium 3). Both isotopes exhibit superfluidity, another property exclusive to helium. ^{3}He enters the superfluid phase at temperatures near 2 mK, which are difficult to reach. Therefore, only a few research groups have been able to investigate ^{3}He superfluidity. By contrast, the critical temperature for superfluid ^{4}He is near 2 K, making it relatively easy to study. Over the last 20 years, interest has been shifted from the well-documented superfluidity of liquid ^{4}He to the possible supersolidity of solid ^{4}He. It has been predicted theoretically that fluid-like behavior and lattice ordering, two seemingly contradictory properties, can coexist in solid ^{4}He. The question of whether supersolidity, the state of solid-state superfluidity, truly exists in solid helium remains one of the most popular and intriguing topics in the field of helium physics. In fact, it was listed among the 125 most significant unanswered questions in the journal *Science*'s commemoration of its 125th anniversary in 2005. (Other questions included "What is the universe made of?," "Are we alone in the universe?," "Can nuclear fusion become our energy source in the future?," "What is the nature of gravity?," and – as was mentioned in Chapter 7 – "What is the structure of water?")

Research on superfluidity in ^{4}He and ^{3}He has yielded several Nobel Prizes, marking its significance in the field. More importantly, the study of helium has significantly improved our understanding of some of the most fundamental and essential issues in physics. It has laid the foundation for modern techniques in low-temperature physics (see Chapter 6 for a brief introduction). Remarkably, many pivotal phenomena in condensed matter physics, such as superconductivity, the quantum Hall effect, the fractional quantum Hall effect, and the anomalous quantum Hall effect, were discovered unexpectedly in experimental conditions involving quantum fluids at low temperatures. In the following discussion of the physical properties of condensed helium, we will briefly review four representative issues that are closely related to full quantum effects. These areas of study have expanded the boundaries of research in helium physics, fostering new frontiers for exploration.

8.1.1 Supersolidity and Mass Flow

The connection between superfluidity and Bose–Einstein condensation was initially established by F. London in 1938. Further, M. Wolfke proposed the possibility of observing superfluidity in solids in the following year. In the 1970s, A. Leggett and his colleagues at the University of Illinois Urbana-Champaign theoretically predicted that certain solids, referred to as "supersolids," could exhibit superfluid behavior. Since then, the search for supersolids has been ongoing. Leggett devised an experiment with the following steps: (1) preparing a sample cell filled with solid ^{4}He; (2) placing the sample cell at one end of a torsion bar; and (3) measuring the period of the torsional oscillator as a function of temperature. In this experimental setup, any alteration in rotational inertia caused by ^{4}He atoms breaking away from the lattice to form a superfluid would result in a change in the torsional oscillator's period. Although similar experimental methods have been used to observe superfluidity in liquid ^{4}He, the supersolid phase of helium remained undiscovered in the following 40 years of experiments. It was only after 2004, when M. H. W. Chan from Pennsylvania State University and E. Kim and his colleagues from the Korea Institute of Science and Technology reported evidence supporting the existence of supersolidity through measurements of the rotational inertia of solid ^{4}He (see Figure 8.1) that this research field again attracted interest [Kim E.2004a, b, 2008, Chan M.H.W.2008, West2009]. This kind of rotational inertia measurement may be explained with a trivial explanation of mechanical effects.

To further investigate the occurrence of superfluid mass flow in the ^{4}He supersolid, M. Ray and R. Hallock conducted an experiment in which they placed the solid material

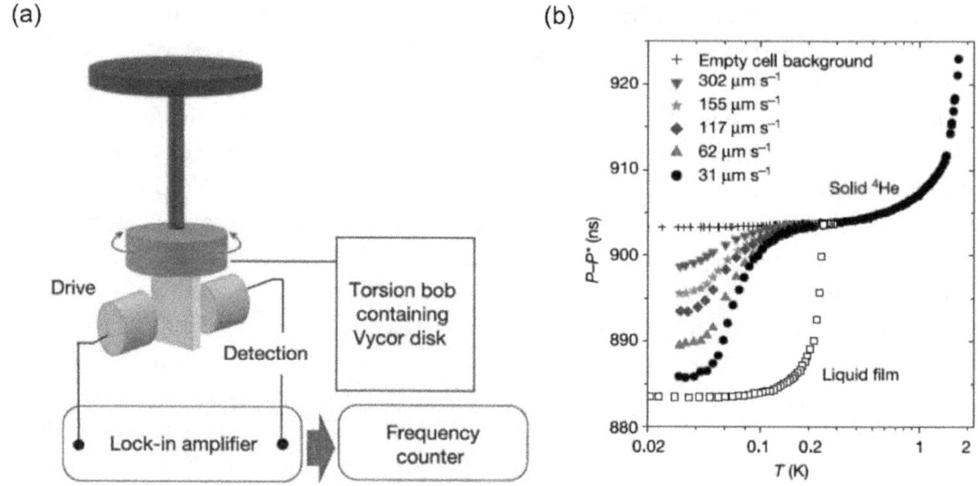

Figure 8.1 (a) Schematic of the resonant torsional oscillator used for measuring the rotational inertia of solid helium. (b) Anomalous decrease in the resonant period, possibly associated with the supersolid phase. From Kim[2004b]. Reproduced with permission from SNCSC.

between two superfluid regions. By applying pressure to one of the superfluid regions, they observed mass flow through the solid material to the other liquid region, as depicted in Figure 8.2 [Ray2011]. Subsequently, J. Beamish and Z. G. Cheng, along with their colleagues, employed a similar approach in which unidirectional mechanical stress was applied to the solid material. They also reported the presence of anomalous mass flow in solid ^{4}He [Cheng Z.G.2015, 2016]. Based on their previous work, Beamish and Z.G. Cheng carried out studies comparing ^{3}He and ^{4}He (see Figure 8.3). The experimental results supported the existence of superfluidity in solid ^{4}He [Cheng Z.G.2018a]. Such experiments have led to continual advancements in the relevant research.

Although the study of supersolidity is still in its early stages, it is evident that this phenomenon cannot be comprehended within the framework of classical physics. Because of helium's extremely low atomic mass, an accurate theoretical understanding can only be achieved by considering full quantum effects. This entails describing both the electrons and nuclei of helium in terms of wavefunctions. At low temperatures, the zero-point motion leads to increased overlap of wavefunctions among helium atoms in solids. As a result, the spatial positions of the quantized helium atoms become indistinguishable, and their frequent exchanges occur without any energy cost. In summarizing the history of supersolid research, Hallock states,

> The story of solid ^{4}He is still unfolding. Can the material exist as a supersolid? If so, the supersolid is nothing like the state originally envisioned by Chester and other theorists more than four decades ago. The pathway to understanding has been convoluted; the real story is unlike the historical summaries in textbooks or review articles that describe in a more direct way our final understanding of a topic. But the journey along a real scientific pathway to discovery can be exciting, especially when one travels with open and friendly colleagues. [Hallock2015]

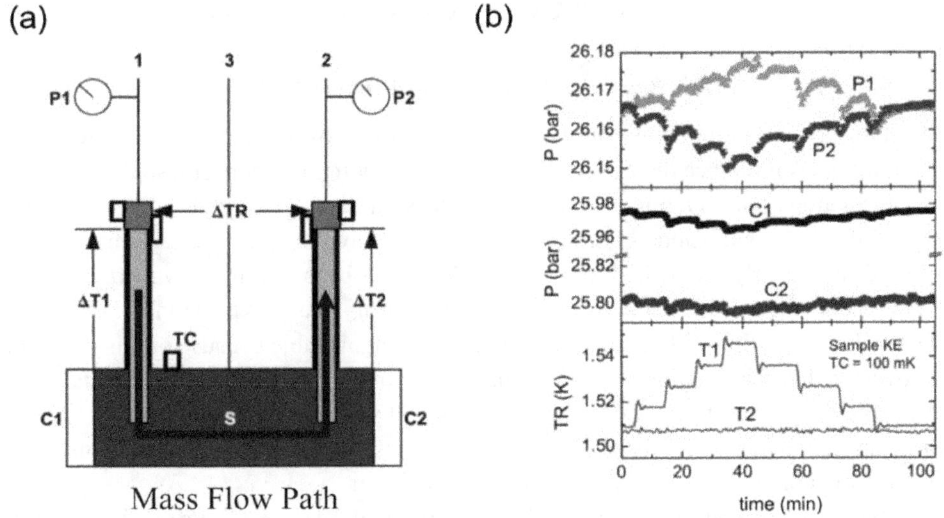

Figure 8.2 (a) Schematic of the experiment carried out by Ray and Hallock on unidirectional mass flow in solid helium. (b) The experimentally observed unidirectional mass flow. Reprinted figure with permission from Ray[2011]. Copyright 2024 by the American Physical Society. The legend "mass flow path" in the image was added based on the description in the original text.

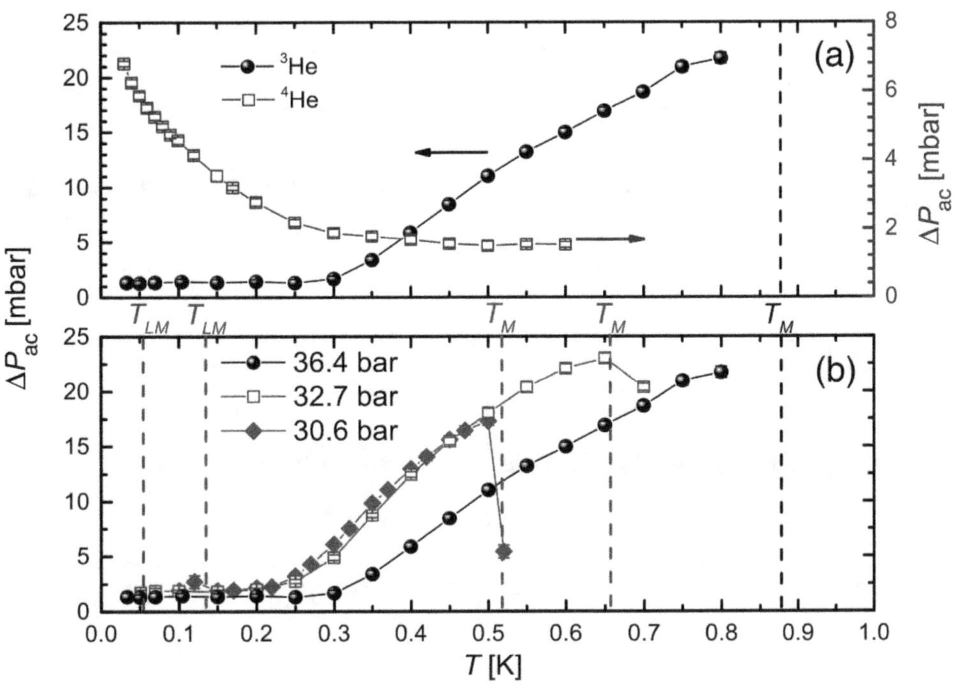

Figure 8.3 (a) Mass flow rate as a function of temperature in solid phases of ^{3}He and ^{4}He. (b) Pressure variation as a function of temperature at different pressures. Reprinted figure with permission from Cheng Z. G.[2018a]. Copyright 2024 by the American Physical Society.

8.1.2 Plastic Deformation and Defect Mobility

In Section 2.1, we commented on the shear modulus measurement of solid ^{4}He. Just as quantum effects influence the flow of liquid helium, making it superfluid, so they influence the deformation of solid helium. In particular, the significant zero-point motion caused by full quantum effects in solid helium results in a large wavefunction overlap among helium atoms. Consequently, solid helium must be described by a complete wavefunction that considers both the electrons and nuclei (i.e., by the Born–Huang expansion). Unlike classical particles, atoms in solid helium are no longer distinguishable because of this wavefunction overlap. Instead, they are free to exchange positions with each other, leading to high defect mobility, quasi-free motion of impurities and vacancies, and dislocation lines that can be approximated as freely vibrating strings. These phenomena cannot be understood when the helium atom is treated classically.

Studying full quantum effects in solid helium allows us to derive laws for the energy-free motion of defects in the elastic or plastic deformation region, which provides a scientific basis for relevant studies in materials science and metallurgy. For example, Beamish and Z.G. Cheng experimentally observed apparent stress release caused by

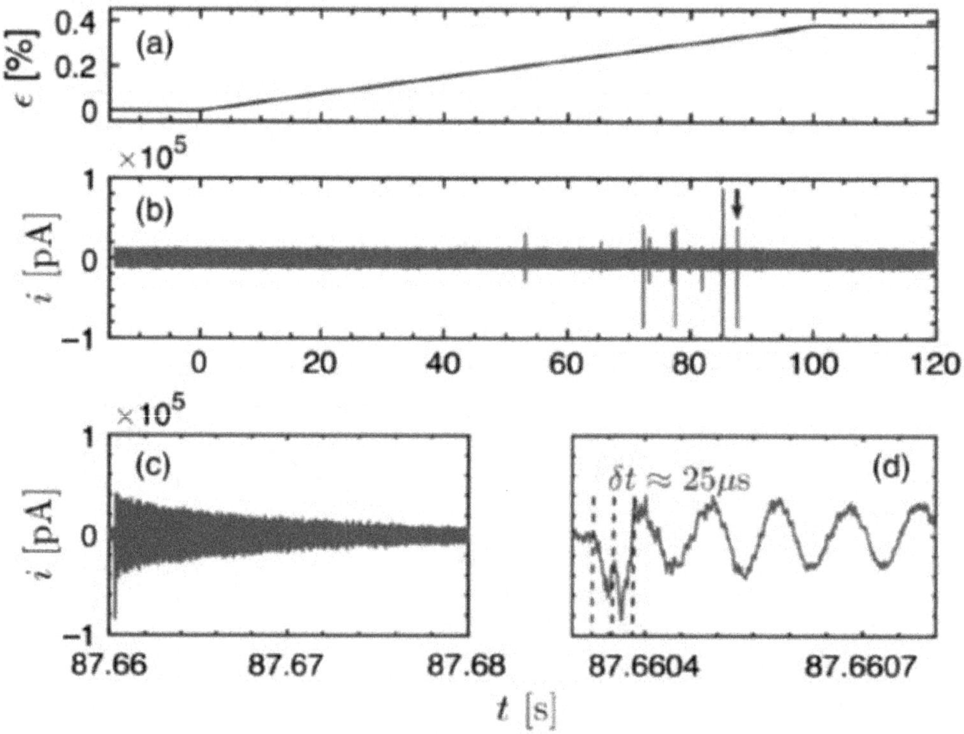

Figure 8.4 Stress release in solid helium due to lattice slip after application of tangential strain: (a) applied strain versus time; (b), (c), and (d) successively enlarged views of the instantaneous current (proportional to the differential shear modulus) generated under applied strain as a function of time. Reprinted figure with permission from Cheng Z.G.[2018b]. Copyright 2024 by the American Physical Society.

lattice slip, which they attributed to the avalanche effect of dislocation lines generated under large tangential stresses. Figure 8.4 shows dislocation lines moving at the speed of sound when the avalanche occurs. The lattice slips have been measured by two current preamplifiers: one with a high gain but long rise time, the other with a relatively lower gain but a prompt response time. The first setup can reveal the total stress release of a phase slip, and the second can measure the time scale of the slip (see the lower right panel of Figure 8.4). The strongest slips have length scales estimated as 5 mm from the stress drop and time scales estimated as 25 μs from the initial dip of the current response. Therefore, the dislocation velocity is around 200 m/s, close to the sound speed of solid helium. This is the first time that a phenomenon similar to acoustic emission has been observed in quantum solids [Cheng Z.G.2018b].

8.1.3 Heat Capacity

Heat capacity measurement is generally used for studying phase transitions. Because the shape of the heat capacity peak near the phase transition point resembles the shape of the letter λ, the superfluid phase transition of liquid helium is also known as the λ phase transition. The superfluid phase transition of liquid helium provides an important experimental method to test the theory of second-order phase transition, and it is one of the few experiments that have been intentionally carried out in a gravity-free environment in space. The objective of this study was to measure the critical exponent [Lipa2003].

The earliest measurements of specific heat in solid helium can be traced back to 1934 [Kaischew1934]. Because of the significance and difficulty of this experiment, continuous efforts were made to measure the specific heat of solid helium over the last 70 years, and it remains a topic for ongoing research today.

Measuring the heat capacity of solid helium can be quite challenging at extremely low temperatures. ^{4}He can only solidify at a pressure greater than 25 GPa. Therefore, a container is required to host solid helium. This container is typically made of metal, but the heat capacity of the metal container at low temperatures is much greater than that of the solid helium. In addition, thermal experiments at extremely low temperatures tend to encounter difficulties that do not apply to conventional experiments, such as boundary thermal resistance, strict thermal insulation conditions, long equilibrium times, and the need for extremely accurate temperature measurements. Until the 1980s, heat capacity measurements of solid helium could only be made with limited accuracy above 0.1 K, making it difficult to determine whether the phase transition from conventional solid to supersolid had occurred. For conventional solid helium, the heat capacity is proportional to the cube of the temperature (T^3). Nevertheless, an experiment in the 1990s found that the heat capacity of solid helium below 0.1 K involves an additional heat-capacity term [Clark2005]. In 2007, M. H. W. Chan and X. Lin successfully measured the heat capacity of solid helium near 20 mK using a silicon sample cell. They also detected a weak heat-capacity signal on the order of μJ/K and a specific heat peak in solid helium (shown in Figure 8.5). In 2009, they established that this peak did not vary with the amount of ^{3}He impurity [Lin X.2007, 2009].

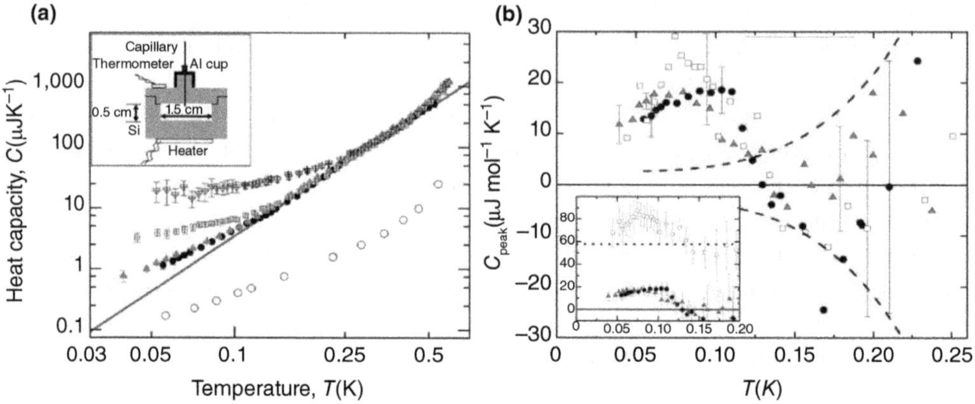

Figure 8.5 (a) Heat capacity of solid helium at extremely low temperatures; the inset shows the diagram of the calorimeter used for heat capacity measurements; (b) specific heat peaks after subtracting the phonon contribution. From Lin X.[2007]. Reproduced with permission from SNCSC.

The characteristic temperature of the specific heat peak found by Chan and X. Lin in solid ^{4}He is similar to the characteristic temperature of the period variation in torsional oscillator experiments. The physics behind this is not fully understood, and no new experimental measurements of the specific heat of solid helium have been reported since the last decade. Currently, the presence of this specific heat peak suggests the potential existence of new states of solid helium at low temperatures (referred to as supersolid) and unexplained physics. All of these are clearly related to full quantum effects and may even lead to new quantum mechanical states. Despite the challenges, further measurement of specific heat is crucial to the study of full quantum effects.

8.1.4 Surface Electronic State

In addition to being the only natural quantum liquid and quantum solid, helium provides a platform for studying other full quantum phenomena. For example, a two-dimensional electron gas can be realized not only in a semiconductor heterojunction but also on the surface of liquid ^{4}He. Using these readily available bosons and simple fermions, we can easily construct a new system that exhibits full quantum effects (in other words, a new two-dimensional full quantum electron gas).

When electrons are adsorbed onto the surface of liquid helium at low temperatures, they are attracted to the induced surface mirror charge, resulting in a vertical effective potential trap that binds the electrons to the ground state energy level. Consequently, one can obtain an electron gas that is unable to move vertically but is free to move in the planar directions. Since this electron gas is weakly coupled to the external environment, high electron mobility can be achieved, making it an ideal system for studying two-dimensional many-body problems. Electrons on the surface of liquid helium (see Figure 8.6) can be generated by the photoelectric effect, a hot cathode tube, or an

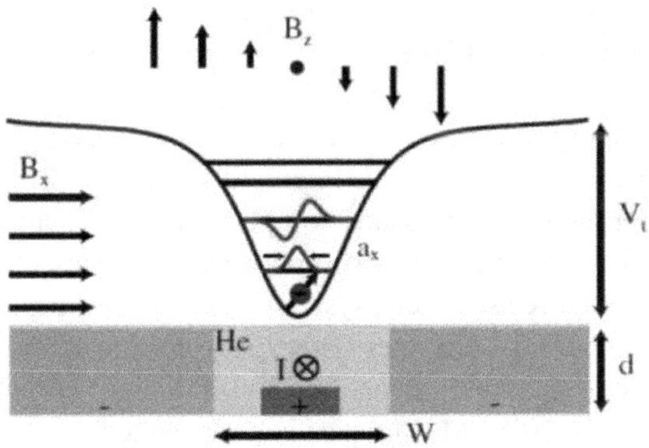

Figure 8.6 Scheme for constructing a spin qubit using the surface-electron system of liquid helium. Reprinted figure with permission from Schuster[2010]. Copyright 2024 by the American Physical Society.

arc discharge. This research direction was first explored in 1971 by D. Thanner at the University of Florida [Sommer1971]; research evolving from his work continues to provide us with new physical discoveries almost 50 years later [Schuster2010, Rees2016].

8.2 Lithium (Li)

Lithium, the third element in the periodic table and the lightest alkali metal, undergoes a series of solid-state phase transitions at high pressures, leading to a remarkably intricate phase diagram. The melting behavior of lithium under high pressure has garnered significant interest due to recent experimental breakthroughs. Because of its exceptionally small atomic mass of 6.941, lithium undoubtedly demonstrates robust nuclear quantum effects, including zero-point fluctuation and quantum tunneling. These effects exert a substantial influence on the high-pressure phase diagram and melting curve of lithium.

In addition to existing as a bulk metal, lithium can form atomic clusters of different sizes. Strong nuclear quantum effects significantly influence the physical behavior of these clusters at low temperatures, as reflected by quantum fluctuations of bond length and the variations of the spectrum features. The mechanism of lithium ion migrations has also attracted significant research interest, especially in the context of lithium batteries. It is increasingly clear that full quantum effects play a major role in their application.

8.2.1 High-Pressure Phase Transition

As the lightest metal at atmospheric pressure, bulk lithium is often considered to be one of the simplest metals. At different pressures and temperatures, it exhibits different crystal structures, and its high-pressure phase diagram has become a popular research topic in recent years [Shimizu2002, Marqués2011, Schaeffer2012, Feng Y.X.2015]. As the pressure increases, bulk lithium undergoes a series of solid phase transitions with gradually decreasing structural symmetry; these correspond to an intriguing metal–semiconductor–metal phase-transition sequence. In addition, reports of the superconductivity of lithium at around 20 K have also attracted significant attention.

Similar to the developments of computational physics, molecular dynamics simulations of metal systems at atmospheric pressure treated atomic nuclei as classical particles in the early stage. However, recent reports have highlighted that, as the spacing between metal atoms decreases under increasing pressure, soft phonon modes start to emerge in the solid phase of certain metals, such as calcium. This observation suggests the significant involvement of nuclear quantum effects in metal systems. Because of the high chemical activity of lithium, the element's melting behavior at high pressures has long been difficult to study experimentally. Initially, the melting curve of bulk lithium could only be measured at a pressure below 8 GPa. In 2010, M. Marqués et al. extended the melting-curve measurement range to 15 GPa using the multi-anvil cell technique [Marqués2011]. A year after this breakthrough, C. L. Guillaume et al. further increased the upper pressure limit to 60 GPa [Guillaume2011]. Through a series of experiments, they discovered that the melting temperature of bulk lithium reaches its maximum value at around 15 GPa; above this pressure, it rapidly decreases. When the pressure reaches ~45 GPa, the melting temperature reaches a minimum of ~190 K. Among all monolithic metallic materials examined experimentally to date, bulk lithium exhibits the lowest melting temperature at this pressure.

When the effect of zero-point energy (ZPE) is considered, it can be observed that nuclear quantum effects play a crucial role in explaining the melting curve of bulk lithium. Theorists have predicted the high-pressure melting curve by using the first-principles molecular-dynamics two-phase method based on the classical approximation of atomic nuclei [Hernandez2010]. At low pressures, this simulation result agreed well with the experimental results obtained by Guillaume et al. However, between 40 and 60 GPa, the experimental value (~190 K) was ~100 K lower than the simulation (~300 K). This discrepancy may be attributed to the nuclear quantum effects that lower the melting point.

In 2012, A. M. J. Schaeffer et al. measured the melting curve of bulk lithium below 64 GPa [Schaeffer2012]. They found melting temperatures of ~300 K. This is generally in line with the first-principles molecular dynamics simulation results for classical nuclei obtained by E. R. Hernandez et al. [Hernandez2010], whose theoretical results also agreed with the low-pressure experimental results of Guillaume et al. This suggested (but did not prove) that the melting temperature of lithium is not significantly changed by nuclear quantum effects. The matter was not settled until 2015, when theoretical studies based on first-principles path-integral dynamics simulations and free-energy calculations (see Figure 8.7) were performed [Feng Y.X.2015].

Y.X. Feng et al. showed that at ~50 GPa, nuclear quantum effects reduced the melting temperature of the *cI*16 phase by ~15 K ($T_m - T'_m = 15\text{K}$). The absolute values of the melting temperatures from the simulations with quantum nuclei were close to the experimental values reported by Schaeffer et al. [Schaeffer 2012]. Therefore, the conflict between theory and experiment was resolved, and the influence of nuclear quantum effects on the melting temperature of lithium had been quantified (a reduction of ~15 K) [Schaeffer2012]. Previous classical calculations [Hernandez2010] that did not account for nuclear quantum effects overestimated this temperature by more than ~100 K.

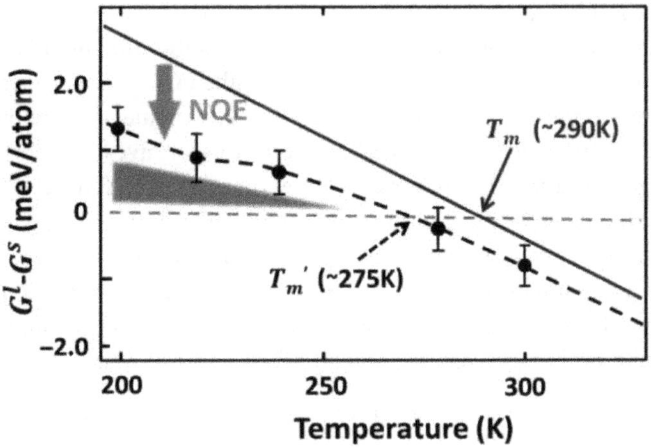

Figure 8.7 Gibbs free energy difference as a function of temperature for liquid (G^L) and solid (G^S) phases of lithium. When ($G^L - G^S$) = 0, the temperature is T_m. The solid and dashed lines in the figure indicate the results obtained by classical molecular dynamics and the quantum PIMD method that considers nuclear quantum effects (NQEs), respectively. Reprinted from Feng Y. X.[2015], with the permission of AIP Publishing.

This study of nuclear quantum effects on the phase transition facilitates further research into quantum-dynamics simulations. In the last decade, more detailed results have been obtained using the path-integral Liouville dynamics method.

In addition to their effect on the melting curve of the *cI*16 phase, nuclear quantum effects also have a significant impact on the phase diagram of metallic lithium at high pressures. At 200 K, the *cI*16 phase of lithium is observed to transform into the *oC*88 (*C2mb*) phase near 60 GPa. However, theoretical studies reveal that the *oC*40 phase at 0 K has a much lower energy in the corresponding pressure range. Combining Raman spectroscopy and first-principles calculation, F. A. Gorelli et al. found that the simulation result was only consistent with the experiment when the quantum properties of the lithium atoms were taken into account [Gorelli2012]. However, in their calculations, the nuclear quantum effects were described by the phonon free energy, which is not very accurate (the consideration of anharmonic effects is incomplete). Computational methods with higher accuracy (e.g., the path-integral Liouville dynamics method) are required. Nuclear quantum effects on the phase diagram of metallic lithium (phases *oC*40, *oC*24, etc.) are expected to be more prominent at higher pressures (>80 GPa); however, this needs to be verified in future studies.

Nuclear quantum effects influence the physical properties of bulk lithium under both low and high pressures. In 1947, Charles discovered that the BBC phase of ^{7}Li at atmospheric pressure transforms into a closest-packed structure (later identified as the 9R structure [Overhauser1984, Smith1987]) below 77 K. It has since been recognized that metallic lithium, previously considered to be simple, possesses a complex ground-state crystal structure [Barrett1947]. In 2017, G. J. Ackland et al. systematically studied the structure of lithium at low pressure using synchrotron X-ray diffraction in diamond anvil cells and multi-scale molecular dynamics simulations based on density functional theory.

After considering the nuclear quantum effects, they found that the closest-packed structure is thermodynamically unstable and, therefore, not a 9R structure [Ackland2017]. The ground state structure for both ^{6}Li and ^{7}Li should be the FCC phase. This in turn raises the question of whether the 9R structure observed in other elements (sodium, samarium) or alloys is thermodynamically stable, which warrants further investigation.

In addition to revealing the influence of pressure on crystal structure, recent studies have predicted that some of the high-pressure phases of bulk lithium have a topologically nontrivial electronic structure; the Z_2 topological variant is nonzero, which is reflected in the presence of a one-dimensional nodal ring formed by the crossing of energy bands near the Fermi energy level. According to these studies, the sequence of high-pressure phase transitions of bulk lithium between 80 and 500 GPa is as follows:

$$Pbca \rightarrow Cmca-24 \rightarrow Cmca-56 \rightarrow P4_2/mbc \rightarrow R\bar{3}m \rightarrow Fd\bar{3}m$$

[Pickard2009, Lv J.2011]. S. A. Mack et al. found that the topologically trivial insulator *Aba*2 phase transitions to a *Pbca* phase with semimetallic Dirac nodal ring at 80 GPa. The nodal ring is protected by a nonzero Berry phase and is therefore topologically nontrivial (as shown in Figure 8.8) [Mack2019]. At 220 GPa, a Lifshitz phase transition occurs between the two *Cmca* phases, and the Dirac nodal ring of the *Cmca-24* phase at the Fermi energy level evolves into two mutually perpendicular *Cmca-56* nodal rings. At 350 GPa, the P4$_2$/*mbc* phase of lithium exhibits a nodal ring near 0.25 eV below the Fermi energy level. Finally, they also computed the two predicted phases for bulk lithium: $R\bar{3}m$ (450 GPa) and $Fd\bar{3}m$ (500 GPa). Both phases exhibit high dispersion bands and metallicity at the Fermi energy level, and the $Fd\bar{3}m$ phase has a distorted hexagonal honeycomb lattice structure

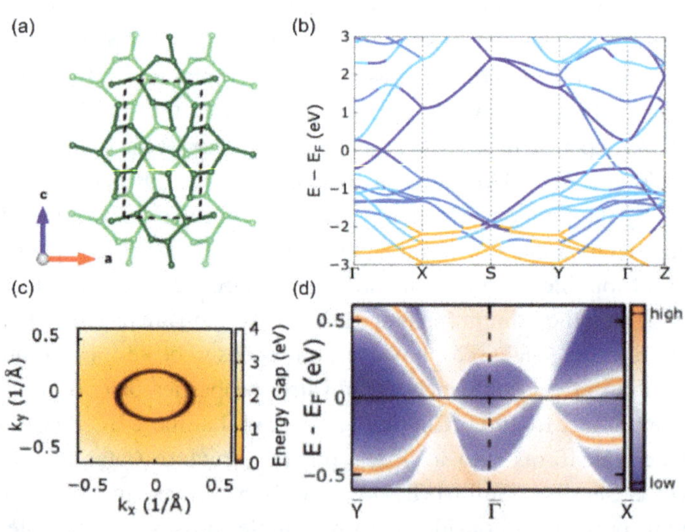

Figure 8.8 *Pbca* phase of lithium at 80 GPa. (a) Single-cell structure of the *Pbca* phase. (b) The energy band structure of the *Pbca* phase at 80 GPa. (c) Projection of the nodal ring on the two-dimensional K-plane. (d) The projected energy band structure along the [001] direction. Reprinted from Mack[2019].

that is very similar to that of graphene, with a Dirac crossing at the Fermi energy of 1 eV. This study suggests that similar electronic structure features may be observed in other light elements under high pressure. In general, high pressure has the potential to induce topological characteristics in the electronic structures of various materials.

8.2.2 Quantum Fluctuation of Bonding Lengths in Clusters

Lithium can also form a variety of cluster structures with different sizes (as shown in Figure 8.9). The optical properties and electronic and atomic structures of lithium clusters Li_n ($n = 2, 4, 5, 6, 8, 20 \ldots$) have been thoroughly studied both experimentally and theoretically [Ellert2002, Wheeler2014]. As the atomic weight of lithium is only about seven times that of hydrogen, its corresponding nuclear quantum effects not only affect the high-pressure phase diagram of metallic lithium but also impact various physical properties of lithium clusters. For example, in subsection 2.1.4, we have discussed the impact on the optical properties of the lithium clusters caused by the quantum effect of the nucleus. In fact, the potential energy surface of lithium clusters is extremely flat, and there are many isomers with similar energies (see Figure 8.9).

The rich structures of lithium clusters have made the study of their dynamical stability a popular research topic. However, in the initial theoretical studies, lithium clusters were simulated using typical first-principles molecular dynamics methods, in which the lithium nuclei are treated as classical particles. In 1992, P. Ballone et al. used the path-integral Monte Carlo method to simulate the physical properties of Li_n ($n = 20$, 40, 92) at temperatures of 30–600 K [Ballone1992]. Their results indicate that the zero-point vibrations

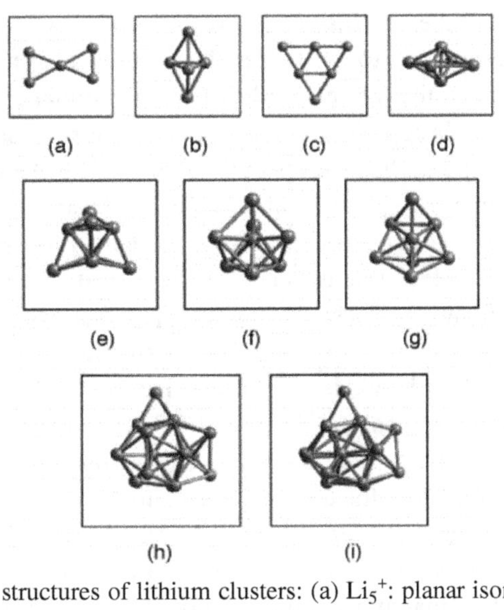

Figure 8.9 Atomic structures of lithium clusters: (a) Li_5^+: planar isomer I (D_{2h}); (b) Li_5^+: isomer II (D_{3h}); (c) Li_6: planar isomer I (D_{3h}); (d) Li_6: isomer II (D_{4h}); (e) Li_8: isomer I (T_d); (f) Li_8: isomer II (C_{3v}); (g) Li_8: isomer III (C_s); (h) Li_{20}: isomer I (C_1); and (i) Li_{20}: isomer II (C_1). Reprinted from Rousseau[1999], with the permission of AIP Publishing.

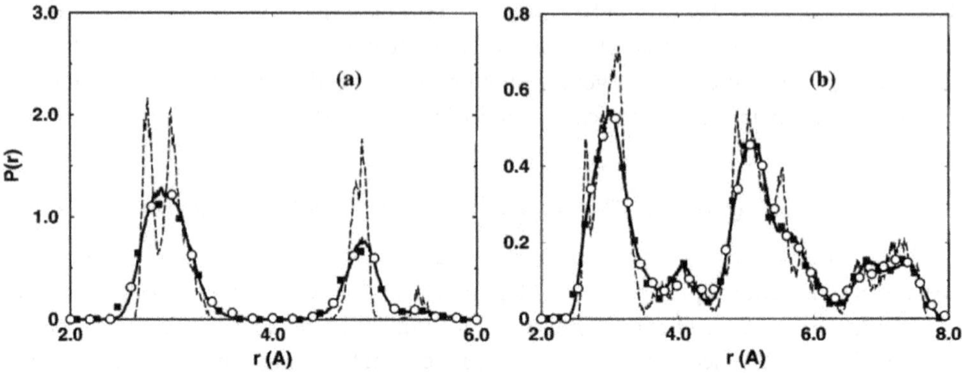

Figure 8.10 Bond length distribution of Li_8 (a) and Li_{20} (b) clusters at 10 K. The solid and dashed curves correspond to the quantum and classical simulation results, respectively; the results of the classical simulations at 100 K are indicated by circles. Reprinted figure with permission from Rousseau[1998]. Copyright 2024 by the American Physical Society.

and quantum tunneling caused by the full quantum effects have a significant impact on the structural and thermodynamic properties of the lithium clusters. Their work also demonstrates that the lithium clusters can exhibit an interesting fluid-like behavior throughout the simulated temperature range (30–600 K).

Later, R. Rousseau and D. Marx used the first-principles path integral molecular dynamics method to study nuclear quantum effects on the atomic structure and electrical properties of lithium clusters [Rousseau1998, 1999]. They reported strong nuclear quantum effects on the Li-Li bond length distribution of Li_8 and Li_{20} clusters. As shown in Figure 8.10(a), the bond length distribution of the Li_8 cluster is characterized by two well-separated peaks near 3.0 Å and 4.8 Å, corresponding to the first and second nearest-neighbor distances (indicated by the solid curves in the figure). In the classical simulation at 10 K, a pronounced splitting of both peaks is observable. In other words, each peak splits into two additional sharp peaks (denoted by the dashed curves in the figure). The classical simulation result is consistent with the presence of a short bond (around 2.8 Å) and a long bond (around 3.1 Å) after optimizing the static atomic structure of the cluster. In Figure 8.10(b), bond-length-peak splitting of the third-nearest neighbor of the Li_{20} cluster can also be observed in the case of the classical simulation. These findings suggest that even at a low temperature of 10 K, classical calculations considering only thermal fluctuation do not disrupt the characteristics of the long and short bonds in Li clusters. However, as shown by the solid curve in Figure 8.10, the splitting of peaks at 3.0 Å and 4.8 Å predicted by classical simulations is not observable at 10 K after nuclear quantum effects are considered. This indicates that at low temperatures, the quantum fluctuations caused by nuclear quantum effects obliterate the distinction between long and short bonds in the lithium clusters. The influence of this phenomenon on its optical properties is very obvious (see discussions in subsection 2.1.4).

Figure 8.11 shows the quantum configurations of the Li_8 and Li_{20} clusters at 10 K. Here, we can clearly see the quantum fluctuation of the lithium nuclei. This indicates that lithium atoms can no longer be described in terms of classical point-like particles, and the

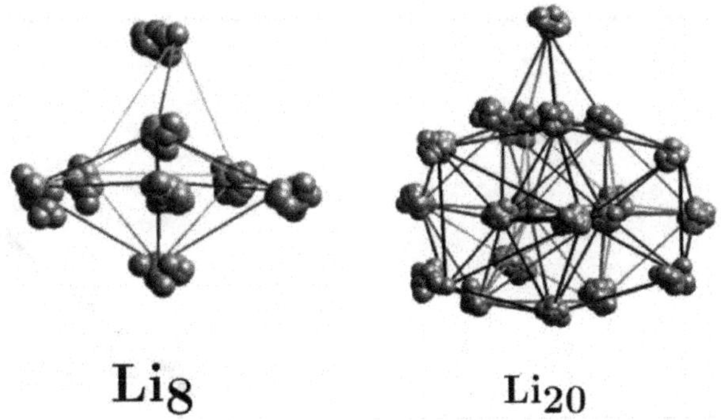

Figure 8.11 Quantum simulation of the atomic structures for Li_8 and Li_{20} clusters at 10 K. Reprinted from Rousseau[1999], with the permission of AIP Publishing.

zero-point vibrations are non-negligible. To investigate the nuclear quantum effects on thermal excitation, Rousseau and Marx calculated the bond length distribution at 100 K using classical simulation. The results of the classical simulations are marked by circles in Figure 8.10. For Li_8 and Li_{20} clusters, the effects of thermal fluctuation obtained by 100 K classical simulations are consistent with the 10 K calculation result that considers both the thermal and quantum fluctuation. In summary, both the combined effects of quantum and thermal fluctuations at 10 K and the thermal fluctuation at 100 K are sufficient to destroy the peak splitting of the static Li cluster structure. Despite the thermal fluctuation is depressed under low temperature (around 10 K), the quantum fluctuation caused by the nuclear quantum effect is strengthened. Their combined effect makes the alternating phenomenon of bond length disappear completely. At higher temperatures (around 100 K), the classical simulation reveals that the disappearance of the alternating bonds in lithium clusters is caused by thermal fluctuation alone.

8.2.3 Effect of Zero-Point Energy on Diffusion

Lithium-ion migrations in materials have recently become a popular research topic because of its role in applications such as lithium-ion batteries. Despite many theoretical studies [Wan W.H.2010, Tritsaris2013, Li W.2015, Mukherjee2018], a unified understanding of the diffusion mechanism has not been achieved, especially at the quantum level. Until recently, the study on the impact of full quantum effects on lithium-ion diffusion behavior has received more and more attention.

K. Toyoura et al. studied the discrepancy between the classically predicted diffusion of lithium ions in LiC_6 and the results of a quantum simulation based on the harmonic approximation [Toyoura2008]. The classical calculation overestimates the diffusion potential in the region below room temperature. This is because full quantum effects involving the zero-point vibrational energy reduce the free potential barrier to diffusion in the lower-temperature range. Above room temperature, the classical value gradually converges to the quantum value, indicating that the full quantum effects are suppressed by thermal fluctuations at high temperatures (Figure 8.12).

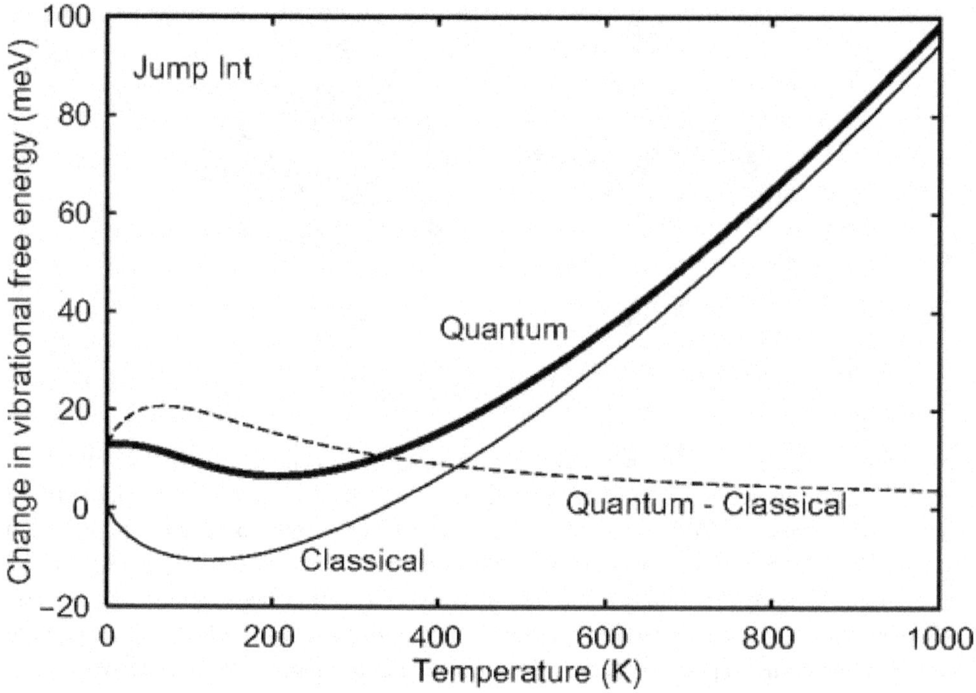

Figure 8.12 Vibrational free energy of lithium as a function of temperature. The thick and the thin curves denote the results obtained using quantum and classical statistics, respectively. The discrepancy between them is indicated by the dashed curve. Reprinted figure with permission from Toyoura[2008]. Copyright 2024 by the American Physical Society.

Later, S. Karmakar et al. examined the diffusion of lithium ions in group IV–VI compounds. They observed that quantum simulations considering ZPE reduce the diffusion potential of lithium ions in SiS and SiSe by 20 meV and 14 meV, respectively. This reduction approximately corresponds to the difference between the top line (circle marker) and the bottom line (pentagon marker) in Figure 8.13. This increases the lithium-ion diffusion rate by 10% at room temperature and ~60% at low temperatures [Karmakar2016]. Figure 8.13 indicates that the result obtained with the inclusion of the Wigner correction (diamond/triangle-marked line) falls between the classical simulation result (circle-marked line) and the quantum ZPE-corrected result (pentagon-marked line). In particular, it converges to the quantum simulation result when the temperature is further reduced, and it approaches the classical simulation result when the temperature increases. More recently, S. Guo et al. discovered that the ZPE correction reduced the lithium-ion diffusion barrier in monolayers of P, As, and Sb by 10 meV at room temperature. Moreover, the influence of nuclear quantum effects became more pronounced at lower temperatures, as depicted in Figure 8.14 [Guo S.2018]. This finding was subsequently corroborated by A. A. Kistanov et al. [Kistanov2019].

Although nuclear quantum effects on lithium-ion diffusion can vary among materials, it is non-negligible at low temperatures. It is evident that this research field holds

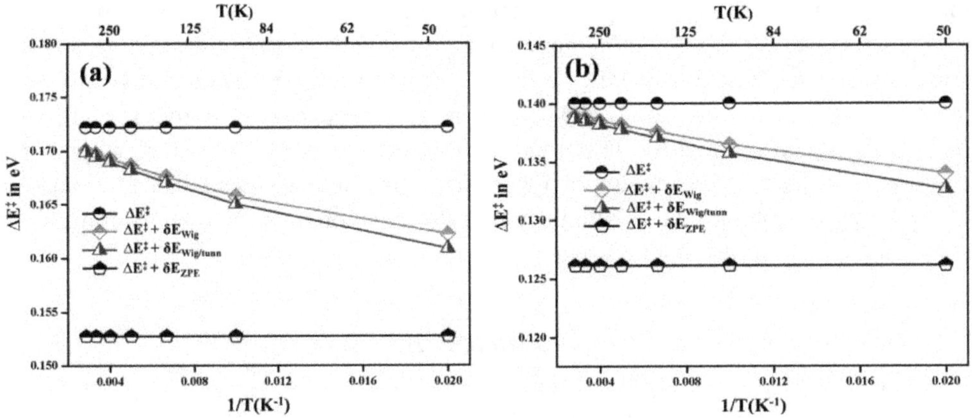

Figure 8.13 Inverse-temperature dependence of Li-ion diffusion potential barriers in (a) SiS and (b) SiSe. The line with circle markers, line with square markers, line with triangle markers, and line with pentagon markers denote the classical, Wigner ZPE-corrected, Wigner ZPE-tunneling-corrected, and quantum ZPE-corrected potential barriers, respectively. Reprinted with permission from Karmakar[2016]. Copyright 2024 American Chemical Society.

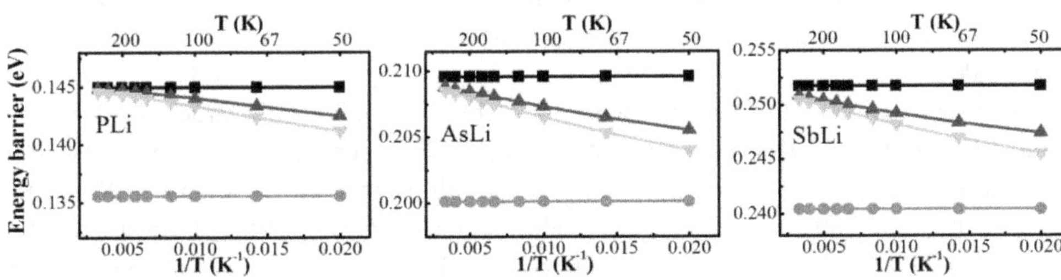

Figure 8.14 Inverse-temperature dependence of the Li-ion diffusion potential barriers in monolayer P, As, and Sb, including the classical potential barrier (black line with square markers), the Wigner-ZPE-corrected potential barrier (light gray line with triangle markers), the Wigner-ZPE-tunneling-corrected potential barrier (black line with triangle markers), and the ZPE-corrected potential barrier (black line with circle markers). Reprinted with permission from Guo S.[2018]. Copyright 2024 American Chemical Society.

significant potential for practical applications, particularly in advancing the understanding of lithium-ion diffusion in electrode materials for the development of high-capacity and high-performance lithium batteries.

8.3 Boron (B)

Boron is in the second period of the periodic table. It is the lightest element in Group III, with a relative atomic mass of 10.8. It exhibits high reactivity and has the capability to form bonds with multiple centers and electrons. Boron can adopt various isomeric structures of different dimensions, including zero-dimensional caged fullerenes, one-dimensional nanotubes, and two-dimensional quasi-planar molecules (borophenes). Moreover, it can combine with numerous other elements, resulting in the formation of stable compounds with a wide range of physical properties. An example of such a compound is magnesium diboride (MgB_2), known for its superconductivity. The presence of distinct bonding formations in boron atoms renders them sensitive to changes in the confined environment surrounding neighboring atoms, leading to distinct full quantum effects. This section introduces full quantum effects in boron and boride materials, classified on the basis of the atomic structure.

8.3.1 Bulk Boron

First, let us discuss the case of bulk elemental boron. In this phase, the number of boron atoms within a single unit cell is large, and the quantum properties of each nucleus must be taken into consideration. In other words, the effect of zero-point vibrations on the crystal structure and physical properties of boron is non-negligible. There are two main structures of bulk boron: α- and β-rhombohedral boron. In 2007, by using density functional theory (DFT) and the generalized gradient approximation, R. A. de Groot's group calculated various possible bulk boron structures that possessed interstitial atoms and partially occupied sites. By adding a correction term for the zero-point vibrational energy, they successfully obtained stable β-rhombohedral structures of bulk boron [van Setten2007].

Their calculation resulted in the two most stable β-rhombohedral structures: a hexagonal structure with 320 atoms per unit cell and a triclinic structure with 106 atoms. Both structures exhibited semiconducting properties, with a bandgap of ~0.35 eV. When the zero-point vibrational energy effect was neglected, the energies of these two β-boron structures were found to be 23 meV/B lower than that of the 105-atom framework and 1 meV/B higher than that of α-boron. However, after including the zero-point vibrational energy correction, de Groot et al. found that the energy of β-boron was 4 meV/B lower than that of α-boron, making it the most stable bulk-boron structure.

In particular, as evident in Figure 8.15, as the temperature increases from 0 K, the Helmholtz free energy difference between α- and β-boron becomes larger. This implies that β-boron is a thermodynamically stable isomer. At temperatures considerably above room temperature, the entropy change due to defects is more pronounced, resulting in a more stable β-rhombohedral structure.

8.3.2 Borophene

The bulk material of boron is usually a three-dimensional structure constructed with cage-like B_{12} molecules as the structural units, and it has multiple allotropes. Boron also possesses a two-dimensional (2D) allotrope, namely borophene, with a simpler atomic arrangement than three-dimensional (3D) bulk boron. With a reduced number of boron atoms in its unit cell, borophene offers a convenient platform for investigating the nuclear quantum effects of boron materials. It exhibits strong electron-phonon coupling and

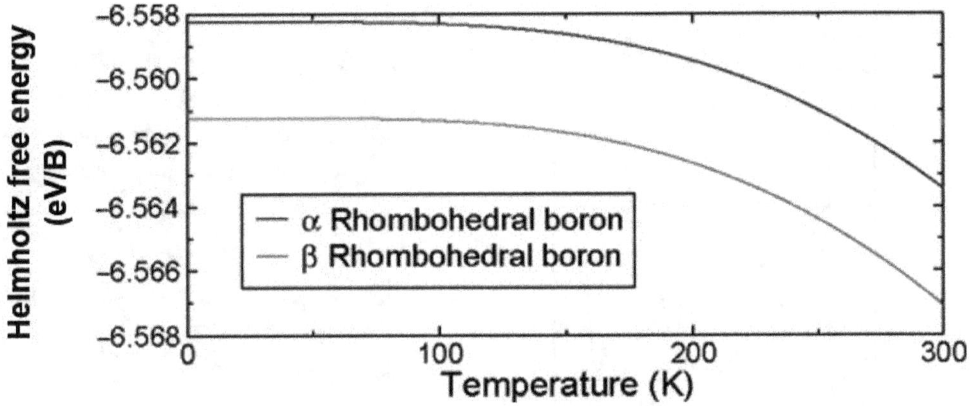

Figure 8.15 Helmholtz free energy as a function of temperature for α (upper curve) and β (lower curve) rhombohedral bulk boron. The volume is fixed in the calculation and the value is taken at 0 K equilibrium. Reprinted with permission from van Setten[2007]. Copyright 2024 American Chemical Society.

could potentially have a higher superconducting temperature than liquid helium. Its other optical, mechanical, and thermal properties have also contributed to making it a popular 2D material in research.

Whether boron can form a 2D planar structure has long been a theoretical concern. Early computational work proposed a variety of different 2D boron structures, including triangular lattices and honeycomb lattices. Important theoretical progress occurred in 2007, when the Ismail-Beigi research group at Yale University [Tang H.2007] and the J. Ni research group at Tsinghua University [Yang X. B. 2008] independently discovered that if hexagonal vacancies are generated in the triangular lattice, the planar boron lattice can be significantly stabilized. They proposed a class of boron sheets featuring triangular and hexagonal motifs. The properties of these sheets were later found to agree with those of experimentally synthesized 2D borophene structures containing periodic vacancies (as depicted in Figure 8.16). Incorporating the zero-point vibrational energy effect into their calculations, Ismail-Beigi et al. compared various potential structures of triangular and hexagonal boron to find the most convincing configuration.

In general, in-plane σ-bonds are stronger than π-bonds derived from p_z orbitals. In borophene, the hexagonal structure lowers its energy by accepting electrons. Because of the presence of antibonding states, the in-plane triangular structure can provide excess electrons. Therefore, by appropriately mixing these two phases, the system can achieve enhanced energetic stability. More specifically, the Fermi energy level of the most stable "hybrid" hexagonal–triangular boron sheet should be located at the zero point of the in-plane projected density of states (PDOS). In this configuration, all unoccupied in-plane bonding states (instead of antibonding states) are filled; the remaining electrons fill the low-energy p_z states, leading to a metallic system. This theoretical finding aligns well with subsequent experimental measurements, which confirm the metallic nature of borophene.

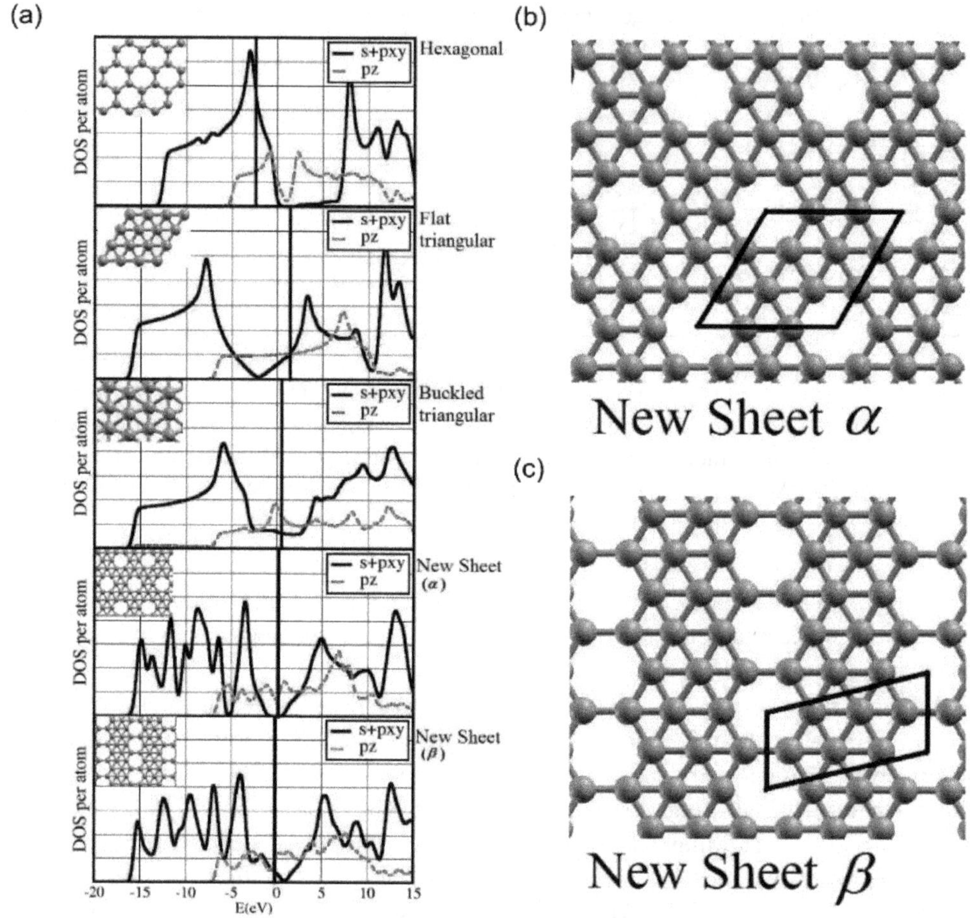

Figure 8.16 (a) Electron density of states (DOS) for five types of borophene: hexagonal, flat triangular, buckled triangular, α-phase, and β-phase. Projections are onto the in-plane $(s + P_x + P_y)$ and out-of-plane (P_z) orbitals. The thick solid line denotes the Fermi energy level E_F. Top view of the (b) α-phase and (c) β-phase borophene structure. The solid line outlines a unit cell. Reprinted figure with permission from Tang H.[2007]. Copyright 2024 by the American Physical Society.

Unlike graphene, which has a honeycomb structure, borophene exhibits a mixed triangular lattice with periodic holes. Because of the infinite possible variation in the density and arrangement of these periodic holes, borophene exhibits a highly polymorphic nature. In other words, it has a large number of isomers. Although there have been several theoretical attempts to predict the structure of borophene, significant advances in this field were achieved independently by Guisinger's group [Mannix2015] at Argonne National Laboratory in 2015, and by K. H. Wu's group [Feng B.J.2016] at the Institute of Physics, Chinese Academy of Sciences, in 2016. Both research groups successfully synthesized single-layer borophene on an Ag(111) surface. However, while Guisinger et al. believed their borophene had a buckled triangular lattice structure, K. H. Wu et al.

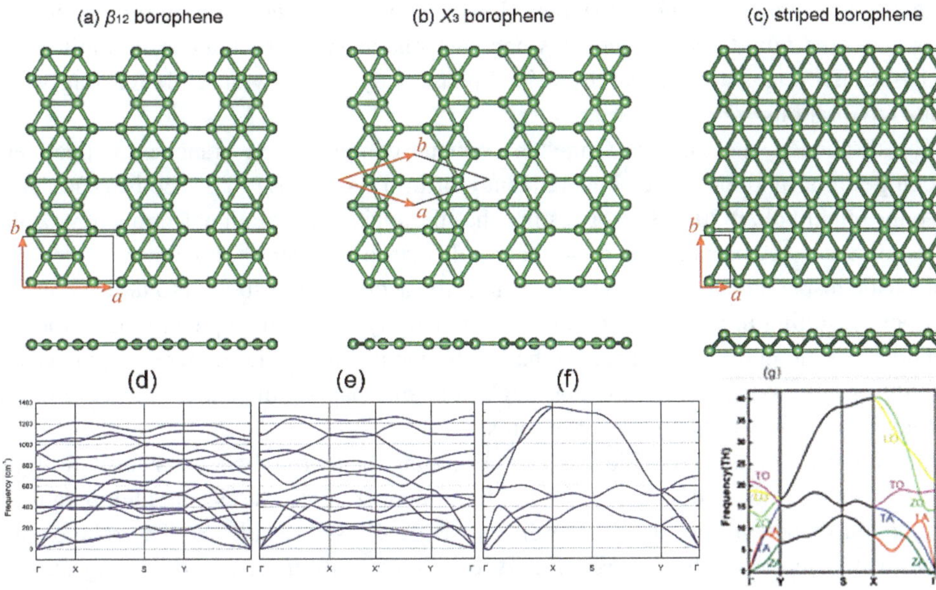

Figure 8.17 Top view and side view of (a) β_{12} borophene, (b) χ_3 borophene, and (c) striped borophene. (d)–(g) Phonon dispersions for (d) β_{12} borophene, (e) χ_3 borophene, and (f) striped borophene along different symmetry lines. (g) An alternative version of (f) after rearranging the K points in reciprocal space. The different branches in the figure are colored differently. Panels (a)–(f) created by Peng B.[2017] licensed under CC BY; panel (g) used with permission of Physical Chemistry Chemical Physics, from Sun H.[2016], Copyright 2024; permission conveyed through Copyright Clearance Center, Inc.

argued that borophene on the Ag(111) surface consisted of β_{12} and χ_3 borophene with holes; subsequent theoretical and experimental studies have confirmed the β_{12} and χ_3 structures as standard structures for borophene.

Once borophene had been successfully grown in the laboratory, the study of its quantum-mechanical properties (such as thermal conductivity, phonon lifetime, thermal expansion, elastic modulus, and particularly lattice dynamics) attracted significant research interest. In 2016, the thermodynamic properties of striped borophene were systematically discussed by X. G. Wan's group. They calculated the phonon dispersion spectrum in borophene, revealing its temperature-dependent structural stability [Sun H.2016]. In 2017, H. Zhang's group comprehensively analyzed the dynamical, thermodynamical, and mechanical stability of the three main borophene structures (β_{12}, χ_3, and the striped phase) using DFT [Peng B.2017].

Two of these structures – β_{12} (Figure 8.17(a)) and χ_3 (Figure 8.17(b)) borophene – have a planar configuration with periodic holes (vacancies), whereas striped borophene has a buckled triangular configuration (Figure 8.17(c)). The most significant difference between the planar and buckled configurations is the absence of holes in the latter; holes further lower the total energy of the system. No vibrational modes with imaginary frequencies are observed in the phonon dispersion spectra of β_{12} and χ_3 borophene, indicating that these structures are dynamically stable. However, in the dispersion spectrum of striped borophene, the vibrational frequencies become imaginary in the long-wave limit along the $\Gamma - X$ direction, showing that the striped phase is dynamically unstable for long-wave acoustic vibrations.

The dispersion spectrum of striped borophene comprises six branches, collectively reaching a total phonon dispersion frequency of up to 40.3 THz. The acoustic branches are only found below 20 THz, suggesting that the maximal vibrational frequency of the acoustic modes in borophene is only about half that in graphene. This limited extension of acoustic dispersion could potentially restrict the acoustic contribution to lattice thermal transport, resulting in a decreased thermal conductivity κ_l. For the optical modes, the top two branches along the $\Gamma - X$ direction have a frequency of ~25 THz; therefore, the group velocity of some optical modes is non-negligible and may significantly contribute to the thermal conductivity κ_l at high temperatures.

Phonons play an indispensable role in determining the thermodynamic stability of light elements such as boron both at 0 K and at finite temperatures. At elevated temperatures, the phonon modes are occupied according to Bose–Einstein statistics. The relationship between the Helmholtz free energy F and temperature is crucial for measuring nuclear quantum effects (see Figure 8.18). In the temperature range of 0 K to 1,000 K, χ_3 borophene exhibits the lowest Helmholtz free energy, while β_{12} borophene also demonstrates significantly lower free energy than striped borophene. This suggests that β_{12} and χ_3 borophene have better thermodynamic stability than striped borophene in a wide temperature

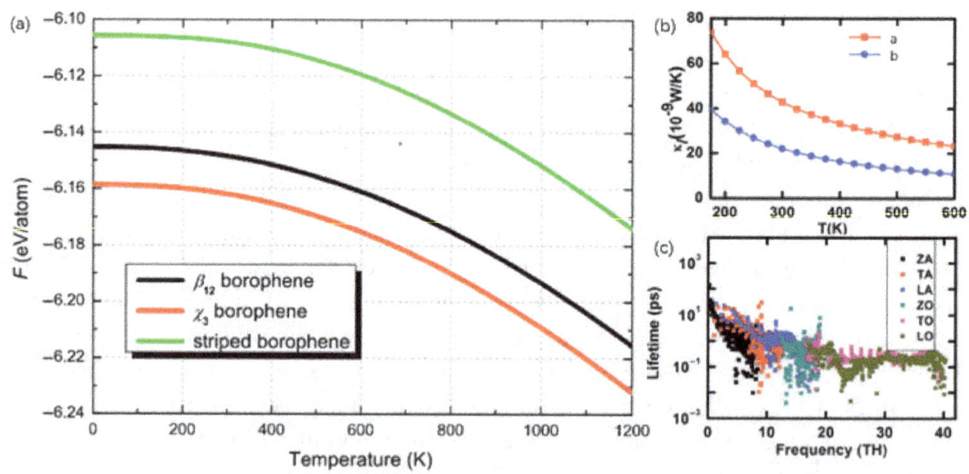

Figure 8.18 (a) Helmholtz free energy versus temperature for β_{12}, χ_3, and striped borophene. (b) Temperature dependence of thermal conductivity κ_l for striped borophene along the (a) and (b) directions. (c) Phonon lifetime distribution of each branch of striped borophene when the temperature is 300 K. Panel (a) created by Peng B.[2017] licensed under CC BY; panels (b) and (c) used with permission of Physical Chemistry Chemical Physics, from Sun H.[2016], Copyright 2024; permission conveyed through Copyright Clearance Center, Inc.

range. Compared with striped borophene, β_{12} and χ_3 borophene have better thermodynamical stability because their phonon modes are softened. In other words, the phonon frequencies of β_{12} and χ_3 borophene are relatively lower at high temperatures, which leads to an increase in entropy and, therefore, a more stable structure.

As it is difficult to define the thickness of a 2D monolayer material, we can use WK^{-1} as the unit of lattice thermal conductivity κ_l in 2D monolayer materials. For striped borophene, κ_l shows a high degree of anisotropy. Moreover, the strong phonon–phonon scattering of borophene leads to a further decrease in the lattice thermal conductivity (as shown in Figure 8.18). Throughout the temperature range considered, the value of κ_l along the b-direction is approximately half of that along the a-direction. The low thermal conductivity along the b-direction is mainly attributable to the short phonon lifetime. The lifetime of the acoustic modes is almost always below 100 ps, which is four orders of magnitude smaller than that of graphene's ZA modes.

The mechanical properties of β_{12} and χ_3 borophene are similar because the two phases have similar in-plane bonding and atomic mass density. Figures 8.19(a) and (b) depict the Young's moduli and Poisson's ratios of the two borophene structures. The stiffness of β_{12} and χ_3 borophene (180–210 N/m) is lower than that of graphene and monolayer boron nitride but higher than that of silicene. Striped borophene exhibits higher stiffness, likely because of strong phonon–phonon scattering. The stiffness of striped borophene along the a-direction (382.3 N/m) is much higher than that along the b-direction; it is even comparable to that of graphene (342.2 N/m). High stiffness implies high thermodynamic instability. This is because increased vibrational frequency can result in increased Helmholtz free energy. Furthermore, the highly buckled structure of striped borophene results in a negative Poisson's ratio. Thus, nuclear quantum effects have a considerable influence on the thermodynamic properties of borophene.

Figure 8.19(c) shows the temperature dependence of Young's modulus E for striped borophene. Evidently, both E_a and E_b initially decrease with increasing temperature but then rapidly increase after reaching room temperature. The temperature enhancement effect differs between borophene and graphene: In graphene, the planar modulus becomes softer as the temperature increases. This discrepancy may be attributed to a substantial thermal contraction effect in borophene, which significantly increases the curvature of the potential energy surface. Calculations using the quasi-harmonic approximation (QHA) reveal that the Poisson's ratios of striped borophene are positive at 0 K. However, as shown in Figure 8.19(d), negative Poisson's ratios are observed within the temperature range of 70–145 K in both the a and b directions, with $V_a = -0.05$ and $V_b = -0.02$ representing the most negative values. These results are consistent with the order of magnitude of DFT calculations.

The linear thermal expansion coefficient (LTEC) of striped borophene has also been calculated using DFT after considering the zero-point vibration. The result shows that the zero-point vibration expands the lattice of striped borophene at 0 K by 0.35% and 0.72% in the (a) and (b) directions, respectively. For diamond, hexagonal boron nitride, and graphene, the lattice is expanded by 0.4%, 0.3%, and 0.3%, respectively (see Section 8.4 for a discussion of carbon). The zero-point vibrational expansion results from

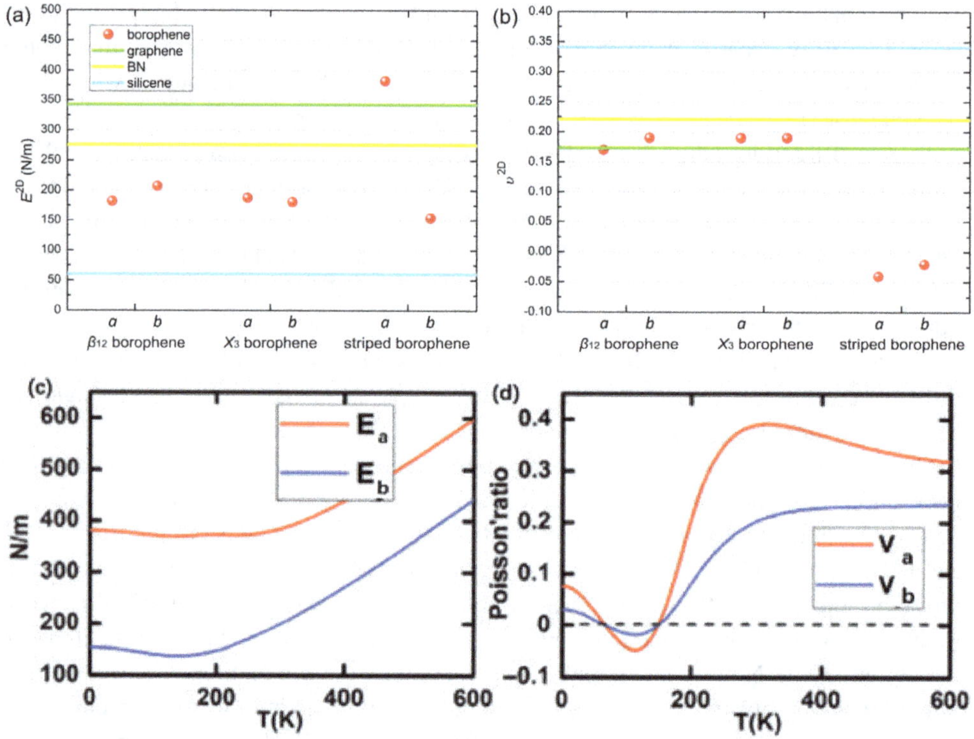

Figure 8.19 (a) 2D Young's modulus E^{2D} and (b) Poisson's ratio v^{2D} for β_{12}, χ_3, and striped borophene along the (a) and (b) directions. Both plots also show graphene, monolayer BN, and silicene. (c) and (d) temperature dependence of (c) Young's modulus and (d) Poisson's ratio along the (a) and (b) directions for striped borophene. Panels (a) and (b) created by Peng B.[2017] licensed under CC BY; panels (c) and (d) used with permission of Physical Chemistry Chemical Physics, from Sun H.[2016], Copyright 2024; permission conveyed through Copyright Clearance Center, Inc.

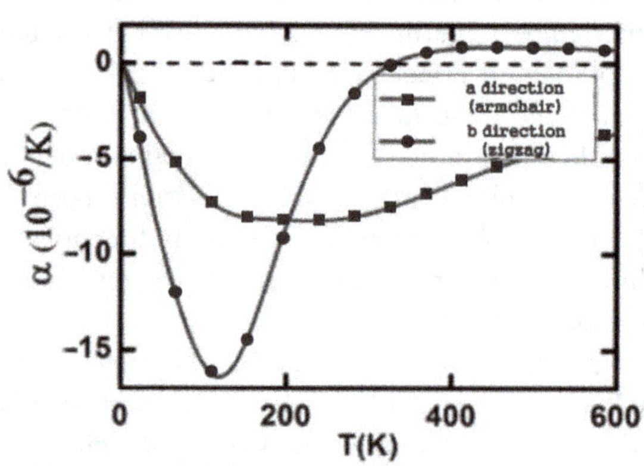

Figure 8.20 Temperature dependence of linear thermal expansion coefficients (LTECs) along the (a) and (b) directions for striped borophene. Used with permission of Physical Chemistry Chemical Physics, from Sun H.[2016], Copyright 2024; permission conveyed through Copyright Clearance Center, Inc.

the competition between phonon modes with positive and negative Grüneisen parameters. This phenomenon is closely related to anharmonic effects and is generally dominated by high-frequency modes.

At finite temperatures, borophene has an anisotropic negative LTEC (Figure 8.20). The LTEC of striped borophene along the a-direction, like the LTEC of graphene, takes negative values over a wide temperature range. However, the most negative value for borophene (-8×10^{-6}) is twice that for graphene (-3.8×10^{-6}). Along the b-direction, the LTEC of borophene decreases drastically in the temperature range of 0 K to 120 K; the LTEC of hexagonal boron nitride (h-BN) exhibits a similar decrease. However, the most negative value for borophene (-16×10^{-6}) is ~2.5 times that for h-BN (-6.5×10^{-6}).

8.3.3 Boron Nitride Nanotubes

So far, we have discussed only elemental boron materials. Boron can also easily form compounds with other elements because of its high reactivity and various ways of bonding. The most interesting of these compounds are those containing nitrogen atoms, such as boron nitride nanotubes and hexagonal boron nitrides. Boron nitride nanotubes (BNNTs), shown in Figure 8.21, were first predicted theoretically in 1995 by S. G. Louie and M. L. Cohen at the University of California, Berkeley [Chopra1995]. Similar to carbon nanotubes (CNTs), they are one-dimensional (1D) structures. In addition to possessing the excellent mechanical properties of CNTs, they exhibit high thermodynamic stability and oxidation resistance (up to 900 °C). They also have excellent insulating characteristics, with a wide bandgap (5–6 eV) that is independent of the nanotube diameter and chiral

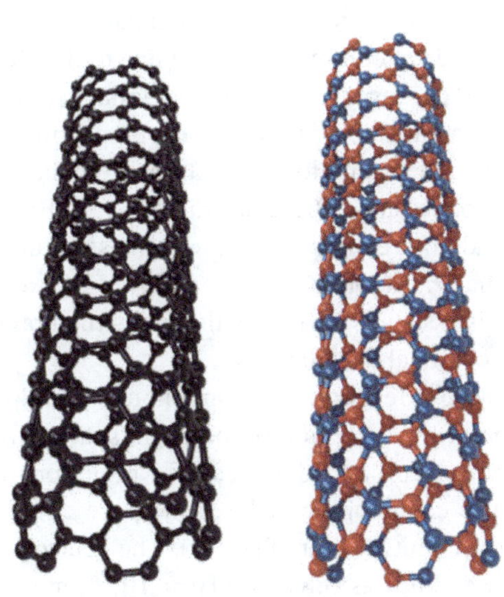

Figure 8.21 Comparison of the carbon nanotube (CNT) and boron nitride nanotube (BNNT) structures. Figure created by Kim J. H.[2018] licensed under CC BY 4.0.

information. These distinctive attributes make BNNTs suitable for a wide range of applications, especially in electronic and optical devices. Nonetheless, the lack of an effective synthesis method hinders their widespread utilization.

Studies of full quantum effects on this type of material primarily concentrate on electron–phonon interactions. In 2005, J. S. Lauret et al. initially reported the optical transition properties of single-walled BNNTs at room temperature and suggested that the splitting of the low-energy line (4.45 eV) might be associated with strong electron–phonon coupling [Lauret2005]. Subsequently, Y. Bando's group utilized cathodoluminescence (CL) spectra to measure the bandgap and the radiative transition spectra of BNNTs and isotopically purified ^{10}BNNTs [Han W.Q.2008]. They reported a direct band gap of 5.38 eV for BNNTs, independent of the tube size and isotope substitution. However, the details of the CL spectra in the low energy region revealed a clear isotope dependence. The purified ^{10}BNNTs had rich phonon replicas between 3.0 and 4.2 eV, reflecting the electron–phonon coupling mechanism; these peaks were significantly attenuated in natural BNNTs. More recently, F. Azizi et al. conducted a study on plasmons in BNNTs, considering the presence of electron–phonon coupling [Azizi2021]. Utilizing the Holstein model Hamiltonian and the Green's function approach, they calculated the charge susceptibility and found that the electron–phonon interaction does not affect the dynamic part of the susceptibility function. However, it does influence the static susceptibility function, which leads to alterations in the system's charge ordering.

Another interesting study was conducted by J. Q. Li's group at Institute of Physics, Chinese Academy of Sciences. They employed ultrafast TEM to investigate the evolution of significant structural features and electronic properties of multi-walled BNNTs during a complete reversible cycle after femtosecond laser excitation. This was done using pulsed electron diffraction and femtosecond-resolved electron energy-loss spectroscopy (EELS) [Li Z.W.2019]. Analysis of the experimental data suggested that initially, a 15 ps nonthermal anisotropic lattice alteration took place due to the electron–phonon coupling. This was followed by a 180 ps Auger recombination. Finally, phonon-driven thermal transients occurred along the radial and axial directions. First-principles calculations support these findings and show that the excitation of electrons from the π to π^* orbitals in BNNTs weakens the intralayer bonds along the axial direction of BNNTs but strengthens the interlayer bonds in the radial direction. As a result, the structure of the BNNTs expands axially and contracts radially. Importantly, time-resolved EELS measurements show that the bandgap contraction is associated with nonthermal structural transients. These results are in agreement with theoretical calculations.

Besides the full quantum effects on electron–phonon coupling of BNNTs, the other notable example is its neutron absorption property. In space exploration, a major safety hazard arises from neutrons produced by cosmic radiation interacting with matter. These neutrons can cause catastrophic failures in space equipment, posing threats to the lives and health of astronauts (Figure 8.22(a)). The neutron absorption cross section of ^{10}B atoms can reach as high as 3,840 b (1b = 10^{-28} m^2). Therefore, ^{10}B-enriched BNNT composites can effectively shield against neutrons produced in the cosmic environment. NASA has explored the possibility of neutron shielding using ^{10}B-enriched

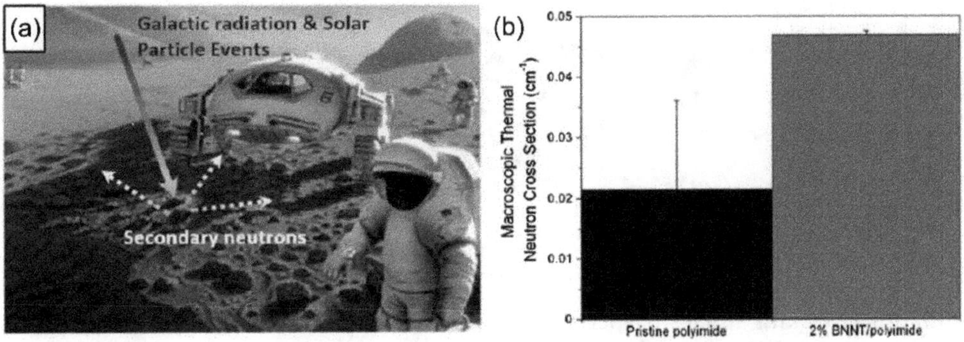

Figure 8.22 BNNT shielding against cosmic rays. Reprinted from Li[2022], with permission of the author.

BNNT composites [Kang J.H.2015]. The neutron cross section of ordinary polyimide with neutron-moderator properties is 0.021 cm^{-1}. However, 2 wt% BNNT/polyimide film has a neutron cross section of 0.047 cm^{-1}, which is 120% more than that of pristine polyimide (Figure 8.22(b)). Recently, hydrogen-enriched nanotubes have attracted considerable research interest because the addition of hydrogen can further enhance shielding against space radiation, including space particle events (SPEs) and galactic cosmic rays (GCRs). Boron nitride nanotubes were chosen due to their large surface area and higher hydrogen binding energy, making them more capable of loading hydrogen atoms than 0D and 2D materials.

8.3.4 Hexagonal Boron Nitride

Another compound composed of boron and nitrogen atoms is hexagonal boron nitride (h-BN), which is a wide-gap semiconductor. It has an atomically smooth surface containing equal numbers of boron and nitrogen atoms. Both h-BN and graphite, a material consisting entirely of carbon atoms, are isoelectronic. The lattice constant (atomic spacing a = 0.2504 nm) and interlayer distance (0.33 nm) of h-BN are also similar to those of graphite because of its graphite-like layer structure. For this reason, h-BN is commonly known as white graphite [Geick1966]. Within each h-BN monolayer, the boron and nitrogen atoms form a planar hexagonal honeycomb mesh by covalent bonding. Each h-BN layer is bonded by weaker van der Waals interactions, and this particular structural orientation results in strong anisotropy (as shown in Figure 8.23). However, compared to those of graphite, the h-BN layers are stacked differently. The AB stacking of graphite refers to the stacking method that provides optimized energy, where adjacent atomic layers are translated by one atomic distance along the edges of the in-plane hexagonal cells. By contrast, the stable structure of h-BN has an AA′ stacking, that is, the B and N atoms in the out-of-plane direction (c-direction) are arranged sequentially. Theoretical studies have confirmed that AA′ stacking is the stacking mode with the lowest energy in h-BN. However, the van der Waals interaction between each layer is relatively weak, which means that it is possible for h-BN to exist in metastable phases

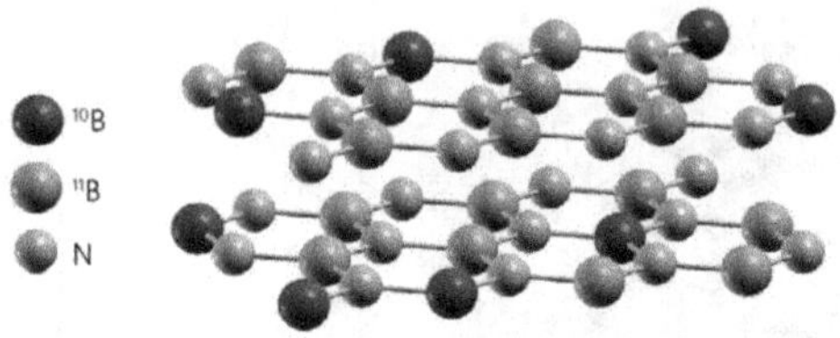

Figure 8.23 Schematic of the h-BN crystal structure. From Vuong[2018]. Reproduced with permission from SNCSC.

with other stacking modes. Experiments have already demonstrated that the AB stacking of h-BN layers is one such metastable stacking configuration.

Hexagonal boron nitride has unique and optimal optoelectronic properties, thermal stability, and chemical inertness. Furthermore, it can weaken Coulomb shielding and protect other 2D materials from environmental pollution and oxidation, exhibiting superior interface characteristics. Therefore, it is commonly used to encapsulate other 2D materials. Some atmospherically unstable 2D materials can be encapsulated between two layers of h-BN to form sandwich structures. The carrier mobility and other properties of 2D materials can also be modulated using h-BN as a dielectric layer. For example, the mobility of h-BN-encapsulated graphene can even reach the theoretical limit of phonon scattering, and its average free range is solely dependent on the sample size. Encapsulation can increase the mobility of black phosphorus from hundreds to thousands of $cm^2/V.s$. In addition, artificial van der Waals heterojunctions, such as graphene/boron-nitride junctions, provide a new platform for studying quantum states such as correlation and topology, demonstrating their potential for device applications in external field modulation. These technical advantages make h-BN the most promising candidate for integration with other 2D materials to produce multifunctional optoelectronic devices.

In addition, several studies have shown that h-BN possesses unique optical properties that offer the possibility of building novel optoelectronic devices. For example, h-BN has a highly anisotropic crystal structure, polar chemical, and other physical properties, and its hyperbolic structure facilitates the formation of polaritons, making it suitable for applications in IR nanoelectronics. Moreover, the h-BN point defects are capable of producing single-photon emission. Therefore, h-BN can be used as a quantum light source, covering the wavelength range from visible light to near-IR at room temperature. Hexagonal boron nitride is an indirect-gap semiconductor with a forbidden bandwidth of ~6 eV. Nevertheless, the internal quantum efficiency of UV radiation can be as high as ~40%, making it possible for h-BN to be used in UV sources and detectors. Very interestingly, K. H. Liu's group at Peking University produced a second-harmonic generation with conversion efficiency of ~8% and facile polarization controllability by a twisted BN films within a thickness of only 3.2 μm [Hong H.2023]. Furthermore, this kind of "twist-phase-matching" nonlinear optical crystals shows isotopic effect. In summary, the unique and abundant physical and chemical properties of h-BN indicate its potential for diverse future applications.

8.3.4.1 Thermal Stability and Bandgap Renormalization

In 2016, F. Calvo and Y. Magnin used the path-integral Monte Carlo (PIMC) and self-consistent phonon (SCP) methods to calculate the thermodynamic properties of an h-BN monolayer. The calculation results were then compared with those of the semi-classical QHA to reveal the nuclear quantum effects on h-BN thermodynamic properties [Calvo2016].

The Monte Carlo (including MC and PIMC) methods calculate the lattice constant of h-BN directly, whereas the SCP and QHA methods obtain it indirectly by minimizing the free energy. However, the SCP method considers the quantum motion of the nuclei, while the QHA method is semiclassical. Figure 8.24 shows the lattice-constant dependence of the free energies ΔF (a;T) obtained by Calvo and Magnin using each method at T = 300 K. The QHA results, represented by the dashed curve in Figure 8.24, were continuous. The SCP results, indicated by the solid curve, were obtained by interpolating a fourth-order polynomial. In the diagram, the classical treatment involves fixing the atoms at static lattice points. Taking the phonon vibration into account, the lattice parameter increased from 2.495 Å (static structure) to 2.498 Å (QHA), and the SCP calculation further increased it by ~0.0015 Å. At zero pressure, the increase in the lattice parameter is attributable to the expansion of the nuclear wavefunction. The strongest stretching modes occur in the monolayer plane, which predominantly affects the in-plane thermomechanical behavior, thereby altering the equilibrium lattice parameter.

Calvo and Magnin also calculated the lattice parameter a as a function of temperature (T) using the same methods. In the classical description, all approaches converged to a = 2.498 Å at T = 0 K (as shown in Figure 8.25(b)). This low-temperature-limiting behavior is consistent with the results expected from the standard model of the static atomic structure. In the classical case, all three methods predicted rather similar

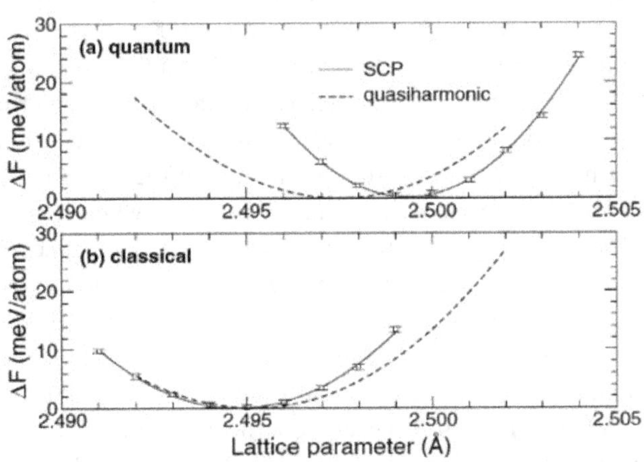

Figure 8.24 Free energy ΔF of an h-BN monolayer at T = 300 K as a function of the lattice parameter, as obtained from QHA (dashed curve) and SCP (solid curve). (a) Quantum-mechanical description of nuclei; (b) Classical description of nuclei in the static configuration. From Calvo[2016]. Reproduced with permission from SNCSC.

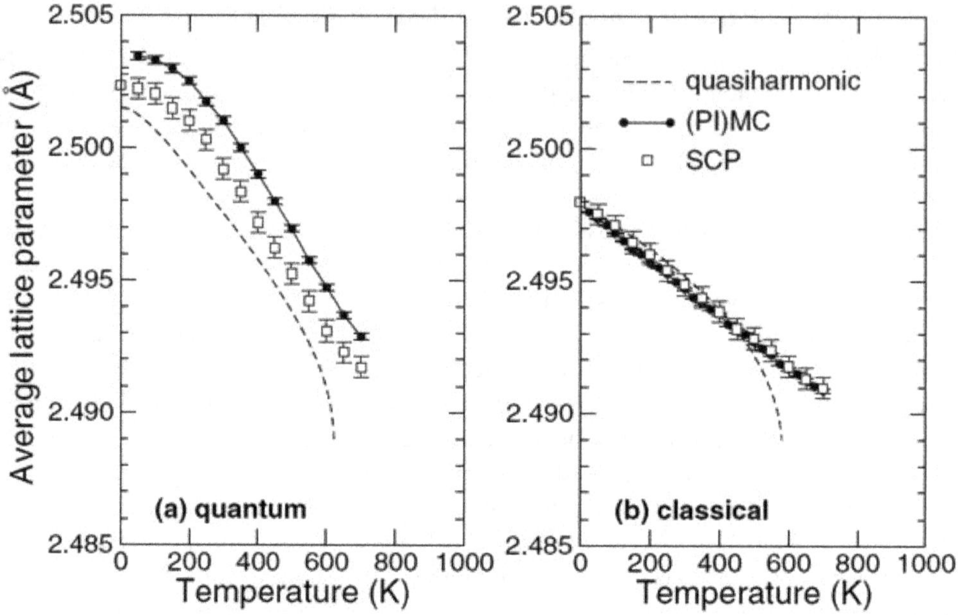

Figure 8.25 Lattice parameter of h-BN monolayer as a function of temperature, as obtained from QHA (dashed curve), SCP (squares), and PIMC (dots). (a) Quantum-mechanical description of atomic nuclei; (b) Classical description of atomic nuclei in the static configuration. From Calvo[2016]. Reproduced with permission from SNCSC.

variations of the lattice parameter with temperature, with an essentially constant slope at finite temperatures. Only when the temperature is higher, the increasing anharmonic effect makes the QHA results deviate significantly.

When the nuclei were treated quantum mechanically, the predictions of the three methods no longer converged to the same value even at T = 0 K (as shown in Figure 8.25(a)). However, in the range 0–400 K, the PIMC and SCP results showed a greater variation than the QHA, probably due to anharmonicity. The lattice parameter predicted by SCP lies between the results predicted by the other two methods because of its partial consideration of anharmonicity. These results indicate that the quantum-wavefunction and zero-point vibrations were prone to anharmonicity even when the system was in the ground state. The zero-temperature limit of the lattice parameter increased from 2.5015 Å (QHA) to 2.5025 Å (SCP) and nearly 2.5030 Å (PIMC). The fact that by two anharmonic methods (SCP and PIMC) the lattice parameters were increased testifies that anharmonic effects induced by the full quantization at low temperature play a greater role in the vibrational delocalization than that solely from the thermal expansion.

Figure 8.26 depicts the thermal expansion coefficient (TEC) as a function of temperature obtained using the same three methods described earlier. All the methods indicate that the TEC of h-BN is negative throughout the entire temperature range. Although results similar to those reported by Sevik et al. were obtained near room temperature [Sevik2014], the magnitude of quantum vibrational delocalization was significantly enhanced.

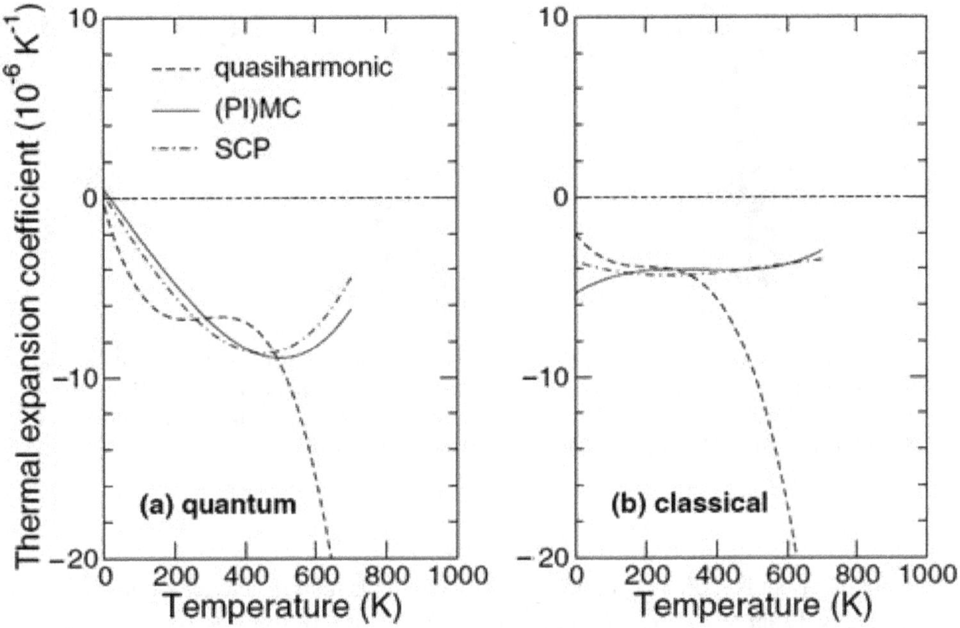

Figure 8.26 In-plane thermal expansion coefficient of an h-BN monolayer obtained from polynomial fits of the lattice parameter, calculated by QHA (dashed curve), (PI)MC (solid curve), and SCP (dashed-dotted curve). (a) Quantum-mechanical description of nuclei; (b) Classical description of nuclei in the static configuration. From Calvo[2016]. Reproduced with permission from SNCSC.

The nuclear quantum effects on the electronic structure of h-BN are mainly reflected in the bandgap renormalization caused by electron–phonon coupling, which originates from full quantum effects. H. Mishra and S. Bhattacharya used density functional perturbation theory (DFPT) and many-body perturbation theory (MBPT) to study the direct bandgap renormalization induced by electron–phonon coupling in an h-BN monolayer [Mishra2019]. As shown in Figure 8.27(a) and (b), the spectral functions of the top valence band and the bottom conduction band have different temperature dependencies. As the temperature increases, the zero-point shift of the occupied state at the top valence band is 161 meV (i.e., the valence band V1 crosses the 0 eV line and shifts 161 meV upward), whereas the zero-point shift of the unoccupied state at the conduction band is –112 meV (i.e., the bottom conduction band C1 reaches 4.596 eV and shifts 112 meV downward).

Evidently, when the nuclei are treated quantum mechanically, the bandgap shrinks by 273 meV (i.e., 112 meV + 161 meV) in total. Mishra and Bhattacharya also observed a remarkable resemblance between the temperature dependence of the direct bandgap at the K point of the Brillouin zone in an h-BN layer and the experimental findings of BN nanotubes with a diameter of 60 nm. As seen in Figure 8.27(c), the bandgap for h-BN monolayer shrinks at a rate of –0.53 meV K^{-1} as the temperature increases. This result confirms the presence of strong electron–phonon interactions in h-BN.

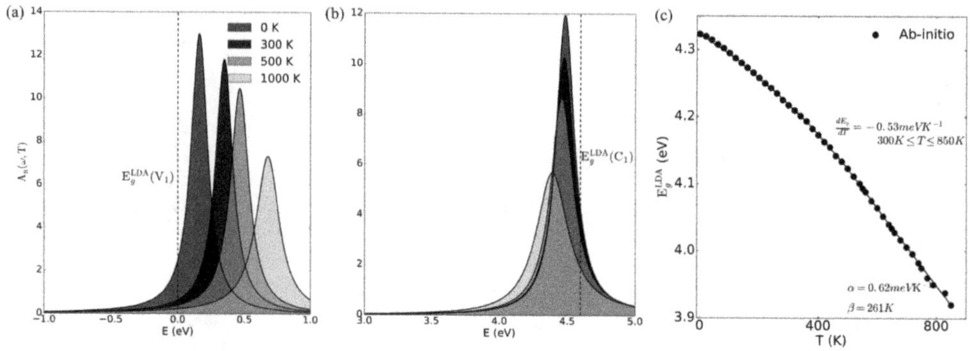

Figure 8.27 Spectral distributions at (a) the top valence band and (b) the bottom conduction band at K point in the Brillouin zone at different temperatures. (c) Bandgap as a function of temperature; the slope is –0.53 meV K^{-1}. Reprinted figure with permission from Mishra[2019]. Copyright 2024 by the American Physical Society.

8.3.4.2 Isotope Effect

The isotopic effect is more pronounced in light elements than in heavier elements because of the relatively large differences in the nuclear masses. Boron has 14 isotopes, out of which two are stable: ^{10}B (with a relative atomic mass of ~10) and ^{11}B (with a relative atomic mass of ~11), with natural abundances of 19.9% and 80.1%, respectively. Nitrogen has 17 isotopes, out of which two are stable: ^{14}N and ^{15}N, with natural abundances of 99.6% and 0.4%, respectively. Therefore, the isotope substitution of boron can significantly change the physical properties of h-BN in natural h-BN crystals.

a Spectroscopy

Because the frequency of a harmonic oscillator is $\sqrt{k/m}$, it decreases with increasing atomic mass. The atomic mass of ^{10}B is smaller than that of ^{11}B. The change in phonon frequency after isotopic substitution can be used to characterize the isotopic purity of h-BN, which has four lattice vibrational modes: E_{1u}, A_{2u}, B_{1g}, and E_{2g}. The modes E_{1u} and A_{2u} are IR active, B_{1g} is neither IR nor Raman active, and the in-plane vibrational mode E_{2g} is Raman active. R. Cuscó et al. have computed the phonon dispersion and phonon density of states for ^{10}B and ^{11}B enriched h-BN, respectively (see Figure 8.28) [Cuscó2018]. The low-energy E_{2g} mode$\left(E_{2g}^{low}\right)$ represents the interlayer vibrational shear mode; the high-energy E_{2g} mode (E_{2g}^{high}) corresponds to the relative vibrational modes of the B and N atoms within the layer. The high-energy intra-layer E_{2g}^{high} mode is best suited for studying isotopic effects, which can be observed by the variation in Raman scattering spectra.

T. Q. P. Vuong et al. measured the Raman spectra of both ^{10}B and ^{11}B enriched h-BN (h-^{10}BN/h-^{11}BN) and natural h-BN (h-^{Na}BN) at room temperature [Vuong2018]. As shown in Figure 8.29, the E_{2g}^{high} vibrational mode corresponds to a Raman shift, that is, it corresponds to the boron and nitrogen atoms undergoing in-plane vibration. As expected, when the mass of the boron atom increases, the vibrational frequencies of the E_{2g}^{high} phonons gradually decrease. The frequencies for h-^{10}BN, h-^{Na}BN, and h-^{11}BN

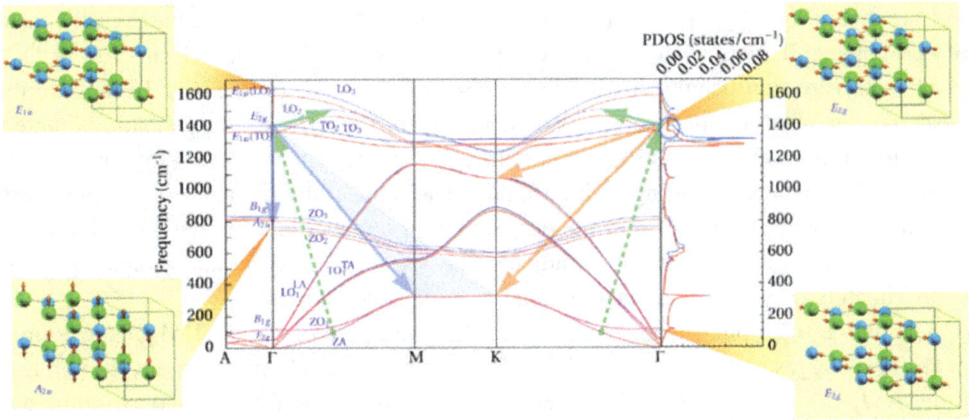

Figure 8.28 Phonon dispersion and corresponding phonon density of states for h-^{10}BN (blue curve) and h-^{11}BN (red curve). The arrows in the figure indicate several major phonon decay processes. The insets on the side show atomic models of the four vibrational modes E_{1u}, A_{2u}, E_{2g}^{low}, and E_{2g}^{high}. Reprinted figure with permission from Cuscó[2018]. Copyright 2024 by the American Physical Society.

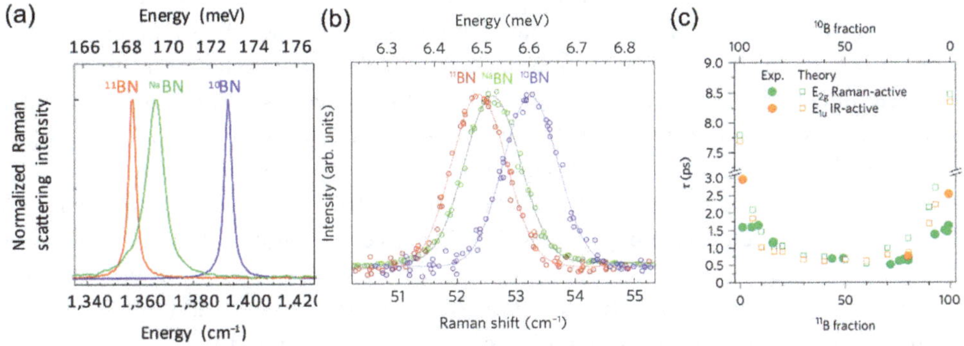

Figure 8.29 (a) Raman spectra in h-^{10}BN, h-^{Na}BN, and h-^{11}BN vibrational modes and (b) Raman mode E_{2g}^{low} at room temperature. (c) E_{2g} (green) and E_{1u} (orange) phonon lifetimes as a function of ^{11}B fraction. The circles are experimental data and the squares denote the theoretical calculation. Figure 8.29 (a) and (b) from Vuong[2018]. Reproduced with permission from SNCSC; Figure 8.29 (c) from Giles[2018]. Reproduced with permission from SNCSC.

are 1,393 cm^{-1}, 1,366 cm^{-1}, and 1,357 cm^{-1}, respectively. In addition to the frequency difference, the full width at half maximum (FWHM) of the Raman peaks also changes drastically. The FWHM of the h-^{10}BN and h-^{11}BN Raman peaks are both 3 cm^{-1}, which is significantly smaller than that of the h-^{Na}BN Raman peak (5 cm^{-1}). The broadening of the Raman peaks in h-^{Na}BN is attributed to the disordered distribution of natural isotopes, which breaks the translational symmetry of the crystal structure and causes elastic phonon scattering [Cardona2005].

Like that of the E_{2g}^{high} mode, the frequency of the E_{2g}^{low} mode also decreases as the mass of boron nuclei increases. However, the energy offsets of E_{2g}^{low} mode (~1 cm^{-1} for h-^{10}BN and h-^{11}BN) are significantly smaller than those of E_{2g}^{high} mode (~36 cm^{-1}). Because of the weak van der Waals coupling between adjacent layers, the energy of E_{2g}^{low} is approximately one-thirtieth that of E_{2g}^{high}. As the E_{2g}^{low} mode has very low vibrational energy and phonon density of states, the spreading of the Raman peak induced by the natural isotopic distribution is almost negligible. From the perspective of lattice vibration, the E_{2g}^{low} vibrational mode involves a rigid movement of adjacent layers, so each layer can be regarded as an approximately homogeneous plane with an average isotopic mass.

The above-mentioned isotope effect can lead to a decreased FWHM of the Raman peak. This suggests that isotope-enriched h-BN can effectively increase the phonon lifetime. A. J. Giles et al. explored the relationship between the E_{2g}^{high} phonon lifetime and isotopic purity, finding good agreement between theory and experiment [Giles2018]. In general, phonon lifetimes are limited by phonon scattering and isotopic disorder scattering, which are characteristic of the material. Giles et al. represented phonon–isotope scattering as mass defects and used DFT calculations, considering anharmonic effects, to calculate the phonon lifetime in h-BN as a function of ^{11}B concentration. The calculation showed that the phonon lifetimes in ^{11}B and ^{10}B enriched h-BN samples (~8 ps) were significantly longer than in natural h-^{Na}BN (~0.5 ps) because of the reduction in mass defects. However, for highly concentrated isotope samples, the experimental measurements still differed significantly from the calculated values. For instance, the experimentally determined phonon lifetimes saturated near 2 ps. This saturation can be attributed to the influence of impurities, such as carbon atoms or other defects, which introduce additional phonon scattering and subsequently decrease the phonon lifetime.

Giles et al. also used Fourier transform infrared (FTIR) reflectance spectroscopy to investigate the relationship between the IR active E_{1u} phonon lifetime and isotopic purity [Giles2018]. The damping constants of the transverse optical phonon were derived from the dielectric function fit of the IR reflectance spectra, and the E_{1u} phonon lifetime was extracted from the damping constants. It was found that the E_{1u} phonon lifetimes in ^{10}B and ^{11}B enriched h-BN samples were longer than those in the natural samples, which is consistent with the theoretical expectations.

The isotope effect is also reflected in the optical properties of h-BN. Natural h-^{Na}BN crystals are hyperbolic materials in the mid-IR band because of their strong in-plane and out-of-plane anisotropy. This allows the existence of hyperbolic phonon polaritons (HPPs) with large momentum [Caldwell2014]. Compared to graphene plasmons, HPPs have features that can provide unique physical properties to h-BN, such as a high-quality factor, longer directional propagation, and ultraslow group velocity.

As the optical loss of the HPPs in h-BN is closely related to the E_{1u} phonon scattering rate, increasing phonon lifetimes can also increase the HPP propagation length. Using s-SNOM, Giles et al. examined the relationship between the propagation length of HPPs and isotopic purity (Figure 8.30) [Giles2018]. The HPP wavelengths were obtained by observing the oscillations of SNOM-tip-launched and h-BN-flake-edge-reflected HPPs.

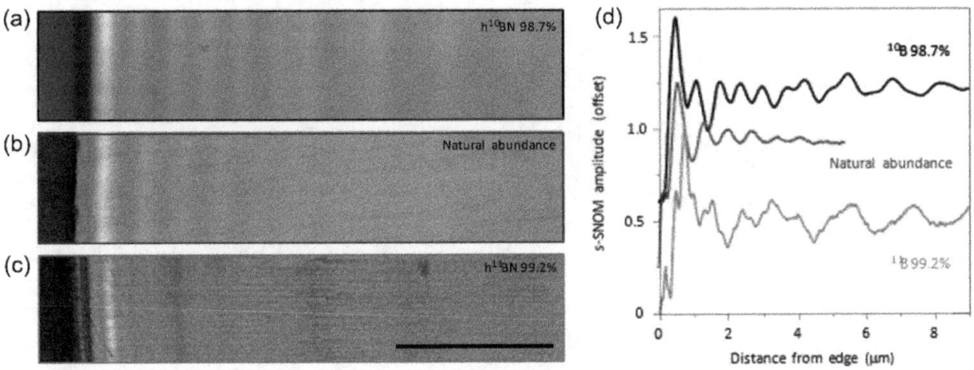

Figure 8.30 Propagation length of hyperbolic phonon polaritons (HPPs) as a function of isotopic purity. Scattering-type scanning near-field optical microscopy (s-SNOM) measurements were collected from 120-nm-thick samples of (a) h-^{10}BN 98.7%, (b) h-^{Na}BN, and (c) h-^{11}BN 99.2%. The scale bar is 5 μm. The HPPs were stimulated with 1,510, 1,480, and 1,480 cm^{-1} incident laser sources, respectively. (d) Linescans extracted from (a)–(c) demonstrating that HPP propagation lengths were significantly longer in the ^{10}B and ^{11}B enriched h-BN samples than in h-^{Na}BN. From Giles[2018]. Reproduced with permission from SNCSC.

It is evident from the figure that HPPs exhibit larger propagation lengths in ^{10}B and ^{11}B enriched h-BN samples than in natural-abundance ones.

In addition to affecting lattice vibrations, the mass change in isotopic substitution also affects the electronic structure because of electron–phonon coupling. An important consequence of this is bandgap renormalization [Cardona2005].

Vuong et al. experimentally studied the bandgap variation in h-BN materials at 8 K using photoluminescence (PL) spectroscopy [Vuong2018] (shown in Figure 8.31). Because h-BN is a semiconductor with an indirect bandgap, the compound luminescence process requires the participation of phonons to satisfy momentum conservation [Cassabois2016a, b, Vuong2017a, b]. As shown in Figure 8.31(a), the energies of the four luminescence peaks were 5.76, 5.79, 5.86, and 5.89 eV, which respectively correspond to the LO, TO, LA, and TA phonon-assisted photoluminescence processes in the PL spectrum of h-BN. The PL spectrum of h-^{10}BN was red-shifted in comparison to that of h-^{Na}BN, while the PL spectrum of h-^{11}BN was blue-shifted. In this case, higher phonon-replica energy corresponded to a larger energy shift, indicating that phonons are directly affected by isotope substitution. After plotting the energy of the luminescence peak as a function of phonon energy and separating the phonon isotopic shift from the isotopic dependence of the bandgap energy, the indirect bandgap energies of h-^{11}BN, h-^{Na}BN, and h-^{10}BN were determined to be 5960.0, 5958.5, and 5954.0 meV, respectively. It is clear that the change in the effective mass of the isotope can lead to a change in the indirect bandgap energy. ^{10}B enrichment of h-BN resulted in a decrease in the indirect bandgap such that $\delta E_g^{10} - \delta E_g^{Na} = -4.4 \pm 0.2$ meV, whereas ^{11}B enrichment of h-BN caused an increase in the indirect bandgap, such that $\delta E_g^{11} - \delta E_g^{Na} = 1.6 \pm 0.2$ meV.

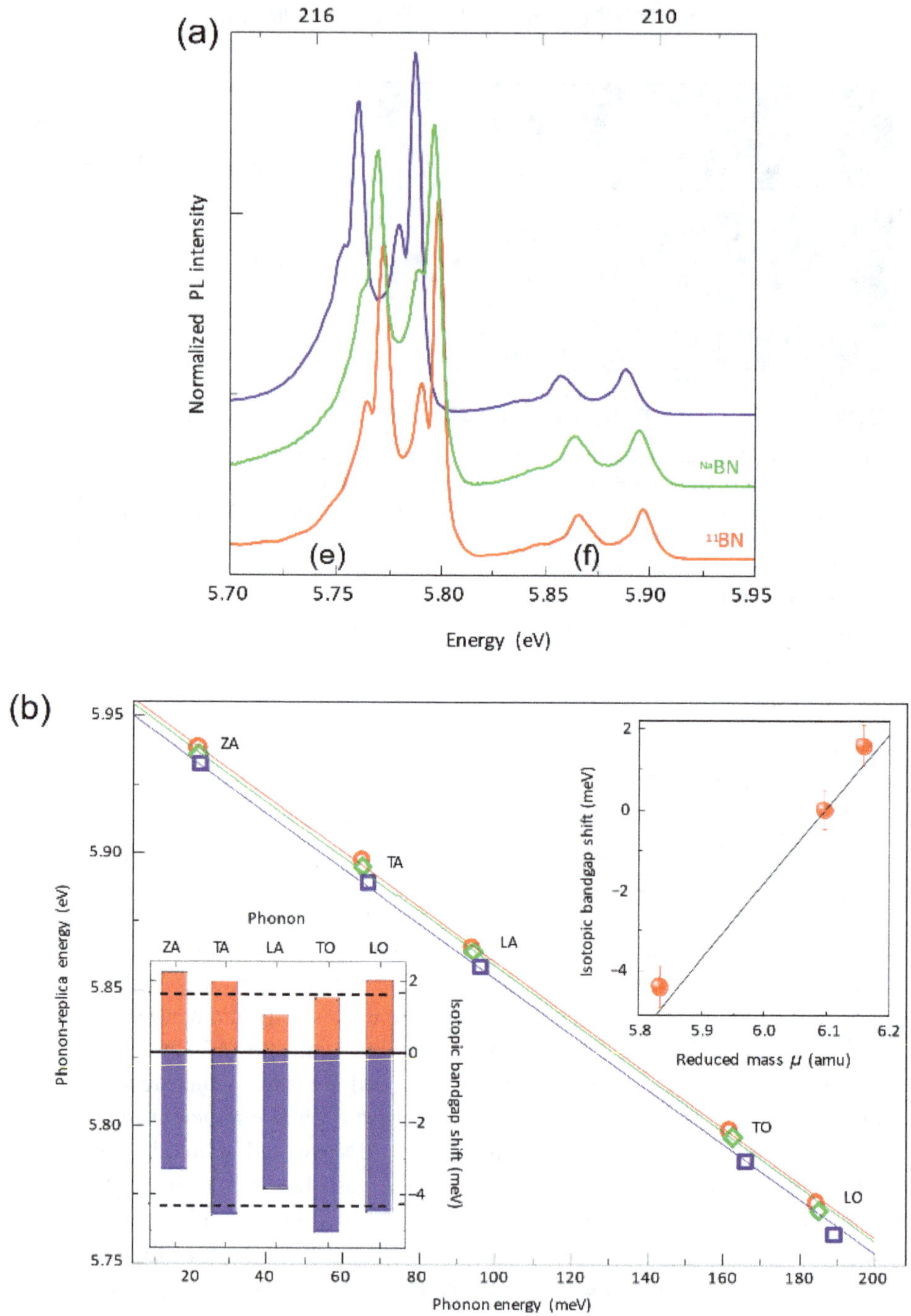

Figure 8.31 Isotope dependence of bandgaps. (a) Photoluminescence (PL) spectra of h-^{10}BN (blue curve), h-^{Na}BN (green curve), and h-^{11}BN (red curve) at 8 K. (b) Energy of the phonon replica in the PL spectra at 8 K versus phonon energy. The inset on the left gives the energy shift of the phonon replica in h-^{10}BN (blue bars) and h-^{11}BN (red bars) compared to h-^{Na}BN, after subtracting the isotopic shift of phonon modes. The inset on the right shows the isotopic dependence of the bandgap energy shift. From Vuong[2018]. Reproduced with permission from SNCSC.

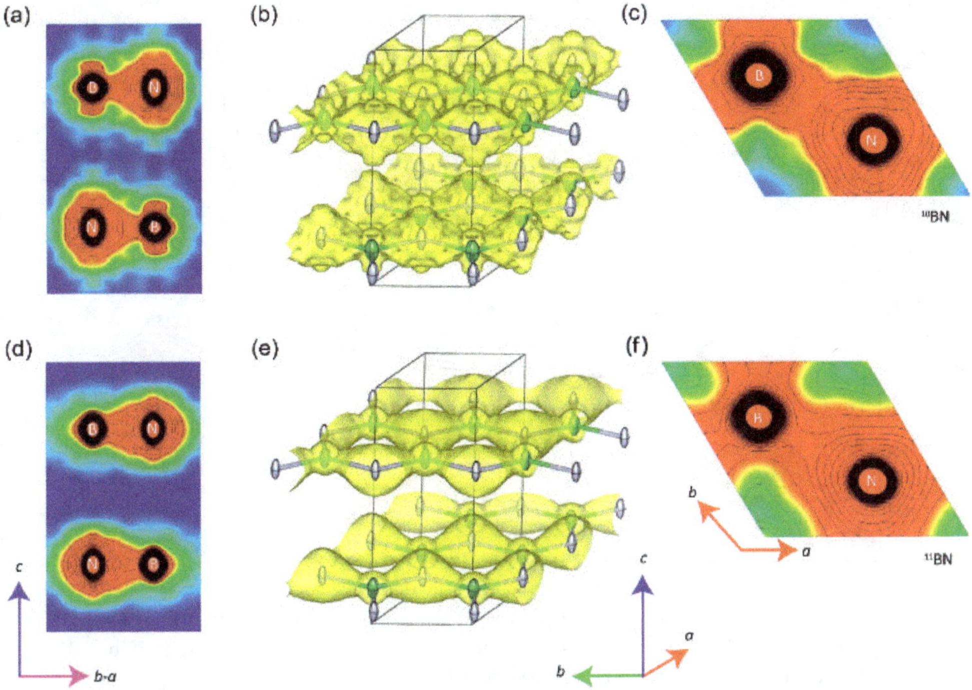

Figure 8.32 Electron density distribution of (a)–(c) h-^{10}BN and (d)–(f) h-^{11}BN. (a) and (d) are 2D contour maps parallel to the c-axis; (b) and (e) are side views of the 3d contour maps; (c) and (f) are 2D contour maps perpendicular to the c-axis. From Vuong[2018]. Reproduced with permission from SNCSC.

The isotope fraction also affects the electron density distribution. Combining theory and experiment, Vuong et al. determined the electron density distributions of h-^{11}BN and h-^{10}BN, and their results are presented in Figure 8.32 [Vuong2018]. The electron density distributions around the boron and nitrogen nuclei of h-^{10}BN, both in the X–Y plane and along the out-of-plane directions, are more dispersed than those of h-^{11}BN. This study successfully demonstrates the isotopic effect on the electron density distribution in layered crystals [Yamanaka1996].

b Electron Energy Loss Spectroscopy

The scanning-transmission electron-microscopy electron-energy-loss spectroscopy (STEM-EELS) method, which can resolve isotopes at the atomic scale, has also been used to examine the isotope effect in h-BN [Li N.2022]. The phonon dispersion and the corresponding phonon density of states for h-^{10}BN and h-^{11}BN based on DFT calculations are shown in Figure 8.33. In the spectrum of phonon density of states, the peaks dominated by in-plane (160–200 meV) and out-of-plane (60–100 meV) modes are displaced. An energy shift of 3.95 meV (sufficiently large to be resolved by STEM-EELS) is observed for both the LO and TO modes in the h-BN isotope samples.

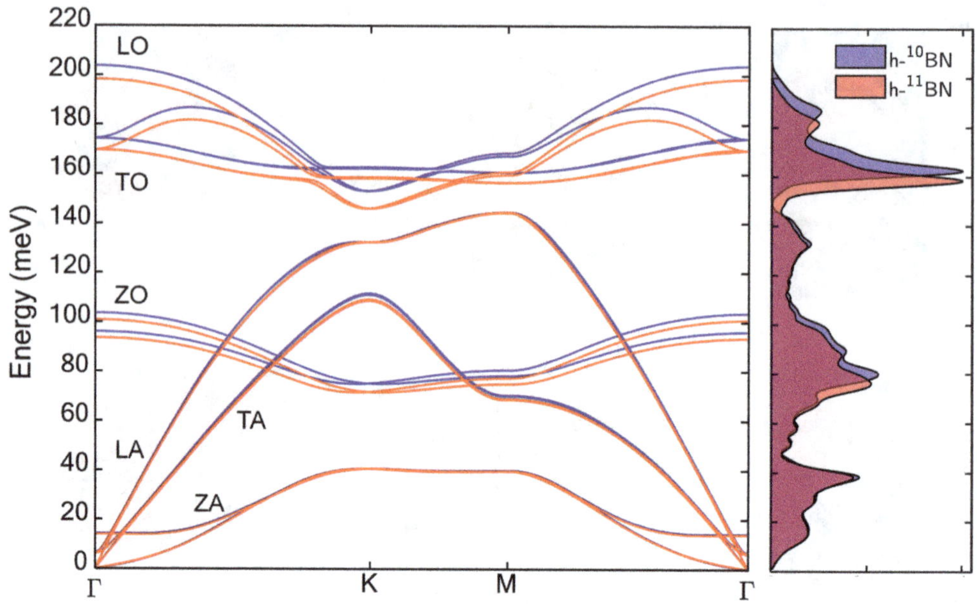

Figure 8.33 Phonon dispersion and phonon density of states distribution for two h-BN isotope samples calculated by the first-principles method. Blue: h-^{10}BN; red: h-^{11}BN. Reprinted from Li N.[2022], with permission of the author.

Y.F. Li et al. replaced the B source with ^{10}B (97.18 at%) and ^{11}B (99.69 at%) sources of ultrahigh isotopic purity, respectively, and successfully prepared isotopically pure h-^{10}BN and h-^{11}BN samples at high temperature [Li Y.F.2021]. The enriched h-BN was then mechanically exfoliated to produce an h-^{10}BN/h-^{11}BN/SiO_2 heterojunction. Atomic-level vibrational spectroscopy was conducted for the ~2.2 nm region on both sides of the heterojunction interfaces. Figures 8.34(a) and (b) show the high-angle annular dark-field (HAADF) images of the data-acquisition region, the experimentally acquired EELS of the out-of-plane vibration signal, and the result of the first-principles calculation based on the density functional approximation using the same parameters as the experiment. Figure 8.34(c) shows the spatial distribution of the two out-of-plane phonon modes. The arrows indicate the direction and magnitude of the eigenvectors. The h-^{10}BN/h-^{11}BN interfaces in these plots are labelled by dashed lines, separating the two different isotope samples. The h-^{10}BN region is shown on the left and the h-^{11}BN region on the right. The upper and lower panels of Figure 8.34(c) respectively show the spatial distribution of phonon modes $ZO_{\text{low q}}$ ($q = \Gamma$, $\omega = 102.468$ meV) and $ZO_{\text{high q}}$ ($q = K$, $\omega = 71.201$ meV and $\omega = 74.691$ meV). The calculations show that the $ZO_{\text{low q}}$ mode had little intensity variation at the interface, while the $ZO_{\text{high q}}$ mode was more concentrated near the interface. In order to quantitatively investigate the experimental EELS vibrational spectrum, the energy variations of the $ZO_{\text{high q}}$ and $ZO_{\text{low q}}$ modes at both sides of the interface were captured using a multi-Gaussian peak-fitting method, as shown in Figure 8.34(d). The cyan and orange regions in Figure 8.34(d) represent the $ZO_{\text{low q}}$ and

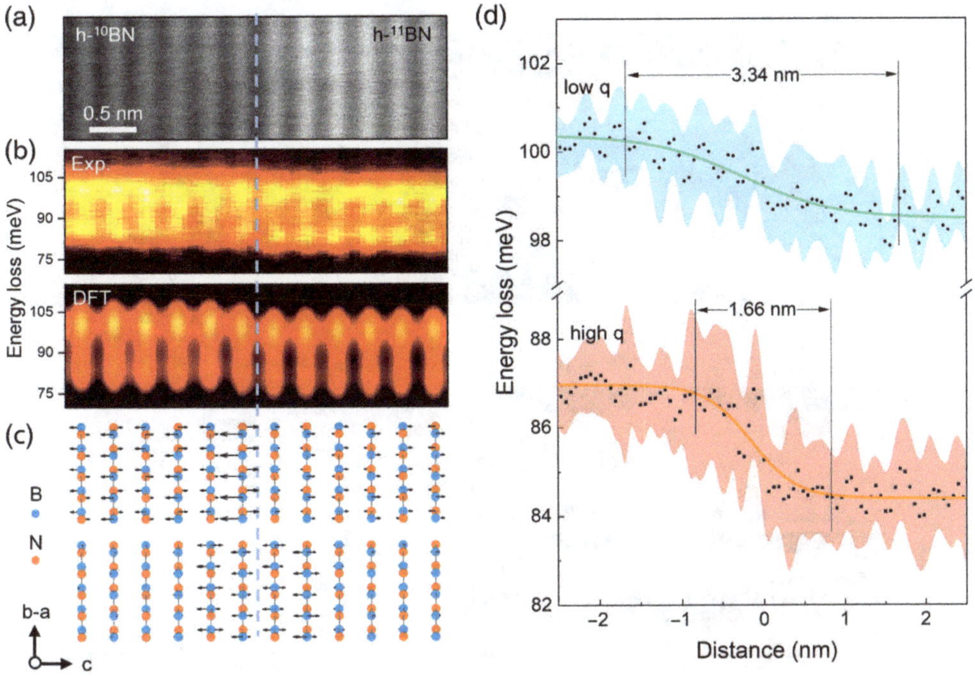

Figure 8.34 Quantitative analysis of the atomically resolved vibrational spectrum at the isotope interface: (a) Atomically resolved HAADF map at the h-^{10}BN/h-^{11} BN interface. (b) Experimentally determined EELS of out-of-plane vibrational signal (top), and first-principles calculation results using the same parameters as the experiment (bottom). (c) Calculated spatial distribution of out-of-plane phonon modes $ZO_{\text{low q}}$ (top) and $ZO_{\text{high q}}$ (bottom) near the h-^{10}BN/h-^{11}BN interface (dashed line); the arrows indicate the direction and magnitude of the eigenvectors. (d) Experimentally measured energy variation of out-of-plane phonons at the interface. Quantitative data are extracted from the experimental spectral lines using the multi-Gaussian peak-fitting method. The black dots are the phonon energies after averaging multiple spectral lines, and the light-colored area is the error range. A logistic function is used to obtain the transition length of each phonon mode. The orange and cyan solid curves are the fitted results for the black dots. Figure created by Li N.[2023] licensed under CC BY 4.0.

$ZO_{\text{high q}}$ modes, respectively. The black dots denote the phonon energies after averaging over multiple spectral lines, and the light-colored region shows the error range.

In addition to nanoscale spatial resolution, 4D-EELS is also capable of resolving the momentum. Using 4D-EELS with the unification of ultrahigh resolution capability in the spatial and momentum space, G. Peng's group at Peking University collected the dispersion spectral lines of isotopic h-BN materials using ~10 nm electron probes. Figure 8.35(a) shows the phonon dispersion in h-^{11}BN; the gray curves correspond to the values obtained using the DFT calculation. The phonon bandgap ω_{Gap} measured in the experiment was subsequently examined at the M point in the second Brillouin zone (q ~ 4.3 Å^{-1}). The phonon bandgap is formed between the LA and LO branches, and its position is marked in Figure 8.35(a) by a white box. Enlarged versions of the same h-^{11}BN and h-^{10}BN regions

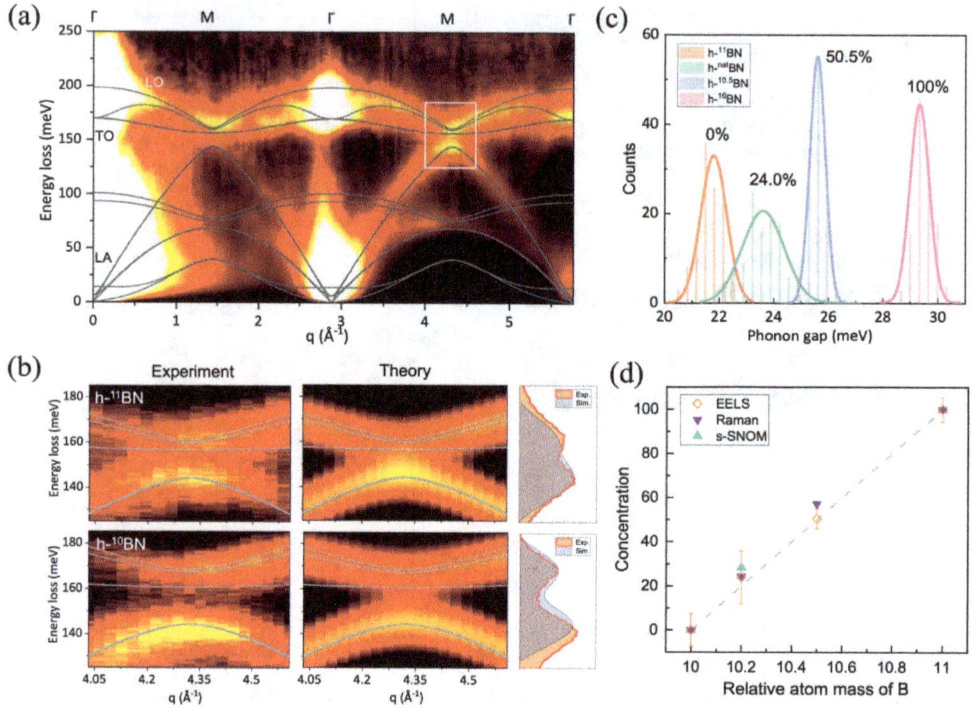

Figure 8.35 Quantitative analysis of isotope composition at the nanometer scale: (a) Phonon dispersion in the $\Gamma - \mathrm{M} - \Gamma$ direction measured by 4D-EELS (color) and calculated by DFT (gray curves) for h-^{11}BN. (b) LA/LO phonon bandgap at the M point in the second Brillouin zone for h-^{11}BN and h-^{10}BN; the measurement range corresponds to the white box in panel (a). A comparison between the experimental and computational results is also shown. (c) Distribution of phonon bandgap widths in the four samples with different isotopes. The number indicates the percentage concentration of ^{10}B in each sample, calculated from the bandgap width. (d) Comparison between EELS measurements and other optical methods in terms of accuracy. Reprinted from Li N.[2022], with permission of the author.

are shown on the left side of Figure 8.35(b), with a comparison between the experimental and theoretical line profiles at q = 4.35 ± 0.05 Å^{-1} on the right. In Figure 8.35(b), the difference between the h-^{11}BN and h-^{10}BN phonon bandgaps becomes more pronounced. Four samples with different isotope doping were studied: isotopically pure h-^{11}BN, natural h-BN, 1:1 h-^{11}BN/h-^{10}BN hybrid, and isotopically pure h-^{10}BN. The width of the phonon bandgap was quantified using the multi-Gaussian peak-fitting method. With statistical analysis, 120 points were uniformly sampled over the range of 100 × 120 nm for each sample, and the distribution of the phonon bandgap is shown in Figure 8.35(c). The short bars of different colors in the figure represent the statistical histograms of different samples, and the solid curves represent the Gaussian fit envelopes of the histograms.

There was a strong positive correlation between the phonon bandgap widths and the material composition of the four samples. Using the bandgaps of isotopically pure samples as reference values, it was even possible to experimentally measure the ^{10}B concentration in mixed samples with arbitrary ratios by comparing the phonon bandgap widths. For example, according to Figure 8.35(d), the experimentally determined phonon bandgap widths for the four samples were 21.76 ± 0.47 meV, 23.63 ± 0.78 meV, 25.61 ± 0.27 meV, and 29.37 ± 0.34 meV, respectively. The corresponding ^{10}B concentrations were found to be (0 ± 6.17)%, (24.0 ± 10.24)%, (50.5 ± 3.54)%, and (100 ± 4.48)%, respectively. Comparing the 4D-EELS measurements with Raman and s-SNOM results showed that this method achieved very high accuracy, with a spatial resolution of 10 nm [Li N.2022].

c Thermal Conductivity

As mentioned previously, h-BN and graphite have similar crystal structures, lattice constants, unit cell masses, and phonon dispersion relationships. However, the room-temperature thermal conductivity of h-BN is approximately five times lower than that of pyrolytic graphite. One important explanation is that the isotope mixture region of h-BN (19.9% ^{10}B, 80.1% ^{11}B) is much larger than that of graphite (98.9% ^{12}C, 1.1% ^{13}C), leading to stronger phonon–isotope scattering. Isotopic mass disorder breaks the translational invariance and induces extra elastic scattering of phonons, depressing the thermal conductance. Lindsay et al. examined the isotope effect on h-BN thermal conductivity through theoretical calculations [Lindsay2011]. According to Figure 8.36, the thermal conductivity peaks at around 100–150 K, then decreases with increasing temperature for both h-^{Na}BN and h-^{11}BN materials. This suggests that intrinsic phonon–phonon scattering resulting from lattice anharmonicity is the primary mechanism that restricts thermal conductivity, as observed in most semiconductors and insulators. Compared to that of h-^{Na}BN, the thermal conductivity of h-^{11}BN exhibits an approximate 30% increase at 300 K and an approximate 70% increase at 100 K, attributed to reduced phonon scattering in the high-purity h-^{11}BN material. The calculations also reveal that there is a significant increase in the thermal conductivity for single-layer h-^{Na}BN and h-^{11}BN, as shown in Figure 8.36 (upper dashed and solid curves). The thermal conductivity enhancement observed in 2D crystals such as single-layer h-BN and graphene is primarily attributed to the symmetry selection rule, which effectively suppresses phonon–phonon scattering [Lindsay2010, Seol2010].

In addition to the isotopic effects in h-BN that affect its thermal conductivity, some other isotopic phenomena are gradually being discovered. For example, because ^{10}B has a large thermal neutron cross section similar to that of BN nanotubes, the detection efficiency of a ^{10}B-enriched h-BN neutron detector is expected to be significantly enhanced. Additionally, phonon scattering due to mass disorder is significantly reduced in isotopically pure h-BN crystals, which increases the phonon lifetime. This also opens up the possibility of reducing optical loss in hyperbolic polarized phonon devices, extending the polarized-phonon lifetime until it is comparable to the optical-phonon lifetime [Low2017]. In addition to direct modulation of phonon scattering, isotope purification can also indirectly modulate the interlayer van der Waals interactions of h-BN and adjust the electron density distribution between adjacent layers via nuclear quantum

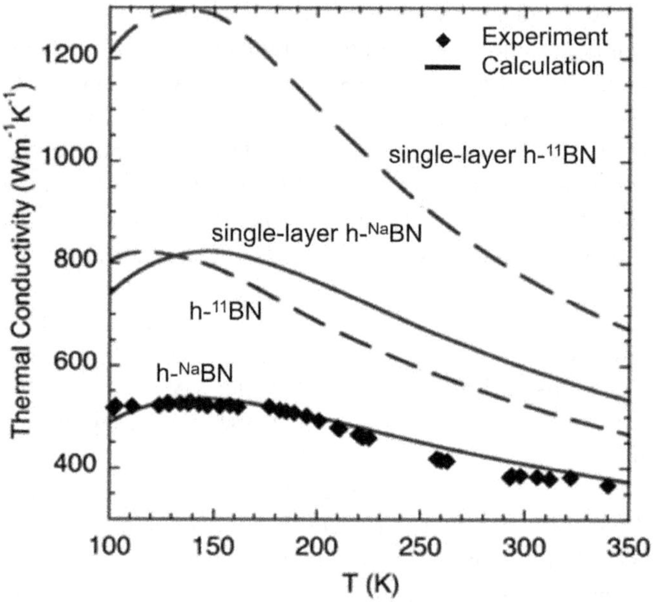

Figure 8.36 Thermal conductivity as a function of temperature for h-^{Na}BN and h-^{11}BN. The bottom solid curve and the black diamonds are the calculated and measured values for h-^{Na}BN, respectively. The middle dashed curve represents the calculated values for h-^{11}BN. The calculated values for single-layer h-^{Na}BN are shown by the middle solid curve and the calculated values for single-layer h-^{11}BN by the upper dashed curve. Reprinted figure with permission from Lindsay[2011]. Copyright 2024 by the American Physical Society.

effects [Vuong2018]. Thus, nuclear quantum effects allow modulation of the electronic structure, providing new research directions for the study of isotopic purification, heterostructures, superlattices, and defects, and so forth.

8.3.5 Magnesium Diboride

Magnesium diboride (MgB_2) is another significant boron-based material that exhibits strong full quantum effects. In 2001, Nagamatsu et al. discovered that MgB_2's superconducting transition temperature can go up to 39 K [Nagamatsu2001], which generated considerable interest and led to extensive research on practical superconducting materials. After more than two decades of study, the normal and superconducting states of MgB_2 are relatively well understood. Studies on the isotope effect and pressure experiments have confirmed that MgB_2 is a phonon-mediated superconductor, and its superconductivity can be at least roughly explained within the framework of BCS theory. However, the superconductivity mechanism of MgB_2 is not entirely consistent with the BCS theory, which is based on a near-free-electron model with weak electron–phonon interaction. This discrepancy arises because the full quantum effects of the electron–phonon interaction in MgB_2 are non-negligible.

8.3.5.1 Crystal Structure and Electronic Structure

MgB_2 crystallizes in a simple hexagonal structure with the space group P6/mmm. It has lattice constants a = 0.3086 nm and c = 0.3524 nm and a crystal density of 2.605 g/cm^3 [Buzea2001]. The boron atoms are arranged in a 2D honeycomb structure, and each magnesium atom is located between two adjacent layers of boron, directly above the boron hexagonal center. The distances between the boron layers are significantly larger than the atomic distances between boron atoms within a single layer, resulting in noticeable anisotropy in MgB_2 (see Figure 8.37). In each atom of the outer boron layer of MgB_2, the p_x, p_y, and s orbitals of the four valence electrons undergo sp^2 hybridization, forming three in-plane σ-covalent bonds at angles of 120° to each other. The p_z orbitals, on the other hand, remain uninvolved in hybridization and are perpendicular to the plane of the sp^2 hybridization orbitals, forming delocalized π bonds at certain intervals shoulder to shoulder.

The magnesium atom in MgB_2 is highly ionized, giving both of the donated outermost electrons to the boron atom. As a result, the boron atom provides most of the electrons near the Fermi surface of MgB_2 [Choi2002]. Although the magnesium ions rarely interact with the σ-electrons of boron, they are strongly attracted to the π-electrons, leading to a significant reduction in the energy band of the π-electrons. This overlap of the σ- and π-energy bands causes the metallization of the σ-energy band [Kortus2001; An J.M.2001], as shown in Figure 8.38. As the σ-band is not fully occupied, a cavity-type conduction band is formed, similar to that in high-temperature copper oxide superconductors.

8.3.5.2 Phonon Structure

The phonon frequency distribution of MgB_2 is shown in Figure 8.39. No significant gaps were observed in the phonon density-of-states spectrum [Bohnen2001]. The in-plane boron phonon vibrations at the band center are nonlinear. The metallic σ-electrons

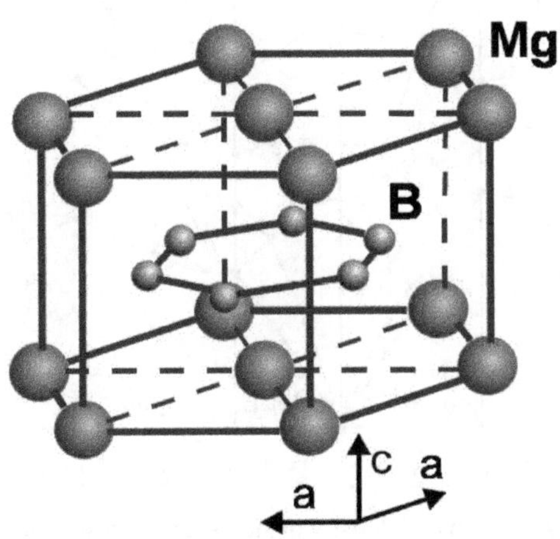

Figure 8.37 Crystal structure of MgB_2. Used with permission of Supercond. Sci. Technol, from Buzea[2001], Copyright 2024; permission conveyed through Copyright Clearance Center, Inc.

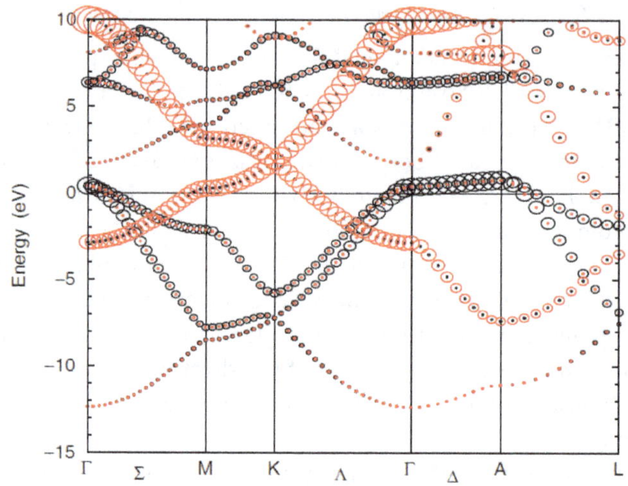

Figure 8.38 Schematic of the p-orbit energy band structure in MgB_2. The red and black circles represent the p_z and $p_{x,y}$ energy bands of boron, respectively. Reprinted figure with permission from Kortus[2001]. Copyright 2024 by the American Physical Society.

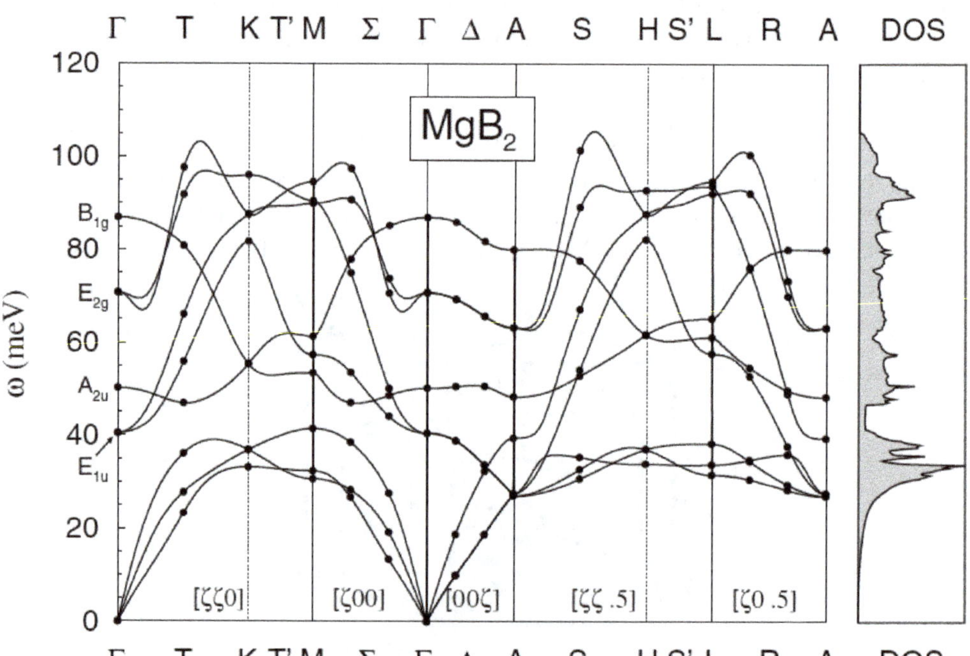

Figure 8.39 Phonon spectrum and density of states (DOS) distribution for MgB_2. Reprinted figure with permission from Bohnen[2001]. Copyright 2024 by the American Physical Society.

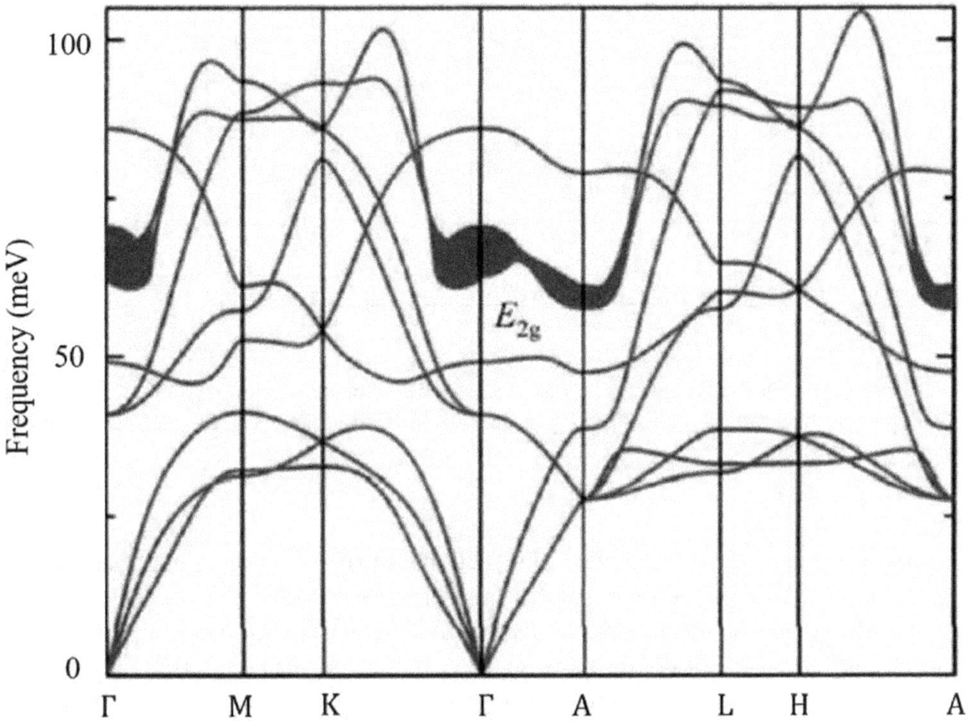

Figure 8.40 Phonon spectrum of MgB_2. The thickness of the curve represents the strength of the electron–E_{2g} phonon coupling. Reproduced from Gao M.[2015].

are additionally coupled to the E_{2g} phonon modes, corresponding to the boron atomic displacement in the 2D plane, due to nuclear quantum effects. These anharmonic phonon modes contribute to strong nonlinear electron–phonon coupling (as shown in Figure 8.40), which leads to superconductivity [Gao M.2015].

According to the theory of electron–phonon coupling in superconductivity, the interaction strength is determined by the electron–phonon coupling coefficient λ, which itself is determined by the effective phonon spectrum $\alpha^2F(\omega)$ of the material. The effective phonon spectrum of MgB_2 [Liu A.Y.2001] has been calculated and is illustrated in Figure 8.41. The high peak in $\alpha^2F(\omega)$ is attributed to the strong coupling between electrons and E_{2g} phonons in the optical branch.

8.3.5.3 Electron–Phonon Coupling and Nonadiabatic Effects

MgB_2 has a superconducting transition temperature as high as 39 K, which is close to the McMillan limit (40 K) predicted by BCS theory. Therefore, it was initially thought that the conventional BCS theory could not explain the superconductivity of MgB_2. However, subsequent experiments on the isotope effect and pressure proved that electron–phonon coupling plays a vital role in superconductivity. Further theoretical

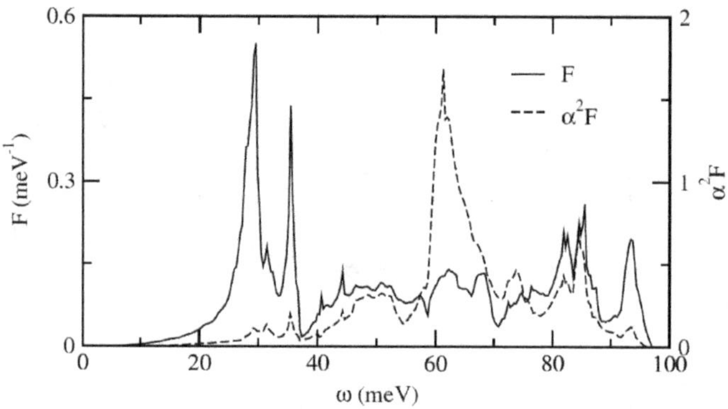

Figure 8.41 Phonon density of states F(ω) and Eliashberg function $\alpha^2F(\omega)$ of MgB_2. Reprinted figure with permission from Liu[2001]. Copyright 2024 by the American Physical Society.

calculations and experiments have established that MgB_2 is a typical multiband superconductor with two hole-type σ-bands, one hole-type π-band, and one electron-type π-band. Moreover, the electron wave vectors in different energy bands are orthogonal to each other. Because of the destabilization of the Fermi surface caused by electron–phonon coupling, superconducting energy gaps can be generated at the Fermi surfaces of the two energy bands. The larger and smaller energy gaps have widths of 5.5–8 meV and 1.5–3.5 meV, respectively.

Another aspect of studying the superconductivity of MgB_2 through the full quantum effect is considering the impact of nonadiabatic processes. Under the adiabatic approximation, the simulation tends to overestimate the superconducting transition temperature by 50% because of electron–phonon coupling. However, the superconducting transition temperature can be correctly evaluated when nonadiabatic effects are considered. Cappelluti et al. conducted a systematic study on the nonadiabatic effects in MgB_2 and concluded that the low Fermi energy σ-electrons can couple with phonons to form new channels; this is known as the small Fermi energy effect [Boeri2005]. MgB_2 is a multiband system consisting of two adiabatic π-bands and two nonadiabatic σ-bands, as shown in Figure 8.42. These energy bands can lead to the formation of nonadiabatic Cooper pairs, explaining the high transition temperature of MgB_2 superconductors and the high isotope exponent α_B of the boron atoms [Cappelluti2002]. K.-P. Bohnen et al. further pointed out that the electron–phonon coupling is related to the high-frequency in-plane vibrations of the boron atom [Bohnen2001].

There are several factors that affect the uncertainties in the Fermi energy level, including the deformation potential, electron energy band renormalization due to lattice vibration, and quantum fluctuation caused by zero-point lattice motion. As a result of these factors, the Born–Oppenheimer adiabatic approximation no longer applies to superconducting MgB_2 [Boeri2005]. This is especially true when the Fermi energy E_F is small and comparable to the phonon ω or the electron–phonon coupling parameter g, where the adiabatic assumption completely fails. In these cases, not only is the Born–Oppenheimer approximation completely unreliable but also the Migdal theorem is invalid [Cappelluti2006].

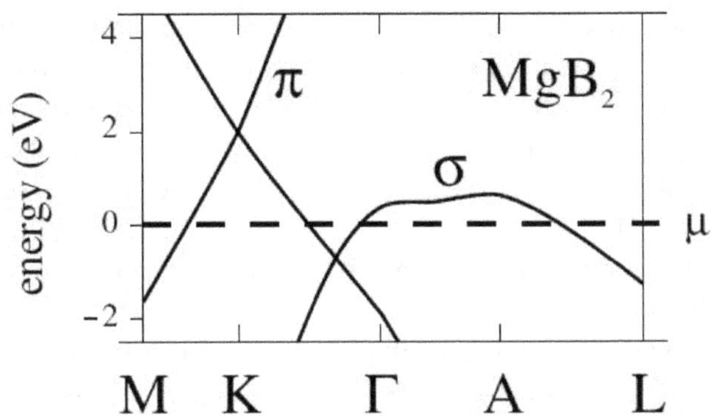

Figure 8.42 Energy band structure of MgB_2. Reprinted figure with permission from Cappelluti[2002]. Copyright 2024 by the American Physical Society.

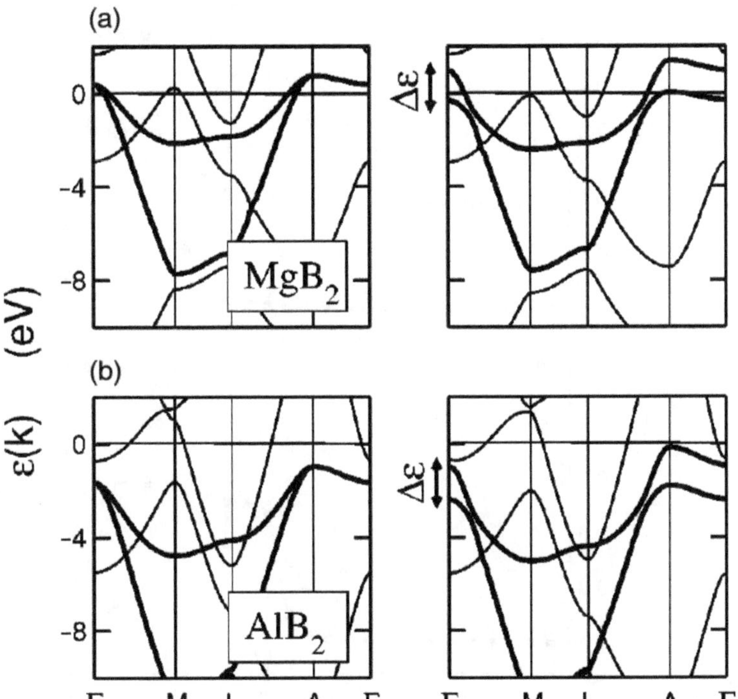

Figure 8.43 Panel (a) shows the energy band of MgB_2 obtained with (right) and without (left) considering the E_{2g} phonon distortion; panel (b) shows the corresponding energy band of AlB_2 under the same conditions. The σ energy band is shown using the thick curve. Reprinted figure with permission from Boeri[2002]. Copyright 2024 by the American Physical Society.

The E_{2g} phonon in MgB_2 also exhibits strong anharmonicity due to the small Fermi energy effect in the σ-band with low hole doping [Boeri2002]. Figure 8.43 shows simulated energy bands before and after lattice distortion (with amplitude u = 0.05 Å) to examine the anharmonicity of the E_{2g} mode. Comparing the results for MgB_2 and AlB_2, it is evident that the σ energy band of AlB_2 is near but never above the Fermi energy level, and its electron distribution remains unchanged; by contrast, the top of the initial σ energy band in MgB_2 is above the Fermi energy level. Once the distortion occurs, some of the energy bands in the Brillouin zone split; this causes the lower σ energy band to fall below the Fermi surface, thus altering the electron distribution of MgB_2. The anharmonicity of the E_{2g} mode is determined by various factors, including the number of energy band shifts and splitting, as well as the precise position of the σ-band before and after distortion.

To study nonadiabatic effects, carrier doping and chemical substitution can be utilized as well. Unfortunately, it is impossible to achieve a higher T_c than that of pure MgB_2 using electron or hole doping. Chemical substitution inevitably results in disorder, which lowers T_c even further. Because nonmagnetic ion substitution preserves time-reversal

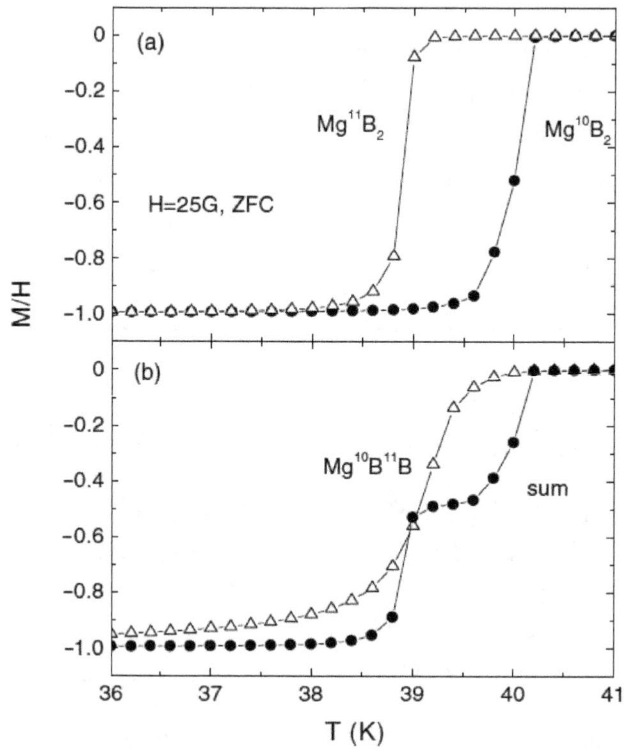

Figure 8.44 Magnetization intensity as a function of temperature. The curves in the figure represent (a) $Mg^{11}B_2$ and $Mg^{10}B_2$ and (b) $Mg^{10}B^{11}B$ and a sum of $Mg^{10}B_2$ and $Mg^{11}B_2$. The data are normalized. Reprinted figure with permission from Bud'ko[2001]. Copyright 2024 by the American Physical Society.

symmetry, according to Anderson's theorem in ME theory, nonadiabatic pairing remains the only plausible explanation [Cappelluti2002].

8.3.5.4 Isotope Effect in MgB_2

Following the discovery of MgB_2 superconductivity, an experiment conducted by S. L. Bud'ko et al. showed that replacing ^{11}B with ^{10}B increased the T_c of a sample from 39.2 K to 40.2 K [Bud'ko2001], as shown in Figure 8.44. Figure 8.45 indicates that the T_c of MgB_2 is inversely related to the square root of the atomic mass for each isotope [Hinks2001]. The corresponding boron isotope exponent α_B is larger, with an estimated value of 0.26. This suggests that the phonons associated with the boron atomic vibrations play a crucial role in the superconductivity of MgB_2. By comparison, the isotopic effect of magnesium is insignificant, with an associated isotope exponent of only 0.02; this indicates that the vibrational frequency of magnesium atoms has little impact on the transition temperature.

Recent research has revealed that using the estimated value of $\alpha^2F(\omega)$ does not produce reliable results that align with the measured T_c and α. To achieve accurate T_c and α values that match the experimental data, it is necessary to take into account the anharmonic effects of phonon modes and the anisotropy of the electron–phonon coupling. By incorporating these full quantum effects, researchers have obtained a close match between the experimental and theoretical values of T_c and α [Hinks2003]. The relevant results are summarized in Table 8.1.

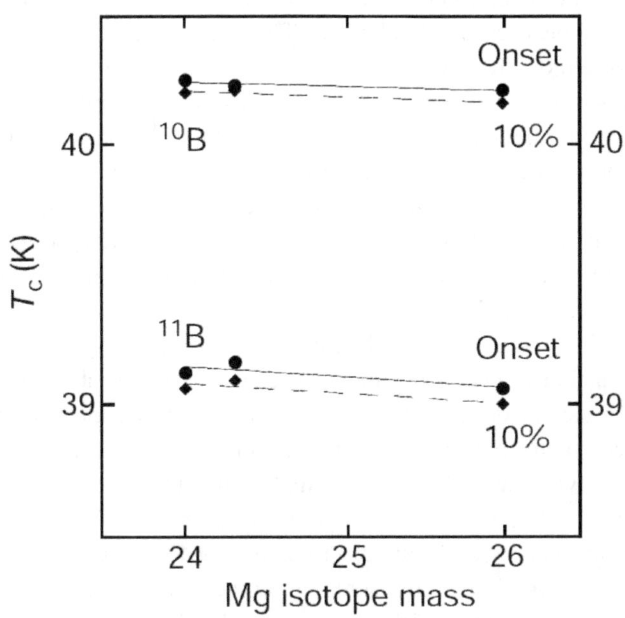

Figure 8.45 Superconducting transition temperatures of MgB_2 samples with different boron isotopes as a function of Mg isotope mass. The solid and dashed lines represent the sample measurements for the onset and 10% transition temperatures, respectively. From Hinks[2001]. Reproduced with permission from SNCSC.

Table 8.1 *Superconducting transition temperature T_c, isotope exponent α, electron–phonon coupling coefficient λ, and phonon frequency ω_{ph} calculated using adiabatic and nonadiabatic approximation after considering both the phonon anharmonicity on the Fermi surface and the anisotropy of the electron–phonon coupling [Hinks2003]. Here ω_{ph} is the energy of the E_{2g} mode. T_c and α_B values are calculated using $\mu^*(\omega_c) = 0.12$; for α_B, the value $\mu^*(\omega_c)$ scales with T_c.*

	Harmonic		Anharmonic	
Parameter	Isotropic	Anisotropic	Isotropic	Anisotropic
$T_c(K)$	28	55	19	39
α_B	0.42	0.46	0.25	0.32
	(0.46)	(0.48)	(0.27)	(0.33)
α_{Mg}	0.04	0.02	0.05	0.03
λ	0.73		0.61	
$\omega_{ph}(meV)$	62.7		75.9	

8.4 Carbon (C)

Carbon belongs to the second period of the periodic table. It has a relative atomic mass of ~12 and is the lightest group IV element. The *s* and *p* orbitals in the outer shell of a carbon atom have almost the same energy. When a carbon atom bonds to its neighbors, its *s* and *p* orbitals can have *sp*, sp^2, or sp^3 hybridization; the bond angle and bond length can vary depending on the chemical environment. The different hybridizations influence how the carbon atom is affected by the surrounding atoms. In some hybridizations, carbon atoms exhibit strong nuclear quantum effects caused by zero-point vibration; in other cases, especially in some carbon materials where a Dirac-cone-type electronic structure exists, nonadiabatic effects are more pronounced than in other semiconductors because of strong electron–phonon coupling.

The discussion in Chapter 2 of physical properties induced by full quantum effects in condensed matter included several examples related to carbon. In subsection 2.1.2, we briefly looked at how full quantum effects can affect the bandgap through electron–phonon coupling, using bandgap renormalization in a diamond as an example. These effects can change the bandgap by ~10%; evidence of this can be found in the optical absorption spectrum of diamonds. In subsection 6.1.4, we learned that graphene has a linear Dirac cone in its electronic energy band structure. As a result, the electrons do not follow the nucleus closely during atomic vibrations. Instead, they prefer to remain in the electronic ground state of the stable atomic structure. This "inertness" of the electrons is closely tied to the quantum coupling of the electronic and nuclear wavefunctions, which results in strong nonadiabatic effects.

In this section, we will discuss the full quantum phenomena exhibited by various materials composed of carbon atoms with different types of structures and bonding.

8.4.1 Diamond

Diamond is the hardest bulk material in nature. This implies that the carbon atoms composing it are joined by strong chemical bonds. At low temperatures, diamond is a typical quasi-harmonic crystal [Ceriotti2009b], and nuclear quantum effects have a significant impact on its physical properties, including its crystal structure, elastic modulus, and electronic structure. Recent theoretical work on nuclear quantum effects in diamond has attempted to provide a reasonable explanation for the experimental observations. From these theoretical and experimental studies, we can clearly observe the influence of full quantum effects on the physical properties of diamond.

8.4.1.1 Thermodynamics and Thermal Conductivity

Condensed matter theory has been very successful in predicting the physical properties of solids. The harmonic approximation (HA) serves as a basic starting point for condensed matter theory. It treats lattice vibrations as phonons but ignores the effective interactions of quantized nuclei and therefore cannot be used to describe phenomena caused by anharmonic effects. The QHA is an improved scheme that can, to a certain extent, consider anharmonic effects by including the thermal excitation of phonons. However, nuclear quantum effects are still not fully reflected in this approach.

Path-integral methods based on the quantization of the nucleus allow for more accurate characterization of nuclear quantum effects at finite temperatures, and they are particularly suitable for studying problems related to anharmonicity. R. Ramírez's group used the PIMC method to investigate the crystal structure and thermodynamic properties of diamond [Herrero2000]. Because diamond only has one type of bonding, the interatomic interaction can be accurately approximated by empirical potentials. Ramírez et al. used a modified Tersoff potential in their calculations. The PIMC results were compared with the classical HA results, the semiclassical QHA results, and the experimentally determined values to evaluate nuclear quantum effects on the crystal structure and thermodynamic stability of diamond.

The calculated and observed temperature dependence of the internal energy and specific heat in diamond at atmospheric pressure are depicted in Figure 8.46. The temperature dependence of the internal energy given by Tersoff potential PIMC was more consistent with the QHA than with the HA results. The cohesive energy predicted by PIMC was 7.31 eV at room temperature and atmospheric pressure, close to the experimental value of 7.36 eV. The ZPE in the zero-temperature limit was calculated to be 0.21 eV/C. As methods using the Tersoff potential tend to overestimate the optical phonon frequency, the specific heat obtained by Tersoff potential PIMC was still off by 70 K from the experimental value. However, at temperatures above 1,000 K, the anharmonic effects caused by the lattice thermal vibrations were more pronounced, and the bulk elastic modulus B obtained by the above PIMC method was significantly larger than the QHA result, as shown in Figure 8.46(c). Elastic energy calculations showed that the zero-temperature classical limit of B was 4.37 Mbar; this value dropped to 4.18 Mbar after the ZPE correction was considered, which indicates that nuclear quantum effects

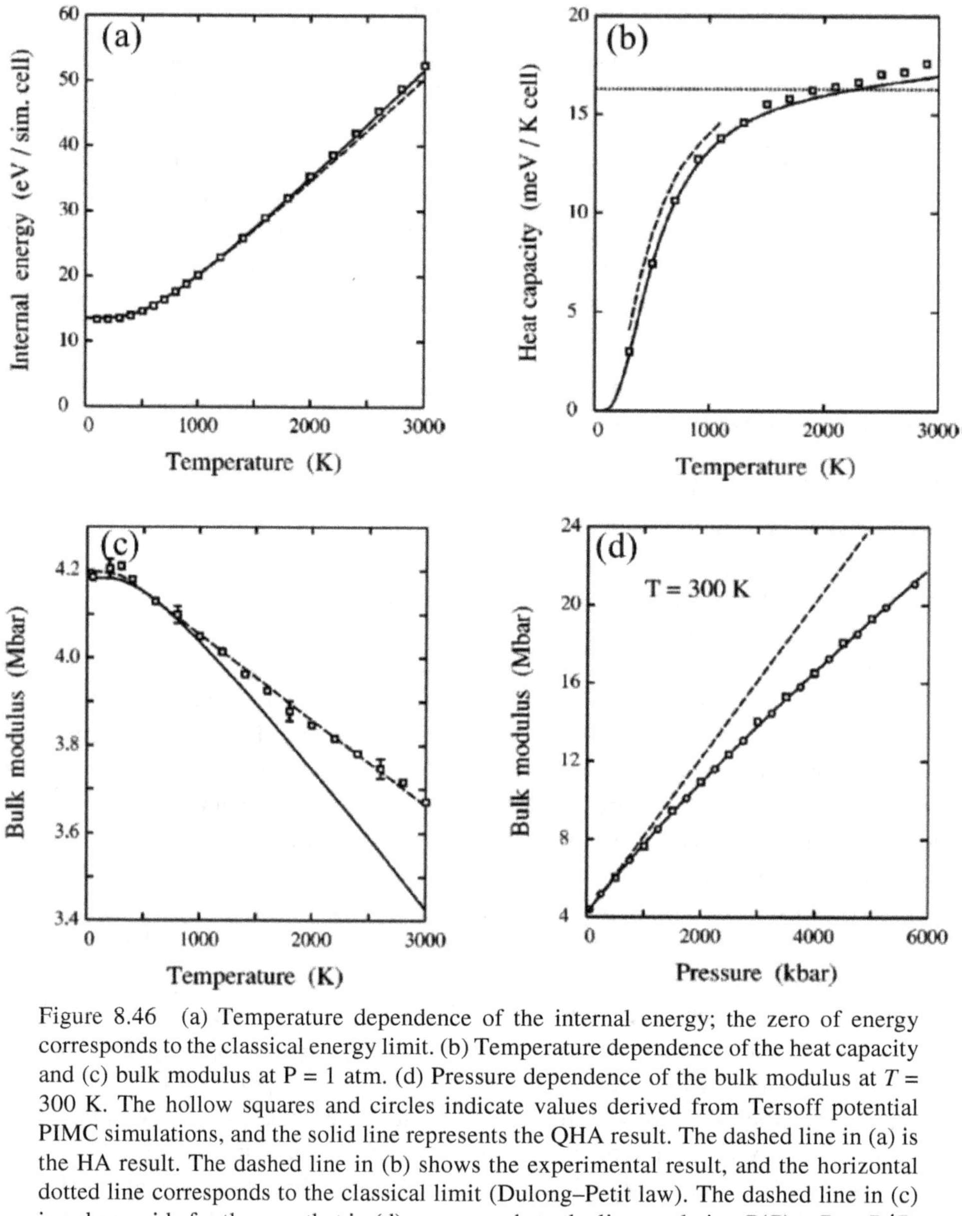

Figure 8.46 (a) Temperature dependence of the internal energy; the zero of energy corresponds to the classical energy limit. (b) Temperature dependence of the heat capacity and (c) bulk modulus at P = 1 atm. (d) Pressure dependence of the bulk modulus at T = 300 K. The hollow squares and circles indicate values derived from Tersoff potential PIMC simulations, and the solid line represents the QHA result. The dashed line in (a) is the HA result. The dashed line in (b) shows the experimental result, and the horizontal dotted line corresponds to the classical limit (Dulong–Petit law). The dashed line in (c) is only a guide for the eye; that in (d) corresponds to the linear relation $B(P) = B_0 + B_0'P$. Reprinted figure with permission from Herrero[2000]. Copyright 2024 by the American Physical Society.

reduce the bulk modulus. By contrast, the pressure dependence of B given by Tersoff potential PIMC was consistent with the QHA results, and both departed appreciably from the linear relationship $B(P) = B_0 + B_0'P$ assumed in the Murnaghan equation of state at high pressures (Figure 8.46(d)).

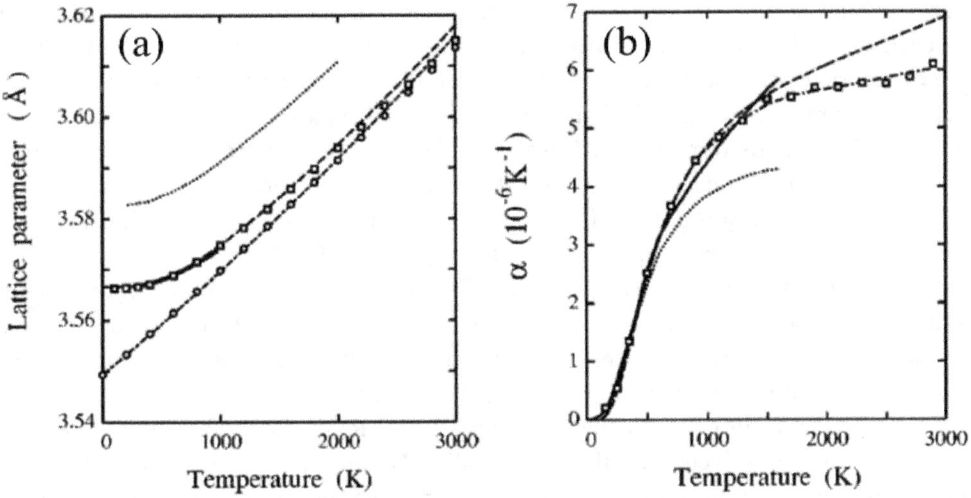

Figure 8.47 (a) Temperature dependence of the lattice parameter in diamond at atmospheric pressure. The hollow squares denote the PI simulation results and the circles correspond to the results derived in the classical limit. Dashed and dashed-dotted lines were obtained using quantum and classical QHA, respectively. The experimental results are given as a solid line [Madelung1982]. The dotted line was calculated using PIMC method with the Tersoff potential. (b) The thermal expansion coefficient of diamond as a function of temperature at atmospheric pressure. The hollow squares are results derived from PIMC simulations with the Tersoff potential. The dashed line is the result of QHA calculation. The dashed-dotted line is only a guide for the eye. The solid line shows the experimental results, and the dotted line represents the DFT calculation in a QHA [Pavone1993]. Reprinted figure with permission from Herrero[2000]. Copyright 2024 by the American Physical Society.

Figure 8.47(a) shows the diamond lattice parameter as a function of temperature T at atmospheric pressure. The Tersoff potential PIMC results agree well with the experimental values, while the values calculated by QHA are slightly higher at high temperatures. The zero-temperature-limit values of the lattice constant were predicted to be 3.5663 and 3.5667 Å by the PIMC and QHA simulations, respectively; both are larger than the classical value of 3.549 Å. The lattice expansion due to ZPE and anharmonicity was equivalent to the effect of a negative pressure of – 66 kbar; it was almost constant in the 0–1,500 K temperature range. At room temperature, the PIMC simulations yielded a lattice constant of 3.5666 Å, which is very close to the experimental value of 3.5668 Å [Madelung1982]. In addition, the pressure dependence of the volume and the lattice constant a given by Tersoff potential PIMC were almost the same as those given by QHA. This implies that the QHA is sufficiently accurate to describe the structure of diamond materials.

Diamond's linear coefficient of thermal expansion $\alpha = \frac{1}{a}\left(\frac{\partial a}{\partial T}\right)_P$ is depicted as a function of temperature at atmospheric pressure in Figure 8.47(b). The TEC obtained from PIMC simulations was in good agreement with the experimental data, while the QHA results had a significantly larger error at higher temperatures. Evidently, anharmonic effects can be characterized by perturbation theory at low temperatures, indicating that QHA is capable of describing the mechanical properties of solids. However, at high temperatures, the values given by QHA for physical quantities such as bulk modulus, TEC, and heat capacity deviated significantly from the PIMC results. This suggests that the QHA based on perturbation theory can no longer describe the effects of thermal fluctuations.

Besides using Tersoff potential, Ramírez's group also studied the vibrational properties of diamond by the path-integral molecular dynamics (PIMD) method using the electronic structure derived from a tight-binding Hamiltonian [Ramírez2006]. Figure 8.48(a) shows the temperature dependence of the vibrational energy of the diamond supercell. In the figure, HA denotes the result predicted by the harmonic approximation, where the vibrational frequencies were derived by diagonalizing the dynamic matrix; LR corresponds to the result given by the linear response (LR) method, where the vibrational frequencies were derived from the tight-binding Hamiltonian PIMD results using the linear response relation. At 100 K, the vibrational energy given by HA deviated from the PIMD result by 0.1 eV, which amounts to 0.8% of the total vibrational energy. This error was a consequence of ignoring the anharmonicity of the phonon vibrations; the LR method, which can adequately describe the anharmonicity of the phonon vibrations, had an error that was only 0.2% of the PIMD result at the same temperature (100 K).

The elastic constant c_{44} can be derived from the phonon velocity along the TA branch by the formula $c_{44} = \rho v_{\mathrm{TA}}^2$. At 200 K, the values of c_{44} given by HA and LR were 551 and 545 GPa, respectively. This suggests that anharmonic effects reduce the elastic constant. Although the c_{44} obtained from LR was lower than the experimental value of 576 GPa, the predicted shift in c_{44} with temperature exhibited a trend consistent with the experimental data (see Figure 8.48(b)).

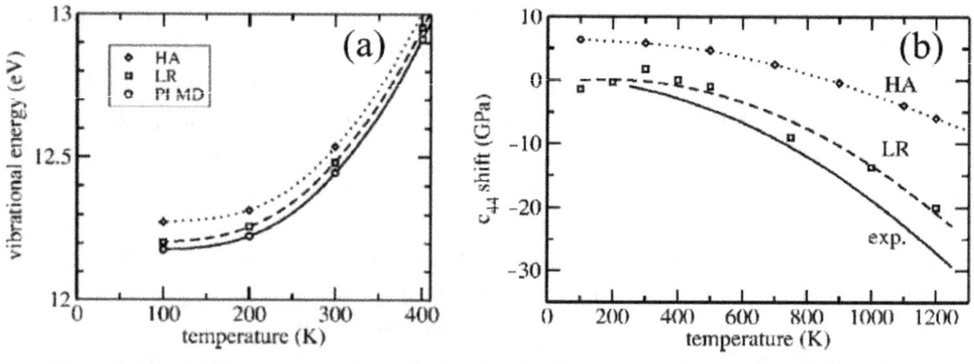

Figure 8.48 (a) Temperature dependence of the vibrational energy of diamond supercell obtained by different methods. The zero of energy corresponds to a crystal with fixed atomic positions and lattice parameter a = 3.5665 Å. (b) Temperature shift of the elastic constant c_{44} of diamond. The zero of the theoretical value is taken as the extrapolated LR value in the zero-temperature limit. The zero of the experimental data corresponds to the extrapolated experimental constant in the zero-temperature limit. Reprinted figure with permission from Ramírez[2006]. Copyright 2024 by the American Physical Society.

Carbon has two stable isotopes, ^{12}C and ^{13}C, with natural abundances of 98.9% and 1.1%, respectively. Isotope substitution can directly affect phonon vibrations in a material and, consequently, thermal conductivity, a physical quantity closely related to electron–phonon coupling. Of all solids, diamond has the highest thermal conductivity. Therefore, isotope substitution would be expected to affect its thermal conductivity significantly. In 1993, L. Wei et al. measured the thermal conductivities of a natural-abundance diamond (98.9% ^{12}C) and an isotopically enriched diamond sample (99.9% ^{12}C) [Wei L.1993]. As shown in Figure 8.49(a), the thermal conductivity increases more rapidly with decreasing temperature in the isotopically enriched diamond than in the natural-abundance diamond. At 104 K, the enhanced thermal conductivity of the diamond due to the isotope effect reaches 410 W/cm·K.

In 2009, A. Ward et al. calculated the temperature-dependent thermal conductivity of diamond samples with different isotopic compositions based on the solution of the Boltzmann transport equation of phonons [Ward2009]. As shown in Figure 8.49(b), the calculation results were in good agreement with the experimental values. Analysis of the phonon spectrum showed that the strong covalent bonding and the low atomic mass of carbon produce high phonon frequencies and optical phonon velocities, leading to extremely high thermal conductivity. The acoustic phonons can carry heat and are responsible for most of the thermal current. The scattering by optical phonons provides an essential scattering channel for the acoustic phonons. In the ^{12}C sample with higher isotopic purity, phonon scattering due to isotopic impurities is reduced, resulting in higher thermal conductivity.

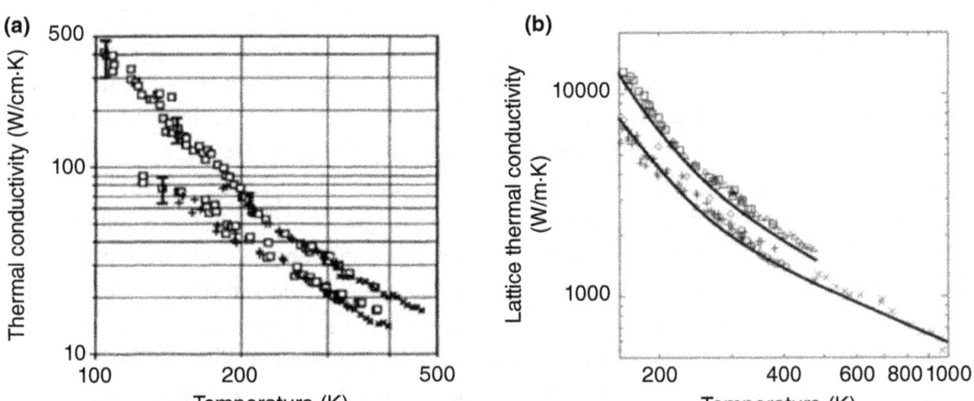

Figure 8.49 (a) Calculated thermal conductivity as a function of temperature for natural-abundance diamond (98.9% ^{12}C) and isotopically enriched diamond (99.9% ^{12}C), in comparison with the experimental measurements (plusses and crosses). The upper (lower) curve corresponds to squares with high (low) isotopic purity. (b) Thermal conductivity of diamond as a function of temperature for different isotopic impurity concentrations, calculated by Ward et al. The upper (lower) solid curve corresponds to the diamond with high (low) isotopic purity. The other symbols in the figure indicate experimentally measured values. Reprinted Figure 8.49 (a) with permission from Wei L.[1993]. Copyright 2024 by the American Physical Society; Reprinted Figure 8.49 (b) with permission from Ward[2009]. Copyright 2024 by the American Physical Society.

In addition, the zero-temperature lattice expansion caused by the zero-point vibration cannot be ignored. For diamond, this effect leads to a 0.5% increase in the lattice constant and a 5% decrease in the bulk elastic modulus. These differences are comparable to the errors of DFT calculations. Therefore, for light elements such as carbon, it is not sufficient to improve the accuracy of electronic structure calculations; nuclear quantum effects must also be considered to obtain reliable results. Unlike path-integral methods, HA and QHA can only provide results that agree with experimental measurements when the anharmonicity is small. When anharmonic effects are too large to be described by perturbation theory (i.e., at high temperatures and pressures), full quantum simulations must be used to provide comprehensive and accurate answers to problems concerning the crystal structure and thermodynamic properties of diamond.

8.4.1.2 Electronic Structure Renormalization

Full quantum effects also have crucial implications for the electronic structure of diamond. This is mainly reflected in the renormalization of the quasiparticle energy level and the bandgap. In Chapter 2, we discussed renormalization from the perspective of electron–phonon coupling. Here we present a detailed discussion of recent research on full quantum effects in calculations of the electronic structure of diamond.

The temperature dependence of the bandgap was experimentally investigated long ago [Logothetidis1992]. From these experimental data, we can easily obtain the bandgap renormalization correction resulting from zero-point vibration, which is around 0.45 eV. Early theoretical studies mainly aimed at explaining the bandgap behavior of diamond as a function of temperature. A notable example is the work of R. Ramírez et al., who combined the PIMD method with a tight-binding Hamiltonian to examine the renormalization correction of the electronic structure in diamond caused by nuclear quantum effects [Ramírez2006]. Figures 8.50(a) and (b) show the electronic density of states at the valence-band top and conduction-band bottom in diamond, respectively. The vertical solid lines correspond to the energy expectation values given by PIMD. With the atom fixed at its equilibrium position, the energy expectation values are indicated by the dashed lines in the figure. Evidently, the electron–phonon coupling resulting from full quantum effects shifts the energy expectation value at the top of the valence band upward. For instance, the energy expectation value is shifted by 0.11 eV upward at 100 K and by 0.14 eV at 1,000 K. Moreover, the electron–phonon coupling shifts the energy expectation value at the bottom of the conduction band downward, with a magnitude larger than the valence band shift. The change in band-edge energy occurs as a result of the renormalization correction of the diamond electronic structure due to nuclear quantum effects. As can be seen in Figure 8.50(c), the quantum effects due to zero-point vibrations are strongest at 0 K, with the bandgap renormalization correction reaching 0.7 eV. This value drops to 0.45 eV at room temperature. Overall, the bandgap renormalization correction of diamond can reach more than 10% of the original bandgap via electron–phonon coupling as a result of nuclear quantum effects. It is worth noting that the temperature dependence of the bandgap given by the tight-binding Hamiltonian PIMD method is in good agreement with previous experimental measurements [Logothetidis1992] and

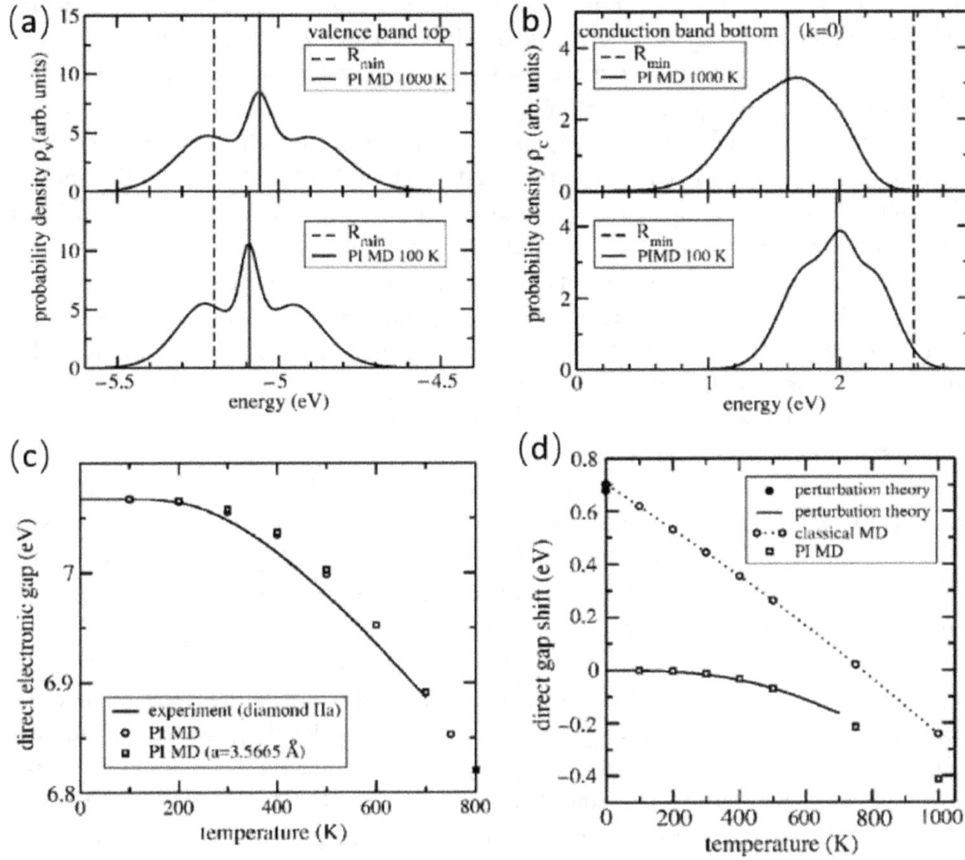

Figure 8.50 Electron probability density at (a) the top of the valence band and (b) the bottom of the conduction band of diamond at 100 K and 1,000 K, derived from tight-binding Hamiltonian path-integral molecular dynamics (PIMD) simulations. The vertical solid line represents the expectation value of the energy of the electronic state given by PIMD, and the dashed line is the expectation value corresponding to the fixed atomic structure. (c) and (d) diamond gap energy and gap offset value as a function of temperature. The solid curve in (c) is a fit to the experimental results [Logothetidis1992]. The hollow circles in (d) are the classical MD results and the hollow squares are the quantum PIMD results. The black circles and the solid line at T = 0 are the results based on perturbation theory. Reprinted figure with permission from Ramírez[2006]. Copyright 2024 by the American Physical Society.

perturbation-theory-based calculations [Zollner1992]. By contrast, the results of classical molecular dynamics, which ignores nuclear quantization, deviate significantly from the experimental data. This suggests that nuclear quantum effects cannot be neglected when studying the bandgap renormalization correction.

Although Ramírez et al. utilized PIMD to quantize the nuclei, their electronic structure calculation remained at the level of the tight-binding Hamiltonian approximation, and the errors in this model left some uncertainty in their final results. For example, they reported a ZPE renormalization correction of 0.7 eV, which is clearly higher than the experimentally determined value of 0.45 eV [Logothetidis1992]. Since 2010, several research teams have been working on the diamond renormalization problem and have carried out more accurate electronic structure calculations. For example, F. Giustino, S. Louie, M. Cohen, and their coworkers developed a first-principles electron–phonon-coupling calculation method based on MBPT. They used this method to systematically study the renormalization of quasiparticle energy levels and bandgaps in a series of semiconductors, including diamond [Giustino2010, Zacharias2016]. First, they calculated the direct bandgap of diamond as a function of temperature (Figure 8.51(a)). We see that the temperature dependence of the direct gap is consistent with the experiment [Logothetidis1992]. This indicates that the calculation method of Giustino et al. is effective in describing the leading contribution of the full quantum effects from the perspective of electron–phonon coupling.

In their study, the electronic structure calculations were carried out using the local density approximation (LDA) of DFT. Although the bandgap derived from the LDA contained some errors, these had only a limited effect on the description of zero-point vibrational renormalization. Giustino et al. introduced the *GW* approximation to quantitatively estimate this error and found that the *GW* correction on the zero-point vibrational renormalization was only 0.01 eV.

Subsequently, Giustino et al. further analyzed the different physical contributions to the electron energy band renormalization. According to Allen–Heine theory, the electron–phonon correction to the single-electron energy level ϵ_{nk} under the adiabatically harmonic approximation is dominated by two terms (see Eq. (2.1)) [Giustino2010]: the self-energy (SE) term $\Delta^{SE}\epsilon_{nk}$ and the Debye–Waller (DW) term $\Delta^{DW}\epsilon_{nk}$, representing the first- and second-order phonon corrections to the electron energy, respectively.

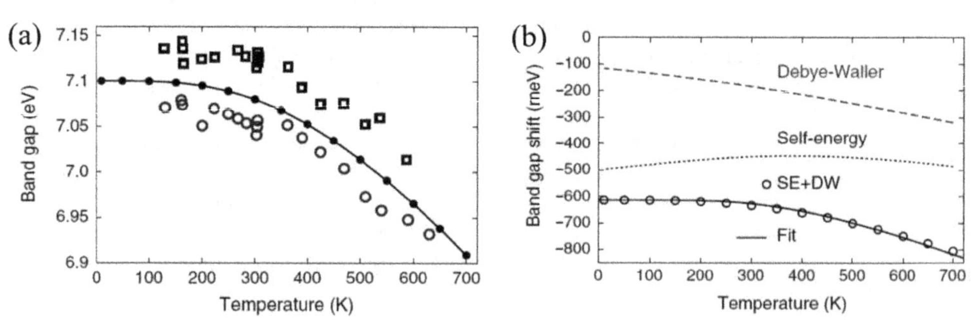

Figure 8.51 (a) Direct bandgap of diamond as a function of temperature after considering nuclear quantum effects. The solid curve shows the result of theoretical calculations; the open circles and squares are the experimental measurements [Logothetidis1992]. (b) Bandgap shift of diamond as a function of temperature. The bandgap shift is decomposed into the self-energy and Debye–Waller components. The solid curve corresponds to the fitting result using the Bose–Einstein law $\Delta E_g(T) = -a\left[1+2\left(e^{\Theta/T}-1\right)^{-1}\right]$, where a = 615 meV and $k_B\Theta = 118$ meV. Reprinted figure with permission from Giustino[2010]. Copyright 2024 by the American Physical Society.

As seen in Figure 8.51(b), both terms cause the bandgap to shrink, but they are different in nature. At low temperatures, the SE contribution dominates, while the contribution of the DW term increases at high temperatures. Using the Bose–Einstein law $\Delta E_g(T) = -a\left[1+2\left(e^{\Theta/T}-1\right)^{-1}\right]$, the fitted model yielded an effective temperature $k_B\Theta =$ 118 meV. This suggests that optical phonons in diamond are the main factor contributing to the bandgap renormalization.

It is worth pointing out that the zero-point renormalization of the bandgap was initially found to be 0.6 eV by Giustino et al. [Giustino2010], which is higher than the experimental measurements. The calculation was later improved by the same research group, and the recalculated temperature dependence of the direct bandgap is presented in Figure 8.52(a). The improved method yielded a zero-point bandgap-renormalization value of 0.45 eV, which agrees with the experimental results [Zacharias2016]. Using the improved method, a corrected renormalization value of 0.345 eV was obtained for the indirect bandgap of diamond. This is also very close to the experimental results, proving that a first-principles calculation based on MBPT can be used to describe the full quantum effects in the electronic structure of a material when anharmonic effects are more pronounced.

In the last few years, several other research groups have studied the renormalization of the electronic structure in diamond utilizing different computational methods that consider nuclear quantum effects [Cannuccia2011, Poncé2015, Monserrat2016, Ishii2021]. At least for diamond bandgap renormalization, the results obtained by different computational methods converge with satisfactory accuracy, and the differences between the theoretical and experimental results remain within the margin of experimental error. Similar studies have been conducted on other semiconductors in an attempt to obtain a whole picture of full quantum effects on electronic structure renormalization (see the discussion in subsection 2.1.2).

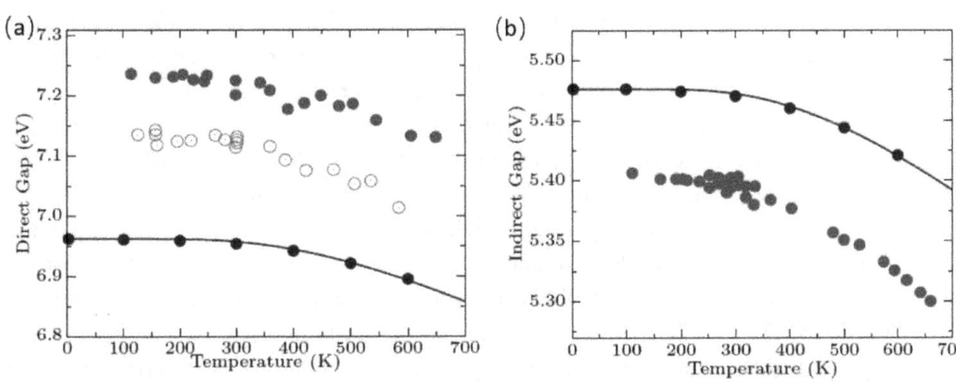

Figure 8.52 Bandgap of diamond as a function of temperature: (a) direct and (b) indirect. The solid lines denote the calculation results, and the others are the experimental values [Logothetidis1992]. Reprinted figure with permission from Zacharias[2016]. Copyright 2024 by the American Physical Society.

8.4.1.3 Optical Properties

Next, we discuss light absorption in diamond. Diamond is a typical indirect-bandgap semiconductor, that is, its indirect bandgap is smaller than its direct bandgap. According to our previous discussion of the electronic structure, an accurate description of the renormalization correction of the electronic structure caused by nuclear quantum effects is essential for a thorough understanding of the optical absorption process. When we introduced full quantum effects on the optical properties of condensed systems in Chapter 2, we reviewed the work of E. Cannuccia and A. Marini, who studied the effect of zero-point vibration on absorption in diamond using the first-principles method [Cannuccia2011]. As shown in Figure 2.19, the zero-point vibrations induced by nuclear quantum effects have a strong dynamic influence on the absorption spectrum of diamond. Some important spectral features in the figure directly reflect the dynamic effects of electron–phonon coupling. These phenomena cannot be properly understood using the conventional harmonic theory, which assumes a static lattice structure under the quasiparticle approximation.

In 1954, to explain phonon-assisted emission or absorption processes in diamond, L. H. Hall, J. Bardeen, and F. J. Blatt proposed a theory of optical absorption involving phonons, subsequently called the HBB (Hall–Bardeen–Blatt) theory [Hall1954]. This theory assumes that the electron transitions, in a time-dependent potential field generated by atomic vibrations, can be calculated by time-dependent perturbation theory. However, the HBB theory does not account for the effect of temperature on the electronic structure; this was considered in the Allen–Heine (AH) theory [Allen1976].

Recently, F. Giustino's group proposed a unified method for calculating direct and indirect optical-transition absorption spectra in solids by combining the Williams–Lax (WL) quasiparticle method [Williams1951, Lax1952], the HBB theory, and the AH theory [Zacharias2015]. The WL formalism treats electrons and phonons equally, and optical excitations correspond to transitions between the electronic and nuclear quantum states. In the calculation, the quantized vibrations of the final state were replaced by a classical continuum, leading to an expression for the optical absorption, which only involves the nuclear wavefunction of the initial state. The temperature dependence of the optical absorption was then obtained as a canonical average over randomly sampled initial states of the system. Furthermore, they demonstrated that at a given temperature, a special configuration of nuclei can be chosen based on the intrinsic vibrations of the atoms. For a sufficiently large supercell, configurational averaging can be simplified to a calculation based on this special atomic configuration, yielding the dielectric function and the optical absorption spectrum of the condensed matter system [Zacharias2016]. With this improved WL method, Giustino's group calculated the optical absorption coefficients of diamond [Zacharias2016]. As shown in Figure 8.53, they successfully predicted the indirect absorption component of the diamond absorption spectrum (solid curve). After considering full quantum effects, the threshold energy of light absorption was reduced from above 7 eV to below 5.5 eV. Comparing this to the results of a previous calculation with a static atomic configuration (dashed curve), the good agreement of the full quantum calculation results with the experimental measurements is evident.

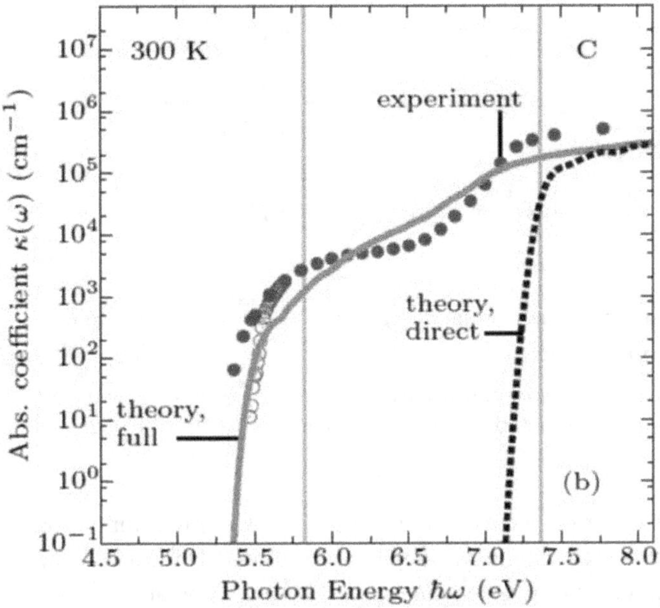

Figure 8.53 Optical absorption coefficient of diamond at room temperature. The solid curve is the result calculated by the WL method. The dashed curve is the result calculated with a static atomic configuration. The gray discs and circles are experimental data points [Phillip1964] and [Clark1964]. The vertical thin lines mark the indirect and direct bandgaps corresponding to the static atomic positions. Reprinted figure with permission from Zacharias[2016]. Copyright 2024 by the American Physical Society.

8.4.2 Graphene

Graphene is a two-dimensional material composed of carbon atoms with sp^2 hybridization. It has a honeycomb lattice structure and contains unique Dirac-cone-like features in the electronic band structure. As the atoms of graphene vibrate, the electrons exhibit a certain "inertness," which leads to more pronounced nonadiabatic effects (see subsection 2.2.1). However, the sp^2 hybridized C–C single bond in graphene is stronger than the sp^3 hybridized C–C single bond in diamond. Therefore, nuclear quantum effects on the physical properties of graphene are not expected to be as significant as in diamond.

8.4.2.1 Thermodynamics

T. Shao et al. calculated the TEC and lattice constant of graphene by combining the first-principles DFT method with QHA [Shao T.2012]. Figure 8.54(a) shows the temperature dependence of the TEC. In the figure, the red solid curve represents the result of a calculation that considers nuclear quantum effects, while the green dashed curve ignores them. The two curves almost overlap, indicating that the influence of nuclear quantum effects on the TEC of graphene is very weak.

According to the red solid curve in Figure 8.54(a), the predicted TEC of graphene is negative below 430 K, with a minimum value of ~-2×10^{-6} K^{-1}. This result can be

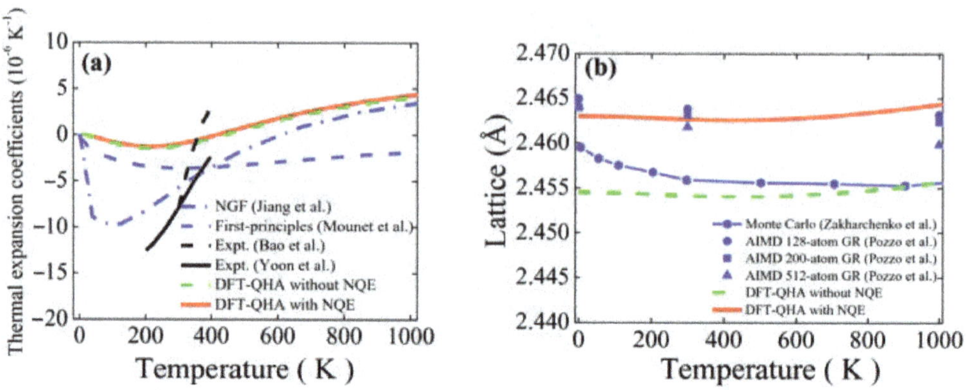

Figure 8.54 Temperature dependence of (a) thermal expansion coefficients and (b) lattice parameter of graphene. Reprinted from Shao T.[2012], with the permission of AIP Publishing.

attributed to the out-of-plane atomic vibrations. The minimum expansion coefficients measured by two experimental groups can be found from the dashed and solid black curves in the figure; both values are $\sim -1 \times 10^{-5}$ K^{-1}. The absolute values of the experimentally measured TECs are significantly greater than the theoretical results. This may be due to the presence of ripples in the real 2D graphene films or to the effect of the substrate on the experimental samples. Neither of these factors was fully considered in the theoretical calculations.

Figure 8.54(b) shows the lattice parameter a of graphene as a function of temperature. It can be seen that the lattice parameters obtained using different methods are relatively close in values, with a difference of ~0.01 Å. The red solid and green dashed curves are the results obtained by T. Shao et al. using the QHA in conjunction with the density functional theory (DFT-QHA) [Shao T.2012] with and without nuclear quantum effects, respectively. By comparing these two curves, it can be observed that the lattice parameter increases by 0.345% and 0.347% at 0 K and 300 K, respectively, when nuclear quantum effects are included. Therefore, as expected, the lattice parameter of graphene is not significantly affected by nuclear quantum effects compared to that of diamond.

In Figure 8.55(a), c_{11} is the elastic constant obtained by deforming the hexagonal graphene lattice along the [n,0] direction. When nuclear quantum effects are not considered, the value of c_{11} is 360.83 N/m at 0 K (corresponding to 1208.78 GPa at a monolayer thickness of 0.335 nm) and 359.93 N/m at 300 K (1205.76 GPa). Similarly, c_{12} is 62.33 N/m (208.81 GPa) at 0 K and 61.36 N/m (205.56 GPa) at 300 K. If nuclear quantum effects are considered, the value of c_{11} decreases by 2.72% at 300 K, a correction comparable to the amount by which it changes when the temperature increases from 0 K to 1,000 K. A similar result is obtained for c_{12}. Clearly, nuclear quantum effects are not negligible when examining the temperature dependence of elastic constants.

Since the elastic constants are the second derivative of the energy density with respect to strain, the effect of nuclear quantum effects on the elastic constants can be estimated

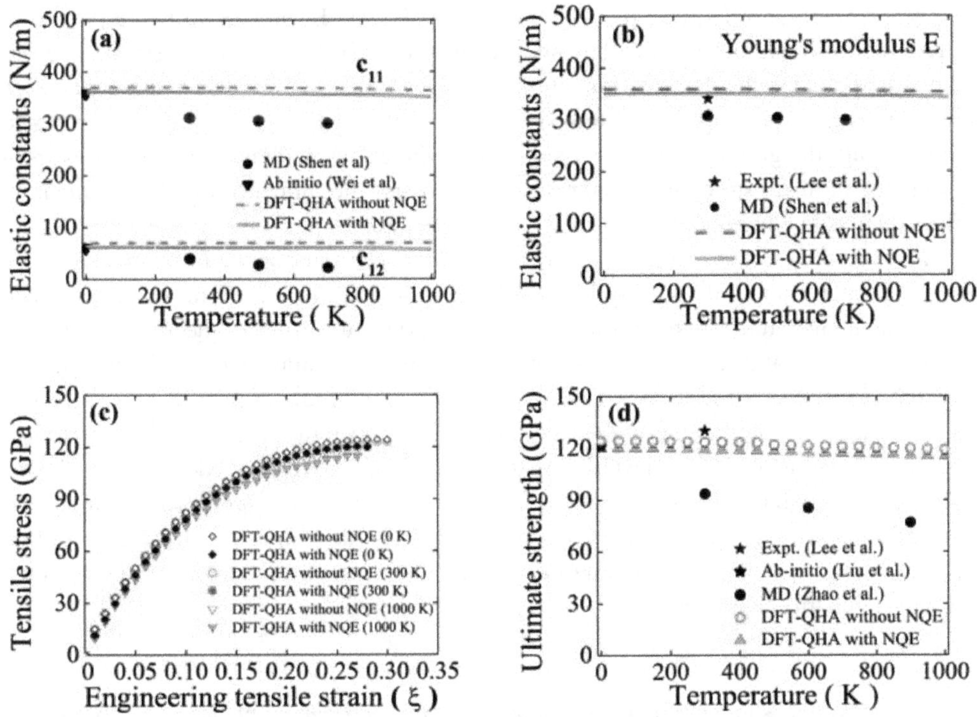

Figure 8.55 Mechanical properties of graphene as a function of temperature: (a) elastic constant; (b) Young's modulus; (c) stress-strain curve; (d) ultimate strength. The data marked "DFT-QHA with NQE" are the results obtained by Shao T. et al. after considering nuclear quantum effects. Reprinted from Shao T.[2012], with the permission of AIP Publishing.

by the second derivative of the ZPE with respect to strain. Although the first derivative of the ZPE with respect to strain is negative, the sign of the second derivative of the ZPE with respect to strain is not determined. Therefore, the influence of nuclear quantum effects on the elastic constants remains undetermined.

In Figure 8.55(b), the value of Young's modulus is 350.07 N/m (1172.73 GPa) at 0 K and 350.01 N/m (1170.72 GPa) at 300 K if nuclear quantum effects are not considered (as indicated by the dashed curve in Figure 8.55(b)). When nuclear quantum effects are considered (as indicated by the solid curve in Figure 8.55(b)), Young's modulus decreases by 2.34% at 300 K. Compared to the case without nuclear quantum effects, Young's modulus decreases by 0.017% ((350.07–350.01)/350.07) as the temperature increases from 0 K to 300 K. When the temperature reaches 1,000 K, Young's modulus is 342.4 N/m, which is only 2.2% ((350.07–342.4)/350.07) lower than its value at 0 K. Evidently, nuclear quantum effects have limited influence on the temperature dependence of Young's modulus. It is worth noting that the calculated results with and without nuclear quantum effects are slightly larger than the experimental value of 340 N/m (1,020 GPa) measured by Lee et al. [Lee2008]; this is likely due to the bias of LDA employed in DFT calculation.

The stress–strain relationships of graphene at 0 K, 300 K, and 1,000 K are plotted in Figure 8.55(c). Without considering nuclear quantum effects, the stress–strain relationship is linear when the deformation strain is below 0.10; when the strain surpasses 0.10, the stress–strain relationship exhibits significant nonlinearity. As the strain continues to increase until the graphene structure is damaged, the corresponding maximum stress represents the ultimate strength of graphene. In summary, the influence of nuclear quantum effects begins to emerge only after the strain exceeds 0.10.

Figure 8.55(d) shows the ultimate strength of graphene as a function of temperature. In the temperature range of 0–1,000 K, the ultimate strength decreases monotonically with increasing temperature. Without considering nuclear quantum effects, it decreases by 4.03% from 120.23 GPa at 0 K to 115.38 GPa at 1,000 K. This softening tendency in the ultimate strength is caused by the weaker interaction between different atomic layers that results from strengthened atomic thermal vibrations as the temperature increases. When nuclear quantum effects are included, the ultimate strength of graphene at 300 K decreases by 3.37% relative to the case not considering nuclear quantum effects. This correction is comparable to the ultimate-strength discrepancy caused by increasing the temperature from 0 K to 1,000 K. This shows that nuclear quantum effects cannot be neglected in temperature-dependent ultimate-strength calculations.

8.4.2.2 Isotope Effect

Although isotope substitution does not affect the electrical or chemical properties of graphene, it does influence physical properties related to atomic vibrations, such as thermal conductivity and phonon transport.

S. Chen et al. experimentally investigated isotope effects on the thermal properties of graphene [Chen S.2012]. In this experiment, graphene samples containing various percentages of ^{13}C were prepared by chemical vapor deposition (CVD), and the temperature dependence of the thermal conductivity of suspended graphene films was measured with micro-Raman spectroscopy. Figure 8.56 depicts the thermal conductivity of graphene grown with pure ^{12}C (0.01% ^{13}C). At 320 K, the thermal conductivity of this sample was greater than 4,000 W(m·K)$^{-1}$, which is twice that of graphene composed of a 50:50 mixture of ^{12}C and ^{13}C. These results show that isotopic scattering has a significant influence on thermal conductivity.

8.4.3 Carbon in Organic Molecules

Carbon can be found in a wide range of organic matter, making it essential to investigate the nuclear quantum effects on carbon in organic molecules. As mentioned previously, carbon atoms are much more massive than hydrogen atoms, so their nuclei would be expected to exhibit weaker quantum tunneling. Nevertheless, studies have found that the nuclear quantum effects of carbon atoms are more pronounced in certain organic molecules.

When we discussed the theory of chemical reaction transition states in Section 3.1, we introduced a reaction rate. Based on similar considerations, let's study a square potential

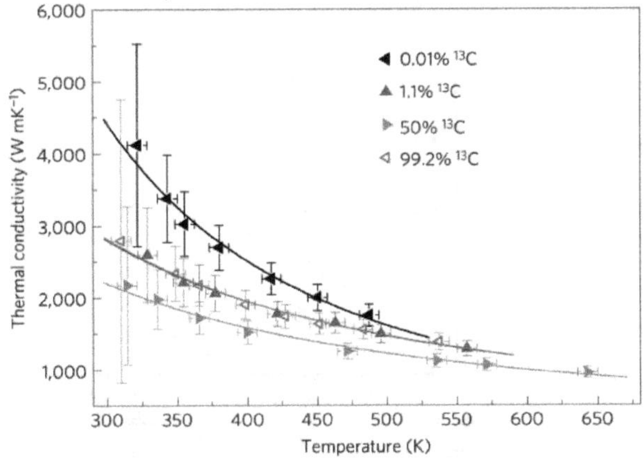

Figure 8.56 Thermal conductivity of suspended graphene films as a function of temperature, measured by micro-Raman spectroscopy. The isotopic concentrations of ^{13}C are 0.01%, 1.1% (natural abundance), 50%, and 99.2%, respectively. The solid curves are a guide to the eye only. The experimental errors were estimated using square-root-sum and error propagation, considering the following sources of error: Raman peak position calibration, temperature resolution of the Raman measurement method, and uncertainty of the laser absorption. From Chen S.[2012]. Reproduced with permission from SNCSC.

barrier of width w and height V. For a particle with energy E and mass m, the probability of transmission through the barrier can be simply expressed as

$$D = D_0 e^{-4\pi w\sqrt{2m(V-E)}/\hbar}, \tag{8.1}$$

where D_0 is a constant of order one. From Eq. (8.1), we see that in some extreme cases, such as when w or $(V–E)$ is small, quantum tunneling of carbon atoms is not negligible.

In a recent study [Lin C.2019], low-temperature scanning tunneling microscopy (STM) was used to observe the flipping reaction occurring between symmetric absorption states of formaldehyde molecules on a metal surface. The reaction rate involving carbon atoms below 10 K was found to be temperature independent, indicating a strong quantum tunneling effect. Density functional theory calculations revealed that the potential barrier for the flipping reaction between symmetric absorption structures was relatively small. This further supports the experimental finding that the flipping rate below 10 K is dominated by the quantum tunneling effect [Lin C.2019].

Theoretical calculations have played a key role in examining the nuclear quantum effects on carbon nuclei in organic molecules. One of the earliest theoretical model systems was the automerization of cyclobutadiene. Cyclobutadiene is an antiaromatic annulene with a strictly rectangular structure. It is composed of four carbon atoms, with two separated C=C double bonds and two C–C single bonds. As a result, two equivalent degenerate states with a 90° difference in spatial rotation are formed. These two degenerate states can be interconverted by shifting between a C=C double bond and a C–C single bond, as shown in Figure 8.57.

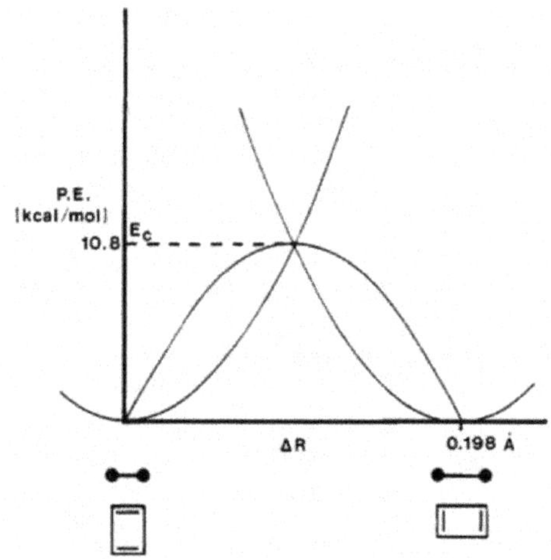

Figure 8.57 Model potential energy functions for the automerization of cyclobutadiene. Reprinted with permission from Carpenter[1983]. Copyright 2024 American Chemical Society.

The automerization reaction pathway between the two degenerate states can be simplified as a simple stretching motion. This means treating the cyclobutadiene molecule (the four carbon atoms and four hydrogen atoms) as a macromolecule composed of two identical atom clusters. Each cluster contains two carbon atoms and two hydrogen atoms. In Figure 8.57, there are two harmonic potential energy curves; their minima represent the two degenerate ground-state energies. Theoretical calculations suggest that the distance between them is $\Delta R = 0.198$ Å. (This value can also be inferred from the typical bond-length difference of 0.18 Å between the C=C double bond and the C–C single bond.) The intersection of these two curves is approximately the highest point of the transition state, whose value depends on ΔR and the force constants of the harmonic vibrations. In accordance with results obtained by higher-order quantum chemical methods (in the range of 8.3–14.0 kcal/mol), the deformation frequency (wave number) of the rectangular molecule is assumed to be 1,000 cm^{-1} and the calculated force constant is 1.532×10^6 dyn/cm, which corresponds to an assumed potential barrier height of 10.8 kcal/mol.

To simplify the tunneling calculation, B. K. Carpenter chose to represent the corresponding potential barrier as a parabola opening downward (see Figure 8.57). The parabola that passes through the assumed transition state ($\Delta R = 0.099$ Å, potential energy $E_c = 10.8$ kcal/mol) and the minima of the two harmonic potential energy curves is much wider than the actual potential barrier; this significantly reduces the tunneling probability.

Using the Bell formula, the probability G(E) of transmission through this potential barrier is

$$G(E) = \frac{1}{1 + \exp\left[\frac{2\pi(E_c - E)}{\hbar \nu_t}\right]}, \tag{8.2}$$

where $\nu_t = \frac{1}{\pi a}\left(\frac{E}{2\mu}\right)^{1/2}$, μ is the reduced mass of the harmonic oscillator, a is half of the width of the potential barrier, and E is the harmonic vibrational energy of the equivalent diatomic molecule [Carpenter1983]. The calculation assumes a temperature of 0 K. If the temperature is to be taken into consideration, the above equation can be multiplied by the Boltzmann distribution and then summed over all energies to estimate the total tunneling rate at different temperatures.

The quantum tunneling rates calculated by Carpenter were 8.08×10^4 /s and 4.65×10^5 /s at 223 K and 263 K, respectively. The corresponding classical reaction rates were 1.01×10^2 /s and 4.82×10^3 /s, respectively. Thus, based on the simplified theoretical model mentioned previously, nuclear quantum effects account for more than 97% of the total reaction rate even when the temperature is just below 273 K [Carpenter1983].

By adding the previously calculated quantum tunneling rate and the classical reaction rate, the activation parameters of the reaction process can be determined, with $\Delta H = 4.6$ kcal / mol and $\Delta S = -15$ cal / (mol • K). In general, experiments suggest that ΔH ranges from 1.6 to 10.0 kcal/mol and ΔS ranges from –32 to 17 cal/(mol • K) [Whitman1982]. Despite the simplicity of the theoretical model used here, it is clear that the consideration of nuclear quantum tunneling effects is necessary for studying this reaction.

Although the physical significance of the theoretical model used by Carpenter et al. is well-established, their model considerably underestimates the contribution of tunneling to the reaction rate [Carpenter1983]. To more quantitively compare the calculation with experimental data, it is necessary to further refine the theoretical model.

The two symmetrical potential energy minima of this 1D tunneling system are degenerate, and there will occur some splitting of the vibrational energy levels, such that $\Delta E_i = E_i^- - E_i^+$, with the – and + symbols representing different symmetries of the coupled wavefunctions. (Here, i is the quantum number of a particular vibrational level; $E_i^{+,-} < E_{i+1}^{+,-}$; and $\Delta E_i < \Delta E_{i+1}$. The symbol 0^+ denotes the ground state.)

Assume that the potential energy satisfies the simple relation $V(R) = AR^2 + B\exp(-CR^2)$, where A, B, and C are undetermined parameters that can be accurately computed from the potential energy of the system [Huang Q.1984]. The R in this equation represents the configuration of the molecule; the corresponding E_i and ΔE_i can be calculated numerically. In this manner, the tunneling rate can be obtained by the equation $k_i = 2\Delta E_i / \hbar$ (i = 0, 1, 2…), and only the statistical average needs to be considered at finite temperatures. Studies have shown that, at 273 K, the main contribution to the tunneling rate comes from the ground state, that is, the vibrational ground state with i = 0. The contribution from the i = 1 excited state is small, and the contribution from higher excited states is negligible. If the values of ΔE_0 are taken from the 0.009–2.0 cm^{-1} range, the tunneling rate can reach 5×10^8 /s to 2×10^{11} /s at –10 °C. And under the same model, the corresponding classical reaction rate is around 10^8 /s. Evidently, when the symmetric double energy splitting occurs, the calculated tunneling rate can increase by several orders of magnitude. (Note the result of the calculation from the previous simple model, the quantum tunneling rate is between 10^4 /s and 10^5 /s.)

Early computational studies were generally based on simple potential energy models in which the selection of parameters was slightly arbitrary, adding uncertainty to the results. Therefore, it is necessary to correct these calculations with more accurate numerical methods. Several researchers have reported results computed by the MINDO/3/HE method [Dewar1984], which uses a semiempirical approach to effectively calculate the force constant matrix and the vibrational energy levels. The HE in the MINDO/3/HE method stands for "half-electron." The MINDO/3/HE method is better suited than first-principles calculations for addressing potential barrier issues related to vibrations. Additionally, when using the first-principles method, it is necessary to incorporate a zero-point vibrational energy correction of ~3.6 cal/mol into the calculated results. The dynamic calculations of the automerization reaction of cyclobutadiene were carried out using the periodic-orbital-based method proposed by J. Bicerano [Bicerano1983]. Normal, deuterated, and ^{13}C enriched cyclobutadiene molecules were compared. The energy-level splitting of each system was calculated first, then the rate of transmission was derived from $\omega_T = 2E/\hbar$, with a value of $\sim 10^{11}$ /s. This result was compared with the reaction rate derived from classical calculations ($\sim 10^8$ /s). The classical reaction rate and the quantum tunneling rate differed by three orders of magnitude even when the temperature reached 350 K.

However, upon further analysis, it becomes clear that this problem is not as simple as it is described previously. The cyclobutadiene molecule has four carbon atoms and four hydrogen atoms, giving rise to multiple vibrational degrees of freedom. Moreover, all the vibrational degrees of freedom are coupled together. Therefore, the quantum tunneling between two vibrational energy levels can be affected by other vibrational degrees of freedom, making it difficult to find an accurate solution to the problem. In addition, external environmental factors have a significant impact on quantum tunneling. For example, the quantum tunneling rates of cyclobutadiene measured in different chemical solutions are different, introducing further complication to the theoretical calculation. Interested readers can refer to a recent report by Schoonmaker et al. for a more detailed discussion [Schoonmaker2018].

When dealing with quantum tunneling in systems with multiple degrees of freedom, new problems may arise in addition to the complexity of the computation. For example, consider the definition of the minimum potential energy path (MEP): In other chemical reaction problems, the MEP represents the optimal reaction path, but this is not necessarily the case when tunneling is involved. The reason for this is simple. As we can see from Eq. (8.1) or Bell's formula, the quantum tunneling rate is related to the energy, the potential barrier width w, and the tunneling frequency v. Upon closer examination of the exponential function, we see that its exponent is directly proportional to w or v, but is only proportional to the square root of the energy. Therefore, the tunneling rate is more dependent on the potential barrier width and the tunneling frequency than on the energy. In general, the reaction path with the greatest quantum tunneling rate does not necessarily cross the transition-state MEP with the lowest potential energy. Therefore, instead of simply finding the classical reaction path with the lowest potential barrier on the Born–Oppenheimer potential energy surface, it is also necessary to calculate the possible transition-state path on the extended potential energy surface. This certainly increases the difficulty of the calculation. The related theoretical methods were discussed in Chapter 3.

In some problems, we can assume that the main quantum tunneling process follows the MEP. This scenario is referred to as the zero-curvature tunneling approximation. There is also a small curvature tunneling (SCT) approximation, which assumes that the main tunneling occurs through a potential energy pathway that is close enough to the MEP for the MEP to serve as a starting point for determining the actual path. On the other hand, the large curvature tunneling (LCT) approximation assumes that the tunneling occurs through a potential energy path that is far from the MEP. Generally, if carbon atoms are involved in the quantum tunneling, the SCT calculation can yield satisfactory results within a relatively short period.

Several programs, such as POLYRATE [Zheng J.J.2010], have been developed to deal effectively with the problems described above. Numerous studies have pointed out that, when the SCT approximation is valid, the results obtained using a simple 1D quantum tunneling model are reasonably close to those calculated with a complex multidimensional potential energy surface model [Doubleday2017]. Interested readers are encouraged to consult additional literature in this field [Borden2016].

8.4.4 Amorphous Carbon

The carbon materials we previously discussed were all crystalline solids, characterized by atoms arranged periodically on a regular lattice. By contrast, amorphous carbon lacks this long-range order. It is crucial to understand the impact of full quantum effects on the physical properties of amorphous materials. This topic is highly intriguing.

Recently, G. Galli's group investigated nuclear quantum effects in diamond-like amorphous carbon using first-principles molecular dynamics with a quantum thermostat [Kundu2022]. Their study showed that the quantum vibronic coupling induced by the nuclear quantum effects significantly changed the physical properties of the material, leading to a renormalization of the electronic and mobility gaps of diamond-like amorphous carbon.

8.4.4.1 Electronic Bandgap Renormalization

Figure 8.58(a)–(c) show the respective electronic densities of states for three representative samples of diamond-like amorphous carbon at 100 K, computed by classical and quantum methods. It can be seen that the quantum simulation led to a wider distribution of the density of states. The top of the valence band was shifted up, while the bottom of the conduction band was shifted down, overall reducing the electronic bandgap. It is also worth noting that the positions of the defect states in Figure 8.58(a) and (b) were largely unaffected by quantum effects, despite their broadened distributions of electronic density of states.

Figure 8.59(a)–(c) shows the HOMO–LUMO electronic bandgap renormalization caused by electron–phonon coupling at different temperatures for the same three amorphous carbon samples as in Figure 8.58. Remarkably, even at 0 K, the quantum vibronic coupling caused a renormalized gap ranging from 80 to 400 meV, which corresponds to 8-12% of the static gap. By contrast, the classical simulations underestimated the extent of gap renormalization (e.g., by 100 to 350 meV at 300 K).

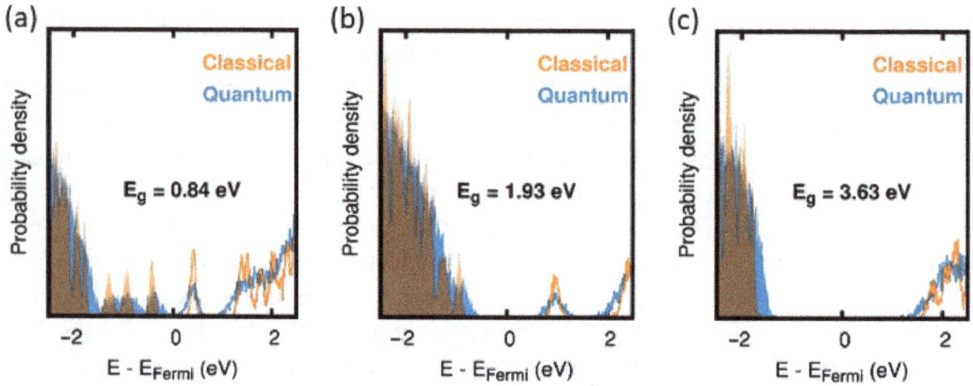

Figure 8.58 Electron density of states at 100 K calculated using the classical and quantum thermostat methods for three representative diamond-like amorphous carbon samples with static HOMO–LUMO gaps of (a) 0.84 eV, (b) 1.93 eV, and (c) 3.63 eV, respectively. The shaded regions indicate the valence bands and unshaded regions represent the conduction bands. Reprinted from Kundu[2022].

Figure 8.59 also provides a comparison between the quantum simulation results of the zero-point renormalization method and those of the frozen-phonon harmonic (FPH) method. The FPH method incorporates nuclear quantization effects, but only within the harmonic approximation. It was observed that the FPH method is more consistent with the quantum simulation results for amorphous carbon samples with a low concentration of defect states. However, significant discrepancies between the two methods were observed for other samples, indicating that the frozen-phonon method lacks robustness. To gain a better understanding of the FPH's limited accuracy in describing vibrational coupling in the diamond-like amorphous carbon system, further analysis was conducted. It revealed that the discrepancy can be attributed to two main factors: (1) the neglect of higher-order terms in the description of the electron–phonon coupling by the FPH method and (2) the inadequate accuracy in describing the strong phonon–phonon coupling interaction.

8.4.4.2 Effect on Orbital Localization and Mobility Gap

The study conducted by Kundu et al. [Kundu2022] delved further into the impact of full quantum effects on orbital localization and the mobility gap in diamond-like amorphous carbon near the Fermi energy level. Specifically, they focused on analyzing the amorphous carbon sample with the smallest HOMO–LUMO energy gap (0.84 eV), which exhibited numerous defect-like states within the mobility gap. Figure 8.60 depicts the joint probability distribution of the electronic energy and the inverse participation ratio (IPR) near the Fermi energy level. The IPR describes the localization of the wavefunction; its value for the ith orbital is calculated from $\int |\psi_i|^4 d^3r / \left(\int |\psi_i|^2 d^3r\right)^2$. A higher IPR value signifies a more localized single-particle wavefunction. The mobility gap is determined by computing the energy difference between occupied and empty states possessing an IPR below a specific threshold (~0.0012).

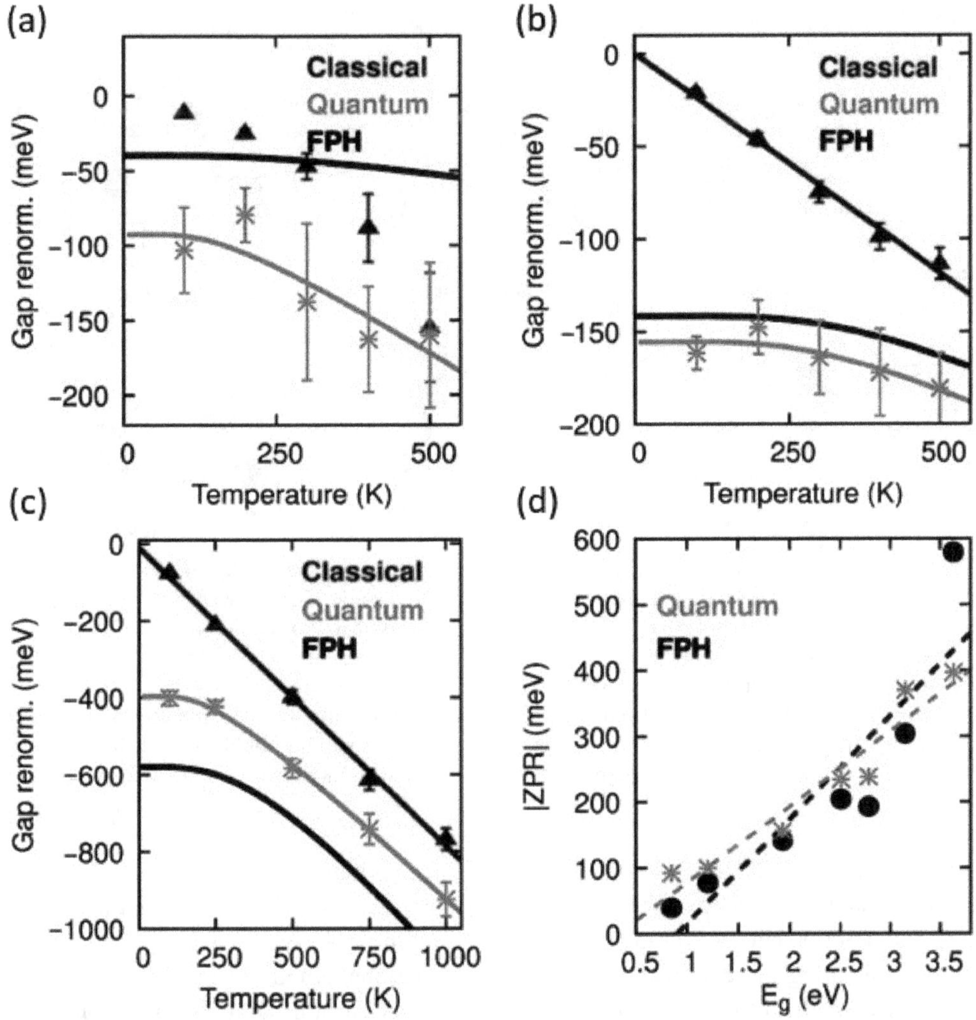

Figure 8.59 (a)–(c) Temperature dependence of HOMO–LUMO energy-gap renormalization due to electron–phonon coupling for amorphous carbon samples with static energy gaps of (a) 0.84 eV, (b) 1.93 eV, and (c) 3.63 eV. The cross and triangle dots are the results obtained from the model fits of quantum and classical molecular dynamics simulations. The deep gray solid curves are the results obtained from the frozen-phonon simple-harmonic approximation (FPH) method. (d) Absolute values of the zero-point renormalization (ZPR) of the HOMO–LUMO gap for the seven amorphous carbon samples with respect to the static HOMO–LUMO energy gap E_g. The light and deep gray dashed lines are linear fits of the quantum simulation and FPH results, respectively. Reprinted from Kundu[2022].

When the results of classical and quantum simulations at the same temperature are compared, it is evident that the presence of full quantum effects leads to a higher degree of delocalization for the electronic states (Figure 8.60). However, it is worth noting that localization is enhanced in a small fraction of electronic states. The influence of quantum

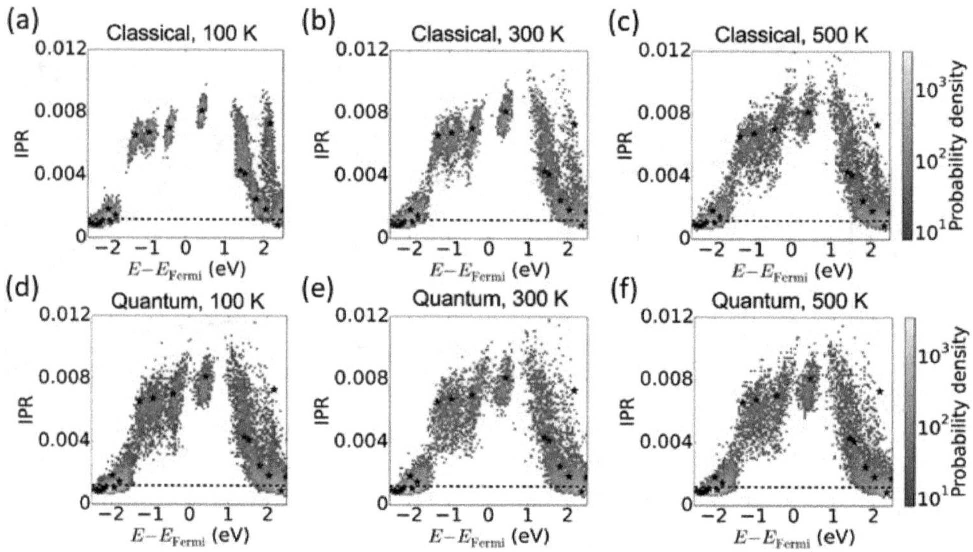

Figure 8.60 Joint probability distribution of electronic energy levels and inverse participation ratio (IPR) for amorphous carbon with a static HOMO–LUMO energy gap of 0.84 eV. (a)–(c) and (d)–(f) show the results of classical and quantum first-principles molecular dynamics simulations, respectively; left (a) and (d), middle (b) and (e), and right (c) and (f) show the respective results at 100 K, 300 K, and 500 K. The black stars represent the IPRs and electronic energies of the configuration that is the closest local minimum to the amorphous carbon sample. The black dashed lines indicate the IPR cutoff used to calculate the mobility-gap threshold (0.0012). Reprinted from Kundu[2022].

fluctuation is particularly pronounced at 100 K (as evidenced by the more delocalized distribution of electronic states in (d) when compared to (a)). These effects remain significant at 300 K but have gradually diminished in importance by 500 K.

By comparing the classical and quantum simulations, Kundu et al. observed a large difference in the mobility edges of the amorphous carbon samples. As depicted in Figure 8.61(a), the classical simulations exhibited static mobility gaps ranging from 4.27 to 4.92 eV at $T = 0$, whereas the quantum simulations yielded smaller values within the narrower range of 4–4.2 eV. Furthermore, Figure 8.61(b)–(d) illustrates a notable disparity in the temperature dependence of the amorphous carbon mobility gap between the classical and quantum simulations. In conclusion, it was observed that the influence of nuclear quantum effects on the mobility gaps of the amorphous carbon systems is prominent even at temperatures significantly higher than room temperature.

8.4.5 Superconductivity

The electrons in a carbon atom exhibit various hybridizations (*sp*, *sp*^{2}, *sp*^{3}) resulting in diverse crystal structures and physical properties in carbon-based compounds. One of the intriguing properties of carbon compounds is their superconductivity, which has

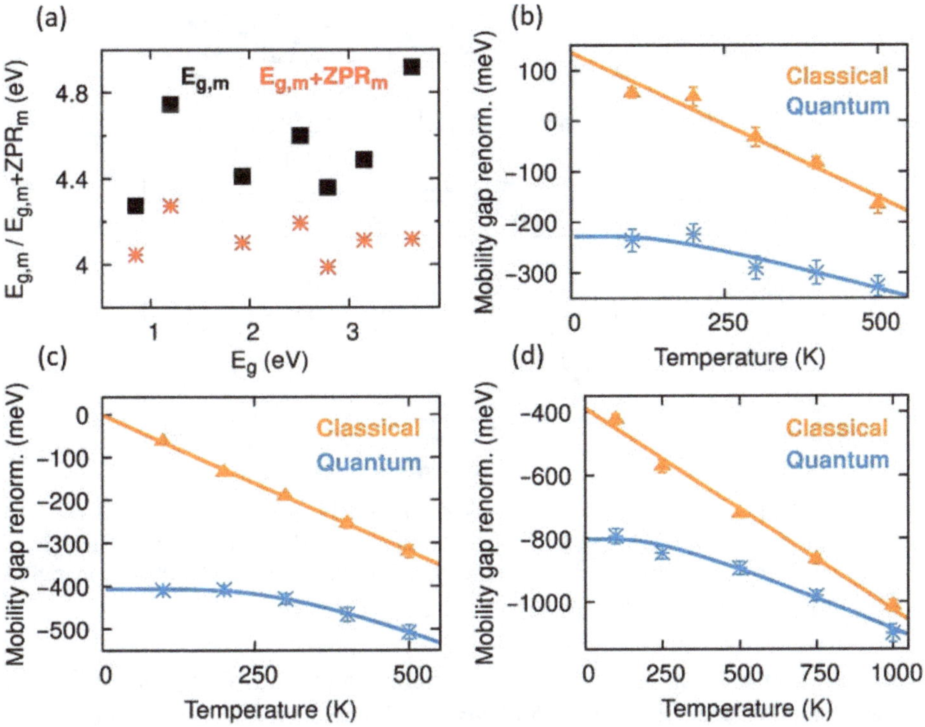

Figure 8.61 (a) Classical static mobility gaps (black squares) and quantum mobility gaps (red asterisks) as a function of static HOMO–LUMO gaps at 0 K for different amorphous carbon samples. (b)–(d) Electron–phonon renormalization of the mobility gaps for the three amorphous carbon samples as a function of temperature. The classical static mobility gaps for the three samples are 4.27, 4.60, and 4.92 eV, respectively. Reprinted from Kundu[2022].

received significant attention in experimental and theoretical studies. From the previous introduction, we know that high pressure can alter superconductivity in some materials, and the physical mechanism for this is primarily achieved by enhancing the strength of the electron–phonon coupling interaction. Recent studies have shown that in addition to pressure, doping with carbon compounds can also lead to superconductivity. Researchers have made extensive efforts to understand the superconducting properties of alkali-metal-doped fullerenes and graphites, as well as boron-doped diamonds.

Graphite is the most commonly found form of carbon. Under high pressures, the low-density arrangement of amorphous carbon gradually changes from carbon pairs to carbon chains. This eventually forms polymers composed of layered graphene. It has been found experimentally that superconductivity can occur after a metal dopant is inserted between graphite layers to form graphite intercalation compounds (GICs). Superconducting carbon compounds include alkali-metal-doped (K, Rb, Cs) graphite, yttrium dicarbide (YC_2), and calcium–graphite compound (CaC_6). Theoretical calculations for YC_2 and CaC_6 show that new superconducting phases appear in a pressurized environment, according to Y. M. Ma's group [Zhang L.J.2006, Feng X.L.2018].

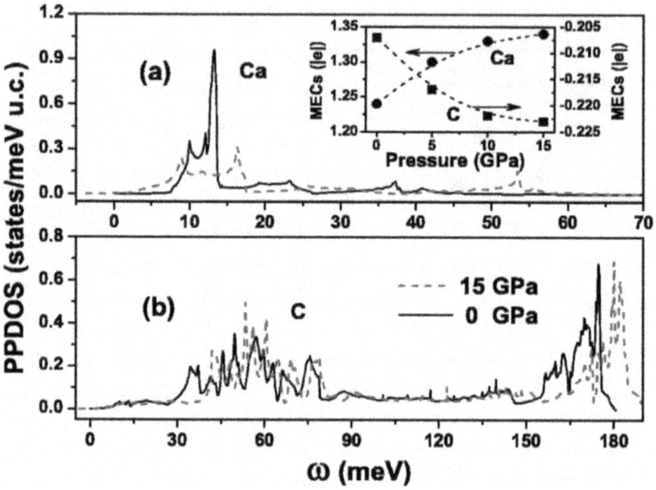

Figure 8.62 Phonon density of states of (a) calcium and (b) carbon in CaC_6 at 0 GPa (solid lines) and 15 GPa (dashed lines). The inset in (a) shows the Mulliken effective charge as a function of pressure. Reprinted figure with permission from Zhang L.J.[2006]. Copyright 2024 by the American Physical Society.

L.J. Zhang et al. combined a global structure search with first-principles calculations and demonstrated that YC_2 stabilizes to a layered graphene structure separated by yttrium atoms under high pressure [Zhang L.J.2006]. They found that the four high-pressure phases of this new structure show conventional BCS superconductivity. For CaC_6, incomplete ionization of dopant Ca in the graphite layer results in a large number of 3d electronic states near the Fermi surface of the original graphite crystal. These electronic states are strongly coupled to the low-frequency phonon modes of the calcium atoms, causing CaC_6 to exhibit superconductivity (see Figure 8.62).

Recent studies have shown that superconductivity can be induced in graphite and other carbon compounds through high pressure, which enhances electron–phonon coupling, and chemical doping. Alkali-metal-doped fullerene superconductors are a prime example, with some exhibiting maximum superconducting transition temperatures of up to 33 K at atmospheric pressure (such as $RbCsC_{260}$) and significant isotope effects. Both experimental and theoretical studies have suggested that electron–phonon coupling is the mechanism behind superconductivity in these fullerenes. While fullerene is an insulator, the introduction of doped alkali metals provides electrons that couple to phonons with specific symmetries. It is widely believed that the main source of superconductivity in alkali-metal-doped fullerenes is the phonons produced by internal molecular vibrations.

The research on the superconductivity of diamond doping is also a topic of particular interest. Another example is doped diamond, discovered by E. A. Ekimov et al. in 2004; they showed that boron-doped diamond bulk materials have a superconducting transition temperature of 4 K [Ekimov2004], as depicted in Figure 8.63. Subsequently, Y. Takano

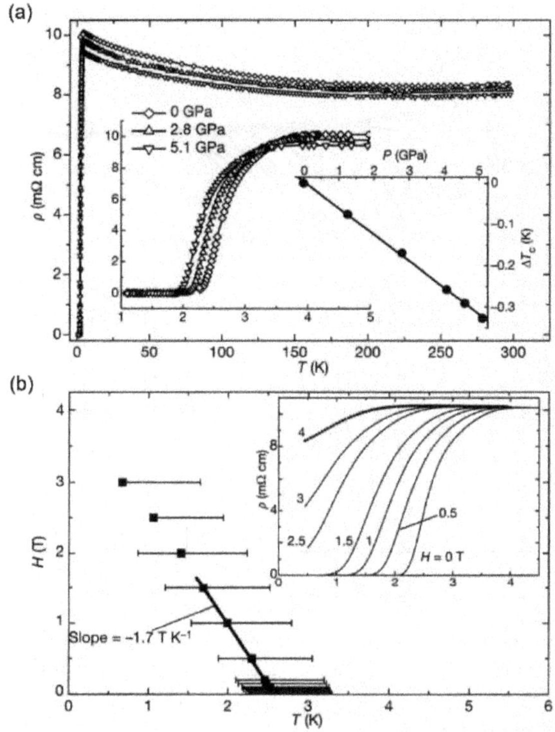

Figure 8.63 Electrical resistivity and upper critical field of boron-doped diamond. (a) Temperature dependence of resistivity at normal and high pressures. The insets show details of the resistivity change below 5 K and the pressure-induced shift ΔT of the midpoint of the resistive transition. (b) Temperature dependence of the upper critical field. The resistive midpoint is used to define H_{c2}. The inset shows the evolution of the resistivity around T_c at 0–4 T magnetic field. From Ekimov[2004]. Reproduced with permission from SNCSC.

et al. found that boron-doped diamond thin films also exhibit superconductivity, with a transition temperature of 7 K [Takano2004] (Figure 8.64). This system shares some similarities with magnesium diboride (MgB_2) in electronic and phonon vibrations. Both systems involve doping in a material with strong covalent (C–C or B–B) bonds. However, boron-doped diamond is hole-doped, while MgB_2 is self-doped. According to theoretical calculations [Lee2004], the enthalpy of formation resulting from C–C bond stretching vibrations in doped diamond is ~60% higher than the enthalpy of B–B bond stretching vibrations in MgB_2. The low enthalpy of formation plays a significant role in determining the superconducting transition temperature. Therefore, as the doping concentration increases, the superconducting transition temperature of boron-doped diamond gradually approaches that of MgB_2.

Chapter 11 covered the yttrium-based hydrides YH_6 and YH_{10}, which exhibit superconductivity at high pressures, with transition temperatures close to room temperature. In this chapter, we delved into the superconducting properties of the yttrium-based

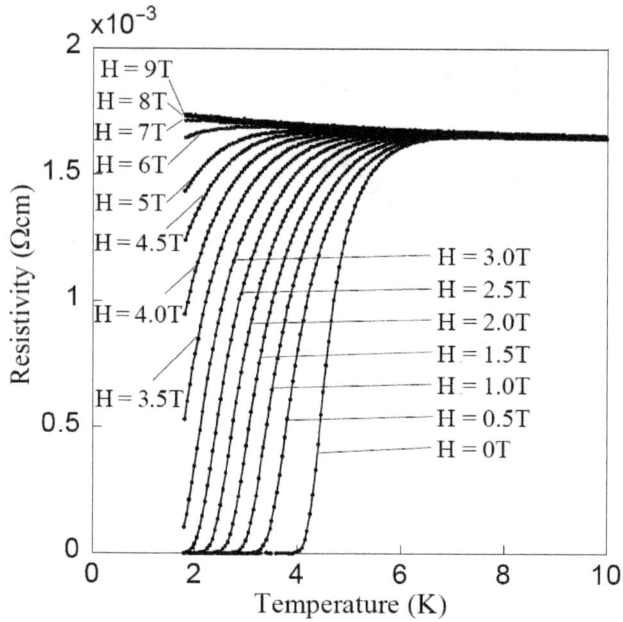

Figure 8.64 Temperature dependence of resistivity of boron-doped diamond at different magnetic field strengths. In the absence of a magnetic field, the resistivity begins to decrease at around 7 K and drops to zero at around 4.2 K. Reprinted from Takano[2004], with the permission of AIP Publishing.

carbide YC_2. The investigation of quantum-mechanical properties, particularly superconductivity, and the behavior of compounds composed of metallic yttrium and other light elements under high pressures requires further attention.

8.5 Oxygen (O)

Oxygen, with a relative atomic mass of 16, is the third most abundant element in the universe by mass, following hydrogen and helium. Because of its high reactivity, oxygen is commonly found in compound form. Although, when studying the physical properties of water, our previous focus was on the full quantum effects of hydrogen atoms (see Section 7.2), the quantum properties of oxygen atoms are also of great significance. This is evident in the diffusion of water dimers on metal surfaces at low temperatures. If we were to solely consider the quantum properties of hydrogen atoms and treat oxygen atoms as classical particles, it would be challenging to explain why the diffusion of a water-molecule pair is faster than that of a single water molecule at low temperatures. We will save the detailed discussion of this for Chapter 10 when we examine full quantum effects in environmental studies.

There is a type of ferroelectric material in oxides that is fully regulated by quantum effects. These are quantum paraelectric oxide ferroelectrics that are unaffected by heat changes at ultralow temperatures. Here we will focus on two typical oxide ferroelectrics that exhibit quantum paraelectricity.

8.5.1 Strontium Titanate

Perovskites are a large class of widely used ferroelectric materials. Generally, they exhibit a paraelectric phase at high temperatures, transitioning to a ferroelectric phase as the temperature decreases. This is accompanied by a structural change. Unpolarized perovskites have a cubic structure. After polarization, symmetry is broken in the direction of polarization, and the structure of the perovskite system can become tetragonal, orthorhombic, or rhombic. Strontium titanate follows the typical ABO_3 formula of perovskites: A central titanium atom is enclosed within each cubic cell, while strontium atoms occupy the eight vertices of the cube, and oxygen atoms are positioned at the adjacent face centers, forming a small octahedron around the central titanium atom. Each strontium atom is surrounded by twelve nearest-neighbor oxygen atoms. At lower temperatures, this cubic structure may undergo a transition to a tetragonal or other phase. However, unlike other perovskites, strontium titanate has a high dielectric constant at low temperatures, and the value of the dielectric constant remains unchanged even as the temperature approaches absolute zero. This phenomenon is known as quantum paraelectricity.

In Chapter 2, when we discussed the influence of full quantum effects on physical properties, Barrett's formula (Eq. (2.7)) was introduced; it considers all titanium, strontium, and oxygen atoms in strontium titanate quantum mechanically [Barrett1952]. From this formula, it is evident that at high temperatures ($T \gg T_1$), the quantum effect diminishes, and the Barrett formula simplifies to the classical Curie–Weiss formula.

In 1979, K. A. Müller and H. Burkard performed detailed low-temperature measurements on strontium titanate. They observed that initially, as the temperature (T) decreases, the dielectric constant of the material increases rapidly, which aligns with the classical Curie–Weiss law. However, below 10 K, the dielectric constant reaches a stable value and no longer varies with temperature, as indicated by Figure 2.20. This experiment confirms the existence of quantum paraelectricity in strontium titanate, although further analysis of the low-temperature data reveals certain limitations of Barrett's formula.

In 1995, D. Vanderbilt's group developed a first-principles effective Hamiltonian method that can reflect the atomic structure and polarization states of ferroelectric materials at finite temperatures [Zhong W.1995]. They concluded that the influence of nuclear quantum effects is not only affected by the nuclear wavefunction but also by the uncertainty of the corresponding atomic positions in space. Combining the model Hamiltonian with the PIMC method, W. Zhong and Vanderbilt conducted further theoretical simulations of strontium titanate; their results are shown in Figure 8.65. The research found that in this type of material, different relative movement patterns between titanium atoms and oxygen atoms lead to two completely different polarization situations. Specifically, under the influence of nuclear quantum effects, they exhibit different physical properties. Compared to the classical Monte Carlo simulation, when considering antiferrodistortive (AFD) modes, the presence of nuclear quantum effects leads to a quantitative reduction in the phase transition temperature. On the other hand, for ferroelectric (FE) modes, the influence of nuclear quantum effects completely eliminates the expected ferroelectric phase transition that typically exists in the low-temperature region. This finding suggests

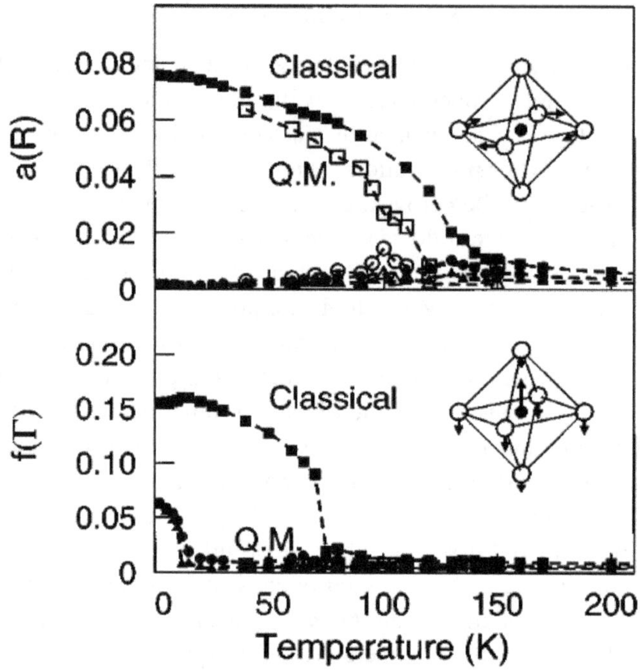

Figure 8.65 Temperature dependence of ferroelectric and structural order parameters obtained by the first-principles effective Hamiltonian method. The $a(R)$ and $f(\Gamma)$ denote the soft modes obtained from the strontium titanate phonon spectrum. The vibrational modes of the atoms, shown in the insets, correspond respectively to the rotation of the titanium–oxygen octahedron around the axis (AFD modes) and the polarization resulting from the relative motion of the titanium and oxygen atoms (FE modes). The results of the path-integral Monte Carlo simulation considering nuclear quantum effects and the classical Monte Carlo calculation are compared in the figure. For AFD modes, the nuclear quantum effects quantitatively reduce the phase transition temperature; for the FE mode, the nuclear quantum effects completely suppress the ferroelectric phase transition that should have occurred in the low-temperature region. Reprinted figure with permission from Zhong W.[1996]. Copyright 2024 by the American Physical Society.

that nuclear quantum effects not only quantitatively change the phase transition temperature as the system structure shifts from the cubic to the tetragonal phase but also qualitatively alter the ferroelectricity of the system [Zhong W.1996].

Understanding the quantum paraelectricity of strontium titanate is important because it suggests that nuclear quantum effects are not exclusive to light elements. In certain physical processes involving coordinate uncertainty, the influence of quantum fluctuations may be more significant than commonly perceived. Moreover, it is possible to prepare new quantum states at higher temperatures using materials similar to strontium titanate, as long as their characteristic scales are effectively reduced and modulated. For ferroelectric materials with perovskite structures, the energy difference between the different symmetric structures is small. Therefore, when the temperature drops, quantum

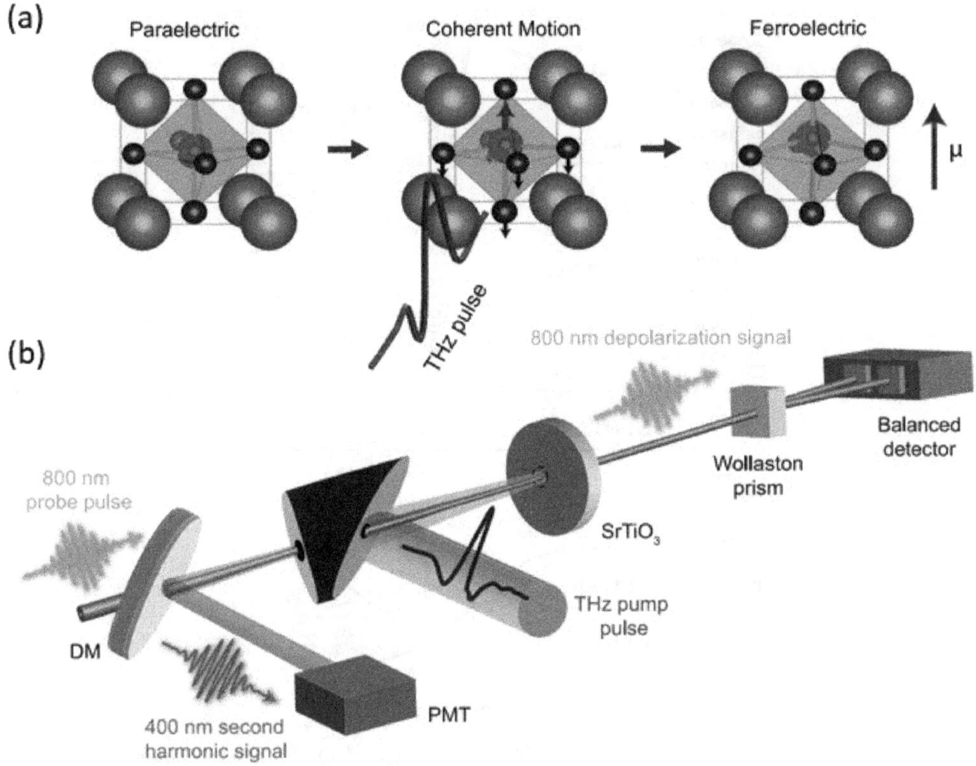

Figure 8.66 Coherent control over material structure: (a) Inducing a paraelectric–ferroelectric transition in $SrTiO_3$; (b) Schematic of terahertz electric-field excitation device. Reprinted figure with permission from Li X.[2019]. Reprinted with permission from AAAS.

fluctuations significantly alter the spontaneous polarization of the material, affecting other physical properties such as the dielectric constant [Höchli1979, Carlson2000].

Having understood the basic principles of quantum paraelectricity, we are in a position to exploit full quantum effects for paraelectric phase modulation. Recently, K. A. Nelson's group used terahertz electric-field excitation to drive the directional migration of $SrTiO_3$ ions, producing a hidden ferroelectric phase that was previously suppressed by quantum fluctuation. The experimental procedure is illustrated in Figure 8.66. The hidden phase has less symmetry, significantly affecting the spectrum of phonon excitations. This confirms the existence of the ferroelectric phase [Li X.2019]. This work demonstrated that external field modulation can be used to control the relative strength of ferroelectric interactions and quantum fluctuation, thus enabling access to the hidden ferroelectric phase.

H. Wu et al. investigated the feasibility of using doping to control nuclear quantum fluctuation and regulate the ferroelectric effect. For instance, changing the doping concentration of Ba in $Ba_xSr_{1\text{-}x}TiO_3$ (BSTO) was found to induce a transition from a quantum paraelectric to a quantum ferroelectric state; this was followed by a phase transition to a classical ferroelectric phase [Wu H.2006]. Similar transitions are also observed when BSTO is doped with other elements such as Pb.

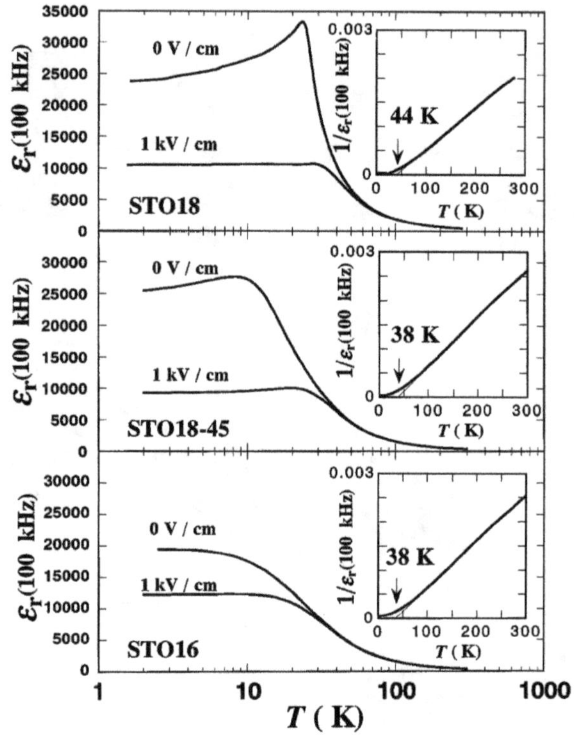

Figure 8.67 Replacement of lighter ^{16}O with heavier ^{18}O in $SrTiO_3$ suppresses the quantum fluctuation because of nuclear quantum effects. Reprinted figure with permission from Itoh[1999]. Copyright 2024 by the American Physical Society.

From the previous discussion, it is evident that isotope substitution serves as a crucial approach for modulating nuclear quantum effects. By substituting the ordinary oxygen atom with one of its isotopes, quantum fluctuation can effectively be suppressed [Kvyatkovskiĭ2001]. Such isotopic substitution of oxygen atoms has been employed in attempts to modulate the properties of ferroelectric materials. For example, replacing ^{16}O with heavier ^{18}O in $SrTiO_3$ allows suppression of the quantum fluctuation without changing the strength of the ferroelectric interactions. Consequently, ferroelectric polarization can be induced in $SrTi(^{16}O_{1-x}{}^{18}O_x)_3$ samples without external electric field excitation (as shown in Figure 8.67, where the polarization is marked by the clear peak in the ^{18}O sample (STO18)) [Itoh1999, Wang R.2001].

8.5.2 Barium Ferrite

Another typical example of an oxide ferroelectric that exhibits quantum paraelectricity is barium ferrite ($BaFe_{12}O_{19}$), an M-type hexaferrite. Because of its ferrimagnetic properties, it shows potential in significant commercial and technological applications. Despite its importance, the crystal structure of barium ferrite has been a subject of controversy for many years. While the commonly held view describes the structure of barium ferrite

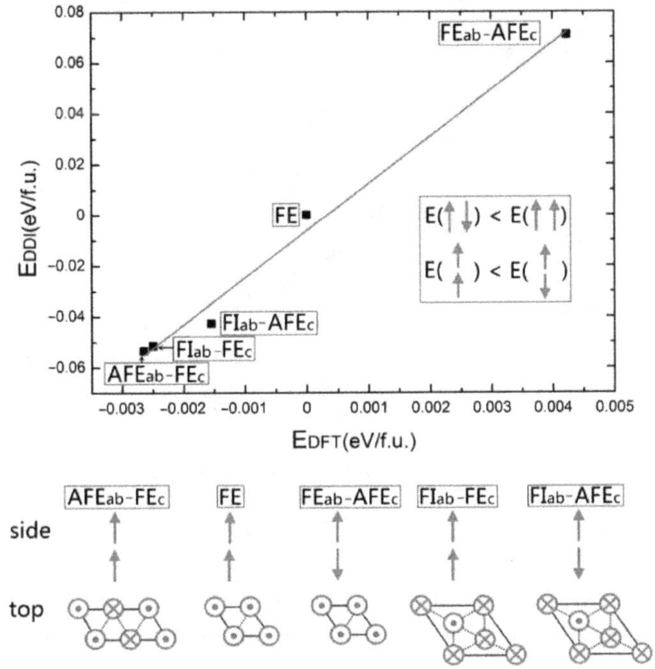

Figure 8.68 Comparison of total energies calculated from the DFT method with those from DDI models for different local dipole configurations of barium ferrite. Figure created by Wang P. S.[2014] licensed under CC BY 3.0.

using space group P63/mmc, certain experiments have observed a symmetry change when the iron ions deviate from the center of the P63/mmc structure along the c-axis. In an attempt to resolve this controversy, P. S. Wang and H. J. Xiang used first-principles calculations to investigate the structure of barium ferrite [Wang P.S.2014].

Figure 8.68 compares the calculated total energy of barium ferrite in different dipole configurations using DFT and the dipole-dipole interaction (DDI) model. It can be seen that the total energy is lowest when the crystal configuration is antiferroelectric in the ab-plane and ferroelectric along the c-direction. Based on this result, P. S. Wang and H. J. Xiang predicted that the crystal structure of barium ferrite exhibits frustrated antiferroelectric behavior. Using this model, they provided a theoretical explanation for the off-centered iron ions observed in experiments [Wang P.S.2014]. Furthermore, considering the ferrimagnetic properties of barium ferrite, they proposed $BaFe_{12}O_{19}$ as a novel multiferroic material, presenting opportunities for investigating frustrated antiferroelectricity and serving as a foundation for the realization of multistate memory devices.

Their DFT calculation did not consider nuclear quantum effects, which are generally assumed negligible for materials composed of relatively heavy elements, such as those in $BaFe_{12}O_{19}$. However, there has been ongoing debate regarding whether nuclear quantum effects should be ignored in this case. To gain further insights into this system, S. P. Shen et al. performed detailed measurements of its dielectric constant [Shen S.P.2016]. The

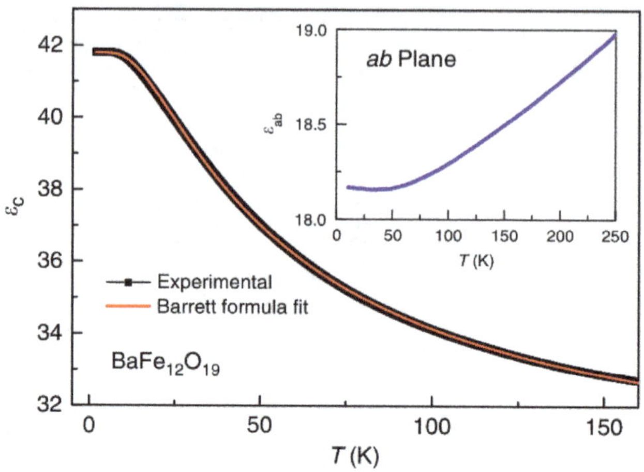

Figure 8.69 Temperature dependence of the dielectric constant of $BaFe_{12}O_{19}$. The black curve is the experimental result, and the red curve is the theoretical result fitted using Barrett's formula. Figure created by Shen S. P.[2016] licensed under CC BY 4.0.

results showed no evidence of ordered antiferroelectricity at low temperatures. Instead, the dielectric constant exhibited a plateau, indicating temperature independence in the low-temperature range (see black curve in Figure 8.69). From our discussion of strontium titanate in subsection 8.5.1, we learned that a plateaued dielectric constant curve signifies the dominant role of nuclear quantum effects in the polarization of barium ferrite at low temperatures. More specifically, it means that it is possible for barium ferrite to exhibit quantum paraelectricity at low temperatures. This contradicts the theoretical prediction made by P.S. Wang et al. By fitting the dielectric-constant-temperature curve of barium ferrite with the temperature-dependent Barrett's formula (Eq. (2.7), which accurately describes the dielectric constant of the quantum paraelectric strontium titanate as a function of temperature), S. P Shen et al. found that the model was in perfect agreement with experimental observations, as indicated by the red curve in Figure 8.69. This provides compelling evidence that the antiferroelectric order of barium ferrite disappears at low temperatures as a result of nuclear quantum effects, yielding a quantum paraelectric state. $BaFe_{12}O_{19}$ thus joins strontium titanate as an example of a quantum paraelectric material.

Furthermore, S. P. Shen et al. made another significant observation in their research. They detected unusual low-energy excitations at low temperatures in both heat capacity and thermal conductivity measurements. Taking into account this phenomenon along with the previously mentioned frustrated antiferroelectric coupling and nuclear quantum effects in $BaFe_{12}O_{19}$, S. P. Shen et al. introduced the concept of "quantum electric–dipole liquids." This term is analogous to the term "quantum spin liquids" systems that have garnered considerable attention recently. The emergence of quantum electric-dipole liquids paves the way for new avenues in physics research.

Although the experiments from S. P. Shen et al. show possible quantum paraelectricity in $BaFe_{12}O_{19}$, theoretical studies confirming this possibility from a microscopic point of view are still lacking. Therefore, X. F. Zhang et al. developed a model Hamiltonian based on first-principles calculations and performed PIMC simulations to study the atomistic details of the quantum paraelectricity of $BaFe_{12}O_{19}$ [Zhang X.F.2020]. As in the discussion in subsection 2.1.5, they compared the dielectric constants obtained by the classical (squares in Figure 2.21(a)) and PIMC (circles in Figure 2.21(a)) simulations. They found that nuclear quantum effects can suppress the antiferroelectric order in $BaFe_{12}O_{19}$ and induce a dielectric-constant plateau at low temperatures. Their results could be fitted well by the Barrett formula (dashed curve in Figure 2.21(a)) and were consistent with the experimental results of S. P. Shen et al. (squares in the inset panel of Figure 2.21(a)). In Figure 2.21(b), they also presented the relationship between the reciprocal of the electric polarization and the temperature, further supporting the above discussion.

Moreover, using PIMC simulations, X. F. Zhang et al. decomposed the nuclear fluctuations (black squares in Figure 2.21(c)) into thermal and quantum contributions (respectively labeled by triangles and circles in Figure 2.21(c)). They found that the proportion of the quantum contribution decreases gradually with increasing temperature until the region where thermal fluctuation becomes dominant, which corresponds to the classical paraelectric phase in the conventional Curie-Weiss theory. The quantum contribution crosses over the classical one at ~10 K, below which the quantum fluctuation is dominant. At such temperatures, although $BaFe_{12}O_{19}$ is still paraelectric, the fluctuations that induce paraelectricity originate in nuclear quantum effects. As a result, the fluctuation-induced properties show no evident temperature dependence in this temperature regime. This is the reason for the plateau of susceptibility observed in the experiments. These calculations provide further confirmation of the quantum paraelectricity of $BaFe_{12}O_{19}$.

Based on the discussion presented in this section, it is evident that nuclear quantum effects exert a significant influence on the behavior of barium ferrite materials at low temperatures. Consequently, $BaFe_{12}O_{19}$ deviates from the predictions of classical first-principles calculations and adopts a quantum paraelectric state instead of an antiferroelectric state. This work once again demonstrates that in some specific cases, even though the material is not composed entirely of light element atoms, the increase in overlap of nuclear quantum states caused by the uncertainty principle related to spatial position makes the effect of nuclear quantum effects more prominent in low-temperature regions, thereby having a decisive impact on the crystal structure and physical properties of the material. Additionally, full quantum effects can produce novel quantum states (such as the quantum electric–dipole liquid) and provide new research platforms for discovering materials with special properties.

8.6 Silicon (Si)

Silicon holds immense significance as the fundamental material for various semiconductor devices and integrated circuits in the current Information Age. Positioned in the third period of the periodic table, it possesses a relatively large atomic mass. Nuclear quantum effects are commonly assumed to be negligible in silicon, but considering that silicon is an indirect bandgap semiconductor, the other type of full quantum effect must also be considered: the nonadiabatic property, specifically the quantum coupling between nuclear and electron motion. In fact, both factors demand careful consideration when investigating light absorption and emission in silicon-based materials. This section will delve into the comprehensive impact of full quantum effects on the physical properties of silicon semiconductors, exploring these aspects in detail.

8.6.1 Anomalous Thermal Expansion

At low temperatures, silicon exhibits anomalous thermal expansion, characterized by a negative TEC. Early efforts to elucidate this phenomenon relied on the QHA [Biernacki1989b]. In general, the QHA model assumes that the frequency of each harmonic oscillator (phonon) depends on volume V and defines the Grüneisen parameter $\gamma_\omega = -\partial(\ln \omega) / \partial(\ln V)$, where ω is the phonon frequency. A negative Grüneisen parameter can lead to a negative TEC. For silicon, the QHA predicts that the low-energy acoustic phonon TA modes have negative Grüneisen parameters, which is presumed to be the main cause of the anomalous lattice expansion at low temperatures. With increasing temperature, more phonons are excited to higher-energy phonon modes. The positive Grüneisen parameters of these higher-energy modes cause the TEC to change sign.

In 2018, researchers at Oak Ridge National Laboratory measured the phonon dispersions of silicon semiconductors at different temperatures using inelastic neutron scattering. The results, depicted in Figure 8.70, revealed that the energies of most phonon modes, including the TA modes at low temperatures, decreased or appeared

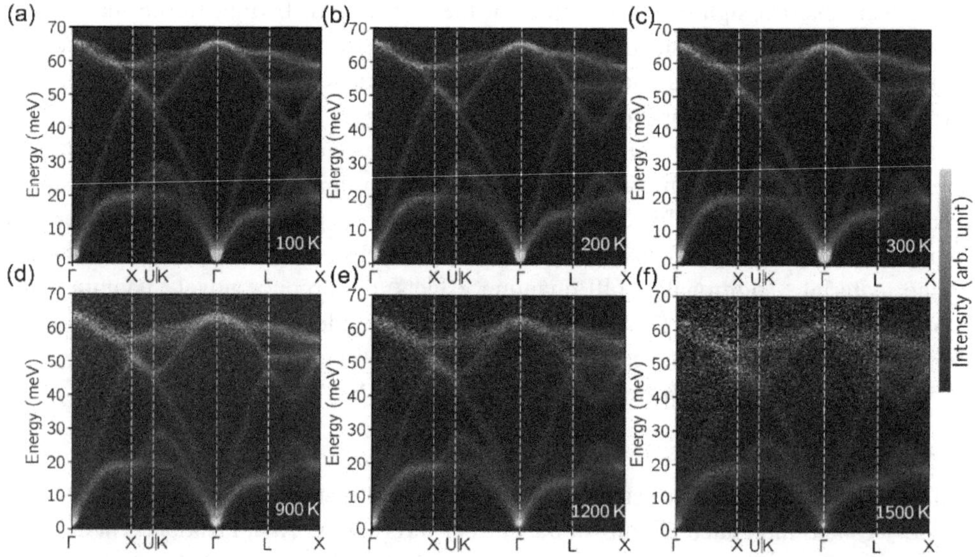

Figure 8.70 Phonon dispersions of silicon semiconductors at different temperatures, measured by inelastic neutron scattering at (a) 100 K, (b) 200 K, (c) 300 K, (d) 900 K, (e) 1,200 K, and (f) 1,500 K. Reprinted from Kim[2018].

softened as the temperature increased [Kim D.S.2018]. Interestingly, these findings contradicted the prediction of QHA calculations that the TA acoustic phonon energy would increase with temperature, resulting in a negative Grüneisen parameter. This discrepancy suggests that the QHA method alone is insufficient to fully explain the anomalous expansion of the silicon lattice at low temperatures. Thus, it is necessary to consider higher-order anharmonic effects.

To address this problem, D. S. Kim et al. adopted the stochastically initialized temperature-dependent effective potential method (s-TDEP). This method retains the higher-order contributions of the lattice dynamics and thus intrinsically includes phonon–phonon interactions and nuclear quantum effects. Using s-TDEP, Kim et al. accurately predicted the softening of phonon energy as the temperature increased. They also predicted positive and negative shifts in phonon energy as the temperature varied. The average predicted phonon energy shift at low temperatures was close to 0, similar to the QHA results. At high temperatures, s-TDEP calculation showed that the average phonon energy of the TA modes softened as the volume and the temperature increased. Compared to the QHA result, the average phonon energy shift of the TA modes exhibited the opposite sign. This indicates that QHA results are consistent with experimental results at low temperatures because the contributions of different phonons cancel each other out. It also indicates that the QHA does not accurately capture the underlying physical mechanism in silicon.

The real physical reason behind the negative TEC is the nuclear quantum effect (see Figure 8.71). The phonon anharmonicity, that is, the phonon–phonon interaction, is important in the entire temperature range; it changes the effective interatomic potential and causes positive and negative shifts in the phonon energies. The nuclear quantum effect leads to a nonzero anharmonic coupling between phonons. At low temperatures, even if the high-energy phonon modes are not excited, the ZPE of the high-energy vibrational modes can still affect the self-energies of the low-energy phonons through anharmonic

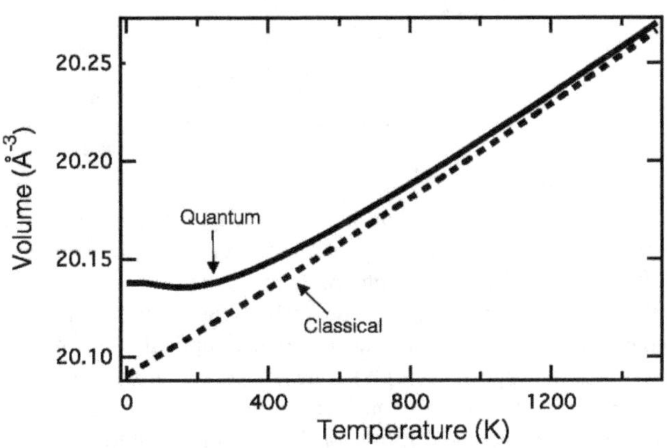

Figure 8.71 Average volume per atom in silicon semiconductor as a function of temperature, calculated by classical and quantum methods. Reprinted from Kim[2018].

coupling, which in turn changes the volume dependence of the free energy in silicon. Calculations using s-TDEP predict that silicon semiconductors exhibit anomalous thermal expansion at low temperatures, which is consistent with the experimental observation. This example emphasizes the importance of considering full quantum effects when studying the thermal properties of materials.

8.6.2 Phonon-Assisted Optical Absorption

Silicon is used not only in integrated circuits and semiconductor devices but also in solar cells. Therefore, the study of the optical absorption in silicon has important practical implications. Similar to diamond, silicon is a semiconductor with an indirect bandgap of ~1.1–1.2 eV, depending on the temperature [Zacharias2015]. At temperatures below 100 K, silicon's indirect bandgap value no longer changes linearly with decreasing temperature; instead, it tends to stabilize at 1.17 eV (as shown in Figure 8.72(a)). This phenomenon is mainly attributed to the action of nuclear quantum effects in silicon and can be explained by the zero-point renormalization of the electronic structure.

Silicon possesses a direct bandgap of 3.4 eV, which exceeds the energy of visible light. Consequently, direct absorption of visible light by electron transitions is not feasible in silicon. Instead, the absorption of visible light in silicon is achieved through indirect processes involving phonon-assisted electron transitions. This necessitates the consideration of nonadiabatic effects resulting from the quantum coupling between electrons and nuclei. Accurate prediction of optical absorption in silicon,

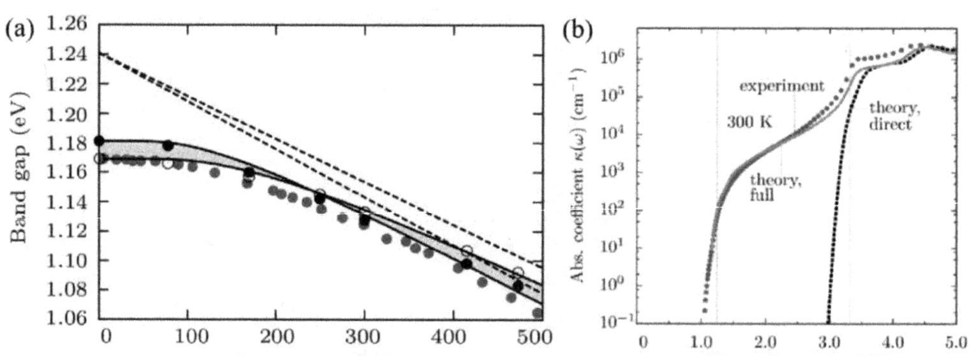

Figure 8.72 (a) Temperature-dependent indirect bandgap of silicon. The light gray dots represent the experimental data, the deep gray dots represent the values of the bandgap fitted by the calculated optical absorption spectrum (solid and hollow dots correspond to fits made in different energy ranges), and the dashed lines represent the high-temperature linear asymptotes. (b) Comparison of the calculated optical absorption spectrum with the experimental values (upper gray dots). The lower gray dots represent the calculated results for the direct optical transition considering only the static atomic configuration; the solid curve represents the calculated result for the phonon-assisted electron transition after considering full quantum effects. Reprinted figure with permission from Zacharias[2015]. Copyright 2024 by the American Physical Society.

as depicted in Figure 8.72(b), requires the description of electrons and nuclei using quantum wavefunctions [Zacharias2015, 2016]. Nonadiabatic effects can significantly reduce the threshold energy for light absorption from ~3 eV to less than 1.0 eV. The numerical results presented in Figure 8.72(b) (solid curve) align perfectly with experimental measurements (upper gray dots in Figure 8.72(b)). By contrast, the results with only direct absorption (lower gray dots in Figure 8.72(b)), obtained by neglecting the contribution of electron–nucleus coupling and relying solely on the static atomic configuration of silicon, fail to explain the experimentally observed optical absorption between 1.1 and 3.3 eV.

9
Full Quantum Effects in Energy Applications

From stellar luminescence to geothermal energy generation, the major sources of energy in the universe are inextricably linked to nuclear quantum effects. For example, stars release energy through internal nuclear fusion, and the nuclear chain reactions can only occur when light elements such as hydrogen and helium react with each other through quantum tunneling.

As human society develops, its energy demands continue to increase. Historically, reliance on fossil fuels has led to environmental problems and undeniable impacts on climate change. Indeed, dependence on fossil fuels has become a barrier hindering the establishment of a sustainable and thriving society and economy. How to use fossil energy efficiently is a new common challenge faced by the scientific and industrial communities. Existing energy sources based on fossil fuels, such as coal, natural gas, and petroleum, predominantly consist of hydrogen, carbon, and other light elements. Thus, full quantum effects undoubtedly play a crucial role in energy conversion, processing, and utilization.

On the other hand, extensive research indicates that traditional fossil fuels are being depleted and will be insufficient to meet the energy demand of industrial development and daily needs in the next 20–30 years. According to the current plan, the next decade will be critical for the global transition from fossil fuels to renewable energy sources. Next-generation clean energy technologies, including solar energy, hydrogen energy, fuel cells, and ion batteries, have attracted significant attention and are gradually becoming indispensable for sustainable development. These new energy technologies often use light elements such as hydrogen and lithium as their primary carriers. The physical and chemical properties of light elements, influenced by full quantum effects, inherently impact the processes of energy generation and conversion in new energy systems, with both nuclear quantum effects and nonadiabatic effects playing substantial roles.

In terms of clean energy research, one of the most attractive topics is how to use the rich water resources. Water, which is inexhaustible on Earth, can serve as a potential reserve of renewable energy. It can be decomposed into hydrogen and oxygen gas, and burning hydrogen with oxygen gas produces water again. Thus, water can serve as both a carrier and a working medium for clean energy, and water decomposition could serve as the foundation for sustainable industrial development. However, the widespread application of water as a renewable energy source is hindered by the lack of cost-effective and efficient catalysts. Without these catalysts, the production of hydrogen releases large amounts of environmentally harmful CO_2. To address this issue, urgent fundamental research into water–material interactions is required, particularly from the perspective of full quantum effects.

In a word, it can be seen from the previous introduction that either traditional energy sources, new-generation clean energy sources, or even future energy sources are all closely associated with light elements. Undeniably, research into full quantum effects is essential for addressing today's energy challenges.

9.1 Energy in the Universe

9.1.1 Stellar Energy

A stable energy supply is required to sustain and drive the formation and evolution of Earth's biosphere. Of the various available sources of energy, sunlight is the most important. It supports photosynthesis, which generates large amounts of oxygen needed for aerobic respiration and thus allows the formation of complex and diverse multicellular organisms. The sunlight-driven evolution of life has been taking place on Earth for billions of years. According to current research, the oldest bacteria can be traced back 3.4 billion years, and the oldest eukaryotes appeared 1.8 billion years ago. The oldest fossil evidence of eumetazoans dates from 630 million years ago [Wacey2011].

The production of sunlight ultimately requires the generation of high-energy photons through light-element fusion inside the Sun. In stars, which are made of plasma, there is Coulomb repulsion between the positively charged atomic nuclei. This necessitates the use of high energy to overcome the Coulomb repulsion and reduce the distance between nuclei to a very small value ($r_c \leq 10^{-15}$ m). Consequently, the nuclei enter the regime of the strong interaction, effectively binding them together. In a star, the primary energy source for overcoming the Coulomb barrier is the kinetic energy of atomic nuclei due to thermal motion. Because of gravitational collapse, the temperature at the center of a star can reach 10^7 K, causing a large number of atoms to undergo vigorous motion. The average kinetic energy of each nucleus can be expressed as

$$E_{kin} = \frac{3}{2} k_B T, \tag{9.1}$$

where k_B denotes the Boltzmann constant, and T denotes the temperature. For a temperature of 10^7 K at the center of the star, the average nuclear kinetic energy is ~1 keV. The Coulomb barrier that needs to be overcome is given by

$$E_{cb} \propto \frac{Z_1 Z_2 e^2}{r_c}, \tag{9.2}$$

where Z_i is the number of charges carried by the i-th nucleus. Based on this equation, the Coulomb barrier is approximately on the order of 1 MeV (corresponding to a temperature of 10^{10} K). Evidently, the kinetic energy of the nuclei due to thermal motion is much lower than the Coulomb repulsion barrier, which is not sufficient to trigger fusion reactions.

However, this intense thermal motion can bring the nuclei close enough for quantum tunneling to occur. In this case, there exists a certain probability for the nuclei to interact with each other and form a more stable bound state, then release nuclear energy, even if they do not possess enough energy to overcome the Coulomb barrier. This probability can be expressed as

$$P_t(v) \propto \exp\left(\frac{-4\pi^2 Z_1 Z_2 e^2}{h}\frac{1}{v}\right), \tag{9.3}$$

where v is the nuclear velocity and h is Planck's constant. Therefore, reducing Z_i increases the probability of quantum-tunneling-induced fusion; for hydrogen nuclei (protons), $Z_i = 1$, this corresponds to the hydrogen nucleus fusion. In the case of stars with low internal temperatures (e.g., the Sun), the tunneling probability leading to the proton–proton chain reaction is $\sim 10^{-20}$. Meanwhile, other nuclear physics processes related to nuclear reactions, such as β^+ decay that transforms the proton into the neutron, can also affect the probability of proton–proton chain reactions.

Theoretically, the probability of a reaction is mainly determined by the nuclear reaction cross section and the velocity distribution of the nucleus. The reaction cross section can be expressed as

$$\sigma(E) = S(E)E^{-1}\exp(-2G), \tag{9.4}$$

where G is the Gamow factor, S(E) is the astrophysical S-factor (which depends on other factors in nuclear physics), and E is the nuclear energy [Trixler2013]. The exp(–2G) term, which is directly proportional to the quantum tunneling probability, exhibits a strong temperature dependence. The reaction cross sections in stars with masses less than 1.5 times the mass of the Sun are smaller than those in massive stars because of the lower core temperatures and pressures. As a result, fusion in these stars occurs mainly through proton–proton chain reactions. It is the extremely low reaction probability that allows the Sun to sustain stable "burning." Because of the large number of protons inside the Sun, fusion reactions will continue to occur until the end of the Sun's lifespan [Trixler2013].

In summary, in smaller stellar objects, the high core temperatures resulting from gravitational contraction are insufficient to trigger nuclear fusion. The combined effect of the high temperature and the quantum tunneling of the nuclei allows the fusion reactions in stars to continue for billions of years. This long timescale is crucial for forming various complex life forms on Earth. Current studies have shown that in the next billions of years, sunlight will still be the most stable source of energy on Earth.

9.1.2 Planetary Energy

Interstellar dust particles can accumulate over millions of years, resulting in the formation of celestial objects such as asteroids, moons, or planets. The habitability of these objects relies on the occurrence of quantum tunneling effects, which can be observed on a scale ranging from several hundred to a few thousand kilometers.

When compared to Saturn's other moon, Mimas, Enceladus is positioned farther away from Saturn and possesses a lower orbital eccentricity. These factors suggest that it should experience lower tidal heating and gravitational energy levels. However, contrary to these expectations, Enceladus exhibits surprisingly intense geological activity and remarkably high internal energy [Kargel2006]. This unexpected behavior can be attributed to the fact that Enceladus contains a higher proportion of rocks in its composition than does Mimas. These rocks are rich in radioisotopes such as ^{235}U, ^{238}U, and ^{232}Th, which undergo decay processes and generate the energy responsible for the observed geological activity.

For heavy isotopes of uranium (U) and thorium (Th), radioactive alpha decay is an exothermic reaction that can be described as

$$ {}^{A}_{Z}X \rightarrow {}^{A-4}_{Z-2}Y + {}^{4}_{2}He, \tag{9.5} $$

where X is the parent isotope and Y is the daughter isotope. The final energy released by the reaction is then given by

$$ E = \left(M_X - M_Y - M_{He}\right)c^2. \tag{9.6} $$

This radiative heating provides sufficient energy to make Enceladus a potentially habitable world. However, the kinetic energy of the released He^{2+} alone is insufficient to overcome the energy barrier between the initial and final states. Like thermonuclear reactions, this process can only occur through quantum tunneling.

One of Saturn's other moons, Titan, is relatively Earth-like. It has an atmosphere composed mainly of nitrogen and experiences periodic precipitation, which are all similar to Earth. In fact, a methane cycle exists on Titan as well. Under the influence of intense radiation and Saturn's strong magnetic field, methane molecules dissociate and react to form organic hydrocarbons such as acetylene, ethane, and ethylene. These processes inevitably involve full quantum dynamics, especially the nonadiabatic chemical reactions. The resulting products can further combine with nitrogen-containing hydrocarbons to form even more complex hydrocarbons. These prebiotic aerosols are then deposited in the lower atmosphere, eventually reaching Titan's surface in the form of precipitation and becoming nitrided aromatic hydrocarbons like nucleotides and amino acids.

However, if the described reaction pathway were the only mechanism, the methane on the surface would be depleted relatively quickly, which contradicts the actual age of Titan. Therefore, there appears to be an additional pathway that replenishes methane from the interior. Current observations suggest that the main source of methane is the planet's oceans. Although methane has a low solubility in water, Titan has vast surface and subsurface oceans, providing large reserves of methane. The presence of liquid water on Titan is sustained by the energy released from isotopic decay. This indicates that full quantum effects also play a significant role in the prebiotic chemistry of Titan [Hörst2012].

Returning to our planet Earth, 50% to 70% of geothermal energy originates from isotopic decay. This energy creates the necessary environment for biological life beneath the ocean, where sunlight cannot reach. Therefore, some studies have suggested that the He^{2+} tunneling in alpha decay is responsible for the chemical evolution of deep-sea life.

On a more practical level, geothermal energy offers an alternative to the use of fossil fuels. In the case of Iceland, a country abundant in geothermal energy, 90% of households rely on it for heating. Iceland has demonstrated to the world that achieving 100% renewable energy generation is possible; nearly all of the country's electricity is derived from hydroelectric or geothermal power plants.

9.2 Hydrogen Energy

In Chapter 7, we discussed the influence of full quantum effects on the physical properties of hydrogen. However, we did not include practical applications of hydrogen in our previous discussion. Hydrogen is considered a promising candidate for new, clean-energy sources. Consequently, investigating the production, storage, and utilization of hydrogen energy has emerged as a crucial interdisciplinary frontier. In terms of fundamental science, comprehending hydrogen's production and storage necessitates a study of the interaction between hydrogen and various materials, and especially the adsorption and diffusion behaviors of hydrogen atoms on material surfaces.

9.2.1 Adsorption and Diffusion of Hydrogen Atoms

9.2.1.1 Carbon Surface

Carbon is one of the most common elements in Earth's environment. The distribution of carbon elements includes carbon substances from natural organic molecules, graphite and diamond, and a series of low-dimensional artificial nano-materials, such as fullerene, nanotube and graphene, and so forth. In addition, hydrogen also exists in the natural environment and is the most common impurity on sample surface in laboratory vacuum chamber. Moreover, it frequently interacts with carbon; thus, carbon-based substrates are ideal for studying the adsorption and diffusion of hydrogen atoms on material surfaces. Given that both hydrogen and carbon are light elements, the influence of full quantum effects is expected to be substantial, particularly at low temperatures. Benzene rings and graphene are commonly employed as substrate materials to explore the adsorption and diffusion of hydrogen atoms on carbon surfaces.

In 2009, C. P. Herrero et al. calculated the adsorption and diffusion of H and D atoms on graphene monolayers using path-integral molecular dynamics based on a tight-binding Hamiltonian [Herrero2009]. As shown in Figure 9.1(a), the vibrations of the adsorbed atoms displayed distinct anharmonicity because of nuclear quantum effects. At low temperatures, the vibrational energies of H and D are lower than the results predicted by the harmonic approximation because of zero-point vibrations, whereas the opposite holds true at high temperatures. In the entire temperature range considered here, the ratio between the vibrational energies of H and D is smaller than the result predicted by the harmonic approximation. After separating the potential contribution of the vibrational energy, its temperature dependence deviates significantly from the results obtained under harmonic approximation. This indicates that the hydrogen atom adsorbed on graphene cannot be treated as a classical particle moving in a harmonic potential field.

The diffusion of H atoms on graphene involves the breaking and recombination of C–H bonds. As a result, the interaction between H and the nearest C atoms needs to be taken

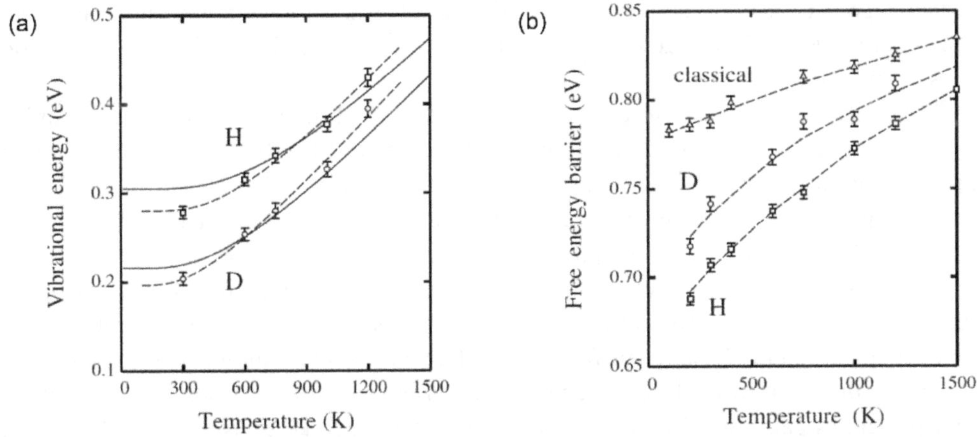

Figure 9.1 (a) Vibrational energy of hydrogen atoms adsorbed on graphene as a function of temperature. Squares and circles represent the results of path-integral molecular dynamics simulations based on the tight-binding model for H and D, respectively. The dashed curves are the reference lines corresponding to the data points for H and D. The solid curves are the vibrational energies obtained using the harmonic approximation for H and D atoms. (b) Effective free-energy barrier for H and D diffusion on graphene as a function of temperature. The top dashed curve and triangles are the results of the classical approximation. Reprinted figure with permission from Herrero[2009]. Copyright 2024 by the American Physical Society.

into consideration. Nuclear quantum effects, especially zero-point vibrations, lead to the renormalization of the diffusion barrier. The zero-point vibrations of C atoms can also enhance phonon-assisted quantum tunneling. The free energy barriers for H and D atoms diffusing on graphene were calculated by Herrero et al. using quantum transition-state theory. As shown in Figure 9.1(b), the diffusion-free energy barrier of the D atom was higher than that of the H atom because of its heavier mass. For the entire temperature range, the calculated quantum diffusion barriers for D and H were smaller than the classical results. At 300 K, the barriers were found to be 0.74 and 0.71 eV for D and H atoms, respectively, whereas the classical limit was 0.79 eV. In other words, nuclear quantum effects renormalized the diffusion free-energy barrier by ~6% and 10% for D and H atoms at room temperature, respectively. Accordingly, the diffusion rate of H atoms was 11.6 s^{-1} at room temperature, ~20 times larger than the value found in the classical calculation. As the temperature increases, the difference between the D and H diffusion free energy potentials and the classical limit decreases, indicating that the nuclear quantum effects are gradually weakened. In other words, the "classicality" of the nucleus increases at high temperatures.

The zero-point energy can further affect the adsorption sites of the atoms. K. Suzuki et al. combined a semiclassical force field model with the path integral molecular dynamics method to simulate the adsorption of H and D atoms on zeolite-templated carbon (ZTC) structures [Suzuki2011]. They found that D atoms exhibited stability on five adsorption sites, whereas H atoms displayed stability on only four sites. The absence of stability for H atoms at the fifth adsorption site was attributed to the impact of full quantum effects, causing spontaneous diffusion to other positions. Through the harmonic approximation, Suzuki et al. determined that zero-point vibrations influenced the stability of adsorption sites for the isotope atoms.

Although both studies described earlier employed path integral molecular dynamics methods and considered nuclear quantum effects, they still relied on the Hamiltonian approximation under the tight-binding or force field models. These studies could only provide qualitative explanations because they did not use the exact potential energy surface calculated by first-principles calculation. Recent advances in first-principles electronic structure calculations allow more precise descriptions of the nuclear quantum effects of atoms adsorbed on material surfaces based on more accurate Born–Oppenheimer potential energy surface. E. R. M. Davidson et al. studied nuclear quantum effects on the adsorption of hydrogen atoms on graphene and coronene surfaces using a first-principles path integral molecular dynamics approach [Davidson2014]. As shown in the inset of Figure 9.2, the hydrogen atom experiences significantly different nuclear quantum effects at different adsorption heights from a surface. Near the surface, the hydrogen atom is in a stable chemisorbed state, corresponding to a localized distribution of path integral replicas. As the hydrogen atom gradually moves away from the surface, its distribution along the direction normal to the surface rapidly broadens. When this occurs, the hydrogen atom gradually approaches a new transition state. On the other side of this transition state is a very shallow valley where the hydrogen atom can undergo physisorption. As the hydrogen atom moves further away from the surface, the adsorption energy gradually decreases, and the distribution of hydrogen atom replicas increasingly resembles that of a free hydrogen atom.

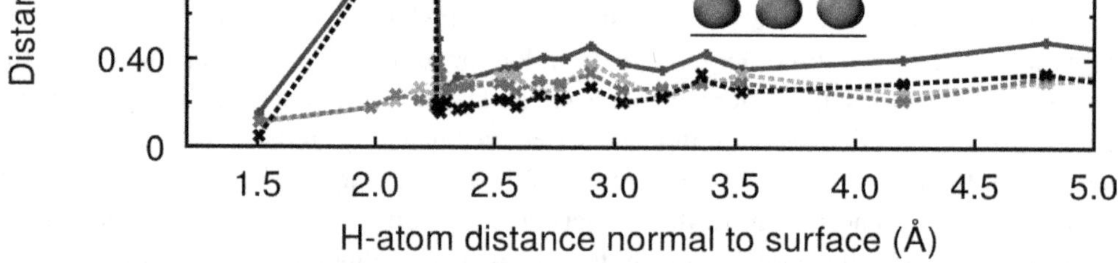

Figure 9.2 Nuclear quantum fluctuation of hydrogen as a function of hydrogen atom distance from the graphene surface. The solid lines denote the size of the distribution described by the different replicas of the path integral. The dashed lines are the three Cartesian components of the path-integral replica distribution. Reprinted with permission from Davidson[2014]. Copyright 2024 American Chemical Society.

On the basis of these studies, Davidson et al. utilized the instanton algorithm to compute the quantum tunneling rate through this newly discovered transition state. The results suggest that quantum tunneling increases the adsorption rate of hydrogen atoms by several orders of magnitude. In addition, Davidson et al. conducted an in-depth investigation of the influence of the electronic potential on nuclear quantum effects. Initially, they observed significant variations in the nuclear quantum effects for different potential energy surfaces derived from density functional theory. Notably, the van der Waals force emerged as a crucial factor in hydrogen adsorption. Neglecting the van der Waals force would eliminate the physisorption state on the potential energy surface; no quantum tunneling can be observed from the physisorption state to the chemisorption state. Their work once again illustrates the importance of using an accurate electronic structure calculation for the study of full quantum effects. We will encounter a similar situation later when we discuss the adsorption of water molecules on material surfaces.

9.2.1.2 Metal Surfaces

The adsorption and diffusion of hydrogen atoms on metal (in addition to carbon) surfaces are also important research topics. They directly relate to hydrogen embrittlement, corrosion, and other challenges encountered in hydrogen storage and transportation and thus significantly influence the use of hydrogen energy and environmental protection. Many notable experimental studies have been conducted to investigate the adsorption of hydrogen atoms on metal surfaces. For instance, real-space imaging and scattering techniques have been employed to measure the diffusion rates and temperature dependence of hydrogen and deuterium atoms, allowing for the determination of nuclear quantum effects through isotope substitution.

In the early 1990s, T. S. Lin and R. Gomer used field emission microscopy to measure the diffusion rates of hydrogen atoms and their isotopes on metal surfaces, such as tungsten and nickel [Lin T.-S.1991]. Their measurements revealed that when the temperature falls below 130 K, the behavior of hydrogen atoms deviates from the classical Arrhenius relation which describes the temperature dependence of the diffusion rate. From 30 K to 100 K, the hydrogen diffusion rate remains almost constant with temperature, indicating that at 100 K, hydrogen atoms enter a diffusion regime dominated by quantum tunneling. In 1992, X.D. Zhu et al. conducted similar measurements on the Ni(100) surface using linear optical diffraction techniques, and their results were comparable to those of Lin et al. [Zhu X.D.1992]. The experimental results of X.D. Zhu et al. are depicted in Figure 9.3.

Another effective experimental method for studying the diffusion of hydrogen atoms is the quasi-elastic helium atom scattering technique. For instance, in 1999, A. P. Graham et al. reported the measurement of hydrogen and deuterium diffusion on the Pt(111) surface using this method [Graham1999]. Since they conducted their measurements in a relatively high temperature range (140–250 K), no deviation from the Arrhenius relation was observed. However, based on the Arrhenius plot, they concluded that the activation energy of H in this temperature range is ~10% lower than that of D. Under these circumstances, full quantum effects are mainly reflected in the differences in zero-point energy and vibrational excitation energy caused by the different masses of the two isotopes. From

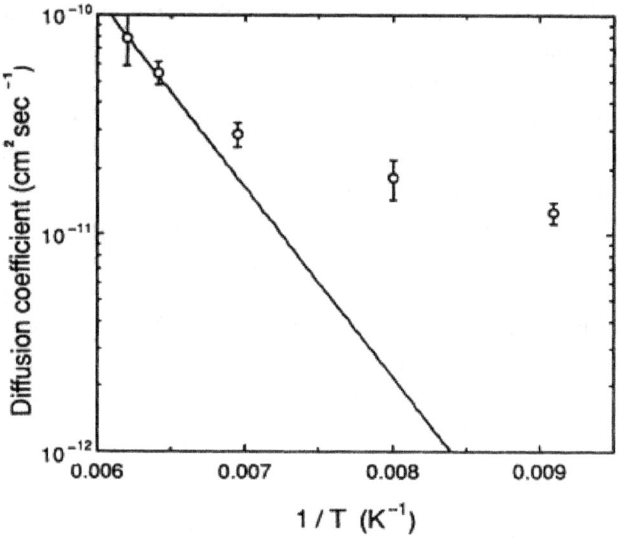

Figure 9.3 Temperature dependence of hydrogen-atom diffusion rate. The data points correspond to the experimentally measured values. The solid line represents the Arrhenius relation obtained from the first two data points. Reprinted figure with permission from Zhu X. D.[1992]. Copyright 2024 by the American Physical Society.

the measurements, it can be inferred that even at room temperature, the zero-point energy and quantum vibrations induced by full quantum effects can still contribute to hydrogen diffusion. However, quantum tunneling only plays a major role under extremely low temperature conditions.

In 2000, L. J. Lauhon et al. pioneered the use of scanning tunneling microscopy to study the diffusion of hydrogen on surfaces [Lauhon2000]. They measured the diffusion rate of hydrogen and deuterium atoms on the Cu(001) surface. In the relatively high temperature range of 65–80 K, the fast diffusion rate made it possible to achieve single-atom tracking using a scanning probe. In the low-temperature region (9–63 K), the diffusion rate of the atoms slowed down, and the experiment was conducted using repeated imaging. The results clearly showed that quantum tunneling rapidly begins to dominate the diffusion process as the temperature decreases to 60 K. This is reflected in the fact that the H-atom diffusion rate no longer decreases with temperature (Figure 1.27(b)). By contrast, for D atoms, the diffusion rate already approaches the lowest measurement limit of the experimental setup at 65 K. At this temperature, the D-atom diffusion is still governed by classical diffusion, which explains why D-atom tunneling was not observed in their experiments.

These experiments mainly focused on the measurement of the diffusion process. Full quantum effects were probed by comparing the experimental measurements with the classical Arrhenius relation, and simplified models were used to provide some phenomenological explanations. Consequently, some phenomena cannot be explained by simple physical models in these experiments. For example, R. Gomer et al. reported observing hydrogen-atom isotope effects smaller than those predicted theoretically [Lin T.S.1991]. Additionally, the temperature dependence of hydrogen diffusion rates varied significantly

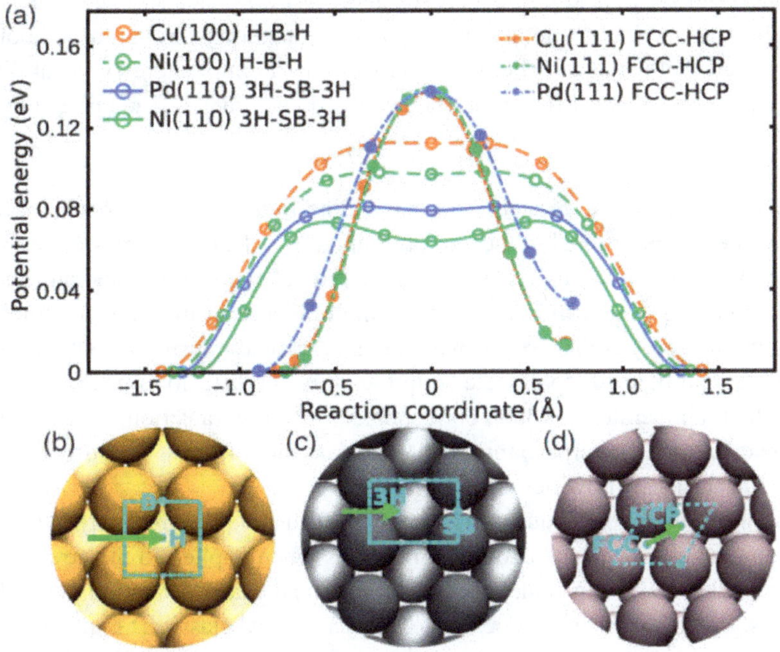

Figure 9.4 Diffusion barriers and reaction pathways of hydrogen atoms on metal surfaces: (a) diffusion barriers on different surfaces; (b)–(d) the (100), (110), and (111) surfaces and their special adsorption sites. Reprinted figure with permission from Fang W.[2017]. Copyright 2024 by the American Physical Society.

from one experiment to another. In some cases, the transition from classical to quantum diffusion occurred rapidly, while in others, it took place gradually over an extended temperature range. It is worth noting that the experimental measurement of hydrogen atom diffusion poses significant challenges, which can result in quantitative differences between different experimental measurements.

In response to these conflicting experimental results, G. Wahnström's group conducted several targeted theoretical studies in the 1990s [Mattsson1993, 1995, 1997, Sundell2004]. Recently, W. Fang et al. applied more advanced instanton methods to systematically study nuclear quantum effects in these systems [Fang W.2017]. We will use their work as an example to discuss hydrogen-atom diffusion on metal surfaces in detail.

Through extensive calculations, they found that the diffusive behavior of hydrogen atoms on a metal surface is mainly determined by two types of potential barriers. As shown in Figure 9.4, these barriers can be classified as either parabolic-top or broad-top. Parabolic-top barriers are predominantly observed on the (111) surface of face-centered cubic metals (Figure 9.4(d)), while broad-top barriers are more commonly found on the (100) (Figure 9.4(b)) or (110) (Figure 9.4(c)) surfaces of metals. The occurrence of broad-top barriers on the latter two surfaces is attributed to the larger spacing between adsorption sites. The hydrogen adsorption energy on the metal surfaces is relatively insensitive to changes in its in-plane position, eventually leading to a plateaued barrier curve instead of a parabolic one. The phenomenon of quantum tunneling through a parabolic barrier has been extensively studied. The transition from high-temperature classical diffusion to low-temperature quantum diffusion is characterized by a transition temperature given by

$$T_c = \hbar\omega_b / 2\pi k_B, \tag{9.7}$$

where ω_b is the imaginary frequency corresponding to the top of the potential barrier; its absolute value is proportional to the curvature there. As the curvature becomes larger, the barrier becomes narrower, indicating a higher critical transition temperature from classical diffusion to quantum diffusion. Below the critical transition temperature, quantum tunneling dominates. In other words, with a higher critical transition temperature, the true diffusion rate can be made to deviate from the classical diffusion rate and can plateau at relatively high temperatures, which corresponds to a slow transition from classical diffusion to the quantum counterpart; this is analogous to the diffusion of hydrogen atoms on a graphene surface in the previous example.

The simple description of quantum tunneling diffusion presented earlier is only applicable to parabolic-top barriers and has limitations when applied to the newly discovered broad-top barriers. W. Fang et al. conducted theoretical analyses and found that for broad-top barriers, hydrogen diffusion does not exhibit a gradual classical-to-quantum transition as temperature decreases. Instead, the classical-to-quantum transition of hydrogen diffusion is sharp.

Figure 9.5(a) compares the two types of potential barriers. For the parabolic-top barrier, the quantum diffusion rate gradually deviates from the classical diffusion rate at the very beginning as the temperature decreases. By contrast, for the broad-top barrier, the quantum diffusion rate remains in good agreement with the classical rate until it reaches the transition temperature. When the temperature falls below the transition temperature, the quantum diffusion rate rapidly deviates from the classical rate and plateaus. In real experiments, a wide variety of surfaces may exist, allowing for the coexistence of both parabolic and broad-top barriers. A comprehensive understanding of quantum tunneling through both parabolic and broad-top barriers is essential for accurately interpreting the full quantum effects of atomic diffusion on metal surfaces at low temperatures.

9.2.2 Production and Diffusion of Hydrogen Gas

In Section 9.2.1, we covered the adsorption and diffusion of hydrogen atoms. In this section, we discuss the production and diffusion of hydrogen molecules (in the gas state). In general, because of the relatively high activation energy, hydrogen gas is not produced by direct collision of two hydrogen atoms; rather, the production of hydrogen gas typically involves the use of catalysts. Therefore, as described in the previous subsection, it is important to study hydrogen's interactions with various materials, particularly its adsorption on graphene, aromatic hydrocarbons, and metal surfaces. Basic research findings in this area have significant implications for practical fields such as controlled nuclear fusion, hydrogen storage, petroleum refining, and hydrogen gas production in space.

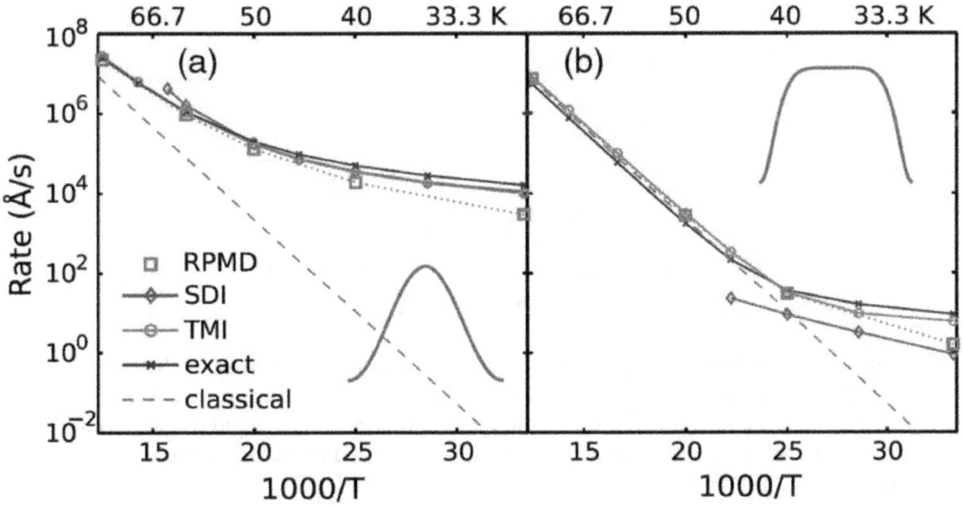

Figure 9.5 Diffusion rates of hydrogen atoms as a function of temperature. (a) Rates on a parabolic-top barrier. (b) Rates on a broad-top barrier. RPMD (Ring-Polymer Molecular Dynamics), TMI (Thermalized Microcanonical Instanton), and SDI (Steepest-Descent approximation to TMI rate) are the results calculated by three different path-integral methods. The black solid line denotes the analytical result, and the dashed line represents the classical diffusion rate. Reprinted figure with permission from Fang W.[2017]. Copyright 2024 by the American Physical Society.

Although industrial hydrogenation of benzene occurs at high temperatures where the quantum tunneling effect is weak, this is not typical of natural hydrogen gas production. In most hydrogen-gas-rich astrophysical environments, temperatures are low. Molecular clouds, for instance, have a temperature of ~10 K, while diffuse nebulae have a temperature of around 100 K, and the photon-dominated region reaches several hundred kelvins. Significant amounts of hydrogen gas have been observed in these environments despite the apparent absence of ideal conditions for its production and the destructive effect of cosmic rays and photons. This phenomenon can be explained by the hypothesis that the adsorption of hydrogen atoms on carbon-based dust is relatively easier in the temperature range above. We can understand interstellar hydrogen enrichment by further considering the dependence of quantum tunneling probability on both the barrier height and width.

To deal with polyatomic systems and asymmetric potential barriers, T. P. M. Goumans and J. Kastner calculated the rate of hydrogen gas formation based on harmonic quantum transition-state theory, also known as instanton theory. They adopted a quantum statistical approach involving discrete closed Feynman path integrals [Goumans2010] to calculate the rate of the hydrogen formation process. In detail, they calculated all possible tunneling paths for hydrogen adsorbed on benzene surfaces at low temperatures, as well as the hydrogen and deuterium quantum tunneling probabilities along these reaction paths, providing a basic model for studying hydrogen adsorption on carbon-based dust (Figure 9.6). Their study demonstrated that harmonic quantum transition-state theory can accurately calculate the quantum tunneling rate during reaction. When applied to the case of hydrogen

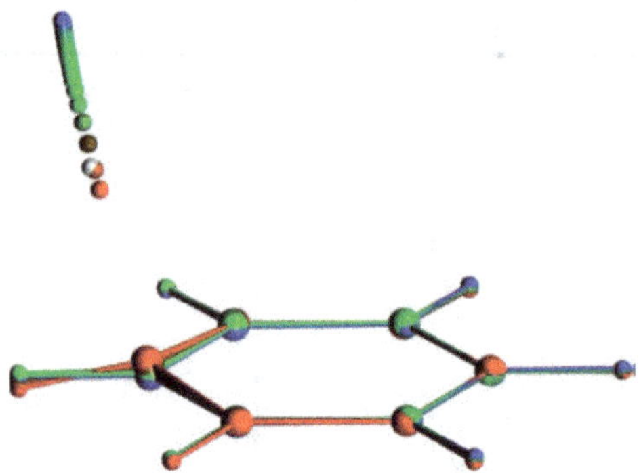

Figure 9.6 The quantum transition state of hydrogen atoms adsorbed on benzene. The white spheres are hydrogen atoms in the classical transition state. The green and red spheres represent the positions of carbon atoms in the shallow and deep potential energy minima on either side of the asymmetric potential barrier. From Goumans[2010]. Reproduced with permission from John Wiley and Sons.

adsorption on benzene systems, the theory predicts that hydrogen atoms are chemisorbed on the edge of benzene in the 100–200 K temperature range; this is the key condition for the next step of hydrogen gas generation. Previous studies have found that after the adsorption of a single hydrogen atom, the adsorption potential for a second hydrogen atom is almost zero. This implies that for hydrogen gas formation, the adsorption of the first hydrogen atom is critical.

From the previous discussion, we know that the adsorption of the first hydrogen atom is crucial to the formation of hydrogen gas. In fact, the quantum motions of the carbon atoms can also quantitatively affect the tunneling result of the hydrogen atoms. After considering the nuclear quantum effects of the entire system (including hydrogen and carbon atoms), T. P. M. Goumans et al. performed more accurate tunneling rate calculations using a simple model consisting only of a hydrogen atom and a benzene ring. They combined first-principles calculations with multidimensional quantum transition state theory to derive the tunneling rate as a function of temperature; the results are presented in Figure 9.7 [Goumans2010]. The classical rates follow the typical Arrhenius relation, indicated by a linear decrease of the logarithm of the rate with the reciprocal of temperature in Figure 9.7. The curves marked by solid symbols represent the adsorption of the hydrogen atom due to quantum tunneling. When the temperature is high, their values are consistent with the classical rates. However, the quantum reaction rates deviate more significantly from the classical results as the temperature decreases. In fact, the logarithm of the quantum tunneling rate converges to a constant value below a specific temperature. In this so-called deep tunneling region, the reaction rate no longer varies with temperature, and the thermal fluctuation is negligible.

Recently, J. Chen's group at Peking University combined the first-principles instanton method with kinetic Monte Carlo simulations and studied the formation of hydrogen gas on graphene using full quantum methods [Han E.X.2022]. The quantum tunneling rates obtained by the first-principles instanton approach are shown in Figure 9.8(a). In contrast

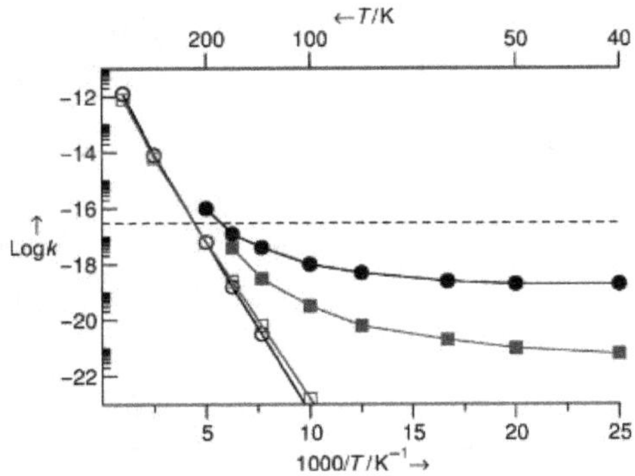

Figure 9.7 Temperature dependence of the quantum tunneling rate k of hydrogen adsorption on benzene. The curves marked with solid symbols are the results obtained from quantum transition-state theory, which considers nuclear quantum effects. The hollow symbols denote the classical reaction rates. From Goumans[2010]. Reproduced with permission from John Wiley and Sons.

to the classical result, the adsorption rate of the hydrogen atoms exceeds the desorption rate at low temperatures (i.e., in the deep tunneling region below 167 K) after quantum tunneling has been taken into consideration. Further analysis of the hydrogen recombination rate reveals that quantum tunneling can enhance ortho-recombination by tens of orders of magnitude at low temperatures. Classically, the ortho-recombination of hydrogen atoms was considered impossible because of the high reaction barrier. However, after including nuclear quantum tunneling in the calculation, ortho-recombination of hydrogen atoms can occur even at low temperatures, competing with meta- and para-recombination.

The first-principles instanton approach searches for the optimal reaction path of the system on a high-dimensional potential energy surface. This allows more detailed information to be extracted from the reaction process. E. X. Han et al. performed an in-depth analysis of the high-dimensional instanton pathways. Figure 9.8(a) indicates significant corner-cutting effects during hydrogen recombination, where the quantum tunneling pathway deviates from the classical pathway. This effect is more prominent in the ortho pathway, suggesting that the recombination of hydrogen atoms does not completely follow the pathway with the lowest energy. Nuclear quantum effects drive the reaction along the pathway of least action, even if the reaction energy is slightly higher.

In addition, E. X. Han et al. found a synchronous–asynchronous reaction mechanism for the ortho-recombination process. The classical ortho pathway involves asynchronous motion of the hydrogen atoms; in the quantum scenario, their motion is synchronous (see Figure 9.8(b) and (c)). This quantum–classical symmetry breaking is related to the complex potential energy surface of the hydrogen-atom adsorption system; it can also be attributed to the classical transition state having two imaginary frequencies corresponding to the C_2 symmetry, whereas the quantum transition state only has one. However, the underlying mechanism of this quantum-protected symmetry is not yet fully understood and requires further investigation.

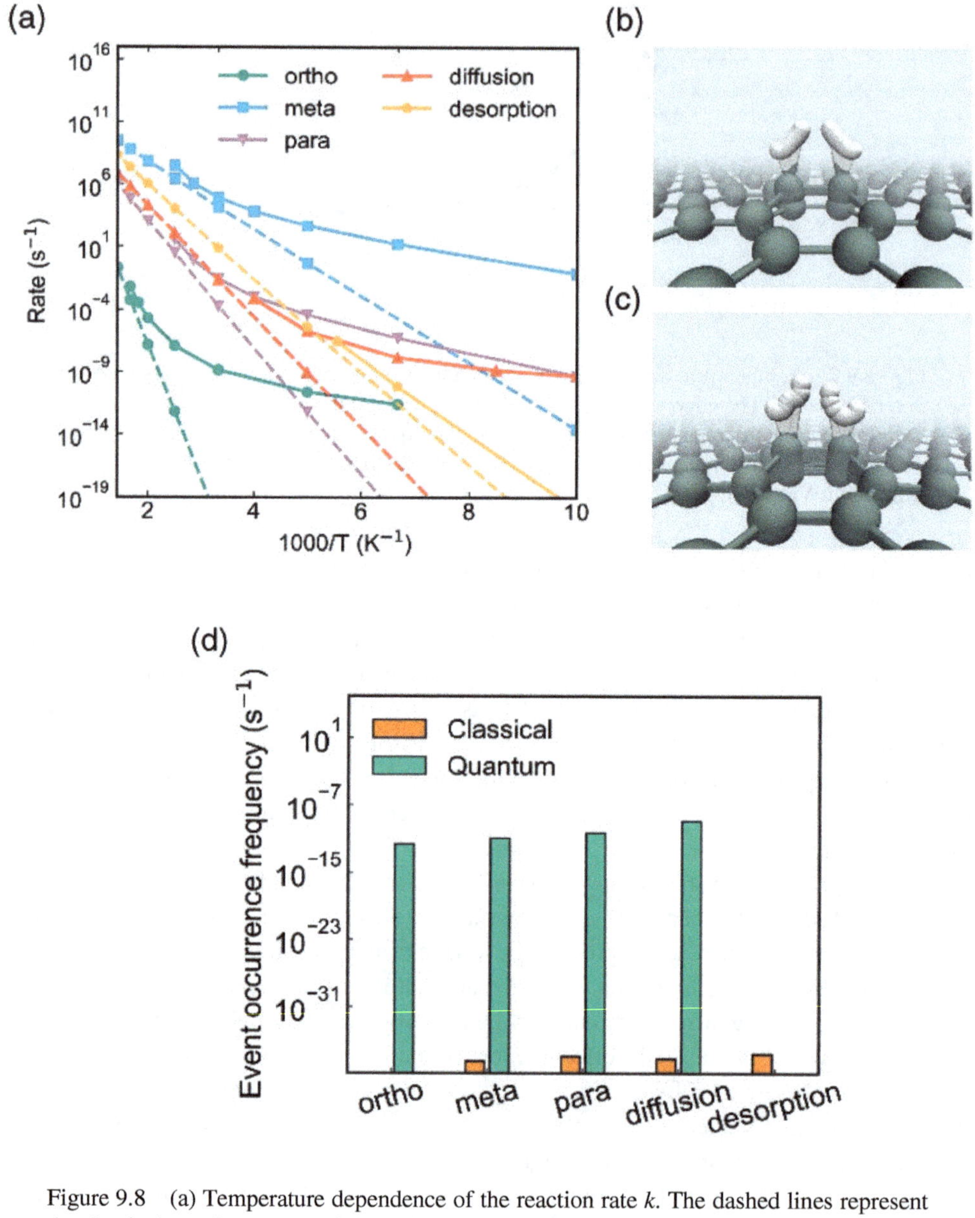

Figure 9.8 (a) Temperature dependence of the reaction rate *k*. The dashed lines represent the classical reaction rates, and the solid curves indicate the quantum reaction rates. (b) The quantum tunneling pathway and (c) the classical reaction pathway for the ortho-recombination of hydrogen atoms (white spheres) at 150 K. (d) Kinetic Monte Carlo simulation of the frequency of desorption, diffusion, and various types of recombination at 100 K. Reprinted with permission from Han E. X.[2022]. Copyright 2024 American Chemical Society.

Figure 9.8(d) depicts the stepwise reaction results calculated by kinetic Monte Carlo simulations at 100 K. These results corroborate the conclusion that "quantum tunneling substantially enhances the reaction rate." Moreover, the desorption of hydrogen atoms is suppressed, contradicting the classical prediction that hydrogen-atom desorption dominates the reaction. These studies clearly demonstrate that full quantum effects influence the entire hydrogen production process, from single-atom diffusion to double-atom recombination.

Because hydrogen gas molecules are chemically stable, their diffusion has received more attention than that of individual hydrogen atoms. A vital question is whether full quantum effects can influence the motion of hydrogen molecules.

H. Nagashima et al. examined the quantum characteristics of hydrogen molecular motion using the centroid molecular dynamics (CMD) method. The motion of hydrogen molecules was simulated over a large temperature range, and the diffusion coefficient was calculated using the Green–Kubo method [Nagashima2017]. The results obtained from quantum calculations were compared with those from classical calculations based on the principle of corresponding states. As shown by the thermodynamic saturation lines of hydrogen gas in Figure 9.9, the predictions, which take into account nuclear quantum effects agree well with the experiment. By contrast, the results obtained from classical molecular dynamics significantly deviate from experimental values. Nevertheless, the difference between the diffusion coefficients calculated by the two methods is small, and the difference in the reduced diffusion coefficients is less than 0.005 (see Figure 9.10).

However, further analysis points out that it does not mean nuclear quantum effects can be ignored at any temperature when discussing hydrogen molecule diffusion. In fact, as shown in Figures 9.9 and 9.10, the full quantum effect changes the action potential between hydrogen molecules: The repulsive region of the action expands while the depth of the potential well decreases. While both effects can influence the diffusion coefficient, they offset each other. As a result, the difference between the diffusion coefficients of quantum and classical simulations is minimal. It is evident that the manifestation of full quantum effects

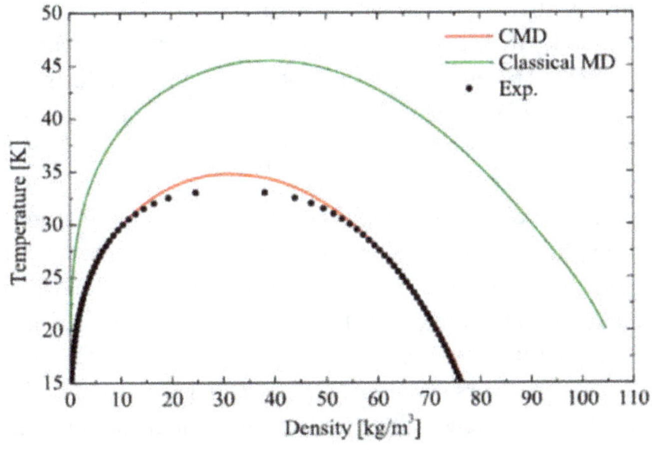

Figure 9.9 Thermodynamic saturation curves for hydrogen gas. The green curve (top line) is the classical molecular dynamics (MD) result, and the red curve (middle line) is the centroid path-integral molecular dynamics (CMD) result. The black dots denote the experimental values. Reprinted from Nagashima[2017], with the permission of AIP Publishing.

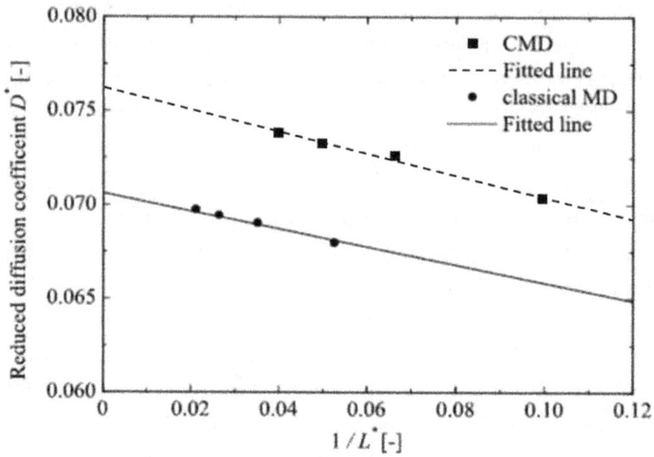

Figure 9.10 Reduced diffusion coefficients of hydrogen obtained by quantum CMD and classical MD methods, respectively. L is the size of the computational model. Note that the size of the computational model has little effect on the difference between the two diffusion coefficients. Reprinted from Nagashima[2017], with the permission of AIP Publishing.

can sometimes be subtle. Without realizing this, our understanding of certain physical phenomena may deviate from the true essence of the problem, even leading to erroneous interpretations.

9.2.3 Dense Hydrogen

Hydrogen, particularly "dense hydrogen" (high-density hydrogen obtained under high pressure), can be used as a propellant for aircraft, rockets, and spacecraft. Therefore, hydrogen is an important energy fuel that is widely used in the field of aerospace. Under high-pressure conditions, hydrogen can exhibit a wide range of quantum behavior due to full quantum effects. Therefore, the study of full quantum effects is crucial for better utilization of hydrogen energy in the aerospace industry. (For information on hydrogen phase transitions and physical properties under high pressures, please refer to the discussion in Chapter 7).

Dense hydrogen is a special form of hydrogen produced under high-pressure conditions. The dynamical properties of dense-hydrogen molecules depend on large-angle scattering and many-body collisions among particles. These properties are fundamental in studying stellar evolution and inertial confinement fusion. Recently, D. D. Kang, J. Y. Dai, J. M. Yuan, and others studied the transport behavior of dense hydrogen using first-principles path integral molecular dynamics simulations, taking nuclear quantum effects into account [Kang D.D.2014]. They concluded that although nuclear quantum effects have only a small impact on the static structure of dense hydrogen, they cause the diffusion coefficient to increase by 20% to 146% at 10 g/cm^3 density in the 3,500–12,000 K temperature range (see Figure 9.11). They also lower the electrical and thermal conductivity of dense hydrogen, as shown in Figure 9.12. The main reason for these changes is the influence of

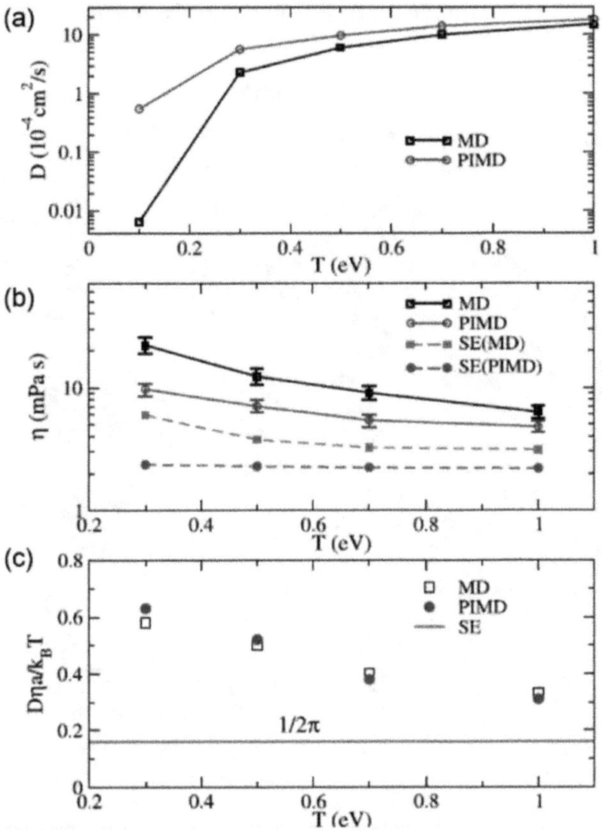

Figure 9.11 Self-diffusion coefficients (a) and shear viscosity (b) of dense hydrogen obtained with path-integral molecular dynamics simulations (PIMD) and classical molecular dynamics simulations (MD). (c) The viscosity obtained from PIMD and MD as a function of temperature versus that obtained from the Stokes–Einstein relation (SE). From Kang D.D.[2014]. Reproduced with permission from SNCSC.

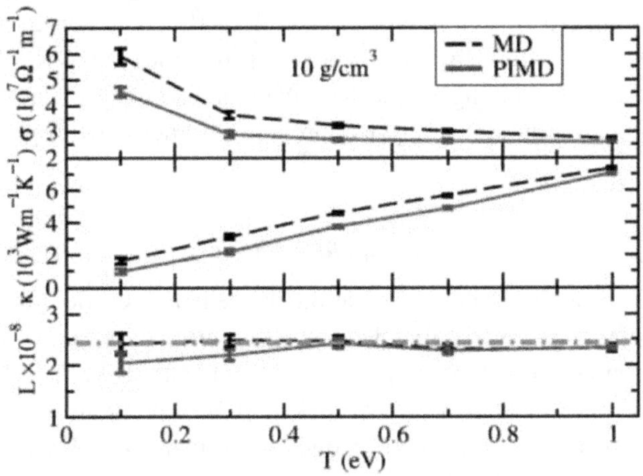

Figure 9.12 Electrical conductivity σ, thermal conductivity κ, and Lorenz number L of dense hydrogen calculated by path-integral molecular-dynamics simulations (PIMD) and classical molecular-dynamics simulations (MD). From Kang D.D.[2014]. Reproduced with permission from SNCSC.

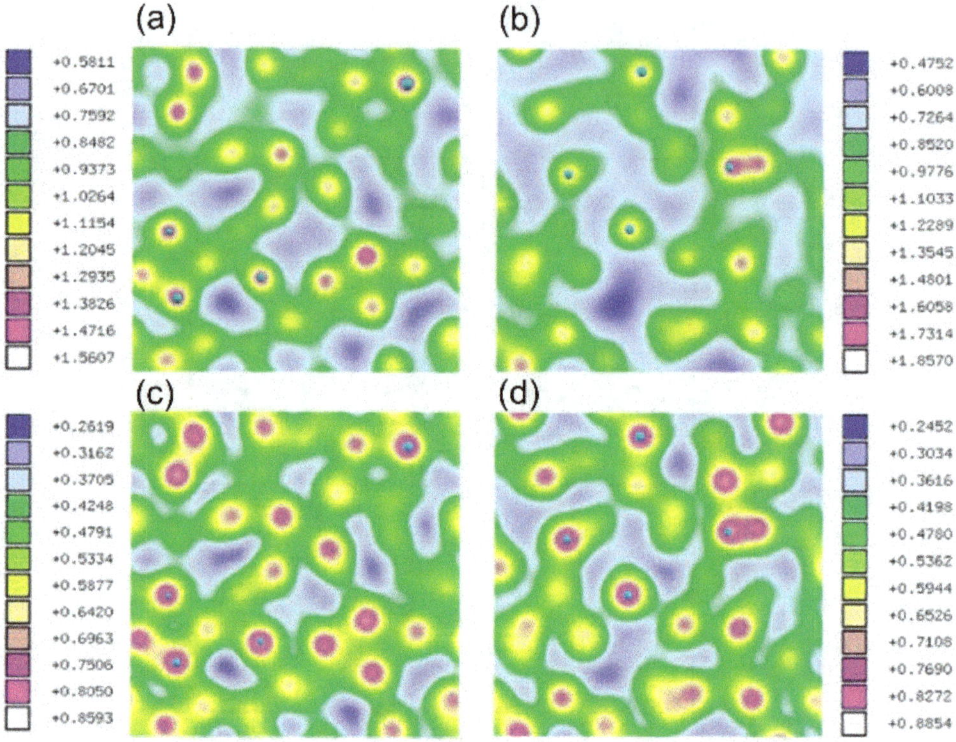

Figure 9.13 Density of electronic states on the cross sections of dense hydrogen: (a) and (b) charge densities; (c) and (d) charge-localization functions. Panels (a) and (c) show classical MD results, and (b) and (d) show path-integral MD results. From Kang D.D.[2014]. Reproduced with permission from SNCSC.

quantum fluctuations, as depicted in Figure 9.13, which modify the charge density distribution and scattering cross section. Nuclear quantum effects in dense hydrogen also lead to significant deviations from the Stokes–Einstein relation, the Wiedemann–Franz law, and isotope effects. These findings show that full quantum effects cannot be excluded from calculations when determining the physical properties of dense hydrogen.

9.2.4 Surface Catalysis

Catalytic processes such as water dissociation are generally carried out at high temperatures, while phenomena related to nuclear quantum effects, such as quantum tunneling, are often more pronounced at low temperatures. As a result, the influence of quantum tunneling on catalytic reactions has been overlooked in most previous discussions. However, recent studies have revealed that nuclear quantum effects can play a significant role at room temperature or higher. Furthermore, the temperature required for catalytic reactions has been substantially reduced with the development and application of new catalysts. Consequently, full quantum effects have the potential to introduce a new degree of control in catalytic reactions. They not only lower the required temperature but also enhance the efficiency of the catalyst.

In practice, it is also worth noting that nuclear quantum effects on surface chemistry reactions are mainly observed in elementary reactions occurring at or below room temperature. However, in industrial applications, most surface reactions occur at high temperatures and involve multiple intermediate steps. Determining whether and how full quantum effects are involved in these catalytic reactions poses a significant challenge for researchers.

As an elementary reaction, water splitting is critical to the study of multiphase catalysis, such as the photolysis and electrolysis of water in hydrogen production, as well as the water-gas shift reaction. The ability of water to decompose and produce hydrogen gas on a solid surface is a much-debated topic in both fundamental and applied energy research. It is generally accepted that water decomposes spontaneously on oxides and highly reactive metal surfaces but not on most precious metal surfaces. However, the existing understanding of water dissociation on solid surfaces is mainly based on the classical picture, with little consideration given to the zero-point motion and quantum tunneling of hydrogen and oxygen nuclei [Meng S.2002, 2004, Feibelman2002].

Through theoretical calculations, researchers at the Technion–Israel Institute of Technology have determined that the dissociation of H_2O and OH on the (111) surfaces of several metals (Pt, Cu, Ru, Rh) is subject to full quantum effects at high temperatures (>500 K). Researchers at the University of Liverpool observed significant isotope effects when studying the dissociation of water on the Ru(0001) surface at low temperatures (~150 K). H_2O molecules on Ru(0001) dissociate to form a mixed OH/H_2O/H hydrogen bond network, whereas D_2O molecules do not dissociate because of weak nuclear quantum effects. Using DFT calculations, Donadio et al. have demonstrated that water molecules adsorb preferentially at the step edge of the stepped Pt(221) surface, forming a one-dimensional hydrogen-bonded chain. After considering the zero-point energy (ZPE) correction, they predicted that the water molecules in this one-dimensional structure would partially dissociate, producing a more stable mixed hydroxyl/water hydrogen-bonded structure (Figure 9.14). Recently, Y. Jiang's group systematically investigated the hydrogen bonding network of the first water layer on the Pt(111) surface using high-resolution STM/AFM and X-ray photoelectron spectroscopy (XPS) [Tian Y.2022]. Their theoretical calculations showed that the hydrogen bonding network formed on the metal surface changed with the water coverage. Indeed, water dissociation is an incredibly complex phenomenon that merits further detailed study.

In addition, full quantum effects of hydrogen nuclei affect the dissociation of CH_4 on Pt(111) and Ni(111) surfaces. Quantum tunneling plays an important role in synthesizing NH_3; it accelerates the reaction rate of NH synthesis on the Ru(0001) surface at 340 K. Based on the previous discussions, it is evident that full quantum effects have an important influence on surface catalytic reactions such as water dissociation, hydrocarbon bond activation, and NH synthesis, and offer new ways to control them.

According to the widely used Bronsted–Evans–Polanyi (BEP) relation in the field of multiphase catalysis [Bronsted1928, Evans1938], the potential barrier between the initial and final states of a chemical reaction is directly related to the energy difference between them.

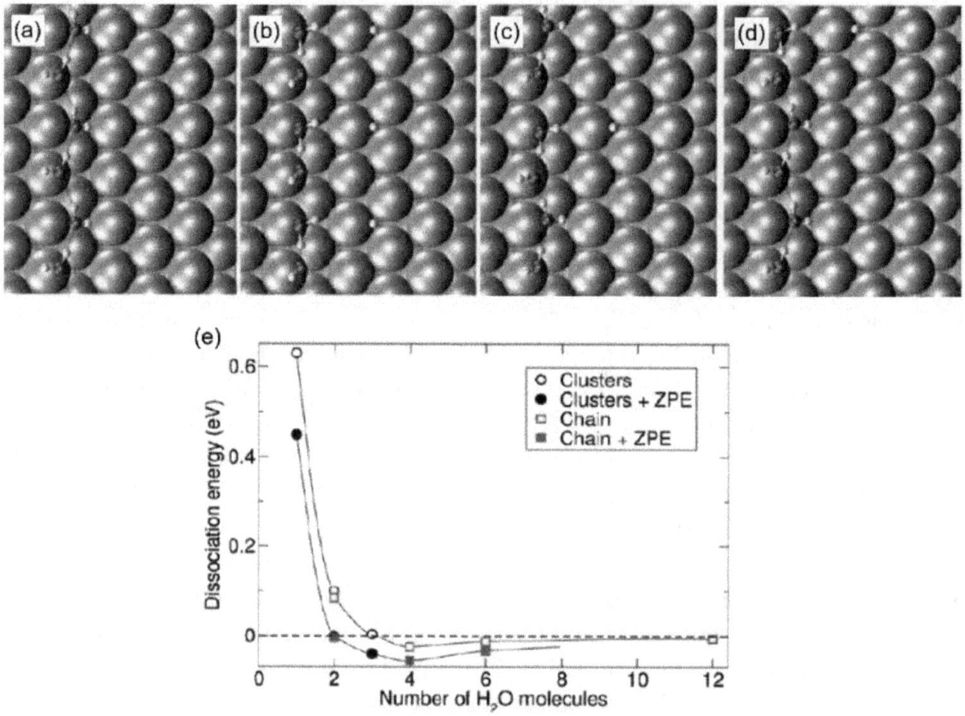

Figure 9.14 Partial dissociation of a 1D water-molecule chain on the Pt(221) surface due to nuclear quantum effects: (a) adsorption configuration of the water chain on the Pt(221) surface; (b)–(d) hydrogen bonding network of partially dissociated (H_2 O+OH); (e) dissociation energies after zero-point-energy (ZPE) correction for different water-molecule configurations on the Pt(221) surface. Reprinted with permission from Donadio[2012]. Copyright 2024 American Chemical Society.

A smaller energy difference corresponds to a larger potential barrier (see Figure 9.15(b)); conversely, a larger energy difference corresponds to a smaller potential barrier (see Figure 9.15(a)). As shown in Figure 9.15, the potential energy difference between the initial and final states can be adjusted by applying an overpotential to the system, allowing us to modify the potential barrier of the reaction. The process shown in Figure 9.15 (a) clearly has a smaller potential barrier than the one shown in (b). If the exact value of the potential barrier is known, the magnitude of the quantum effect contribution to the transmission probability can be easily derived. This result can be quantified through isotope substitution (assuming that the potential barrier remains constant for different isotopes).

K. Sakaushi et al. investigated the isotope effect in oxygen reduction on a Pt electrode by replacing the hydrogen in water with deuterium [Sakaushi2018]. During this reaction, OH is obtained by combining a proton with oxygen through proton transfer. Proton transfer requires crossing a certain potential barrier; calculating the probability of this process requires the use of both temperature-dependent classical transition state theory and temperature-independent quantum tunneling theory.

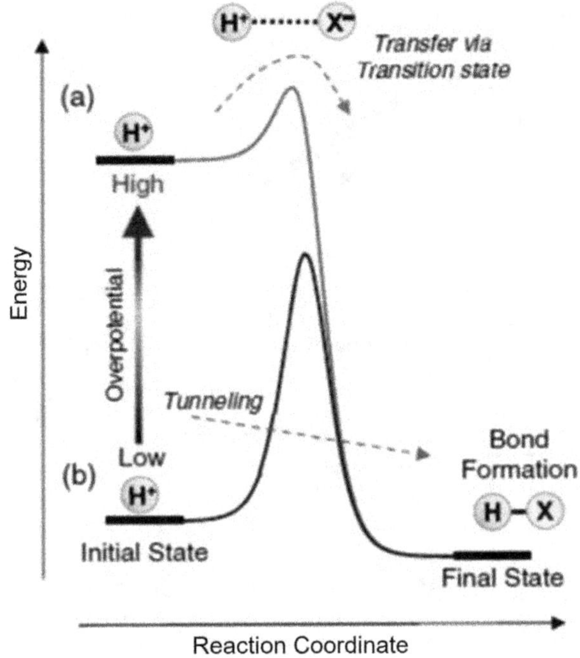

Figure 9.15 Schematic for two different paths of proton transport: (a) the classical transition-state path; and (b) the quantum tunneling path. In the electrochemical system, the contribution of the two mechanisms can be tuned by the applied potential. Reprinted figure with permission from Sakaushi[2018]. Copyright 2024 by the American Physical Society.

In experiments, Sakaushi et al. found that the isotope effect is small at high overpotential, indicating a small contribution from quantum tunneling. However, the isotope effect is more pronounced at low overpotential, indicating a significant contribution from quantum tunneling (as shown in Figure 9.16). This suggests that the proton migration is enhanced at a lower overpotential. As a result, the probability of proton transfer through the classical transition state is reduced, and quantum tunneling becomes more pronounced.

The conclusion drawn from this experiment can be generalized. Full quantum effects are most easily observed in electrochemical processes involving hydrogen, given its smaller atomic mass than heavier elements. It has also been suggested that quantum tunneling can be observed in electrochemical processes involving other atoms, such as oxygen, and even molecules, such as hydroxide. Although the effects are less pronounced because of the relatively larger masses involved, full quantum effects are expected to be present in all types of reactions. The probability of quantum tunneling depends not only on the mass of the tunneling atom or molecule but also on the width of the tunneling barrier. The height and the width of the potential barrier are the key parameters for tunneling. This implies that if we can experimentally modify the width of the potential barrier, we can change the reaction rate by controlling nuclear quantum effects. In summary, future chemical reaction and catalyst design will benefit from the incorporation of full quantum-mechanical principles.

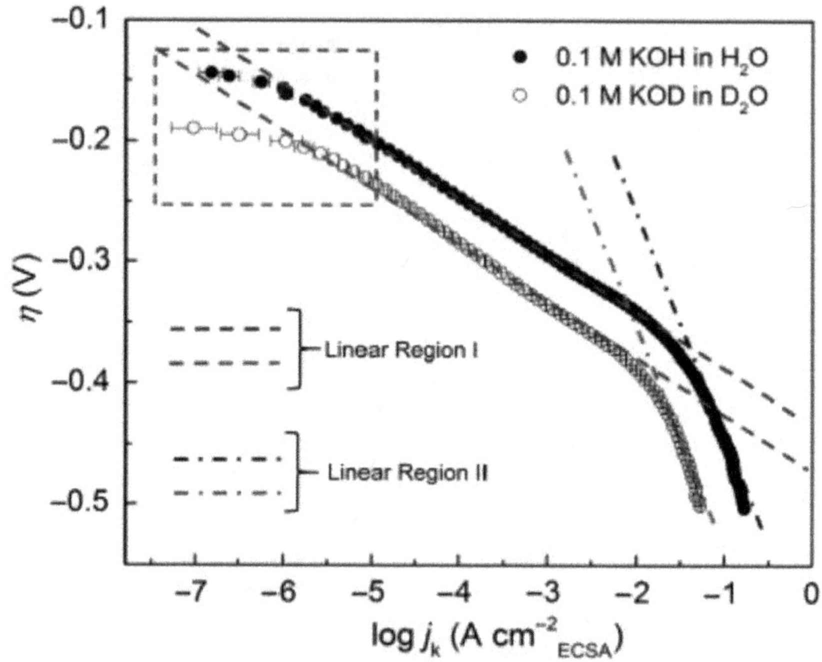

Figure 9.16 Change in overpotential η with current density j_k. Low η corresponds to the transmission region, where the contribution of nuclear quantum effects is more prominent. Reprinted figure with permission from Sakaushi[2018]. Copyright 2024 by the American Physical Society.

9.3 Solar Energy Conversion

Among the various types of new energy, solar energy has attracted widespread attention for its abundant reserves, clean and pollution-free characteristics, and relative lack of geographical restrictions. Solar cells directly convert solar energy into easily accessible electrical energy. High-efficiency and stable solar cells can support grid-connected power generation and provide electricity for small appliances, offering prospects for wide applications. Although silicon-based solar cells currently dominate the solar cell market because of their high conversion efficiency and stability, their complex fabrication process and high manufacturing cost still limit their widespread practical use. Reducing the manufacturing cost, improving the photoelectric conversion efficiency, and enhancing the stability of solar cells are the key factors for the large-scale application of solar cell technology. To address these practical challenges, improving the photovoltaic conversion efficiency of solar cells is of paramount importance. During photoelectric conversion, nonadiabatic effects can influence the charge transfer efficiency of solar cells. This was discussed in detail in Chapter 2.

However, there exists a theoretical upper limit to the efficiency of photovoltaic (PV) cells. According to the band theory of semiconductors, only a portion of the solar spectrum can be used by PV cells. This is because photons with energy lower than the bandgap of the solar cell material are not absorbed by the cell but rather reflected back into space. For photons with energy exceeding the bandgap, only the portion of the energy equal to the bandgap is utilized; the excess energy is dissipated as light or heat.

Nuclear quantum effects renormalize the electronic structure of silicon crystals and alter their indirect absorption spectrum (see Chapter 8). However, the comprehensive influence of full quantum effects (both nuclear quantum effects and nonadiabatic effects) still requires further in-depth research. Unlike other materials, silicon PV cells can utilize up to 45% of the solar spectrum, which is the highest efficiency ever achieved. If we consider the efficiency loss due to other factors, the actual efficiency of single-junction solar cells is ~32% at best [Shockley1961].

PV cells made of single-crystal silicon can be classified into three types. The first type is the passivated emitter and rear cell (PERC) with a mainstream p-type silicon-based emitter. In mass production, it can achieve an efficiency of 23.0%. The second type is the *n*-type tunnel oxide passivated contact (TOPC) PV cell. The TOPC cell is being rapidly developed, and its mass production efficiency ranges between 24.0% and 24.5%. The third type is the *n*-type heterojunction technology (HJT) cell, which is currently the most efficient solar cell in mass production, reaching an efficiency of 24.5%–25.0%. PV cells utilizing black-silicon–aluminum rear cells with polycrystalline silicon have an efficiency of around 19.5%, while polycrystalline PERC PV cells have an efficiency of around 20.5%. Therefore, there is not much room for improvement in the efficiency of crystalline silicon PV cells.

Recently, D. M. Bierman et al. from the Massachusetts Institute of Technology proposed a method to modify the energy distribution of light incident on the surface of a solar cell. This method directs all the energy into the above-bandgap portion of the spectrum and adjusts it to ensure complete absorption by the PV cell [Bierman2016]. Based on this principle, they developed a solar thermophotovoltaic system (depicted in Figure 9.17) that significantly enhances the theoretical efficiency of the PV cell to over 60%. The system comprises an absorber, a thermal emitter, and an optical filter. The incident sunlight must pass through all three components before reaching the PV cell. First, sunlight enters the outer absorber, which is composed of carbon nanotubes. This absorber efficiently captures and converts sunlight into thermal energy without any reflection. The thermal energy is then utilized to heat the inner-layer photonic crystal made of silicon and silicon dioxide. The photonic crystal acts as a "selective emitter," "squeezing" the above-bandgap portion of the solar spectrum to match the bandgap of the PV cell. At this stage, 50% of the solar spectrum is still below the bandgap. Finally, an optical filter is used to reflect all radiation below the bandgap into the absorber, where it is converted into thermal energy to heat the emitter, and then directed towards the PV cell. After this step, the solar spectrum below the bandgap is completely utilized. As a result, a significant portion of the energy is concentrated slightly above the bandgap, as the remaining energy in the high-energy solar spectrum is limited. In this way, all light incident on the PV cell can be used to excite the electron-hole pairs, generating photocurrent (as shown in Figure 9.17(b)). An advantage of this approach is that it minimizes the waste in the high-energy solar spectrum. All these designs are based on the renormalized electronic structures obtained by considering full quantum effects. Although further research is needed in this area, it is reasonable to assert that full quantum effects play a significant role in photoabsorption in solar materials such as Si.

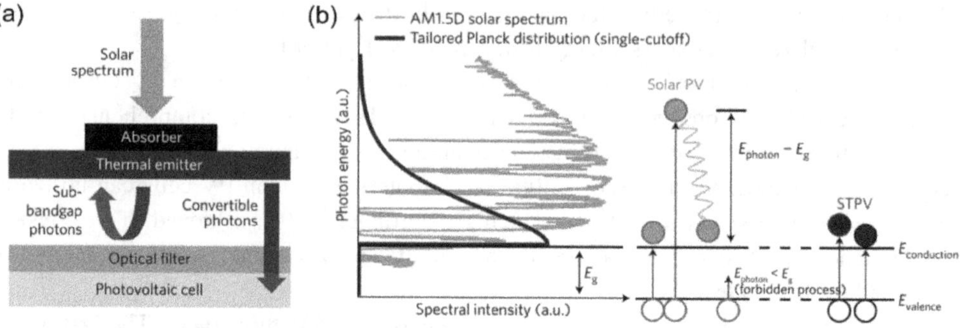

Figure 9.17 (a) Schematic of thermophotovoltaic cell. (b) Energy conversion mechanism of solar thermophotovoltaic cell (STPV) and photovoltaic cell (PV). From Bierman[2016]. Reproduced with permission from SNCSC.

The significance of this invention lies in the ability of thermophotovoltaic cells to generate electricity even in the absence of sunlight, thereby addressing the stability issue associated with solar power generation. Integrating PV cells with a thermal-energy storage system allows power output to be decoupled from sunlight input. When sunlight is available, the absorber is heated, and the resulting thermal energy is stored rather than converted into electricity. When electricity is needed, the stored thermal energy can be released to generate power, regardless of the weather conditions. However, it should be noted that this technology is still far from reaching the stage of practical application. For example, ensuring the stability of photonic crystals (which typically break at high temperatures) and cells in a high-temperature (1,000–2,000 K) environment under intense sunlight is still a challenge. Compared with conventional PV cells, thermophotovoltaic cells still have significantly high manufacturing costs, which consequently restrict their potential applications.

Research on conventional PV technology has progressed through several developmental stages, with interest shifting from mainstream crystalline silicon to group II–VI cadmium telluride (CdTe) and mixed crystal semiconductor (CIGS) thin-film cells and, more recently, to perovskite thin-film cells. This subject is very broad, and surveying it exceeds the scope of this book. To demonstrate that full quantum effects influence the physicochemical properties of solar-cell materials, we will focus more narrowly on a specific organic semiconducting material with photosensitive properties.

According to recent studies, the efficiency of organic solar cells made from P3HT [poly(3-hexylthiophene), poly(3-hexylthiophene)] and PCBM (phenyl-C_{61}-butyric acid methyl ester) has reached 5.4% [Berger2018]. The crystallinity of P3HT polymers is one of the key factors affecting their PV performance. Researchers at Oak Ridge National Laboratory and the University of South Carolina found that selectively replacing the H atom at the thiophene ring of the P3HT backbone with a D atom can significantly reduce the crystallinity. By contrast, replacing the H atom on the side chain has little effect on the crystallinity [Jakowski2017].

To explain this phenomenon, the researchers studied the nuclear quantum effects using a second-order perturbation method [Jakowski2018]. They started by computing physical

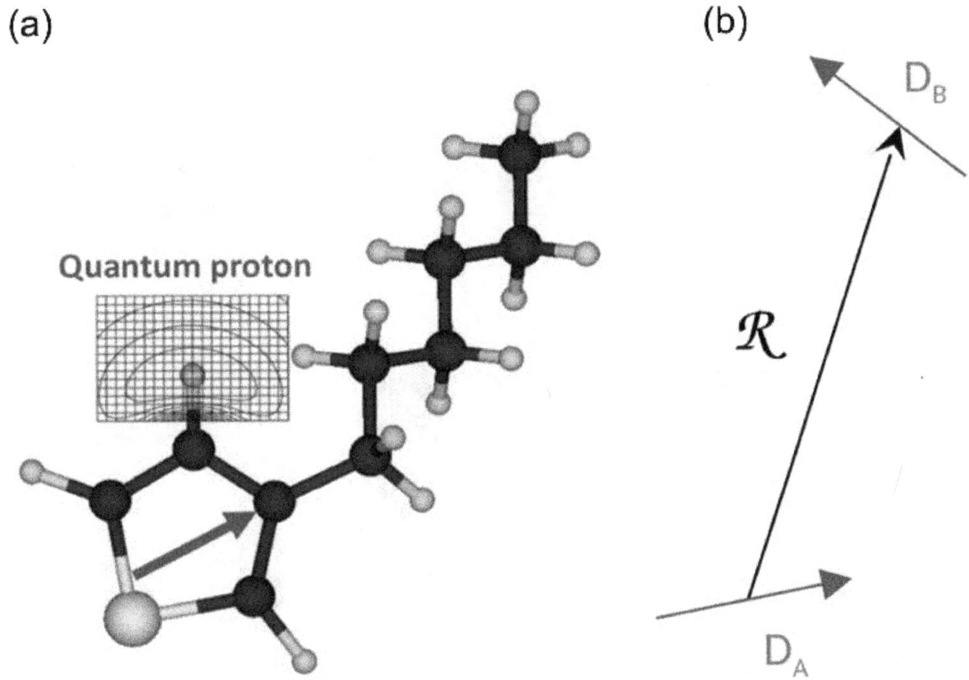

Figure 9.18 (a) P3HT monomer structure. Big white ball: sulphur atom; black balls: carbon atoms; small white balls: hydrogen atoms. The thiophene-bound proton is treated quantum mechanically. The contour of the potential energy surface computed with DFT is plotted on the grid. The arrow indicates the dipole moment localized on the thiophene ring. The side chain exhibits a negligible dipole moment. (b) A schematic of the two interacting dipoles separated by the distance R on two adjacent chains of the P3HT crystal. From Jakowski[2018]. Reproduced with permission from John Wiley and Sons.

quantities such as dipole–dipole interactions and polarization rates. To simplify the calculations, the researchers only considered the quantum effect of the replaced H or D atom on the thiophene ring; the other atoms were still treated as classical point charges in fixed positions, as shown in Figure 9.18(a). The dipole moment of each P3HT monomer is localized on the thiophene ring, and there is a dipole–dipole interaction between two monomers separated by the distance R (see Figure 9.18(b)). The isotope substitution initially affects the ZPE, which in turn changes the stability of P3HT. Calculations showed that the ZPE of the P3HT backbone is smaller than that of the side chain, making the crystal more stable. This effect is more prominent in the H atom than in the D atom, with a ZPE difference of 0.68 meV.

As shown in Figure 9.19, nuclear quantum vibrations also lead to changes in the magnitude of the dipole moment, affecting the dipole–dipole interaction. This long-range electrostatic interaction is the primary factor that determines the crystallization of P3HT. According to perturbation theory, the first-order term of the dipole–dipole interaction depends only on the average dipole moment. Because of the anharmonicity of the potential energy surface and the asymmetry of the vibrational wavefunction [see Figure 9.19(a)],

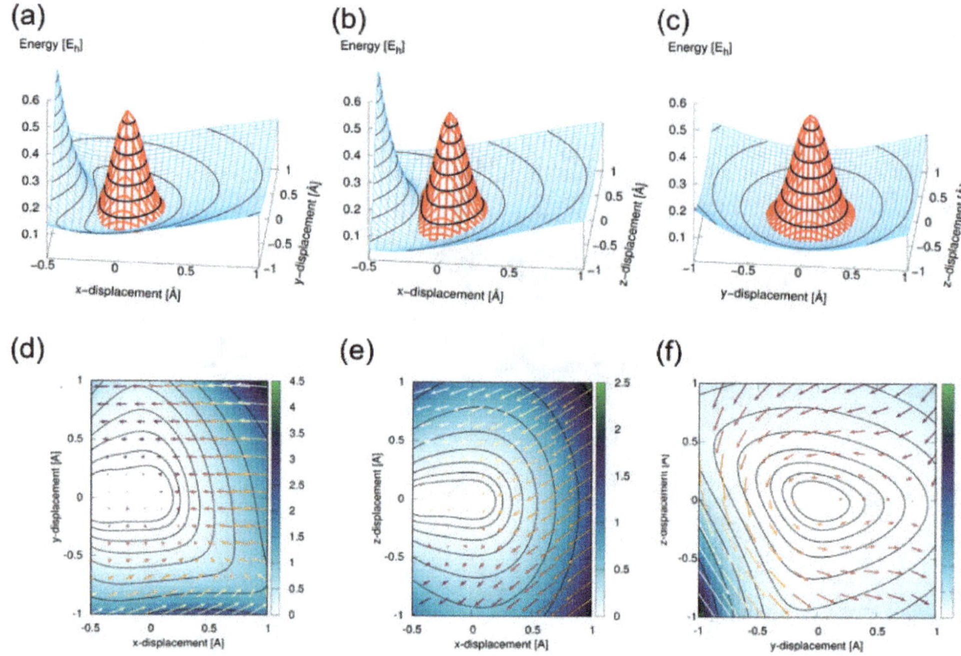

Figure 9.19 Projection of the vibrational ground state and the dipole moment of the H atom bound to the thiophene ring of P3HT. (a)–(c) show the ground state wavefunction (red, thick line) and potential energy surface (blue, thin line) of H in the xy, xz, and yz planes. (d)–(f) show the absolute magnitude of the dipole-moment variance due to the vibration of H, projected on the xy, xz, and yz planes. The arrows indicate the direction of the dipole moment in the projection plane. The displacement along the x-axis corresponds to the CH bond stretching mode, while the displacements along the y- and z-axes correspond to the bending modes. From Jakowski[2018]. Reproduced with permission from John Wiley and Sons.

the average of the local dipole moment over the vibrational wavefunction deviates from its minimum value at the potential energy surface. This deviation is larger for H than for D. As a result, the hydrogen-containing P3HT is more stable than the deuterium-containing P3HT, and the corresponding energy is lowered by 0.33 meV. The second-order term of the dipole–dipole interaction can be divided into two parts: the induction interaction, which results from the distortion of the charge distribution due to the change in dipole moment of neighboring molecules, and the dispersion interaction, which describes the interaction between the instantaneous induction dipole moments of the molecules. Both terms contribute to attractive interactions. Further calculations showed that the total dipole–dipole interaction is slightly greater for the H system compared to the D system (only by 0.27 μeV). This suggests that the polarization rate of the H atom is higher than that of the D atom.

In summary, the crystallinity of P3HT decreased after D replaced H on the main chain because of the differences in ZPE and interchain dipole–dipole interaction strength between the isotopes. These findings highlight the substantial influence of isotopic purity on the structural stability and physical properties of polymer crystals and other systems with long-range interactions. By considering the full quantum effects, it becomes possible to effectively manipulate the crystallinity of the material, thereby enhancing the solar cell efficiency of organic semiconductors.

9.4 Fuel Cells

In simple terms, a fuel cell is a chemical device that directly converts the chemical energy of fuel into electrical energy. It is often referred to as the fourth form of power generation, following hydroelectric, thermal, and nuclear power. Like a regular battery, a fuel cell consists of an anode, a cathode, an electrolyte, and supporting components. However, the active material of a conventional battery is stored internally, thus limiting its capacity. In a fuel cell, the anode and cathode do not contain the active material and only act as catalytic conversion elements. During the operation of the fuel cell, the fuel and oxidant are supplied externally to carry out the reaction. In principle, a fuel cell can continuously generate electricity as long as the reactants are provided constantly and the reaction products are removed continuously.

The dynamic properties of proton transport and the proton exchange membrane are crucial in fuel cell technology. Considering full quantum effects is essential for understanding the interactions among the protons, the exchange membrane, and the reaction potential.

In Chapter 7, we discussed proton migration through 2D materials and in bulk materials. Contrary to the theoretical prediction, Geim et al. found that monolayer graphene and hexagonal boron nitride are highly permeable to protons [Hu S.2014]. At room temperature, hexagonal boron nitride exhibits better proton transport ability than graphene (Figure 7.29). This is primarily due to the different electron density distributions in these two 2D materials. It is worth noting that after nuclear quantum effects are included in theoretical calculations, the ZPE on the activation energy barrier of the proton transport is predicted to reach the order of tens of meV. It has further been demonstrated that the hydrogenation of graphene leads to lattice distortion, which is more favorable for proton transport. Irradiating graphene films modified by Pt nanoparticles with low-intensity visible light can further enhance the proton-transport efficiency [Lozada-Hidalgo2018].

Proton migration in bulk materials plays a crucial role in fuel cell performance. Using perovskite oxides as an example, Q.F. Zhang et al. explored the decomposition of long-range proton transport into two fundamental processes: a transfer process (T-process) between adjacent oxygen atoms and a reorientation process (R-process) around a single oxygen atom [Zhang Q.F.2008]. Previous theoretical simulations based on static crystal structures had suggested that the T-process is slower than the R-process, thus impeding long-range proton migration. However, a redshift due to a strong O–H bond stretching mode was observed in far-infrared spectral measurements. This implies that significant breaking and recombination of hydrogen bonds are involved in proton migration. According to this result, the T-process is faster than the R-process because it breaks and recombines O–H bonds, and hydrogen bonding plays a more active role during the process under the influence of nuclear quantum effects. Such experimental and theoretical observations have generated considerable disagreement and debate. To resolve this controversy, a comprehensive consideration of the full quantum effects of hydrogen migration is necessary.

In perovskite systems (e.g., $BaZrO_3$), both quantum vibrations and quantum tunneling effects play substantial roles in the hydrogen transport process. Using full quantum molecular dynamics simulations (PIMD), Q.F. Zhang et al. studied the quantum vibration and tunneling of protons [Zhang Q.F.2008]. They found that although both the T-process and R-process are fundamental for proton transport in $BaZrO_3$, the full quantum effects involved in these processes are very different. The T-process consists of the breaking of one O–H bond and the formation of another. In this process, the stretching mode of the O–H bond near the transition state is significantly softened, considerably reducing the ZPE in the transition state. However, the R-process, which involves the rotation of the H atom around the O atom, does not entail bond breaking or recombination and, therefore, exhibits minimal ZPE change throughout the process. As a result, the quantum vibrational effect is canceled by the R-process. Quantum vibration is mainly reflected in the change of bonding between atoms. The influence of quantum effects on proton transport is primarily manifested through the transfer process rather than the reorientation process, as predicted by the classical approximation. This suggests that quantum effects primarily affect the bonding changes between atoms in proton migration.

Sundell et al. [Sundell2007] conducted calculations considering the ZPE correction. They determined that the potential barrier of protons in perovskite oxides was reduced by 0.12 and 0.04 for the T-process and R-process, respectively. These findings align with experimental observations. Thus, it is clear that nuclear quantum effects on potential barriers can alter the proton-transport rate and paths in both thin film materials and bulk materials. This area of research will undoubtedly be helpful for the development of fuel cells.

In addition to hydrogen–oxygen fuel cells, other fuel cells involve another common element, carbon, in energy conversion and chemical reaction. Carbon has an atomic number of 6, and its atomic mass is several times larger than that of hydrogen. Nonetheless, if the potential barrier of the reaction is small, the full quantum effects can completely dominate at low temperatures.

Recently, C. F. Lin et al. studied the adsorption and transfer of formaldehyde (CH_2O) on a Cu(110) surface using low-temperature STM. Their findings revealed that the molecule exhibited frequent flipping behavior between two symmetrical adsorption configurations (Figure 9.20). Particularly, the flipping rate remained constant below 10 K, indicating that temperature had no impact. The transfer process was attributed to concerted carbon and hydrogen tunneling. C.F. Lin et al. also demonstrated that the transfer rate significantly changed when a hydrogen atom in the formaldehyde molecule was replaced with a deuterium atom [Lin C.F.2019].

This example again illustrates that even heavier elements, such as carbon, can display significant full quantum effects in specific scenarios, particularly when the atomic or molecular hopping distance is very short (~0.1 Å) during the reaction. Despite the existence of a substantial barrier, notable quantum tunneling can still be observed. A comprehensive knowledge of full quantum effects is necessary to understand and predict these phenomena correctly.

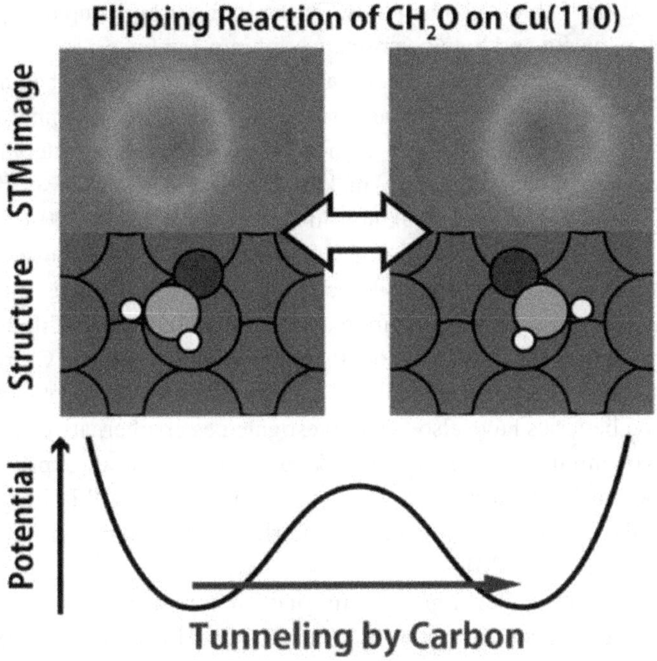

Figure 9.20 Two equivalent configurations of formaldehyde molecules adsorbed on the Cu(110) surface, with corresponding STM images. Deep grey ball: oxygen atom; light grey ball: carbon atom; while balls: hydrogen atoms. The potential energy curves of the flipping reaction are also shown. Below 10 K, the interchange between the two configurations can occur through quantum tunneling. In this process, the carbon atoms are displaced by ~0.1 Å. The black ball, gray ball, and white ball represent carbon atom, oxygen atom, and hydrogen atom, respectively. Figure created by Lin C. F.[2019] licensed under CC BY.

9.5 Ion Cells

Currently, the most widely used type of ion battery is the lithium-ion battery, which was initially commercialized by Sony Corporation in Japan. The cathode material of the lithium-ion battery consists of a lithium-containing compound, and the anode material is made of carbon. Lithium ions are transported between the cathode and anode during charging and discharging through the electrolyte. Therefore, the lithium-ion migration and storage are critical to the performance of the lithium-ion batteries. As lithium is the lightest metal, with a relative atomic mass of only approximately seven times that of hydrogen, the behaviors of lithium-ion batteries are expected to be influenced by full quantum effects. For example, S. Karmakar et al. studied the diffusion of lithium ions in group IV sulfur compounds. After considering the quantum statistics of the ZPE, they reported a 10% increase in the lithium-ion diffusion rate at room temperature and more than 60% at low temperatures [Karmakar2016]. The ZPE correction has been reported to reduce the diffusion potential of lithium ions by 0.01 eV in monolayer P, As, and Sb at room temperature [Guo G.C.2015]. Consequently, incorporating the ZPE correction has significant implications for lithium transport, particularly in achieving better performance of electrode materials for lithium-ion batteries. Furthermore, it opens up new possibilities for developing high-capacity and high-performance lithium-ion batteries.

Because of the limited resources of lithium, the price of lithium-ion batteries is expected to remain high, prompting ongoing research into alternative battery technologies. One such technology that has gained considerable attention is the sodium-ion battery [Zhao C.2020]. Sodium reserves are abundant in nature, making them a viable and cost-effective alternative. However, sodium-ion batteries exhibit lower energy density (100–165 Wh/kg) than lithium-ion batteries (150–300 Wh/kg) and also significantly differ in cycle life.

The study of sodium migration is crucial to further improving the efficiency of sodium-ion batteries. Sodium ions require more energy to move than lithium ions because of sodium's larger atomic mass (~23, more than triple that of lithium). Currently, the exact influence of full quantum effects on sodium-ion transport remains unknown.

Aluminum-ion batteries have also been investigated as an alternative solution. Because the atomic mass of aluminum is ~27, any nuclear quantum effects are expected to be weaker than those in sodium-ion batteries. As research on this topic is still in its infancy, we will not elaborate on the details of aluminum-ion batteries. Interested readers are encouraged to consult the literature [Lin M.2015].

In future clean energy development, the use of renewable energy, such as wind and solar power, is expected to continue increasing. However, the storage and efficient utilization of these forms of energy still pose significant challenges. In the case of stationary electrochemical energy storage, such as home-based wind and solar energy storage, the primary obstacle does not lie in the size or weight of the battery but in ensuring the system's safety. Large-scale energy storage necessitates stacking numerous batteries together. The risk of fire and explosion is ever-present because of the inherent high reactivity and other safety issues associated with lithium-ion battery materials. Typically, commercially available lithium-ion batteries utilize organic solvents as electrolytes to achieve high voltage and energy density. Although lithium-ion battery adopting organic electrolyte have superior energy performance, they also have unresolved safety issues: Its cathode is too reactive, its anode is prone to oxygen release, and the organic solvent in the electrolyte has a low flash point. Therefore, lithium-ion batteries are prone to thermal runaway, making them challenging to salvage in accidents. Fires caused by lithium-ion batteries may destroy the entire system, leading to catastrophic consequences. These problems limit the applicability of lithium-ion batteries in large-scale energy storage.

In all ion batteries, the electrolyte facilitates the transportation of ions (such as lithium, sodium, or aluminum) between the cathode and anode. Hence, selecting the solvent used in the electrolyte plays a critical role in determining battery performance. Solvents can be classified as inorganic or organic. While various organic solvents are available, most are volatile and combustible. On the other hand, viable options for inorganic solvents are limited. Water is one such solvent. Researchers have developed aqueous lithium-ion batteries that utilize water as an electrolyte to improve safety. Recently, S. Xin's group at the Institute of Chemistry, Chinese Academy of Sciences, investigated nuclear quantum effects on the electrolytic properties of hydrogen during aqueous lithium transport [Chou J.2022].

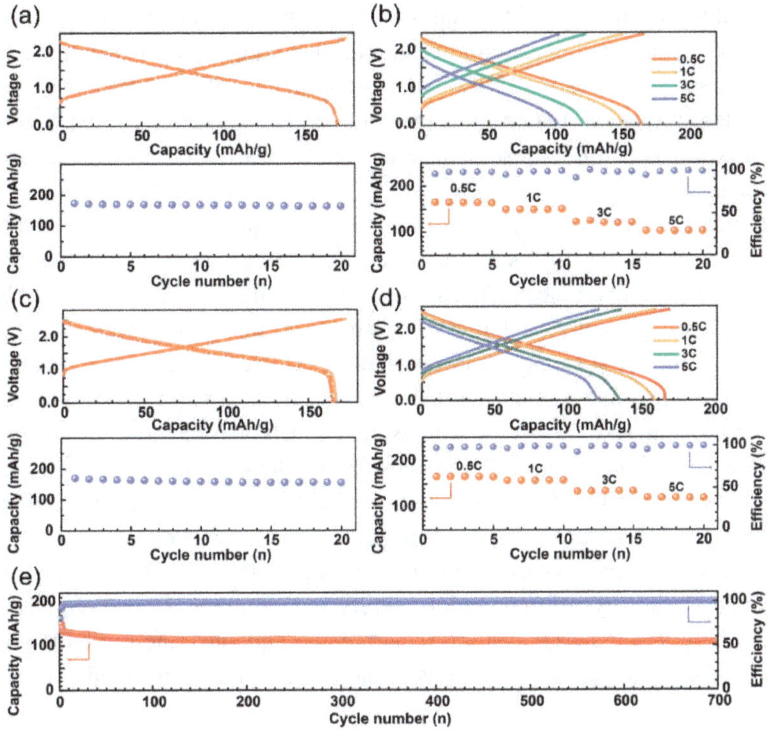

Figure 9.21 Cycle performance and the charge-discharge curves of aqueous lithium-ion batteries at different C-rates in the presence of D_2O/CAN electrolyte. From Chou J.[2022]. Reproduced with permission from John Wiley and Sons.

Their study demonstrated the isotope effect by highlighting the significant differences in electrochemical properties between heavy water (D_2O) and water (H_2O). Compared to H_2O-based electrolytes, the D_2O-based electrolytes have a wider electrochemical window, a higher percentage of coordinated water, and longer hydrogen-bond lifetimes. They also exhibit higher anode stability when operating with high-voltage layered-oxide cathode materials, including $LiCoO_2$ and $LiNi_{0.8}Co_{0.1}MnO_{0.12}$. This leads to improved cycle life and excellent rate performance (see Figure 9.21). For example, a high-capacity retention of 81.5% and an average Coulombic efficiency of 99.6% were obtained after 700 cycles, as shown in Figure 9.21(e).

Aqueous lithium-ion batteries are still in the research stage. Another more mature aqueous electrolyte-based technology is the aqueous zinc-ion battery. Using a relatively inert electrode immersed in an aqueous solution ensures the safety of zinc-ion batteries. In addition, rechargeable aqueous zinc-ion batteries can achieve energy densities comparable to those of $LiFePO_4$-based batteries and are equipped with a wide operating temperature range and excellent rate performance. Furthermore, the production process also offers certain advantages. Whereas lithium-ion battery production must meet strict requirements, specifically concerning the drying conditions, aqueous zinc-ion batteries can be manufactured in a general production environment. Because water is used as the electrolyte solvent, there is no need for specific ambient humidity conditions, simplifying the production process and reducing associated costs. Additionally, the inherent safety of aqueous zinc-ion batteries facilitates convenient battery transportation and installation. These advantages pave the way for extensively utilizing aqueous zinc-ion batteries in large-scale electrochemical energy storage.

In recent years, several research teams have made significant breakthroughs in increasing the stability of anodes, high-capacity cathodes, and electrolyte additives for aqueous zinc-ion batteries. For example, C. Y. Zhi's group from the City University of Hong Kong mass-produced 50 Ah zinc-ion batteries at the Songshan Lake Materials Laboratory, demonstrating promising application prospects [Liang G.J.2021, Zhao Y.W.2022]. Although zinc atoms have a relatively large atomic mass (over 65), isotopic substitution in the water nevertheless inevitably affects zinc-ion transport through the ion hydrates formed in the aqueous environment (see subsection 10.2.4). This also directly affects battery performance. The role of the full quantum effect in this is still unclear. Undoubtedly, fundamental research is crucial to enhancing battery performance.

Among the various aqueous-ion batteries, the aqueous zinc system offers higher efficiency and energy density. With its unique safety advantages and excellent energy performance enabled by the material system, rechargeable aqueous zinc-ion batteries have the potential to make significant contributions to large-scale electrochemical energy storage.

10
Full Quantum Effects in Environmental Applications

Full quantum effects play a crucial role in many processes that occur in deep space, including galaxy formation, interstellar chemistry, and possibly even the emergence and evolution of life. The universe is primarily composed of light elements, particularly hydrogen, and has an average temperature of ~3 K. Consequently, quantum tunneling and quantum fluctuations, which are phenomena associated with nuclear quantum effects, are expected to become more prominent as hydrogen atoms interact with their surroundings. In addition, nonadiabatic processes related to photochemical reactions are also significant in such environments.

On Earth, we are surrounded by phenomena arising from full quantum effects, which often have significant implications for human life and civilization. The ozone layer in the atmosphere can be depleted by halogen atoms such as those produced by the photolysis of chlorofluorocarbons (CFCs) and halons, which can even lead to the formation of ozone holes in the polar regions. Understanding this phenomenon requires comprehensive knowledge of the full quantum effects in oxygen atoms. In addition, the interaction of light atoms/molecules with other materials at the molecular level plays a crucial role in effectively removing impurities and toxic components in water, addressing atmospheric pollution caused by photochemical smog, and mitigating global warming. Although these interactions occur at noncryogenic temperatures, it is imperative to acknowledge the significance of full quantum effects when studying them.

10.1 Space Environment

In Chapter 9, we discussed the process of interstellar hydrogen production. Nebulae, which are cloudy objects in interstellar space, consist mainly of dust and gas, particularly H_2. The temperature in molecular clouds and diffuse nebulae (~10–100 K) does not meet the required value for conventional H_2 generation. Furthermore, H_2 is decomposed and destroyed by ultraviolet radiation and cosmic rays. Thus, one possible explanation for the abundance of H_2 in nebulae is that quantum tunneling enhances the diffusion rate of hydrogen adsorbed on the dust. Quantum tunneling enhances the synthesis of H_2 molecules, which are then released into space in the form of gas.

Full quantum effects are not only related to H_2 production but are also associated with the formation of water molecules, which are essential for life. From 5 K to 20 K, H_2O is the most abundant component of the ice-mantle surface of interstellar dust particles. Its formation is strongly related to the adsorption of H_2 molecules on the dust surface. The process of water formation can be expressed as:

$$OH + H_2 \rightarrow H_2O + H. \tag{10.1}$$

When OH reacts with H_2 at such low temperatures, a high chemical barrier of the process must be overcome via quantum tunneling. Numerous isotopic experiments have shown that replacing H_2 with D_2 reduces the reaction rate by an order of magnitude, demonstrating the importance of nuclear quantum effects.

CO is the second most abundant molecule on the surfaces of interstellar dust particles. It undergoes continuous hydrogenation via quantum tunneling on a water-rich surface. This process further results in the formation of organic molecules such as formaldehyde (CH_2O), which are precursors of amino acids, sugars, and other complex molecules of biological significance. The continuous hydrogenation process is described by the following equation:

$$CO \rightarrow HCO \rightarrow CH_2O \rightarrow CH_3O \rightarrow CH_3OH. \tag{10.2}$$

The above discussion demonstrates that full quantum effects are closely associated with many fundamental evolutionary processes in cosmic environments.

10.2 Water Problem in Nature

Water treatment is a central issue for ecological change and conservation. In Chapter 7, we focused on the fundamental science of water under cryogenic and ultrahigh-vacuum conditions. By contrast, the discussion on environmental science research is often conducted under ambient conditions (i.e., near room temperature and atmospheric pressure). Therefore, in this section, we focus on the anomalous properties of water resulting from full quantum effects under such ambient conditions and explore their impact on the environment. To gain a deeper understanding of the principles governing these intricate issues, we approach these topics from a fundamental science perspective.

10.2.1 Adsorption and Diffusion of Water Molecules

In its natural state, water generally exists in a molecular form. Therefore, in environmental and water sciences, the diffusion of water molecules on material surfaces and their mutual interactions are crucial aspects. Full quantum effects must be considered for a comprehensive understanding of these phenomena.

10.2.1.1 Carbon Surface

Since 2009, using carbon surfaces as an ideal substrate material, E. G. Wang's group, in collaboration with Michaelides' group, has been conducting systematic and in-depth theoretical studies of the adsorption and diffusion of water on carbon surfaces [Ma J.2009, 2011]. The adsorption of water molecules is considerably more complicated than the hydrogen-adsorption process discussed previously. There are two main adsorption configurations of water molecules on carbon surfaces: the two-leg and one-leg structures, as shown in Figure 10.1(a) and (b), respectively.

Researchers have compared adsorption energies calculated by DFT with more accurate diffusion Monte Carlo (DMC) calculations, as shown in Figure 10.2. They found that an accurate calculation of the adsorption energy of water molecules requires a rigorous description of long-range van der Waals interactions. For example, when the ordinary PBE functional was used for the DFT calculations, the adsorption energy curve of the water molecules exhibited only a very shallow valley with a depth of ~30 meV. If the zero-point energy of the water molecule and the finite-temperature vibrational excitation are calculated from this adsorption potential energy surface, it can be concluded that there are no stable adsorption sites for water molecules on the graphene surfaces. In more extreme cases, if electronic structure calculations are performed using a BLYP-like functional in DFT, the results may even indicate repulsion between water molecules and graphene surfaces. However, in exact DMC calculations, the adsorption energy of water molecules on graphene was 80 meV, which implies that water molecules can be securely adsorbed on the graphene surface even at room temperature. J. Ma et al. found that in the DFT framework, the results of some functionals overestimated the adsorption capacity of water, making it difficult to characterize the full quantum effects in the adsorption of water molecules accurately. For example, the calculated zero-point energy for the adsorption of water molecules is excessively large.

J. Ma et al. further analyzed and tested different calculation methods that combined the DFT functional and van der Waals correction; the PBE-D approach yielded the best results (as shown in Figure 10.2). At the same time, Brandenburg et al. analyzed other relevant systems using the PBE method [Brandenburg2019]. Studies conducted by J. Ma et al. laid a solid foundation for future investigations into the full quantum effects of water adsorption [Ma J.2009, 2011]. Once the precise electronic structure is known, the following study on full-quantum-effect contributions may be considered.

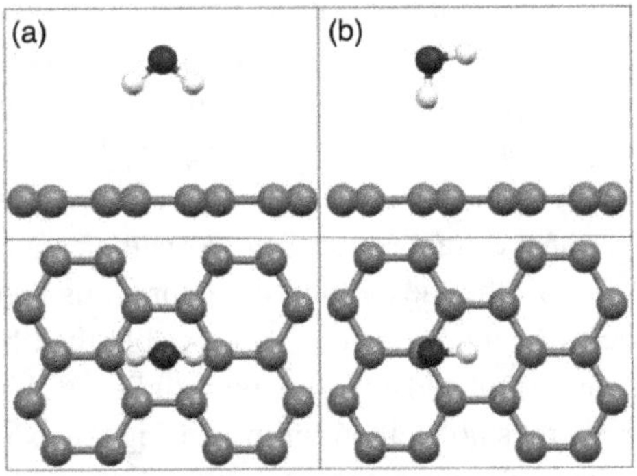

Figure 10.1 Adsorption configuration of water molecules on a graphene surface: (a) two-leg configuration; (b) one-leg configuration. The upper panels show the view from the side, while the lower panels show the view from above. Reprinted figure with permission from Ma J.[2011]. Copyright 2024 by the American Physical Society.

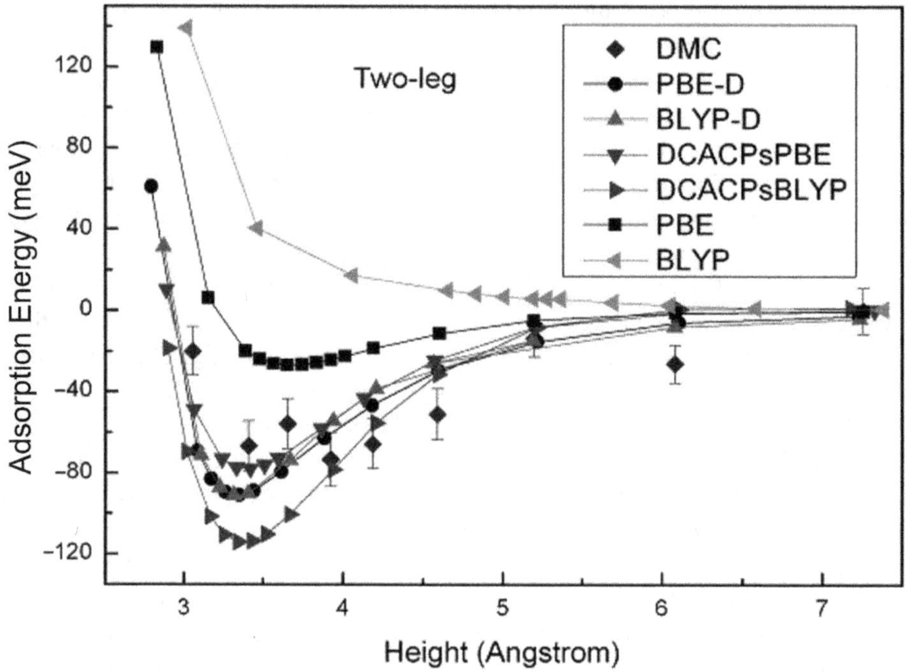

Figure 10.2 Adsorption energy for two-leg configuration of a water molecule on a graphene surface. DMC means diffusion Monte Carlo; the other results were obtained by various DFT methods. Reprinted figure with permission from Ma J.[2011]. Copyright 2024 by the American Physical Society.

First, in the results of theoretical calculation, the water molecules in the most stable adsorption structure contain hydrogen atoms oriented toward the graphene surface. In this configuration, a weak hydrogen bond is formed between the water molecules and graphene. According to the experimental results reported by J. Guo et al. [Guo J.2016a], nuclear quantum effects further weaken this already weak hydrogen bond. Therefore, the adsorption energy of water molecules on graphene is expected to be even lower. Moreover, in contrast to the one-leg configuration, the two-leg configuration has two weak hydrogen bonds, which are both weakened by nuclear quantum effect. Consequently, nuclear quantum effects must have a more significant impact on the adsorption energy of the two-leg structure. Therefore, the most reasonable inference is that the one-leg structure is a more stable adsorption configuration for water molecules on the graphene surface. Nonetheless, it is important to acknowledge the possible existence of other adsorption configurations that are minimally affected by nuclear quantum effects and are, therefore, more stable than both one-leg and two-leg configurations. Detailed theoretical and experimental studies of this possibility have not yet been reported. Moreover, there is no method for measuring the exact adsorption energies of these systems. It is clear that nuclear quantum effects will play a crucial and sensitive role in future research on adsorption energy comparison between experiment and theory.

In 2014, S. Hu et al. reported the fabrication of graphene membranes that were highly permeable to protons [Hu S.2014]. Their results demonstrated that protons can effectively pass through the centers of the hexagonal ring structure in defect-free graphene at room temperature. However, these experimental results disagree with theoretical predictions of the potential barrier to proton permeation through graphene. To understand this phenomenon, Y.X. Feng et al. performed further calculations and found that the solution environment around the graphene film significantly lowers the potential barrier, thereby facilitating proton permeation [Feng Y.X.2017]. Notably, the protons were connected to water or other protons and pointed in the direction of the hexagonal graphene ring. This is closely related to the issue of how hydrogens are oriented relative to the graphene surface in water adsorption, discussed earlier in this section. If a hydrogen atom in water is oriented toward the graphene surface, it can act as a proton in the environment of a proton-conductive membrane and penetrate the graphene monolayer, as predicted by Y.X. Feng et al.

Clearly, the proton-permeation process is affected by nuclear quantum effects. Y.X. Feng et al. used path-integral molecular dynamics to calculate that the effective penetration barrier of this process is reduced by ~0.4 eV because of nuclear quantum effects. In Figure 10.3, the two solid lines with square and triangle markers show the barrier energy variation as a function of the displacement of the proton from the graphene plane for the classical (MD) and quantum (PIMD) calculations, respectively. The center part of the inset shows the instantaneous configurational distribution of the path-integral molecular dynamics simulation as the proton approaches the center of the hexagonal ring of the graphene plane (horizontal coordinates from left to right in the figure). When a proton reaches the

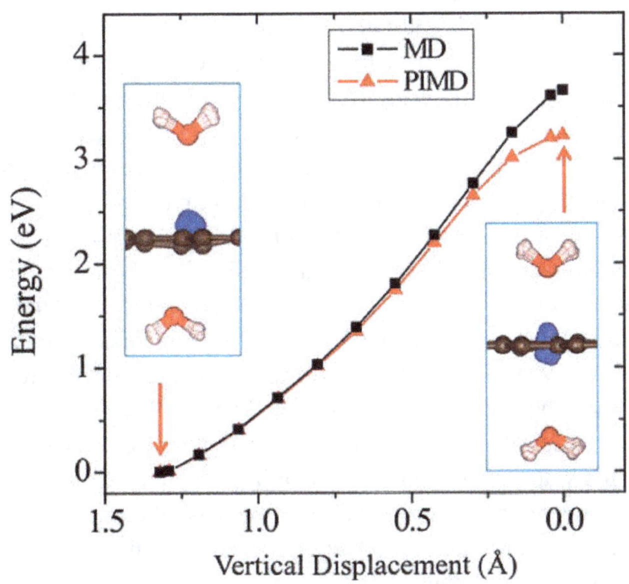

Figure 10.3 Free energy as a function of the distance from the proton to the graphene surface (brown) obtained by two molecular-dynamics methods based on DFT electronic structure calculation. MD is the classical molecular dynamics simulation, and PIMD is the path-integral molecular dynamics simulation. Reprinted with permission from Feng Y. X.[2017]. Copyright 2024 American Chemical Society.

surface, the spherical distribution of the proton is extended in the direction perpendicular to the surface (see the right inset of the figure). Such extended distribution clearly demonstrates the quantum nature of atomic nuclei. It is worth pointing out that in the calculation by Y.X. Feng et al., the electronic structure was provided by DFT based on the van der Waals correction; therefore, the numerical results are relatively reliable.

In environmental research, many systems are more complex than simple adsorption of water onto graphene surfaces. These systems are governed by hydrogen bonds and van der Waals forces. Studying the full quantum effects in these practical systems relies heavily on the accurate calculation of their Born–Oppenheimer potential energy surfaces. Therefore, it is essential to obtain an accurate description of the electronic structure by using the DMC method.

10.2.1.2 Metal Surface

The adsorption and diffusion of water molecules on metal surfaces have also frequently been studied in environmental science research. Based on experimental measurements, it is generally accepted that nuclear quantum effects have a non-negligible influence on the diffusion of water molecules and clusters on surfaces. Because of the coupling between hydrogen bonding among water molecules and the interaction between the water molecules and substrates, the diffusion of the water molecules occurs in various modes. These distinct modes compete and are highly susceptible to nuclear quantum effects.

The nuclear quantum effect in the pairwise diffusion of two water molecules on a metal surface is particularly interesting. Hydrogen bonds are formed between water molecules on metal surfaces, resulting in the formation of dimers. In a well-known experiment, T. Mitsui et al. found that the diffusion rate of water dimers on a Pd(111) surface exceeded that of individual water molecules at low temperatures [Mitsui2002]. This is puzzling, since it is often true that water dimers have larger adsorption energies than monomers. This is also the case for adsorptions of water molecules on Pd(111). As such, for the conventional translational diffusion mechanism, there is simply no reason that a dimer can diffuse faster than a monomer. One mechanism, that can circumvent this problem, is to make use of the high-dimensional nature of the water dimer diffusion process. In addition to the weak water-metal interactions, hydrogen bond can also form between the water molecules in dimer. And the role of the donor and acceptor can switch if the water dimer can "dance" on the metal surface. In so doing, the barrier of the dimer diffusion can be lowered.

Motivated by this, V.A. Ranea et al. performed a systematic theoretical study of the different diffusion mechanisms of water dimer on Pd(111). They proposed one mechanism that is analogous to the waltz dance (as depicted in Figure 10.4). They found that the diffusion barrier of the water dimer can be substantially decreased by this mechanism when compared to the conventional translational one. Because the mass of hydrogen is small, it was expected that nuclear quantum effects may influence the barrier of donor-acceptor switching during this process [Ranea2004].

Sixteen years later, this problem was further analyzed by W. Fang et al. [Fang W.2020]. After performing systematic theoretical calculations, they found that water dimer waltzing diffusion can occur on many different surfaces. Whether it is the dominant diffusion mechanism indeed depends on the role played by nuclear quantum effects [Fang W.2020].

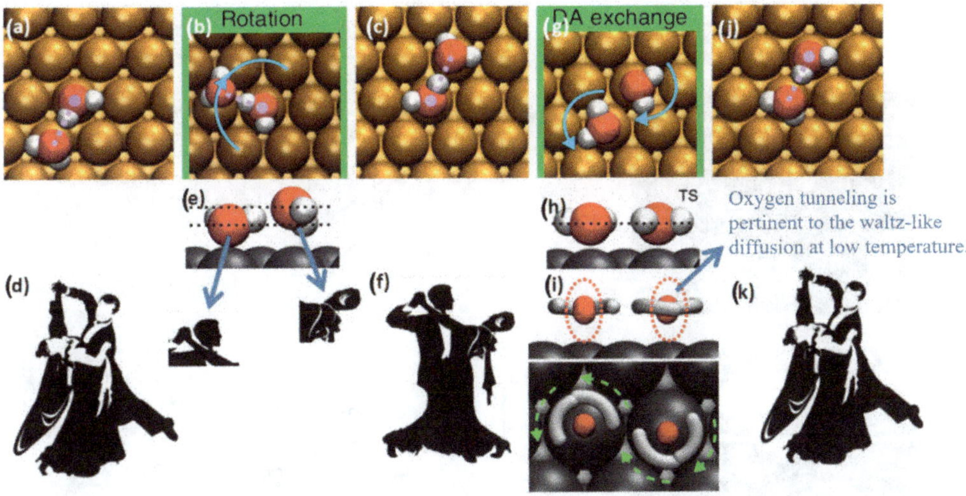

Figure 10.4 Schematic of water-dimer diffusion on a metal (Pd(111)) surface at low temperatures. This process can be explained by a "waltz-like" diffusion mechanism. The initial configuration of the water dimer is shown in (a). At the beginning of the diffusion process, the water molecule at the edge rotates around the water molecule at the center; the system changes to the configurations shown in (b) and (c). The waltz dance moves corresponding to these three configurations are depicted in (d)–(f). The water molecule at the center is strongly adsorbed on the metal surface, resulting in a small rotation barrier for the water molecule near the edge. After the rotation, the two water molecules exchange roles to continue diffusion. In this case, in a water dimer connected by a hydrogen bond, one of the water molecules makes the transition from donor to acceptor, while the other makes the opposite transition. This process is illustrated in (g). The transition state of the hydrogen-bond exchange is depicted in (h). At low temperatures, this process must be accomplished through oxygen tunneling, as shown in (i). The hydrogen tunneling must be coordinated with the oxygen tunneling to complete the hydrogen-bond exchange. After this step, the waltz-like rotation can continue (j)–(k). Figure created by Fang W.[2020] licensed under CC BY 4.0. The material is modified.

As shown in Figure 10.5, the most critical step in waltzing diffusion is when the hydrogen bond between two water molecules undergoes donor-acceptor exchange. Considering nuclear quantum effects, W. Fang et al. observed significant quantum tunneling in the hydrogen bond exchange between water dimers as the temperature decreased below 40 K. For water monomers, where hydrogen bond exchange was absent, the diffusion rate followed the Arrhenius relationship and decreased accordingly (Figure 10.5). Similarly, the diffusion rate of water dimers decreased with decreasing temperature, but at a different rate owing to atomic tunneling. Below 30 K, the diffusion rate of the dimers decreased more slowly than that of the monomers. Consequently, the diffusion of the water dimers became more prominent. Specifically, the diffusion of water dimers below 30 K surpassed that of water monomers, whereas the reverse held above 30 K. Examining the distribution along the calculated quantum tunneling pathways, it is evident that tunneling plays a predominant role during the temperature transition from high to low.

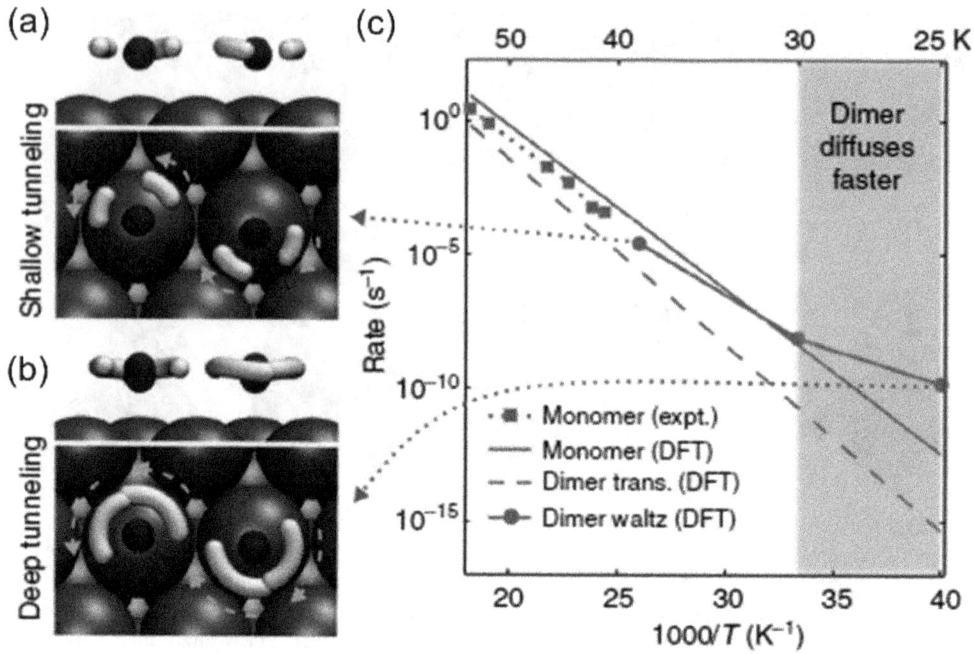

Figure 10.5 Diffusion rate of a dimer consisting of two water molecules on a surface. The molecular dynamics (MD) results are represented by the gray dashed line and the path-integral molecular dynamics (PIMD) results correspond to the solid line with circle markers. The theoretical results (solid line with no markers) and the experimental results (square dots) of a water monomer are also compared in this figure. Both MD and PIMD are molecular dynamics simulations based on DFT electronic structure. Figure created by Fang W.[2020] licensed under CC BY 4.0.

Interestingly, it is not only the nuclear quantum effects of the hydrogen atoms that contribute to the hydrogen bond exchange in the water dimer but also the quantum properties of the oxygen atoms. In the stable adsorption structure of a water dimer, the two molecules are not at the same height above the surface (see Figure 10.4(e)). Typically, the hydrogen-bond donor is positioned closer to the surface than the acceptor. However, during the hydrogen-bond exchange, the roles of the donor and acceptor are reversed. To return to the most stable adsorption configuration, the oxygen atoms in these water molecules undergo a displacement perpendicular to the surface with a magnitude of up to 0.7 Å. Further calculations revealed that purely classical motion cannot account for this displacement of the oxygen atoms. Instead, the oxygen atom and the hydrogen atom undergo quantum tunneling motion together. At 25 K, the tunneling distance of an oxygen atom can be extended up to 0.22 Å (as depicted in Figure 10.4)

10.2.2 Bulk Water

As discussed in the previous examples, nuclear quantum effects can often be probed using isotope substitution. However, the nuclear quantum effects in water observed under natural environmental conditions differ from those measured at low temperatures and high pressures in the laboratory. At room temperature or low pressure in an ambient environment, the isotopic effects in water are mostly attributable to the zero-point energy associated with O–H stretching, which corresponds to an energy of ~2,000 K along the stretching coordinate and varies with the structure of the local hydrogen bonds. Therefore, the nuclear quantum effects at room temperature or low pressure are mainly reflected by the changes in the hydrogen bond strength and configuration. In contrast, the uncertainty in the proton position is comparable to the distance between the two minima on the potential energy surface, resulting in quantum tunneling and proton delocalization. Therefore, at low temperatures or high pressures in a lab environment, nuclear quantum effects manifest as quantum tunneling and the delocalization of atomic nuclei.

According to Table 7.1, when the system is in the gas phase, the O–D hydrogen bond is 1.6% stronger than the O–H bond, with a bond energy of 7.5 kJ/mol. Researchers generally attribute this difference to the zero-point energy difference of 5.8 kJ/mol between the O–H and O–D bonds. The fact that D_2O and H_2O have gas-phase dipole moments of the same magnitude indicates that the difference in their bond lengths and angles is insignificant (less than 0.05%). However, the dissociation energy of D_2O dimer is 12.7% higher than that of H_2O, suggesting that D forms stronger hydrogen bonds than H in the gas phase.

Many experimental observations have demonstrated that hydrogen bonds formed by D_2O are stronger than those formed by H_2O in the liquid phase. After deuterium substitution, the melting point of water increases by 3.82 K; this indicates that the D atoms form stronger hydrogen bonds, making deuterium ice more stable than hydrogen ice. In addition, the temperature of maximum density (TMD) is increased by 7.21 K after deuteration, meaning that the difference between the melting point and TMD increases from 3.98 K to 7.37 K. This implies that the universal scaling law is insufficient to explain the isotope effect.

As shown in Table 7.1, the heat capacity of D_2O is significantly greater than that of H_2O, suggesting that nuclear quantum effects remain important for water, even at ambient temperatures. By comparing the densities at TMD and 298 K, it can be observed that D_2O and T_2O undergo a slight expansion with respect to H_2O. In addition, $H_2{}^{18}O$ undergoes a small contraction with respect to H_2O. In the past, some researchers claimed that this isotopic phenomenon is caused by stronger hydrogen bonding in H_2O. The opposite is true, as strong hydrogen bonds lead to a decrease in density owing to their directionality. The observed values in Table 7.1 indicate that, to a large extent, nuclear quantum effects weaken the hydrogen bond strength in water and strengthen it after deuteration.

However, some of the other properties listed in the table are not consistent with this interpretation. For example, the critical temperatures of D_2O and T_2O are lower than those of H_2O. This indicates that nuclear quantum effects strengthen the hydrogen bonds of H_2O at high temperatures, in contrast to their influence at low temperatures. The liquid–gas surface tension of D_2O is also anomalously low compared with that of H_2O. The volume of liquid water and hexagonal ice Ih increases by 0.1% after replacing H with D atoms; this is also an anomaly, because the volume of liquids and solids tends to decrease after hydrogen atoms are replaced with heavier isotopes. Full quantum effects are responsible for numerous anomalous phenomena in natural environments.

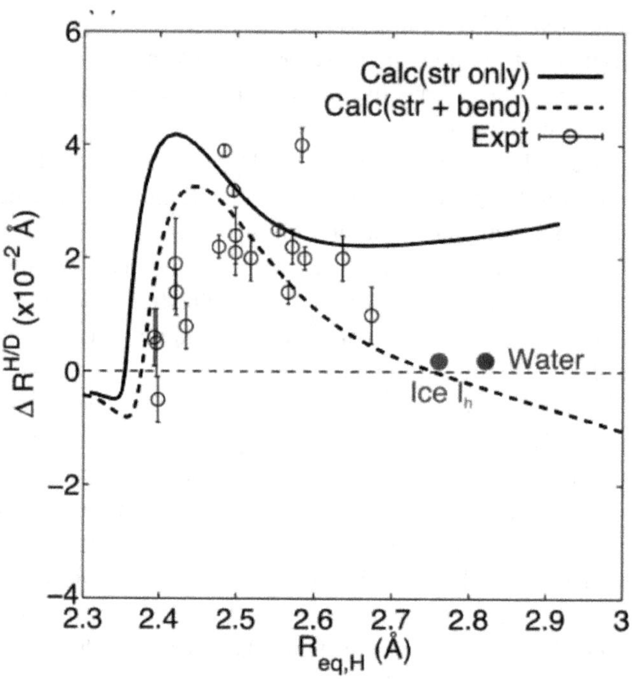

Figure 10.6 Calculated and experimental change in hydrogen bond length as a function of the original hydrogen bond length after H/D substitution in water. "str" refers to stretching, and "bend" refers to bending. Reprinted from McKenzie[2014], with the permission of AIP Publishing.

Figure 10.6 illustrates the change in the O–O distance of water and ice upon H/D substitution ($\Delta R^{H/D} = R_{eq,H} - R_{eq,D}$); this change is called the secondary geometric isotope effect (the primary effect is the alteration in the O–H distance). The horizontal coordinate in the figure denotes the O–O distance of water, and the vertical coordinate represents the change after replacing H with D. The values of the vertical coordinates reflect the strength of the isotope effect. The solid curve in the figure represents the calculated result of the zero-point energy contribution considering only O–H stretching, the dashed curve denotes the calculated result considering both O–H stretching and bending, and the scattered data points correspond to the experimental values [Sokolov1988]. The secondary geometric isotope effects in hexagonal ice Ih and liquid water are relatively minor. Quantum effects compete with each other and cancel each other out [McKenzie2014].

XRS (X-ray Raman Scattering), XAS (X-ray Absorption Spectroscopy), and XES (X-ray Emission Spectroscopy) have been used to probe the structure of water, providing substantial evidence that D_2O has stronger hydrogen bonds than H_2O at room temperature. The hydrogen bond configuration in liquid water is often discussed in terms of two features: the local tetrahedral shell structure and the symmetry of the hydrogen bonds. For instance, the characteristic features in the XRS spectra reflect the enhanced symmetry of the hydrogen bonds of H_2O compared to those of D_2O. Furthermore, X-ray, γ-ray, and

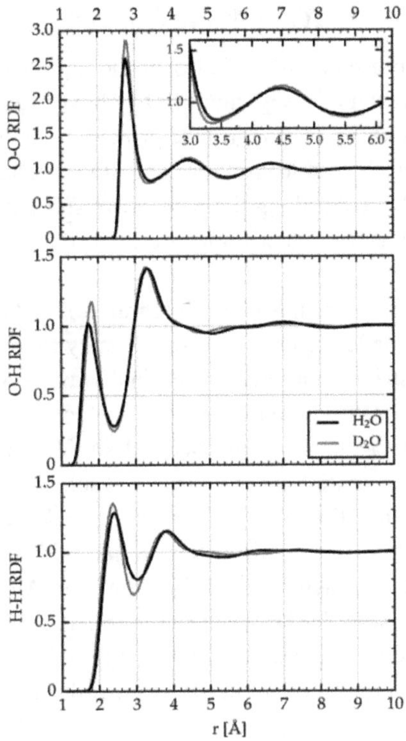

Figure 10.7 Radial distribution function (RDF) of water (H_2O and D_2O).

neutron scattering experiments have confirmed that the shell structure features are more pronounced in D_2O than in H_2O at room temperature.

Figure 10.7 shows the radial distribution functions (RDF) of O–O, O–H, and H–H in both light and heavy water obtained by combined neutron and X-ray scattering at 296 K. We see that the first peak of the O–O RDF in D_2O is higher than that in H_2O, indicating that D_2O has a more prominent shell structure. In contrast, the O–H covalent bond is ~3% longer than the O–D bond in water, implying that the hydrogen bonds in H_2O are shorter. Additionally, the distance between the first H–H molecule in H_2O is ~2% longer than in D_2O. Overall, the hydrogen bond structure of H_2O exhibits higher symmetry than that of D_2O.

The isotope effects on the dynamic properties of water are more prominent than those on its static properties. One possible reason for this is that the dynamics of water are mainly driven by fluctuations in the hydrogen-bond network, which are generally affected by nuclear quantum effects rather than by average water properties. Another reason is that replacing H with D increases the mass, effectively slowing down the dynamic processes even in the classical limit. The magnitude of this decrease ranges from 0.707 to 0.949, where $\left(m(H)/m(D)\right)^{1/2} = 0.707$ for purely elastic collisions by isotopic substitution and $\left(m(H_2O)/m(D_2O)\right)^{1/2} = 0.949$ for elastic collisions of protons in water molecules.

10.2.3 Water Clusters

As acidic solvents and representative moieties in the aqueous environments, protonated and deprotonated water clusters ($H_5O_2^+$ and $H_3O_2^-$) have been extensively studied, both theoretically and experimentally. Through first-principles molecular orbital calculations, researchers have found that the stable structure of $H_3O_2^-$ becomes asymmetric because of the presence of a potential barrier along the proton transfer pathway. In the $H_5O_2^+$ cation, the C_2 symmetry distorts the two water molecules, and the proton is located between the two oxygen atoms.

Tuckerman et al. used first-principles path-integral molecular dynamics to simulate this system. After considering nuclear quantum effects, the hydrogen nuclei of the corresponding $H_3O_2^-$ system were distributed approximately halfway between the two oxygen atoms (with only one minimum on the potential energy surface) [Tuckerman2002]. This conclusion differs significantly from those of classical simulations, which typically contain two minima on a classical potential energy surface. Tachikawa et al. later found that OO bond of $D_3O_2^-$ was longer than that of $H_3O_2^-$ and that the deuterium nuclei were localized around the hydrogen bonds of the two oxygen atoms, with no sign of delocalization as hydrogen nuclei do in $H_3O_2^-$ [Tachikawa2005]. However, McCoy et al., who conducted further theoretical studies using the DMC and vibrational configuration interaction methods, reported completely contrary results [McCoy2005]. These disagreements highlight the need for careful consideration of influence from full quantum effects, even in seemingly simple hydrogen-bond systems.

To gain a deeper understanding of the delocalization of hydrogen nuclei by isotopic effects, Tachikawa's group performed a detailed simulation considering the full quantum effects [Suzuki2008]. The two-dimensional distribution of the hydrogen-bond structural parameters is shown in Figure 10.8, where δ_{OH^*} is defined as the deviation of the proton position from the R_{OO} center. The distribution obtained from the quantum simulation exhibits a single peak at low temperatures. This indicates that the hydrogen, deuterium, and tritium nuclei were delocalized (O...X...O), where X represents a hydrogen, deuterium, or tritium nucleus. Because of the difference in the zero-point energy, the δ_{OH^*} of the H atoms was broader than that of D and T. At high temperatures, there were two peaks for deuterium and tritium, whereas hydrogen retained only one peak. This indicates that the deuterium and tritium nuclei were localized (O–X–H), whereas the hydrogen nucleus remained delocalized. In addition, the OO distance increased with increasing temperature. At high temperatures, δ_{OH^*} and R_{OO} were strongly correlated. $D_3O_2^-$ had a longer OO bond than did $H_3O_2^-$. However, at low temperatures, where the isotope effects were more pronounced, the average OO distance of H was the longest among all the isotopic water clusters. This elongation of the OO distance was caused by the zero-point energy of the two OH^* vibrational modes under the anharmonic effect.

The two-dimensional potential energy surface of $H_3O_2^-$ is plotted in Figure 10.9 [Suzuki2018]. As R_{OO} increases, the potential barrier between the two oxygen atoms gradually emerges and increases. Consequently, the OO bond elongates because of the increase of thermal excitation. However, in a high temperature regime, after isotopic substitution, the delocalization of the deuterium and tritium nuclei is less pronounced; instead, they

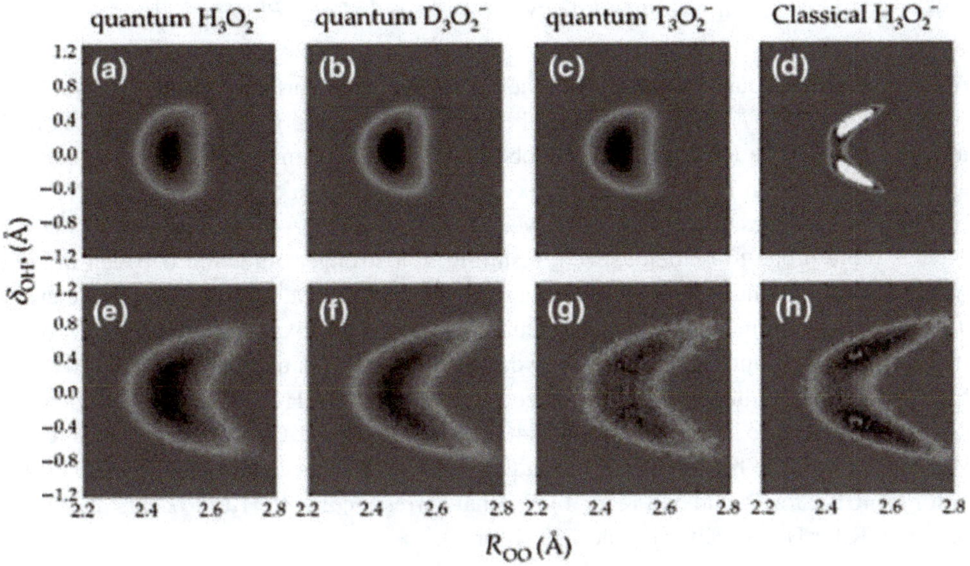

Figure 10.8 Two-dimensional distribution of hydrogen bond structural parameters in water clusters. (a) Quantum $H_3O_2^-$ at 50 K; (b) Quantum $D_3O_2^-$ at 50 K; (c) Quantum $T_3O_2^-$ at 50 K; (d) Classical $H_3O_2^-$ at 50 K; (e) Quantum $H_3O_2^-$ at 600 K; (f) Quantum $D_3O_2^-$ at 600 K; (g) Quantum $T_3O_2^-$ at 600 K; (h) Classical $H_3O_2^-$ at 600 K. Reprinted from Suzuki[2008], with the permission of AIP Publishing.

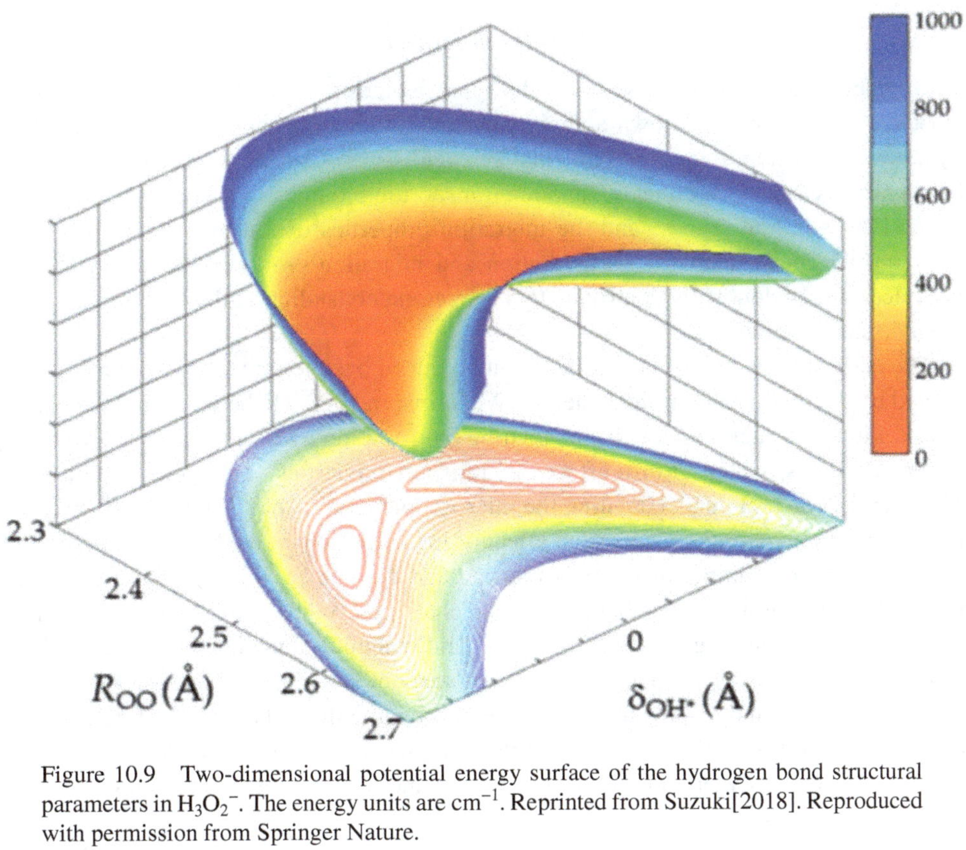

Figure 10.9 Two-dimensional potential energy surface of the hydrogen bond structural parameters in $H_3O_2^-$. The energy units are cm^{-1}. Reprinted from Suzuki[2018]. Reproduced with permission from Springer Nature.

remain localized around the minima of the potential energy surfaces, situated between the oxygen atoms. $H_5O_2^+$ exhibits the same behavior. The change in OO bond length at high temperatures can be attributed to the Ubbelohde effect, assuming that H/D mediates the attractive interaction between the two oxygen atoms. From a chemical perspective, it is highly likely that resonance occurs between O–H···O and O···H–O; compared to D and T, H is more prone to be delocalized, resulting in a stronger attraction between the two oxygen atoms and thus a shorter OO distance. At high temperatures, the tritium nucleus in $T_3O_2^-$ localizes around the two potential minima between the oxygen atoms, which loosens hydrogen bonding and increases the OO distance. A similar situation occurs for $H_5O_2^+$ at high temperatures, although the difference there is small. For $H_3O_2^-$, the quantum delocalization characteristic is still significant at 400 K, whereas for $H_5O_2^+$, it does not. The reason is that the resonance frequency corresponding to $H_3O_2^-$ occurs at a lower temperature (464 K for H, 401 K for D, and 343 K for T), but that corresponding to $H_5O_2^+$ occurs at a higher one (898 K for H, 847 K for D, and 763 K for T).

In conclusion, the temperature dependence of the hydrogen bond structure for $H_3O_2^-$ and $H_5O_2^+$ can be qualitatively understood in the low-temperature case as a result of the isotopic effect on the zero-point energy under an anharmonic potential field. However, at high temperatures, the proton distribution and the coupling between the OO vibrational modes determine the isotope effect.

10.2.4 Ion Hydrates

As a solvent, water not only dissolves salts but also binds to dissolved ions to form various clusters. This process is known as ion hydration, and the clusters formed are called ion hydrates. Many important environmental science issues, such as aerosol formation, salt dissolution, seawater desalination, and ion transport in biological ion channels, are related to the ion hydration process and the migration properties of ion hydrates at interfaces. Studying the mechanisms of ion hydration and migration properties of ion hydrates at interfaces will undoubtedly contribute to a deeper understanding of these important issues in environmental science.

The earliest research on ion hydrates can be traced back to transport experiments conducted by the renowned physical chemist W. Nernst in 1900. He believes that the various kinds of ions in water do not exist in isolation but form ion hydrates with the surrounding water molecules, which in turn affect the transport properties of the ions. Despite over 100 years of research, understanding the microscopic factors that determine the transport properties of ion hydrates, such as the effect of ion hydration on the hydrogen bond structure of water, the number of hydration shells, and the number and distribution of water molecules on each hydration shell, remains a challenge. When water interacts with ions, full quantum effects (e.g., zero-point energy and quantum tunneling) play a crucial role in ion hydration at low temperatures, or even at room temperature, because of the small mass of hydrogen and oxygen atoms. Before introducing the full quantum effects, we will discuss the stability of the structure and dynamic behavior of ion hydrates confined to a surface. We will only provide an overview of some recent experimental and theoretical results; for a more in-depth discussion, interested readers can refer to D. Y. Cao's PhD dissertation [Cao D.Y.2020].

10.2.4.1 Ion Hydrates on Surface

Despite the significant advances in high-resolution scanning probe techniques over the past three decades, obtaining detailed molecular – or even sub-molecular – level hydrate structures in real space remains an extremely challenging task. This is because most previous studies utilized scanning tunneling microscopy (STM), which inevitably suffers from interference from the metal substrate, particularly electronic states near the Fermi level, significantly reducing the imaging resolution for water molecules. Another experimental challenge is the lack of a controlled method for preparing single-ion hydrates on a surface. This dual challenge was not addressed until a major scientific breakthrough was achieved in 2018. Y. Jiang's group at Peking University successfully prepared single sodium ion (Na^+) hydrates with varying numbers of water molecules for the first time and achieved atomic-level imaging of hydrates using spatially enhanced scanning probe techniques [Peng J.B.2018a].

a Sodium-ion Hydrate Preparation and Imaging

Y. Jiang's group devised an ingenious experiment. First, they grew a bilayer NaCl single crystal on the surface of an Au substrate. This insulating layer significantly reduced the influence of the substrate on the experimental signal and improved the STM resolution for imaging water molecules. Thus, they successfully obtained high-resolution STM images of the internal structures of water molecules. Figure 10.10 illustrates the process of preparing Na^+ hydrates in the STM experiment. Initially, a single Cl-ion (Cl^- was extracted from the NaCl(100) surface using an STM tip and an interaction-enhanced STM Cl^- tip (Cl-STM) was prepared. Next, using this enhanced STM Cl^- tip, Na^+ was extracted from the surface, and water molecules were continuously moved and adsorbed around the Na^+ one by one. By allowing the Na^+ to acquire different numbers of water molecules, controlled preparation of Na^+ hydrates could be achieved.

Using this method, Na^+ hydrates of different sizes were prepared on the NaCl(100) surface, and images at the atomic level were obtained using a spatially enhanced atomic force microscope (AFM) based on a CO tip [Peng J.B.2018a] (Figure 10.11). Figure 10.11(a)–(e)

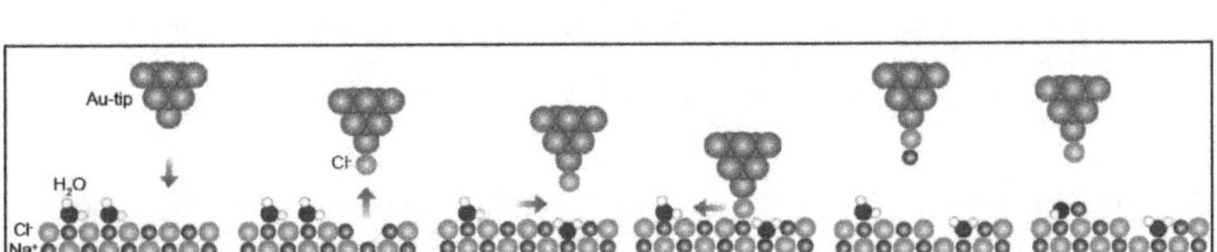

Figure 10.10 Experimental procedure for the preparation of $Na^+{\bullet}nH_2O$ hydrates (n = 1–5) by manipulating individual water molecules using STM tips. Reprinted from Cao D.Y.[2020], with permission of the author.

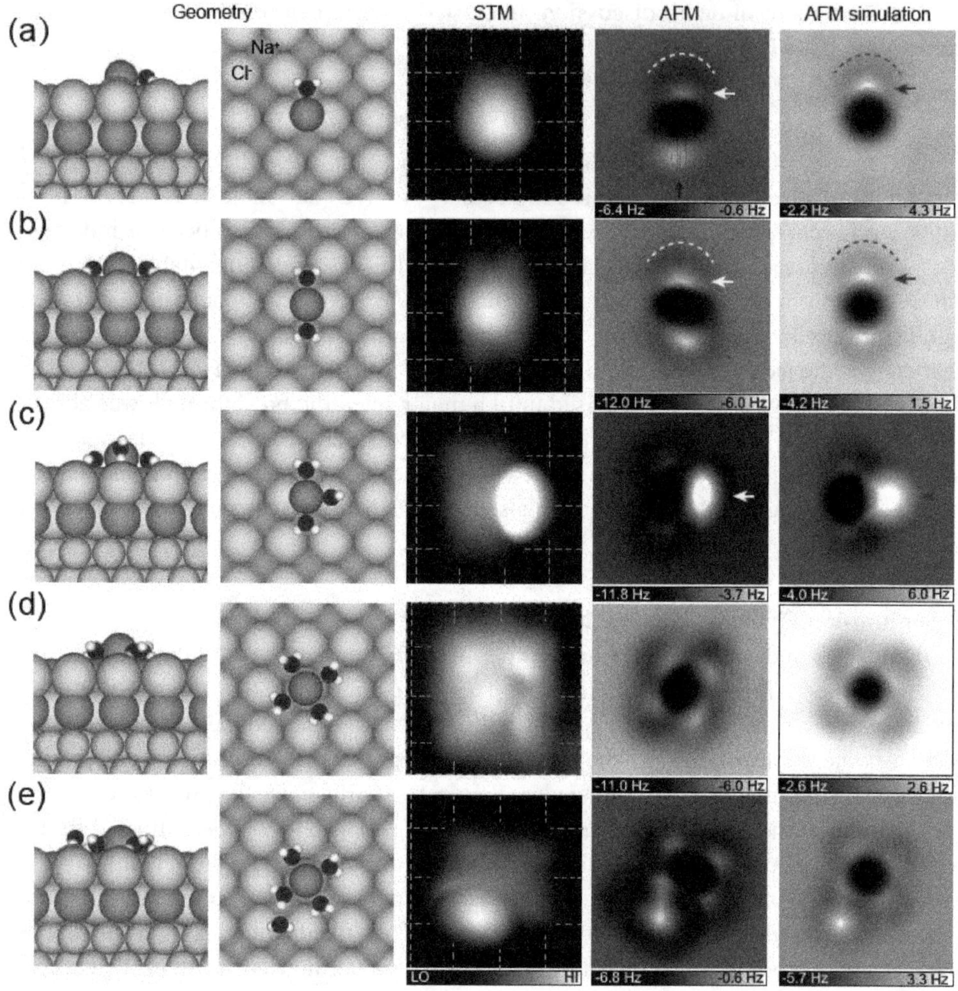

Figure 10.11 Structural diagrams, STM images, and AFM images of Na^+ hydrates supported by Au/NaCl(001) substrate. (a)–(e) Side views of the atomic model (first column), top views of the atomic model (second column), experimental STM images (third column), experimental AFM images (fourth column), and simulated AFM images (fifth column) for the $Na^+{\bullet}nH_2O$ cluster. From (a) to (e), each row corresponds to $Na^+{\bullet}nH_2O$ with $n = 1, 2, 3, 4, 5$, respectively. The H, O, and Na atoms in the hydrates are represented by white, deep gray, and gray spheres in the second column, and Na and Cl of the base are labelled directly in the diagrams. In the first column, the first layer is Cl and second layer is Na, and the third and forth layers are Au. Reprinted from Cao D.Y.[2020], with permission of the author.

shows the side and top views of the most stable Na^+ hydrate configurations obtained using DFT calculations, the experimental STM and AFM images, and the simulated AFM images. In the STM, AFM, and simulated AFM images, the dashed square grid represents the Cl atomic lattice on the NaCl(001) surface. In panels (a) and (b), the bright protrusions in the AFM images, labeled with white and gray left arrows, correspond to the O nuclei in

Table 10.1 *Lennard–Jones (LJ) parameters for AFM simulations. [Cao D.Y.2020].*

Element	ε (meV)	r (Å)
H	0.680	1.487
O	9.106	1.661
Cl	11.491	1.948
Na^+	10.0	1.40

the water molecules. The dark depressions, marked by dashed curves, correspond to the H nuclei in the water molecule. The fuzzy feature indicated by the up arrow in the AFM image in (a) reflects the shift of the water molecule in $Na^+{\bullet}H_2O$ due to the perturbation of the $Na^+{\bullet}H_2O$ structure by the tip during scanning. The white arrow in the AFM image in (c) corresponds to the "standing" water molecule in $Na^+{\bullet}3H_2O$. Using their STM/AFM scanning probe system, Y. Jiang et al. precisely controlled the specific experimental parameters. In Figure 10.11(a)–(e), the set points of the STM imaging are as follows: (a) V = 100 mV, I = 20 pA; (b) V = 150 mV, I = 30 pA; (c) V = 100 mV, I = 30 pA; (d) V = 100 mV, I = 50 pA; and (e) V = 100 mV, I = 15 pA. The AFM tip heights (both experimental and simulated) were set to be 130 pm (7.90 Å), 80 pm (8.10 Å), 100 pm (7.95 Å), 100 pm (8.10 Å), and 100 pm (7.99 Å). The tip height of the AFM image in the experiments was referenced to the STM setpoint on the NaCl surface (100 mV, 50 pA). The tip height of the simulated AFM image was defined as the vertical distance between the apex atom of the metal tip and the Na^+ in the hydrate. The AFM amplitude for both the experimental and simulated AFM images was set to 100 pm. AFM simulations were performed using the probe-particle model method with a d_{z^2} tip (k = 0.75 N/m, Q = –0.2e). The LJ parameters for each atom used in the simulations are listed in Table 10.1. The size of each image in panels (a)–(e) is 1.5 nm × 1.5 nm.

The most stable Na^+ hydrate structures obtained from DFT (shown in Figure 10.11(a)–(e)) were determined by comparing the total energies of different candidate configurations and selecting the one with the lowest total energy. For $Na^+{\bullet}H_2O$, $Na^+{\bullet}2H_2O$, and $Na^+{\bullet}3H_2O$, Na^+ was adsorbed on the bridge site, whereas for $Na^+{\bullet}4H_2O$ and $Na^+{\bullet}5H_2O$, it was adsorbed on the Cl top site. These theoretical results agree with the experimental measurements. In the hydrate structure and high-resolution AFM images, the Na^+ was imaged as a dark depression owing to the electrostatic attraction between the negatively charged CO tip apex and the Na^+. The negatively charged O atoms and positively charged H atoms in the water molecules were imaged as bright protrusions (Figure 10.11(a) and (b), marked by white arrows) and dark rings (Figure 10.11(a) and (b), marked by white dashed curves), respectively. For $Na^+{\bullet}3H_2O$, the positions of the "standing" water molecules were higher, resulting in stronger Pauli repulsion between the tip and the water molecules. As a result, the "standing" water molecules in the $Na^+{\bullet}3H_2O$ were imaged particularly clearly as bright protrusions in the AFM images (as indicated by the white arrow in Figure 10.11(c)). These results confirmed that the AFM simulation images of the most stable structures obtained from the DFT calculations aligned perfectly with the high-resolution AFM images obtained in the experiments. By analyzing and comparing the AFM simulation and experimental images, the atomic structure of $Na^+ \bullet nH_2O$ (n = 1–5) on the NaCl(001) surface could be determined with confidence.

b Dynamics of Sodium-Ion Hydrate

Furthermore, Y. Jiang et al. developed a unique inelastic electron tunneling (IET) technique to investigate the dynamic properties of different Na^+ hydrates. By applying a bias voltage, hot electrons or holes are injected into the substrate, propagate along the substrate surface, and transfer energy to the hydrates, thereby this method excites the diffusion of the hydrates on the substrate (Figure 10.12(a)). The isotope substitution method was used to determine the influence of the nuclear quantum effects on hydrate diffusion. Figure 10.12(b) shows the relationship between the diffusion probability of $Na^+ \bullet 3D_2O$ and $Na^+ \bullet 3H_2O$ and the bias voltage, with the lateral distance between the tip and Na^+ set to $d = 4$ (where d is in units of the lattice constant of NaCl(001): $d = 1$ corresponds to a lateral distance of 0.39 nm between the tip and Na^+). The figure shows that there was an abrupt increase in the diffusion probability of $Na^+ \bullet 3D_2O$ and $Na^+ \bullet 3H_2O$ at bias voltages of ±150 mV and ±170 mV, respectively. Further analysis reveals that the energies of the hot electron/hole at bias voltages of ±150 mV and ±170 mV correspond to the energies of the bending vibration for D_2O and H_2O, respectively. This indicates that hydrate diffusion is related to the excitation of bending vibrations within the water molecule.

Moreover, the hydrates containing heavy water correspond to a narrower bias window, which may be related to nuclear quantum effects. The inset in Figure 10.12(b) shows the diffusion probability of $Na^+ \bullet 3H_2O$ as a function of current at a bias of 170 mV, with $d = 2$. The dashed line represents the experimental data points, and the solid line in the inset represents the least-squares fit of the experimental data using the R–I^N relation, which yields $N = 1.02 \pm 0.08$. The relationship between the diffusion probability of the hydrates and current is linear, indicating that the diffusion of the hydrates was a single-electron process. Therefore, the difficulty of hydrate diffusion can be determined based on the magnitude of the threshold bias (V_{eff}) required to excite the diffusion of different hydrates. The specific parameters used in the experiment are provided in the figure captions.

In the previous experiment, a CO tip was used to excite the diffusion of the Na^+ hydrate; however, the diffusion direction was almost random, which was not conducive for experimental detection. Density functional theory (DFT) calculations revealed that a Cl tip with a strong negative charge could effectively attract the diffusion of the Na^+ hydrate toward the tip. This makes experimental measurement more convenient and provides an effective method for studying the required V_{eff} for the diffusion of different hydrates. In the experiment, a Cl tip was used with the relative height set to –165 pm (referenced to the STM set point on NaCl: V = 100 mV, I = 10 pA) to induce the diffusion of $Na^+ \bullet nD_2O$ (n = 1–5) of different sizes. The specific experimental process was as follows: first, the lateral distance d between the Cl tip and the hydrate was determined, and then the bias voltage V_{eff} was gradually ramped up while the tip height was kept constant; the tunneling current underwent a sudden increase when V_{eff} was sufficiently high (Figure 10.13(a)). This sharp

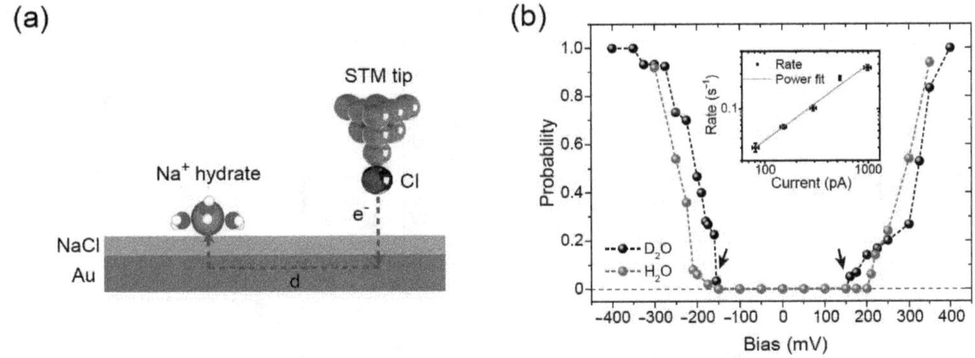

Figure 10.12 (a) Schematic of the diffusion of the experimental process used to excite the diffusion of Na^+ hydrate. Here, *d* denotes the lateral distance between the Na^+ hydrate and the STM tip in units of the lattice constant of NaCl; $d = 1$ corresponds to a distance of 0.39 nm. The gray dashed arrow denotes the direction of hot electron flow. (b) The diffusion probability of $Na^+{\bullet}3D_2O$ and $Na^+{\bullet}3H_2O$ as a function of the bias voltage excited by the CO tip ($d = 4$). The voltage pulse duration was 1.2 s, and the diffusion probability was obtained statistically from 50 excitation events. The threshold bias required to excite D_2O diffusion is indicated by two black arrows. The tip height is –165 pm relative to the STM set point on NaCl: 100 mV, 10 pA. Reprinted from Cao D.Y.[2020], with permission of the author.

increase indicated that the hydrate had moved under or was in direct contact with the tip. Figure 10.13(b) shows the relationship between the V_{eff} required to induce the diffusion of $Na^+{\bullet}3D_2O$ and the lateral distance *d* between the tip and hydrate. The V_{eff} required to excite $Na^+{\bullet}3D_2O$ diffusion gradually increases with increasing *d*, and the difference between the positive and negative V_{eff} is negligible, which again reveals that hydrate diffusion is caused by vibrational excitation.

The dynamics of hydrate diffusion can be determined by studying the V_{eff} values required for the diffusion of different hydrates. Generally, the variation in V_{eff} is positively correlated with *d*. This is because, as *d* increases, so does the proportion of hot electrons/holes scattered into the substrate or flowing in other directions on the surface. Consequently, the number of hot electrons/holes that eventually reach the hydrate decreases, leading to a lower diffusion probability for the hydrate.

Figure 10.13(c) shows the relationship between V_{eff} and *d* required to excite the diffusion of different hydrates. When $d = 2$, the bias voltage satisfied V_{eff} ($Na^+{\bullet}3D_2O$) < V_{eff} ($Na^+{\bullet}4D_2O$) ≈ V_{eff} ($Na^+{\bullet}5D_2O$) < V_{eff} ($Na^+{\bullet}nD_2O$) ($n = 1, 2$). When $d > 2$, the hydrate never reached the tip except in the case of $Na^+{\bullet}3D_2O$, even when the bias voltage was increased to 700 meV. This sufficiently means that the diffusion distances of the other four $Na^+{\bullet}nD_2O$ ($n = 1, 2, 4$, and 5) hydrates were short. The diffusion of $Na^+{\bullet}3D_2O$ could still be excited by a relatively small V_{eff} (~400 mV), even when $d = 7$. Although V_{eff} is not equivalent to the diffusion potential, it can be used as an indicator of the mobility of different hydrates. Based on the previous observations, $Na^+{\bullet}3D_2O$ diffuses more easily than $Na^+{\bullet}4D_2O$ and $Na^+{\bullet}5D_2O$, while $Na^+{\bullet}4D_2O$ and $Na^+{\bullet}5D_2O$ diffuse more easily than $Na^+{\bullet}D_2O$ or $Na^+{\bullet}2D_2O$. In summary, in $Na^+{\bullet}nD_2O$ hydrates, the ion hydrate is most

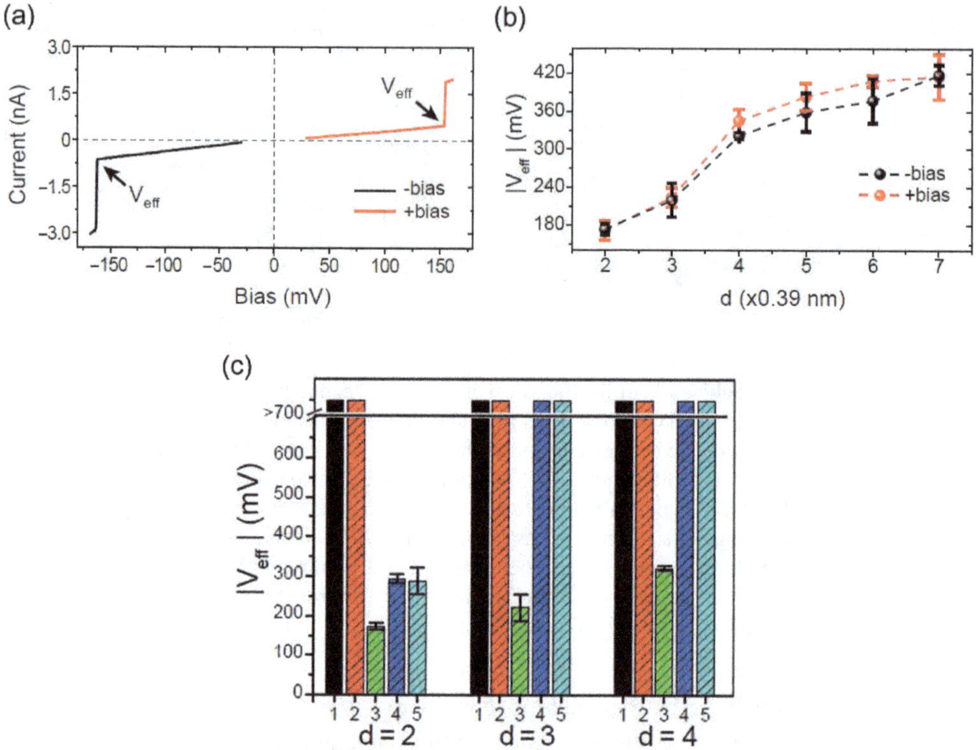

Figure 10.13 (a) Current–bias relationship of $Na^+{\bullet}3D_2O$ with a Cl tip at d = 2, where the arrows point to the jump in current that occurs when the bias voltage reaches V_{eff}. (b) V_{eff} required to induce $Na^+{\bullet}3D_2O$ diffusion as a function of d, the lateral distance between the tip and the hydrate as defined in Figure 10.12. The red (+bias) and black dots (-bias) correspond to the positive and negative bias voltages, respectively. (c) Comparison of V_{eff} required to induce $Na^+{\bullet}nD_2O$ diffusion (n = 1–5) for d = 2, 3, and 4. Reprinted from Cao D.Y.[2020], with permission of the author.

susceptible to diffusion when n = 3, followed by n = 4 or 5, and finally n = 1 or 2. The different diffusion abilities of these hydrates may be related to the differences in their symmetry match with the surface lattice and their quantum tunneling.

c First-Principles Study

To gain a deeper understanding of the dynamic properties of different Na^+ hydrates observed experimentally, D.Y. Cao et al. calculated the classical $Na^+{\bullet}nH_2O$ (n = 1–5) diffusion pathways and the corresponding potential barriers using the climbing image nudged elastic band (cNEB) method within the framework of DFT [Cao D.Y.2020]. Figure 10.14 shows the energy profiles of different Na^+ hydrates corresponding to diffusion with the lowest energy barriers (Figure 10.14(a)) and snapshots of the hydrates along the transition paths (Figure 10.14(b)). As shown in the figure, the diffusion of $Na^+{\bullet}nH_2O$ (n = 1, 2, 3) followed the rotational mode, in which the diffusion of Na^+ was accompanied by rotation of the water molecules. By contrast, the diffusion of $Na^+{\bullet}nH_2O$ (n = 4 and 5) followed

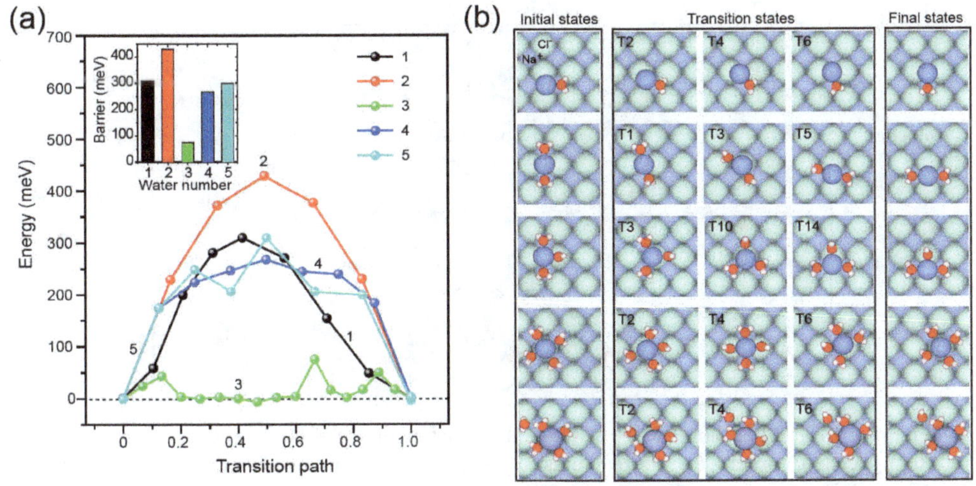

Figure 10.14 (a) DFT-calculated energy profiles of the diffusion of $Na^{+}{\bullet}nH_2O$ (n = 1–5) with the lowest energy barrier; the inset depicts the diffusion energy barriers for different hydrates. (b) DFT-calculated diffusion pathway of $Na^{+}{\bullet}nH_2O$ (n = 1–5) with the lowest energy barrier; the first column, the middle three columns, and the last column represent the initial state, transition state, and final state of $Na^{+}{\bullet}nH_2O$ (n = 1–5) diffusion, respectively. The number "n" in the "Tn" labels at the top-left corners corresponds to the (n+1)st data point in (a). Reprinted from Cao D.Y.[2020], with permission of the author.

a translational mode, in which the water molecules underwent local rearrangements. Notably, $Na^{+}{\bullet}3H_2O$ had the lowest diffusion barrier among all the ion hydrates discussed. In particular, during the diffusion process from the Cl bridge to the Cl top site and back, the energy variation of $Na^{+}{\bullet}3H_2O$ was relatively flat, with a diffusion energy barrier of less than 80 meV. In contrast, the diffusion energy barriers of the other hydrates exceeded 200 meV. These DFT results qualitatively explain the experimental observations and clarify why the diffusion of $Na^{+}{\bullet}3H_2O$ occurs more readily. At low temperatures, nuclear quantum effects may further reduce the corresponding diffusion energy barrier; however, the exact magnitudes of these effects require further investigation.

Among the $Na^{+}{\bullet}nH_2O$ (n = 1–5) hydrates, the diffusion barrier of $Na^{+}{\bullet}3H_2O$ was the lowest, indicating the presence of a "magic number" effect in the diffusion of Na^+ hydrates; an ion hydrate containing a certain "magic" value of the number of water molecules has distinct diffusion properties and a significantly higher diffusion capacity compared with hydrates having other numbers of water molecules. In particular, its mobility is one to two orders of magnitude higher than that of other hydrates and considerably higher than that of bulk-phase ions. The magic number varies depending on the ion species and substrate type; in the present case, it is 3. This dependence can be proved in the specific examples provided later.

To investigate why the diffusion of Na^+ hydrates on the NaCl(001) surface has a magic number $n = 3$, D.Y. Cao et al. conducted detailed computational studies. The results suggest that this effect is primarily associated with a specific metastable configuration of $Na^{+}{\bullet}3H_2O$

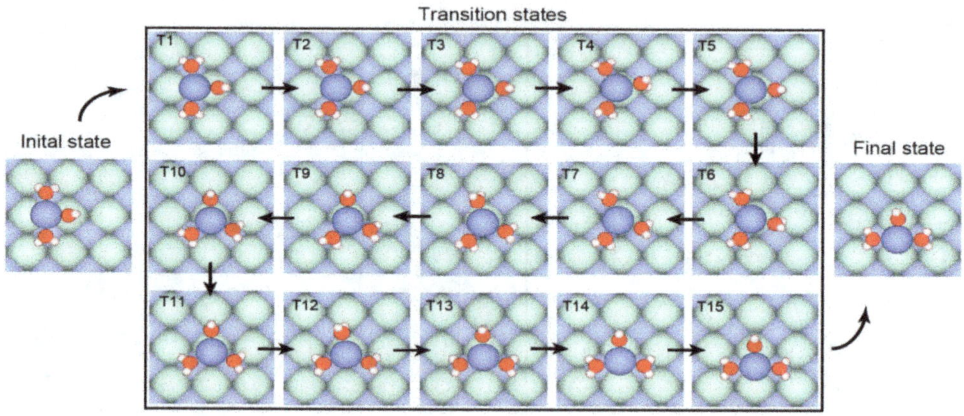

Figure 10.15 Snapshots of the rotational diffusion of $Na^+{\bullet}3H_2O$. The number "n" in the upper-left corner "Tn" corresponds to the (n+1)st data point of $Na^+{\bullet}3H_2O$ in Figure 10.14(a). Reprinted from Cao D.Y.[2020], with permission of the author.

(the "T3" state in Figure 10.15). Unlike the most stable state (the initial state of diffusion) of $Na^+{\bullet}3H_2O$, in which Na^+ is located at the Cl bridge site of the NaCl substrate, Na^+ is located at the Cl top site in the "T3" metastable state. Because of the symmetrical mismatch between $Na^+{\bullet}3H_2O$ and the tetragonal NaCl(001) surface, Na^+ on both the Cl bridge site and the Cl top site cannot make the surrounding water molecules meet the optimal adsorption configuration, resulting in a small energy difference of only a few meV between the most stable state and the state "T3". In addition, the conversion barrier between the stable state and the metastable state "T3" state is only ~50 meV (Figure 10.14(a)). After diffusion from the most stable state to the "T3" state, a certain barrier must be overcome to reach the "T10" state for the rotational diffusion of the hydrate. In the motion near the "T10" state, the three water molecules within the hydrate rotate collectively around the Na^+ adsorbed on the Cl top site (from "T6" to "T11"). The water molecule in the bottom-left corner of the "T9" state needs to first break the hydrogen bond with the Cl^- on the NaCl surface ("T10" state), and then form a new hydrogen bond with another Cl^- on the surface ("T11" state), while the other two water molecules only need to make minor adjustments without breaking any bonds with the surface. This implies that only one hydrogen bond is broken during the diffusion process. To diffuse from the "T11" state to the final (most stable) state, $Na^+{\bullet}3H_2O$ just needs to adjust the orientation of the water molecules slightly and then undergo the reverse process from the most stable state to the "T3" state, finally reaching the final state (the most stable state) and completing the diffusion of the entire hydrate. Therefore, the rotational diffusion of $Na^+{\bullet}3H_2O$ has a very small barrier (~80 meV) to overcome, and the collective rotation of the three water molecules around the Na^+ adsorbed on the Cl top site is easier to be affected by stepwise quantum tunnelling, which can facilitate $Na^+{\bullet}3H_2O$ diffusion in multiple directions with a relatively low energy barrier.

Unlike $Na^+{\bullet}3H_2O$, the structures of the other $Na^+{\bullet}nH_2O$ ($n = 1, 2, 4, 5$) hydrates match well with the symmetry of the NaCl(001) surface, allowing them to remain more stable at specific positions (with Na^+ at the Cl bridge or Cl top sites) and be less prone to diffusion.

Removing one H_2O molecule from Na^+•$3H_2O$ significantly increases the travel distance of water during rotation. On the other hand, adding one H_2O molecule results in the formation of Na^+•$4H_2O$, which exhibits a perfect symmetry match with the NaCl(001) surface. Either of these modifications would impede the diffusion of ion hydrates. Therefore, the degree of symmetry matching between the ion hydrate and the surface determines the diffusion barrier of the hydrate and is the main cause of the magic number effect. Owing to its symmetrical mismatch with the tetragonal NaCl(001) surface, Na^+•$3H_2O$ exhibits a small energy difference between the metastable and most stable states, allowing interconversion with a small potential barrier. This promotes the collective rotation of the three water molecules around Na^+ within Na^+•$3H_2O$. Furthermore, nuclear quantum tunneling effects may cause an additional reduction of the energy barrier for the stepwise rotational diffusion of Na^+•$3H_2O$; this effect is more prominent at low temperatures. Related study has important significance for fully understanding the surface diffusion of ion hydrates.

d Molecular Dynamics Simulation

The STM/AFM experiments described above were conducted at a low temperature (5 K), whereas DFT calculations only provide results for the case at 0 K. To simulate and study the diffusion behavior at finite temperatures, especially the dynamics of ion hydrates on the NaCl surface at temperatures near room temperature, Y. Q. Gao's group at Peking University employed classical molecular dynamics (MD) simulations to clarify the diffusion of hydrates of different sizes [Cao D.Y.2020]. Classical MD simulations were performed using the AMBER program package, and a polarizable force field was developed based on the electronic continuum theory. This force field can reproduce a range of physical and chemical properties of NaCl, including the density, surface tension, viscosity, and dielectric constant of the crystal, as well as the diffusion coefficient of NaCl in solution. In this force field, E_{ij} – the potential energy between unbound atoms i and j – can be expressed as the sum of the Lennard–Jones (LJ) potential (discussed in subsection 1.3.1) and the Coulomb potential. This is expressed as

$$E_{ij} = 4\varepsilon_{ij}\left[\left(\frac{\sigma_{ij}}{r_{ij}}\right)^{12} - \left(\frac{\sigma_{ij}}{r_{ij}}\right)^{6}\right] + \lambda_i\lambda_j \frac{q_i q_j}{4\pi\epsilon_0 r_{ij}}, \tag{10.3}$$

where r_{ij} is the distance between atom i and atom j; q_i and q_j are the charges of the atoms i and j; ϵ_0 is the permittivity of vacuum; λ_i and λ_j describe the Coulombic polarizable effects of atoms i and j; σ_{ij} are the distances at which the interparticle potential is zero; and ε_{ij} is the well depth of the LJ potential. Lorentz–Berthelot combination rules were used to describe the LJ interactions between two different types of atoms. Tables 10.2 and 10.3 show the parameters of the force field used in the classical MD simulations. To test the effect of the force field parameters on the simulation results, a nonpolarizable force field with the SPC/E water model was used to study the diffusion properties of ion hydrates in the MD simulations. The conclusions obtained using the nonpolarizable force field were consistent with those obtained using the polarizable force field, demonstrating the reliability of the polarized force field simulations.

Table 10.2 *Parameters of polarizable force field of NaCl*

Model	q [e]	λ_c	σ [Å]	$(\varepsilon / k_B)[K]$
Na	+1	0.885	2.52	17.44
Cl	−1	0.885	3.85	192.45

Table 10.3 *Force field parameters for polarizable water model based on SPC/ε model*

Model	r_{OH} [Å]	Θ [deg]	$q_H(q_O)$ [e]	σ [Å]	(ε / k_B)[K]
SPC	1	109.47	0.445 (−0.890)	3.1785	84.9

To simulate the diffusion behavior of hydrates on the NaCl(001) surface, the actual system used in the MD simulations was constructed using a slab model, which is commonly used in surface physics. The computational system consisted of four layers of NaCl(001) substrate and supported Na^+ hydrates. This system differs from experiments in several ways, such as neglecting the gold substrate beneath the NaCl thin film and increasing the thickness of the NaCl layer. However, considering that the interaction between ion hydrates and the topmost NaCl layer mainly affects the diffusion of hydrates and that the influence of the gold substrate on hydrates is weak, the above simplification can still ensure that the MD simulation provides reasonable results for the diffusion law of Na^+ hydrates. The simulation used periodic boundary conditions, the bottom layer of NaCl was fixed, all hydrogen-containing bonds were constrained using the SHAKE algorithm, and the cutoff radius for the LJ interaction was set to be 1.0 nm. After the initial equilibration, each system underwent a 200-ns dynamic simulation with a time step of 2 fs.

Owing to the limitations of the available computational capacity and the inherently stochastic nature of diffusion, it was difficult to obtain an accurate diffusion coefficient (D) for the atoms within the hydrate. Therefore, mean-square displacement (MSD) was used to compare the diffusion capabilities of different hydrates. Taking Na^+ as the center of the hydrate, the equilibrium distribution of the hydrate at different positions was assumed to follow the Boltzmann distribution, and the equilibrium fraction of the ith state of the hydrate was

$$\frac{N_i}{N_{total}} = \frac{e^{-E_i/RT}}{\sum_{k=1}^{N_{total}} e^{-E_k/RT}}, \tag{10.4}$$

where E_i is the free energy of the ith state, R is the molar ideal gas constant, and T is temperature. From this, the relative free energy of the different states can be calculated as:

$$\Delta G = -RT \ln \frac{N_i}{N_{total}}. \tag{10.5}$$

Classical MD simulations of the diffusion of different Na^+ hydrates on the NaCl(001) surface not only obtained the relative magnitude of the diffusion rate but also revealed some diffusion information that could not be obtained from 0 K DFT calculations or low-temperature experiments. Figure 10.16(a) shows the x–y diffusion trajectory of $Na^+{\bullet}nH_2O$ (n = 1–5) at 300 K over 200 ns. It can be seen from the figure that the diffusion range of $Na^+{\bullet}3H_2O$ (green) is much greater than that of the other hydrates. Figure 10.16(b) depicts the zoomed-in trajectory of Figure 10.16(a), illustrating the details of diffusion behavior of different hydrates. Figure 10.16(c) shows the MSD of hydrate diffusion at different temperatures over a certain period of time obtained from the MD calculations. This result indicates that the mobility of $Na^+{\bullet}3H_2O$ remained high even near room temperature. For example, at 225 K, the mobility of $Na^+{\bullet}3H_2O$ was more than one order of magnitude higher than that of the other clusters. The results of these MD simulations confirm that $Na^+{\bullet}3H_2O$ diffuses faster than the other hydrates, which is consistent with the findings from the DFT calculations and experimental observations that $Na^+{\bullet}3H_2O$ exhibits a higher level of diffusivity.

Further analysis of the arrangement of the diffusion trajectory points in Figure 10.16(b) was conducted by D.Y. Cao et al. They found that for $Na^+{\bullet}nH_2O$ (n = 1 or 2), the lines connecting the diffusion trajectory points were slanted, indicating that the hydrates hopped between the Cl bridge sites, whereas for $Na^+{\bullet}nH_2O$ (n = 4 or 5), the lines were either horizontal or vertical, indicating that the hydrates hopped between the Cl top sites. These observations are consistent with the DFT-calculated diffusion pathways of the Na^+ hydrates with the lowest energy barriers.

By contrast, hybrid behavior consisting of two types of hopping patterns was observed in the MD results for $Na^+{\bullet}3H_2O$. This is slightly different from the DFT-calculated $Na^+{\bullet}3H_2O$ diffusion path with the lowest energy barrier, where $Na^+{\bullet}3H_2O$ diffuses between the Cl bridge sites. This difference may be attributed to the metastable state of $Na^+{\bullet}3H_2O$, namely the "T3" state obtained from DFT calculations, which has a small energy difference from the stable state. As mentioned in subsection 10.2.3, $Na^+{\bullet}3H_2O$ can diffuse from the stable state to the "T3" state (Na^+ is located at the Cl top site) after crossing a small potential barrier. Moreover, the three water molecules in the "T3" state can rotate collectively around the Na^+, allowing for diffusion in multiple directions with a small energy barrier. Therefore, the hybrid behavior of $Na^+{\bullet}3H_2O$ observed in the MD simulations was a combination of the two hopping patterns.

D.Y. Cao et al. investigated the properties of the most stable and metastable states of Na^+ hydrates of different sizes, including their free energy and particle density distributions. Figures 10.16(d)–(f) depict the free-energy distributions of $Na^+{\bullet}3H_2O$ obtained from the MD simulations and the particle density distributions of the most stable and metastable states of $Na^+{\bullet}3H_2O$. Figures 10.17(a)–(h) show the free-energy distribution and the most stable state particle density distributions of Na^+ hydrates other than $Na^+{\bullet}3H_2O$. It can be seen from Figure 10.17 that the free-energy distribution of $Na^+{\bullet}nH_2O$ (n = 1, 2, 4, 5) contains a single minimum as a result of the perfect symmetry match with the NaCl(001). These free energy minima of Na^+ are located at the Cl bridge (n = 1, 2) and Cl top sites (n = 4, 5). By contrast, the free energy distribution of $Na^+{\bullet}3H_2O$ shows two minimum points, with the Na^+ located at the Cl top site in the metastable state and the Cl bridge sites in the most stable state, owing to the symmetry mismatch between $Na^+{\bullet}3H_2O$ and the substrate (Figures 10.16(d)–(f)).

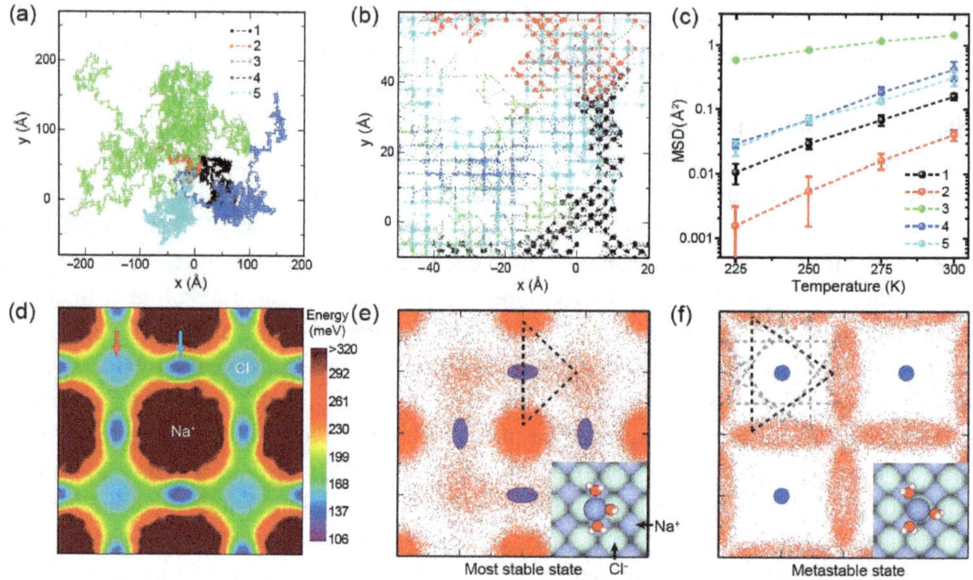

Figure 10.16 Diffusion behavior of Na^+ hydrate on the NaCl(001) surface obtained from MD simulations. (a) Diffusion trajectory of $Na^+{\bullet}nH_2O$ (n = 1–5) in x–y plane at 300 K over 200 ns. (b) Zoomed-in image of (a) showing the diffusion behavior of different hydrates. (c) MSD of $Na^+{\bullet}nH_2O$ (n = 1–5) diffusion at different temperatures; error bars reflect the standard deviation from 10 different datasets. (d) Free-energy distribution of $Na^+{\bullet}3H_2O$ on the NaCl surface at 300 K: the hydrate position is labeled by the position of Na^+; the positions of Na^+ and Cl^- on the underlying NaCl(001) surface are marked as "Na^+" and "Cl^-"; the blue arrow indicates the Na^+ at the Cl bridge site in the most stable $Na^+{\bullet}3H_2O$ state; and the red arrow denotes the Na^+ at the Cl top site in the metastable state. (e) and (f) Density distribution of the most stable and metastable $Na^+{\bullet}3H_2O$ at 300 K: the blue and red dots represent the positions of Na^+ and O in the hydrate; the three water molecules within a $Na^+{\bullet}3H_2O$ molecule are connected with black or gray dashed lines for clarity. The surface areas in (e)–(f) are the same as those in (d). The Na^+ position in (e) is constrained within an elliptical area centered at the Cl bridge site (semimajor axis: 0.5 Å; semiminor axis: 0.25 Å); the Na^+ position in (f) is constrained within a circular region centered at the Cl top site (radius: 0.25 Å). Insets of (e) and (f) show the MD structures of $Na^+{\bullet}3H_2O$ at the Cl bridge site and the Cl top site, respectively, with an image size of 1.2 nm × 1.2 nm. The size of the images in (d)–(f) is 0.8 nm × 0.8 nm. Reprinted from Cao D.Y.[2020], with permission of the author.

According to the particle density distributions, the most stable high-temperature structures of the different hydrates, as determined by MD simulations, are the same as those obtained from DFT calculations at 0 K (see Figure 10.11). In particular, in the density distributions of Na^+ and O in the most stable and metastable $Na^+{\bullet}3H_2O$ states obtained from the MD simulations (Figure 10.16(e) and (f)), two characteristic triangular structures can be identified (see black dashed triangles and insets in Figures 10.16(e) and (f)). These two structures are very similar to the initial (final) and "T3" transition states calculated by DFT. Notably, the triangular structures of the metastable state are distributed in four equivalent configurations (black and gray dashed triangles in Figure 10.16(f)), indicating a stepwise rotation diffusion of the three water molecules around Na^+ within the hydrate. Therefore, the structural information on the most stable structures of different hydrates and metastable Na^+•$3H_2O$ obtained from MD simulations is in good agreement with the results obtained from DFT, revealing that the contribution of adsorption energy is more significant than that of entropy, even at high temperatures.

In-depth analysis of the energy changes along the lowest diffusion barrier pathway obtained from MD simulation reveals that the energy variation during diffusion of Na^+•$3H_2O$ has an "M-shaped" pattern, which differs from the single-peak energy variation of other Na^+ hydrates (see Figure 10.17(i)). This suggests that the diffusion of Na^+•$3H_2O$ is a multistep process with a metastable state and a reduced diffusion barrier. As a result, the migration of Na^+•$3H_2O$ is accelerated, which is consistent with the DFT calculations.

However, a further comparison of the energy changes along the lowest diffusion barrier pathway of the different hydrates reveals some inconsistencies between the MD simulations (Figure 10.17(i)) and DFT calculations (Figure 10.14(a)). For example, the energy changes observed during Na^+•nH_2O (n = 3–5) diffusion differ significantly between the two methods. Additionally, the diffusion barriers of Na^+•$4H_2O$ and Na^+•$5H_2O$ in the MD simulations are only slightly higher than those of Na^+•$3H_2O$, in dramatic contrast to the large differences between them in the DFT calculations. There are two reasons for the discrepancy between the results obtained from the MD simulations and DFT calculations. The DFT considers only one specific pathway when computing the potential barrier, whereas MD simulations calculate the diffusion barrier by statistically sampling multiple pathways at high temperatures. However, the contribution of entropy at high temperatures may alter the diffusion mode of hydrates in MD simulations from that in DFT calculations at low temperatures. In conclusion, the energy variations during the diffusion of hydrates observed from MD simulations at finite temperatures differed quantitatively from those obtained from DFT calculations at 0 K. Apart from the specific potential energy landscape, the contributions of thermodynamic and conformational entropy may also have an important effect on the magic number phenomenon of Na^+ hydrates at finite temperatures.

Certain internal structural changes are inevitable during the collective diffusion of clusters composed of sodium ions and water molecules. To investigate the correlation between the structural changes to water molecules in hydrates and cluster diffusion, D.Y. Cao et al. used MD simulations to study the rotational relaxation of water molecules around Na^+ during cluster diffusion. The rotation relaxation was represented by the rotation time correlation function, which is defined as:

$$C_2(\mathrm{t}) = \left\langle P_2\left[\mu_{\mathrm{WAT}}(0)\cdot\mu_{\mathrm{WAT}}(t)\right]\right\rangle, \qquad (10.6)$$

where P_2 is a second-order Legendre polynomial, and $\mu_{\mathrm{WAT}}(0)$ and $\mu_{\mathrm{WAT}}(\mathrm{t})$ are the direction vectors of the water dipole at times 0 and t, respectively. Figure 10.17(j) shows the rotational relaxation of water molecules in the hydrates as a function of time according to the MD simulations. From this figure, it can be seen that the rotational relaxation rate of the water molecules within the hydrates followed the same trend as the diffusion rate of the hydrate (Figure 10.16(c)) when the number of water molecules changed. This indicates that the migration of hydrates is related to the rotational motion of water molecules. The rotational relaxation rate of Na^+•$3H_2O$ was the fastest, indicating that the

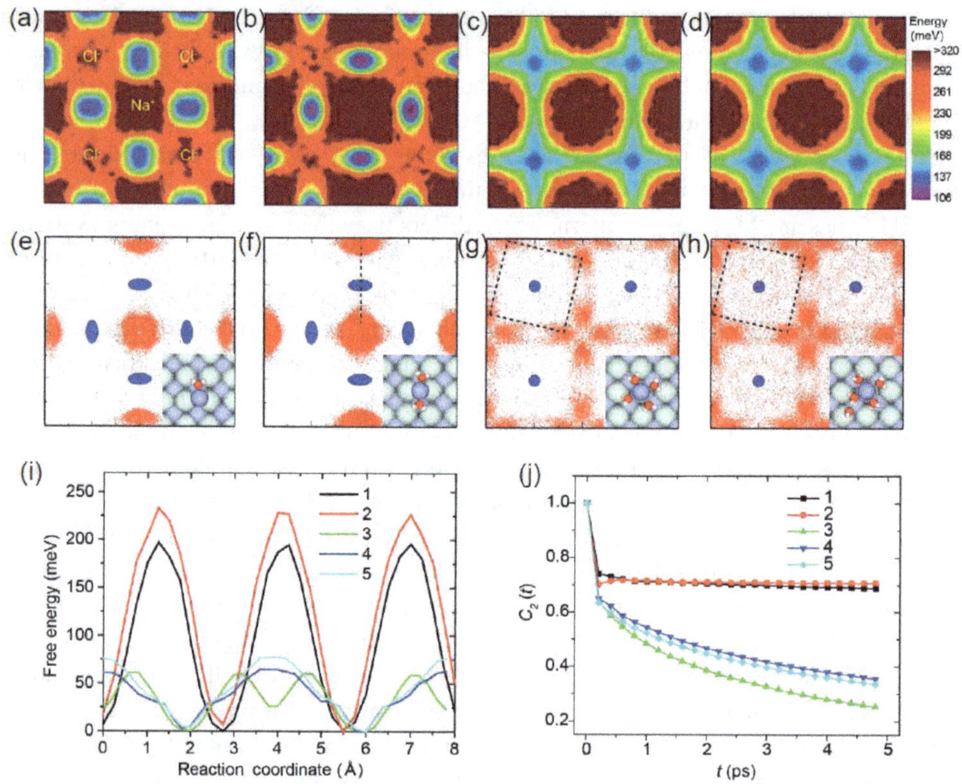

Figure 10.17 Diffusion behavior of $Na^+ \bullet nH_2O$ (n = 1–5) obtained from MD simulations at 300 K. (a)–(d) Relative free energy distribution of $Na^+ \bullet nH_2O$ (n = 1, 2, 4, 5) on the NaCl(001) surface. (e)–(h) Density distribution of the most stable $Na^+ \bullet nH_2O$ (n = 1, 2, 4, 5) on the NaCl(001) surface; blue and red dots represent Na^+ and O; for clarity, water molecules in hydrates are connected by black dashed lines; insets of (e)–(h) show the most stable structures of Na^+ hydrates. The positions of Na^+ and Cl^- on the NaCl(001) are marked as "Na^+" and "Cl^-" respectively, in panel (a). The positions of Na^+ and Cl^- on the NaCl(001) surface are the same in figures (a)–(h). (i) Relative free energy variation of different hydrates along the pathway with the lowest diffusion barrier. (j) Time correlation function C(t) for the rotational relaxation of water molecules in the Na^+ hydration shell for different hydrates. Image sizes (a)–(h): 0.8 nm × 0.8 nm; (e)–(h) insets: 1.2 nm × 1.2 nm. Reprinted from Cao D.Y.[2020], with permission of the author.

water molecules in $Na^+ \bullet 3H_2O$ were the most susceptible to change. This is consistent with the DFT prediction, stepwise diffusion of $Na^+ \bullet 3H_2O$ through collective rotation of the three H_2O molecules around Na^+ (Figure 10.15). In summary, MD simulations qualitatively confirmed the conclusion obtained from DFT calculations, that is, symmetry matching between different hydrates and the substrate is the main factor affecting hydrate diffusion properties. Finally, it should be emphasized that this rotational diffusion process may evolve into a fast diffusion dominated by stepwise quantum tunneling after considering the full quantum effects. These theoretical inferences align with experimental observations.

e Other Ion Hydrates

To verify the generality of the magic number effect in the dynamic behavior of ion hydrates on salt surfaces, D. Y. Cao et al. conducted classical MD simulations of the diffusion of other ion hydrates on the NaCl(001) surface at 300 K [Cao D.Y.2020]. To compare the dynamic properties of different cation hydrates on the surface, MD simulations were conducted to calculate the migration rates of K^+•$n$$H_2O$ (n = 1–5) and Li^+•$n$$H_2O$ (n = 1–5). Figure 10.18 shows the MSDs of the different hydrates over a period of 1 ns obtained from the MD simulations. It can be observed from the figure that the diffusion of K^+ hydrate on the NaCl(001) surface still exhibits a magic number effect, with a specific magic number of 3. Similar to the Na^+ hydrates, the K^+ hydrates with three water molecules showed significantly higher migration rates than the other K^+ hydrates. For the Li^+ hydrate, the magic number is 2. The change in the migration rate of Li^+ hydrates with the number of water molecules was observed to differ from those of Na^+ and K^+. The significantly smaller radius of Li^+ compared to those of Na^+ and K^+ could lead to changes in factors such as structural matching, thus affecting the diffusion behavior of the Li^+ hydrate. However, it should also be noted that the accuracy of the force field of Li^+ has not been verified as reliably as those of Na^+ and K^+. Therefore, further studies of Li^+ hydrates are required.

The discussion thus far has focused mainly on the structural and dynamic properties of cation hydrates on surfaces. However, it is important to consider anion hydrates as well.

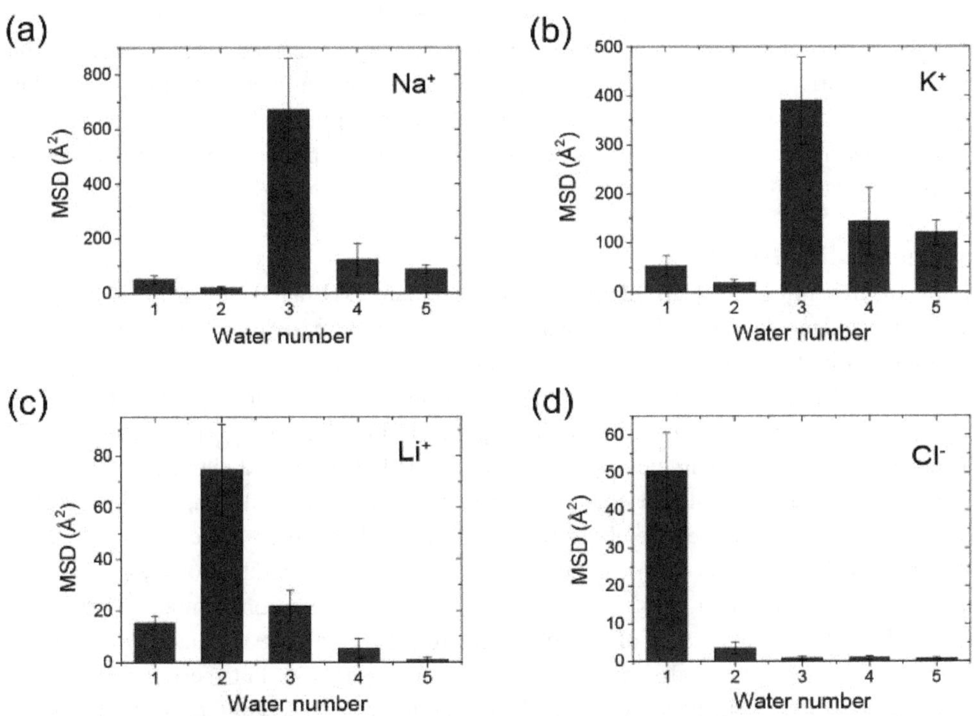

Figure 10.18 MSD of X•$n$$H_2O$ over the period of 1 ns (X = Na^+, K^+, Li^+, Cl^-, n = 1–5) obtained from MD calculations at 300 K. Reprinted from Cao D.Y.[2020], with permission of the author.

D. Y. Cao et al. investigated the migration rate of $Cl^{-}\bullet nH_2O$ (n = 1–5) using classical MD simulations. The results suggest that the dynamic properties of Cl^- hydrate on the NaCl(001) surface are significantly different from those of Na^+ hydrate. As shown in Figure 10.18(d), the MSD result of the Cl^- hydrates indicates that the diffusion capability of Cl^- hydrate is weaker than that of Na^+ hydrate. When n = 1, the diffusion capability of $Cl^{-}\bullet H_2O$ is much greater than that of other Cl^- hydrates. Therefore, it is reasonable to conclude that Cl^- hydrates have a magic number of 1, rather than 3 as observed for Na^+ and K^+ hydrates. This difference in the specific magic numbers can be explained by the distinct interactions of Cl^- and Na^+ with water/NaCl(001). The significantly higher diffusion capability of $Cl^{-}\bullet H_2O$ can also be attributed to the symmetry mismatch between the $Cl^{-}\bullet nH_2O$ and the NaCl lattice.

The previous work provides answers for understanding the anomalously ultrafast ion transport in ion channels, that is, certain hydrated states will increase the migration rate of ions [Sahu2019]. Especially, it would be natural to assume that full quantum effects, which can affect the formation and strength of hydrogen bonds in hydrates, can also affect the thermodynamic stability and dynamic behavior of surface-confined ion hydrates. This could be elucidated by considering the behavior of free ion hydrates in bulk water. However, because of the complex competition of different interactions among ions, water molecules, and substrate materials, the specific influence of full quantum effects remains unclear, and further investigation is needed in this regard.

10.2.4.2 Ion Hydrates in Bulk

The previous examples show that surface-confined ion hydrates tend to have specific structures. In addition, their dynamic behavior exhibits a magic-number effect owing to their matching relationship with the surface lattice. This also applies to free ion hydrates in bulk water under natural conditions. We now explore the intriguing question of how these systems demonstrate full quantum effects.

a Fluoride-Ion Hydrate

Using the first-principles path-integral molecular dynamics (PIMD) method, Y. Kawashima et al. investigated the nuclear quantum and thermal effects on hydrogen bonding in various fluoride-ion hydrates, $F^{-}\bullet nH_2O$ (n = 1–3). Based on specific numerical studies, they identified several important structural parameters in fluoride ion hydrates: OH^* bond length R_{OH^*}, F^-...H length R_{FH^*}, F^-...O length R_{FO}, and H^*FO angle θ_{H^*FO} [Kawashima2013]. The variations in these parameters were also recorded. They found significant discrepancies between the classical and quantum simulation results. In addition, the mean values of the structural parameters for both the simulations changed as the coordination number increased. For the covalent bond length R_{OH^*}, quantum simulations yielded larger average R_{OH^*} values than classical simulations for all coordination numbers of the water molecules. The average value of R_{FO} obtained from the quantum simulation was smaller than that from the classical simulation in the case of $F^-(H_2O)$, but the reverse was true in the case of $F^-(H_2O)_3$, because the nuclear quantum effects strengthened the formation of hydrogen bonds between water molecules. Overall, nuclear quantum effects increased R_{FH^*}, R_{FO}, θ_{H^*FO}, but decreased R_{OH^*} with increasing coordination number. Evidently, the patterns of the full quantum effects in these systems are not entirely consistent.

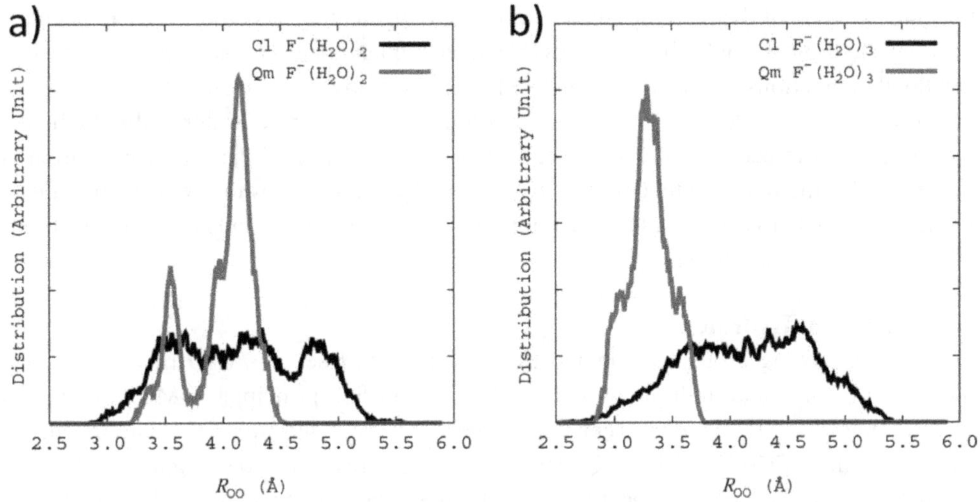

Figure 10.19 Distribution of OO distance for F^{-}•$2H_2O$ and F^{-}•$3H_2O$. The gray and black curves denote the quantum and classical simulation results, respectively. Reprinted with permission from Kawashima[2013]. Copyright 2024 American Chemical Society.

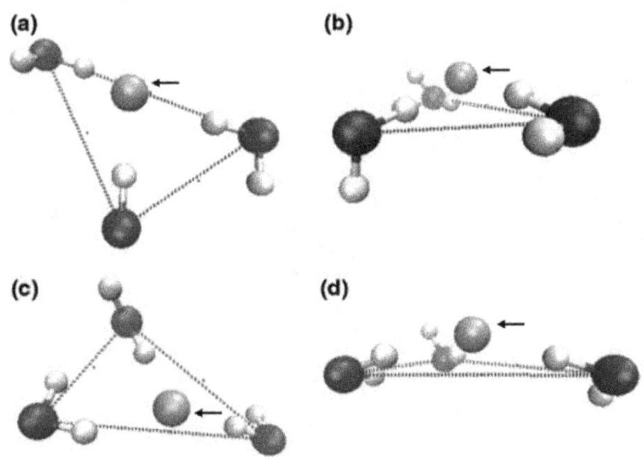

Figure 10.20 Geometric configuration of the F^{-} ion hydrate F^{-}•$3H_2O$; (a) and (b) are the results of classical simulations and (c) and (d) are the results of quantum simulations. The isolated gray sphere represents the F^{-} ion (indicated by arrow). Reprinted with permission from Kawashima[2013]. Copyright 2024 American Chemical Society.

Subsequently, Kawashima et al. analyzed some special phenomena caused by full quantum effects in this system. Figure 10.19 depicts the one-dimensional distribution of the OO distance (R_{OO}). For the F^{-}•nH_2O (n = 2, 3) system, R_{OO} ranged from 3.0 Å to 5.3 Å in the classical simulation. In the quantum simulation, it was mainly localized around two values (4.2 Å and 3.4 Å), which indicates hydrogen bond formation between the two water molecules due to nuclear quantum effects in this system. This is an important characteristic of ion hydrates in bulk water.

Figure 10.20 illustrates the hydrogen-bond configuration. In the classical simulation, the hydrogen atoms of water are all oriented toward the F ion, whereas in the quantum simulation, some of them are oriented toward the oxygen atoms of other water molecules, forming a hydrogen bond. This indicates that nuclear quantum effects strengthen the hydrogen bond interactions between the water molecules. Moreover, the nuclear quantum effects on the water–water hydrogen bonds suppress those on the fluorine–water hydrogen bonds and compete with each other. Further calculations have indicated that this competition is the reason for the opposite tendencies observed in R_{OH^*} and R_{FO} as the coordination number increases. Clearly, the nuclear quantum effects on hydrogen bonds affect the stability of the ion-hydrate structures.

b Chloride-Ion Hydrate

Similarly, Q. Wang et al. investigated the structures of free $Cl^{-}\bullet nH_2O$ (n = 2–4) complexes with single and multiple hydration shells using first-principles PIMD simulations to determine the difference between these free complexes and the previously discussed surface-confined $Cl^{-}\bullet nH_2O$ [Wang Q.2013]. The results are presented in Figure 10.21. From this research, we can compare the differences between the free movement of hydrated chloride ions in bulk water and the restricted movement on the material surfaces. Firstly, free $Cl^{-}\bullet nH_2O$ hydrates in bulk water are composed of multishell structures. In a single-shell structure, water molecules are directly connected to chloride ions via hydrogen bonds. In contrast, in the multishell structure, the outer-layer water molecules form hydrogen bonds

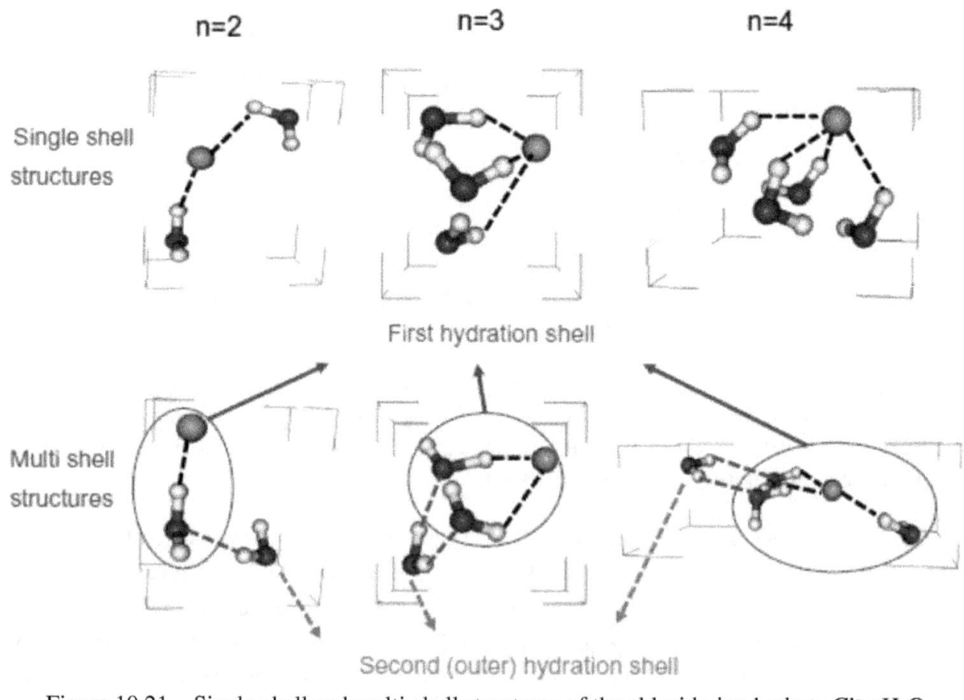

Figure 10.21 Single-shell and multi-shell structures of the chloride-ion hydrate $Cl^{-}\bullet nH_2O$ (n = 2–4). The gray, black, and white spheres in the figure represent the chlorine, oxygen, and hydrogen atoms, respectively. Reprinted from Wang Q.[2013]. Copyright 2024, with permission from Elsevier.

with those in the first hydration layer. Because the hydrogen bonds formed between water molecules and Cl^- ions are weaker, the rearrangement of single- and multi-shell structures is common in $Cl^-{\bullet}nH_2O$ ($n = 2$–4) hydrates.

Owing to the flexible structures of Cl ion hydrates in bulk water, the bond arrangement within them frequently changes. Based on previous research, Q. Wang et al. proposed the following geometric rules for the characteristic radial and angular sizes of ion–hydrogen-bond structures:

$$r(\text{Cl}\ldots\text{O}) < r_c(\text{Cl}\ldots\text{O}),$$

$$r(\text{Cl}\ldots\text{Y}^*) < r_c(\text{Cl}\ldots\ \text{Y}^*),$$

$$\theta(\text{Cl}\ldots 0 - \text{Y}^*) < \theta_c(\text{Cl}\ldots\text{O-Y}^*),$$

$$\delta > \delta_c,$$

where Y^* represents a bound atom of hydrogen or deuterium, and the subscript c represents the cutoff of the upper limit of ionic hydrogen bonds (e.g., $r_c(\text{Cl}\ldots\text{O})$ represents the longest value of Cl ion hydrogen bond). In the final rule, $\delta = |r(\text{Cl}\ldots.\text{Y}_1) - r(\text{Cl}\ldots\text{Y}_2)|$, with Y_1 and Y_2 representing hydrogen atoms within the same water molecule. When $\delta = 0$, the water molecules are symmetrically bonded to the Cl^- ion. Larger values of δ correspond to stronger ionic hydrogen bonds. Using these geometric rules, Q. Wang et al. calculated the probability distribution function of r(Cl….O) in $Cl^-{\bullet}nH_2O$ when $n = 1$ (shown in Figure 10.22). No water–water hydrogen bonds were observed because only one water molecule was present in the cluster. Figure 10.22 shows the change in r(Cl….O) for quantum simulations of hydrogen (H(qu)) and deuterium (D(qu)), and for a classical simulation of

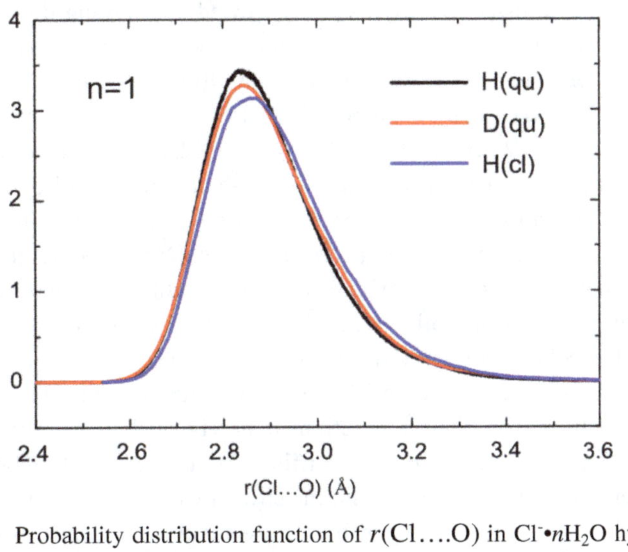

Figure 10.22 Probability distribution function of r(Cl….O) in $Cl^-{\bullet}nH_2O$ hydrate when $n = 1$. The black (top), red (middle), and blue lines (bottom) in the figure represent the simulation results for quantum hydrogen (H(qu)), quantum deuterium (D(qu)), and classical hydrogen (H(cl)), respectively. Reprinted from Wang Q.[2013]. Copyright 2024, with permission from Elsevier.

hydrogen (H(cl)). These results suggest that Cl^- ion–water stretching is enhanced when quantum effects are considered. Because of their heavier masses, the D atoms form relatively weak hydrogen bonds with the ions.

To study the rearrangement pattern of single-layer or multilayer ionic hydrate structures, the longest value of r(Cl....O) at each time step, denoted as r^1(Cl....O), was chosen for $n > 1$. When r^1(Cl....O) $< r_c$(Cl....O), all water molecules are bound to Cl ions through ion–water hydrogen bond interactions, thus forming a single-shell structure. Conversely, if r^1(Cl....O) $> r_c$(Cl....O), a multishell structure is formed, and the outer water molecules are bound to ions through water–water hydrogen bond interactions. Probability distribution plots of chloride ion hydrates as functions of $r^1\left(\text{Cl....O}\right)$ distances for $n > 1$ are shown in Figure 10.23 and the corresponding typical structural features (S1–S15) are shown in Figure 10.24.

The observed proportion of multishell structures increases with the number of water molecules. This indicates that as the number of water molecules increases, the probability of water–water hydrogen bond interactions also increases. As shown in Figure 10.23(a), for $n = 2$, the proportions of single-shell structures (S1 and S2) for H and D are smaller in the quantum simulations than in the classical simulations. In contrast, the proportion of multishell structures (S3) is larger in the quantum simulations. Interestingly, D(qu) exhibits a higher probability of multishell structures (S3) than H(qu). In general, when a proton is replaced by a deuteron, both ion–water and water–water hydrogen bond interactions are weakened because of the heavier mass of D. Nevertheless, the ion–water hydrogen bond remains weaker than the water–water hydrogen bond. Therefore, the water–water hydrogen bond network is preferred in the D(qu) clusters, leading to an increased S3 probability. Moreover, because the zero-point energy of D(qu) is smaller than that of H(qu), the behavior of hydrogen bond clusters in D(qu) can be viewed as "cooling down" with respect to H(qu). As shown in Figure 10.23, the probability difference between H(qu) and H(cl) was significantly smaller than that between H(qu) and D(qu). This implies that the effective potential is more sensitive to H/D isotope substitution.

For $n = 3$ (Figure 10.23(b)) and $n = 4$ (Figure 10.23(c)), the probability distribution functions for H(cl), D(qu), and H(qu) in the single-shell structure followed the same trends as those for $n = 2$ because of the isotope effect. However, this was not the case for the multi-shell structures. For $n = 3$, the probabilities of D(qu) and H(qu) were similar in magnitude in structures S6 and S7, whereas the probability of D(qu) was slightly higher in the S8 structure. It is worth noting that S8 had more/fewer water–water hydrogen bond configurations than S7/S6. For $n = 4$, the probability of D(qu) was larger than that of H(qu) in the S11–S13 structures but lower in S14 and S15. Evidently, the S11–S13 structures had more water–water hydrogen bond interactions than S14 or S15. The differences observed between the $n = 3$ and $n = 4$ cases may be attributed to the competition between ion–water and water–water hydrogen bonding within the structures. These results indicate that structures with more water–water hydrogen bonds are favored in D(qu), which is consistent with the findings for $n = 2$. This study highlights the critical influence of full quantum effects on the structure of ion hydrates, which must have important implications for reaction processes in environmental science.

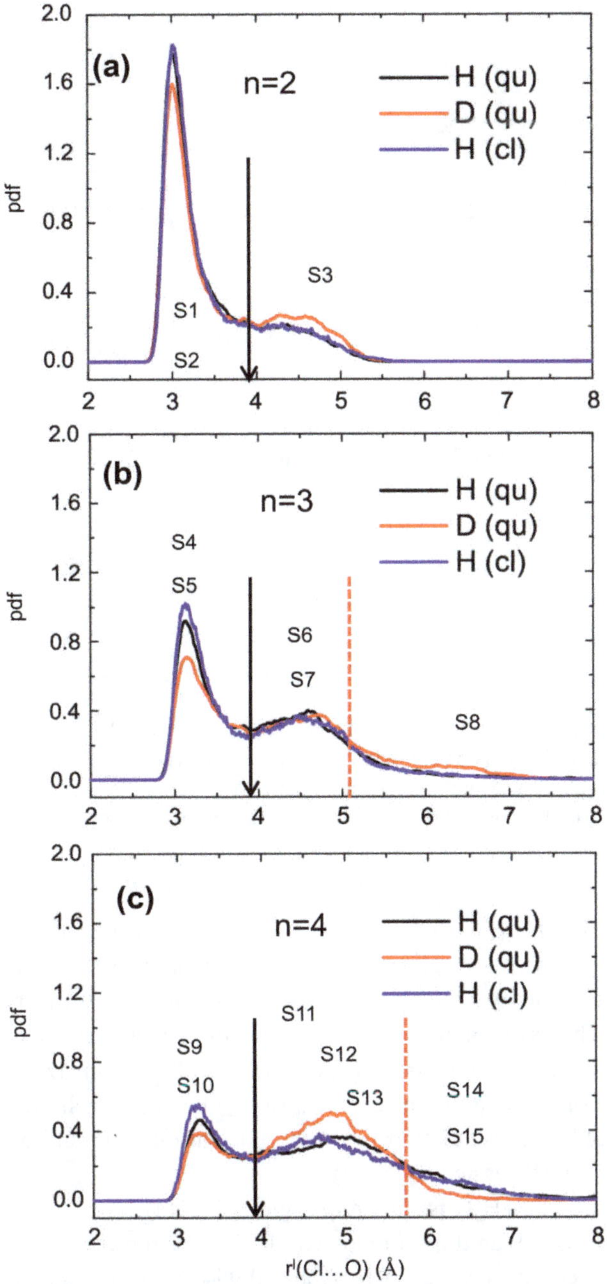

Figure 10.23 Probability distribution function (pdf) of chloride-ion hydrate as a function of r^1(Cl....O) distance, with (a) $n = 2$, (b) $n = 3$, and (c) $n = 4$. The black, red, and blue curves correspond to the quantum simulations of hydrogen (H(qu)), quantum simulations of deuterium (D(qu)), and classical simulations of hydrogen (H(cl)), respectively. Reprinted from Wang Q.[2013]. Copyright 2024, with permission from Elsevier.

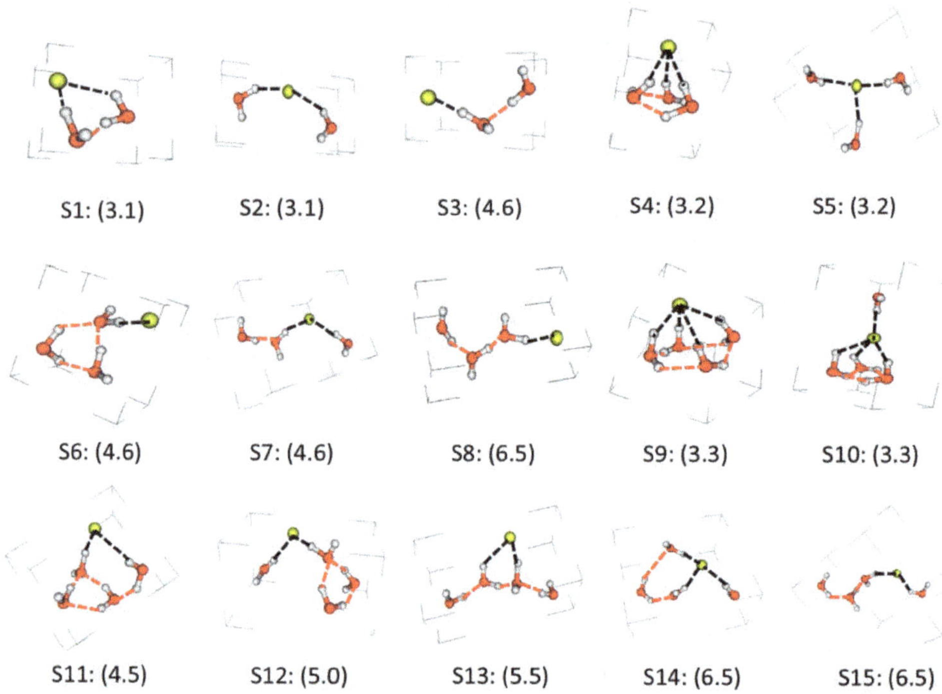

Figure 10.24 Schematic illustrations of $Cl^-(H_2O)_{2\text{-}4}$ structures (S1–S15). The r^1(Cl….O) distance of each structure is presented in parentheses (in Å). The black and red dashed lines represent ion–water and water–water hydrogen bond interactions, respectively. Reprinted from Wang Q.[2013]. Copyright 2024, with permission from Elsevier.

10.2.5 Fluorine–Water Reaction

Studying the reaction $F+(H_2O)_n \rightarrow FH+(H_2O)_{n-1}OH$ in bulk water can aid in understanding dissolution processes relevant to environmental research. T. Udagawa et al. calculated the potential energy profiles and molecular geometries of this reaction using DFT [Udagawa2016]. They systematically studied the nuclear quantum effects of hydrogen and deuterium atoms (referred to as the "quantum H" and "quantum D" cases, respectively) using a multicomponent quantum mechanical (MC_QM) approach. The results were subsequently compared with calculations in which nuclear quantum effects were not considered (referred to as the "classical H" case).

The reaction path of $F+H_2O$ is shown in Figure 10.25(a). The covalent bond lengths of quantum H were longer than those of classical H. For example, the O–H_2 and F–H_1 bond lengths of the exit complex were 0.996 Å and 0.954 Å in the quantum H case, respectively, whereas in the classical H case, they were 0.972 Å and 0.930 Å. However, hydrogen bonds showed the opposite behavior: The O–H distance of the exit complex (1.755 Å) with quantum H was shorter than that with classical H (1.802 Å). The calculated covalent and hydrogen bond lengths in the quantum D case fell between the results for the quantum H and classical H cases because the nuclear wave function of deuterium is more localized than that of hydrogen.

The potential energy profiles of the reaction show that nuclear quantum effects decreased the relative energy of the transition state. The Mulliken charges of hydrogen in the entrance complex obtained by the multicomponent quantum mechanical calculations were 0.255 for quantum H, 0.250 for quantum D, and 0.237 for classical H. Thus, the electrostatic interactions weakened in the order quantum H > quantum D > classical H. As the electrostatic interaction is the second-largest component of the total interaction energy in the entrance complex, the results that considered nuclear quantum effects were more stable than the classical H results.

The nuclear quantum effects of hydrogen and deuterium atoms also enhance hydrogen bond interactions. Owing to the presence of two hydrogen bonds (F–H(D) and O…H(D)) in both the transition state and exit complex, significant stabilization occurs when nuclear quantum effects are included. The relative energies of quantum H and quantum D in the transition state were 3.4 kcal/mol and 2.5 kcal/mol lower than that of classical H, respectively, while in the exit complex, the relative energies of quantum H and quantum D were 1.9 kcal/mol and 1.4 kcal/mol also lower than that of classical H. In addition, the activation energy of the transition state was 2.8 kcal/mol when nuclear quantum effects were neglected, indicating a positive potential barrier for this reaction. However, after

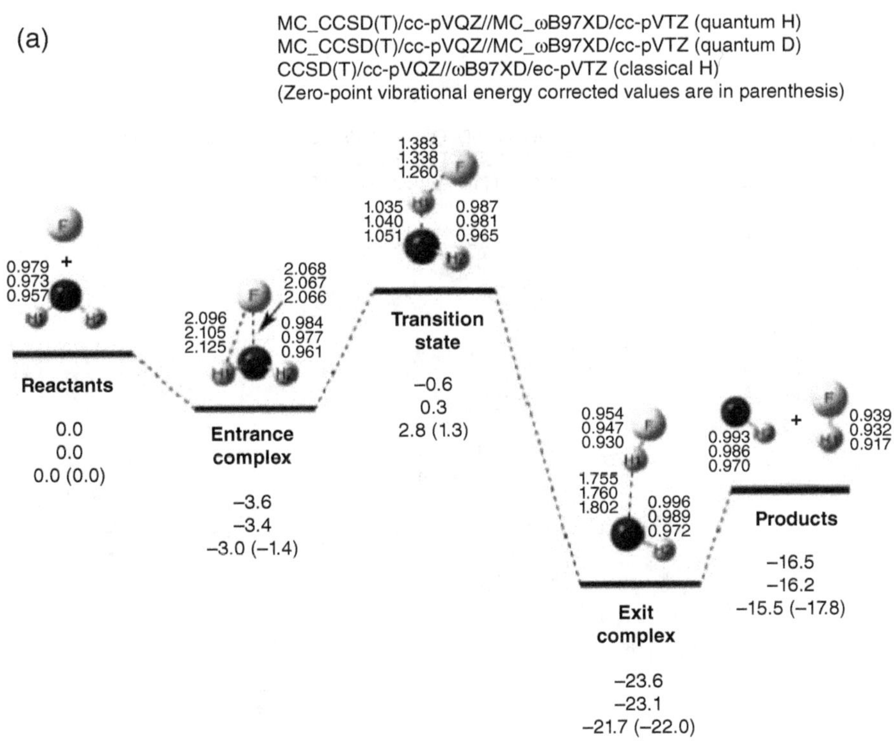

Figure 10.25 Potential energy profiles (kcal/mol) and molecular geometries (stationary point structure) for the reaction $F + (H_2O)_n \rightarrow FH + (H_2O)_{n-1}OH$ for (a) $n = 1$, (b) $n = 2$, and (c) $n = 3$. Black ball represents oxygen. Reprinted from Udagawa[2016], with the permission of AIP Publishing.

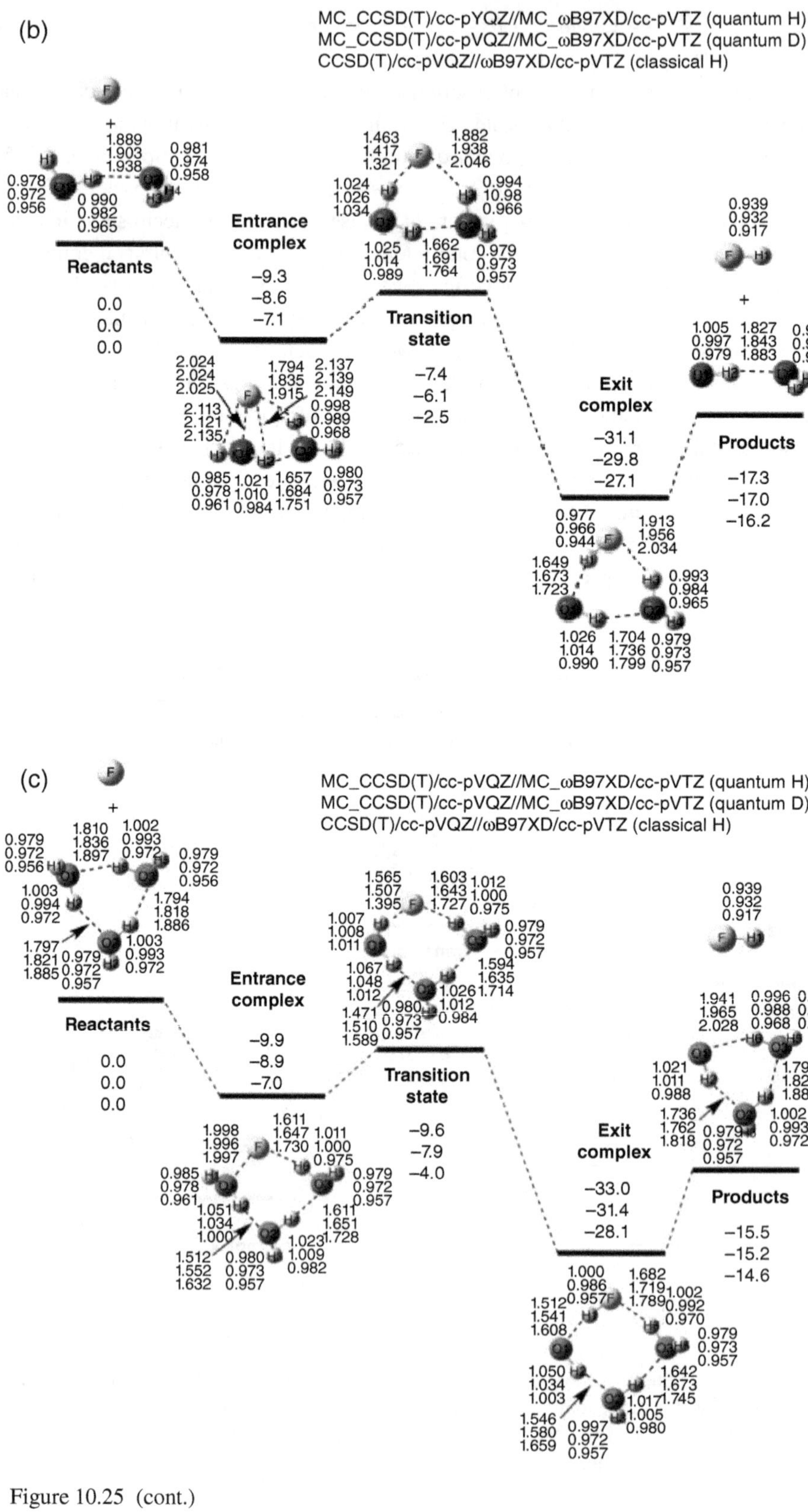

Figure 10.25 (cont.)

considering the nuclear quantum effect, the relative energy of the transition state in quantum H was 0.6 kcal/mol lower than that of the reactants. Moreover, the relative energies of the entrance complex in quantum H and quantum D were 0.6 kcal/mol and 0.4 kcal/mol lower than the classical-H result, respectively. By contrast, although the zero-point energy correction reduced the relative energy of the transition state to 1.3 kcal/mol, it increased the relative energy of the entrance complex by 1.6 kcal/mol, effectively destabilizing it. This suggests that, in addition to simple zero-point energy corrections, the geometric relaxation due to nuclear quantum effects and its influence on the potential energy profile of the $F + H_2O \rightarrow FH + OH$ reaction must also be considered.

The reaction $F + (H_2O)_2 \rightarrow FH + (H_2O)OH$ is shown in Figure 10.25(b). After considering the nuclear quantum effects, the relative energies of all the states were found to be lower than the classical H results. When nuclear quantum effects were considered, more hydrogen bonds were present in the transition state and exit complex than in the entrance complex, significantly increasing the stability (i.e., lowering the energy). Correspondingly, the relative energy difference between the entrance complex and the transition state decreased to 1.9 kcal/mol for quantum H and 2.5 kcal/mol for quantum D. When the nuclear quantum effects were not considered, the relative energy difference was 4.6 kcal/mol. Compared to the $F+H_2O$ reaction, the relative energy of the transition state of the second water molecule was reduced in the $F+(H_2O)_2$ reaction, which in turn lowered the potential barrier. In the $F+(H_2O)_2$ reaction, the calculated transition-state energy was significantly lower than the reactant energy for all three methods, indicating that the second water molecule catalyzed the forward reaction.

Subsequently, the reaction $F + (H_2O)_3 \rightarrow FH + (H_2O)_2 OH$ was investigated, and the results are shown in Figure 10.25(c). The inclusion of nuclear quantum effects results in a more pronounced energy change in the transition state and exit complex than in the entrance complex. In the transition state, the third water molecule reduced the relative energies by 2.2 kcal/mol and 1.8 kcal/mol for quantum H and D, respectively. In the classical H case, the relative energy is reduced by only 1.5 kcal/mol. As a result, the relative energies of the transition state in quantum H and quantum D are only 0.3 kcal/mol and 1.0 kcal/mol higher than those of the entrance complexes. This implies that a third water molecule can also catalyze the reaction and lower the relative energy of the transition state, again demonstrating that nuclear quantum effects can significantly lower the reaction barrier.

10.3 Atmospheric Chemistry

10.3.1 Ozone Formation and Stability

A significant amount of ozone (O_3) is present in the Earth's atmosphere. Ozone is an allotrope of oxygen (O_2). When ultraviolet radiation from the sun passes through the atmosphere and is absorbed by O_2, the molecule undergoes photolysis to produce oxygen atoms. These oxygen atoms can then combine with O_2 molecules to form ozone. In the stratosphere, ozone absorbs large amounts of potentially harmful ultraviolet radiation, forming a protective shield between the outer space and the Earth's surface. However, ozone is unstable and dissociates into oxygen atoms and O_2 molecules after absorbing ultraviolet radiation. In this way, ozone–oxygen cycling is established in the atmosphere.

As early as 40 years ago, scientists discovered that the ozone layer in the stratosphere can be depleted by halogens. Halogens are produced by the photolysis of compounds such as CFCs and halons, which were commonly used in industry during the twentieth century and released into the atmosphere. The resulting depletion of ozone could lead to the loss of its protective effect against UV radiation, which the ozone layer prevents from reaching the Earth's surface. This environmental concern has sparked extensive experimental and theoretical research. On the other hand, anthropogenic emissions can increase the concentration of ozone at ground level, posing significant human health risks. Therefore, the formation and transformation of ozone are important research topics in environmental protection.

Although the ozone molecule is small, the complexity of its electronic structure makes the study of its quantum chemistry challenging. Ozone has an isomer known as cyclic ozone (shown in Figure 10.26), with the symmetry of an equilateral triangle (D_{3h} point group). Without resonance, the cyclic configuration satisfies the condition for a stable Lewis structure (a perfect octet of three oxygen atoms). According to early theoretical calculations, the energy of a cyclic ozone molecule is ~30 kcal/mol higher than that of "normal" (or "open") ozone. The energy barrier for the isomerization reaction from cyclic to open ozone is ~24 kcal/mol. In addition, the energy of cyclic ozone is 6–7 kcal/mol higher than the sum of the energies of a ground-state O_2 molecule and an oxygen atom. The reaction barrier for the dissociation of cyclic ozone is ~47 kcal/mol. Thus, the intrinsic thermal stability of cyclic ozone is closely related to its isomerization rate. This also implies that, in ozone production starting from O_2+O, it is possible to avoid the high-energy potential barrier by using cyclic ozone as an intermediate state before transforming into the more stable open isomer.

Theoretically, cyclic ozone is a dynamically stable molecule that can be observed under suitable experimental conditions. However, despite considerable efforts over the last two decades, no experimental evidence of its existence has been obtained. One possible reason for the absence of evidence could be that under experimental conditions, the topology of the O_3 potential energy surface (PES) leads to the absence of energetically favorable paths connecting open ozone or the ground state O_2+O to cyclic ozone. Another possible reason is that the isomerization reaction from cyclic to open ozone occurs rapidly, and the lifetime of cyclic ozone is too short to be detected using spectroscopic techniques. Although the isomerization barrier is as high as ~24 kcal/mol, the movement of oxygen nuclei during the reaction is relatively small. Therefore, the width of the corresponding energy barrier is also small. Quantum tunneling of oxygen could thus make a significant contribution to the dynamic process of ozone production.

To better understand the transformation between these conformations, researchers have recently calculated the temperature dependence of the reaction rate constant during single-molecule isomerization using variational transition state theory with microcanonical optimized multidimensional tunneling correction at quantized reactant states

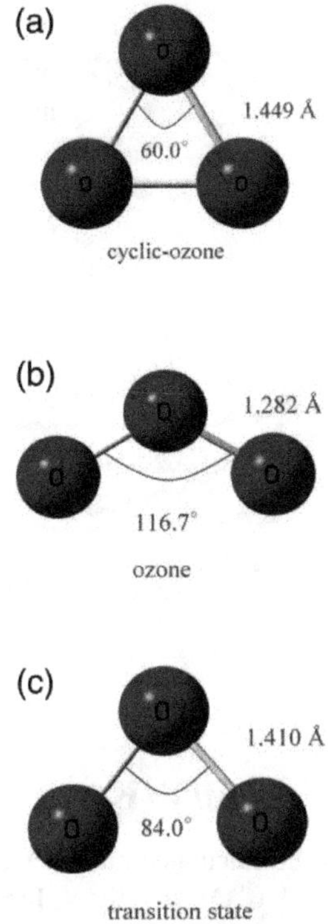

Figure 10.26 Calculated structure of the ozone molecule. From (a) to (c): cyclic ozone, open ozone, and the transition state. Reprinted with permission from Chen J. L.[2011]. Copyright 2024 American Chemical Society.

(μOMT – QRST). The results effectively demonstrate the importance of quantum tunneling and provide an estimate of the lifetime of the elusive cyclic ozone to the correct order of magnitude [Chen J.L.2011]. Figure 10.27 shows the Arrhenius plot of the calculated reaction rate as the temperature increased from 25 K to 500 K. The results obtained from the μOMT and μOMT-QRST methods are also given. The rate constants were found to span 165 orders of magnitude in this temperature range owing to the relatively large energy barrier (21.1 kcal/mol). Taking the quantum tunneling correction into account, the temperature dependence of the reaction rate is greatly weakened at low temperatures, as indicated by the curvature of the Arrhenius plots.

Quantum tunneling leads to significant changes in the motion of oxygen atoms during chemical reactions. In this system, the overall structural changes were minimal when the barrier was high, indicating a narrow energy barrier and an increased tunneling probability. After quantum tunneling was included, the reaction rate calculated by μOMT-QRST was three times larger than that predicted by classical transition state theory (TST) at 500 K. This result suggests that, at this temperature, quantum tunneling and the over-the-barrier process are equally important. When the temperature decreased below 500 K, tunneling

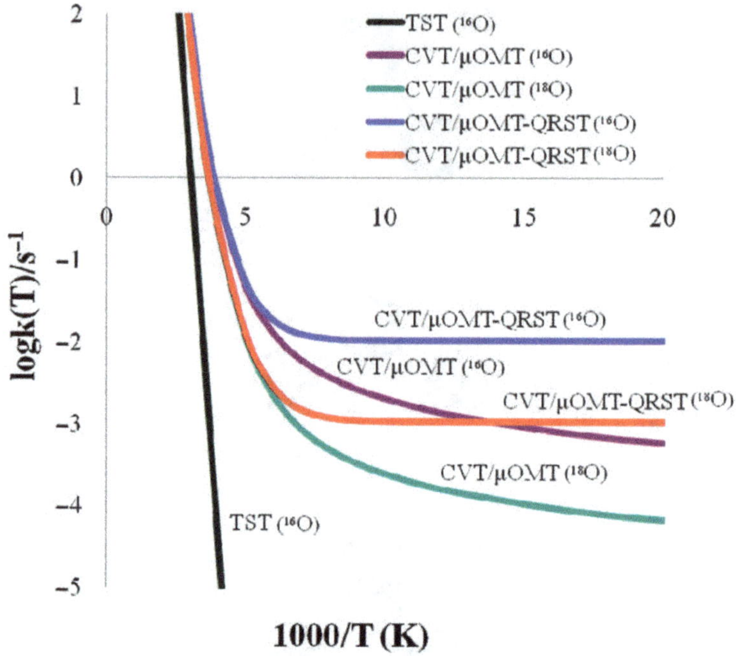

Figure 10.27 Arrhenius plot of the calculated rate for the cyclic-to-open ozone reaction. The curves for $(^{16}O)_3$ are labeled (^{16}O) and those for $(^{18}O)_3$ are labeled (^{18}O). The transition state theory (TST) curve of (^{18}O) is not shown because it almost entirely overlaps that of (^{16}O), which is consistent with the Arrhenius relation. CVT, canonical variational theory. μOMT, microcanonical optimized multidimensional tunneling correction. μOMT-QRST, microcanonical optimized multidimensional tunneling correction at quantized reactant states. Reprinted with permission from Chen J. L.[2011]. Copyright 2024 American Chemical Society.

became the dominant factor in determining the reaction rate. For instance, the rates calculated considering oxygen tunneling were 200, 10^7, and 10^{27} times higher than the classical reaction rates at 300, 200, and 100 K, respectively. The QRST rate constant levelled off below 150 K, entering the reaction regime dominated by deep quantum tunneling process with an asymptotic value of $\sim 1 \times 10^{-2}$ / s. At this temperature, essentially all reactant molecules were in the vibrational ground state; thus, the overall tunneling contribution did not change significantly as the temperature decreased further. For common and important reactions in atmospheric chemistry, such as reactions between ozone molecules and alkenes to form Criegee intermediates and their subsequent decay products, significant quantum tunneling and zero-point energy effects are expected to occur. However, extensive research in this area has yet to be conducted.

10.3.2 Reaction of HOSO with NO_2

Mitigation of acid rain is another pressing issue in environmental governance. Acid rain is a severe environmental pollutant. Sulfur dioxide (SO_2) and nitrogen oxides are the primary triggers of acid rain formation. Experimental measurements have shown that the atmosphere contains trace amounts of the relatively stable hydroxysulfinyl radical (HOSO), formed by the recombination of SO_2 and hydrogen atoms. Atmospheric chemical reactions involving HOSO, such as hydrogen transfer reactions between HOSO and NO_2, are of great importance in the study of acid rain, and the nuclear quantum effect of hydrogen atoms has a critical impact on these reactions. The nuclear quantum effects of hydrogen atoms cannot be ignored in several other important atmospheric chemical processes involving radicals, such as the reaction between OH radicals and NO_2 to form HNO_3. However, in previous discussions, atomic nuclei have often been treated as classical point charges without consideration of full quantum effects.

To include full quantum effects, Sugimoto et al. employed a multicomponent quantum mechanical (MC_QM) approach that described not only electrons but also hydrogen nuclei in terms of wave functions [Sugimoto2019]. Therefore, the MC_QM method allows for a more accurate representation of the physical properties of hydrogen bond systems and proton-transfer-reaction processes. Moreover, researchers have investigated the H/D isotope effect using the MC_QM method and directly assessed the different nuclear quantum effects of hydrogen and deuterium atoms.

Figure 10.28 shows the reaction energy profiles of the HOSO + NO_2 reaction obtained by CCSD(T) / cc-pVTZ // B3LYP / 6-31G** calculations based on the MC_QM method. The optimized geometric parameters for each stationary point structure during the reaction are shown in Figure 10.29. In Figure 10.29, the hydrogen bond H...O distances in

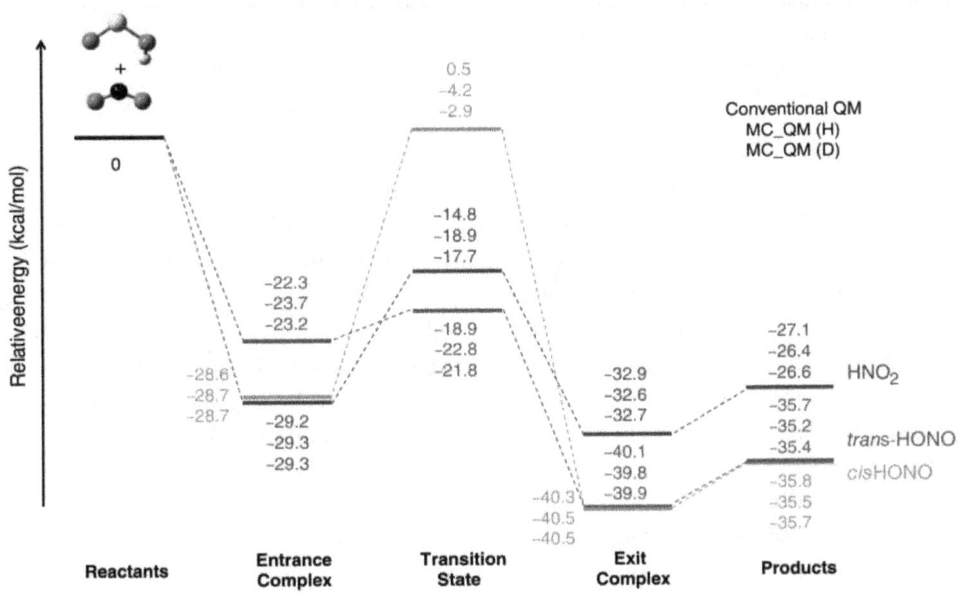

Figure 10.28 Reaction energy profiles of the HOSO + NO_2 reactions computed by multicomponent quantum mechanical (MC_QM) method. From Sugimoto[2019]. Reproduced with permission from John Wiley and Sons.

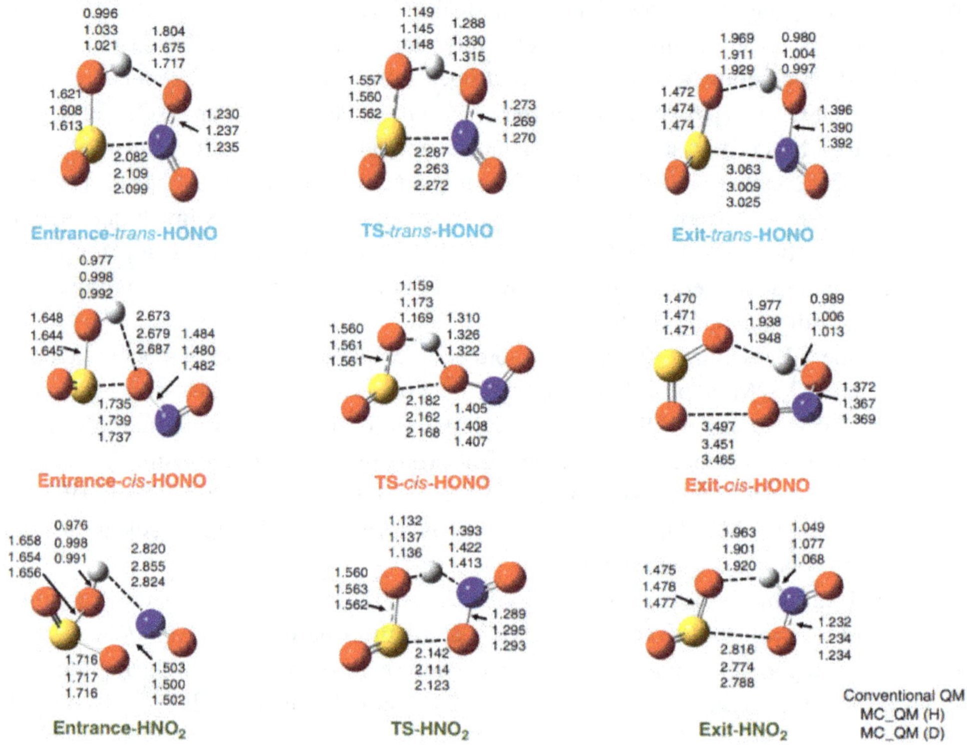

Figure 10.29 Optimized structures and representative interatomic distances (Å) of the stationary point structures in the HOSO + NO_2 reaction computed by multicomponent quantum mechanical (MC_QM) method. Here the red, gray, yellow, and blue balls are oxygen, hydrogen, sulfur, and nitrogen, respectively. From Sugimoto[2019]. Reproduced with permission from John Wiley and Sons.

the Entrance-*trans*-HONO point structure were much shorter than those in Entrance-*cis*-HONO and Entrance-HNO_2. This implies that the proton in the HOSO radical was more attracted to the oxygen atoms of the NO_2 molecule. As a result, longer O–H covalent bonds were observed in Entrance-*trans*-HONO. In the transition state (TS) structures, the length of the O–H covalent bond was slightly elongated (ranging from 1.132 Å to 1.159 Å), whereas the H...O hydrogen bonds were remarkably shortened (ranging from 1.288 Å to 1.393 Å). In the exit complexes, the migrant hydrogen atoms formed covalent O–H or N–H bonds with NO_2 molecules and hydrogen bonds with SO_2 molecules. Therefore, hydrogen bonds were important in all stationary point structures during the reaction.

Figure 10.28 shows the relative energies of the structures at each stationary position obtained from the multicomponent quantum mechanical calculations. Nuclear quantum effects were particularly pronounced in the TS structures for all reactions. The differences between the relative energies of TS structures obtained by conventional DFT calculations and MC_QM(H) calculations were 3.9 kcal/mol, 4.7 kcal/mol, and 4.1 kcal/mol for the reactions leading to *trans*-HONO, *cis*-HONO, and HNO_2, respectively. By contrast, nuclear quantum effects barely affected the relative energies of the entrance and exit complexes, except for that of Entrance-*trans*-HONO.

To investigate the nuclear quantum effects on the relative energies, Sugimoto et al. focused on the optimized geometric parameters obtained by using multicomponent quantum mechanical calculations (Figure 10.29). In Entrance-*cis*-HONO and Entrance-HNO_2, the (OSO)H...O(NO) and (OSO)H...N(O_2) distances were almost equivalent to the sum of the van der Waals radii of the hydrogen and oxygen or nitrogen atoms. Thus, these hydrogen bond interactions are likely to be weak, and the nuclear quantum effect of the hydrogen atom barely affects the relative energies of these complexes. Nuclear quantum effects also barely affected the relative energies of the exit complexes, and the hydrogen-bonded H...O distances were less than 2 Å. In fact, in Exit-*trans*-HONO and Exit-*cis*-HONO, the hydrogen-bonded H...O distance decreased after the nuclear quantum effects of the hydrogen atoms were incorporated into the calculation. The interatomic distances between atoms with the same charge (S and N atoms in Exit-*trans*-HONO and O atoms in Exit-*cis*-HONO) also become shorter. By contrast, the S...N distance in Exit-*trans*-HONO became longer after the nuclear quantum effect was included. Thus, nuclear quantum effects only slightly lowered the relative energy of Exit-*trans*-HONO.

After further considering nuclear quantum effect, Figure 10.28 also shows the MC_QM(D) calculation results for deuterium. As deuterons are twice as heavy as protons, their nuclear quantum effects are less significant. When the nuclear quantum effects of deuterons were considered, the relative energy of the TS structures was reduced. However, the magnitude of reduction was smaller than that observed in the hydrogen calculations. For reactions leading to *trans*-HONO, *cis*-HONO, and HNO_2, the differences between the relative energies of TS structures obtained by conventional DFT and MC_QM(D) calculations were 2.9 kcal/mol, 3.4 kcal/mol, and 2.9 kcal/mol, respectively.

The differences between the nuclear quantum effects of hydrogen and deuterium atoms were reflected in the optimized geometric parameters. After taking the nuclear quantum effects of hydrogen into account, the X–H (X = O or N) covalent bond length increased because of the anharmonicity of the potential energy curve along the covalent bond direction when using the multicomponent quantum mechanical method. Because the nuclear quantum effect of deuterons is less pronounced than that of protons, the covalent bond length in X–D is shorter than in X–H. By contrast, deuterium substitution leads to the elongation of the hydrogen-bonded H(D)...O distances, as seen in various hydrogen bond systems. The TS structure is relaxed by including the nuclear quantum effects of hydrogen and deuterium. Although the difference between the zero-point vibrational energies of hydrogen and deuterium atoms can be considered with the aid of conventional DFT methods, the PES is invariant in density functional theory calculations, thus geometric changes due to deuterium substitution are essentially neglected in DFT calculations. By representing the H/D isotope effect on the geometric parameters using the MC_QM method, the H/D isotope effect on the calculated activation barrier can be analyzed in detail.

10.3.3 Reaction of HOSO + NO_2 with H_2O

We next consider the reaction of HOSO + NO_2 with H_2O. Sugimoto et al. [Sugimoto2019] analyzed two types of hydrogen ion transfer mechanisms: (a) a reaction in which H_2O bridges HOSO and NO_2 at a site opposite to the proton-transfer site, and the proton of HOSO directly migrates to NO_2 and (b) a hydrogen-relay reaction in which H_2O mediates proton transfer from HOSO to H_2O, and then from H_2O to NO_2.

The reaction energy profiles and optimized structures at each stationary position are shown in Figures 10.30 and 10.31, respectively. All optimized structures in reaction (a) have energies less than the sum of the energies of the reactant molecules (HOSO, NO_2, and H_2O). The activation barrier for the hydrogen atoms in reaction (a) was calculated to be 2.3 kcal/mol using conventional DFT calculations, −0.1 kcal/mol using the quantum MC_QM(H) method, and 0.5 kcal/mol using the MC_QM(D) method. Evidently, the nuclear quantum effects of the hydrogen atom reduced the activation barrier of the reaction, regardless of the presence of H_2O. In particular, the activation barrier vanished in the MC_QM(H) results when the nuclear quantum effects of hydrogen atoms were included. The result is in agreement with the CBS-QB3 energy profile corrected by the zero-point vibrational energy renormalization, in which the activation barrier for the reaction with one H_2O molecule was only 0.1 kcal/mol. It is worth noting that the previous CBS-QB3

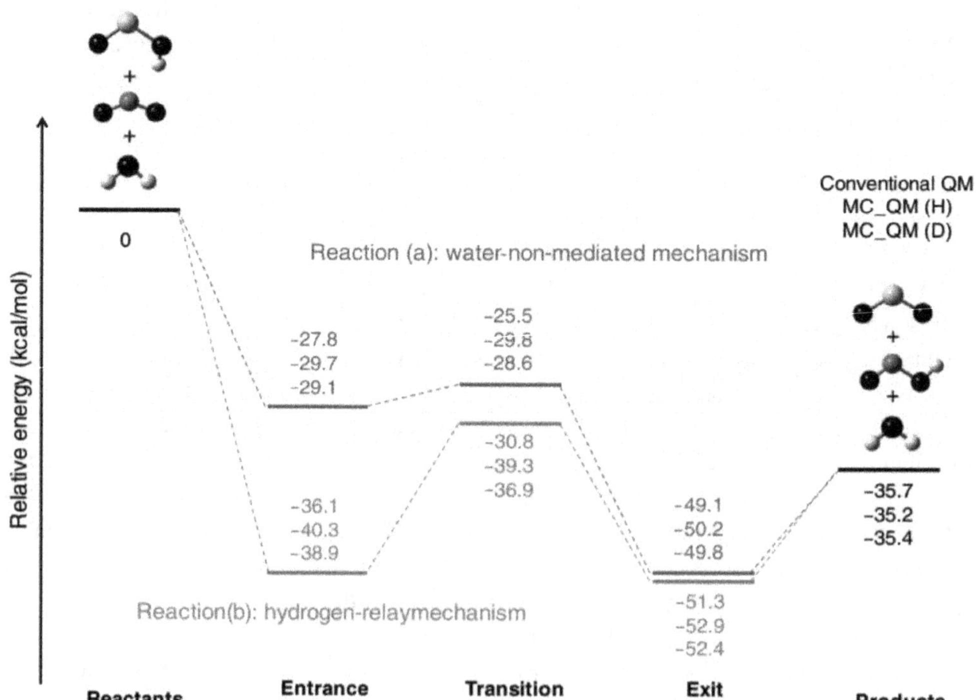

Figure 10.30 Reaction energy profiles of the HOSO + NO_2 + H_2O reaction obtained by multicomponent quantum mechanical (MC_QM) method. From Sugimoto[2019]. Reproduced with permission from John Wiley and Sons.

calculations included only zero-point vibrational energies, whereas the MC_QM calculations considered not only the kinetic energies of the hydrogen atoms but also the electronic structure and the renormalization of geometric parameters.

Figure 10.31 shows that in reaction (b), water molecules mediate proton transfer from HOSO to NO_2, and the reaction proceeds in a concerted manner. The relative energies of the entrance complex, TS structure, and exit complex in reaction (b) were found to be lower than those in reaction (a) [Sugimoto2019]. In particular, the relative energies of Entrance-W1-relay (reaction (b)) were 8.3 kcal/mol to 10.6 kcal/mol lower than those of Entrance-W1-direct (reaction (a)). As a result, the TS structure in the hydrogen-relay-type reaction (TS-W1-relay) was more stable than that in TS-W1-direct, and its activation barrier was larger than that in reaction (a). The nuclear quantum effects of the hydrogen atom lowered the relative energy of TS-W1-relay by 8.5 kcal/mol; this value is approximately twice that resulting from the nuclear quantum effects observed in reaction (a). Therefore, the nuclear quantum effects of the hydrogen atoms are more important in the TS structure of the water-mediated reaction pathway. According to the MC_QM(H) results, the activation barrier decreased from 5.3 kcal/mol to 1.0 kcal/mol. Geometric relaxation and geometric H/D isotope effects induced by the nuclear quantum effects of hydrogen were also observed in the optimized stationary point structures, as shown in Figure 10.31.

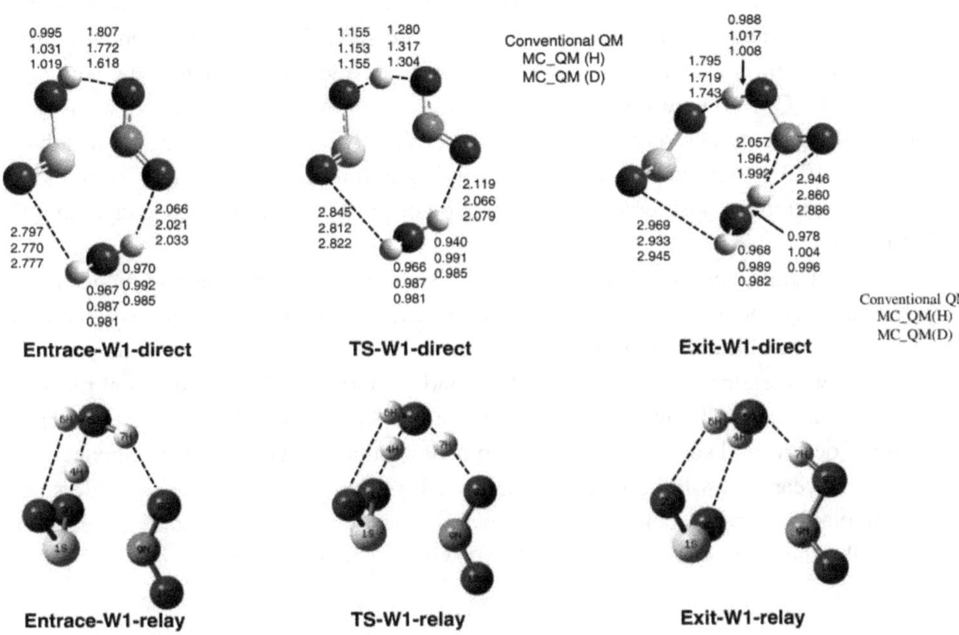

Figure 10.31 Optimized structure and interatomic relative distances (Å) of the stationary point structures in the HOSO + NO_2 + H_2O reaction obtained by multicomponent quantum mechanical (MC_QM) method. Here the dark grey (labeled "*O"), light grey (labeled "*H"), medium-light grey (labeled "*S"), medium grey (labeled "*N") balls are oxygen, hydrogen, sulfur, and nitrogen, respectively. From Sugimoto[2019]. Reproduced with permission from John Wiley and Sons.

The reaction profile of HOSO + NO_2 with H_2O can be summarized as follows. In addition to the previously reported proton-transfer reaction (reaction (a)), the water-mediated proton-relay reaction (reaction (b)) was analyzed. The multicomponent quantum mechanical calculations revealed that the structures at each stationary position in the proton-relay mechanism had lower energies than the corresponding structures in the proton-transfer mechanism. However, the activation barrier in the direct proton-transfer mechanism was smaller and vanished in the (MC_QM) (H) calculation results. Because the energies of all stationary point structures in the proton-relay mechanism are much lower than the sum of the energies of the reactant molecules (HOSO, NO_2, and H_2O), the activation barrier is only 1.0 kcal/mol for hydrogen atoms in multicomponent quantum mechanical calculations (2.0 kcal/mol for deuterium atoms). These results suggest that the reaction is more likely to proceed via a proton-relay mechanism.

In conclusion, the multicomponent quantum mechanical method can be employed to analyze the nuclear quantum effects on proton transfer in the HOSO + NO_2 system, both in the presence and absence of H_2O. The numerical results indicate that nuclear quantum effects of hydrogen and deuterium atoms reduce the relative energies of the stationary point structures. The most significant stabilization is found in the TS structure of the water-mediated proton-relay reaction as a result of nuclear quantum effects. Interestingly, after the nuclear quantum effects of the hydrogen atoms are included, the activation barrier of the direct hydrogen-transfer mechanism in (MC_QM) (H) calculations vanishes. Although the entrance complex and transition state have lower relative energies than those in direct proton transfer when the full quantum effect is considered, the activation barriers in MC_QM (H) are much smaller for the proton-relay-type reaction with water. Consequently, proton-relay-type reactions are more likely to occur. The results in subsections 10.3.1–10.3.3 clearly demonstrate that full quantum effects are significant factors that must be seriously considered in the protection of the stratospheric ozone layer and the mitigation of acid rain.

Given the authors' research background, the discussion of environmental science in this chapter primarily focuses on fundamental research in the fields of water resource management and atmospheric environmental governance. The processes previously considered involve light elements, particularly H, N, and O, and the relevant chemical reactions exhibit pronounced full quantum effects. Many other environmental problems, such as the pollution of oceans and soils with plastics and heavy metals, have gained significant attention in recent years. Resolving these problems will require a comprehensive understanding of complex processes such as photodissociation and nonadiabatic chemical reactions. Undoubtedly, the study of full quantum effects will play a crucial role.

11
Full Quantum Effects in Device Applications

Chapters 1–10 introduce full quantum effects in terms of the crucial role in various physical and chemical phenomena, current experimental and theoretical methods, and applications in the field of condensed matter physics.Typical full-quantum condensed-matter systems and issues related to energy and environmental concerns are also demonstrated. In summary, Chapters 1–10 aim to uncover the nature of full quantum effects and study their impact on physical and chemical properties.

The study of full quantum effects represents the forefront of research in theoretical and experimental condensed matter physics and is currently in a phase of rapid development. Such fundamental research on full quantum effects is essential for exploring new quantum states and materials and lays the foundation for the application of full-quantum devices.

One objective of research is to develop next-generation devices with better performance and more comprehensive functionality. In contrast to the theoretical and experimental studies, device research on full quantum effects remains in an early stage of development. Apart from the existing full quantum applications in the field of electronics, such as superconductivity and magnetism, the other potential device application areas of full quantum effects are not well defined yet. The extensive applications to superconductors, semiconductors, and other quantum devices are not the focus of this book; they have been well described in many other monographs.

Our previous discussion clarifies that full quantum effects provide a new and effective means of control. For example, the efficiency of charge transfer in $MoSe_2/WSe_2$ van der Waals heterojunctions is extremely sensitive to the strength of the interlayer electronic coupling. By making a small modification to the interlayer electronic coupling through full-quantum band-structure renormalization, the charge transfer efficiency can be substantially improved by nuclear quantum effects in such materials. By modifying the parameters of electronic structures, ultrasensitive quantum-control applications, such as novel, high-efficiency optoelectronic and energy devices, may be realized. By twist-phase matching 2D thin-films, novel nonlinear optical crystals with isotopic effect can also be obtained. In addition, the nonadiabatic electron–phonon coupling in the excited states depends on the photocarrier dynamics in the presence of optical fields. Utilizing this characteristic allows real-time measurements of electron–phonon coupling and precise manipulation through optical techniques. Therefore, exploring the topological properties of photocarriers (e.g., chirality or valley selectivity) may lead to new physical discoveries and device applications for materials with optically modulated electron–phonon coupling.

In Sections 11.2–11.6, we present certain ideas regarding the development and application of full quantum devices. Our goal is to demonstrate the advantages of refined, diverse, and synergistic ways in which the device performance may be controlled through full quantum effects. We aim to inspire and assist readers in comprehending and using full quantum effects to develop innovative devices. It is expected that breakthroughs in this field would lead to developments of novel quantum devices that could surpass traditional technologies and prove indispensable to the next generation of industrial advances.

11.1 Devices for Hydrogen-Isotope Separation

In Chapter 7, the proton penetrability of 2D materials is introduced. Different 2D materials, e.g., monolayer graphene, monolayer hexagonal boron nitride (h-BN), and monolayer molybdenum disulfide, exhibit significant differences in their proton-penetrating behavior. Table 11.1 summarizes the proton conductivities of these different materials. Theoretical studies have shown that the high proton-penetrating rates of monolayer graphene and monolayer h-BN films are governed primarily by nuclear quantum effects.

Furthermore, numerous studies have shown that the introduction of transition metals (e.g., nickel or platinum) as modifiers in 2D materials has a significant activating effect on proton transport [Xin Y.B.2017]. This is because transition metals can catalyze protons or promote hydrogen synthesis from atoms to molecules. Consequently, the presence of transition metals leads to the rapid transfer of incoming hydrogen away from the surface of 2D materials, further accelerating proton penetrating rates of the 2D material. Table 11.2 reveals that proton tunneling through graphene or monolayer h-BN is significantly enhanced when the surface is modified by Pt. When hydrogen passes through a catalytically activated h-BN monolayer, the proton conductivity per unit area reaches 3,000 mS/cm^2 at room temperature, indicating unimpeded proton penetrability.

Monolayer graphene and monolayer h-BN are promising materials for hydrogen technology due to their high proton conductivities. For example, proton-conducting membranes, which serve as the core components of fuel cells, require both high proton conductivity and the ability to block fuel permeation. Thus, the design of commercial Nafion film [Mauritz2004] makes a trade-off between membrane thickness and proton conductivity, directly affecting the performance of the fuel cell. Conversely, graphene and monolayer h-BN are expected to revolutionize the field of proton-conducting membrane materials by achieving both high proton conductivity and extremely low thickness of conducting-membrane in fuel cells.

Table 11.1 *Proton conductivity per unit area of different thin-film materials. [Hu S.2014, Xin Y.B.2017]*

Film type	Nafion	h-BN	Graphene	MoS_2	Metal
Proton conductivity/$mS \cdot cm^{-2}$	5,000	100	5	$<10^{-8}$	$<10^{-8}$

Table 11.2 *Proton conductivity per unit area of Pt catalytically activated 2D materials [Hu S.2014, Xin Y.B.2017].*

	Proton conductivity / mS·cm^{-2}		Barrier height converted from conductivity / eV	
2D Material	No activation	Pt-activated	No activation	Pt-activated
Graphene	5	90	0.78	0.24
Bilayer h-BN	6	100	0.61	0.24
Monolayer h-BN	100	3,000	0.30	Approx. $3k_BT/2$

The aforementioned studies on proton transport provide an important insight; that is, the high proton permeability of 2D materials, together with the influence of full quantum effects, could potentially enable isotope separation. For example, separating hydrogen isotopes efficiently is generally difficult because protium, deuterium, and tritium atoms have similar chemical properties. In 2016, researchers combined graphene or h-BN with Nafion films to construct a quantum filter for hydrogen isotope separation (Figure 11.1) [Lozada-Hidalgo2016]. The conductivities of the hydrogen and deuterium nuclei in the same kind of graphene and monolayer h-BN were found to differ by an order of magnitude under the influence of nuclear quantum effects. This difference in conductivity enables the separation of deuterium atoms from water, which could play an important role in applications such as nuclear wastewater treatment. This isotope separation method also offers significant advantages in terms of energy conservation and reduced processing cost.

Further research reveals an activation barrier difference of ~60 meV between protons and deuterons in different 2D materials. This difference can be understood by considering that the protons and deuterons in the experiment move not in vacuum but rather along the hydrogen-bonded networks provided by sulfonate (SO_3^-) groups and water in Nafion. It is reasonable to expect that the proton and deuteron nuclei remain transiently bonded to the sulfonate and water groups (the initial state before hydrogen transfer) before penetrating through the 2D crystals. The zero-point energies of these hydrogen-bonded networks are estimated to be ~0.2 eV for protons and ~0.14 eV for deuterons. The zero-point oscillations effectively increase the initial state energy and reduce the activation barriers by 0.2 eV (for protons) and 0.14 eV (for deuterons). This explanation is consistent with experimental observation as all these 2D materials exhibit lower penetration activation energies, with a smaller reduction ($\Delta E = 60$ meV) observed for deuterons. Consequently, the results show that different activation energies can lead to an order of magnitude difference in the conductivities of protons and deuterons in graphene or monolayer h-BN. By exploiting this difference in the penetration barriers, a highly efficient quantum filter (depicted in Figure 11.1) can be designed to effectively separate deuterium from water, thus realizing hydrogen isotope separation.

For example, S. Zhang et al. from Tianjin University, in collaboration with Geim's group in Manchester University, found that the separation factor of hydrogen isotopes can be as high as 8 in experiments using electrochemical pumping [Lozada-Hidalgo2017].

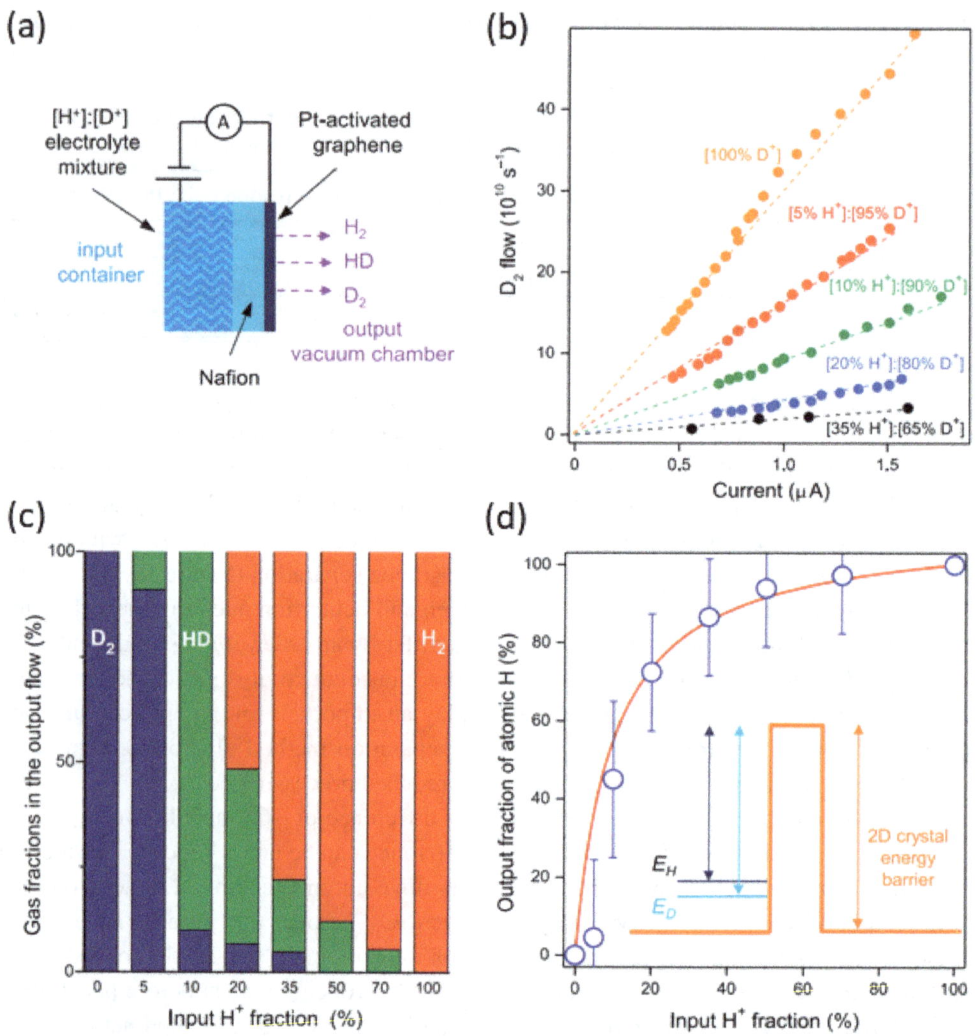

Figure 11.1 Schematic of the quantum filter for hydrogen-isotope separation and its working principle. From Lozada-Hidalgo[2016]. Reprinted with permission from AAAS.

Their method has an extremely low energy cost and a small environmental footprint. Because no toxic substances are involved, it is suitable for industrial-scale applications. In later experiments, the same group reveal that irradiating a Pt-modified graphene film with low-intensity visible light could further enhance the proton transport efficiency [Lozada-Hidalgo2018].

11.2 Ultrasensitive Quantum Tunneling Devices

Quantum tunneling is the transport of microscopic particles (including electrons and nuclei) or clusters through a potential barrier with a certain probability, even when the barrier height is greater than the kinetic energy. Although this phenomenon is impossible within the framework of classical physics, it occurs frequently in our daily life. For instance, as mentioned earlier, quantum tunneling effectively controls the rate of burning in the sun, allowing it to sustain static burning over a long time period. Furthermore, research has shown that quantum tunneling of electrons and protons is one of the crucial factors for redox reactions and the spontaneous mutation process of DNA. Quantum tunneling is widely applied in semiconductor physics, superconductor physics, particle physics, biology, and chemistry. Many functional devices and precision-measuring instruments have been developed based on the principle of quantum tunneling, including tunnel diodes based on the principle of electron tunnelling, Josephson junctions, magnetic tunnel junctions, and scanning tunneling microscopes. These devices and instruments play substantial roles in daily life and scientific research.

Consider a particle (e.g., electron or atom) with mass *m* and kinetic energy *E* with a simple square potential barrier of height *V* and width *w*. The wave function of a particle has a certain probability of being reflected by the potential barrier and a certain probability of tunneling through it. By solving the Schrödinger equation using appropriate boundary conditions, the probability *D* that the particle will pass from one side of the barrier to the other can be calculated using Eq. (8.1):

$$D = D_0 e^{-4\pi w\sqrt{2m(V-E)}/h}.$$

From this equation, it can be seen that the particle tunneling probability is highly sensitive to the width and height of the potential barrier. This is the working principle of many precision instruments (e.g., the scanning tunneling microscope). By utilizing the high sensitivity of quantum tunneling to the barrier height, quantum tunneling can be engineered and controlled.

Hexagonal boron nitride (h-BN) is a wide-bandgap (~5.97 eV) insulator that can be used as a barrier material for electron tunneling. As a naturally layered material, it maintains excellent insulating properties even when reduced to a single layer by mechanical exfoliation or chemical vapor deposition [Wang L.2019]. Boron has two stable isotopes, ^{10}B and ^{11}B; naturally occurring boron is a mixture of these isotopes with fractions of ~19.9% and 80.1% [Szegedi1990]. Because of nuclear quantum effects, ^{10}BN and ^{11}BN exhibit different electronic structures and energy barriers. When designing such devices, the isotopic composition can be precisely adjusted to obtain different heights of energy barrier. For example, L. Liu's group at Peking University prepared $WS_2/^{10}BN$ and $WS_2/^{11}BN$ heterostructures with ultrahigh isotopic purities, enabling effective control of the energy barriers and interlayer electron–phonon coupling strength (see Figure 11.2) in order to meet special device requirements [Li Y.F.2022].

In addition, L. Liu's group prepared isotopic junction $^{10}BN/^{11}BN$ or isotopic superlattice $^{10}BN/^{11}BN$ by selectively controlling the purity of isotopes in hexagonal boron nitride (h-BN) and alternately changing the thickness of layered structures with different isotopic atomic ratios. These structures made of light elements have ideal lattice-matched interfaces, and their bandgap is sensitive to isotopic effects. The corresponding potential barrier modulation can be restricted to a very small energy range. Such precise barrier control is difficult to achieve by other experimental means.

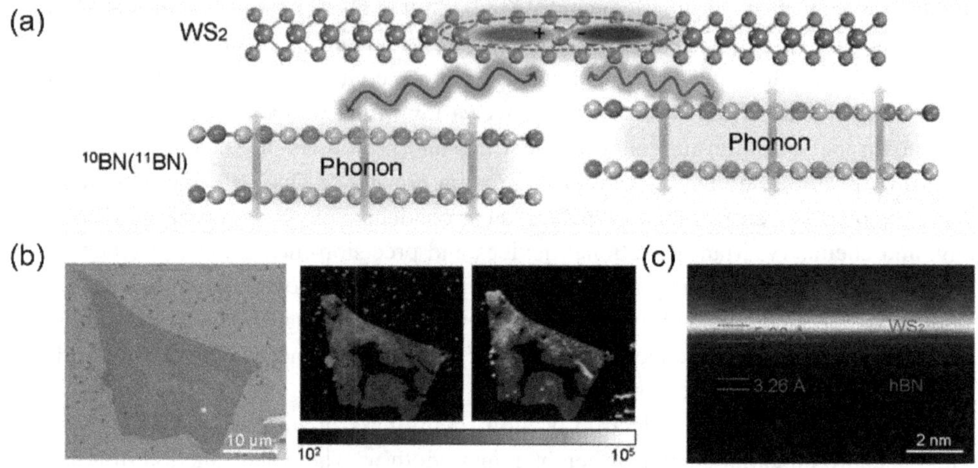

Figure 11.2 Modulation of the electron–phonon coupling between the layers of a WS_2/BN van der Waals heterostructure. (a) Illustration of the coupling between the electrons in the WS_2 layer and the phonons in the BN layer. The energy barriers and interlayer coupling strengths can be modulated by isotope, temperature, and pressure in isotopically engineered materials. (b) Left panel shows WS_2 layered islands on BN film prepared using CVD. Raman (middle panel) and PL (right panel) mappings were conducted to confirm the high quality and uniformity of the heterostructure sample. (c) Cross-sectional STEM image. Reprinted with permission from Li Y. F.[2022]. Copyright 2024 American Chemical Society.

Similarly, H. Watanabe et al. fabricated isotopically enriched diamond superlattices composed of ^{12}C and ^{13}C (Figure 11.3(a)). By modulating the diamond energy gap, they achieved carrier confinement in different spatial regions [Watanabe2009]. The corresponding cathodoluminescence (CL) spectra are shown in Figure 11.3(b). When the ^{13}C layer is thin (i.e., for samples A and B in the figure), the carriers generated in the layer migrate into the lower-energy ^{12}C layer and are confined there. When the thickness of the ^{13}C layer exceeds the mean free path of the carriers (corresponding to samples C and D), luminescence emanating from the ^{13}C layer is also observed. This indicates that diamond superlattices composed of different isotopes are capable of realizing arranged carrier confinement in space. A bandgap energy difference between ^{12}C and ^{13}C was measured to be 17 meV.

In 2013, Watanabe and collaborators studied the dependence of the diamond bandgap on the isotopic composition of homoepitaxial diamond-film samples $^{12}C_{1-x}{}^{13}C_x (0 \le x \le 1)$ [Watanabe2013]. They found that the relationship between the indirect bandgap and isotopic composition can be expressed as $E_g(x_f) = E_{ex}(x_f) + E_x + \hbar\omega_{TO}(x_f)$, where x_f is the ^{13}C composition; E_{ex} is the free-exciton energy of the transverse optical (TO) phonons for inducing annihilation; ω_{TO} is the frequency of the TO phonons; and the free-exciton binding energy $E_x = 80$ meV is almost independent of isotope substitution. When ^{12}C was replaced with ^{13}C, the bandgap changed by 15.4 meV. Si and Ge belong to the same group as carbon but have larger atomic masses. Therefore, the full quantum effects on

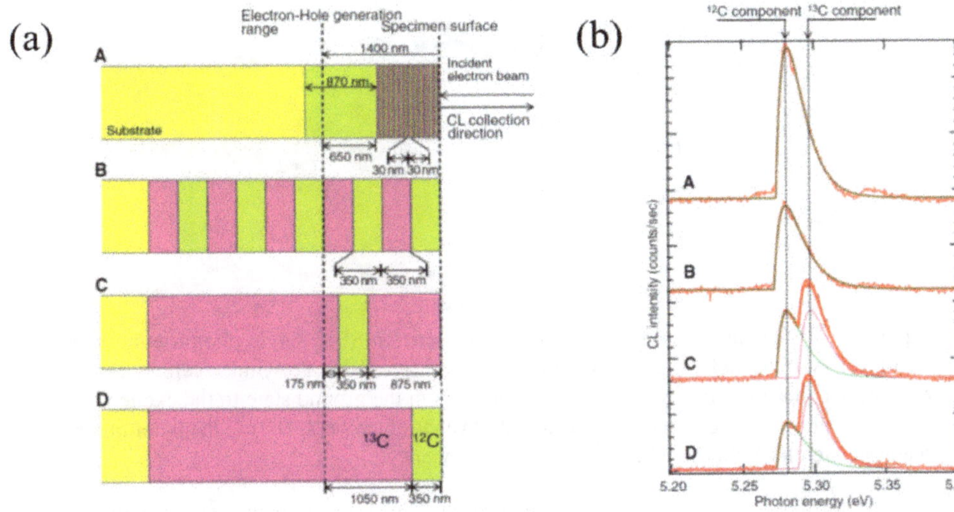

Figure 11.3 (a) Schematic of an isotopically enriched diamond superlattice structure. Green and purple represent the ^{12}C and ^{13}C layers, respectively. (b) CL spectra of different superlattices measured at 80 K. The thick red line shows one fitting of an experimental spectrum. From Watanabe[2009]. Reprinted with permission from AAAS.

their electronic states are expected to be one to two orders of magnitude smaller than on carbon's, possibly on the scale of a few meV [Cardona2005].

Given that the quantum tunneling is particularly sensitive to potential barriers, the barrier differences generated by manipulating isotopic band engineering are expected to be a new and effective approach to fabricating precise and ultrasensitive quantum tunneling devices. For example, we can take the advantage of full quantum effects to develop full quantum materials and devices, such as isotope-engineered compounds, isotope heterostructures, and isotope superlattices for magnetic resonance spectroscopy. All these studies illustrate once again that full quantum effects provide us with new design directions at a time when experimental instrument designers constantly strive for the highest accuracy and electronic device performance depends on extreme precision.

11.3 Single-Photon Emitters and Nano-cavities

Single-photon emitters (SPEs), which emit only one photon per excitation cycle, can be prepared using quantum optics technology. The mechanism of SPE is based on the discrete energy levels of quantum systems. When these systems are excited, electrons (generally speaking) can transition from the ground state to the excited state, and then return to the ground state through spontaneous emission, releasing a photon during the process.

Point defects in solids could serve as potential single-photon sources. A high-quality SPE material should have a wide bandgap to ensure that the ground and excited states within the bandgap are not influenced by thermal fluctuation, allowing the stable presence of the multiple particle state that can be maintained at room temperature. Hexagonal

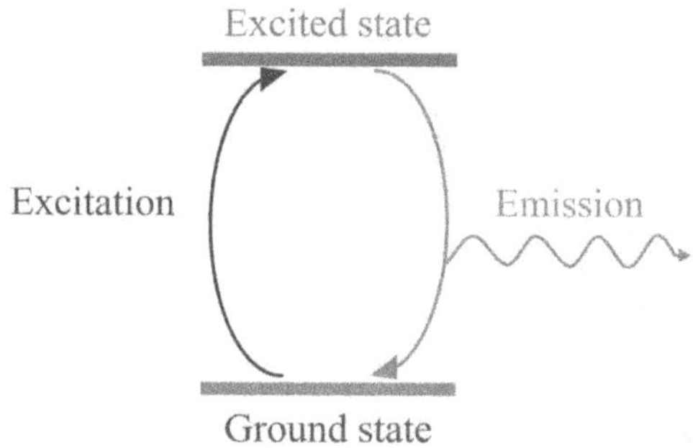

Figure 11.4 Schematic of a single-photon emitter. An ideal single-photon emitter is a two-level system, including a ground state and an excited state. Under certain external excitation conditions, the system can be excited from the ground state to the excited state, and then return to the ground state through de-excitation process. During the latter process, the system emits a photon with a controllable frequency.

boron nitride (h-BN) is such a wide bandgap material. In 2016, researchers first reported the development of h-BN SPEs [Tran2016a]. These single-photon emitters not only have high quantum efficiency and brightness (typically exceeding 1 × 106 counts/second) but also have very good optical stability [Tran2016b]. The significant advantages of h-BN have stimulated research enthusiasm for further development of its single-photon emission sources. Currently, in experiments, various point defects are introduced into h-BN through techniques such as ion beam, electron irradiation, plasma treatment, or thermal annealing (Figure 11.5(a)), to fabricate single-photon emitters.

Most current SPE sources use the narrowband emission technology. At low temperatures, the h-BN SPE source exhibits a narrow emission linewidth close to the Fourier transform limit. However, this emission is limited by the intrinsic dipole moments of the emission defects (Figure 11.5(b)) [Tran2018]. To reduce the influence of spectral broadening, the Stark effect can be used to stabilize the emission linewidth. The upper panel of Figure 11.5(c) shows a device prepared using h-BN and graphene, and the lower panel shows the observed Stark shift in the SPE sources [Noh2018]. Due to the excellent SPE properties of h-BN, it can even serve as a quantum light source covering from the visible to near-IR band at room temperature.

h-BN SPE sources can also be integrated into dielectric optical cavities and waveguides to achieve more efficient optical functions. By adopting the extraction-placement method, h-BN flakes are precisely integrated into optical waveguides. The advantage of this method is that the optical bandgap of h-BN is independent of its size, which allows for the direct fabrication of photonic crystal cavities, microring resonators, and planar lenses on bulk h-BN. For example, Figure 11.5(d) shows a 1D photonic crystal cavity made of h-BN [Kim S.2018]. In addition, owing to their inherent hyperbolic structure characteristics, h-BN materials can be applied in IR nano-optoelectronic devices by forming hyperbolic phonon polaritons.

Considering that the base materials of these quantum optoelectronic devices, such as BN and graphene, are composed of light elements, their electronic energy band

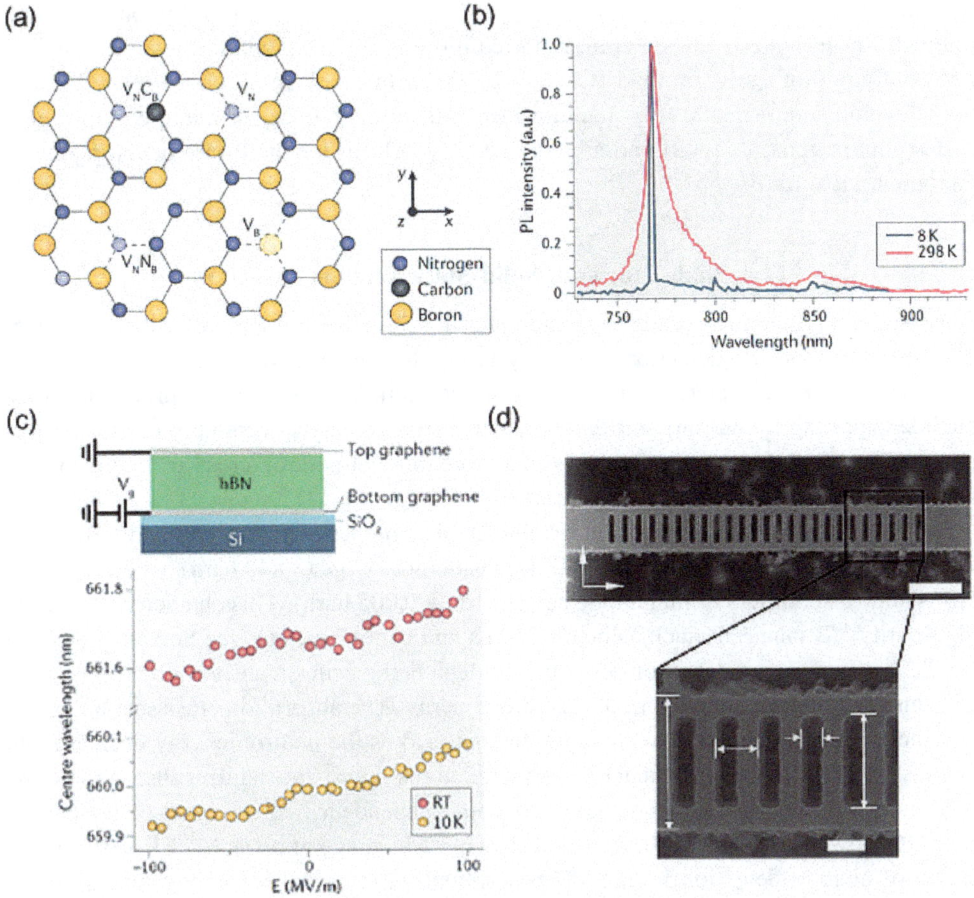

Figure 11.5 (a) Typical point defects in h-BN: nitrogen vacancy (VN), boron vacancy (VB), nitrogen vacancy plus substituted carbon (VNCB), and nitrogen vacancy plus anti-site nitrogen (VNNB). (b) Typical PL spectra of h-BN SPE at room temperature (red) and low temperature (blue). (c) Stark effect in h-BN SPE device, where graphene sheets are used as the bottom and top electrodes; the data points show a significant wavelength shift (i.e., Stark effect) at both room temperature and cryogenic temperature (10 K). (d) 1D photonic crystal cavity made of h-BN with high-quality factor. Figure 11.5 (b) reprinted with permission from Tran[2018]. Copyright 2024 American Chemical Society; Figure 11.5 (c) reprinted with permission from Noh[2018]. Copyright 2024 American Chemical Society; Figure 11.5 (d) Figure created by Kim[2018] licensed under CC BY 4.0.

structures can be accurately controlled through full quantum effects (e.g., isotopic substitutions). Especially in the presence of optical field, precise control of the light transition process, the nonadiabatic dynamics process of the electron–phonon coupling and photocarriers, can introduce new device design parameters, such as the selectivity of chirality and valley, thereby achieving ultra-sensitive operation of optoelectronic devices for specific needs. The foundation of this research lies in accurate characterization of the electronic ground states, excited states, and defect energy levels in materials. In this field, theoretical studies of point defects have been a subject of especially long-standing interest. Recently, Y. L. Chen et al. reported more reliable calculation results using the full-configuration QMC method [Chen Y.L.2023]. In-depth research on related issues may develop some special full quantum-mechanical techniques for manipulating light–matter interactions; this will undoubtedly be very helpful for the future development of nanophotonic circuits.

11.4 High-Efficiency Solid-State Neutron Detectors

Our discussion of neutron scattering techniques in subsection 6.1.2 clarifies that in nuclear reaction detection, the detection efficiency is highly sensitive to the neutron absorption cross section of the material. The neutron cross section represents the probability of a nuclide capturing a neutron, and this parameter significantly affects the detector performance. Therefore, finding a material with a large thermal-neutron cross section is crucial for improving the sensitivity of detectors.

An important distinction between the two stable isotopes of boron, ^{10}B and ^{11}B, is that the thermal-neutron cross section of ^{10}B is considerably large (3,840 barn), whereas that of ^{11}B is almost six orders of magnitude smaller (only 0.005 barn). This characteristic makes high-purity ^{10}B materials such as h-^{10}BN films and nanotubes, etc (see Section 8.3) ideal candidates for neutron detection. Figure 11.6 depicts the neutron absorption cross section of common materials as a function of neutron energy. The absorption cross section generally increases exponentially (with some fluctuations) as the neutron energy decreases. In contrast to some of the other materials depicted in the figure, the neutron absorption cross section of ^{10}B demonstrates a stable exponential dependence across most of the energy range [Chadwick2006, Doan2017]. Another major advantage of using h-^{10}BN in the application of neutron detection is that ^{10}B has a small atomic number and is insensitive to γ-rays, which allows for higher n/γ resolving capability. Moreover, h-BN is a wide-bandgap (E_g ~ 5.97 eV) semiconductor with an ultrahigh resistivity and an extremely low leakage current. These characteristics are critical for improving the signal-to-noise ratio of neutron detectors. Therefore, h-^{10}BN is considered a highly promising material for thermal neutron detection [Maity2017]. Compared with commercially available ^{3}He-based proportional counters, h-^{10}BN-based devices offer advantages such as low manufacturing cost, solid-state properties, and high efficiency.

In h-^{10}BN, a nuclear reaction occurs when a ^{10}B atom captures a thermal neutron. The resultant α-particle and ^{7}Li ion have large kinetic energies, and their corresponding mean free ranges are ~5 μm and ~2 μm, respectively [Knoll2010]. The kinetic energy of these particles dissipates as they travel through the material, and the energy is converted into a cluster of electron–hole pairs along the particle trajectories within the h-BN. Under an applied electric field, charges with different signs will drift toward the two opposing electrodes, resulting in detectable charge pulses that serve as thermal-neutron detection signals. Previous studies have shown that h-BN neutron detectors based on ^{10}B enrichment (99.9%) exhibit the highest detection efficiency (up to 51.4%) among solid-state detectors to date [Maity2016].

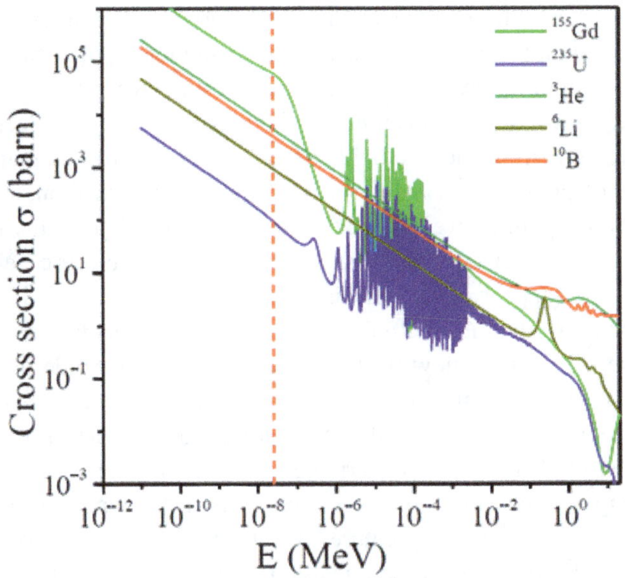

Figure 11.6 Neutron absorption cross section as a function of the neutron energy in common materials. Reprinted from Doan[2016], with permission of the author.

11.5 Functional Ferroelectric Devices

Ferroelectric materials have two or more spontaneous polarization states, which can be reversed by applying an external electric field [Fatuzzo1967, Zhong W.1996]. Spontaneous polarization arises from the relative displacement of the positive and negative charge centers within the unit cell in the absence of an external field, resulting in the formation of an electric dipole moment. Consequently, the symmetry of the crystal structure is restricted. Of the 32 crystallographic point groups, only 10 meet this requirement and are known as polar point groups: $1(C_1)$, $2(C_2)$, $m(C_s)$, $mm2(C_{2v})$, $4(C_4)$, $4mm(C_{4v})$, $3(C_3)$, $3m(C_{3v})$, $6(C_6)$, and $6mm(C_{6v})$. Because the polar point groups are all non-centrosymmetric and satisfy the condition for piezoelectric crystals, ferroelectric materials are inherently piezoelectric [Martin1972].

11.5.1 Ferroelectric Phase Transition

A ferroelectric domain is a small region within which the electric dipoles are aligned in the same direction. In ferroelectric materials, multiple domains form spontaneously, thereby reducing the energy of the system. The interface that separates the two domains is known as the domain wall. Under an applied electric field, the motion of the domain walls can lead to changes in the relative domain size. Hysteresis occurs during this phase transition, as shown in Figure 11.7(a). As the applied electric field slowly increases from zero, the polarization direction of the electric dipoles inside the ferroelectric material gradually aligns with the direction of the external electric field. Meanwhile, the domains with spontaneous polarization orientation aligned with the electric field expand as the domain wall moves outward, and those with spontaneous polarization opposite to the electric field gradually diminish until the crystal reaches a state of saturated polarization and becomes a single-domain structure. In the saturated polarization state, when the applied electric field strength gradually decreases, the magnitude of the polarization also declines, but it does not return to zero. When the electric field strength reaches zero, the material still retains a certain degree of polarization, known as the remnant polarization, whose strength is denoted by P_r. At this point, if we reverse the electric field and gradually increase its magnitude from zero, the polarization continues to decrease. The electric field E_c that causes the polarization to disappear completely is called the coercive field. If the electric field becomes more negative than E_c, the polarization eventually becomes saturated again (but with the opposite sign). If we now begin to decrease the magnitude of the electric field until it returns to zero, reverse the field such that it points in the original direction, and then gradually increase it, the *P–E* curve will form a closed loop known as a hysteresis loop.

Another important characteristic of ferroelectric materials is the loss of ferroelectricity above certain temperature. This high-temperature phase is known as the paraelectric phase. The transition between the low-temperature ferroelectric and high-temperature paraelectric phases is called the ferroelectric phase transition, and the transition temperature is known as the Curie temperature [Damodaran2016]. The symmetry of the ferroelectric phase is always lower than that of the paraelectric phase. For example, $BaTiO_3$ crystals have a Curie temperature of 394 K. During the ferroelectric phase transition, the structure changes from a cubic lattice ($Pm\bar{3}m$) to a tetragonal lattice (*P4mm*), accompanied by spontaneous polarization along the *c* direction of the tetragonal phase. As the temperature decreases, $BaTiO_3$ undergoes additional ferroelectric phase transitions at 273 K and 203 K, but these temperatures are not Curie temperatures.

During the ferroelectric phase transition, these materials exhibit anomalous behavior in their dielectric, piezoelectric, elastic, optical, and thermal properties. For example, most ferroelectric materials have a large dielectric constant (on the order of 10^4–10^5) near the Curie point [Zhong W.L.1996]. Above the Curie temperature, the low-frequency dielectric constant (ε_r) of the ferroelectric materials obeys the Curie–Weiss law [Damjanovic1998]:

$$\varepsilon_r(0) = \varepsilon_r(\infty) + \frac{C}{T - T_c} \approx \frac{C}{T - T_c}, \tag{11.1}$$

where $\varepsilon_r(0)$ and $\varepsilon_r(\infty)$ are the low-frequency and optical-frequency relative permittivities, respectively. C is the Curie constant, and T_c is the Curie–Weiss temperature. Since $\varepsilon_r(\infty)$ is much smaller than $\varepsilon_r(0)$ and is temperature-independent, it is often neglected in calculations.

Ferroelectric phase transitions can be categorized as first or second order, depending on the behavior of the thermodynamic function near the critical point. For instance, the thermodynamic function of a first-order phase transition has a discontinuity in its first-order derivative near the critical point. The thermodynamic function of a second-order phase transition, on the other hand, has a discontinuity in its second-order derivative. In a first-order ferroelectric phase transition, the two phases coexist, and latent heat is released during the transition. The relationship between the free energy and the polarization of a first-order transition is shown in Figure 11.7(b). By contrast, a second-order phase transition does not produce latent heat and the two phases cannot coexist.

11.5.2 Ferroelectric Switching

Many applications of ferroelectric materials rely on ferroelectric polarization switching or the movement of domain walls. Nonvolatile ferroelectric memories, for example, take advantage of the fact that spontaneous polarization in ferroelectrics can be reversed under an external electric field, and the two polarization states are used to denote the "0" and "1" states in data storage [Scott1998, Hong S.2016]. Next, we discuss the effects of temperature, defects, electric fields, and stress on polarization switching in ferroelectrics.

When the temperature is below the Curie temperature, the depth of the ferroelectric double-well potential curve is altered. In particular, the energy corresponding to the lowest points of the potential well can be adjusted. In ferroelectrics, the potential well of the free-energy polarization curve deepens as the temperature decreases, implying that a larger external field is required for polarization switching. Conversely, as the temperature increases, the well depth decreases, suggesting that the polarization can be switched using a smaller external field. Although a larger electric field is required for the ferroelectric transition in a deeper potential well, the reversed polarization can better maintain its performance, which is beneficial for the stability of ferroelectric devices. With a shallow potential well, ferroelectric polarization can be reversed under a weaker external electric field, which reduces the energy consumption of the ferroelectric device. However, the polarization stability will decrease as a result, and the polarization is easier to be destroyed. Therefore, temperature is often used to adjust the depth of a potential well according to different application scenarios.

Defects, such as grain boundaries, point defects and dislocations within a crystal, also significantly affect ferroelectric switching. In early studies, the experimentally measured switching electric-field strength was found to be much smaller than the theoretically predicted value, which is known as Landauer's paradox [Landauer1957]. Subsequent investigations revealed that one of the reasons for Landauer's paradox is the presence of various defects in practical ferroelectric materials. The effects of these defects were not adequately considered then in the theoretical calculations.

Generally, defects can act as nucleation centers for polarization switching. They can reduce the external electric-field strength required for switching [Jesse2008], which in turn affects the thermal stability of the ferroelectric phase. In addition, these defects tend to favor polarization in a certain direction, breaking the symmetry of the free-energy–polarization curve so that the two polarization states are no longer equal. As indicated by the asymmetric gray curve in Figure 11.7(b), the potential well on the left side becomes shallower because of the presence of defects. If the influence of the defects is significant enough, it may even lead to the disappearance of this potential well, resulting in a "frozen" region where the polarization can no longer be reversed. Advances in material preparation techniques and characterization methods have enabled researchers to achieve the epitaxial growth of various high-quality ferroelectric thin film structures, such as $PbTiO_3$ and $BiFeO_3$ films.

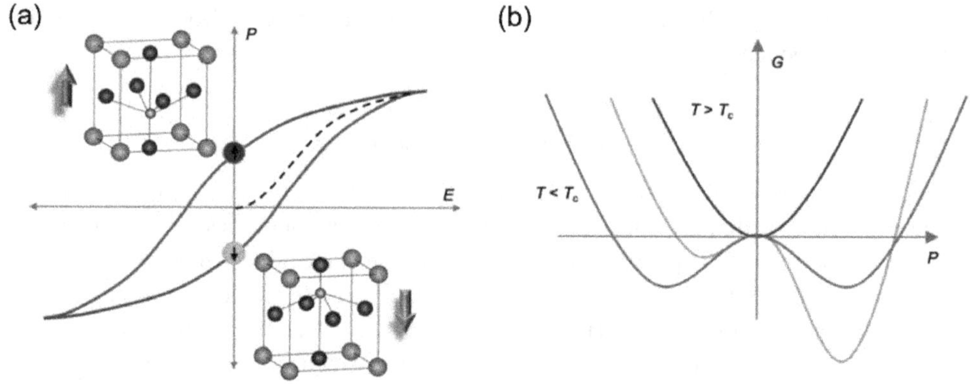

Figure 11.7 (a) Hysteresis loop of ferroelectrics and the atomistic structure corresponding to the tetragonal phase of $BaTiO_3$. (b) Free-energy–polarization curves, including $T > T_c$, $T < T_c$ and the gray curve is the asymmetric potential well due to the presence of external fields or defects. Reprinted from Zhong W.L.[1996] (in Chinese).

By manipulating the boundary conditions of the stress and electrostatic fields, the domain structures can be controlled, and novel topological structures can be created in the thin films [Zeches2009, Jia C.L.2011, Yu P.2012, Tang Y.2015, Damodaran2016, Martin2016, Yadav2016]. Moreover, the development of aberration-corrected transmission electron microscopy allows for direct observation of defect structures in ferroelectric materials. This has provided insights into how defects influence the distribution of ferroelectric polarization and affect the ferroelectric phase transition [Scott2000, Chu M.W.2004, Jia C.L.2009, 2011, Su D.2011, Gao P.2013, 2018, Li L.2015, Li M.2019, Li X. M.2019]. By utilizing these techniques to control the defect type and distribution, designing nanoscale devices with desired performance characteristics and effectively controlling their ferroelectric properties are possible.

The unique properties of ferroelectrics, such as their high dielectric constants [Ramirez2000], spontaneous polarization [Ederer2005], and polarization-switching capability under an external electric field [Gao P.2011], make them highly suitable for developing various devices, including storage elements of dynamic random-access memories, pyroelectric devices, and piezoelectric sensors. Some of these devices have found widespread practical applications, for example, as ultrasound probes in medical equipment.

11.5.3 Full Quantum Ferroelectric Functional Devices

The physical properties of ferroelectric materials can be profoundly affected by full quantum effects. When such effects are considered, the nuclei within the material are no longer treated as classical particles. Their positions are no longer deterministic but instead follow a probability distribution described by a quantum-mechanical wavefunction, as shown in Figure 11.8. This is equivalent to lowering the potential-well depth for polarization switching or increasing the system temperature. If full quantum effects are sufficiently strong, the atomic nucleus is no longer confined to the bottom of the potential well associated with a specific position (and polarization). Rather, it becomes delocalized and can even tunnel through the energy barrier to reach a reversed polarization state, resulting in a transition from the ferroelectric to the paraelectric phase. Therefore, full quantum effects are crucial in ferroelectric materials, affecting their properties and applications.

Generally, if a Curie–Weiss temperature T_c is observed in a material, nuclear quantum effects are not sufficient to alter its ferroelectricity. Conversely, the absence of a Curie–Weiss temperature may suggest that the Born–Oppenheimer potential energy surface of the material lacks inherent ferroelectric characteristics. Alternatively, this could also indicate that the ferroelectricity of the Born–Oppenheimer potential energy surface is overwhelmed by nuclear quantum effects. Using this principle, we can design a device that induces a phase transition between the ferroelectric and paraelectric phases by controlling nuclear quantum effects. Such a device is known as a full-quantum ferroelectric functional device.

Because nuclear quantum effects originate in the wave–particle duality of atomic nuclei or quasiparticles in materials, smaller effective masses or lower temperatures are associated with more significant nuclear quantum effects. Furthermore, as nuclear quantum effects primarily manifest themselves in the low-temperature region, the quantum fluctuations are overwhelmed by thermal fluctuations if the characteristic temperature of the physical process (T_c in the case of the ferroelectric phase transition) is high (Figure 11.9). In previous studies, when addressing practical problems, an equivalent correction was often applied to the potential energy surface instead of including an accurate description of the nuclear quantum effects. This approach aims to obtain numerical results consistent with experimental observations. The main reason for using it was that existing DFT calculations typically rely on static atomic structures rather than real-time dynamic structures. Consequently, in practice, adjustment of the polarization-switching barrier height due to nuclear quantum effects could only be achieved by modifying the Born–Oppenheimer potential energy surface, which is derived from static atomic structures. Moreover, the

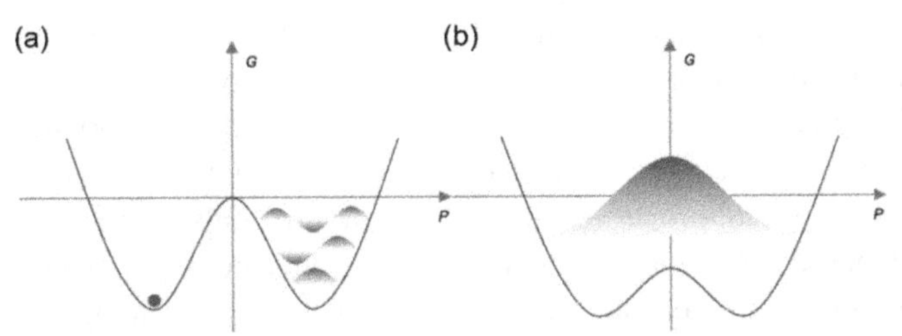

Figure 11.8 Physical images of ferroelectric properties under full quantum effects. (a) Illustration of full quantum effects on ferroelectricity. Left: the solid point indicates that the atom (classical particle) is in a stationary position (the bottom of the potential well) when nuclear quantum effects are not taken into account; right: the positions of the quantized nuclei can no longer be determined after considering nuclear quantum effects. Instead, nuclei evolve into a "cloud" and their probability distributions are described by nuclear wavefunctions. (b) With significant nuclear quantum effects, the nuclei are not confined in the potential well of a single polarization state, leading to a ferroelectric–paraelectric phase transition.

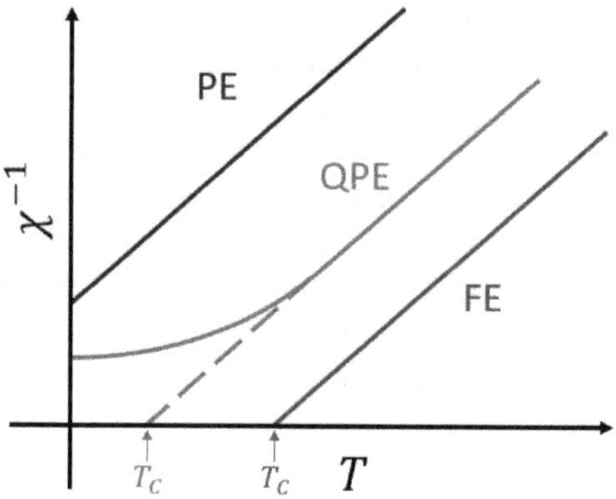

Figure 11.9 Polarizability as a function of temperature. Both the ferroelectric (FE) and paraelectric (PE) material follow the Curie–Weiss law. The FE material undergoes a phase transition at a finite temperature. The quantum paraelectric (QPE) material follows the Curie–Weiss law at higher temperatures and should have undergone the phase transition at the extrapolated T_c, where the gray dashed line crosses the T axis. However, it remains in the paraelectric phase and follows the solid gray curve because nuclear quantum effects dominate in the low-temperature region.

results obtained by adjusting the absolute value of the Born–Oppenheimer potential energy surface seemed to agree with those obtained by considering the nuclear quantum-effect-induced quantum fluctuations, and both methods seemed to yield free-energy surfaces consistent with the experimental observations. However, the physics underlying these two approaches is fundamentally different. Full quantum effects are crucial in ferroelectric systems. Therefore, full quantum simulations that extend beyond the scope of the Born–Oppenheimer approximation are essential to accurately uncover the physical mechanisms of the system based on precise electronic structures. In this context, it is important to note that the positional uncertainty caused by the quantum-mechanical properties of nuclei ensures the presence of nuclear quantum effects, irrespective of whether the material is composed of light elements. Hence, nuclear quantum effects are bound to have a significant impact on the physical properties of ferroelectric materials.

In view of applications, nuclear quantum effects are also important for modulating the performance of ferroelectric materials and devices. For example, they can significantly change the physical properties of ferroelectric materials with perovskite structures, including their phase transition temperatures and spontaneous polarizations. It has been seen that quantum fluctuations are closely related to nuclear quantum effects. For some ferroelectric materials with perovskite structures, the energy difference between the different symmetric structures is very small. Therefore, when the temperature decreases, quantum fluctuations can significantly alter material properties such as the spontaneous polarization and the dielectric constant [Höchli1979, Carlson2000]. Evidently, full quantum effects are critical for modulating ferroelectric device properties [Zhong W.1996]. For example, in $SrTiO_3$, thermal fluctuations are rapidly suppressed as the temperature decreases; thus, quantum fluctuations gradually dominate. The presence of these quantum fluctuations prevents the occurrence of a ferroelectric phase transition for $SrTiO_3$ even at low temperatures, resulting in the low-temperature quantum paraelectric phase shown in Figure 11.9 [Müller1979, Zhong W.1996]. In a different material $BaTiO_3$, nuclear quantum effects reduce the phase transition temperature by ~35–50 K while significantly suppressing spontaneous polarization.

The elemental doping has been identified as another effective means for modulating quantum effects. For example, Ba and Pb doping in $SrTiO_3$ can weaken quantum fluctuations. Similarly, in $Ba_xSr_{1-x}TiO_3$ (BST), adjusting the Ba-doping concentration leads to a transition from the quantum paraelectric phase to the quantum ferroelectric phase, and ultimately to the classical ferroelectric phase.

The isotope effect, which is another manifestation of the full quantum effect, can also be utilized to manipulate material properties and device performances. For example, in Section 8.1, by replacing lighter ^{16}O with heavier ^{18}O in $SrTiO_3$, quantum fluctuations can be suppressed while the potential energy surface remains unchanged [Kvyatkovskiĭ2001]. Thus, ferroelectric polarization has been induced in $SrTi(^{16}O_{1-x}{}^{18}O_x)_3$ samples even in the absence of an external electric field, expanding the application of $SrTiO_3$ in electronics [Itoh1999, Wang R.2001]. In the hydrogen-containing ferroelectric system KH_2PO_4, the Curie temperature can be increased from ~122 K to ~229 K by replacing the hydrogen atoms with heavier deuterium atoms; the spontaneous polarization is also enhanced after such isotopic substitution [Blinc1960, Koval2002, 2005]. Therefore, we can tune the Curie temperature and polarization magnitude of ferroelectrics using isotope substitution, enabling them to better meet practical device requirements.

In summary, full quantum effects offer a theoretical explanation of ferroelectricity and the mechanisms underlying ferroelectric phase transitions. Full quantum effects significantly influence the physical properties of both perovskite ferroelectrics containing heavier elements and ferroelectrics containing hydrogen. Because full quantum effects can be tuned using techniques such as isotope substitution, it is possible to modulate important properties such as the Curie temperature, ferroelectric polarization, and hysteresis-loop parameters. Controlling the operational temperature range can also change the strength of the full quantum effects, which are more pronounced at low temperatures. These studies open new directions in material research beyond conventional modulation methods, such as doping, strain, and defects. These results offer valuable insights for the design of novel full-quantum ferroelectric functional devices.

11.6 Stability of Low-Dimensional Semiconductor Devices

Low-dimensional materials (e.g., 0D quantum dots, 1D nanotubes, 2D thin films and heterojunctions) are especially sensitive to small changes in external fields (e.g., optical, electric, and stress fields). This is due to the confinement of electrons within a limited space, leading to a reduced total charge quantity. This characteristic makes these materials particularly useful for the development of new devices, hence the saying "the interface is the device." Due to the spatial confinement, electrons are evidently more sensitive to the quantum behavior of nuclei. In many cases, conclusions drawn by ignoring full quantum effects are invalid.

Over the past few decades, the study of 2D quantum materials has given rise to completely new physical concepts and research areas (e.g., semiconductor heterostructure/superlattice physics, twistronics, and valleytronics), significantly expanding the scope of condensed matter physics. The atomically flat surfaces, atomically sharp and clean interfaces, and extremely high specific surface areas in these 2D materials make them ideal for carrier transport. 2D semiconductors are considered excellent candidates for next-generation transistor-channel materials [Liu Y.2019]. In addition, they exhibit clear size effects, allowing their physical properties to be tuned for different application scenarios. For example, in 2006, it was demonstrated that the bandgap of a graphene nanoribbon (GNR) can be tuned using the nanoribbon width. GNRs can be categorized into three families based on their variation patterns: $3m$, $3m + 1$, and $3m + 2$, where m is a positive integer. Materials such as 2D chalcogenides and phosphenes also exhibit similar behavior, that is, their transport properties can be continuously tuned by changing their sizes.

The tunable physical properties of 2D semiconductors clearly suggest flexible and diverse application possibilities. However, such tunability also presents a great challenge which is associated with the instability of device performance. For example, because the semiconducting properties of a GNR are primarily determined by its boundaries, any structural defect present at an edge can alter its energy band structure and transport properties. Phosphene is also extremely sensitive to water and oxygen molecules in air. If these materials are used to construct field-effect transistors (FETs), the variability of their physical properties will have a significant impact on the uniformity of device performance. Compared to bulk materials, 2D semiconductors possess much larger specific surface areas, which are mainly responsible for this instability. The planar surface lacks longitudinal constraints, and isolated electron pairs perpendicular to it exist in a metastable state. Although 2D semiconductors exhibit excellent electrical conductivities, they are sensitive to changes in their own structures or external conditions, posing a serious threat to the stability of their electrical properties. These challenges make it difficult to find device processing applications for 2D materials in industry.

Considerable efforts have been made to investigate the atomic vibration effects on the physical and chemical properties, synthesis processes, morphological evolution, and defect formation mechanisms of 2D semiconductors. However, research concerning the influence of atomic vibrations on electrical properties and device performance is lacking. Molecular dynamics (MD) simulations are the primary tools for studying atomic vibrations. Previously, most MD methods treated electrons quantum mechanically and nuclei classically owing to the significant mass difference between electrons and atoms. The most commonly used approach is the BOMD method, which combines the Born–Oppenheimer approximation with the Car–Parrinello MD simulation, as described in Chapter 4. However, as previously discussed, this approximation introduces unavoidable inaccuracy, particularly for condensed matter systems comprising light elements [Markland2018].

An open question is how such approximations affect the device design. To address this question, we use GNR as an example and systematically investigate the impact of nuclear quantum effects on its electrical properties and FET performance. Numerical

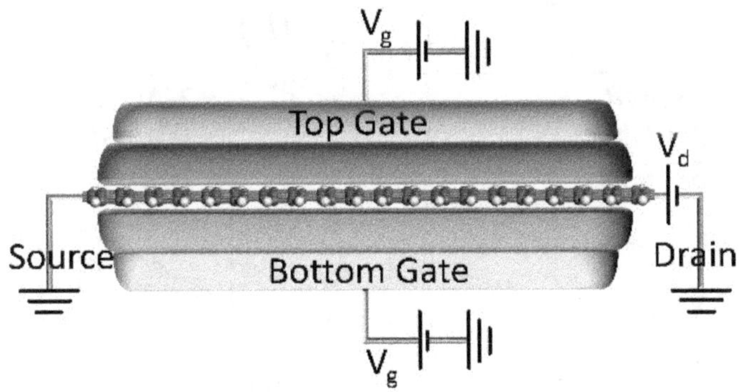

Figure 11.10 Schematic of a double-gate FET constructed from graphene nanoribbon (GNR). V_g denotes the gate voltage and V_d corresponds to the (equal) source and drain voltages.

simulations are conducted using the full-quantum path-integrated MD (PIMD) method combined with first-principles electronic structure calculations and quantum transport simulations. A double-gate FET device model is employed, as shown in Figure 11.10. The top and bottom gate voltages are kept the same, whereas the source and drain voltages are maintained at 0.1 V.

The open-source software package I-PI and the first-principles software QUANTUM ESPRESSO [Giannozzi2017] are utilized in the PIMD simulations. The I-PI is configured as the main server port responsible for studying the quantum motion of atomic nuclei, whereas QUANTUM ESPRESSO serves as the access point using the atomic structure calculated by the main server port to determine the electronic structure and atomic forces through DFT calculations. For GNRs with varying widths, an increase in both the average value and standard deviation of the bond lengths is observed with increasing temperature after the PIMD simulations for a total duration of 10^4 fs (with a step of 1 fs). These results indicate that the GNR structure expanded with increasing temperature (see Figure 11.11). Notably, this expansion effect in GNRs becomes more pronounced after including the nuclear quantum effects in the calculation.

Altered atomic structures inevitably result in altered electrical properties. The most crucial change lies in the bandgap E_g, which depends on temperature-induced atomic displacements and becomes more prominent after considering bandgap renormalization due to nuclear quantum effects. During the PIMD simulations, E_g is no longer constant because uncertainty of the atomic configuration is introduced by quantum consideration. Consequently, a certain degree of broadening is observed relative to the bandgap (E_g) of the static atomic structure, indicating quantum fluctuation of bandgap due to full quantum effect. As depicted in Figure 11.11(c), upon utilizing the initial value of $E_g \pm 0.2$ eV as the reference standard, it is observed that the GNRs with 7, 10, and 13 six-membered rings deviated from this energy range for 19%, 10%, and 6% of the PIMD simulation, respectively, after including the nuclear quantum effects. This result, which is induced by quantum fluctuations, does not keep the electrical properties of the GNR stable. It can also be seen that the incorporation of full quantum effects at any given temperature leads

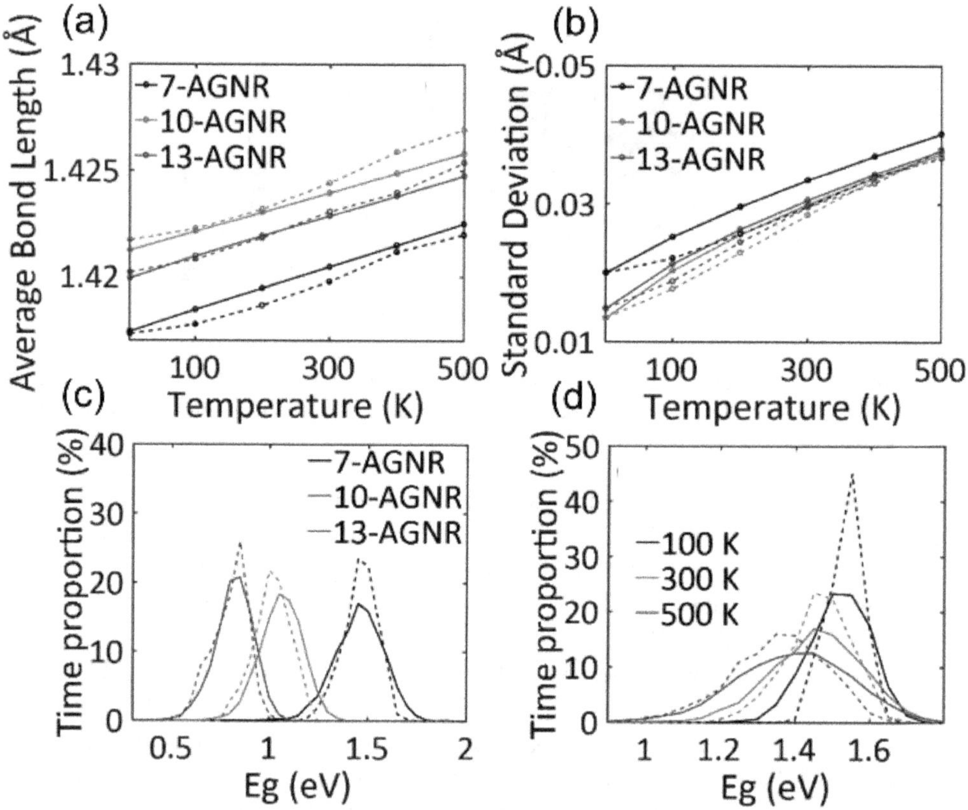

Figure 11.11 PIMD simulations of the (a) average and (b) standard deviation of the C–C bond lengths in three kinds of GNR. The relationship between the time proportion and the bandgap E_g is given in (c) for the different kinds of GNR and in (d) for different temperatures. The solid lines represent the quantum PIMD simulation results, and the dashed lines correspond to the classical BOMD simulation results.

to significant corrections in E_g; this is an unavoidable outcome of the electronic structure renormalization resulting from the quantum treatment of the nucleus.

In the atomic model used for the simulations, the GNRs are H-terminated at the edges. To reveal the full quantum effects on the renormalized electronic structure and bandgap of the GNRs, we employ the PIMD method to calculate E_g for various C–C and C–H bond lengths. The results are presented in Figure 11.12(a). From this figure, we can see that when the strain is within the –5% to 5% range, the C–H bond length in the GNR changes, but no effect on E_g is observed. However, increasing the C–C bond length reduces E_g significantly. By comparing the energy band structure of 7-AGNR at the maximum and minimum bandgaps (Figure 11.12(b)), it is observed that the most significant changes occurred near the conduction band minimum and valence band maximum. These energy levels are the result of π-bond coupling along the width of the GNR. In light of these observations, it is reasonable to conclude that the quantum vibrations of the GNR's atoms along the width direction account for the fluctuations in their E_g values. This conclusion comes

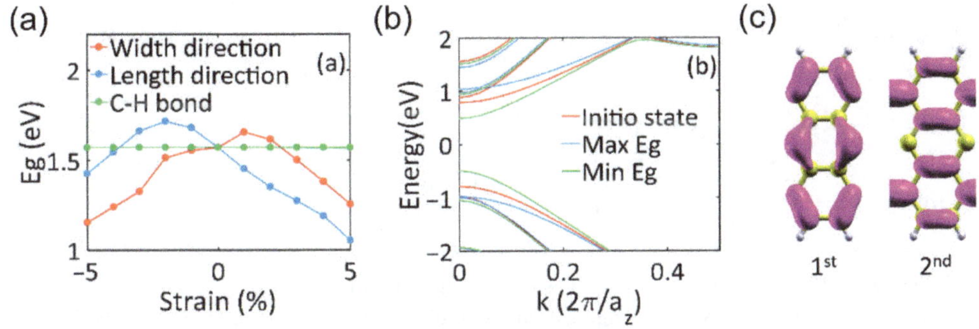

Figure 11.12 PIMD simulations of (a) 7-AGNR bandgap E_g as a function of stress and (b) 7-AGNR maximum and minimum energy band E_g diagram. (c) The real-space density distributions of states at the first and second conduction-band minimums.

naturally, as there is not much space left for the C atoms to vibrate along the length of the nanoribbon. In the width direction, however, C atom chains are truncated, leading to more vibrational degrees of freedom.

The PIMD simulation results for the characteristics of the GNR-FET are shown in Figure 11.13. Here, 10 instantaneous configurations obtained from PIMD simulations are manually selected, each of which represented five large and five small E_g values. The electronic wave functions are calculated from these instantaneous configurations using DFT and then transformed into a tight-binding model using the maximally localized Wannier function. The tight-binding Hamiltonian is also computed. As the transformation is unitary, the transformation matrix could completely preserve the transient characteristics of the GNR. By incorporating this matrix into the nonequilibrium Green's function, the relationship between the current and gate voltage for the corresponding instantaneous configuration of the GNR-FET is revealed.

In Figure 11.13(a), the transfer characteristics of the device exhibit significant deviations for different instantaneous configurations. Curves with large (small) E_g values correspond to smaller (larger) drain currents, which agrees with the theoretical predictions for semiconductor devices. Figures 11.12(b) and (c) display the off-state leakage current as a function of temperature and ribbon width, respectively. From these figures, it is clear that the leakage current tended to increase with temperature. In extreme cases, this is amplified by five orders of magnitude; this is sufficient to cause severe leakage in narrow-gap semiconductors, leading to destructive degradation of the FET switching performance.

As typical 2D semiconductors, GNRs offer additional vibrational freedom in the transverse direction. However, owing to nuclear quantum effects, they undergo atomic vibrations along their edges, leading to increased device instability. This is because the involvement of full quantum effects results in the renormalization of their electronic structures, effectively enhancing the impact of temperature on the GNR lattice structure. Consequently, the electrical performance of the device is also affected. PIMD simulations predict a different bandgap for each instantaneous atomic structure; this introduces

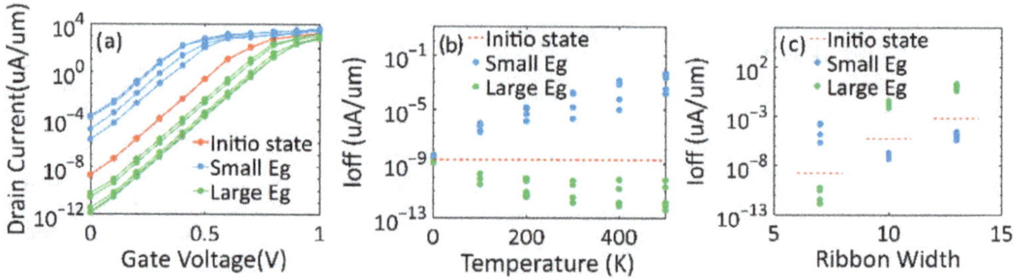

Figure 11.13 (a) Relationship between drain current and gate voltage, (b) temperature dependence of the off-state leakage current, and (c) off-state leakage current as a function of ribbon width for GNR-FETs with different instantaneous configurations. Five of the configurations correspond to large E_g and five to small E_g.

instability and increases the probability of current leakage in transient devices, which ultimately degrades the switching performance of the transistors. Through the above discussion of PIMD, it is evident that such perturbations can have a significant effect on the performance of narrow-gap semiconductor devices.

To reduce these negative effects, we propose several possible solutions. First, when considering the application of 2D materials in FETs, it is generally advisable to avoid the use of narrow-gap semiconductors. Second, it is necessary to implement spatial confinement of 2D semiconductors, which can be achieved through advanced techniques such as colloidal encapsulation. Finally, the surfaces of 2D semiconductors should be properly passivated, even if this involves a trade-off in their electrical performance.

From the above discussion, it can be seen that full quantum effects substantially affect the performance stability of 2D semiconductor electronic devices. This merits further in-depth research. We hope that professionals in the semiconductor industry will pay close attention to full quantum effects in their practical device designs.

The discussion in this chapter clearly suggest that the full quantum effects are helpful on not only discovering new physical and chemical phenomena in condensed matter systems but also exploring new materials with novel physical and chemical properties. In the research and development (R&D) of such materials for specific devices, it is important to carefully evaluate the advantages and disadvantages of utilizing full-quantum materials to leverage their unique characteristics and technological benefits and ensure that the final design effectively meets specific requirements. In summary, the study of new quantum states in condensed matter will undoubtedly extend our knowledge and enable the realization of full quantum devices with more comprehensive and powerful functions.

Appendix A

Nuclear Quantum Effect and Path-Integral Molecular Dynamics Simulations

Path-integral molecular dynamics (PIMD) is an efficient approach to studying the nuclear quantum effect (see discussion in Section 5.1 for details). According to path-integral theory, the density matrix of a quantum-mechanical many-body system under the adiabatic approximation is given by:

$$\rho(x_{\mathrm{b}},x_{\mathrm{a}};\beta)=\lim_{\varepsilon\to 0}\prod_{i=1}^{P}\prod_{I=1}^{N}\left[\left(\frac{m_I P}{2\pi\beta\hbar^2}\right)^{\frac{1}{2}}\int\frac{dx_I^{(i)}}{A}\right]$$
$$\times exp\left[-\beta\sum_{i=1}^{P}\sum_{I=1}^{N}\left\{\frac{m_I P}{2\beta^2\hbar^2}\left(x_I^{(i)}-x_I^{(i-1)}\right)^2+\frac{1}{2P}\left(V\left(x_I^{(i)}\right)+V\left(x_I^{(i-1)}\right)\right)\right\}\right], \tag{A.1}$$

where each quantized nucleus is equivalent to a necklace-like ring-polymer, and each necklace contains P segments, which can be regarded as P "beads." The neighboring beads are connected through harmonic effect, and the anharmonic Coulomb energy term $(V(x_I^{(i)})+V(x_I^{(i-1)}))/2P$ depends only on the atomic configuration of the ith and i-1th segment. Thus, the effective interaction potential of a complex quantum system can be expressed as:

$$V_{eff}=\sum_{i=1}^{P}\left\{\sum_{I=1}^{N}\left\{\frac{m_I P}{2\beta^2\hbar^2}\left(x_I^{(i)}-x_I^{(i-1)}\right)^2+\frac{1}{2P}\left(V\left(x_I^{(i)}\right)+V\left(x_I^{(i-1)}\right)\right)\right\}\right\}. \tag{A.2}$$

Further introduce the vibration frequency $\omega_P = P^{\frac{1}{2}} / \beta\hbar$, and the Eq. (A.2) can be rewritten as:

$$V_{eff}=\sum_{i=1}^{P}\left\{\sum_{I=1}^{N}\left\{\frac{1}{2}m_I\omega_P^2\left(x_I^{(i)}-x_I^{(i+1)}\right)^2+\frac{1}{2P}\left(V\left(x_I^{(i)}\right)+V\left(x_I^{(i-1)}\right)\right)\right\}\right\}. \tag{A.3}$$

Thus, the dynamics of quantized nuclei can be represented by the classical dynamics of a ring-polymer system under the action of an effective potential. At each simulation time step, the electronic structures in the different path segments were not coupled to each other, allowing self-consistent electronic calculations to be performed independently for each path segment. Therefore, simulations of different "beads" can be efficiently performed using parallel computing. The mutual coupling between the "beads" and the dynamic evolution of each "bead" can be computed after the parallel simulations are completed.

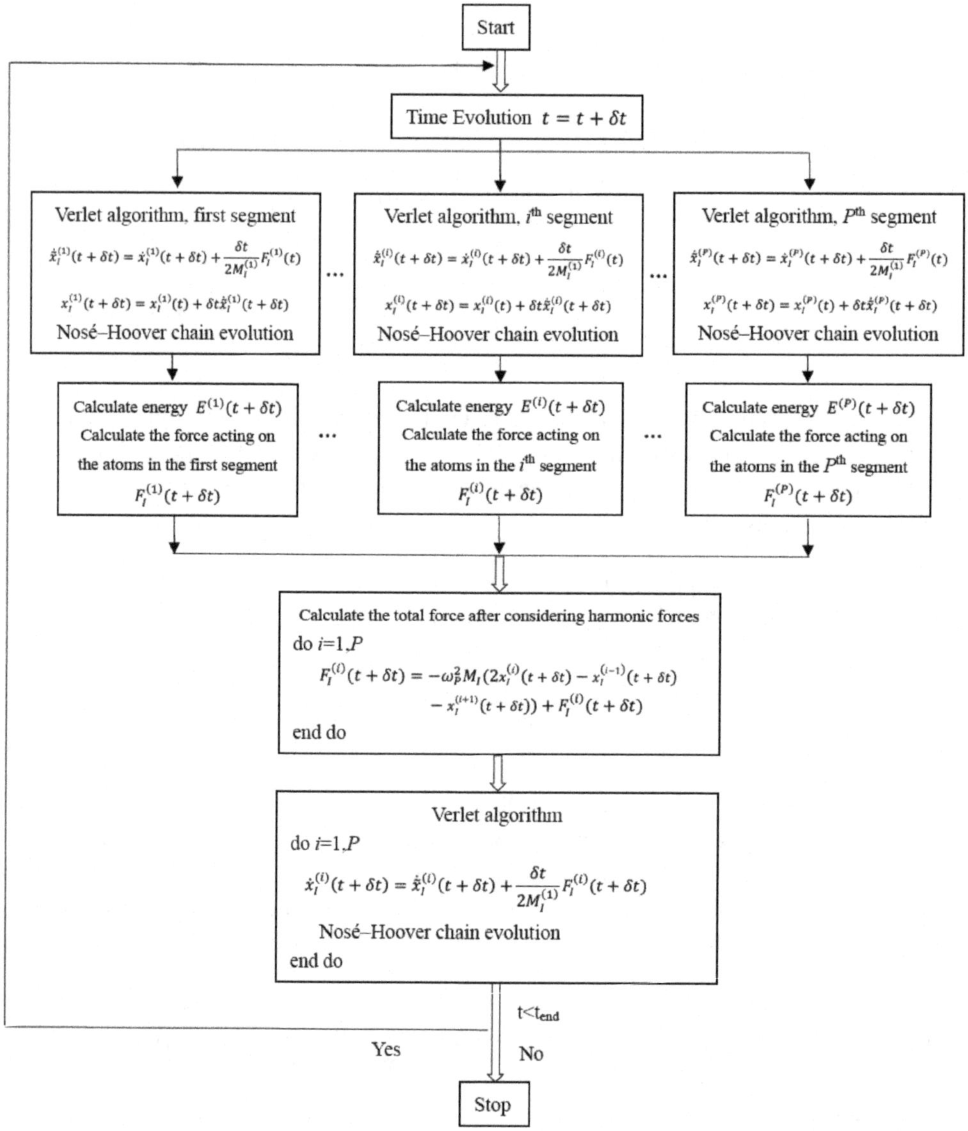

Figure A.1 Flowchart of the full quantum PIMD simulation program

Figure A.1 summarizes the procedure based on the above discussion. The flowchart shows the steps in the PIMD program for a full quantum simulation in Cartesian coordinates. In practice, the Cartesian coordinates and the forces acting on the atom can also be transformed into positions and forces in canonical coordinates. In this process, the first-principles simulation program calculates the self-consistent electronic structure, system energy, and the force on the nucleus. At the same time, the PIMD program is responsible for computing the force between different "beads," coupled coordinates, velocity, and position evolution of different beads.

The full-quantum PIMD simulation program required a series of parameters. Because the program is based on first-principles molecular dynamics simulations, its application requires configuring both the parameters related to the electronic structure calculations (e.g., pseudopotentials, cutoff energy, wavefunction initialization, and self-consistent calculation method) and those related to the molecular dynamics simulation (e.g., simulation temperature, simulation duration, time step, ensemble, and heat bath). In addition, the PIMD program contains several other important parameters including:

(1) Number of beads in each path segment. Because each quantized nucleus is equivalent to a ring-polymer chain composed of beads, the number of beads determines the accuracy of the quantum simulation. In general, nuclear quantum effects are more pronounced when the system contains lighter nuclei (H, He, Li, etc.) and the simulation temperature is relatively low. Consequently, a larger number of beads was required for the simulation.
(2) The coupling motion mode between the beads can be divided into Cartesian and centroidal coordinates. The Cartesian coordinate mode is a general coupling mode, that is, the harmonic interactions between different beads were calculated based on their distances, and determines its acceleration, velocity, and coordinates in the next step. In the centroid mode, the atomic coordinates of all beads undergo a canonical transformation to extract their centroid modes, and approximate them as classical nuclei. At the same time, by setting the "fictitious masses" to ensure rapid motion of the non-centroid modes, the centroid motion can be effectively sampled.

Some of the existing DFT-based computational software packages include full quantum PIMD simulation. Car-Parrinello Molecular Dynamics (CPMD) was the earliest first-principles molecular dynamics software to implement full quantum simulations. It played an important role in the early studies of nuclear quantum effects, particularly in the simulation of molecular and cluster systems. However, the full quantum simulation function in CPMD has significant limitations when applied to condensed matter systems. To overcome these difficulties, E. G. Wang's group collaborated with S. W. Gao and Wahnstrom to optimize the full quantum simulation function in the CPMD software. In particular, they focused on improving the efficiency of k-space sampling in electronic structure calculations for periodic lattice systems and the parallel simulation function of k-space sampling, so that it can be applied to computational simulation of condensed matter systems. In 2008, Q.F. Zhang et al. applied the optimized PIMD method to the study of full quantum effects in $BaZrO_3$ hydrogen nucleus transport (see Section 7.6 for details) [Zhang Q.F.2008]. The centroid PIMD method was used in this study. The quantum paths were generally discretized into $P = 16$ segments ($P = 32$ when the temperature was below 100 K). Each centroid and noncentroid mode was connected into a Nosé–Hoover chain to ensure the ergodicity of mode sampling and enable canonical ensemble simulations. They found that the transport pictures obtained from full quantum molecular dynamics simulations were fundamentally different from the results of classical molecular dynamics (where nuclei are treated as classical particles). The most important discovery was that the potential barrier corresponding to the proton transfer process between oxygen atoms can drop to a maximum of ~0.6 eV because of the nuclear quantum effects of hydrogen. This has revolutionized the understanding of proton transport processes in the classical picture, leading to results that are consistent with experimental measurements.

In addition, CP2K software also includes full quantum PIMD simulation functionality. It employs an atomic-orbital basis set for first-principles simulations; this offers a significant advantage in terms of computational speed over the plane-wave basis set, particularly for the problem of large computational tasks such as full quantum simulations. Consequently, CP2K can be used to some extent to investigate nuclear quantum effects in condensed matter systems. However, because the employment of orbital basis sets can lead to large errors in metal systems or systems with strong metallicity, the application scope of this method is limited.

The Vienna *Ab initio* Simulation Package (VASP) is currently one of the most widely used first-principles electronic structure calculation and molecular dynamics software packages. It is known for its standardized programming, excellent computational reliability, and high computational speed. VASP primarily uses adiabatic molecular dynamics methods to avoid the issues associated with the simultaneous integration of the equations of motion for electrons and nuclei that arise in the CPMD method. This makes VASP quite versatile and suitable for numerous applications. However, the standard version of VASP does not possess the capability to execute full quantum PIMD simulations. Meanwhile, the molecular dynamics functionality of the VASP method is relatively limited, and the simulation function of ensemble and heat-bath simulations is not perfect enough.

To address these problems, in 2011, E. G. Wang's group developed a first-principles PIMD simulation program based on VASP electronic structure calculations. The program follows a quadratic parallel process, where different path segments are computed in parallel. The self-consistent electronic calculations within each path segment are also carried out in parallel, which substantially improves the efficiency of the dynamic evolution. The PIMD function of the program is entirely self-developed, which is fully compatible with VASP in terms of first-principles self-consistent electronic structure calculations, system energy, and atomic force calculations. The program developed by Q. F. Zhang and J. Chen contains the following components:

1. *An efficient parallel computing program for different path segments.* By adding the message-passing interface (MPI) in the electronic structure and the wavefunction calculation of the original VASP code, the CPU parallel-computing structures are optimized from a 2D to 3D topology, enabling parallel computing for different path segments and significantly improving the computational efficiency.
2. *Two types of PIMD methods: the Cartesian coordinate mode and the centroid mode.* This allows for simulations using different PIMD modes.
3. *The canonical ensemble and heat bath of the Nosé–Hoover chain.* A canonical ensemble heat bath function of the Nose-Hoover chain is introduced in PIMD, which replaces the Nosé heat bath of the original program. It ensures the ergodicity of the simulation and realizes the canonical ensemble calculation.

Using this developed VASP-based first-principles PIMD program, J. Chen, Q. F. Zhang, Y. X. Feng, and X. Z. Li et al. conducted a series of studies on nuclear quantum effects for different condensed matter systems. For example, in 2013, they successfully applied this method to investigate the melting point of liquid metallic hydrogen at different pressures (see Section 7.1 for details) [Chen J.2013]. In their calculations, the Nosé–Hoover chain was applied to realize canonical ensemble simulations, and the number of segments of the quantum path was taken as $P = 32$. Simulations were performed using the centroid-PIMD method to simulate the proton migrations of liquid hydrogen. Their findings showed that considering the nuclear quantum effects of atoms, the melting point of liquid metal hydrogen is always below 200 K at different pressures. This is significantly different from the conclusion that the melting point of liquid metal hydrogen is generally ~300 K obtained from classical MD simulations. This work demonstrates the importance of full quantum effects in the study of phase transition of condensed metallic hydrogen.

Appendix B

Nonadiabatic Effect and the Time-Dependent Ab-Initio *Package (TDAP)*

The Born–Oppenheimer approximation, also referred to as the adiabatic approximation, is one of the most fundamental hypotheses in quantum physics and chemistry. Because the mass of the electrons is three to four orders of magnitude smaller than that of the nuclei in most molecules and condensed matter systems, Born and Oppenheimer assumed that the motions of nuclei and electrons are separable; this is the starting point in most quantum-mechanical simulations. However, because it neglects the quantum-mechanical correlations between electrons and nuclei, the approximation fails in many cases, such as physical, chemical, and biological processes involving electron/nucleus transfer with significant tunnelling and/or nonadiabatic effects, chemical bonding interactions involving quantum zero-point motions or electron–nucleus couplings, and hydrogen bond breaking/reforming (inherently a quantum-mechanical process) in complex hydrogen-bonded networks such as water. In these processes, molecules and condensed matter systems have enough wisdom and possibilities to explore unusual regions of the configuration space, the adiabatic potential energy surfaces that drives the time evolution of the whole system cross and split, and the quantum wave packets will also evolve in a wide range of potential energy space and various possible state manifolds. To describe these real-world physical and chemical phenomena, excited-state molecular dynamics calculations that go beyond ground-state simulations and the Born–Oppenheimer approximation are necessary. The Time-Dependent *Ab initio* Package (TDAP) described here has been developed to address precisely such problems.

TDAP is a first-principles software package targeting excited-state calculations and dynamic simulations of condensed matter. In 2008, S. Meng and E. Kaxiras developed the first version of TDAP, with the aim of performing quantum dynamics simulations of electron–nucleus coupling based on first principles. An improved version of the software package is currently maintained by the Institute of Physics and the material simulation platform of Songshan Lake Materials Laboratory, Chinese Academy of Sciences, under S. Meng's leadership [Meng S.2008a, Ma W.2016, Lian C.2018, You P.W.2021b, Zhao R.J.2023]. Readers who are interested in accessing and downloading the TDAP software can visit the links http://tdap.iphy.ac.cn or http://tdap.sslab.org.cn. TDAP employs a real-time, real-space implementation of time-dependent density functional theory based on numerical atomic basis sets, enabling first-principles molecular dynamics simulations of condensed matter beyond the adiabatic approximation.

To accurately characterize the excited-state properties of materials and establish quantum-dynamical time-evolution paths for relevant physical quantities, the theoretical treatment of time-dependent nonadiabatic phenomena is essential and can be challenging on many levels. Currently, approximated full quantum-mechanical solutions can be obtained for realistic macromolecules and large-scale condensed matter systems. Consequently, over the past 50 years, researchers have dedicated considerable effort to developing hybrid methods that combine quantum and semi-classical simulations to address this problem. These methods encompass Ehrenfest dynamics, trajectory potential energy surface hopping, and the linearized semi-classical initial-value representation.

TDAP is a real-time first-principles computational method primarily used to model coupled electron–nucleus quantum dynamic processes in molecules and condensed materials. Within the framework of the mean-field theory of mixed quantum-classical systems, the forces acting on the nuclei are averaged over many possible adiabatic electronic states induced by the nuclear motion. In particular, TDAP follows the semi-classical Ehrenfest dynamic evolution beyond the Born–Oppenheimer approximation.

A flowchart of TDAP is shown in Figure B.1. In practical calculations, TDAP employs local atomic basis sets and real-time propagation of wave functions to solve the time-dependent Kohn–Sham equations. This endows it with several advantages over conventional methods:

(1) Atomic-orbital basis sets, which are small and fast, are adopted for the expansion of electronic wavefunctions. This enables the simulation of a large periodic system or a finite-sized supercell with a large vacuum space without incurring significant computational cost, yet still maintaining relatively high computational accuracy.
(2) Real-time excited-state trajectories are achieved by the simultaneous and direct evolution of the electron density and nuclear configuration. Between each electronic and nuclear evolution step, forces are calculated self-consistently from the mean-field theory. This provides a powerful computational tool for directly modeling the ultrafast dynamics of coupled electrons and nuclei during photoexcitation, offering a more reliable microscopic picture of the quantum state of photoexcited matter.
(3) The method has relatively high parallelization computation efficiency. Because all the occupied electronic orbitals propagate independently, their dynamic evolution can be uniformly distributed over several processors with little mutual communication, thereby facilitating large-scale parallel computing.
(4) The photoabsorption spectra, polarizabilities, and dynamic quantum evolution of the macromolecules and condensed matter systems are calculated within the same scheme on an equal footing. Nonlinear effects and nonperturbative processes can also be rigorously simulated in a nonadiabatic manner.

In recent years, S. Meng's group, in collaboration with E. G. Wang et al., succeeded in calculating the optical absorption spectra of molecules and condensed systems using the TDAP method. They have simulated chemical reactions induced by interface electron injection, electron–hole recombination, and charge transfer. Their research has focused on the dynamic properties of quantum materials under ultrafast and ultraintense photoexcitation.

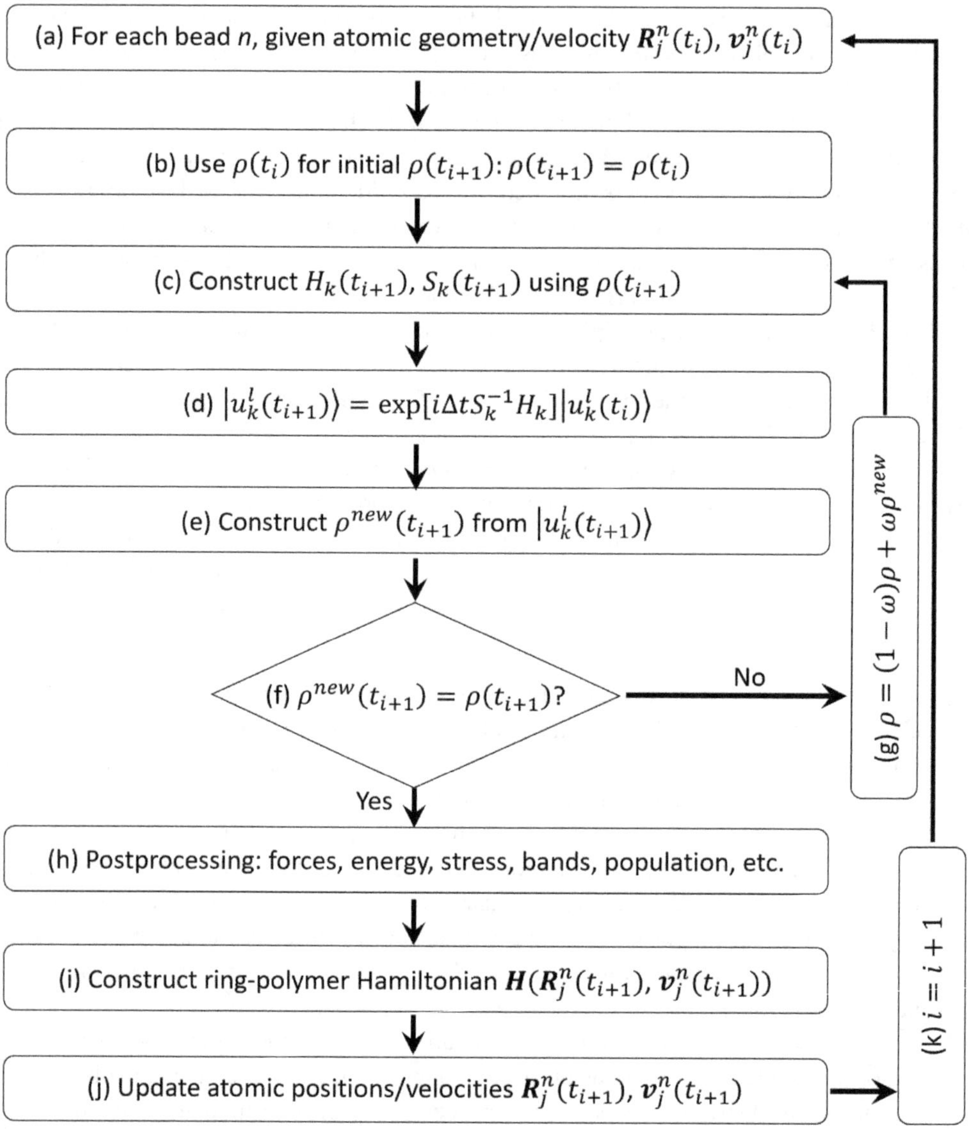

Figure B.1 Flowchart of the TDAP program for full-quantum simulations. It illustrates the simulation framework for time-dependent DFT combined with ring-polymer path-integral molecular dynamics. For ordinary Ehrenfest dynamics, the number of beads n can be taken as 1. Specific details can be found at: http://tdap.iphy.ac.cn or http://tdap.sslab.org.cn.

In particular, they have been successful in understanding the excited-state reaction dynamics when a single path or a large number of similar paths dominate [Lakhotia H.2020, Lian C.2020, Guan M.X.2021, You P.W.2021a-b, Guan M.X.2022, Xu J.Y.2022].

However, under the mean-field approximation, Ehrenfest dynamics still employs a single average trajectory to describe nuclear evolution, even though the nuclear wave function in the excited-state evolution path has been decomposed into different segments.

Therefore, the conventional Ehrenfest method cannot accurately handle cases involving multiple reaction paths in the excited state, particularly when the nuclear trajectories of a specific excited state are of interest. Furthermore, the conventional Ehrenfest method cannot account for the balance between electronic states, leading to deviations from the true behavior in certain long-term dynamic evolutions. For alternative strategies that go beyond the Ehrenfest dynamics approach, readers can refer to semi-classical methods, such as trajectory potential energy surface hopping, initial value representation, and so on, that calculate electronic states in the classical pathway after the transition to simulate nonadiabatic processes in molecules and materials.

Currently, S. Meng et al. are developing new methods to be included in TDAP that go beyond the conventional Ehrenfest mean-field approximation, such as nonadiabatic path-integral simulation methods based on time-dependent DFT. These new approaches aim to provide a more accurate description to address the problems mentioned above, especially for coupled electron–nucleus quantum states and their dynamic evolution. See subsection 5.4.3 for examples of relevant studies.

References

A

Abel, W. R., A. C. Anderson, and J. C. Wheatley, 1966, *Propagation of zero sound in liquid He^3 at low temperatures*, **Phys. Rev. Lett. 17**, 74.

Ackland, G. J., M. Dunuwille, M. Martinez-Canales, I. Loa, R. Zhang, S. Sinogeikin, W. Cai, and S. Deemyad, 2017, *Quantum and isotope effects in lithium metal*, **Science 356**, 1254.

Agrawal, K. V., S. Shimizu, L. W. Drahushuk, D. Kilcoyne, and M. S. Strano, 2017, *Observation of extreme phase transition temperatures of water confined inside isolated carbon nanotubes*, **Nat. Nanotech. 12**, 267.

Alder, B. J., and T. E. Wainwright, 1959, *Studies in molecular dynamics. I. General method*, **J. Chem. Phys. 31**, 459.

Alexandrov, B. S., V. G. Stanev, A. R. Bishop, and K. Ø. Rasmussen, 2012, *Anharmonic dynamics of intramolecular hydrogen bonds driven by DNA breathing*, **Phys. Rev. E 86**, 061913.

Alkauskas, A., Q. Yan, and C. G. van de Walle, 2014, *First-principles theory of nonradiative carrier capture via multiphonon emission*, **Phys. Rev. B 90**, 075202.

Allen, J. F., and H. Jones, 1938, *New phenomena connected with heat flow in helium II*, **Nature 141**, 243.

Allen, L. J., H. G. Brown, S. D. Findlay, and B. D. Forbes, 2018, *A quantum mechanical exploration of phonon energy-loss spectroscopy using electrons in the aloof beam geometry*, **Microscopy 67**, i24.

Allen, P. B., and R. C. Dynes, 1975, *Transition temperature of strong-coupled superconductors reanalyzed*, **Phys. Rev. B 12**, 905.

Allen, P. B., and V. Heine, 1976, *Theory of the temperature dependence of electronic band structures*, **J. Phys. C: Solid State Phys. 9**, 2305.

Allen, P. B., 1978, *New method for solving Boltzmann's equation for electrons in metals*, **Phys. Rev. B 17**, 3725.

Allen, P. B., and M. Cardona, 1981, *Theory of the temperature dependence of the direct gap of germanium*, **Phys. Rev. B 23**, 1495.

Althorpe, S. C., 2011, *On the equivalence of two commonly used forms of semiclassical instanton theory*, **J. Chem. Phys. 134**, 114104.

An, J. M., and W. E. Pickett, 2001, *Superconductivity of MgB_2: Covalent bonds driven metallic*, **Phys. Rev. Lett. 86**, 4366.

Ananth, N., 2013, *Mapping variable ring polymer molecular dynamics: A path-integral based method for nonadiabatic processes*, **J. Chem. Phys. 139**, 124102.

Andersen, H. C., 1983, *Rattle: A "velocity" version of the shake algorithm for molecular dynamics calculations*, **J. Comp. Phys. 52**, 24.

Anderson, J. B., 1975, *A random-walk simulation of the Schrödinger equation: H+3*, **J. Chem. Phys. 63**, 1499.

Anderson, J. B., 1976, *Quantum chemistry by random walk. H2P, H+3 D3h 1A'1, H2 3Σ +u' H4 1Σ +g' Be 1S*, **J. Chem. Phys.** 65, 4121.

Anderson, P. W., 1972, *More Is different*, **Science 177**, 393.

Anderson, P. W., 1987a, *The resonating valence bond state in La_2CuO_4 and superconductivity*, **Science 235**, 1196.

Anderson, P. W., G. Baskaran, Z. Zou, and T. Hsu, 1987b, *Resonating-valence-bond theory of phase transitions and superconductivity in La_2CuO_4-based compounds*, **Phys. Rev. Lett. 58**, 2790.

Andersson, S., G. Nyman, A. Arnaldsson, U. Manthe, and H. Jonsson, 2009, *Comparison of quantum dynamics and quantum transition state theory estimates of the $H+CH_4$ reaction rate*, **J. Phys. Chem. A 113**, 4468.

Andreani, C., D. Colognesi, J. Mayers, G. F. Reiter, and R. Senesi, 2005, *Measurement of momentum distribution of light atoms and molecules in condensed matter systems using inelastic neutron scattering*, **Adv. Phys. 54**, 377.

Andreani, C., G. Romanelli, and R. Senesi, 2013, *A combined INS and DINS study of proton quantum dynamics of ice and water across the triple point and in the supercritical phase*, **Chem. Phys. 427**, 106.

Andreev, A. F., and I. M. Lifshitz, 1969, Quantum theory of defects in crystals, *JETP* 29, 1107.

Andzelm, J., and E. Wimmer, 1992, *Density functional Gaussian-type-orbital approach to molecular geometries, vibrations, and reaction energies*, **J. Chem. Phys. 96**, 1280.

Annaberdiyev, A., G. Wang, C. A. Melton, M. C. Bennett, L. Shulenburger, and L. Mitas, 2018, *A new generation of effective core potentials from correlated calculations: 3d transition metal series*, **J. Chem. Phys. 149**, 134108.

Antončík, E., 1955, *On the theory of temperature shift of the absorption curve in non-polar crystal*, **Czech. J. Phys. 5**, 449.

Aoki, K., H. Yamawaki, M. Sakashita, and H. Fujihisa, 1996, *Infrared absorption study of the hydrogen-bond symmetrization in ice to 110 GPa*, **Phys. Rev. B 54**, 15673.

Arunan, E., G. Desiraju, R. Klein, J. Sadlej, S. Scheiner, I. Alkorta, D. Clary, R. Crabtree, J. Dannenberg, P. Hobza, H. Kjaergaard, A. Legon, B. Mennucci, and D. Nesbitt, 2011, *Definition of the hydrogen bond (IUPAC Recommendations 2011)*, **Pure Appl. Chem. 83**, 1637.

Ashcroft, N. W., 1968, *Metallic hydrogen: A high-temperature superconductor?* **Phys. Rev. Lett. 21**, 1748.

Ashcroft, N. W., and N. D. Mermin, 1976, *Solid state physics,* (Harcourt College Publishers, New York).

Ashcroft, N. W., 2000, *The hydrogen liquids*, **J. Phys.: Condens. Matter 12**, A129.

Ashcroft, N. W., 2004, *Hydrogen dominant metallic alloys: High temperature superconductors?* **Phys. Rev. Lett. 92**, 187002.

Atou, T., M. Hasegawa, L. J. Parker, and J. V. Badding, 1996, *Unusual chemical behavior for potassium under pressure: Potassium-silver compounds*, **J. Am. Chem. Soc. 118**, 12104.

Awaji, S., K. Watanabe, H. Oguro, H. Miyazaki, S. Hanai, T. Tosaka, and S. Ioka, 2017, *First performance test of a 25 T cryogen-free superconducting magnet*, **Supercond. Sci. Tech. 30**, 065001.

Azizi, F., and H. Rezania, 2021, *Collective plasmonic oscillations of zigzag boron-nitride nanotubes in the presence of Holstein phonons*, **Phys. E Low-Dimens. Syst. Nanostruct. 130**, 114687.

B

Babaev, E., A. Sudbo, and N. W. Ashcroft, 2004, *A superconductor to superfluid phase transition in liquid metallic hydrogen*, **Nature 431**, 666.

Badding, J. V., 1998, *High-pressure synthesis, characterization, and tuning of solid state materials*, **Annu. Rev. Mater. Sci. 28**, 631.

Bafile, U., M. Celli, M. Zoppi, and J. Mayers, 1998, *Deep inelastic neutron scattering on liquid hydrogen in the crossover region between the molecular and atomic regimes*, **Phys. Rev. B 58**, 791.

Bai, H. Y., M. D. Bird, L. D. Cooley, I. R. Dixon, K. L. Kim, D. C. Larbalestier, W. S. Marshall, U. P. Trociewitz, H. W. Weijers, D. V. Abraimov, and G. S. Boebinger, 2020, The 40 T superconducting magnet project at the national high magnetic field laboratory, *IEEE Trans. Appl. Supercon.* **30**, 4300405.

Bai, X. D., E. G. Wang, J. Yu, and H. Yang, 2000, *Blue-violet PL from large-scale highly aligned BCN nanofibers*, **Appl. Phys. Lett. 77**, 67.

Bakker, H. J., and H. K. Nienhuys, 2002, *Delocalization of protons in liquid water*, **Science 297**, 587.

Bakó, I., Á. Madarász, and L. Pusztai, 2021, *Nuclear quantum effects: Their relevance in neutron diffraction studies of liquid water*, **J. Mol. Liq. 325**, 115192.

Baldini, E., C. A. Belvin, M. Rodriguez-Vega, I. O. Ozel, D. Legut, A. Kozlowski, A. M. Oles, K. Parlinski, P. Piekarz, J. Lorenzana, G. A. Fiete, and N. Gedik, 2020, *Discovery of the soft electronic modes of the trimeron order in magnetite*, **Nat. Phys. 16**, 541.

Ball, P., 2008, *Water – an enduring mystery*, **Nature 452**, 291.

Ballone, P., and P. Milani, 1992, *Delocalization and tunneling in the ionic structure of lithium microclusters*, **Phys. Rev. B 45**, 11222.

Bandura, A. V., J. D. Kubicki, and J. O. Sofo, 2008, *Comparisons of multilayer H_2O adsorption onto the (110) surfaces of α-TiO_2 and SnO_2 as calculated with density functional theory*, **J. Phys. Chem. B 112**, 11616.

Bang, J., S. Meng, Y.-Y. Sun, D. West, Z. Wang, F. Gao, and S. B. Zhang, 2013, *Regulating energy transfer of excited carriers and the case for excitation-induced hydrogen dissociation on hydrogenated graphene*, **Proc. Natl. Acad. Sci. U. S. A. 110**, 908.

Banhart, F., J. Kotakoski, and A. V. Krasheninnikov, 2011, *Structural defects in graphene*, **ACS Nano 5**, 26.

Baratoff, A., and B. N. J. Persson, 1988, *Theory of the local tunneling spectrum of a vibrating adsorbate*, **J. Vac. Sci. Technol. A 6**, 331.

Barbatti, M., 2011, *Nonadiabatic dynamics with trajectory surface hopping method*, **WIRES Comput. Mol. Sci. 1**, 620.

Bardeen, J., and D. Pines, 1955, *Electron-phonon interaction in metals*, **Phys. Rev. 99**, 1140.

Bardeen, J., L. N. Cooper, and J. R. Schrieffer, 1957, *Theory of superconductivity*, **Phys. Rev. 108**, 1175.

Baroni, S., S. de Gironcoli, A. Dal Corso, and P. Giannozzi, 2001, *Phonons and related crystal properties from density-functional perturbation theory*, **Rev. Mod. Phys. 73**, 515.

Barrett, C., 1947, *A low temperature transformation in lithium*, **Phys. Rev. 72**, 245.

Barrett, J. H., 1952, *Dielectric constant in perovskite type crystals*, **Phys. Rev. 86**, 118.

Bartels, C., and M. Karplus, 1998, *Probability distributions for complex systems: adaptive umbrella sampling of the potential energy*, **J. Phys. Chem. B 102**, 865.

Bartels-Rausch, T., 2013, *Ten things we need to know about ice and snow*, **Nature 494**, 27.

Bartelt, N. C., and K. Thurmer, 2022, *Ionicity and hydrogen affinity of water layers on metals*, **Proc. Natl. Acad. Sci. U. S. A. Nexus 1**, 1.

Bartolomei, M., M. I. Hernández, J. Campos-Martínez, and R. Hernández-Lamoneda, 2019, *Graphene multi-protonation: A cooperative mechanism for proton permeation*, **Carbon 144**, 724.

Barwick, B., H. S. Park, O. H. Kwon, J. S. Baskin, and A. H. Zewail, 2008, *4D imaging of transient structures and morphologies in ultrafast electron microscopy*, **Science 322**, 1227.

Baskaran, G., and P. W. Anderson, 1988, *Gauge theory of high-temperature superconductors and strongly correlated Fermi systems*, **Phys. Rev. B 37**, 580.

Basran, J., S. Patel, M. J. Sutcliffe, and N. S. Scrutton, 2001, *Importance of barrier shape in Enzyme-catalyzed reactions: Vibrationally assisted hydrogen tunneling in tryptophan tryptophylquinone-dependent amine dehydrogenases*, **J. Biol. Chem. 276**, 9.

Bassett, W. A., 2009, *Diamond anvil cell, 50th birthday*, **HPR 29**, 163.

Bastard, G., E. E. Mendez, L. L. Chang, and L. Esaki, 1982, *Exciton binding energy in quantum wells*, **Phys. Rev. B 26**, 1974.

Beck, M. H., A. Jackle, G. A. Worth, and H. D. Meyer, 2000, *The multiconfiguration time-dependent Hartree (MCTDH) method: A highly efficient algorithm for propagating wavepackets*, **Phys. Rep. 324**, 1.

Becke, A. D., 1989, *Basis-set-free density-functional quantum chemistry*, **Int. J. Quantum Chem. 36**, 599.

Bennett, C. H., 1977, *Algorithms for chemical computations* (ACS Symposium Series No.46), (American Chemical Society, Washington, DC).

Ben-Nun, M., and T. J. Martinez, 1998, *Nonadiabatic molecular dynamics: Validation of the multiple spawning method for a multidimensional problem*, **J. Chem. Phys. 108**, 7244.

Benoit, M., D. Marx, and M. Parrinello, 1998, *Tunnelling and zero-point motion in high-pressure ice*, **Nature 392**, 258.

Berger, P. R., and M. Kim, 2018, *Polymer solar cells: P3HT:PCBM and beyond*, **J. Renew. Sustain. Energy 10**, 013508.

Bergmann, U., D. Nordlund, P. Wernet, M. Odelius, L. G. M. Pettersson, and A. Nilsson, 2007, *Isotope effects in liquid water probed by x-ray Raman spectroscopy*, **Phys. Rev. B 76**, 024202.

Berne, B. J., and D. Thirumalai, 1986, *On the simulation of quantum systems: Path integral methods*, **Annu. Rev. Phys. Chem. 37**, 401.

Berne, B. J., G. Ciccotti, and D. F. Coker, 1998, *Classical and quantum dynamics in condensed phase simulations* (World Scientific, New Jersey).

Bianconi, A., S. Agrestini, G. Bianconi, D. Di Castro, and NL Saini, 2001, *A quantum phase transition driven by the electron lattice interaction gives high Tc superconductivity*, **J. Alloys Compd. 317**, 537.

Bicerano, J., H. F. Schaefer III, and W. H. Miller, 1983, *Structure and tunneling dynamics of malonaldehyde. A theoretical study*, **J. Am. Chem. Soc.** 105, 2550.

Bierman, D. M., A. Lenert, W. R. Chan, B. Bhatia, I. Celanovic, M. Soljacic, and E. N. Wang, 2016, *Enhanced photovoltaic energy conversion using thermally based spectral shaping*, **Nat. Energy 1**, 16068.

Biermann, S., D. Hohl, and D. Marx, 1998a, *Proton quantum effects in high pressure hydrogen*, **J. Low Temp. Phys. 110**, 97.

Biermann, S., D. Hohl, and D. Marx, 1998b, *Quantum effects in solid hydrogen at ultra-high pressure*, **Solid State Commun. 108**, 337.

Biernacki, S., and M. Scheffler, 1989b, *Negative thermal expansion of diamond and zinblende semiconductors*, **Phys. Rev. Lett. 63**, 290.

Bikondoa, O., C. L. Pang, R. Ithnin, C. A. Muryn, H. Onishi, and G. Thornton, 2006, *Direct visualization of defect-mediated dissociation of water on TiO_2 (110)*, **Nat. Mater. 5**, 189.

Bircher, M. P., E. Liberatore, N. J. Browning, S. Brickel, C. Hofmann, A. Patoz, O. T. Unke, T. Zimmermann, M. Chergui, P. Hamm, U. Keller, M. Meuwly, H.-J. Woerner, J. Vanicek, and U. Rothlisberger, 2017, *Nonadiabatic effects in electronic and nuclear dynamics*, **Struct. Dyn. 4**, 061510.

Bird, M. D., S. Bole, Y. M. Eyssa, B. J. Gao, and H. J. Schneider-Muntau, 1996a, *The world's first 27 T and 30 T resistive magnets*, **IEEE Trans. Magn. 32**, 2444.

Bird, M. D., S. Bole, Y. M. Eyssa, B. J. Gao, and H. J. Schneider-Muntau, 1996b, *Design of a poly-Bitter magnet at the NHMFL*, **IEEE Trans. Magn. 32**, 2542.

Bird, M. D., I. R. Dixon, and J. Toth, 2004, *Design of the next generation of Florida-Bitter magnets at the NHMFL*, **IEEE Trans. Appl. Supercon. 14**, 1253.

Bitter, F., 1936, *The design of powerful electromagnets – Part I. The use of iron*, **Rev. Sci. Instrum. 7**, 479.

Blanc, J., V. Bonačić-Koutecký, M. Broyer, J. Chevaleyre, P. Dugourd, J. Koutecký, C. Scheuch, J. Wolf, and L. Wöste, 1992, *Evolution of the electronic structure of lithium clusters between four and eight atoms*, **J. Chem. Phys. 96**, 1793.

Blinc, R., 1960, *On the isotopic effects in the ferroelectric behaviour of crystals with short hydrogen bonds*, **J. Phys. Chem. Solids 13**, 204.

Blöchl, P. E., 1994, *Projector augmented-wave method*, **Phys. Rev. B 50**, 17953.

Boeri, L., G. B. Bachelet, E. Cappelluti, and L. Pietronero, 2002, *Small Fermi energy and phonon anharmonicity in MgB_2 and related compounds*, **Phys. Rev. B 65**, 214501.

Boeri, L., E. Cappelluti, and L. Pietronero, 2005, *Small Fermi energy, zero-point fluctuations, and nonadiabaticity in MgB_2*, **Phys. Rev. B 71**, 012501.

Bohm, D., 1952, *A suggested interpretation of the quantum theory in terms of "hidden" variables. I*, **Phys. Rev. 85**, 166.

Bohnen, K. P., R. Heid, and B. Renker, 2001, *Phonon dispersion and electron-phonon coupling in MgB_2 and AlB_2*, **Phys. Rev. Lett. 86**, 5771.

Bohr, N., 1913, *On the constitution of atoms and molecules*, **Lond. Edinb. Dublin Philos. Mag. J. Sci. 26**, 1.

Bok, J. M., J. J. Bae, H. Y. Choi, C. M. Varma, W. T. Zhang, J. F. He, Y. X. Zhang, L. Yu, and X. J. Zhou, 2016, *Quantitative determination of pairing interactions for high-temperature superconductivity in cuprates*, **Sci. Adv. 2**, e1501329.

Bonev, S. A., E. Schwegler, T. Ogitsu, and G. Galli, 2004, *A quantum fluid of metallic hydrogen suggested by first-principles calculations*, **Nature 431**, 669.

Booth, G. H., A. J. W. Thom, and A. Alavi, 2009, *Fermion Monte Carlo without fixed nodes: A game of life, death, and annihilation in slater determinant space*, **J. Chem. Phys. 131**, 054106.

Booth, G. H., A. Grüneis, and A. Alavi, 2013, *Towards an exact description of electronic wavefunctions in real solids*, **Nature 493**, 365.

Borden, W. T., 2016, *Reactions that involve tunneling by carbon and the role that calculations have played in their study*, **WIREs Comput. Mol. Sci. 6**, 20.

Born, M., and R. Oppenheimer, 1927, *Zur quantentheorie der molekeln*, **Ann. Phys. 84**, 457.

Born, M., and V. Fock, 1928, *Beweis des adiabatensatzes*, **Z. Phys. 51**, 165.

Born, M., 1951, *Coupling of electron and nuclear motions in molecules and crystal (in German)*, **Nachr. Akad. Wiss. Göttingen, Math.-Phys. Klasse IIa, Math.-Phys.-Chem. Abt., S. Art. 6**, 1.

Born, M., and K. Huang, 1954, *Dynamical theory of crystal lattices* (Oxford University Press, Oxford).

Borodin, D., N. Hertl, G. B. Park, M. Schwarzer, J. Fingerhut, Y. Wang, J. Zuo, F. Nitz, G. Skoulatakis, A. Kandratsenka, D. J. Auerbach, D. Schwarzer, H. Guo, T. N. Kitsopoulos, and A. M. Wodtke, 2022, *Quantum effects in thermal reaction rates at metal surfaces*, **Science 377**, 394.

Bove, L. E., S. Klotz, A. Paciaroni, and F. Sacchetti, 2009, *Anomalous proton dynamics in ice at low temperatures*, **Phys. Rev. Lett. 103**, 165901.

Boys, S. F., 1950, *Electronic wave functions. I. A general method of calculation for the stationary states of any molecular system*, **Proc. R. Soc. A**, **200**, 542.

Brandenburg, J. G., A. Zen, D. Alfè, and A. Michaelides, 2019, *Interaction between water and carbon nanostructures: How good are current density functional approximations?* **J. Chem. Phys. 151**, 164702.

Brickwedde, F. G. 1982, *Harold Urey and the discovery of deuterium*, **Phys. Today 35**, 34.

Bridgman, P. W., 1931, *The physics of high pressure* (G. Bell and Sons, Ltd., London).

Brillouin, L., 1926, *La mécanique ondulatoire de Schrödinger: une méthode générale de resolution par approximations successives*, **C. R. Acad. Sci. 183**, 24.

Bronsted, J. N., 1928, *Acid and basic catalysis*, **Chem. Rev. 5**, 231.

Brougham, D. F., A. J. Horsewill, and R. I. Jenkinson, 1997, *Proton transfer dynamics in the hydrogen bond: A direct measurement of the incoherent tunnelling rate by NMR and the quantum-to-classical transition*, **Chem. Phys. Lett. 272**, 69.

Brougham, D. F., R. Caciuffo, and A. J. Horsewill, 1999, *Coordinated proton tunnelling in a cyclic network of four hydrogen bonds in the solid state*, **Nature 397**, 241.

Buckingham, R. A., 1938, *The classical equation of state of gaseous helium, neon and argon*, **Proc. Roy. Soc. A. 168**, 264.

Bud'ko, S. L., G. Lapertot, C. Petrovic, C. E. Cunningham, N. Anderson, and P. C. Canfield, 2001, *Boron isotope effect in superconducting MgB_2*, **Phys. Rev. Lett. 86**, 1877.

Bunch, J. S., S. S. Verbridge, J. S. Alden, A. M. van der Zande, J. M. Parpia, H. G. Craighead, and P. L. McEuen, 2008, *Impermeable atomic membranes from graphene sheets*, **Nano Lett. 8**, 2458.

Bünermann, O., A. Kandratsenka, and A. M. Wodtke, 2021, *Inelastic scattering of H atoms from surfaces*, **J. Phys. Chem. A 125**, 3059.

Burkatzki, M., C. Filippi, and M. Dolg, 2007, *Energy-consistent pseudopotentials for quantum Monte Carlo calculations*, **J. Chem. Phys. 126**, 234105.

Buzea, C., and T. Yamashita, 2001, *Review of the superconducting properties of MgB_2*, **Supercond. Sci. Technol. 14**, R115.

C

Cahlik, A., J. Hellerstedt, J. I. Mendieta-Moreno, M. Svec, V. M. Santhini, S. Pascal, D. Soler-Polo, S. I. Erlingsson, K. Vyborny, P. Mutombo, O. Marsalek, O. Siri, and P. Jelinek, 2021, *Significance of nuclear quantum effects in hydrogen bonded molecular chains*, **ACS Nano 15**, 10357.

Caldeira, A. O., and A. J. Leggett, 1983, *Path integral approach to quantum Brownian motion*, **Phys. A: Stat. Mech. Appl. 121**, 587.

Caldwell, J. D., A. V. Kretinin, Y. Chen, V. Giannini, M. M. Fogler, Y. Francescato, C. T. Ellis, J. G. Tischler, C. R. Woods, A. J. Giles, M. Hong, K. Watanabe, T. Taniguchi, S. A. Maier, and K. S. Novoselov, 2014, *Sub-diffractional volume-confined polaritons in the natural hyperbolic material hexagonal boron nitride*, **Nat. Commun. 5**, 5221.

Callen, H. B., and T. A. Welton, 1951, *Irreversibility and generalized noise*, **Phys. Rev. 83**, 34.

Callen, H. B., and R. F. Greene, 1952a, *On a theorem of irreversible thermodynamics*, **Phys. Rev. 86**, 702.

Callen, H. B., and R. F. Greene, 1952b, *On a theorem of irreversible thermodynamics II*, **Phys. Rev. 88**, 1387.

Calvo, F., and Y. Magnin, 2016, *Nuclear quantum effects on the thermal expansion coefficient of hexagonal boron nitride monolayer*, **Eur. Phys. J. B 89**, 56.

Cannuccia, E., and A. Marini, 2011, *Effect of the quantum zero-point atomic motion on the optical and electronic properties of diamond and trans-polyacetylene*, **Phys. Rev. Lett. 107**, 255501.

Cannuccia, E., and A. Marini, 2012, *Zero point motion effect on the electronic properties of diamond, trans-polyacetylene and polyethylene*, **Eur. Phys. J. B 85**, 320.

Cao, D. Y., 2020, *Structures and dynamics of interfacial water*, PhD dissertation, Peking University, Beijing, 2020 (in Chinese).

Cao, J. S., and G. A. Voth, 1993, *A new perspective on quantum time correlation functions*, **J. Chem. Phys. 99**, 10070.

Capaz, R. B., C. D. Spataru, P. Tangney, M. L. Cohen, and S. G. Louie, 2005, *Temperature dependence of the band gap of semiconducting carbon nanotubes*, **Phys. Rev. Lett. 94**, 036801.

Cappelluti, E., S. Ciuchi, C. Grimaldi, L. Pietronero, and S. Strässler, 2002, *High Tc superconductivity in MgB_2 by nonadiabatic pairing*. **Phys. Rev. Lett. 88**, 117003.

Cappelluti, E., and L. Pietronero, 2006, *Electron–phonon interaction and breakdown of the adiabatic principle in fullerides and MgB_2*, **J. Phys. Chem. Solids 67**, 1941.

Car, R., and M. Parrinello, 1985, *Unified approach for molecular dynamics and density-functional theory*, **Phys. Rev. Lett. 55**, 2471.

Carbone, F., O. H. Kwon, and A. H. Zewail, 2009, *Dynamics of chemical bonding mapped by energy-resolved 4D electron microscopy*, **Science 325**, 181.

Cardona, M., and M. L. W. Thewalt, 2005, *Isotope effects on the optical spectra of semiconductors,* **Rev. Mod. Phys. 77**, 1173.

Carlson, C. M., T. V. Rivkin, P. A. Parilla, J. D. Perkins, D. S. Ginley, A. B. Kozyrev, V. N. Oshadchy, and A. S. Pavlov, 2000, *Large dielectric constant ($\varepsilon/\varepsilon_0$> 6000) $Ba_{0.4}Sr_{0.6}TiO_3$ thin films for high-performance microwave phase shifters*. **Appl. Phys. Lett. 76**, 1920.

Carpenter, B. K., 1983, *Heavy-atom tunneling as the dominant pathway in a solution-phase reaction? Bond shift in antiaromatic annulenes*, **J. Am. Chem. Soc. 105**, 1700.

Carr, T. H. G., J. J. Shephard, and C. G. Salzmann, 2014, *Spectroscopic signature of stacking disorder in ice I*, **J. Phys. Chem. Lett. 5**, 2469.

Caruso, F., M. Hoesch, P. Achatz, J. Serrano, M. Krisch, E. Bustarret, and F. Giustino, 2017, *Nonadiabatic Kohn anomaly in heavily boron-doped diamond*, **Phys. Rev. Lett. 119**, 017001.

Cassabois, G., P. Valvin, and B. Gil, 2016a, *Hexagonal boron nitride is an indirect bandgap semiconductor*, **Nat. Photonics 10**, 262.

Cassabois, G., P. Valvin, and B. Gil, 2016b, *Intervalley scattering in hexagonal boron nitride*, **Phys. Rev. B 93**, 035207.

Cederbaum, L. S., 2008, *Born-Oppenheimer approximation and beyond for time-dependent electronic processes*, **J. Chem. Phys. 128**, 1241010.

Ceperley, D. M., and B. J. Alder, 1980, *Ground state of the electron-gas by a stochastic method*, **Phys. Rev. Lett. 45**, 566.

Ceperley, D. M., 1995, *Path integrals in the theory of condensed helium*, **Rev. Mod. Phys. 67**, 279.

Ceriotti, M., G. Bussi, and M. Parrinello, 2009a, *Langevin equation with colored noise for constant-temperature molecular dynamics simulations*, **Phys. Rev. Lett. 102**, 020601.

Ceriotti, M., G. Bussi, and M. Parrinello, 2009b, *Nuclear quantum effects in solids using a colored-noise thermostat*, **Phys. Rev. Lett. 103**, 030603.

Ceriotti, M., D. E. Manolopoulos, and M. Parrinello, 2011, *Accelerating the convergence of path integral dynamics with a generalized Langevin equation*, **J. Chem. Phys. 134**, 084104.

Ceriotti, M., and D. E. Manolopoulos, 2012, *Efficient first-principles calculation of the quantum kinetic energy and momentum distribution of nuclei*, **Phys. Rev. Lett. 109**, 100604.

Ceriotti, M., W. Fang, P. G. Kusalik, R. H. Mckenzie, A. Michaelides, M. A. Morales, and T. E. Markland, 2016, *Nuclear quantum effects in water and aqueous systems: Experiment, theory, and current challenges*, **Chem. Rev. 116**, 7529.

Chadi, D. J., and M. L. Cohen, 1975, *Tight-binding calculations of the valence bands of diamond and zincblende crystals*, **Phys. Status Solidi 68**, 405.

Chadwick, M. B., P. Oblozinsky, M. Herman, N. M. Greene, R. D. McKnight, D. L. Smith, P. G. Young, R. E. MacFarlane, G. M. Hale, S. C. Frankle, A. C. Kahler, T. Kawano, R. C. Little, D. G. Madland, P. Moller, R. D. Mosteller, P. R. Page, P. Talou, H. Trellue, M. C. White, W. B. Wilson, R. Arcilla, C. L. Dunford, S. F. Mughabghab, B. Pritychenko, D. Rochman, A. A. Sonzogni, C. R. Lubitz, T. H. Trumbull, J. P. Weinman, D. A. Brown, D. E. Cullen, D. P. Heinrichs, D. P. McNabb, H. Derrien, M. E. Dunn, N. M. Larson, L. C. Leal, A. D. Carlson, R. C. Block, J. B. Briggs, E. T. Cheng, H. C. Huria, M. L. Zerkle, K. S. Kozier, A. Courcelle, V. Pronyaev, and S. C. van der Marck, 2006, *ENDF/B-VII.0: Next generation evaluated nuclear data library for nuclear science and technology*, **Nucl. Data Sheets 107**, 2931.

Chan, M. H. W., 2008, *Supersolidity*, **Science 319**, 1207.

Chandler, D., 1978, *Statistical mechanics of isomerization dynamics in liquids and the transition state approximation*, **J. Chem. Phys. 68**, 2959.

Chandler, D., and P. Wolynes, 1981, *Exploiting the isomorphism between quantum theory and classical statistical mechanics of polyatomic fluids*, **J. Chem. Phys. 74**, 4078.

Chang, C. Z., J. Zhang, X. Feng, J. Shen, Z. Zhang, M. Guo, K. Li, Y. Ou, P. Wei, L. L., Wang, Z. Q. Ji, Y. Feng, S. Ji, X. Chen, J. Jia, X. Dai, Z. Fang, S. C. Zhang, K. He, Y. Wang, L. Lu, X. C. Ma, and Q. K. Xue, 2013, *Experimental observation of the quantum anomalous Hall effect in a magnetic topological insulator*, **Science 340**, 167.

Chang, K. J., M. M. Dacorogna, M. L. Cohen, J. M. Mignot, G. Chouteau, and G. Martinez, 1985, *Superconductivity in high-pressure metallic phases of Si*, **Phys. Rev. Lett. 54**, 2375.

Chang, Y., Z. Chen, J. Zhou, Z. Luo, Z. He, G. Wu, M. N. R. Ashfold, K. Yuan, and X. Yang, 2019, *Striking isotopologue-dependent photodissociation dynamics of water molecules: The signature of an accidental resonance*, **J. Phys. Chem. Lett. 10**, 4209.

Chao, Y., and Y. W. Ma, 2021, *Superconducting materials: Challenges and opportunities for large-scale applications*, **iScience 24**, 102541.

Chayes, J. T., L. Chayes, and M. B. Ruskai, 1985, *Density functional approach to quantum lattice systems*, **J. Statist. Phys. 38**, 493.

Chen, C., F. Tian, D. Duan, K. Bao, X. Jin, B. Liu, and T. Cui, 2014, *Pressure induced phase transition in MH (M=V, Nb)*, **J. Chem. Phys. 140**, 114703.

Chen, J., X. Z. Li, Q. F. Zhang, M. I. J. Probert, C. J. Pickard, R. J. Needs, A. Michaelides, and E. G. Wang, 2013, *Quantum simulation of low-temperature metallic liquid hydrogen*, **Nat. Commun. 4**, 2064.

Chen, J., 2014, *Computer simulation of nuclear quantum effects of hydrogen*, PhD dissertation, Peking University, Beijing, 2014 (in Chinese).

Chen, J., J. Guo, X. Z. Meng, J. B. Peng, J. M. Sheng, L. M. Xu, Y. Jiang, X. Z. Li, and E. G. Wang, 2014a, *An unconventional bilayer ice structure on a NaCl(001) film*, **Nat. Commun. 5**, 4056.

Chen, J., X. G. Ren, X. Z. Li, D. Alfe, and E. G. Wang, 2014b, *On the room-temperature phase diagram of high-pressure hydrogen: An ab-initio molecular dynamics perspective and a diffusion Monte Carlo study*, **J. Chem. Phys. 141**, 024501.

Chen, J. L., and W.-P. Hu, 2011, *Theoretical prediction on the thermal stability of cyclic ozone and strong oxygen tunneling*, **J. Am. Chem. Soc. 133**, 16045.

Chen, S., Q. Wu, C. Mishra, J. Kang, H. Zhang, K. Cho, W. Cai, A. A. Balandin, and R. S. Ruoff, 2012, *Thermal conductivity of isotopically modified graphene*, **Nat. Mater. 11**, 203.

Chen, W., F. Ambrosio, G. Miceli, and A. Pasquarello, 2016, *Ab initio electronic structure of liquid water*, **Phys. Rev. Lett. 117**, 186401.

Chen, X., P. C. Liu, Z. W. Hu, and L. Jensen, 2019, *High-resolution tip-enhanced Raman scattering probes sub-molecular density changes*, **Nat. Commun. 10**, 2567.

Chen, Y. F., Y. T. Tsai, L. Hirsch, and D. M. Bassani, 2017, *Kinetic isotope effects provide experimental evidence for proton tunneling in methylammonium lead triiodide perovskites*, **J. Am. Chem. Soc. 139**, 16359.

Chen, Y. L., T. Jiang, H. Chen, E. Han, A. Alavi, K. Yu, E. G. Wang, and J. Chen, 2023, *Multiconfigurational nature of electron correlation within nitrogen vacancy centers in diamond*, **Phys. Rev. B 108**, 045111.

Cheng, B., E. A. Engel, J. Behler, C. Dellago, and M. Ceriotti, 2019, *Ab initio thermodynamics of liquid and solid water*, **Proc. Natl. Acad. Sci. U. S. A. 116**, 1110.

Cheng, J. G., J. S. Zhou, J. B. Goodenough, Y. Sui, Y. Ren, and M. R. Suchomel, 2011, *High-pressure synthesis and physical properties of perovskite and post-perovskite* $Ca_{1-x}Sr_xIrO_3$, **Phys. Rev. B 83**, 064401.

Cheng, J. G., K. Matsubayashi, S. Nagasaki, A. Hisada, T. Hirayama, M. Hedo, H. Kagi, and Y. Uwatoko, 2014, *Integrated-fin gasket for palm cubic-anvil high pressure apparatus*, **Rev. Sci. Instrum. 85**, 093907.

Cheng, J. G., K. Matsubayashi, W. Wu, J. P. Sun, F. K. Lin, J. L. Luo, and Y. Uwatoko, 2015, *Pressure induced superconductivity on the border of magnetic order in MnP*, **Phys. Rev. Lett. 114**, 117001.

Cheng, J. P., J. Kono, B. D. McCombe, I. Lo, W. C. Michel, and C. E. Stutz, 1995, *Evidence for a stable excitonic ground state in a spatially separated electron-hole system*, **Phys. Rev. Lett. 74**, 450.

Cheng, Z. G., J. Beamish, A. D. Fefferman, F. Souris, S. Balibar, and V. Dauvois, 2015, *Helium mass flow through a solid-superfluid-solid junction*, **Phys. Rev. Lett. 114**, 165301.

Cheng, Z. G., and J. Beamish, 2016, *Compression-driven mass flow in bulk solid* ^4He, **Phys. Rev. Lett. 117**, 025301.

Cheng, Z. G., and J. Beamish, 2018a, *Mass flow through solid* ^3He *in the bcc phase*, **Phys. Rev. Lett. 121**, 225304.

Cheng, Z. G., and J. Beamish, 2018b, *Plastic deformation in a quantum solid: Dislocation avalanches and creep in helium*, **Phys. Rev. Lett. 121**, 055301.

Chevy, F., 2016, *Bose polarons that strongly interact*, **Physics 9**, 86.

Ch'ng, L. C., A. K. Samanta, G. Czakó, J. M. Bowman, and H. Reisler, 2012, *Experimental and theoretical investigations of energy transfer and hydrogen-bond breaking in the water dimer*, **J. Am. Chem. Soc. 134**, 15430.

Cho, S., S. Kim, J. H. Kim, J. Zhao, J. Seok, D. H. Keum, J. Baik, D.-H. Choe, K. J. Chang, K. Suenaga, S. W. Kim, Y. H. Lee, and H. Yang, 2015, *Phase patterning for ohmic homojunction contact in* $MoTe_2$, **Science 349**, 625.

Choi, H. J., D. Roundy, H. Sun, M. L. Cohen, and S. G. Louie, 2002, *The origin of the anomalous superconducting properties of* MgB_2, **Nature 418**, 758.

Chopra, N. G., R. J. Luyken, K. Cherrey, V. H. Crespi, M. L. Cohen, S. G. Louie, and A. Zettl, 1995, *Boron-nitride nanotubes*, **Science 269**, 966.

Chou, J., Y. Zhao, X. T. Li, W. P. Wang, S. J. Tan, Y. H. Wang, J. Zhang, Y. X. Yin, F. Y. Wang, S. Xin, and Y. G. Guo, 2022, *Hydrogen isotope effects on aqueous electrolyte for electrochemical lithium-ion storage*, **Angew. Chem. Int. Ed. 61**, e202203137.

Choudhury, R. R., and R. Chitra, 2018, *Investigation of nuclear quantum effect on the hydrogen bonds of ammonium dihydrogen phosphate using single-crystal neutron diffraction and theoretical modelling*, **Pramana – J. Phys. 91**, 53.

Chowdhury, S. N., and P. Huo, 2017, *Coherent state mapping ring polymer molecular dynamics for non-adiabatic quantum propagations*, **J. Chem. Phys. 147**, 214109.

Chu, C. W., and B. Lorenz, 2009, *High pressure studies on Fe-pnictide superconductors*, **Physica C 469**, 385.

Chu, M. W., I. Szafraniak, R. Scholz, C. Harnagea, D. Hesse, M. Alexe, and U. Gosele, 2004, *Impact of misfit dislocations on the polarization instability of epitaxial nanostructured ferroelectric perovskites*, **Nat. Mater. 3**, 87.

Chu, W., Q. Zheng, O. V. Prezhdo, J. Zhao, and W. A. Saidi, 2020, *Low-frequency lattice phonons in halide perovskites explain high defect tolerance toward electron-hole recombination*, **Sci. Adv. 6**, eaaw7453.

Clark, A. C., and M. H. W. Chan, 2005, *Specific heat of solid helium*, **J. Low Temp. Phys. 138**, 853.

Clark, C. D., P. J. Dean, and P. V. Harris, 1964, *Intrinsic edge absorption in diamond*, **Proc. R. Soc. Lon. Ser-A 277**, 312.

Cleland, D., and A. Alavi, 2010, *Communications: Survival of the fittest: Accelerating convergence in full configuration-interaction quantum Monte Carlo*, **J. Chem. Phys. 132**, 041103.

Cohen, A. J., P. Mori-Sanchez, and W. T. Yang, 2012, *Challenges for density functional theory*, **Chem. Rev. 112**, 289.

Coleman, S., 1977, *Fate of the false vacuum: Semiclassical theory,* **Phys. Rev. D 16**, 1762.

Condon, E. U., 1928, *Nuclear motions associated with electron transitions in diatomic molecules*, **Phys. Rev. 32**, 858.

Craig, C. F., W. R. Duncan, and O. V. Prezhdo, 2005, *Trajectory surface hopping in the time-dependent Kohn-Sham approach for electron-nuclear dynamics*, **Phys. Rev. Lett. 95**, 163001.

Craig, I. R., and D. E. Manolopoulos, 2004, *Quantum statistics and classical mechanics: Real time correlation functions from ring polymer molecular dynamics*, **J. Chem. Phys. 121**, 3368.

Craig, I. R., and D. E. Manolopoulos, 2005a, *Chemical reaction rates from ring polymer molecular dynamics*, **J. Chem. Phys. 122**, 084106.

Craig, I. R., and D. E. Manolopoulos, 2005b, *A refined ring polymer molecular dynamics theory of chemical reaction rates*, **J. Chem. Phys. 123**, 034102.

Cui, L. J., N. H. Chen, S. J. Jeon, and I. F. Silvera, 1994, *Megabar pressure triple point in solid deuterium*, **Phys. Rev. Lett. 72**, 3048.

Cuscó, R., L. Artús, J. H. Edgar, S. Liu, G. Cassabois, and B. Gil, 2018, *Isotopic effects on phonon anharmonicity in layered van der Waals crystals: Isotopically pure hexagonal boron nitride*, **Phys. Rev. B 97**, 155435.

D

Dacorogna, M. M., M. L. Cohen, and P. K. Lam, 1985, *Self-consistent calculation of the q dependence of the electron-phonon coupling in aluminum*, **Phys. Rev. Lett. 55**, 837.

Dale, H. J. A., A. G. Leach, and G. C. Lloyd-Jones, 2021, *Heavy-atom kinetic isotope effects: Primary interest or zero point?* **J. Am. Chem. Soc. 143**, 21079.

Damjanovic, D., 1998, *Ferroelectric, dielectric and piezoelectric properties of ferroelectric thin films and ceramics*. **Rep. Prog. Phys. 61**, 1267.

Damodaran, A. R., J. C. Agar, S. Pandya, Z. Chen, L. Dedon, R. Xu, B. Apgar, S. Saremi, and L. W. Martin, 2016, *New modalities of strain-control of ferroelectric thin films*. **J. Phys.: Condens. Matter 28**, 263001.

Das, A., S. Pisana, B. Chakraborty, S. Piscanec, S. K. Saha, U. V. Waghmare, K. S. Novoselov, H. R. Krishnamurthy, A. K. Geim, A. C. Ferrari, and A. K. Sood, 2008, *Monitoring dopants by Raman scattering in an electrochemically top-gated graphene transistor*, **Nat. Nanotechnol. 3**, 210.

Datta, S., M. R. Melloch, and R. L. Gunshor, 1985, *Possibility of an excitonic ground state in quantum wells*, **Phys. Rev. B 32**, 2607.

Davidson, E. R. M., J. Klimeš, D. Alfè, and A. Michaelides, 2014, *Cooperative interplay of van der Waals forces and quantum nuclear effects on adsorption: H at graphene and at coronene*, **ACS Nano 8**, 9905.

Day, J., and J. Beamish, 2007, *Low-temperature shear modulus changes in solid ^{4}He and connection to supersolidity*, **Nature 450**, 853.

De Marco, L., G. Valtolina, K. Matsuda, W. G. Tobias, J. P. Covey, and J. Ye, 2019, *A degenerate Fermi gas of polar molecules*, **Science 363**, 853.

de' Medici, L., A. Georges, G. Kotliar, and S. Biermann, 2005, *Mott transition and Kondo screening in f-electron metals*, **Phys. Rev. Lett. 95**, 066402.

de Oteyza, D. G., P. Gorman, Y.-C. Chen, S. Wickenburg, A. Riss, D. J. Mowbray, G. Etkin, Z. Pedramrazi, H.-Z. Tsai, A. Rubio, M. F. Crommie, and F. R. Fischer, 2013, *Direct imaging of covalent bond structure in single-molecule chemical reactions*, **Science 340**, 1434.

Debye, P., 1912, *Zur Theorie der spezifischen Wärmen*, **Ann. Phys. 39**, 789.

Della Sala, F., R. Rousseau, A. Görling, and D. Marx, 2004, *Quantum and thermal fluctuation effects on the photoabsorption spectra of clusters*, **Phys. Rev. Lett. 92**, 183401.

Dellago, C., M. M. Naor, and G. Hummer, 2003, *Proton transport through water-filled carbon nanotubes*, **Phys. Rev. Lett. 90**, 105902.

de-Picciotto, R., M. Reznikov, M. Heiblum, V. Umansky, G. Bunin, and D. Mahalu, 1997, *Direct observation of a fractional charge*, **Nature 389**, 162.

Dewar, M. J. S., K. M. Merz Jr., and J. J. P. Stewart, 1984, *Tunneling dynamics of cyclobutadiene*, **J. Am. Chem. Soc., 148**, 4040.

D'Hendecourt, L. B., L. J. Allamandola, and J. M. Greenberg, 1985, *Time dependent chemistry in dense molecular clouds. I. grain surface reactions, gas/grain interactions and infrared spectroscopy*, **A&A 152**, 130.

Ding, H. F., J. H. Hu, W. W. Liu, Y. Xu, C. X. Jiang, T. H. Ding, L. Li, X. Z. Duan, and Y. Pan, 2012, *Design of a 135 MW power supply for a 50 T pulsed magnet*, **IEEE Trans. Appl. Supercon. 22**, 5400504.

Ding, Z., F. Z., Shi, and J. F. Du, 2020, *Nanoscale magnetic imaging based on quantum sensing with diamond and its applications to condensed matter physics*, **PHYSICS**, **44**, 359 (in Chinese).

Doan, T. C., *Thermal neutron detector based on hexagonal boron nitride*, PhD dissertation, TTU, Lubbock, 2016.

Donadio, D., L. M. Ghiringhelli, and L. Delle Site, 2012, *Autocatalytic and cooperatively stabilized dissociation of water on a stepped platinum surface*, **J. Am. Chem. Soc. 134**, 19217.

Doubleday, C., R. Armas, D. Walker, C. V. Cosgriff, and E. M. Greer, 2017, *Heavy-atom tunneling calculations in thirteen organic reactions: Tunneling contributions are substantial, and Bell's formula closely approximates multidimensional tunneling at ≥250 K*, **Angew. Chem. Int. Ed. 56**, 13099.

Drechsel-Grau, C., and D. Marx, 2014, *Quantum simulation of collective proton tunneling in hexagonal ice crystals*, **Phys. Rev. Lett. 112**, 148302.

Drechsel-Grau, C., and D. Marx, 2017, *Collective proton transfer in ordinary ice: Local environments, temperature dependence and deuteration effects*, **Phys. Chem. Chem. Phys. 19**, 2623.

Drozdov, A. P., M. I. Eremets, I. A. Troyan, V. Ksenofontov, and S. I. Shylin, 2015, *Conventional superconductivity at 203 kelvin at high pressures in the sulfur hydride system*, **Nature 525**, 73.

Drozdov, A. P., P. P. Kong, V. S. Minkov, S. P. Besedin, M. A. Kuzovnikov, S. Mozaffari, L. Balicas, F. F. Balakirev, D. E. Graf, V. B. Prakapenka, E. Greenberg, D. A. Knyazev, M. Tkacz, and M. I. Eremets, 2019, *Superconductivity at 250 K in lanthanum hydride under high pressures*, **Nature 569**, 528.

Drummond, N. D., B. Monserrat, J. H. Lloyd-Williams, P. L. Rios, C. J. Pickard, and R. J. Needs, 2015, *Quantum Monte Carlo study of the phase diagram of solid molecular hydrogen at extreme pressures*, **Nat. Commun. 6**, 7794.

Duan, D., Y. Liu, F. Tian, D. Li, X. Huang, Z. Zhao, H. Yu, B. Liu, W. Tian, and T. Cui, 2014, *Pressure-induced metallization of dense $(H_2S)_2H_2$ with high-Tc superconductivity*, **Sci. Rep. 4**, 6968.

Dubrovinsky L., S. Khandarkhaeva, T. Fedotenko, D. Laniel, M. Bykov, C. Giacobbe, E. L. Bright, P. Sedmak, S. Chariton, V. Prakapenka, A. V. Ponomareva, E. A. Smirnova, M. P. Belov, F. Tasnádi, N. Shulumba, F. Trybel, I. A. Abrikosov, and N. Dubrovinskaia, 2022, *Materials synthesis at terapascal static pressures*, **Nature 605**, 274.

Dupuis, R., J. S. Dolado, M. Benoit, J. Surga, and A. Ayuela, 2017, *Quantum nuclear dynamics of protons within layered hydroxides at high pressure*, **Sci. Rept. 7**, 4842.

Dwyer, C., 2017, *Prospects of spatial resolution in vibrational electron energy loss spectroscopy: Implications of dipolar scattering*, **Phys. Rev. B 96**, 224102.

Dyke, T. R., and J. S. Muenter, 1973, *Electric dipole-moments of low J states of H_2O and D_2O*, **J. Chem. Phys. 59**, 3125.

Dynes, R. C., 1972, *Mcmillan's equation and the Tc of superconductors*, **Solid State Commun. 10**, 615.

E

Ederer, C., and N. A. Spaldin, 2005, *Effect of epitaxial strain on the spontaneous polarization of thin film ferroelectrics*, **Phys. Rev. Lett. 95**, 257601.

Egger, D. A., L, Kronik, and A. M. Rappe, 2015, *Theory of hydrogen migration in organic-inorganic halide perovskites*, **Angew. Chem. Int. Ed. 54**, 12437.

Ehrenfest, P., 1932, *Some of the quantum mechanics questions*, **Z. Fur. Phys. 78**, 555.

Einstein, A., 1911, *Elementare Betrachtungen über die thermische Molekularbewegung in festen Körpern*, **Ann. Phys. 35**, 679.

Einstein, A., 1917, *On the quantum theory of radiation*, **Phys. Z. 18**, 121.

Eisenberg, D. S., and W. Kauzmann, 1969, *The structure and properties of water* (Oxford University Press, Oxford).

Ekimov, E. A., V. A. Sidorov, E. D. Bauer, N. N. Mel'nik, N. J. Curro, J. D. Thompson, and S. M. Stishov, 2004, *Superconductivity in diamond*, **Nature 428**, 542.

Ellert, C., M. Schmidt, H. Haberland, V. Veyret, and V. Bonačić-Koutecký, 2002, *Vibrational structure in the optical response of small Li-cluster ions*, **J. Chem. Phys. 117**, 3711.

Eremets, M. I., V. W. Struzhkin, H. K. Mao, and R. J. Hemley, 2001, *Superconductivity in boron*, **Science 293**, 272.

Eremets, M. I., I. A. Trojan, S. A. Medvedev, J. S. Tse, and Y. Yao, 2008, *Superconductivity in hydrogen dominant materials: Silane*, **Science 319**, 1506.

Eremets, M. I., and I. A. Troyan, 2011, *Conductive dense hydrogen*, **Nat. Mater. 10**, 927.

Errea, I., M. Calandra, C. J. Pickard, J. Nelson, R. J. Needs, Y. Li, H. Liu, Y. Zhang, Y. Ma, and F. Mauri, 2015, *High-pressure hydrogen sulfide from first principles: A strongly anharmonic phonon-mediated superconductor*, **Phys. Rev. Lett. 114**, 157004.

Errea, I., M. Calandra, C. J. Pickard, J. R. Nelson, R. J. Needs, Y. Li, H. Liu, Y. Zhang, Y. Ma, and F. Mauri, 2016, *Quantum hydrogen-bond symmetrization in the superconducting hydrogen sulfide system*, **Nature 532**, 81.

Errea, I., F. Belli, L. Monacelli, A. Sanna, T. Koretsune, T. Tadano, R. Bianco, M. Calandra, R. Arita, F. Mauri, and J. A. Flores-Livas, 2020, *Quantum crystal structure in the 250-kelvin superconducting lanthanum hydride*, **Nature 578**, 66.

Esaki, L., and R. Tsu, 1970, *Superlattice and negative differential conductivity in semiconductors*, **IBM J. Res. Develop. 14**, 61.

Evans, M. G., and M. Polanyi, 1938, *Inertia and driving force of chemical reactions*, **Tran. Faraday Soc. 34**, 11.

Evans, A. C., J. Mayers, D. N. Timms, and M. J. Cooper, 1993, *Deep inelastic neutron scattering in the study of atomic momentum distribution*, **Z. Naturforsch. A 48**, 425.

F

Fan, H. Y., 1951, *Temperature dependence of the energy gap in semiconductors*, **Phys. Rev. 82**, 900.

Fang, W., J. O. Richardson, J. Chen, X. Z. Li, and A. Michaelides, 2017, *Simultaneous deep tunneling and classical hopping for hydrogen diffusion on metals*, **Phys. Rev. Lett. 119**, 126001.

Fang, W., J. Chen, Y. Feng, X.-Z. Li, and A. Michaelides, 2019, *The quantum nature of hydrogen*, **Int. Rev. Phys. Chem. 38**, 35.

Fang, W., J. Chen, P. Pedevilla, X. Z. Li, J. O. Richardson, and A. Michaelides, 2020, *Origins of fast diffusion of water dimers on surfaces*, **Nat. Commun. 11**, 1689.

Fatuzzo, E., and W. J. Merz, 1967, *Ferroelectricity* (North-Holland, Amsterdam).

Fausti, D., R. I. Tobey, N. Dean, S. Kaiser, A. Dienst, M. C. Hoffmann, S. Pyon, T. Takayama, H. Takagi, and A. Cavalleri, 2011, *Light-induced superconductivity in a stripe-ordered cuprate*, **Science 331**, 189.

Feibelman, P. J., 2002, *Partial dissociation of water on Ru(0001)*, **Science 295**, 99.

Feng, B. J., J. Zhang, Q. Zhong, W. B. Li, S. Li, H. Li, P. Cheng, S. Meng, L. Chen, and K. H. Wu, 2016, *Experimental realization of two-dimensional boron sheets*, **Nat. Chem. 8**, 563.

Feng, X., J. Zhang, G. Gao, H. Liu, and H. Wang, 2015, *Compressed sodalite-like MgH_6 as a potential high-temperature superconductor*, **RSC Adv. 5**, 59292.

Feng, X. L., S. Y. Lu, C. J. Pickard, H. Y. Liu, S. A. T. Redfern, and Y. M. Ma, 2018, *Carbon network evolution from dimers to sheets in superconducting ytrrium dicarbide under pressure*, **Commun. Chem. 1**, 85.

Feng, Y. X., J. Chen, D. Alfe, X. Z. Li, and E. G. Wang, 2015, *Nuclear quantum effects on the high pressure melting of dense Lithium*, **J. Chem. Phys. 142**, 064506.

Feng, Y. X., J. Chen, W. Fang, E. G. Wang, A. Michaelides, and X.-Z. Li, 2017, *Hydrogenation facilitates proton transfer through two dimensional honeycomb crystals*, **J. Phys. Chem. Lett. 8**, 6009.

Feng, Y. X., Y. Zhao, W.-K. Zhou, Q. Li, W. A. Saidi, Q. Zhao, and X.-Z. Li, 2018a, *Proton migration in hybrid lead iodide perovskites: From classical hopping to deep quantum tunneling*, **J. Phys. Chem. Lett. 9**, 6536.

Feng, Y. X., Z. Wang, J. Guo, J. Chen, E. G. Wang, Y. Ying, and X.-Z. Li, 2018b, *The collective and quantum nature of proton transfer in the cyclic water tetramer on NaCl(001)*, **J. Chem. Phys. 148**, 102329.

Fermi, E., 1927, *Un metodo statistico per la determinazione di alcune priorieta dell'atome*, **Rend. Accad. Naz. Lincei 6**, 602.

Fermi, E., J. R. Pasta, and S. Ulam, 1955, *Studies of the nonlinear problems, Report LA-1940* (Los Alamos Scientific Laboratory, Los Alamos).

Ferray, M., A. L'Huillier, X. F. Li, L. A. Lompre, G. Mainfray, and C. Manus, 1988, *Multiple-harmonic conversion of 1064 nm radiation in rare gases*, **J. Phys. B: At. Mpl. Opt. Phys. 21**, L31.

Ferriere, K. M., 2001, *The interstellar environment of our galaxy*, **Rev. Mod. Phys. 73**, 1031.

Feynman, R. P., 1949, *Space-time approach to quantum electrodynamics*, **Phys. Rev. 76**, 769.

Feynman, R. P., 1953a, *The λ-transition transition in liquid Helium*, **Phys. Rev. 90**, 1116.

Feynman, R. P., 1953b, *Atomic theory of the λ transition in Helium*, **Phys. Rev. 91**, 1291.

Feynman, R. P., 1953c, *Atomic theory of liquid Helium near absolute zero*, **Phys. Rev. 91**, 1301.

Feynman, R. P, and A. R. Hibbs, 1965, *Quantum mechanics and path integrals* (McGraw-Hill Inc, New York).

Flynn, C. P., and A. M. Stoneham, 1970, *Quantum theory of diffusion with application to light interstitials in metals*, **Phys. Rev. B 1**, 3966.

Fock, V., 1930, *Näherungsmethode zur lösung des quantenmechanischen mehrkörperproblems*, **Z. Physik 61**, 126.

Franck, J., 1926, *Elementary processes of photochemical reactions*, **Trans. Faraday. Soc. 21**, 536.

Fröhlich, H., 1937, *Theory of electrical breakdown in ionic crystals*, **Proc. Roy. Soc. Lond. A 160**, 230.

Fröhlich, H., and J. Odwyer, 1950, *Time dependence of electronic processes in dielectrics*, **Proc. Phys. Soc. Lond. A 63**, 81.

Fujishima, A., and K. Honda, 1972, *Electrochemical photolysis of water at a semiconductor electrode*, **Nature 238**, 37.

Fumagalli, L., A. Esfandiar, R. Fabregas, S. Hu, P. Ares, A. Janardanan, Q. Yang, B. Radha, T. Taniguchi, K. Watanabe, G. Gomila, K. S. Novoselov, and A. K. Geim, 2018, *Anomalously low dielectric constant of confined water*, **Science 360**, 1339.

G

Ganeshan, S., R. Ramírez, and M. V. Fernández-Serra, 2013, *Simulation of quantum zero-point effects in water using a frequency-dependent thermostat*, **Phys. Rev. B 87**, 134207.

Gao, B. J., L. R. Ding, Z. J. Wang, Y. Zhang, J. Li, and J. Su, 2016, *Water-cooled resistive magnets at CHMFL*, **IEEE Trans. Appl. Supercon. 26**, 0600506.

Gao, G., A. R. Oganov, A. Bergara, M. Martinez-Canales, T. Cui, T. Iitaka, Y. Ma, and G. Zou, 2008, *Superconducting high pressure phase of germane*, **Phys. Rev. Lett. 101**, 107002.

Gao, L., Y. Y. Xue, F. Chen, Q. Xiong, R. L. Meng, D. Ramirez, C. W. Chu, J. H. Eggert, and H. K. Mao, 1994, *Superconductivity up to 164-K in $HgBa_2Ca_{m-1}Cu_mO_{2m+2+\delta}$ (m=1, 2, and 3) under quasi-hydrostatic pressures*, **Phys. Rev. B. 50**, 4260.

Gao, M., Z. Y. Lu, and T. Xiang, 2015, *Finding high-temperature superconductors by metallizing the σ-bonding electrons*, **PHYSICS 44**, 421 (in Chinese).

Gao, P., C. T. Nelson, J. R. Jokisaari, S. H. Baek, C. W. Bark, Y. Zhang, E. G. Wang, D. G. Schlom, C. B. Eom, and X. Q. Pan, 2011, *Revealing the role of defects in ferroelectric switching with atomic resolution*, **Nat. Commun. 2**, 1600.

Gao, P., J. Britson, J. R. Jokisaari, C. T. Nelson, S.-H. Baek, Y. Wang, C.-B. Eom, L.-Q. Chen, and X. Pan, 2013, *Atomic-scale mechanisms of ferroelastic domain-wall-mediated ferroelectric switching*, **Nat. Commun. 4**, 3791.

Gao, P., S. Yang, R. Ishikawa, N. Li, B. Feng, A. Kumamoto, N. Shibata, P. Yu, and Y. Ikuhara, 2018, *Atomic-scale measurement of flexoelectric polarization at $SrTiO_3$ dislocations*, **Phys. Rev. Lett. 120**, 267601.

Gao, Y. Q., 2008, *An integrate-over-temperature approach for enhanced sampling*, **J. Chem. Phys. 128**, 064105.

Garrett, B. C., D. G. Truhlar, R. S. Grev, and A. W. Magnuson, 1980, *Improved treatment of threshold contributions in variational transition-state theory*, **J. Phys. Chem. 84**, 1730.

Geick, R., C. H. Perry, and G. Rupprech, 1966, *Normal modes in hexagonal boron nitride*, **Phys. Rev., 146**, 543.

Georges, A., G. Kotliar, W. Krauth, and M. J. Rozenberg, 1996, *Dynamical mean-field theory of strongly correlated fermion systems and the limit of infinite dimensions*, **Rev. Mod. Phys. 68**, 13.

Gerritzen, D., and H. H. Limbach, 1984, *Kinetic isotope effects and tunneling in cyclic double and triple proton transfer between acetic acid and methanol in tetrahydrofuran studied by dynamic proton and deuteron NMR spectroscopy*, **J. Am. Chem. Soc. 106**, 869.

Ghosh, S., P. Verma, C. J. Cramer, L. Gagliardi, and D. G. Truhlar, 2018, *Combining wave function methods with density functional theory for excited states*, **Chem. Rev. 118**, 7249.

Giannozzi, P., S. Baroni, N. Bonini, M, Calandra, R. Car, C. Cavazzoni, D. Ceresoli, G. L. Chiarotti, M. Cococcioni, I. Dabo, A. D. Corso, S. de Gironcoli, S. Fabris, G. Fratesi, R. Gebauer, U. Gerstmann, C. Gougoussis, A. Kokalj, M. Lazzeri, L. Martin-Samos, N. Marzari, F. Mauri, R. Mazzarello, S. Paolini, A. Pasquarello, L. Paulatto, C. Sbraccia, S. Scandolo, G. Sclauzero, A. P. Seitsonen, A. Smogunov, P. Umari1, and R. M. Wentzcovitch, 2009, *Quantum Espresso: A modular and open-source software project for quantum simulations of materials*, **J. Phys.: Condens. Matter 21**, 395502.

Giannozzi, P., O. Andreussi, T. Brumme, O. Bunau, M. B. Nardelli, M. Calandra, R. Car, C. Cavazzoni, D. Ceresoli, M. Cococcioni, N. Colonna, I. Carnimeo, A. D. Corso, S. de Gironcoli, P. Delugas, R. A. DiStasio Jr, A. Ferretti, A. Floris, G. Fratesi, G. Fugallo, R. Gebauer, U. Gerstmann, F. Giustino, T. Gorni, J. Jia, M. Kawamura, H.-Y. Ko, A.

Kokalj, E. Küçükbenli, M. Lazzeri, M. Marsili, N. Marzari, F. Mauri, N. L. Nguyen, H.-V. Nguyen, A. Otero-de-la-Roza, L. Paulatto, S. Poncé, D. Rocca, R. Sabatini, B. Santra, M. Schlipf, A. P. Seitsonen, A. Smogunov, I. Timrov, T. Thonhauser, P. Umari, N. Vast, X. Wu, and S. Baroni, 2017, *Advanced capabilities for materials modelling with Quantum ESPRESSO*, **J. Phys-Condens. Mat. 29**, 465901.

Giessibl, F. J., 2003, *Advances in atomic force microscopy*, **Rev. Mod. Phys. 75**, 949.

Giles, A. J., S. Dai, I. Vurgaftman, T. Hoffman, S. Liu, L. Lindsay, C. T. Ellis, N. Assefa, I. Chatzakis, T. L. Reinecke, J. G. Tischler, M. M. Fogler, J. H. Edgar, D. N. Basov, and J. D. Caldwell, 2018, *Ultralow-loss polaritons in isotopically pure boron nitride*, **Nat. Mater. 17**, 134.

Giuliani, A., M. A. Ricci, and F. Bruni, 2012, *Quantum effects and the local environment of water hydrogen: Deep inelastic neutron scattering study*, **Phys. Rev. B 86**, 104308.

Giustino, F., M. L. Cohen, and S. G. Louie, 2007, *Electron-phonon interaction using Wannier functions*, **Phys. Rev. B 76**, 165108.

Giustino, F., S. G. Louie, and M. L. Cohen, 2010, *Electron-phonon renormalization of the direct band gap of diamond*, **Phys. Rev. Lett. 105**, 265501.

Giustino, F., 2017, *Electron-phonon interactions from first principles*, **Rev. Mod. Phys. 89**, 015003.

Godbeer, A. D., J. S. Al-Khalili, and P. D. Stevenson, 2015, *Modelling proton tunnelling in the adenine-thymine base pair*, **Phys. Chem. Chem. Phys. 17**, 13034.

Golze, D., M. Dvorak, and P. Rinke, 2019, *The GW compendium: A practical guide to theoretical photoemission spectroscopy*, **Front. Chem. 7**, 377.

Goncharov, A. F., I. I. Mazin, J. H. Eggert, R. J. Hemley, and H. K. Mao, 1995, *Invariant points and phase-transitions in deuterium at megabar pressures*, **Phys. Rev. Lett. 75**, 2514.

Goncharov, A. F., V. V. Struzhkin, M. S. Somayazulu, R. J. Hemley, and H. K. Mao, 1996, *Compression of ice to 210 gigapascals: Infrared evidence for a symmetric hydrogen-bonded phase*, **Science 273**, 218.

Goncharov, A. F., V. V. Struzhkin, H. K. Mao, and R. J. Hemley, 1999, *Raman spectroscopy of dense H_2O and the transition to symmetric hydrogen bonds*, **Phys. Rev. Lett. 83**, 1998.

Goncharov, A. F., R. J. Hemley, and H. K. Mao, 2011, *Vibron frequencies of solid H_2 and D_2 to 200 GPa and implications for the P-T phase diagram*, **J. Chem. Phys. 134**, 174501.

Gorelli, F. A., S. F. Elatresh, C. L. Guillaume, M. Marqués, G. J. Ackland, M. Santoro, S. A. Bonev, and E. Gregoryanz, 2012, *Lattice dynamics of dense lithium*, **Phys. Rev. Lett. 108**, 055501.

Gorelov, V., M. Holtzmann, and D. M. Ceperley, 2020, *Energy gap closure of crystalline molecular hydrogen with pressure*, **Phys. Rev. Lett. 124**, 116401.

Goumans, T. P. M., and J. Kästner, 2010, *Hydrogen-atom tunneling could contribute to H2 formation in space*, **Angew. Chem. Int. Ed. 49**, 7350.

Graham, A. P., A. Menzel, and J. P. Toennies, 1999, *Quasielastic helium atom scattering measurements of microscopic diffusional dynamics of H and D on the Pt(111) surface*, **J. Chem. Phys. 111**, 1676.

Green, M. S., 1952, *Markoff random processes and the statistical mechanics of time-dependent phenomena*, **J. Chem. Phys. 20**, 1281.

Green, M. S., 1954, *Markoff random processes and the statistical mechanics of time-dependent phenomena. II. Irreversible processes in fluids*, **J. Chem. Phys. 22**, 398.

Greenwood, D., 1958, *The Boltzmann equation in the theory of electrical conduction in metals*, **Proc. Phys. Soc. 71**, 585.

Grimm, R. C., and R. G. Storer, 1971, *Monte-Carlo solution of Schrödinger's equation*, **J. Comput. Phys. 7**, 134.

Grimvall, G., 1981, *The electron-phonon interaction in metals* (North-Holland, Amsterdam).

Gross, L., F. Mohn, N. Moll, P. Liljeroth, and G. Meyer, 2009, *The chemical structure of a molecule resolved by atomic force microscopy*, **Science 325**, 1110.

Gross, L., F. Mohn, N. Moll, B. Schuler, A. Criado, E. Guitian, D. Pena, A. Gourdon, and G. Meyer, 2012, *Bond-order discrimination by atomic force microscopy*, **Science 337**, 1326.

Guan, M. X., E. Wang, P. W. You, J. T. Sun, and S. Meng, 2021, *Manipulating Weyl quasiparticles by orbital-selective photoexcitation in WTe_2*, **Nat. Commun. 12**, 1885.

Guan, M. X., X. Liu, D. Chen, X. Li, Y. Qi, Q. Yang, P. You, and S. Meng, 2022, *Optical control of multistage phase transition via phonon coupling in $MoTe_2$*, **Phys. Rev. Lett. 128**, 015702.

Guillaume, C. L., E. Gregoryanz, O. Degtyareva, M. I. McMahon, M. Hanfland, S. Evans, M. Guthrie, S. V. Sinogeikin, and H. K. Mao, 2011, *Cold melting and solid structures of dense lithium*, **Nat. Phys. 7**, 211.

Guo, G.-C., D. Wang, X.-L. Wei, Q. Zhang, H. Liu, W. M. Lau, and L.-M. Liu, 2015, *First-principles study of phosphorene and graphene heterostructure as anode materials for rechargeable Li batteries*, **J. Phys. Chem. Lett. 6**, 5002.

Guo, J., X. Z. Meng, J. Chen, J. B. Peng, J. M. Sheng, X. Z. Li, L. M. Xu, J. R. Shi, E. G. Wang, and Y. Jiang, 2014, *Real-space imaging of interfacial water with submolecular resolution*, **Nat. Mater. 13**, 184.

Guo, J., J. T. Lü, Y. X. Feng, J. Chen, J. B. Peng, Z. R. Lin, X. Z. Meng, Z. C. Wang, X. Z. Li, E. G. Wang, and Y. Jiang, 2016a, *Nuclear quantum effects of hydrogen bonds probed by tip-enhanced inelastic electron tunneling*, **Science 352**, 321.

Guo, J., K. Bian, Z. Lin, and Y. Jiang, 2016b, *Perspective: Structure and dynamics of water at surfaces probed by scanning tunneling microscopy and spectroscopy*, **J. Chem. Phys. 145**, 160901.

Guo, J., X. Z. Li, J. B. Peng, E. G. Wang, and Y. Jiang, 2017, *Atomic-scale investigation of nuclear quantum effects of surface water: Experiments and theory*, **Prog. Surf. Sci. 92**, 203.

Guo, J., and Y. Jiang, 2022, *Submolecular insights into interfacial water by hydrogen-sensitive scanning probe microscopy*, **Acc. Chem. Res. 55**, 1680.

Guo, J. Q., 2022, *First-principles study of the structures and dynamics of light-element liquids*, PhD dissertation, Peking University, Beijing, (in Chinese).

Guo, J. Q., B. C. Cheng, L. M. Xu, E. G. Wang, and J. Chen, 2022, *Onset of metallic transition in molecular liquid hydrogen*, DOI: 10.48550/arXiv.2201.03157.

Guo, S., X. Hu, W. Zhou, X. Liu, Y. Gao, S. Zhang, K. Zhang, Z. Zhu, and H. Zeng, 2018, *Mechanistic understanding of two-dimensional phosphorus, arsenic, and antimony high-capacity anodes for fast-charging lithium/sodium ion batteries*, **J. Phys. Chem. C 122**, 29559.

Gutzwiller, M. C., 1963, *Effect of correlation on the ferromagnetism of transition metals*, **Phys. Rev. Lett. 10**, 159.

H

Habershon, S., D. E. Manolopoulos, T. E. Markland, and T. F. Miller, 2013, *Ring-polymer molecular dynamics: Quantum effects in chemical dynamics from classical trajectories in an extended phase space*. **Annu. Rev. Phys. Chem. 64**, 387.

Hachtel, J. A., J. S. Huang, I. Popovs, S. Jansone-Popova, J. K. Keum, J. Jakowski, T. C. Lovejoy, N. Dellby, O. L. Krivanek, and J. C. Idrobo, 2019, *Identification of site-specific isotopic labels by vibrational spectroscopy in the electron microscope*, **Science 363**, 525.

Hack, M. D., and D. G. Truhlar, 2000, *Nonadiabatic trajectories at an exhibition*, **J. Phys. Chem. A 104**, 7917.

Hage, F. S., D. M. Kepaptsoglou, Q. M. Ramasse, and L. J. Allen, 2019, *Phonon spectroscopy at atomic resolution*, **Phys. Rev. Lett. 122**, 016103.

Hage, F. S., G. Radtke, D. M. Kepaptsoglou, M. Lazzeri, and Q. M. Ramasse, 2020, *Single-atom vibrational spectroscopy in the scanning transmission electron microscope*, **Science 367**, 1124.

Hahn, S., K. Kim, K. Kim, X. B. Hu, T. Painter, I. Dixon, S. Kim, K. R. Bhattarai, S. Noguchi, J. Jaroszynski, and D. C. Larbalestier, 2019, *45.5-tesla direct-current magnetic field generated with a high-temperature superconducting magnet*, **Nature 570**, 496.

Haid, S., M. Marszalek, A. Mishra, M. Wielopolski, J. Teuscher, J. E. Moser, R. Humphry-Baker, S. M. Zakeeruddin, M. Grätzel, and P. Bäuerle, 2012, *Significant improvement of dye-sensitized solar cell performance by small structural modification in π-conjugated donor–acceptor dyes*, **Adv. Funct. Mater. 22**, 1291.

Haldane, F. D. M., 1988, *Exact Jastro-Gutzwiller resonating-valence-bond ground state of the spin-1/2 antiferromagnetic Heisenberg chain with 1/r2 exchange*, **Phys. Rev. Lett. 60**, 635.

Hall, G. G., 1951, *The molecular orbital theory of chemical valency. VIII. A method of calculating ionization potentials*, **Proc. R. Soc. A 205**, 541.

Hall, L. H., J. Bardeen, and F. J. Blatt, 1954, *Infrared absorption spectrum of germanium*, **Phys. Rev. 95**, 559.

Hallock, R. B., 2015, *Is solid helium a supersolid?* **Phys. Today 68**, 30.

Hama, T., H. Ueta, A. Kouchi, and N. Watanabe, 2015, *Quantum tunneling observed without its characteristic large kinetic isotope effects*, **Proc. Natl. Acad. Sci. U. S. A. 112**, 7438.

Hamann, D. R., M. Schlüter, and C. Chiang, 1979, *Norm-conserving pseudopotentials*, **Phys. Rev. Lett. 43**, 1494.

Hamm, P., and G. Stock, 2015, *Nonadiabatic vibrational dynamics in the HCO-2·H2O complex*, **J. Chem. Phys. 143**, 134308.

Han, E. X., W. Fang, M. Stamatakis, J. O. Richardson, and J. Chen, 2022, *Quantum tunnelling driven H_2 formation on graphene*, **J. Phys. Chem. Lett. 13**, 3173.

Han, K.-L., 2004, *Nonadiabatic dynamics with split-operator scheme on multiple potential energy surfaces*, **Acta Phys. -Chim. Sin. 20**, 1032.

Han, W. Q., H. G. Yu, C. Zhi, J. Wang, Z. Liu, T. Sekiguchi, and Y. Bando, 2008, *Isotope effect on band gap and radiative transitions properties of boron nitride nanotubes*, **Nano Lett. 8**, 491.

Hao, W. J., Y. N. Zhai, Q. Y. Zhang, and J. M. Zhao, 2021, *Attosecond light source in material science investigation*, **Chin. Sci. Bull. 66**, 856 (in Chinese).

Harada, Y., T. Tokushima, Y. Horikawa, O. Takahashi, H. Niwa, M. Kobayashi, M. Oshima, Y. Senba, H. Ohashi, K. T. Wikfeldt, A. Nilsson, L. G. M. Pettersson, and S. Shin, 2013, *Selective probing of the OH or OD stretch vibration in liquid water using resonant inelastic soft-X-ray scattering*, **Phys. Rev. Lett. 111**, 193001.

Hardy, E. H., A. Zygar, and M. D. Zeidler, 2001, *Isotope effect on the translational and rotational motion in liquid water and ammonia*, **J. Chem. Phys. 114**, 3174.

Hare, D. E., and C. M. Sorensen, 1987, *The density of supercooled water. II. Bulk samples cooled to the homogeneous nucleation limit*, **J. Chem. Phys. 87**, 4840.

Harrison, W. A., 1980, *Electronic structure and the properties of solids: The physics of the chemical bond* (W. H. Freeman and Company, San Francisco).

Hartree, D. R., 1928, *The wave mechanics of an atom with a non-coulomb central field. Part I. Theory and methods*, **Math. Proc. Cam. Phil. Soc. 24**, 89.

Hasan, M. Z., and C. L. Kane, 2010, *Colloquium: Topological insulators*, **Rev. Mod. Phys. 82**, 3045.

Hedin, L., 1965, *New method for calculating the one-particle Green's function with application to the electron-gas problem*, **Phys. Rev. 139**, A796.

Hehre, W. J., R. Ditchfield, and J. A. Pople, 1972, *Self-consistent molecular orbital methods. XII. Further extensions of Gaussian-type basis sets for use in molecular orbital studies of organic molecules*, **J. Chem. Phys. 56**, 2257.

Heinrich, A. J., C. P. Lutz, J. A. Gupta, and D. M. Eigler, 2002, *Molecule cascades*, **Science 298**, 1381.

Hele, T. J., 2011, *Master's Thesis*, Exeter College, Oxford University.

Hele, T. J., and S. C. Althorpe, 2013, *Derivation of a true ($t \to 0+$) quantum transition-state theory. I. Uniqueness and equivalence to ring-polymer molecular dynamics transition-state-theory*, **J. Chem. Phys. 138**, 084108.

Hellmann, S., T. Rohwer, M. Kallane, K. Hanff, C. Sohrt, A. Stange, A. Carr, M. M. Murnane, H. C. Kapteyn, L. Kipp, M. Bauer, and K. Rossnagel, 2012, *Time-domain classification of charge-density-wave insulators*, **Nat. Commun. 3**, 1069.

Hemley, R. J., P. M. Bell, and H. K. Mao, 1987, *Laser techniques in high-pressure geophysics*, **Science 237**, 605.

Henson, A. B., S. Gersten, Y. Shagam, J. Narevicius, and E. Narevicius, 2012, *Observation of resonances in Penning ionization reactions at sub-kelvin temperatures in merged beams*, **Science 338**, 234.

Hentschel, M., R. Kienberger, Ch. Spielmann, G. A. Reider, N. Milosevic, T. Brabec, P. Corkum, U. Heinzmann, M. Drescher, and F. Krausz, 2001, *Arrosecond metrology*, **Nature 414**, 509.

Herman, M. F., and E. Kluk, 1984, *A semiclasical justification for the use of non-spreading wavepackets in dynamics calculations*, **Chem. Phys. 91**, 27.

Herman, M. F., 1997, *Improving the accuracy of semiclassical wavepacket propagation using integral conditioning techniques*, **Chem. Phys. Lett. 275**, 445.

Hernandez, E. R., A. Rodriguez-Prieto, A. Bergara, and D. Alfè, 2010, *First-principles simulations of lithium melting: Stability of the bcc phase close to melting*, **Phys. Rev. Lett. 104**, 185701.

Herrero C. P., and R. Ramírez, 2000, *Structural and thermodynamic properties of diamond: A path-integral Monte Carlo study*, **Phys. Rev. B 63**, 024103.

Herrero, C. P., and R. Ramírez, 2009, *Vibrational properties and diffusion of hydrogen on graphene*, **Phys. Rev. B 79**, 115429.

Hill, P. G., R. D. C. MacMillan, and V. Lee, 1982, *A fundamental equation of state for heavy-water*, **J. Phys. Chem. Ref. Data 11**, 1.

Hinks, D. G., H. Claus, J. D. Jorgensen, 2001, *The complex nature of superconductivity in MgB_2 as revealed by the reduced total isotope effect*, **Nature 411**, 457.

Hinks, D. G., and J. D. Jorgensen, 2003, *The isotope effect and phonons in MgB_2*, **Physica C 385**, 98.

Hirsch, K., and W. Holzapfel, 1986, *Effect of high pressure on the Raman spectra of ice VIII and evidence for ice X*, **J. Chem. Phys. 84**, 2771.

Hladky-Hennion, A. C., G. Allan, and M. de Billy, 2005, *Localized modes in a one-dimensional diatomic chain of coupled spheres*, **J. Appl. Phys. 98**, 054909.

Ho, W., 2002, *Single-molecule chemistry*, **J. Chem. Phys. 117**, 11033.

Höchli, U. T., and L. A. Boatner, 1979, *Quantum ferroelectricity in $K_{1-x}Na_xTaO_3$ and $KTa_{1-y}Nb_yO_3$*. **Phys. Rev. B 20**, 266.

Hohenberg, P., and W. Kohn, 1964, *Inhomogeneous electron gas*, **Phys. Rev. 136**, B864.

Hohenberg, P. C., and P. M. Platzman, 1966, *High-energy neutron scattering from liquid He_4*, **Phys. Rev. 152**, 198.

Holz, M., S. R. Heil, and A. Sacco, 2000, *Temperature-dependent self-diffusion coefficients of water and six selected molecular liquids for calibration in accurate ^1H NMRPFG measurements*, **Phys. Chem. Chem. Phys. 2**, 4740.

Hong, F., L. X. Yang, P. F. Shan, P. T. Yang, Z. Y. Liu, J. P. Sun, Y. Y. Yin, X. H. Yu, J. G. Cheng, and Z. X. Zhao, 2020, *Superconductivity of lanthanum superhydride investigated using the standard four-probe configuration under high pressures*, **Chin. Phys. Lett. 37**, 107401.

Hong, H, C. Huang, C. Ma, J. Qi, X. Shi, C. Liu, S. Wu, Z. Sun, E. G. Wang, and K. H. Liu, 2023, *Twist phase matching in two-dimensional materials*, **Phys. Rev. Lett. 131**, 233801.

Hong, J. N., Y. Tian, T. Liang, X. Liu, Y. Song, D. Guan, Z. Yan, J. Guo, B. Tang, D. Y. Cao, J. Guo, J. Chen, D. Pan, L. M. Xu, E. G. Wang, and Y. Jiang, 2024, *Imaging surface structure and premelting of ice Ih with atomic resolution*, **Nature 630**, 375.

Hong, N.-S., D. Petrović, R. Lee, G. Gryn'ova, M. Purg, J. Saunders, P. Bauer, P. D. Carr, C.-Y. Lin, P. D. Mabbitt, W. Zhang, T. Altamore, C. Easton, M. L. Coote, S. C. L. Kamerlin, and C. J. Jackson, 2018, *The evolution of multiple active site configurations in a designed enzyme*, **Nat. Commun. 9**, 3900.

Hong, S., S. M. Nakhmanson, and D. D. Fong, 2016, *Screening mechanisms at polar oxide heterointerfaces*, **Rep. Prog. Phys. 79**, 076501.

Hong, X. P., J. Kim, S. F. Shi, Y. Zhang, C. Jin, Y. Sun, S. Tongay, J. Wu, Y. Zhang, and F. Wang, 2014, *Ultrafast charge transfer in atomically thin MoS_2/WS_2 heterostructures*, **Nat. Nanotechnol. 9**, 682.

Hoover, W. G., 1985, *Canonical dynamics: Equilibrium phase-space distributions*, **Phys. Rev. A 31**, 1695.

Hoover, W. G., 1986, *Constant-pressure equations of motion*, **Phys. Rev. A 34**, 2499.

Horita, J., and D. J. Wesolowski, 1994, *Liquid-vapor fractionation of oxygen and hydrogen isotopes of water from the freezing to the critical-temperature*, **Geochim. Cosmochim. Acta 58**, 3425.

Horsewill, A. J., and Q. Xue, 2002, *Magnetic field-cycling investigations of molecular tunnelling*, **Phys. Chem. Chem. Phys. 4**, 5475.

Horsewill, A. J., and W. Wu, 2006, *Proton tunneling in a hydrogen bond measured by cross-relaxation field-cycling NMR*, **J. Magn. Reson. 179**, 169.

Hörst, S. M., R. V. Yelle, A. Buch, N. Carrascon, G. Cernogora, O. Dutuit, E. Quirico, E. Sciamma-O'Brien, M. A. Smith, Á. Somogyi, C. Szopa, R. Thissen, and V. Vuitton, 2012, *Formation of amino acids and nucleotide bases in a titan atmosphere simulation experiment*, **Astrobiology 12**, 809.

Howie, R. T., C. L. Guillaume, T. Scheler, A. F. Goncharov, and E. Gregoryanz, 2012a, *Mixed molecular and atomic phase of dense hydrogen*, **Phys. Rev. Lett. 108**, 125501.

Howie, R. T., T. Scheler, C. L. Guillaume, and E. Gregoryanz, 2012b, *Proton tunneling in phase IV of hydrogen and deuterium*, **Phys. Rev. B 86**, 214104.

Hu, C. P., H. Hirai, and O. Sugino, 2007, *Nonadiabatic couplings from time-dependent density functional theory: Formulation in the Casida formalism and practical scheme within modified linear response*, **J. Chem. Phys. 127**, 064103.

Hu, M. G., Y. Liu, D. D. Grimes, Y. W. Lin, A. H. Gheorghe, R. Vexiau, N. Bouloufa-Maafa, O. Dulieu, T. Rosenband, and K. K. Ni, 2019, *Direct observation of bimolecular reactions of ultracold KRb molecules*, **Science 366**, 1111.

Hu, S., M. Lozada-Hidalgo, F. C. Wang, A. Mishchenko, F. Schedin, R. R. Nair, E. W. Hill, D. W. Boukhvalov, M. I. Katsnelson, R. A. W. Dryfe, I. V. Grigorieva, H. A. Wu, and A. K. Geim, 2014, *Proton transport through one-atom-thick crystals*, **Nature 516**, 227.

Hu, S. Q., H. Zhao, C. Lian, X. B. Liu, M. X. Guan, and S. Meng, 2022, *Tracking photocarrier-enhanced electron-phonon coupling in nonequilibrium*, **npj Quantum Mater. 7**, 14.

Hu, X. G., Y. M. Cai, and Q. S. Li, 1991, *The development of transition state theory (II)*, **JXPII 6**, 55 (in Chinese).

Huang, M. J., and M. Wolfsberg, 1984, *Tunneling in the automerization of cyclobutadiene*, **J. Am. Chem. Soc. 106**, 4039.

Huang, Q., D. L. Yu, B. Xu, W. T. Hu, Y. M. Ma, Y. B. Wang, Z. S. Zhao, B. Wen, J. L. He, Z. Y. Liu, and Y. J. Tian, 2014, *Nanotwinned diamond with unprecedented hardness and stability*, **Nature 510**, 250.

Huang, X., L. Wang, K. Liu, L. Liao, H. Sun, J. Wang, X. Tian, Z. Xu, W. Wang, L. Liu, Y. Jiang, J. Chen, E. G. Wang, and X. Bai, 2023, *Tracking cubic ice at molecular resolution*, **Nature 617**, 86.

Huo, P., and D. F. Coker, 2012, *Consistent schemes for non-adiabatic dynamics derived from partial linearized density matrix propagation*, **J. Chem. Phys. 137**, 22A535.

Huynh, M. H. V., and T. J. Meyer, 2007, *Proton-coupled electron transfer*, **Chem. Rev. 107**, 5004.

Hybertsen, M. S., and S. G. Louie, 1986, *Electron correlation in semiconductors and insulators: Band gaps and quasiparticle energies*, **Phys. Rev. B 34**, 5390.

I

Ishii, K., J. Haruyama, and O. Sugino, 2021, *Optical representation of thermal nuclear fluctuation effect on band-gap renormalization*, **Phys. Rev. B 104**, 245144.

Itoh, M. R., R. Wang, Y. Inaguma, T. Yamaguchi, Y. J. Shan, and T. Nakamura, 1999, *Ferroelectricity induced by oxygen isotope exchange in strontium titanate perovskite*. **Phys. Rev. Lett. 82**, 3540.

Iwashita, T., B. Wu, W.-R. Chen, S. Tsutsui, A. Q. R. Baron, and T. Egami, 2017, *Seeing real-space dynamics of liquid water through inelastic x-ray scattering*, **Sci. Adv. 3**, e1603079.

J

Jaccard, D., K. Behnia, and J. Sierro, 1992, *Pressure-induced heavy fermion superconductivity of $CeCu_2Ge_2$*, **Phys. Lett. A 163**, 475.

Jakowski, J., J. Huang, S. Garashchuk, Y. Luo, K. Hong, J. Keum, and B. G. Sumpter, 2017, *Deuteration as a means to tune crystallinity of conducting polymers*, **J. Phys. Chem. Lett. 8**, 4333.

Jakowski, J., J. Huang, B. G. Sumpter, and S. Garashchuk, 2018, *Theoretical assessment of the nuclear quantum effects on polymer crystallinity via perturbation theory and dynamics*, **Int. J. Quantum Chem. 118**, e25712.

Jastrow, R., 1955, *Many-body problem with strong forces*, **Phys. Rev. 98**, 1479.

Jenkinson, R. I., A. Ikram, A. J. Horsewill, and H. P. Trommsdorff, 2003, *The quantum dynamics of proton transfer in benzoic acid measured by single crystal NMR spectroscopy and relaxometry*, **Chem. Phys. 294**, 95.

Jérome, D., A. Mazaud, M. Ribault, and K. Bechgaard, 1980, *Superconductivity in a synthetic organic conductor (TMTSF)$_2$PF$_6$*, **J. Physique Lett. 41**, 95.

Jesse, S., B. J. Rodriguez, S. Choudhury, A. P. Baddorf, I. Vrejoiu, D. Hesse, M. Alexe, E. A. Eliseev, A. N. Morozovska, J. Zhang, L. Q. Chen, and S. V. Kalinin, 2008, *Direct imaging of the spatial and energy distribution of nucleation centres in ferroelectric materials*, **Nat. Mater. 7**, 209.

Ji, S. H., F. Wang, L. L. Zhu, X. G. Xu, Z. P. Wang, and X. Sun, 2013, *Non-critical phase-matching fourth harmonic generation of a 1053-nm laser in an ADP crystal*, **Sci. Rep. 3**, 1605.

Jia, C. L., S. B. Mi, K. Urban, I. Vrejoiu, M. Alexe, and D. Hesse, 2009, *Effect of a single dislocation in a heterostructure layer on the local polarization of a ferroelectric layer*, **Phys. Rev. Lett. 102**, 117601.

Jia, C. L., K. W. Urban, M. Alexe, D. Hesse, and I. Vrejoiu, 2011, *Direct observation of continuous electric dipole rotation in flux-closure domains in ferroelectric Pb(Zr,Ti)O$_3$*, **Science 331**, 1420.

Jiang, F., T. Peng, H. X. Xiao, J. L. Zhao, Y. Pan, F. Herlach, and L. Li, 2014, *Design and test of a flat-top magnetic field system driven by capacitor banks*, **Rev. Sci. Instrum. 85**, 045106.

Jiang, T. H., Y. L. Chen, N. A. Bogdanov, E. G. Wang, A. Alavi, and J. Chen, 2021, *A full configuration interaction quantum Monte Carlo study of ScO, TiO, and VO molecules*, **J. Chem. Phys. 154**, 164302.

Jiang, T. H., W. Fang, A. Alavi, and J. Chen, 2022, *General analytical nuclear force and molecular potential energy surface from full configuration interaction quantum Monte Carlo*, DOI: 10.48550/arXiv.2204.13356.

Jiao, K. J., Y. K. Xing, Q. L. Yang, H. Qiu, and T. S. Mei, 2020, *Site-selective C–H functionalization via synergistic use of electrochemistry and transition metal catalysis*, **Acc. Chem. Res. 53**, 2.

Jokisaari, J. R., J. A. Hachtel, X. Hu, A. Mukherjee, C. H. Wang, A. Konecna, T. C. Lovejoy, N. Dellby, J. Aizpurua, O. L. Krivanek, J. C. Idrobo, and R. F. Klie, 2018, *Vibrational spectroscopy of water with high spatial resolution*, **Adv. Mater. 30**, 1802702.

Jones, H., R. J. Nicholas, and W. J. Siertsema, 2000, *The upgrade of the Oxford high magnetic field laboratory*, **IEEE Trans. Appl. Supercon. 10**, 1552.

Jones, H., P. Frings, O. Portugall, M. von Ortenberg, A. Lagutin, F. Herlach, and L. van Bockstal, 2006, *ARMS: A successful. European program for an 80-T user magnet*, **IEEE Trans. Appl. Supercon. 16**, 1684.

K

Kaczmarek, A., M. Shiga, and D. Marx, 2009, *Quantum effects on vibrational and electronic spectra of hydrazine studied by "on-the-fly" ab initio ring polymer molecular dynamics*, **J. Phys. Chem. A 113**, 1985.

Kaischew, R., and F. Simon, 1934, *Some thermal properties of condensed helium*, **Nature 133**, 460.

Kang, D. D., H. Sun, J. Dai, W. Chen, Z. Zhao, Y. Hou, J. Zeng, and J. Yuan, 2014, *Nuclear quantum dynamics in dense hydrogen*, **Sci. Rep. 4**, 5484.

Kang, J. H., G. Sauti, C. Park, V. I. Yamakov, K. E. Wise, S. E. Lowther, C. C. Fay, S. A. Thibeault, and R. G. Bryant, 2015, *Multifunctional electroactive nanocomposites based on piezoelectric boron nitride nanotubes*, **ACS Nano 9**, 11942.

Kapil, V., C. Schran, A. Zen, J. Chen, C. J. Pickard, and A. Michaelides, 2022, *The first-principles phase diagram of monolayer nanoconfined water*, **Nature 609**, 512.

Kapitza, P., 1924, *A method of producing strong magnetic fields.*, **Proc. Roy. Soc. Lond. A 105**, 691.

Kapitza, P., 1938, *Viscosity of liquid helium below the λ-point*, **Nature 141**, 74.

Kargel, J. S., 2006, *Enceladus: Cosmic gymnast, volatile miniworld*, **Science 311**, 1389.

Karmakar, S., C. Chowdhury, and A. Datta, 2016, *Two-dimensional group IV monochalcogenides: Anode materials for Li-ion batteries*, **J. Phys. Chem. C 120**, 14522.

Kavokine, N., M. L. Bocquet, and L. Bocquet, 2022, *Fluctuation-induced quantum friction in nanoscale water flows*, **Nature 602**, 84.

Kawashima, Y., K. Suzuki, and M. Tachikawa, 2013, *Ab initio path integral simulations for the fluoride ion-water clusters: Competitive nuclear quantum effect between F^--water and water-water hydrogen bonds*, **J. Phys. Chem. A 117**, 5205.

Kell, G. S., 1977, *Effects of isotopic composition, temperature, pressure, and dissolved-gases on density of liquid water*, **J. Phys. Chem. Ref. Data 6**, 1109.

Kendall, R. A., and T. H. Dunning Jr., 1992, *Electron affinities of the first-row atoms revisited. Systematic basis sets and wave functions*, **J. Chem. Phys. 96**, 6796.

Kent, P. R. C., G. Kotliar, 2018, *Toward a predictive theory of correlated materials*, **Science 361**, 348–354.

Keutsch, F. N., J. D. Cruzan, and R. J. Saykally, 2003, *The water trimer*, **Chem. Rev. 103**, 2533.

Kim, D. S., O. Hellman, J. Herriman, H. L. Smith, J. Y. Y. Lin, N. Shulumba, J. L. Niedziela, C. W. Li, D. L. Abernathy, and B. Fultz, 2018, *Nuclear quantum effect with pure anharmonicity and the anomalous thermal expansion of silicon*, **Proc. Natl. Acad. Sci. U. S. A. 115**, 1992.

Kim, D. Y., and M. H. W. Chan, 2014, *Upper limit of supersolidity in solid helium*, **Phys. Rev. B90**, 064503.

Kim, E., and M. H. W. Chan, 2004a, *Observation of superflow in solid helium*, **Science 305**, 1941.

Kim, E., and M. H. W. Chan, 2004b, *Probable observation of a supersolid helium phase*, **Nature 427**, 225.

Kim, E., J. S. Xia, J. T. West, X. Lin, A. C. Clark, and M. H. W. Chan, 2008, *Effect of ^3He impurities on the nonclassical response to oscillation of solid ^4He*, **Phys. Rev. Lett. 100**, 065301.

Kim, J. H., T. V. Pham, J. H. Hwang, C. S. Kim, and M. J. Kim, 2018, *Boron nitride nanotubes: Synthesis and applications*, **Nano Converg. 5**, 17.

Kim, K. H., A. Spah, H. Pathak, F. Perakis, D. Mariedahl, K. Amann-Winkel, J. A. Sellberg, J. H. Lee, S. Kim, J. Park, K. H. Nam, T. Katayama, and A. Nilsson, 2017, *Maxima in the thermodynamic response and correlation functions of deeply supercooled water*, **Science 358**, 1589.

Kim, S., J. E. Froch, J. Christian, M. Straw, J. Bishop, D. Totonjian, K. Watanabe, T. Taniguchi, M. Toth, and I. Aharonovich, 2018, *Photonic crystal cavities from hexagonal boron nitride*, **Nat. Commun. 9**, 2623.

Kirilyuk, A., A. V. Kimel, and T. Rasing, 2010, *Ultrafast optical manipulation of magnetic order*, **Rev. Mod. Phys. 82**, 2731.

Kirkwood, J. G., 1946, *The statistical mechanical theory of transport processes I. General theory*, **J. Chem. Phys. 14**, 180.

Kistanov, A. A., D. R. Kripalani, Y. Cai, S. V. Dmitriev, K. Zhou, and Y.-W. Zhang, 2019, *Ultrafast diffusive cross-sheet motion of lithium through antimonene with 2 + 1 dimensional kinetics*, **J. Mater. Chem. A 7**, 2901.

Kitamura, H., S. Tsuneyuki, T. Ogitsu, and T. Miyake, 2000, *Quantum distribution of protons in solid molecular hydrogen at megabar pressures*, **Nature 404**, 259.

Kleinert, H., 1989, *Gauge fields in condensed matter, Vol. I, Superflow and vortex lines* (World Scientific, Singapore).

Klinman, J. P., and A. R. Offenbacher, 2018, *Understanding biological hydrogen transfer through the lens of temperature dependent kinetic isotope effects*, **Acc. Chem. Res. 51** 1966.

Knoll, G. F., 2010, *Radiation detection and measurement*, 4th edition (John Wiley & Sons, Hoboken, NJ).

Koch, W., and M. C. Holthausen, 2001, *A Chemist's guide to density functional theory* (Wiley-VCH Verlag GmbH, Weinheim).

Koch, M., M. Pagan, M. Persson, S. Gawinkowski, J. Waluk, and T. Kumagai, 2017, *Direct observation of double hydrogen transfer via quantum tunneling in a single porphycene molecule on a Ag(110) surface*, **J. Am. Chem. Soc. 139**, 12681.

Kohn, W., and L. J. Sham, 1965, *Self-consistent equations including exchange and correlation effects*, **Phys. Rev. 140**, A1133.

Kohn, W., 1983, *v-representability and density functional theory*, **Phys. Rev. Lett. 51**, 1596.

Kolesnikov, A. I., G. F. Reiter, N. Choudhury, T. R. Prisk, E. Mamontov, A. Podlesnyak, G. Ehlers, A. G. Seel, D. J. Wesolowski, and L. M. Anovitz, 2016, *Quantum tunneling of water in beryl: A new state of the water molecule*, **Phys. Rev. Lett. 116**, 167802.

Kolesnikov, A. I., G. F. Reiter, T. R. Prisk, M. Krzystyniak, G. Romanelli, D. J. Wesolowski, and L. M. Anovitz, 2018, *Inelastic and deep inelastic neutron scattering of water molecules under ultra-confinement*, **J. Phys: Conf. Series 1055**, 012002.

Konecna, A., F. Iyikanat, and F. J. Abajo, 2021, *Theory of atomic-scale vibrational mapping and isotope identification with electron beams*, **ACS nano 15**, 9890.

Kortus, J., I. I. Mazin, K. D. Belashchenko, V. P. Antropov, and L. L. Boyer, 2001, *Superconductivity of metallic boron in MgB_2*, **Phys. Rev. Lett. 86**, 4656.

Koval, S., J. Kohanoff, R. L. Migoni, and E. Tosatti, 2002, *Ferroelectricity and isotope effects in hydrogen-bonded KDP crystals,* **Phys. Rev. Lett. 89**, 187602.

Koval, S., J. Kohanoff, J. Lasave, G. Colizzi, and R. L. Migoni, 2005, *First-principles study of ferroelectricity and isotope effects in H-bonded KH_2PO_4 crystals*, **Phys. Rev. B 71**, 184102.

Kramers, H. A., 1926, *Wellenmechanik und halbzahlige quantisierung*, **Z. Phys. 39**, 828.

Krivanek, O. L., T. C. Lovejoy, N. Dellby, T. Aoki, R. W. Carpenter, P. Rez, E. Soignard, J. T. Zhu, P. E. Batson, M. J. Lagos, R. F. Egerton, and P. A. Crozier, 2014, *Vibrational spectroscopy in the electron microscope*, **Nature 514**, 209.

Krüger, K., Y. Wang, S. Tödter, F. Debbeler, A. Matveenko, N. Hert, X. Zhou, B. Jiang, H. Guo, A. M. Wodtke, and O. Bünermann, 2022, *Hydrogen atom collisions with a semiconductor efficiently promote electrons to the conduction band*, **Nat. Chem. 15**, 326.

Kuang, G. L., and S. F. Shao, 2018, *The technology and scientific significance of high magnetic field*, **Sci. Tech. Rev. 36**, 93 (in Chinese).

Kubo, R., 1957a, *Statistical-mechanical theory of irreversible processes. 1. General theory and simple applications to magnetic and conduction problems*, **J. Phys. Soc. Jap. 12**, 570.

Kubo, R., 1957b, *Statistical-mechanical theory of irreversible processes. 2. Response to thermal disturbance*, **J. Phys. Soc. Jap. 12**, 1203.

Kudish, A. I., D. Wolf, and F. Steckel, 1972, *Physical properties of heavy-oxygen water. Absolute viscosity of $H_2{}^{18}O$ between 15 and 35°C*, **J. Chem. Soc. Faraday Trans. 68**, 2041.

Kumagai, T., M. Kaizu, S. Hatta, H. Okuyama, T. Aruga, I. Hamada, and Y. Morikawa, 2008, *Direct observation of hydrogen-bond exchange within a single water dimer*, **Phys. Rev. Lett. 100**, 166101.

Kumagai, T., M. Kaizu, H. Okuyama, S. Hatta, T. Aruga, I. Hamada, and Y. Morikawa, 2009, *Tunneling dynamics of a hydroxyl group adsorbed on Cu(110)*, **Phys. Rev. B 79**, 035423.

Kumagai, T., M. Kaizu, H. Okuyama, S. Hatta, T. Aruga, I. Hamada, and Y. Morikawa, 2010, *Symmetric hydrogen bond in a water-hydroxyl complex on Cu(110)*, **Phys. Rev. B 81**, 045402.

Kumagai, T., A. Shiotari, H. Okuyama, S. Hatta, T. Aruga, I. Hamada, T. Frederiksen, and H. Ueba, 2011, *H-atom relay reactions in real space*, **Nat. Mat. 11**, 167.

Kumagai, T., 2015, *Direct observation and control of hydrogen-bond dynamics using low-temperature scanning tunneling microscopy*, **Prog. Surf. Sci. 90**, 239.

Kundu, A., Y. Song, and G. Galli, 2022, *Influence of nuclear quantum effects on the electronic properties of amorphous carbon*, **Proc. Natl. Acad. Sci. 119**, e2203083119.

Kvyatkovskiĭ, O. E., 2001, *Theory of isotope effect in displacive ferroelectrics*. **Solid State Commun. 117**, 455.

L

Labet, V., P. Gonzalez-Morelos, R. Hoffmann, and N. W. Ashcroft, 2012a, *A fresh look at dense hydrogen under pressure. I. An introduction to the problem, and an index probing equalization of H-H distances*, **J. Chem. Phys. 136**, 074501.

Labet, V., R. Hoffmann, and N. W. Ashcroft, 2012b, *A fresh look at dense hydrogen under pressure. III. Two competing effects and the resulting intra-molecular H-H separation in solid hydrogen under pressure*, **J. Chem. Phys. 136**, 074503.

Laio, A., and M. Parrinello, 2002, *Escaping free-energy minima*, **Proc. Natl. Acad. Sci. U. S. A. 99**, 12562.

Lakhotia, H., H. Y. Kim, M. Zhan, S. Hu, S. Meng, and E. Goulielmakis, 2020, *Laser picoscopy of valence electrons in solids*, **Nature 583**, 55.

Lam, P. K., M. M. Dacorogna, and M. L. Cohen, 1986, *Self-consistent calculation of electron-phonon couplings*, **Phys. Rev. B 34**, 5065.

Landau, L. D., 1933, *Über die bewegung der elektronen in kristalgitter*, **Phys. Z. Sowjetunion 3**, 644.

Landauer, R., 1957, *Electrostatic considerations in $BaTiO_3$ domain formation during polarization reversal*. **J. Appl. Phys. 28**, 227.

Landt, L., K. Klünder, J. E. Dahl, R. M. K. Carlson, T. Möller, and C. Bostedt, 2009, *Optical response of diamond nanocrystals as a function of particle size, shape, and symmetry*, **Phys. Rev. Lett. 103**, 047402.

Lascola, R., R. Withnall, and L. Andrews, 1988, *Infrared spectra of hydrazine and products of its reactions with HF, F2, and O3 in solid argon*, **Inorg. Chem. 27**, 642.

Laughlin, R. B., 1983, *Anomalous quantum Hall effect: An incompressible quantum fluid with fractionally charged excitations*, **Phys. Rev. Lett. 50**, 1395.

Lauhon, L. J., and W. Ho, 2000, *Direct observation of the quantum tunneling of single hydrogen atoms with a scanning tunneling microscope*, **Phys. Rev. Lett. 85**, 4566.

Lauret, J. S., R. Arenal, F. Ducastelle, A. Loiseau, M. Cau, B. Attal-Tretout, E. Rosencher, and L. Goux-Capes, 2005, *Optical transitions in single-wall boron nitride nanotubes*, **Phys. Rev. Lett. 94**, 1.

Lawrence, J. E., and D. E. Manolopoulos, 2018, *Analytic continuation of Wolynes theory into the Marcus inverted regime*, **J. Chem. Phys. 148**, 102313.

Lawrence, J. E., and D. E. Manolopoulos, 2019a, *An analysis of isomorphic RPMD in the golden rule limit*, **J. Chem. Phys. 151**, 244109.

Lawrence, J. E., T. Fletcher, L. P. Lindoy, and D. E. Manolopoulos, 2019b, *On the calculation of quantum mechanical electron transfer rates*, **J. Chem. Phys. 151**, 114119.

Lawrence, J. E., and D. E. Manolopoulos, 2020, *A general non-adiabatic quantum instanton approximation*, **J. Chem. Phys. 152**, 204117.

Lax, M., 1952, *The Franck-Condon principle and its application to crystals*, **J. Chem. Phys. 20**, 1752.

Lazzeri, M., and F. Mauri, 2006, *Nonadiabatic Kohn anomaly in a doped graphene monolayer*, **Phys. Rev. Lett. 97**, 266407.

Lee, C., X. Wei, J. W. Kysar, and J. Hone, 2008, *Measurement of the elastic properties and intrinsic strength of monolayer graphene*, **Science 321**, 385.

Lee, K. W., and W. E. Pickett, 2004, *Superconductivity in boron-doped diamond*, **Phys. Rev. Lett. 93**, 237003.

Leggett, A. J., 1970, *Can a solid be "superfluid"?* **Phys. Rev. Lett. 25**, 1543.

Lennard-Jones, J. E., 1924, *On the determination of molecular fields. – II. from the equation of state of a gas*, **Proc. Roy. Soc. Lond. A 106**, 463.

Leuenberger, M. N., and D. Loss, 2001, *Quantum computing in molecular magnets*, **Nature 410**, 789.

Levy, M., 1982, *Electron densities in search of Hamiltonians*, **Phys. Rev. A 26**, 1200.

Lewis, E. S., and L. Funderburk, 1967, *Rates and isotope effects in the proton transfers from 2-nitropropane to pyridine bases*, **J. Am. Chem. Soc. 89**, 2322.

Li, C. Y., J. B. Le, Y. H. Wang, S. Chen, Z. L. Yang, J. F. Li, J. Cheng, and Z. Q. Tian, 2019, *In situ probing electrified interfacial water structures at atomically flat surfaces*, **Nat. Mater. 18**, 697.

Li, L., J. R. Jokisaari, and X. Pan, 2015, *In situ electron microscopy of ferroelectric domains*, **MRS Bull. 40**, 53.

Li, M., B. Wang, H.-J. Liu, Y.-L. Huang, J. Zhang, X. Ma, K. Liu, D. Yu, Y.-H. Chu, and L.-Q. Chen, 2019, *Direct observation of weakened interface clamping effect enabled ferroelastic domain switching*, **Acta Mater. 171**, 184.

Li, N., X. Guo, X. Yang, R. Qi, T. Qiao, Y. Li, R. Shi, Y. Li, K. Liu, Z. Xu, L. Liu, F. J. Abajo, Q. Dai, E. G. Wang, and P. Gao, 2021, *Direct observation of highly confined phonon polaritons in suspended monolayer hexagonal boron nitride*, **Nat. Mater. 20**, 43.

Li, N., 2022, *Study of phonons and phonon polaritons in low-dimensional boron nitride materials by electron energy loss spectroscopy*, PhD dissertation, Peking University, Beijing, (in Chinese).

Li, N., R. Shi, Y. Li, R. Qi, F. Liu, X. Zhang, Z. Liu, Y. Li, X. Guo, K. Liu, Y. Jiang, X. Z. Li, J. Chen, L. Liu, E. G. Wang, and P. Gao, 2023, *Phonon transition across an isotopic interface*, **Nat. Commun. 14**, 2382.

Li, P. E., C. Guo, S. Wang, D. Ma, T. Feng, Y. Wang, and Y. Qiu, 2022, *Facile and general electrochemical deuteration of unactivated alkyl halides*, **Nat. Commun. 13**, 3774.

Li, R. Z., J. Chen, X. Z. Li, E. G. Wang, and L. M. Xu, 2015, *Supercritical phenomenon of hydrogen beyond the liquid–liquid phase transition*, **N. J. Phys. 17**, 063023.

Li, W., Y. Yang, G. Zhang, and Y. Zhang, 2015, *Ultrafast and directional diffusion of lithium in phosphorene for high-performance lithium-ion battery*, **Nano Lett. 15**, 1691.

Li, X., T. Qiu, J. H. Zhang, E. Baldini, J. Lu, A. M. Rappe, and K. A. Nelson, 2019, *Terahertz field-induced ferroelectricity in quantum paraelectric $SrTiO_3$*, **Science 364**, 1079.

Li, X. M., M. Li, X. Li, S. Tian, A. Y. Abid, N. Li, J. Wang, L. Zhang, X. Li, and Y. Zhao, 2019, *Effect of single point defect on local properties in $BiFeO_3$ thin film*, **Acta Mater. 170**, 132.

Li, X. Z., M. I. J. Probert, A. Alavi, and A. Michaelides, 2010, *Quantum nature of the proton in water-hydroxyl overlayers on metal surfaces*, **Phys. Rev. Lett. 104**, 066102.

Li, X. Z., B. Walker, and A. Michaelides, 2011, *Quantum nature of the hydrogen bond*, **Proc. Natl. Acad. Sci. U. S. A. 108**, 6369.

Li, X. Z., B. Walker, M. I. J. Probert, C. J. Pickard, R. J. Needs, and A. Michaelides, 2013, *Classical and quantum ordering of protons in cold solid hydrogen under megabar pressures*, **J. Phys.: Condens. Matter 25**, 085402.

Li, X. Z., and E. G. Wang, 2014, *Computer simulations of molecules and condensed matters: From electronic structures to molecular dynamics* (Peking University Press, Beijing).

Li, Y., R. Qi, R. Shi, N. Li, and P. Gao, 2020, *Manipulation of surface phonon polaritons in SiC nanorods*, **Sci. Bull. 65**, 820.

Li, Y. F., X. Wen, C. J. Tan, N. Li, R. J. Li, X. Y. Huang, H. F. Tian, Z. X. Yao, P. C. Liao, S. L. Yu, S. Z. Liu, Z. J. Li, J. J. Guo, Y. Huang, P. Gao, L. F. Wang, S. L. Bai, and L. Liu, 2021, *Synthesis of centimeter-scale high-quality polycrystalline hexagonal boron nitride films from Fe fluxes*, **Nanoscale 13**, 11223.

Li, Y. F., X. Zhang, J. Wang, X. Ma, J. A. Shi, X. Guo, Y. Zuo, R. Li, H. Hong, N. Li, K. Xu, X. Huang, H. Tian, Y. Yang, Z. Yao, P. C. Liao, X. Li, J. Guo, Y. Huang, P. Gao, L. Wang, X. Yang, Q. Dai, E. G. Wang, K. Liu, W. Zhou, X. Yu, L. Liang, Y. Jiang, X. Z. Li, and L. Liu, 2022, *Engineering interlayer electron-phonon coupling in WS2/BN heterostructures*, **Nano Lett. 22**, 2725.

Li, Y. W., J. Hao, H. Liu, Y. L. Li, and Y. Ma, 2014, *The metallization and superconductivity of dense hydrogen sulfide*, **J. Chem. Phys. 140**, 174712.

Li, Y. W., J. Hao, H. Liu, J. S. Tse, Y. Wang, and Y. Ma, 2015, *Pressure-stabilized superconductive yttrium hydrides*, **Sci. Rep. 5**, 9948.

Li, Z. W., R. J. Xiao, P. Xu, C. Zhu, S. Sun, D. Zheng, H. Wang, M. Zhang, H. Tian, H. X. Yang, and J. Q. Li, 2019, *Lattice dynamics and contraction of energy bandgap in photoexcited semiconducting boron nitride nanotubes*, **ACS Nano 13**, 11623.

Lian, C., M. X. Guan, S. Q. Hu, and S. Meng, 2018, *Photoexcitation in solids: First-principles quantum simulations by real time TDDFT*, **Adv. Theory Simul. 1**, 1800055.

Lian, C., S. J. Zhang, S. Q. Hu, M. X. Guan, and S. Meng, 2020, *Ultrafast charge ordering by self-amplied exciton-phonon dynamics in TiSe2*, **Nat. Commun. 11**, 43.

Liang, G. J., and C. Y. Zhi, 2021, *A reversible Zn-metal battery*, **Nat. Nanotechnol. 16**, 854.

Liao, G. Q., Y. T. Li, H. Liu, G. G. Scott, D. Neely, Y. H. Zhang, B. J. Zhu, Z. Zhang, C. Armstrong, E. Zemaityte, P. Bradford, P. G. Huggard, D. R. Rusby, P. McKenna, C. M. Brenner, N. C. Woolsey, W. M. Wang, Z. M. Sheng, and J. Zhang, 2019, *Multimillijoule coherent terahertz bursts from picosecond laser-irradiated metal foils*, **Proc. Natl. Acad. Sci. 116**, 3994.

Lide, D. R., 1999, *CRC Handbook of Chemistry and Physics*, CRC netBase (Chapman & Hall, London).

Lieb, E. H., 1983, *Density functional for coulomb systems*, **Int. J. Quantum Chem. 24**, 243.

Lin, C. F., E. Durant, M. Persson, M. Rossi, and T. Kumagai, 2019, *Real-space observation of quantum tunneling by a carbon atom: Flipping reaction of formaldehyde on Cu(110)*, **J. Phys. Chem. Lett. 10**, 645.

Lin, M., M. Gong, B. Lu, Y. Wu, D. Wang, M. Guan, M. Angell, C. Chen, J. Yang, B. Hwang, and H. Dai, 2015, *An ultrafast rechargeable aluminium-ion battery*, **Nature 520**, 325.

Lin, T.-S., and R. Gomer, 1991, *Diffusion of ^1H and ^2H on the Ni(111) and (100) planes*. **Surf. Sci. 255**, 41.

Lin, X., A. C. Clark, and M. H. W. Chan, 2007, *Probable heat capacity signature of the supersolid transition*, **Nature 449**, 1025.

Lin, X., A. C. Clark, Z. G. Cheng, and M. H. W. Chan, 2009, *Heat capacity peak in solid ^{4}He: Effects of disorder and ^{3}He impurities*, **Phys. Rev. Lett. 102**, 125302.

Lindsay, R. B., 1924, *On the atomic models of the alkali metals*, **J. Math. Phys. 3**, 191.

Lindsay, L., D. A. Broido, and N. Mingo, 2010, *Flexural phonons and thermal transport in graphene*, **Phys. Rev. B**, **82**, 115427.

Lindsay, L., and D. A. Broido, 2011, *Enhanced thermal conductivity and isotope effect in single-layer hexagonal boron nitride*, **Phys. Rev. B**, **84**, 155421.

Lipa, J. A., J. A. Nissen, D. A. Stricker, D. R. Swanson, and T. C. P. Chui, 2003, *Specific heat of liquid helium in zero gravity very near the lambda point*, **Phys. Rev. B 68**, 174518.

Littlejohn, R. G., 1992, *The Van Vleck formula, Maslov theory, and phase space geometry*, **J. Stat. Phys. 68**, 7.

Liu, A. Y., I. I. Mazin, and J. Kortus, 2001, *Beyond Eliashberg superconductivity in MgB_2: Anharmonicity, two-phonon scattering, and multiple gaps*, **Phys. Rev. Lett. 87**, 087005.

Liu, G. D., G. L. Wang, Y. Zhu, H. B. Zhong, G. C. Zhang, X. Y. Wang, Y. Zhou, W. T. Zhang, H. Y. Liu, L. Zhao, J. Q. Meng, X. L. Dong, C. T. Chen, Z. Y. Xu, and X. J. Zhou, 2008, *Development of a vacuum ultra-violet laser-based angle-resolved photoemission system with a super-high energy resolution better than 1 meV*, **Rev. Sci. Instruments. 79**, 023105.

Liu, H., I. I. Naumov, R. Hoffmann, N. W. Ashcroft, and R. J. Hemley, 2017, *Potential high-Tc superconducting lanthanum and yttrium hydrides at high pressure*. **Proc. Natl. Acad. Sci. U. S. A. 114**, 6990.

Liu, H. Y., H. Wang, and Y. M. Ma, 2012, *Quasi-molecular and atomic phases of dense solid hydrogen*, **J. Phys. Chem. C 116**, 9221.

Liu, H. Y., Y. Yuan, D. Liu, X. Z. Li, and J. R. Shi, 2020, *Superconducting transition temperatures of metallic liquids*, **Phys. Rev. Research 2**, 013340.

Liu, J., and N. Makri, 2006, *Symmetries and detailed balance in forward-backward semiclassical dynamics*, **Chem. Phys. 322**, 23.

Liu, J., and W. H. Miller, 2009, *A simple model for the treatment of imaginary frequencies in chemical reaction rates and molecular liquids*, **J. Chem. Phys. 131**, 074113.

Liu, J., 2014, *Path integral liouville dynamics for thermal equilibrium systems*, **J. Chem. Phys. 140**, 224107.

Liu, J., F. Xia, D. Xiao, F. J. G. de Anajo, and D. Sun, 2020, *Semimetals for high-performance photodetection*, **Nat. Mater. 19**, 830.

Liu, J. H., L. Wang, L. Qin, Q. L. Wang, and Y. M. Dai, 2020, *Design, fabrication, and test of a 12 T REBCO insert for a 27 T all-superconducting magnet*, **IEEE Trans. Appl. Supercon. 30**, 5203006.

Liu, K. H., W. L. Wang, Z. Xu, X. D. Bai, E. G. Wang, Y. G. Yao, J. Zhang, and Z. F. Liu, 2009, *Chirality-dependent transport properties of double-walled nanotubes measured in situ on their field-effect transistors*, **J. Am. Chem. Soc. 131**, 62.

Liu, K. H., J. Deslippe, F. J. Xiao, R. B. Capaz, X. P. Hong, S. Aloni, A. Zettl, W. L. Wang, X. D. Bai, S. G. Louie, E. G. Wang, and F. Wang, 2012, *An atlas of carbon nanotube optical transitions*, **Nat. Nanotechnol. 7**, 325.

Liu, K. H., C. H. Jin, X. P. Hong, J. H. Kim, A. Zettl, E. G. Wang, and F. Wang, 2014, *Van der Waals-coupled electronic states in incommensurate double-walled carbon nanotubes*, **Nat. Phys. 10**, 737.

Liu, L. M., C. J. Zhang, G. Thornton, and A. Michaelides, 2010, *Structure and dynamics of liquid water on rutile TiO_2 (110)*, **Phys. Rev. B 82**, 161415.

Liu, X. D., R. T. Howie, H. C. Zhang, X. J. Chen, and E. Gregoryanz, 2017, *High-pressure behavior of hydrogen and deuterium at low temperatures*, **Phys. Rev. Lett. 119**, 065301.

Liu, X. D., P. Dalladay-Simpson, R. T. Howie, H. C. Zhang, W. Xu, J. Binns, G. J. Ackland, H. K. Mao, and E. Gregoryanz, 2020, *Counterintuitive effects of isotopic doping on the phase diagram of H_2-HD-D_2 molecular alloy*, **Proc. Natl. Acad. Sci. 117**, 13374.

Liu, Y., Y. Huang, and X. Duan, 2019, *Van der Waals integration before and beyond two-dimensional materials*, **Nature 567**, 323.

Lloyd-Williams, J. H., and B. Monserrat, 2015, *Lattice dynamics and electron-phonon coupling calculations using nondiagonal supercells*, **Phys. Rev. B 92**, 184301.

Locher, R., M. Lucchini, J. Herrmann, M. Sabbar, M. Weger, A. Ludwig, L. Castiglioni, M. Greif, M. Hengsberger, L. Gallmann, and U. Keller, 2014, *Versatile attosecond beamline in a two-foci configuration for simultaneous time-resolved measurements*, **Rev. Sci. Instrum. 85**, 013113.

Logothetidis, S., J. Petalas, H. M. Polatoglou, and D. Fuchs, 1992, *Origin and temperature dependence of the first direct gap of diamond*, **Phys. Rev. B 46**, 4483.

Lorenz, U. J., and A. H. Zewail, 2014, *Observing liquid flow in nanotubes by 4D electron microscopy*, **Science 344**, 1496.

Loubeyre, P., R. Letoullec, E. Wolanin, M. Hanfland, and D. Husermann, 1999, *Modulated phases and proton centring in ice observed by X-ray diffraction up to 170 GPa*, **Nature 397**, 503.

Loubeyre, P., F. Occelli., and P. Dumas, 2020, *Synchrotron infrared spectroscopic evidence of the probable transition to metal hydrogen*, **Nature 577**, 631.

Loveday, J., 2012, *High-pressure physics* (CRC Press, New York).

Low, T., A. Chaves, J. D. Caldwell, A. Kumar, N. X. Fang, P. Avouris, T. F. Heinz, F. Guinea, L. Martin-Moreno, and F. Koppens, 2017, *Polaritons in layered two-dimensional materials*, **Nat. Mater., 16**, 182.

Löwdin, P. O., 1963, *Proton tunneling in DNA and its biological implications*. **Rev. Mod. Phys. 35** 724.

Lozada-Hidalgo, M., S. Hu, O. Marshall, A. Mishchenko, A. N. Grigorenko, R. A. W. Dryfe, B. Radha, I. V. Grigorieve, and A. K. Geim, 2016, *Sieving hydrogen isotopes through two-dimensional crystals*, **Science 351**, 68.

Lozada-Hidalgo, M., S. Zhang, S. Hu, A. Esfandiar, I. V. Grigorieva, and A. K. Geim, 2017, *Scalable and efficient separation of hydrogen isotopes using graphene-based electrochemical pumping*, **Nat. Commun. 8**, 15215.

Lozada-Hidalgo, M., S. Zhang, S. Hu, V. G. Kravets, F. J. Rodriguez, A. Berdyugin, A. Grigorenko, and A. K. Geim, 2018, *Giant photoeffect in proton transport through graphene membranes*, **Nat. Nanotechnol. 13**, 300.

Lu, G., 1997, *Fundamentals of physics* (Higher Education Press, Beijing) (in Chinese).

Luo, Y. R., 1983, *Progress in transition state theory*, **Chemistry 10**, 8 (in Chinese).

Lupi, L., A. Hudait, B. Peters, M. Grünwald, R. G. Mullen, A. H. Nguyen, and V. Molinero, 2017, *Role of stacking disorder in ice nucleation*, **Nature 551**, 218.

Lv, J., Y. Wang, L. Zhu, and Y. Ma, 2011, *Predicted novel high-pressure phases of lithium*, **Phys. Rev. Lett. 106**, 015503.

M

Ma, J., 2009, *Van der Waals interaction of water on graphene surfaces*, PhD dissertation, Graduate University of Chinese Academy of Sciences, Beijing, (in Chinese).

Ma, J., A. Michaelides, D. Alfe, L. Schimka, G. Kresse, and E. G. Wang, 2011, *Adsorption and diffusion of water on graphene from first principles*, **Phys. Rev. B 84**, 033402.

Ma, L., K. Wang, Y. Xie, X. Yang, Y. Wang, M. Zhou, H. Liu, X. Yu, Y. Zhao, H. Wang, G. Liu, and Y. Ma, 2022, *High-temperature superconducting phase in clathrate calcium hydride CaH 6 up to 215 K at a pressure of 172 GPa*, **Phys. Rev. Lett. 128**, 167001.

Ma, R. Z., D. Y. Cao, C. Q. Zhu, Y. Tian, J. B. Peng, J. Guo, J. Chen, X. Z. Li, J. S. Francisco, X. C. Zeng, L. M. Xu, E. G. Wang, and Y. Jiang, 2020, *Atomic imaging of the edge structure and growth of a two-dimensional hexagonal ice*, **Nature 577**, 60.

Ma, W., Y. Jiao, and S. Meng, 2013, *Modeling charge recombination in dye-sensitized solar cells using first-principles electron dynamics: Effects of structural modification*, **Phys. Chem. Chem. Phys. 15**, 17187.

Ma, W., Y. Jiao, and S. Meng, 2014, *Predicting energy conversion efficiency of dye solar cells from first principles*, **J. Phys. Chem. C 118**, 16447.

Ma, W., J. Zhang, L. Yan, Y. Jiao, Y. Gao, and S. Meng, 2016, *Recent progresses in real-time local-basis implementation of time dependent density functional theory for electron-nucleus dynamics*, **Comp. Mater. Sci. 112**, 478.

Ma, X., S. Cambré, W. Wenseleers, S. K. Doorn, and H. Htoon, 2017, *Quasiphase transition in a single file of water molecules encapsulated in (6, 5) carbon nanotubes observed by temperature-dependent photoluminescence spectroscopy*, **Phys. Rev. Lett. 118**, 027402.

Ma, X. C., E. G. Wang, W. Zhou, D. A. Jefferson, J. Chen, S. Z. Deng, N. S. Xu, and J. Yuan, 1999, *Polymerized carbon nitrogen nanobells and their field emission*, **Appl. Phys. Lett. 75**, 3105.

Ma, Y. M., M. Eremets, A. R. Oganov, Y. Xie, I. Trojan, S. Medvedev, A. O. Lyakhov, M. Valle, and V. Prakapenka, 2009, *Transparent dense sodium*, **Nature 458**, 182.

Maali, A., T. Cohen-Bouhacina, and H. Kellay, 2008, *Measurement of the slip length of water flow on graphite surface*, **Appl. Phys. Lett. 92**, 2007.

Mack, S. A., S. M. Griffin, and J. B. Neaton, 2019, *Emergence of topological electronic phases in elemental lithium under pressure*, **Proc. Natl. Acad. Sci. U. S. A. 116**, 9197.

Madelung, O., 1982, *Physics of group IV elements and III–IV compounds* (Springer-Verlag, Berlin).

Madeo, J., M. K. L. Man, C. Sahoo, M. Campbell, V. Pareek, E. L. Wong, A. Al-Mahboob, N. S. Chan, A. Karmakar, B. M. K. Mariserla, X. Q. Li, T. F. Heinz, T. Cao, and K. M. Dani, 2020, *Directly visualizing the momentum-forbidden dark excitons and their dynamics in atomically thin semiconductors*, **Science 370**, 1199.

Mahan G., 1998, *Many-particle physics* (Plenum, New York).

Maier, S., and M. Salmeron, 2015, *How does water wet a surface?* **Acc. Chem. Res. 48**, 2783.

Maiman, T. H., 1960, *Stimulated optical radiation in ruby*, **Nature 187**, 493.

Maity, A., T. C. Doan, J. Li, J. Y. Lin, and H. X. Jiang, 2016, *Realization of highly efficient hexagonal boron nitride neutron detectors*, **Appl. Phys. Lett. 109**, 072101.

Maity, A., S. J. Grenadier, J. Li, J. Y. Lin, and H. X. Jiang, 2017, *Toward achieving flexible and high sensitivity hexagonal boron nitride neutron detectors*, **Appl. Phys. Lett. 111**, 033507.

Maksyutenko, P., T. R. Rizzo, and O. V. Boyarkin, 2006, *A direct measurement of the dissociation energy of water*, **J. Chem. Phys. 125**, 181101.

Malard, L. M., D. C. Elias, E. S. Alves, and M. A. Pimenta, 2008, *Observation of distinct electron-phonon couplings in gated bilayer graphene*, **Phys. Rev. Lett. 101**, 257401.

Mallikarjunaiah, K. J., K. C. Paramita, K. P. Ramesh, and R. Damle, 2007, *Study of molecular reorientation and quantum rotational tunneling in tetramethylammonium selenate by 1H NMR*, **Solid State Nucl. Magn. Reson. 32**, 11.

Mallikarjunaiah, K. J., K. Jugeshwar Singh, K. P. Ramesh, and R. Damle, 2008, *1H NMR study of internal motions and quantum rotational tunneling in $(CH_3)_4NGeCl_3$*, **Magn. Reson. Chem. 46**, 110.

Mamin, H. J., M. Kim, M. H. Sherwood, C. T. Rettner, K. Ohno, D. D. Awschalom, and D. Rugar, 2013, *Nanoscale nuclear magnetic resonance with a nitrogen-vacancy spin sensor*, **Science 339**, 557.

Mankowsky, R., A. Subedi, M. Forst, S. O. Mariager, M. Chollet, H. T. Lemke, J. S. Robinson, J. M. Glownia, M. P. Minitti, A. Frano, M. Fechner, N. A. Spaldin, T. Loew, B. Keimer, A. Georges, and A. Cavalleri1, 2014, *Nonlinear lattice dynamics as a basis for enhanced superconductivity in $YBa_2Cu_3O_{6.5}$*, **Nature 516**, 71.

Mannix, A. J., X. F. Zhou, B. Kiraly, J. D. Wood, D. Alducin, B. D Myers., X. L Liu., B. L. Fisher, U. Santiago, J. R. Guest, M. J. Yacaman, A. Ponce, A. R. Oganov, M. C. Hersam, and N. P. Guisinger, 2015, *Synthesis of borophenes: Anisotropic, two-dimensional boron polymorphs*, **Science 350**, 1513.

Mao, H. K., and R. J. Hemley, 1994, *Ultrahigh-pressure transitions in solid hydrogen*, **Rev. Mod. Phys. 66**, 671.

Mao, H. K., X. J. Chen, Y. Ding, B. Li, and L. Wang, 2018, *Solids, liquids, and gases under high pressure*, **Rev. Mod. Phys. 90**, 015007.

Marcus, R. A., and M. E. Coltrin, 1977, *A new tunneling path for reactions such as $H+H_2$àH_2+H*, **J. Chem. Phys. 67**, 2609.

Marcus, R. A., 1993, *Electron transfer reactions in chemistry: Theory and experiment*, **Rev. Mod. Phys. 65**, 599.

Marini, A., 2008, *Ab initio finite-temperature excitons*, **Phys. Rev. Lett. 101**, 106405.

Markland, T. E., and M. Ceriotti, 2018, *Nuclear quantum effects enter the mainstream*, **Nat. Rev. Chem. 2**, 0109.

Marom, N., 2016, *Accurate description of the electronic structure of organic semiconductors by GW methods*, **J. Phys.: Condens. Matter 29**, 103003.

Marqués, M., M. I. McMahon, E. Gregoryanz, M. Hanfland, C. L. Guillaume, C. J. Pickard, G. J. Ackland, and R. J. Nelmes, 2011, *Crystal structures of dense lithium: A metal-semiconductor-metal transition*, **Phys. Rev. Lett. 106**, 095502.

Martin, R. M., 1972, *Piezoelectricity*, **Phys. Rev. B 5**, 1607.

Martin, R. M., 2004, *Electronic structure, basic theory and practical methods* (University of Princeton, Cambridge).

Martin, L. W., and A. M. Rappe, 2016, *Thin-film ferroelectric materials and their applications*, **Nat. Rev. Mater. 2**, 16087.

Martyna, G. J., M. L. Klein, and M. E. Tuckerman, 1992, *Nosé-Hoover chains: the canonical ensemble via continuous dynamics*, **J. Chem. Phys. 97**, 2635.

Marx, D., and M. Parrinello, 1996, *Ab initio path integral molecular dynamics: Basic ideas*, **J. Chem. Phys. 104**, 4077.

Marx, D., 2006, *Proton transfer 200 years after von Grotthuss: Insights from ab initio simulations*, **ChemPhysChem 7**, 1848.

Marx, D., A. Chandra, and M. E. Tuckerman, 2010, *Aqueous basic solutions: Hydroxide solvation, structural diffusion, and comparison to the hydrated proton*, **Chem. Rev. 110**, 2174.

Mason, P. E., H. C. Schewe, T. Buttersack, V. Kostal, M. Vitek, R. S. McMullen, H. Ali, F. Triter, C. Lee, D. M. Neumark, S. Thurmer, R. Seidel, B. Winter, S. E. Bradforth, and P. Jungwirth, 2021, *Spectroscopic evidence for a gold-coloured metallic water solution*, **Nature 595**, 673.

Mastel, S., A. A. Goyyadinov, C. Maisseni, A. Chuvilin, A. Berger, and R. Hillenbrand, 2018, *Understanding the image contrast of material boundaries in IR nanoscopy reaching 5 nm spatial resolution*, **ACS Photonics 5**, 3372.

Matsuoka, T., and K. Shimizu, 2009, *Direct observation of a pressure-induced metal-to-semiconductor transition in lithium*, **Nature 458**, 186.

Mattsson, T. R., U. Engberg, and G. Wahnstrom, 1993, *H-diffusion on Ni(100) – a quantum Monte-Carlo simulation*, **Phys. Rev. Lett. 71**, 2615.

Mattsson, T. R., and G. Wahnstrom, 1995, *Quantum Monte-Carlo study of surface diffusion*, **Phys. Rev. B 51**, 1885.

Mattsson, T. R., and G. Wahnstrom, 1997, *Isotope effect in hydrogen surface diffusion*, **Phys. Rev. B 56**, 14944.

Mauritz, K. A., and R. B. Moore, 2004, *State of understanding of Nafion*, **Chem. Rev. 104**, 4535.

Mazin, I. I., R. J. Hemley, A. F. Goncharov, M. Handfland, and H. Mao, 1997, *Quantum and classical orientational ordering in solid hydrogen*, **Phys. Rev. Lett. 78**, 1066.

Mazzola, G., and S. Sorella, 2015, *Distinct metallization and atomization transitions in dense liquid hydrogen*, **Phys. Rev. Lett. 114**, 105701.

McCoy, A. B., X. Huang, S. Carter, and J. M. Bowman, 2005, *Quantum studies of the vibrations in H_3O_2- and D_3O_2-*, **J. Chem. Phys. 123**, 064317.

McKenzie, R. H., C. Bekker, B. Athokpam, and S. G. Ramesh, 2014, *Effect of quantum nuclear motion on hydrogen bonding*, **J. Chem. Phys. 140**, 174508.

McMahon, J. M., and D. M. Ceperley, 2011, *Ground-state structures of atomic metallic hydrogen*, **Phys. Rev. Lett. 106**, 165302.

McMillan, W. L., 1965, *Ground state of liquid He_4*, **Phys. Rev. 138**, A442.

McMillan, W. L., 1968, *Transition temperature of strong-coupled superconductors*, **Phys. Rev. 167**, 331.

Mead, C. A., and D. G. Truhlar, 1979, *On the determination of Born–Oppenheimer nuclear motion wave functions including complications due to conical intersections and identical nuclei*, **J. Chem. Phys. 70**, 2284.

Medvedev, S., T. M. McQueen, I. A. Troyan, T. Palasyuk, M. I. Eremets, R. J. Cava, S. Naghavi, F. Casper, V. Ksenofontov, G. Wortmann, and C. Felser, 2009, *Electronic and magnetic phase diagram of β-$Fe_{1\cdot01}$Se with superconductivity at 36.7 K under pressure*, **Nat. Mater. 8**, 630.

Meier, T., S. Petitgirard, S. Khandarkhaeva, and L. Dubrovinsky, 2018, *Observation of nuclear quantum effects and hydrogen bond symmetrisation in high pressure ice*, **Nat. Commun. 9**, 2766.

Meisner, J., and J. Kästner, 2016, *Atom tunneling in chemistry*, **Angew. Chem. Int. Ed. 55**, 5400.

Mendoza, M., S. Succi, and H. J. Herrmann, 2014, *Kinetic formulation of the Kohn-Sham equations for ab initio electronic structure calculations*, **Phys. Rev. Lett. 113**, 096402.

Meng, J. Q., G. Liu, W. Zhang, L. Zhao, H. Liu, X. Jia, D. Mu, S. Liu, X. Dong, J. Zhang, W. Lu, G. Wang, Y. Zhou, Y. Zhu, X. Wang, Z. Xu, C. Chen, and X. J. Zhou, 2009, *Coexistence of Fermi arcs and Fermi pockets in a high-Tc copper oxide superconductor*, **Nature 462**, 335.

Meng, S., L. F. Xu, E. G. Wang, and S. W. Gao, 2002, *Vibrational recognition of hydrogen-bonded water networks on a metal surface*, **Phys. Rev. Lett. 89**, 176104.

Meng, S., E. G. Wang, and S. W. Gao, 2004, *Water adsorption on metal surfaces: a general picture from density functional theory studies*, **Phys. Rev. B 69**, 195404.

Meng, S., and E. Kaxiras, 2008a, *Real-time, local basis-set implementation of time-dependent density functional theory for excited state dynamics simulations*, **J. Chem. Phys. 129**, 054110.

Meng, S., J. Ren, and E. Kaxiras, 2008b, *Natural dyes adsorbed on TiO_2 nanowire for photovoltaic applications: Enhanced light absorption and ultrafast electron injection.* **Nano Lett. 8**, 3266.

Meng, S., and E. Kaxiras, 2010, *Electron and hole dynamics in dye-sensitized solar cells: Influencing factors and systematic trends*, **Nano Lett. 10**, 1238.

Meng, S., and E. G. Wang, 2023, *Water: Basic science* (Springer, Singapore).

Meng, X. Z., J. Guo, J. B. Peng, J. Chen, Z. C. Wang, J. R. Shi, X. Z. Li, E. G. Wang, and Y. Jiang, 2015, *Direct visualization of concerted proton tunnelling in a water nanocluster*, **Nat. Phys. 11**, 235.

Menzeleev, A. R., F. Bell, and T. F. Miller, 2014, *Kinetically constrained ring-polymer molecular dynamics for non-adiabatic chemical reactions*, **J. Chem. Phys. 140**, 064103.

Mermin, N. D., 1965, *Thermal properties of the inhomogeneous electron gas*, **Phys. Rev. 137**, A1441.

Metropolis, N., A. W. Rosenbluth, M. N. Rosenbluth, N. Marshall, A. H. Teller, and W. Teller, 1953, *Equation of state calculations by fast computing machines*, **J. Chem. Phys. 21**, 1087.

Michaelides, A., and K. Morgenstern, 2007, *Ice nanoclusters at hydrophobic metal surfaces*, **Nat. Mater. 6**, 597.

Migdal, A. B., 1958, *Interaction between slectrons and lattice vibrations in a normal metal*, **Sov. Phys. JETP 7**, 996.

Miller, D. A. B, D. S. Chemla, T. C. Damen, A. C. Gossard, W. Wiegamn, T. H. Wood, and C. A. Burrus, 1985, *Electric field dependence of optical absorption near the band gap of quantum-well structures*, **Phys. Rev. B 32**, 1043.

Miller, W. H., 1970, *Classical S matrix: Numerical application to inelastic collisions*, **J. Chem. Phys. 53**, 3578.

Miller, W. H., 1974, *Quantum mechanical transition state theory and a new semiclassical model for reaction rate constants*, **J. Chem. Phys. 61**, 1823.

Miller, W. H., 1975, *Semiclassical limit of quantum mechanical transition state theory for nonseparable systems*, **J. Chem. Phys. 62**, 1899.

Miller, W. H., S. D. Schwartz, and J. W. Tromp, 1983, *Quantum mechanical rate constants for bimolecular reactions*, **J. Chem. Phys. 79**, 4889.

Miller, W. H., 2001, *The semiclassical initial value representation: A potentially practical way for adding quantum effects to classical molecular dynamics simulations*, **J. Phys. Chem. A 105**, 2942.

Miller, W. H., Y. Zhao, M. Ceotto, and S. Yang, 2003, *Quantum instanton approximation for thermal rate constants of chemical reactions*, **J. Chem. Phys. 119**, 1329.

Milonni, P. W., and M. L. Shih, 1991, *Zero-point energy in early quantum-theory*, **Am. J. Phys. 59**, 684.

Mishra, H., and S. Bhattacharya, 2019, *Giant exciton-phonon coupling and zero-point renormalization in hexagonal monolayer boron nitride*, **Phys. Rev. B 99**, 165201.

Mitsuhashi, R., Y. Suzuki, Y. Yamanari, H. Mitamura, T. Kambe, N. Ikeda, H. Okamoto, A. Fujiwara, M. Yamaji, N. Kawasaki, Y. Maniwa, and Y. Kubozono, 2010, *Superconductivity in alkali-metal-doped picene*, **Nature 464**, 76.

Mitsui, T., M. K. Rose, E. Fomin, D. F. Ogletree, and M. Salmeron, 2002, *Water diffusion and clustering on Pd(111)*. **Science 297**, 1850.

Miyata, A., H. Ueda, Y. Ueda, H. Sawabe, and S. Takeyama, 2011, *Magnetic phases of a highly frustrated magnet, $ZnCr_2O_4$, up to an ultrahigh magnetic field of 600 T*, **Phys. Rev. Lett. 107**, 207203.

Miyazaki, T., 2004, *Atom tunneling phenomena in physics, chemistry and biology* (Springer, Berlin).

Mogg, L., G.-P. Hao, S. Zhang. C. Bacaksiz, Y.-C. Zou, S. J. Haigh, F. M. Peeters, A. K. Gerim, and M. Lozada-Hidalgo, 2019a, *Atomically thin micas as proton-conducting membranes*, **Nat. Nanotech. 14**, 962.

Mogg, L., S. Zhang, G.-P. Hao, K. Gopinadhan, D. Barry, B. L. Liu, H. M. Cheng, A. K. Gerim, and M. Lozada-Hidalgo, 2019b, *Perfect proton selectivity in ion transport through two-dimensional crystals*, **Nat. Commun. 10**, 4243.

Moller, C., and M. S. Plesset, 1934, *Note on an Approximation Treatment for Many-electron Systems*, **Phys. Rev. 46**, 618.

Monacelli, L., I. Errea, M. Calandra, and F. Mauri, 2021, *Black metal hydrogen above 360 GPa driven by proton quantum fluctuations*, **Nat. Phys. 17**, 63.

Monacelli, L., M. Casula, K. Nakano, S. Sorella, and F. Mauri, 2023, *Quantum phase diagram of high-pressure hydrogen*, **Nat. Phys. 19**, 845.

Monserrat, B., 2016, *Vibrational averages along thermal lines*, **Phys. Rev. B 93**, 014302.

Moore, E. B., and V. Molinero, 2011, *Is it cubic? Ice crystallization from deeply supercooled water*, **Phys. Chem. Chem. Phys. 13**, 20008.

Morales, M. A., C. Pierleoni, E. Schwegler, and D. M. Ceperley, 2010, *Evidence for a first-order liquid-liquid transition in high-pressure hydrogen from Ab initio simulations*, **Proc. Natl. Acad. Sci. U. S. A. 107**, 12799.

Morales, M. A., J. M. McMahon, C. Pierleoni, and D. M. Ceperley, 2013, *Nuclear quantum effects and nonlocal exchange-correlation functionals applied to liquid hydrogen at high pressure*, **Phys. Rev. Lett. 110**, 065702.

Morrone, J. A., and R. Car, 2008, *Nuclear quantum effects in water*, **Phys. Rev. Lett. 101**, 017801.

Mukherjee, S., L. Kavalsky, K. Chattopadhyay, and C. V. Singh, 2018, *Adsorption and diffusion of lithium polysulfides over blue phosphorene for Li-S batteries*, **Nanoscale 10**, 21335.

Müller, K., and A. H. Burkard, 1979, *$SrTiO_3$: An intrinsic quantum paraelectric below 4 K*, **Phys. Rev. B 19**, 3593.

Murakami, M., K. Hirose, K. Kawamura, N. Sata, and Y. Ohishi, 2004, *Post-perovskite phase transition in $MgSiO_3$*, **Science 304**, 855.

N

Nagamatsu, J., N. Nakagawa, T. Muranaka, Y. Zenitani, and J. Akimitsu, 2001, *Superconductivity at 39 K in magnesium diboride*, **Nature 410**, 63.

Nagashima, H., S. Tsuda, N. Tsuboi, A. K. Hayashi, and T. Tokumasu, 2017, *A molecular dynamics study of nuclear quantum effect on diffusivity of hydrogen molecule*, **J. Chem. Phys. 147**, 024501.

Nagata, Y., R. E. Pool, E. H. G. Backus, and M. Bonn, 2012, *Nuclear quantum effects affect bond orientation of water at the water-vapor interface*, **Phys. Rev. Lett. 109**, 226101.

Nakamura, M., K. Tamura, and S. Murakami, 1995, *Isotope effects on thermodynamic properties-mixtures of x(D_2O or H_2O)+(1-x)CH_3CN at 298.15 K*, **Thermochim. Acta 253**, 127.

Narten, A. H., and H. A. Levy, 1969, *Observed diffraction pattern and proposed models of liquid water*, **Science 165**, 447.

Navrotsky, A., 1998, *Energetics and crystal chemical systematics among ilmenite, lithium niobate, and perovskite structures*, **Chem. Mater. 10**, 2787.

Negele, J. W., and H. Orland, 1998, *Quantum many-particle systems (Frontiers in Physics)* (Perseus Books, Reading, MA).

Nellis, W. J., S. T. Weir, and A. C. Mitchell, 1999, *Minimum metallic conductivity of fluid hydrogen at 140 GPa*, **Phys. Rev. B 59**, 3434.

Ni, K.-K., S. Ospelkaus, M. Miranda, A. Pe'er, B. Neyenhuis, J. Zirbel, S. Kotochigova, P. Julienne, D. Jin, and J. Ye, 2008, *A high phase-space-density gas of polar molecules*, **Science 322**, 5899.

Nie, S., P. J. Feibelman, N. C. Bartelt, and K. Thurmer, 2010, *Pentagons and heptagons in the first water layer on Pt(111)*, **Phys. Rev. Lett. 105**, 026102.

Nilsson, A., D. Nordlund, I, Waluyo, N. Huang, H. Ogasawara, S. Kaya, U. Bergmann, L.-Å. Näslund, H. Öström, Ph. Wernet, K. J. Andersson, T. Schiros, and L. G. M. Pettersson, 2010, *X-ray absorption spectroscopy and X-ray Raman scattering of water and ice; an experimental view*, **J. Electron Spectrosc. Relat. Phenom. 177**, 99.

Nilsson, F., and F. Aryasetiawan, 2018, *Recent progress in first-principles methods for computing the electronic structure of correlated materials*, **Computation 6**, 26.

Noble, D. L., A. Aibout, and A. J. Horsewill, 2009, *^{1}H-^{19}F spin-lattice relaxation spectroscopy: Proton tunnelling in the hydrogen bond studied by field-cycling NMR*, **J. Magn. Reson. 201**, 157.

Noh, G., D. Choi, J. H. Kim, D. G. Im, Y. H. Kim, H. Seo, and J. Lee, 2018, *Stark tuning of single-photon emitters in hexagonal boron nitride*, **Nano Letter 18**, 4710.

Nose, S., 1984a, *A molecular dynamics method for simulations in the canonical ensemble*, **Mol. Phys. 52**, 255.

Nose, S., 1984b, *A unified formulation of the constant temperature molecular dynamics methods*, **J. Chem. Phys. 81**, 511.

Nose, S., 1986, *An extension of the canonical ensemble molecular dynamics method*, **Mol. Phys. 57**, 187.

Nova, T. F., A. S. Disa, M. Fechner, and A. Cavalleri, 2019, *Metastable ferroelectricity in optically strained $SrTiO_3$*, **Science 364**, 1075.

Novoselov, K. S., A. K. Geim, S. V. Morozov, D. Jiang, Y. Zhang, S. V. Dubonos, I. V. Grigorieva, and A. A. Firsov, 2004, *Electric field effect in atomically thin carbon films*, **Science 306**, 666.

Novoselov, K. S., A. K. Geim, S. V. Morozov, D. Jiang, M. I. Katsnelson, I. V. Grigorieva, S. V. Dubonos, and A. A. Firsov, 2005, *Two-dimensional gas of massless Dirac fermions in graphene*, **Nature 438**, 197.

Novoselov, K. S., Z. Jiang, Y. Zhang, S. V. Morozov, H. L. Stormer, U. Zeitler, J. C. Maan, G. S. Boebinger, P. Kim, and A. K. Geim, 2007, *Room-temperature quantum hall effect in graphene*, **Science 315**, 1379.

O

Oganov, A. R., and S. Ono, 2004, *Theoretical and experimental evidence for a post-perovskite phase of $MgSiO_3$ in Earth's D'' layer*, **Nature 430**, 445.

Onnes, H. K., 1911, *The superconductivity of mercury*, **Comm. Phys. Lab. Univ. Leiden**, **122**, 122–124.

Osheroff, D. D., R. C. Richardson, and D. M. Lee, 1972, *Evidence for a new phase of solid He^3*, **Phys. Rev. Lett. 28**, 885.

Overhauser, A. W., 1984, *Crystal structure of lithium at 4.2K*, **Phys. Rev. Lett. 53**, 64.

P

Paesani, F., and G. A. Voth, 2009, *The properties of water: Insights from quantum simulations*, **J. Phys. Chem. B 113**, 5702.

Palstra, T. T. M., O. Zhou, Y. Iwasa, P. E. Sulewski, R. M. Fleming, and B. R. Zegarski, 1995, *Superconductivity at 40K in cesium doped C_{60}*, **Solid State Commun. 93**, 327.

Pamuk, B., J. M. Soler, R. Ramirez, C. P. Herrero, P. W. Stephens, P. B. Allen, and M. V. Fernandez-Serra, 2012, *Anomalous nuclear quantum effects in ice*, **Phys. Rev. Lett. 108**, 193003.

Pan, D., L. M. Liu, G. A. Tribello, B. Slater, A. Michaelides, and E. G. Wang, 2008, *Surface energy and surface proton order of ice Ih*, **Phys. Rev. Lett. 101**, 155703.

Papaconstantopoulos, D. A., B. M. Klein, M. J. Mehl, and W. E. Pickett, 2015, *Cubic H_3S around 200 GPa: An atomic hydrogen superconductor stabilized by sulfur*, **Phys. Rev. B 91**, 184511.

Parr, R. G., and W. T. Yang, 1989, *Density-functional theory of atoms and molecules* (Oxford University Press, Oxford).

Patrick, C. E., and F. Giustino, 2013, *Quantum nuclear dynamics in the photophysics of diamondoids*, **Nat. Commun. 4**, 2006.

Patrick, C. E., and F. Giustino, 2014, *Unified theory of electron-phonon renormalization and phonon-assisted optical absorption*, **J. Phys.: Condens. Matter 26**, 365503.

Paul, P. M., E. S. Toma, P. Breger, G. Mullot, F. Auge, Ph. Balcou, H. G. Muller, and P. Agostini, 2001, *Observation of a train of attosecond pulses from high harmonic generation*, **Science 292**, 1689.

Pavone, P., K. Karch, O. Schutt, W. Windl, D. Strauch, P. Giannozzi, and S. Baroni, 1993, *Ab-initio lattice-dynamics of diamond*, **Phys. Rev. B 48**, 3156.

Peng, B., H. Zhang, H. Z. Shao, Z. Y. Ning, Y. F. Xu, G. Ni, H. L. Lu, D. W. Zhang, and H. Y. Zhu, 2017, *Stability and strength of atomically thin borophene from first principles calculations*, **Mater. Res. Lett. 5**, 399.

Peng, F., Y. Sun, C. J. Pickard, R. J. Needs, Q. Wu, and Y. Ma, 2017, *Hydrogen clathrate structures in rare earth hydrides at high pressures: Possible route to room-temperature superconductivity*, **Phys. Rev. Lett. 119**, 107001.

Peng, J. B., D. Y. Cao, Z. L. He, J. Guo, P. Hapala, R. Z. Ma, B. W. Cheng, J. Chen, W. J. Xie, X. Z. Li, P. Jelínek, L. M. Xu, Y. Q. Gao, E. G. Wang, and Y. Jiang, 2018a, *The effect of hydration number on the interfacial transport of sodium ions*, **Nature 557**, 701.

Peng, J. B., J. Guo, P. Hapala, D. Y. Cao, R. Z. Ma, B. W. Cheng, L. M. Xu, M. Ondracek, P. Jelinek, E. G. Wang, and Y. Jiang, 2018b, *Weakly Perturbative imaging of interfacial water with submolecular resolution by atomic force microscopy*, **Nat. Commun. 9**, 122.

Peng, S., H. Zhang, S. Yan, W. Zhang, J. Fu, X. Lin, S. Hao, Z. Jin, Y. Zhang, C. Zhang, F. Miao, S. J. Liang, and G. Ma, 2021, *Observation of negative terahertz photoconductivity in large area type-II Dirac semimetal $PtTe_2$*, **Phys. Rev. Lett. 126**, 227402.

Perdew, J. P., and A. Zunger, 1981, *Self-interaction correction to density-functional approximations for many-electron systems*, **Phys. Rev. B 23**, 5048.

Pérez, A., and O. A. von Lilienfeld, 2011, *Path integral computation of quantum free energy differences due to alchemical transformations involving mass and potential*, **J. Chem. Theory Comput. 7**, 2358.

Persson, B. N. J., and A. Baratoff, 1987, *Inelastic electon tunneling from a metal tip: The contribution from resonant processes*, **Phys. Rev. Lett. 59**, 339.

Peterson, A. A., F. Abild-Pedersen, F. Studt, J. Rossmeisl, and J. K. Norskov, 2010, *How copper catalyzes the electroreduction of carbon dioxide into hydrocarbon fuels*, **Energ. Environ. Sci. 3**, 1311.

Phillip, H. R., and E. A. Taft, 1964, *Kramers-Kronig analysis of reflectance data for diamond*, **Phys. Rev. A 136**, 1445.

Pickard, C. J., and R. J. Needs, 2007, *Structure of phase III of solid hydrogen*. **Nat. Phys. 3**, 473.

Pickard, C. J., and R. J. Needs, 2009, *Dense low-coordination phases of lithium*, **Phys. Rev. Lett. 102**, 146401.

Pietropaolo, A., R. Senesi, C. Andreani, A. Botti, M. A. Ricci, and F. Bruni, 2008, *Excess of proton mean kinetic energy in supercooled water*, **Phys. Rev. Lett. 100**, 127802.

Pietropaolo, A., Roberto Senesi, Carla Andreani, Jerry Mayers, 2009, *Quantum effects in water proton kinetic energy maxima in stable and supercooled liquid*, **Braz. J. Phys. 39**, 319.

Pisana, S., M. Lazzeri, C. Casiraghi, K. S. Novoselov, A. K. Geim, A. C. Ferrari, and F. Mauri, 2007, *Breakdown of the adiabatic Born-Oppenheimer approximation in graphene*, **Nat. Mater. 6**, 198.

Planck, M., 1900a, *On an improvement of Wein's equation for the spectrum*, **Verh. Dtsch. Phys. Ges. Berlin 2**, 202.

Planck, M., 1900b, *On the theory of the energy distribution law of the normal spestrum*, **Verh. Dtsch. Phys. Ges. Berlin 2**, 237.

Polian, A., and M. Grimsditch, 1984, *New high-pressure phase of H_2O: Ice X*, **Phys. Rev. Lett. 52**, 1312.

Pollock, E. L., and D. M. Ceperley, 1984, *Simulation of quantum many-body systems by path-integral methods*, **Phys. Rev. B 30**, 2555.

Pollock, E. L., and D. M. Ceperley, 1987, *Path-integral computation of superfluid densities*, **Phys. Rev. B 36**, 8343.

Poltavsky, L., L. Zheng, M. Mortazavi, and A. Tkatchen, 2014, *Quantum tunneling of thermal protons through pristine graphene*, **arXiv:1605.06341,** DOI: https://doi.org/10.48550/arXiv.1605.06341.

Poltavsky, L., L. Zheng, M. Mortazavi, and A. Tkatchen, 2018, *Quantum tunneling of thermal protons through pristine graphene*, **J. Chem. Phys. 148**, 204707.

Poncé, S., G. Antonius, P. Boulanger, E. Cannuccia, A. Marini, M. Côté, and X. Gonze, 2014, *Verification of first-principles codes: Comparison of total energies, phonon frequencies, electron-phonon coupling and zero-point motion correction to the gap between ABINIT and QE/Yambo*, **Comput. Mater. Sci. 83**, 341.

Poncé, S., Y. Gillet, J. L. Janssen, A. Marini, M. Verstraete, and X. Gonze, 2015, *Temperature dependence of the electronic structure of semiconductors and insulators*, **J. Chem. Phys. 143**, 102813.

Pozzo, M., M. P. Desjarlais, and D. Alfè, 2011, *Electrical and thermal conductivity of liquid sodium from first-principles calculations*, **Phys. Rev. B 84**, 054203.

Prager, M., and A. Heidemann, 1997, *Rotational tunneling and neutron spectroscopy: A compilation*, **Chem. Rev. 97**, 2933.

Price, W. S., H. Ide, Y. Arata, and O. Soderman, 2000, *Temperature dependence of the self-diffusion of supercooled heavy water to 244 K*, **J. Phys. Chem. B 104**, 5874.

Pugnat, P., and H. J. Schneider-Muntau, 2014, *Hybrid magnets-past, present, and future*, **IEEE Trans. Appl. Supercon. 24**, 4300106.

Purvis III, G. D., and R. J. Barlett, 1982, *A full coupled-cluster singles and doubles model: The inclusion of disconnected triples*, **J. Chem. Phys. 76**, 1910.

Q

Qi, R. S., N. Li, J. Du, R. Shi, Y. Huang, X. Yang, L. Liu, Z. Xu, Q. Dai, D. Yu, and P. Gao, 2021a, *Four-dimensional vibrational spectroscopy for nanoscale mapping of phonon dispersion in BN nanotubes*, **Nat. Commun. 12**, 1179.

Qi, R. S., R. Shi, Y. Li, Y. Sun, M. Wu, N. Li, J. Du, K. Liu, C. Chen, J. Chen, F. Wang, D. Yu, E. Wang, and P. Gao, 2021b, *Measuring phonon dispersion at an interface*, **Nature 599**, 399.

Qi, X. L., and S. C. Zhang, 2011, *Topological insulators and superconductors*, **Rev. Mod. Phys. 83**, 1057.

R

Rahman, A., 1964, *Correlations in the motion of atoms in liquid argon*, **Phys. Rev. A 136**, 405.

Ramasesha, K., L. De Marco, A. Mandal, and A. Tokmakoff, 2013, *Water vibrations have strongly mixed intra- and intermolecular character*, **Nat. Chem. 5**, 935.

Ramirez, A. P., M. A. Subramanian, M. Gardel, G. Blumberg, D. Li, T. Vogt, and S. M. Shapiro, 2000, *Giant dielectric constant response in a copper-titanate*, **Solid State Commun. 115**, 217.

Ramirez, R., and L. M. Falicov, 1971, *Theory of Alpha-Gamma phase transition in metallic cerium*, **Phys. Rev. B 3**, 2425.

Ramírez, R., C. P. Herrero, and E. R. Hernández, 2006, *Path-integral molecular dynamics simulation of diamond*, **Phys. Rev. B 73**, 245202.

Ranea, V. A., A. Michaelides, R. Ramírez, P. L. de Andres, J. A. Vergés, and D. A. King, 2004, *Water dimer diffusion on Pd(111) assisted by an H-bond donor-acceptor tunneling exchange*, **Phys. Rev. Lett. 92**, 136104.

Ray, M. W., and R. B. Hallock, 2011, *Mass flow through solid ^{4}He induced by the fountain effect*, **Phys. Rev. B 84**, 144512.

Rayleigh, J. W. S. Baron, 1899, *Scientific papers,* Vol. 6 (Cambridge University Press, Cambridge).

Rees, D. G, N. R. Beysengulov, J.-J. Lin, and K. Kono, 2016, *Stick-slip motion of the Wigner solid on liquid helium*, **Phys. Rev. Lett. 116**, 206801.

Reiter, G. F., J. Mayers, and J. Noreland, 2002, *Momentum-distribution spectroscopy using deep inelastic neutron scattering*, **Phys. Rev. B 65**, 104305.

Reiter, G. F., J. C. Li, J. Mayers, T. Abdul-Redah, and P. Platzman, 2004, *The proton momentum distribution in water and ice*, **Braz. J. Phys. 34**, 142.

Reiter, G. F., C. Burnham, D. Homouz, P. M. Platzman, J. Mayers, T. Abdul-Redah, A. P. Moravsky, J. C. Li, C. K. Loong, and A. I. Kolesnikov, 2006, *Anomalous behavior of proton zero point motion in water confined in carbon nanotubes*, **Phys. Rev. Lett. 97**, 247801.

Reiter, G. F., A. I. Kolesnikov, S. J. Paddison, P. M. Platzman, A. P. Moravsky, M. A. Adams, and J. Mayers, 2012, *Evidence for an anomalous quantum state of protons in nanoconfined water*, **Phys. Rev. B 85**, 045403.

Reiter, G. F., A. Deb, Y. Sakurai, M. Itou, V. G. Krishnan, and S. J. Paddison, 2013, *Anomalous ground state of the electrons in nanoconfined water*, **Phys. Rev. Lett. 111**, 036803.

Ren, W. L., W. Z. Fu, and J. Chen, 2022, *Towards the ground state of molecules via diffusion Monte Carlo on neural networks*, DOI: **10.48550/arXiv.2204.13903**.

Repp, J., G. Meyer, K. H. Rieder, and P. Hyldgaard, 2003, *Site determination and thermally assisted tunneling in homogenous nucleation*, **Phys. Rev. Lett. 91**, 206102.

Richardson, J. O., and S. C. Althorpe, 2009, *Ring-polymer molecular dynamics rate-theory in the deep-tunneling regime: Connection with semiclassical instanton theory*, **J. Chem. Phys. 131**, 214106.

Richardson, J. O., and S. C. Althorpe, 2011, *Ring-polymer instanton method for calculating tunneling splittings*, **J. Chem. Phys. 134**, 054109.

Richardson, J. O., and M. Thoss, 2013, *Communication: Nonadiabatic ring-polymer molecular dynamics*, **J. Chem. Phys. 139**, 031102.

Richardson, J. O., R. Bauer, and M. Thoss, 2015, *Semiclassical Green's functions and an instanton formulation of electron-transfer rates in the nonadiabatic limit*. **J. Chem. Phys. 143**, 134115.

Richardson, J. O., 2016a, *Microcanonical and thermal instanton rate theory for chemical reactions at all temperatures*, **Faraday Discuss. 195**, 49.

Richardson, J. O., C. Pérez, S. Lobsiger, A. A. Reid, B. Temelso, G. C. Shields, Z. Kisiel, D. J. Wales, B. H. Pate, and S. C. Althorpe, 2016b, *Concerted hydrogen-bond breaking by quantum tunneling in the water hexamer prism*, **Science 351**, 1310.

Rocher-Casterline, B. E., A. K. Mollner, L. C. Ch'ng, and H. Reisler, 2011a, *Imaging H_2O photofragments in the predissociation of the HCl-H_2O hydrogenbonded dimer*, **J. Phys. Chem. A 115**, 6903.

Rocher-Casterline, B. E., L. C. Ch'ng, A. K. Mollner, and H. Reisler, 2011b, *Communication: Determination of the bond dissociation energy (D_0) of the water dimer, $(H_2O)_2$, by velocity map imaging*, **J. Chem. Phys. 134**, 211101.

Roothaan, C. C. J., 1951, *New developments in molecular orbital theory*, **Rev. Mod. Phys. 23**, 69.

Rousseau, R., and D. Marx, 1998, *Fluctuations and bonding in lithium clusters*, **Phys. Rev. Lett. 80**, 2574.

Rousseau, R., and D. Marx, 1999, *The role of quantum and thermal fluctuations upon properties of lithium clusters*, **J. Chem. Phys. 111**, 5091.

Roux, B., and M. Karplus, 1991a, *Ion transport in a model gramicidin channel. Structure and thermodynamics*, **Biophys. J. 59**, 961.

Roux, B., and M. Karplus, 1991b, *Ion transport in a gramicidin-like channel: Dynamics and mobility*, **J. Phys. Chem. 95**, 4856.

Roux, B., 1995, *The calculation of the potential of mean force using computer simulations*, **Comput. Phys. Commun. 91**, 275.

Runge, E., and E. K. U. Gross, 1984, *Density-functional theory for time-dependent systems*, **Phys. Rev. Lett. 52**, 997.

S

Sachdev, S., and B. Keimer, 2011, *Quantum criticality,* **Phys. Today 64**, 29.

Sachdev, S., 2012, *Quantum phase transitions of antiferromagnets and the cuprate superconductors, in Modern theories of many-particle systems in condensed matter physics* (**Springer**, Berlin Heidelberg).

Sahu, S, 2019, *Colloquium: Ionic phenomena in nanoscale pores through 2D materials*, **Rev. Mod. Phys. 91**, 021004.

Sai-Halasz, G. A., R. Tsu, and L. Esaki, 1977, *A new semiconductor superlattice*, **Appl. Phys. Lett. 30**, 651.

Sai-Halasz, G. A., L. L. Chang, J. M. Welter, C. A. Chang, and L. Esaki, 1978, *Optical absorption of $In_{1-x}Ga_xAs$-$GaSb_{1-y}As_y$ superlattices*, **Solid State Commun. 27**, 935.

Sakaushi, K., A. Lyalin, T. Taketsugu, and K. Uosaki, 2018, *Quantum-to-classical transition of proton transfer in potential-induced dioygen reduction*, **Phys. Rev. Lett. 121**, 236001.

Sala, F. D., R. Rousseau, A. Gorling, and D. Max, 2004, *Quantum-to-classical transition of proton transfer in potential-induced dioygen reduction*, **Phys. Rev. Lett. 121**, 236001.

Sano, W., T. Koretsune, T. Tadano, R. Akashi, and R. Arita, 2016, *Effect of Van Hove singularities on high-TC superconductivity in H_3S*, **Phys. Rev. B 93**, 094525.

Saparov, D., B. Xiong, Y. Ren, and Q. Niu, 2022, *Lattice dynamics with molecular Berry curvature: Chiral optical phonons,* **Phys. Rev. B. 105**, 064303.

Schaefer, J., E. H. G. Backus, Y. Nagata, and M. Bonn, 2016, *Both inter- and intramolecular coupling of O-H groups determine the vibrational response of the water/air interface*, **J. Phys. Chem. Lett. 7**, 4591.

Schaeffer, A. M. J., W. B. Talmadge, S. R. Temple, and S. Deemyad, 2012, *High pressure melting of lithium*, **Phys. Rev. Lett. 109**, 185702.

Schawlow, A. L., 1981, *Laser Spectroscopy*, **New Scientist 92**, 225.

Schlemmer, S., T. Kuhn, E. Lescop, and D. Gerlich, 1999, *Laser excited N_2^+ in a 22-pole ion trap: Experimental studies of rotational relaxation processes*, **Int. J. Mass Spectrom. 185**, 589.

Schneider-Muntau, H. J., 1981, *Polyhelix magnets*, **IEEE Trans. Magn. 17**, 1775.

Schoonmaker R., T. Lancaster, and S. J. Clark, 2018, *Quantum mechanical tunneling in the automerization of cyclobutadiene*, **J. Chem. Phys.**, 104109.

Schreck, S., and P. Wernet, 2016, *Isotope effects in liquid water probed by transmission mode x-ray absorption spectroscopy at the oxygen K-edge*, **J. Chem. Phys. 145**, 104502.

Schreiner, P. R., H. P. Reisenauer, D. Ley, D. Gerbig, C.-H. Wu, and W. D. Allen, 2011, *Methylhydroxycarbene: Tunneling control of a chemical reaction*, **Science 332**, 1300.

Schreiner, P. R., 2017, *Tunneling control of chemical reactions: The third reactivity paradigm*, **J. Am. Chem. Soc. 139**, 15276.

Schrödinger, E., 1926, *An undulatory theory of the mechanics of atoms and molecules*, **Phys. Rev. 28**, 1049.

Schrödinger, E., 1944, *What is life?* (Cambridge University Press, Cambridge).

Schuster, D. I., A. Fragner, M. I. Dykman, S. A. Lyon, and R. J. Schoelkopf, 2010, *Proposal for manipulating and detecting spin and orbital states of trapped electrons on helium using cavity quantum electrodynamics*, **Phys. Rev. Lett. 105**, 040503.

Scott, J. F., 1998, *The physics of ferroelectric ceramic thin films for memory applications*, **Ferroelectrics Rev. 1**, 1.

Scott, J. F., and M. Dawber, 2000, *Oxygen-vacancy ordering as a fatigue mechanism in perovskite ferroelectrics*, **Appl. Phys. Lett. 76**, 3801.

Secchi, E., S. Marbach, A. Nigues, D. Stein, A. Siria, and L. Bocquet, 2016, *Massive radius-dependent flow slippage in carbon nanotubes*, **Nature 537**, 210.

Seidl, A., A. Görling, P. Vogl, J. A. Majewski, and M. Levy, 1996, *Generalized Kohn-Sham schemes and the band-gap problem*, **Phys. Rev. B 53**, 3764.

Senesi, R., D. Flammini, A. I. Kolesnikov, E. D. Murray, G. Galli, and C. Andreani, 2013a, *The quantum nature of the OH stretching mode in ice and water probed by neutron scattering experiments*, **J. Chem. Phys. 139**, 074504.

Senesi, R., G. Romanelli, M. A. Adams and C. Andreani, 2013b, *Temperature dependence of the zero point kinetic energy in ice and water above room temperature,* **Chem. Phys. 427**, 111.

Senga, R., Y. C. Lin, S. Morishita, R. Kato, T. Yamada, M. Hasegawa, and K. Suenaga, 2022, *Imaging of isotope diffusion using atomic-scale vibrational spectroscopy*, **Nature 603**, 68.

Seol, J. H., I. Jo, A. L. Moore, L. Lindsay, Z. H. Aitken, M. T. Pettes, X. Li, Z. Yao, R. Huang, D. Broido, N. Mingo, R. S. Ruoff, and L. Shi, 2010, *Two-dimensional phonon transport in supported graphene*, **Science, 328**, 213.

Serwatka, T., R. G. Melko, A. Burkov, and P. N. Roy, 2023, *Quantum phase transition in the one-dimensional water chain*, **Phys. Rev. Lett. 130**, 026201.

Sevik, C., 2014, *Assessment on lattice thermal properties of two-dimensional honeycomb structures: Graphene, h-BN, h-MoS2, and h-MoSe2*, **Phys. Rev. B 89**, 035422.

Shakib, F. A., and P. Huo, 2017, *Ring polymer surface hopping: Incorporating nuclear quantum effects into nonadiabatic molecular dynamics simulations*, **J. Phys. Chem. Lett. 8**, 3073.

Shao, J. S., and N. Makri, 1999, *Forward-backward semiclassical dynamics with linear scaling*, **J. Phys. Chem. A 103**, 9479.

Shao, T., B. Wen, R. Melnik, S. Yao, Y. Kawazoe, and Y. Tian, 2012, *Temperature dependent elastic constants and ultimate strength of graphene and graphyne*, **J. Chem. Phys. 137**, 194901.

Shastry, B. S., 1988, *Exact solution of an S=1/2 Heisenberg antiferromagnetic chain with long ranged interactions*, **Phys. Rev. Lett. 60**, 639.

Shen, R., W. L. Guo, and W. Y. Zhong, 2010, *Hydration valve controlled non-selective conduction of Na^+ and K^+ in the NaK channel*, **Biochim. Biophys. Acata 1798**, 1474.

Shen, S. P., J. C. Wu, J. D. Song, X. F. Sun, Y. F. Yang, Y. S. Chai, D. S. Shang, S. G. Wang, J. F. Scott, and Y. Sun, 2016, *Quantum electric-dipole liquid on a triangular lattice*, **Nat. Commun. 7**, 10569.

Shen, Y. R., and V. Ostroverkhov, 2006, *Sum-frequency vibrational spectroscopy on water interfaces: polar orientation of water molecules at interfaces*, **Chem. Rev. 106**, 1140.

Sheng, S. X., J. B. Wu, X. Cong, W. B. Li, J. Gou, Q. Zhong, P. Cheng, P. H. Tan, L. Chen, and K. H. Wu, 2017, *Vibrational properties of a monolayer silicene sheet studied by tip-enhanced Raman spectroscopy*, **Phys. Rev. Lett. 119**, 196803.

Sheng, S. X., R. Z. Ma, J. B. Wu, X. Cong, W. B. Li, L. J. Kong, X. Cong, D. Y. Cao, W. Q. Hu, J. Gou, J. W. Luo, P. Cheng, P. H. Tan, Y. Jiang, L. Chen, and K. H. Wu, 2018, *The pengatonal nature of self-assembled silicon chains and magic clusters on Ag(110)*, **Nano Lett. 18**, 2937.

Shi, C., H. A. Hansen, A. C. Lausche, and J. K. Nørskov, 2014, *Trends in electrochemical CO_2 reduction activity for open and close-packed metal surfaces*, **Phys. Chem. Chem. Phys. 16**, 4720.

Shi, L., and L. W. Wang, 2012, *Ab Initio calculations of deep-level carrier nonradiative recombination rates in bulk semiconductors*, **Phys. Rev. Lett. 109**, 245501.

Shi, L., K. Xu, and L. W. Wang, 2015, *Comparative study of ab initio nonradiative recombination rate calculations under different formalisms*, **Phys. Rev. B 91**, 205315.

Shi, Q., and E. Geva, 2003, *Semiclassical theory of vibrational energy relaxation in the condensed phase*, **J. Phys. Chem. A 107**, 9059.

Shi, R. C., Y. Li, Q. Luo, and P. Gao, 2020, *Atomic resolution vibrational EELS acquired from an annular aperture*, **Microsc. Microanal. 26**, 2640.

Shi, Y. G., Y. F. Guo, X. Wang, A. J. Princep, D. Khalyavin, P. Manuel, Y. Michiue, A. Sato, K. Tsuda, S. Yu, M. Arai, Y. Shirako, M. Akaogi, N. L. Wang, K. Yamaura, and A. T. Boothroyd, 2013, *A ferroelectric-like structural transition in a metal*, **Nat. Mater. 12**, 1024.

Shi, Z. W., C. H. Jin, W. Yang, L. Ju, J. Horng, X. B. Lu, H. A. Bechtel, M. C. Martin, D. Y. Fu, J. Q. Wu, K. Watanabe, T. Taniguchi, Y. B. Zhang, X. D. Bai, E. G. Wang, G. Y. Zhang, and F. Wang, 2014, *Gate-dependent pseudospin mixing in graphene/boron nitride moiré superlattices*, **Nat. Phys. 10**, 743.

Shimizu, K., H. Ishikawa, D. Takao, T. Yagi, and K. Amaya, 2002, *Superconductivity in compressed lithium at 20 K*, **Nature 419**, 597.

Shiotari, A., and Y. Sugimoto, 2017, *Ultrahigh-resolution imaging of water networks by atomic force microscopy*, **Nat. Commun. 8**, 14313.

Shockley, W., and H. J. Queisser, 1961, *Detailed balance limit of efficiency of p-n junction solar cells*, **J. Appl. Phys. 32**, 510.

Shushkov, P., R. Li, and J. C. Tully, 2012, *Ring polymer molecular dynamics with surface hopping*, **J. Chem. Phys. 137**, 22A549.

Si, Q., and F. Steglich, 2010, *Heavy fermions and quantum phase transitions*, **Science 329**, 1161.

Singh D., 1991, *Ground-state properties of lanthanum: Treatment of extended-core states*, **Phys. Rev. B 43**, 6388.

Silvera, I. F., and T. J. Wijngaarden, 1981, *New low-temperature phase of molecular deuterium at ultrahigh pressure*, **Phys. Rev. Lett. 47**, 39.

Sjöstedt, E., L. Nordström, and D. J. Singh, 2000, *An alternative way of linearizing the augmented plane-wave method*, **Solid State Commun. 114**, 15.

Skinner, L. B., C. J. Benmore, J. C. Neuefeind, and J. B. Parise, 2014, *The structure of water around the compressibility minimum*, **J. Chem. Phys. 141**, 214507.

Slater, J. C., and H. C. Verma, 1929, *The theory of complex spectra*, **Phys. Rev. 34**, 1293.

Slater, J. C., 1930a, *Note on Hartree's method*, **Phys. Rev. 35**, 210.

Slater, J. C., 1930b, *Atomic shielding constants*, **Phys. Rev. 36**, 57.

Slater, J. C., 1950, *The Lorentz correction in barium titanate*, **Phys. Rev. 78**, 748.

Slater, J. C., 1951, *A simplification of the Hartree-Fock method*, **Phys. Rev. 81**, 385.

Slocombe, L., M. Sacchi, and J. Al-Khalili, 2022, *An open quantum systems approach to proton tunnelling in DNA*, **Commun. Phys. 5**, 109.

Smith, H. G., 1987, *Martensitic phase transformation of single-crystal lithium from bcc to a 9R-related structure*, **Phys. Rev. Lett. 58**, 1228.

Smith, J. D., C. D. Cappa, K. R. Wilson, B. M. Messer, R. C. Cohen, and R. J. Saykally, 2004, *Energetics of hydrogen bond network rearrangements in liquid water*, **Science 306**, 851.

Sobota, J. A., Y. He, and Z. X. Shen, 2021, *Angle-resolved photoemission studies of quantum materials*, **Rev. Mod. Phys. 93**, 025006.

Soddy, F., 1913, *Intra-atomic charge*, **Nature 92**, 399.

Sokolov, N. D., M. V. Vener, and V. A. Savelev, 1988, *Tentative study of the strong hydrogen bond dynamics: Part I. Geometric isotope effects*, **J. Mol. Struct. 177**, 93.

Somayazulu, M., M. Ahart, A. K. Mishra, Z. M. Geballe, M. Baldini, Y. Meng, V. V. Struzhkin, and R. J. Hemley, 2019, *Evidence for superconductivity above 260 K in lanthanum superhydride at megabar pressures*, **Phys. Rev. Lett. 122**, 027001.

Sommer, W. T., and D. J. Tanner, 1971, *Mobility of electrons on the surface of liquid ^{4}He*, **Phys. Rev. Lett. 27**, 1345.

Song, K., Y. Xu, L. Z. Song, and Q. Shi, 2016, *Quantum dynamical theories for proton transfer reactions in condensed phase*, **Sci Sin Chim 46**, 1039 (in Chinese).

Soper, A. K., 2000, *The radial distribution functions of water and ice from 220 to 673 K and at pressures up to 400 MPa*, **Chem. Phys. 258**, 121.

Stavrou, E., Y. S. Yao, A. F. Goncharov, S. S. Lobanov, J. M. Zaug, H. Y. Liu, E. Greenberg, and V. B. Prakapenka, 2018, *Synthesis of xenon and iron-nickel intermetallic compounds at earth's core thermodynamic conditions*, **Phys. Rev. Lett. 120**, 096001.

Stiopkin, I. V., C. Weeraman, P. A. Pieniazek, F. Y. Shalhout, J. L. Skinner, and A. V. Benderskii, 2011, *Hydrogen bonding at the water surface revealed by isotopic dilution spectroscopy*, **Nature 474**, 192.

Stipe, B. C., M. A. Rezaei, and W. Ho, 1998, *Single-molecule vibrational spectroscopy and microscopy*, **Science 280**, 1732.

Stock, G., and M. Thoss, 1997, *Semiclassical description of nonadiabatic quantum dynamics*, **Phys. Rev. Lett. 78**, 578.

Stolyarova, E., D. Stolyarov, K. Bolotin, S. Ryu, L. Liu, K. T. Rim, M. Klima, M. Hybertsen, I. Pogorelsky, I. Pavlishin, K. Kusche, J. Hone, P. Kim, H. L. Stormer, V. Yakimenko, and G. Flynn, 2009, *Observation of graphene bubbles and effective mass transport under graphene films*, **Nano Lett. 9**, 332.

Stroscio, J. A., and R. J. Celotta, 2004, *Controlling the dynamics of a single atom in lateral atom manipulation*, **Science 306**, 242.

Struzhkin, V. V., R. J. Hemley, H. K. Mao, and Y. A. Timofeev, 1997, *Superconductivity at 10–17 K in compressed sulphur*, **Nature 390**, 382.

Su, D., Q. Meng, C. A. F. Vaz, M.-G. Han, Y. Segal, F. J. Walker, M. Sawicki, C. Broadbridge, and C. H. Ahn, 2011, *Origin of 90° domain wall pinning in $Pb(Zr_{0.2}Ti_{0.8})O_3$ heteroepitaxial thin films*, **Appl. Phys. Lett. 99**, 102902.

Sugimoto, H., M. Tachikawa, and T. Udagawa, 2019, *Multicomponent QM study on the reaction of HOSO + NO_2 with H_2O: Nuclear quantum effect on structure and reaction energy profile*, **Int. J. Quantum Chem. 119**, e25895.

Sun, B., S. Niu, R. P. Hermann, J. Moon, N. Shulumba, K. Page, B. Zhao, A. S. Thind, K. Mahalingam, J. Milam-Guerrero, R. Haiges, M. Mecklenburg, B. C. Melot, Y. Jho, B. M. Howe, R. Mishra, A. Alatas, B. Winn, M. E. Manley, J. Ravichandran, and A. J. Minnich, 2020, *High frequency atomic tunneling yields ultralow and glass-like thermal conductivity in chalcogenide single crystals*, **Nat. Commun. 11**, 6039.

Sun, H., Q. Li, and X. G. Wan, 2016, *First-principles study of thermal properties of borophene*, **Phys. Chem Chem. Phys. 18**, 14927.

Sun, Z. R., D. Pan, L. Xu, and E. G. Wang, 2012, *Role of proton ordering in adsorption preference of polar molecule on ice surface*, **Proc. Natl. Acad. Sci. U. S. A. 109**, 13117.

Sun, Z. R., L. X. Zheng, M. H. Chen, M. L. Klein, F. Paesani, and X. F. Wu, 2018, *Electron-hole theory of the effect of quantum nuclei on the X-Ray absorption spectra of liquid water*, **Phys. Rev. Lett. 121**, 137401.

Sundell, P. G., and G. Wahnström, 2004, *Activation energies for quantum diffusion of hydrogen in metals and on metal surfaces using delocalized nuclei within the density-functional theory*, **Phys. Rev. Lett. 92**, 155901.

Sundell, P. G., M. E. Björketun, and G. Wahnström, 2007, *Density-functional calculations of prefactors and activation energies for H diffusion in $BaZrO_3$*, **Phys. Rev. B 76**, 094301.

Suo, L., O. Borodin, T. Gao, M. Olguin, J. Ho, X. Fan, C. Luo, C. Wang, and K. Xu, 2015, *"Water-in-salt" electrolyte enables high-voltage aqueous lithium-ion chemistries*, **Science 350**, 938.

Sutcliffe, M. J., and N. S. Scrutton, 2002, *A new conceptual framework for enzyme catalysis. Hydrogen tunneling coupled to enzyme dynamics in flavoprotein and quinoprotein enzymes*, **Eur. J. Biochem. 269**, 3096.

Suzuki, K., M. Shiga, and M. Tachikawa, 2008, *Temperature and isotope effects on water cluster ions with path integral molecular dynamics based on the fourth order Trotter expansion*, **J. Chem. Phys. 129**, 144310.

Suzuki, K., M. Tachikawa, H. Ogawa, S. Ittisanronnachai, H. Nishihara, T. Kyotani, and U. Nagashima, 2011, *Isotope effect of proton and deuteron adsorption site on zeolite-templated carbon using path integral molecular dynamics*, **Theor. Chem. Acc. 130**, 1039.

Suzuki, K., Y. Kawashima, and M. Tachikawa, 2018, Nuclear quantum effect and H/D isotope effect on hydrogen-bonded system with path integral simulation, **Frontiers of Quantum Chemistry,** Springer Nature Singapore.

Svishchev, I. M., and P. G. Kusalik, 1994, *Dynamics in liquid water, water-d2, and water-t2: A comparative simulation study*, **J. Phys. Chem. 98**, 728.

Syage, J. A., R. B. Cohen, and J. Steadman, 1992, *Spectroscopy and dynamics of jet-cooled hydrazines and ammonia. I. Single-photon absorption and ionization spectra.* **J. Chem. Phys. 97**, 6072.

Szado, A., and N. S. Ostlund, 1996, *Modern quantum chemistry: Introduction to advanced electronic structure theory* (Dover Publications Inc., Mineola, New York).

Szegedi, S., Váradi, M., Buczkó, Cs. M., Várnagy, M., Sztaricskai, T., 1990, *Determination of boron in glass by neutron transmission method.* **J. Radioanal. Nucl. Chem. 146**, 177.

T

Tachikawa, M., and M. Shiga, 2005, *Geometrical H/D isotope effect on hydrogen bonds in charged water clusters*, **J. Am. Chem. Soc. 127**, 11908.

Takahashi, H., K. Igawa, K. Arii, Y. Kamihara, M. Hirano, and H. Hosono, 2008, *Superconductivity at 43 K in an iron-based layered compound* $LaO_{1-x}F_xFeAs$, **Nature 453**, 376.

Takano, Y., M. Nagao, I. Sakaguchi, M. Tachiki, T. Hatano, K. Kobayashi, H. Umezawa, and H. Kawarada, 2004, *Superconductivity in diamond thin films well above liquid helium temperature*, **Appl. Phys. Lett. 85**, 2851.

Takayanagi, T., and Y. Kurosaki, 1998, *van der Waals resonances in cumulative reaction probabilities for the* $F + H_2$, D_2, *and HD reactions*, **J. Chem. Phys. 109**, 8929.

Tang, H., and S. Ismail-Beigi, 2007, *Novel precursors for boron nanotubes: The competition of two-center and three-center bonding in boron sheets*, **Phys. Rev. Lett.**, **99**, 115501.

Tang, Y., Y. Zhu, X. Ma, A. Y. Borisevich, A. N. Morozovska, E. A. Eliseev, W. Wang, Y. Wang, Y. Xu, and Z. Zhang, 2015, *Observation of a periodic array of flux-closure quadrants in strained ferroelectric* $PbTiO_3$ *films*. **Science 348**, 547.

Tao, X., P. Shushkov, and T. F. Miller, 2018, *Path-integral isomorphic Hamiltonian for including nuclear quantum effects in non-adiabatic dynamics*, **J. Chem. Phys. 148**, 102327.

Tao, X., P. Shushkov, and T. F. Miller, 2019, *Simple flux-side formulation of state-resolved thermal reaction rates for ring-polymer surface hopping*, **J. Phys. Chem. A. 123**, 3013.

Tao, Z. S., C. Chen, T. Szilvasi, M. Keller, M. Mavrikakis, H. Kapteyn, and M. Murnane, 2016, *Direct time-domain observation of attosecond final-state lifetimes in photoemission from solids*, **Science 353**, 62.

Tennyson, J., P. F. Bernath, L. R. Brown, A. Campargue, A. G. Csaszar, L. Daumont, R. R. Gamache, J. T. Hodges, O. V. Naumenko, O. L. Polyansky, L. S. Rothman, A. C. Vandaele, N. F. Zobov, N. Denes, A. Z. Fazliev, T. Furtenbacher, I. E. Gordon, S. M. Hu, T. Szidarovszky, and I. A. Vasilenko, 2014, *IUPAC critical evaluation of the rotational-vibrational spectra of water vapor. Part IV. Energy levels and transition wavenumbers for* $D_2{}^{16}O$, $D_2{}^{17}O$, *and* $D_2{}^{18}O$, **J. Quant. Spectros. Radiat. Transfer. 142**, 93.

Terada, Y., S. Yoshida, O. Takeuchi, and H. Shigekawa, 2010, *Real-space imaging of transient carrier dynamics by nanoscale pump-probe microscopy*, **Nat. Photon. 4**, 869.

Thapa, M. J., W. Fang, and J. O. Richardson, 2019, *Nonadiabatic quantum transition-state theory in the golden-rule limit. I. Theory and application to model systems*, **J. Chem. Phys. 150**, 104107.

The LNCMP-team, 2004, *The LNCMP: A pulsed-field user-facility in Toulouse*, **Physica B 346**, 668.

Thomas, T. H., 1927, *The calculation of atomic fields*, **Proc. Cambridge Phil. Soc. 23**, 542.

Thompson, J. D., 1984, *Low-temperature pressure variations in a self-clamping pressure cell*, **Rev. Sci. Instrum. 55**, 231.

Tian, Y., C. Zhou, X. Fu, S. Ji, Y. Leng, and R. Li, 2021, *Research progress of generation and control of ultrafast and coherent electron sources based on optical fields*, **Acta Photon. Sin. 50**, 0850202.

Tian, Y., J. N. Hong, D. Y. Cao, S. F. You, Y. Z. Song, B. W. Cheng, Z. C. Wang, D. Guan, X. M. Liu, Z. P. Zhao, X. Z. Li, L. M. Xu, J. Guo, J. Chen, E. G. Wang, and Y. Jiang, 2022, *Visualizing Eigen/Zundel cations and their interconversion in monolayer water on metal surfaces*, **Science 377**, 315.

Tian, Y. C., He Tian, Y. L. Wu, L. L. Zhu, L. Q. Tao, W. Zhang, Y. Shu, D. Xie, Y. Yang, Z. Y. Wei, X. H. Lu, T.-L. Ren, C.-K. Shih, J. M. Zhao, 2015, *Coherent generation of photo-thermo-acoustic wave from graphene sheets*, **Sci. Rep. 5**, 10582.

Tian, Y. C., W. H. Zhang, F. S. Li, Y. L. Wu, Q. Wu, F. Sun, G. Y. Zhou, L. L. Wang, X. C. Ma, Q. K. Xue, and J. M. Zhao, 2016, *Ultrafast dynamics evidence of high temperature superconductivity in single unit cell FeSe on* $SrTiO_3$, **Phys. Rev. Lett. 116**, 107001.

Tian, Y. J., B. Xu, D. L. Yu, Y. M. Ma, Y. B. Wang, Y. B. Jiang, W. T. Hu, C. C. Tang, Y. F. Gao, K. Luo, Z. S. Zhao, L. M. Wang, B. Wen, J. L. He, and Z. Y. Liu, 2013, *Ultrahard nanotwinned cubic boron nitride*, **Nature 493**, 385.

Tian, Z. Y., Q. Y. Zhang, Y. W. Xiao, G. A. Gamage, F. Tian, S. Yue, V. G. Nadjiev, J. M. Bao, Z. F. Ren, E. J. Liang, and J. M. Zhao, 2022, *Ultraweak electron-phonon coupling strength in cubic boron arsenide unveiled by ultrafast dynamics*, **Phys. Rev. B 105**, 174306.

Toth, J., and S. T. Bole, 2018, *Design, construction, and first testing of a 41.5 T all-resistive magnet at the NHMFL in Tallahassee*, **IEEE Trans. Appl. Supercon. 28**, 4300104.

Toyoura, K., Y. Koyama, A. Kuwabara, F. Oba, and I. Tanaka, 2008, *First-principles approach to chemical diffusion of lithium atoms in a graphite intercalation compound*, **Phys. Rev. B 78**, 214303.

Trail, J. R., and R. J. Needs, 2017, *Shape and energy consistent pseudopotentials for correlated electron systems*, **J. Chem. Phys. 146**, 204107.

Tran, T. T., K. Bray, M. J. Ford, M. Toth, and I. Aharonovich, 2016a, *Quantum emission from hexagonal boron nitride monolayers*, **Nat. Nanotechnol. 11**, 37.

Tran, T. T., C. Elbadawi, D. Totonjian, C. J. Lobo, G. Grosso, H. Moon, D. R. Englund, M. J. Ford, I. Aharonovich, and M. Toth, 2016b, *Robust multicolor single photon emission from point defects in hexagonal boron nitride*, **ACS Nano 10**, 7331.

Tran, T. T., M. Kianinia, M. Nguyen, S. Kim, Z. Q. Xu, A. Kubanek, M. Toth, and I. Aharonovich, 2018, *Resonant excitation of quantum emitters in hexagonal boron nitride*, **ACS Photonics 5**, 295.

Tritsaris, G. A., E. Kaxiras, S. Meng, and E. G. Wang, 2013, *Adsorption and diffusion of lithium on layered silicon for Li-ion storage*, **Nano Lett. 13**, 2258.

Trixler, F., 2013, *Quantum tunneling to the origin and evolution of life*, **Current Org. Chem. 17**, 1758.

Tromp, J. W., and W. H. Miller, 1986, *New approach to quantum mechanical transition-state theory*, **J. Phys. Chem. 90**, 3482.

Truhlar, D. G., W. L. Hase, and J. T. Hynes, 1983, *Current status of transition-state theory*, **J. Phys. Chem. 87**, 2664.

Truhlar, D. G., B. C. Garrett, and S. J. Klippenstein, 1996, *Current status of transition-state theory*, **J. Phys. Chem. 100**, 12771.

Tse, J. S., Y. Yao, and K. Tanaka, 2007, *Novel superconductivity in metallic SnH_4 under high pressure*, **Phys. Rev. Lett. 98**, 117004.

Tsui, D. C., H. L. Stormer, and A. C. Gossard, 1982, *Two-dimensional magnetotransport in the extreme quantum limit*, **Phys. Rev. Lett. 48**, 1559.

Tuckerman, M. E., D. Marx, M. L. Klein, and M. Parrinello, 1996, *Efficient and general algorithms for path integral Car-Parrinello molecular dynamics*, **J. Chem. Phys. 104**, 5579.

Tuckerman, M. E., D. Marx, M. L. Klein, and M. Parrinello, 1997, *On the quantum nature of the shared proton in hydrogen bonds*, **Science 275**, 817.

Tuckerman, M. E., D. Marx, and M. Parrinello, 2002, *The nature and transport mechanism of hydrated hydroxide ions in aqueous solution*, **Nature 417**, 925.

Tully, J. C., 1990, *Molecular dynamics with electronic transitions*, **J. Chem. Phys. 93**, 1061.

Tully, J. C., 1998, *Mixed quantum-classical dynamics*, **Faraday Discuss. 110**, 407.

U

Ubbelohde, A. R., and K. J. Gallagher, 1955, *Acid-base effects in hydrogen bonds in crystals*, **Acta Crystallogr. 8**, 71.

Udagawa, T., and M. Tachikawa, 2016, *Nuclear quantum effect and H/D isotope effect on $F + (H_2O)_n \rightarrow FH + (H_2O)_{n-1}OH$ (n = 1–3) reactions*, **J. Chem. Phys. 145**, 164310.

Ulmer, G. C., and H. L. Barnes, 1987, *Hydrothermal experimental techniques* (Wiley, New York).

Unke, O. T., J. C. Castro-Palacio, R. J. Bemish, and M. Meuwly, 2016, *Collision-induced rotational excitation in $N_2^+(^2\Sigma_g^+, v=0)-Ar$: Comparison of computations and experiment*, **J. Chem. Phys. 144**, 224307.

V

Vanderbilt, D., 1990, *Soft self-consistent pseudopotentials in a generalized eigenvalue formalism*, **Phys. Rev. B 41**, 7892.

van Leeuwen, R., 1996, *The Sham-Schlüter equation in time-dependent density-functional theory*, **Phys. Rev. Lett. 76**, 3610.

van Setten, M. J., M. A. Uijttewaal, G. A. de Wijs, and R. A. de Groot, 2007, *Thermodynamic stability of boron: The role of defects and zero point motion*, **J. Am. Chem. Soc. 129**, 2458.

van Vleck, J. H., 1928, *On dielectric constants and magnetic susceptibilities in the new quantum mechanics Part III – Application to dia- and paramagnetism*, **Phys. Rev. 31**, 587.

Vardi-Kilshtain, A., N. Nitoker, and D. T. Major, 2015, *Nuclear quantum effects and kinetic isotope effects in enzyme reactions*, **Arch. Biochem. Biophys. 582**, 18.

Venkatraman, K., B. D. Levin, K. March, P. Rez, P. A. Crozier, 2019, *Vibrational spectroscopy at atomic resolution with electron impact scattering*, **Nat. Phys. 15**, 1237.

Vértes, A., S. Nagy, Z. Klencsár, R. Q. Lovas, and F. Rösch, 2011, *Handbook of Nuclear Chemistry* (Springer Science + Business Media B.V., Dordrecht, the Netherlands).

Vestergaard, C. L., M. B. Mikkelsen, W. Reisner, A. Kristensen, and H. Flyvbjerg, 2016, *Transition state theory demonstrated at the micron scale with out-of-equilibrium transport in a confined environment*, **Nat. Commun. 7**, 10227.

Vojta, M., 2003, *Quantum phase transitions*, **Rep. Prog. Phys. 66**, 2069.

Vol'kenshtein, M. V., R. R. Dogonadze, A. K. Madumarov, Z. D. Urushadze, and Y. I. Kharkats, 1972, *The theory of enzyme catalysis*. **Mol. Biol.** (Moscow) **6**, 431.

Von Klitzing, K., G. Dorda, and M. Pepper, 1980, *New method for high-accuracy determination of the fine-structure constant based on quantized Hall resistance*, **Phys. Rev. Lett. 45**, 494.

Vosko, S. H., L. Wilk, and M. Nusair, 1980, *Accurate spin-dependent electron liquid correlation energies for local spin-density calculations – A critical analysis*, **Can. J. Phys. 58**, 1200.

Voth, G. A., D. Chandler, and W. H. Miller, 1989, *Rigorous formulation of quantum transition state theory and Its dynamical corrections*, **J. Chem. Phys. 91**, 7749.

Voth, G. A., 1993, *Feynman path integral formulation of quantum mechanical transition-state theory*, **J. Phys. Chem. 97**, 8365.

Voth, G. A., 1996, *Path-integral centroid methods in quantum statistical mechanics and dynamics*, **Adv. Chem. Phys. 93**, 135.

Vuong, T. Q. P., G. Cassabois, P. Valvin, V. Jacques, A. Van Der Lee, A. Zobelli, K. Watanabe, T. Taniguchi, and B. Gil, 2017a, *Phonon symmetries in hexagonal boron nitride probed by incoherent light emission*, **2D Materials 4**, 011004.

Vuong, T. Q. P., G. Cassabois, P. Valvin, V. Jacques, R. Cusco, L. Artus, and B. Gil, 2017b, *Overtones of interlayer shear modes in the phonon-assisted emission spectrum of hexagonal boron nitride*, **Phys. Rev. B 95**, 045207.

Vuong, T. Q. P., S. Liu, A. Van der Lee, R. Cusco, L. Artus, T. Michel, P. Valvin, J. H. Edgar, G. Cassabois, and B. Gil, 2018, *Isotope engineering of van der Waals interactions in hexagonal boron nitride*, **Nat. Mater. 17**, 152.

W

Wacey, D., M. R. Kilburn, M. Saunders, J. Cliff, and M. D. Brasier, 2011, *Microfossils of sulphur-metabolizing cells in 3.4-billion-year-old rocks of Western Australia*, **Nature Geosci. 4**, 698.

Walker, D., M. A. Carpenter, and C. M. Hitch, 1990, *Some simplifications to multianvil devices for high-pressure experiments*, **Am. Mineral. 75**, 1020.

Walsh, J. P. S., S. M. Clarke, Y. Meng, S. D. Jacobsen, and D. E. Freedman, 2016, *Discovery of $FeBi_2$, ACS Central* **Sci. 2**, 867.

Walsh, J. P. S., and D. E. Freedman, 2018, *High-pressure synthesis: A new frontier in the search for next-generation intermetallic compounds*, **Acc. Chem. Res. 51**, 1315.

Wan, W. H., Q. F. Zhang, Y. Cui, and E. G. Wang, 2010, *First principles study of lithium insertion in bulk silicon*, **J. Phys. Condens. Matter 22**, 415501.

Wang, B. Y., 2016, *Review on Bose-Einstein condensation*, **Highl. Sci. Eng. Technol. 38**, 19.

Wang, E. G., Y. Zhou, C. S. Ting, J. Zhang, T. Pang, and C. Chen, 1995, *Excitons in spatially separated electron-hole systems: A quantum Monte Carlo study*, **J. Appl. Phys. 78**, 7099.

Wang, E. G., 1997, *Research on carbon nitrides*, **Prog. Mater. Sci. 41**, 241.

Wang, E. G., 2016, *Observation of full quantum effect in water clusters*, **Inter. Conf. on Quantum Physics**, Shenzhen, China (Nov.6–9).

Wang, E. G., 2022, *Full quantum nature of water: A study at atom level*, ***Forum on Beauty of Science***, Tsinghua University (Feb.24).

Wang, H., J. S. Tse, K. Tanaka, T. Iitaka, and Y. Ma, 2012, *Superconductive sodalite-like clathrate calcium hydride at high pressures*, **Proc. Natl. Acad. Sci. U. S. A. 109**, 6463.

Wang, H. B., X. Sun, and W. H. Miller, 1998, *Semiclassical approximations for the calculation of thermal rate constants for chemical reactions in complex molecular systems*, **J. Chem. Phys. 108**, 9726.

Wang, H. C., H. W. Liu, Y. A. Li, Y. J. Liu, J. F. Wang, J. Liu, J. Y. Dai, Y. Wang, L. Li, J. Q. Yan, D. Mandrus, X. C. Xie, and J. Wang, 2018, *Discovery of log-periodic oscillations in ultraquantum topological materials*, **Sci. Adv. 4**, eaau5096.

Wang, L., X. Z. Xu, L. N. Zhang, R. X. Qiao, M. H. Wu, Z. C. Wang, S. Zhang, J. Liang, Z. H. Zhang, W. Chen, X. D. Xie, J. Y. Zong, Y. W. Shan, Y. Guo, M. Willinger, H. Wu, Q. Y. Li, W. L. Wang, P. Gao, S. W. Wu, Y. Zhang, Y. Jiang, D. P. Yu, E. G. Wang, X. D. Bai, Z. J. Wang, F. Ding, and K. H. Liu, 2019, *Dual-coupling-guided Epitaxial growth of 100-square-centimetre single-crystal hexagonal boron nitride monolayer on copper*, **Nature 570**, 91.

Wang, P. S., and H. J. Xiang, 2014, *Room-temperature ferrimagnet with frustrated antiferroelectricity: Promising candidate toward multiple-state memory*, **Phys. Rev. X 4**, 011035.

Wang, Q., K. Suzuki, U. Nagashima, M. Tachikawa, and S. W. Yan, 2013, *Geometric isotope effects on small chloride ion water clusters with path integral molecular dynamics simulations*, **Chem. Phys. 426**, 38.

Wang, R., and M. Itoh, 2001, *Suppression of the quantum fluctuation in ^{18}O-enriched strontium titanate*, **Phys. Rev. B 64**, 174104.

Wang, W. L., X. D. Bai, K. H. Liu, Z. Xu, D. Golberg, Y. Bando, and E. G. Wang, 2006, *Direct synthesis of B-C-N single-walled nanotubes by bias-assisted hot filament chemical vapor deposition*, **J. Am. Chem. Soc. 128**, 6530.

Wang, X., W. Dong, C. Xiao, L. Che, Z. Ren, D. Dai, X. Wang, P. Casavecchia, X. Yang, B. Jiang, D. Xie, Z. Sun, S. Lee, D. Zhang, H. Werner, and M. Alexander, 2008, *The extent of non–Born-Oppenheimer coupling in the reaction of Cl (^2P) with para-H_2*, **Science**, **322**, 573.

Wang, X. F., R. H. Liu, Z. Gui, Y. L. Xie, Y. J. Yan, J. J. Ying, X. G. Luo, and X. H. Chen, 2011, *Superconductivity at 5 K in alkali-metal-doped phenanthrene*, **Nat. Commun. 2**, 507.

Ward, A., D. A. Broido, D. A. Stewart, and G. Deinzer, 2009, *Ab initio theory of the lattice thermal conductivity in diamond*, **Phys. Rev. B 80**, 125203.

Watanabe, H., C. E. Nebel, and S. Shikata, 2009, *Isotopic homojunction band engineering from diamond*, **Science 324**, 1425.

Watanabe, H., T. Koretsune, S. Nakashima, S. Saito, and S. Shikata, 2013, *Isotope composition dependence of the band-gap energy in diamond*, **Phys. Rev. B 88**, 205420.

Watkins, M., D. Pan, E. G. Wang, A. Michaelides, J. Vande Vondele, and B. Slater, 2011, *Large variation of vacancy formation energies in the surface of crystalline ice*, **Nat. Mater. 10**, 794.

Wech, P. F., and N. Balakrishnan, 2005, *Quantum dynamics of the Li + HF → H + LiF reaction at ultralow temperatures*, **J. Chem. Phys. 122**, 154309.

Wei, L., P. K. Kuo, R. L. Thomas, T. R. Anthony, and W. F. Banholzer, 1993, *Thermal conductivity of isotopically modified single crystal diamond*, **Phys. Rev. Lett. 70**, 3764.

Wei, S. H, and H. Krakauer, 1985a, *Local-density-functional calculation of the pressure-induced metallization of BaSe and BaTe*, **Phys. Rev. Lett. 55**, 1200.

Wei, S. H., H. Krakauer, and M. Weinert, 1985b, *Linearized augmented-plane wave calculation of the electronic structure and total energy of tungsten*, **Phys. Rev. B 32**, 7792.

Weickert, F., B. Meier, S. Zherlitsyn, T. Herrmannsdorfer, R. Daou, M. Nicklas, J. Haase, F. Steglich, and J. Wosnitza, 2012, *Implementation of specific-heat and NMR experiments in the 1500 ms long-pulse magnet at the Hochfeld-Magnetlabor Dresden*, **Meas. Sci. Technol. 23**, 105001.

Weir, S. T., A. C. Mitchell, and W. J. Nellis, 1996, *Metallization of fluid molecular hydrogen at 140 GPa*, **Phys. Rev. Lett. 76**, 1860.

Weir, S. T., 1998, *Metallization of fluid hydrogen by multiple shock compression*, **J. Phys. Condens. Mat. 10**, 11147.

Wentzel, G., 1926, *Eine Verallgemeinerung der Quantenbedingungen für die Zwecke der Wellenmechanik*, **Z. Phys. 38**, 518.

West, J. T., X. Lin, Z. G. Cheng, and M. H. W. Chan, 2009, *Supersolid behavior in confined geometry*, **Phys. Rev. Lett. 102**, 185302.

Wheeler, S. E., K. W. Sattelmeyer, P. R. Schleyer, and H. F. Schaefer, 2014, *Binding energies of small lithium clusters (Li(N)) and hydrogenated lithium clusters (Li(N)H)*, **J. Chem. Phys. 120**, 4683.

Whitman, D. W., and B. K. Carpenter, 1982, *Limits on the activation parameters for automerization of cyclobutadiene-1,2-d2*, **J. Am. Chem. Soc.**, 6473.

Wien, W., 1893, *Die obere Grenze der Wellenlangen, welche in der Warmestrahlung fester Korper vorkemmen konnen; Folgerungen aus dem zweiten Hauptsatz der Warmetheorie*, **Annalan der Physik**, **285** 633.

Wigner, E., 1932, *On the quantum correction for thermodynamic equilibrium*, **Phys. Rev. 40**, 749.

Wigner, E., and H. B. Huntington, 1935, *On the possibility of a metallic modification of hydrogen*, **J. Chem. Phys. 3**, 764.

Williams, F., 1951, *Theoretical low temperature spectra of the thallium activated potassium chloride phosphor*, **Phys. Rev. 82**, 281.

Wipf, H., A. Magerl, S. M. Shapiro, S. K. Satija, and W. Thomlinson, 1981, *Neutron-spectroscopic evidence for hydrogen tunneling states in niobium*, **Phys. Rev. Lett. 46**, 947.

Witt, A., S. D. Ivanov, M. Shiga, H. Forbert, and D. Marx, 2009, *On the applicability of centroid and ring polymer path integral molecular dynamics for vibrational spectroscopy*, **J. Phys. Chem. A 130**, 194510.

Wittig, C., 2005, *The Landau-Zener formula*, **J. Phys. Chem. B 109**, 8428.

Wolfke, M., 1939, *Can a quantum crystal be superfluid?* **Ann. Acad. Sci. Techn. Varsovie 6**, 14.

Wu, D., Z. P. Zhao, B. Lin, Y. Song, J. Qi, J. Jiang, Z. Yuan, B. Cheng, M. Zhao, Y. Tian, Z. Wang, M. H. Wu, K. Bian, K. H. Liu, L. M. Xu, X. C. Zeng, E. G. Wang, and Y. Jiang, 2024, *Probing structural superlubricity of two-dimensional water transport with atomic resolution,* **Science 384**, 1254.

Wu, H., and W. Z. Shen, 2006, *Dielectric and pyroelectric properties of $Ba_xSr_{1-x}TiO_3$: Quantum effect and phase transition*, **Phys. Rev. B 73**, 094115.

Wu, H. L., R. Sato, A. Yamaguchi, M. Kimura, M. Haruta, H. Kurata, and T. Teranishi, 2016, *Formation of pseudomorphic nanocages from Cu_2O nanocrystals through anion exchange reactions*, **Science 351**, 1306.

Wu, J.-Z., M. Jie, and S.-T. Jia, 2018, *Production of ultracold molecules and their manipulation by external fields based on laser cooled atoms*, **Physics 47**, 162.

Wu, K. H., E. G. Wang, S. Liu, and X. R. Wang, 2001, *Bistable characteristic and current jumps in field electron emission of nanocrystalline diamond films*, **J. Appl. Phys. 89**, 4810.

Wu, Q., H. X. Zhou, Y. L. Wu, L. L. Hu, S. L. Ni, Y. C. Tian, F. Sun, F. Zhou, X. L. Dong, Z. X. Zhao, and J. M. Zhao, 2020, *Ultrafast quasiparticle dynamics and electron-phonon coupling in* $(Li_{0.84}Fe_{0.16})OHFe_{0.98}Se$, **Chin. Phys. Lett. 37**, 097802.

Wu, R., Y. L. Zhang, S. C. Yan, F. Bian, W. L. Wang, X. D. Bai, X. H. Lu, J. M. Zhao, and E. G. Wang, 2011, *Purely coherent nonlinear optical response in solution dispersions of graphene sheets*, **Nano Lett. 11**, 5159.

Wu, W., J. G. Cheng, K. Matsubayashi, P. P. Kong, F. K. Lin, C. Q. Jin, N. L. Wang, Y. Uwatoko, and J. L. Luo, 2014, *Superconductivity in the vicinity of antiferromagnetic order in CrAs*, **Nat. Commun. 5**, 5508.

Wu, Y. L., Q. O. Wu, F. Sun, C. Cheng, S. Meng, and J. M. Zhao, 2015, *Emergence of electron coherence and two-color all-optical switching in* MoS_2 *based on spatial self-phase modulation*, **Proc. Natl. Acad. Sci. U. S. A. 112**, 11800.

Wu, Y. L., X. Yin, J. Hasaien, Y. Ding, and J. M. Zhao, 2020, *High-pressure ultrafast dynamics in* Sr_2IrO_4*: Pressure-induced phonon bottleneck effect*, **Chin. Phys. Lett. (Express Letter) 37**, 047801.

Wu, Y. L., X. Yin, J. Z. L. Hasaien, Z. Y. Tian, Y. Ding, and J. M. Zhao, 2021, *On-site in situ high-pressure ultrafast pump-probe spectroscopy instrument*, **Review of Scientific Instruments 92**, 113002.

X

Xia, X., X. M. Chen, and J. J. Quinn, 1992, *Magnetoexcitons in a GaSb-AlSb-InAs quantum-well structure*, **Phys. Rev. B 46**, 7212.

Xiao, C., X. Xu, S. Liu, T. Wang, W. R. Dong, T. Yang, Z. Sun, D. Dai, X. Xu, D. H. Zhang, and X. Yang, 2011, *Experimental and theoretical differential cross sections for a four-atom reaction:* $HD + OH \rightarrow H_2O + D$, **Science 333**, 440.

Xie, Y., H. Zhao, Y. Wang, Y. Huang, T. Wang, X. Xu, C. Xiao, Z. Sun, D. H. Zhang, and X. Yang, 2020, *Quantum interference in* $H + HD \rightarrow H_2 + D$ *between direct abstraction and roaming insertion pathways*, **Science 368**, 767.

Xie, Z. J., S. He, C. Chen, Y. Feng, H. Yi, A. Liang, L. Zhao, D. Mou, J. He, Y. Peng, X. Liu, Y. Liu, G. Liu, X. Dong, L. Yu, J. Zhang, S. Zhang, Z. Wang, F. Zhang, F. Yang, Q. Peng, X. Wang, C. Chen, Z. Xu, and X. J. Zhou, 2014, *Orbital-selective spin texture and its manipulation in a topological insulator*, **Nat. Commun. 5**, 3382.

Xin, Y. B., Q. Hu, D. H. Niu, X. H. Zheng, H. L. Shi, M. Wang, Z. S. Xiao, A. P. Huang, and Z. B. Zhang, 2017, *Research progress of hydrogen tunneling in two-dimensional materials*, **Acta. Physica. Sinica. 66**, 056601.

Xu, J., D. Chen, and S. Meng, 2022, *Decoupled ultrafast electronic and structural phase transitions in photoexcited monoclinic* VO_2. **Sci. Adv. 8**, eadd2392.

Xu, J. J., and C. Wang, 2022, *Perspective-electrolyte design for aqueous batteries: From ultra-high concentration to low concentration*, **J. Electrochem. Soc. 169**, 030530.

Xu, J. H., E. G. Wang, C. S. Ting, and W. P. Su, 1993, *Tight-binding theory of the electronic structures for rhombohedral semimetals*, **Phys. Rev. B 48**, 17271.

Xu, J. Y., H. Y. Jiang, Y. T. Shen, X. Z. Li, E. G. Wang, and S. Meng, 2019, *Transparent proton transport through a two-dimensional nanomesh material*, **Nat. Commun. 10**, 3971.

Xu, J. Y., D. Q. Chen, and S. Meng, 2022, *Decoupled ultrafast electronic and structural phase transitions in photoexcited monoclinic VO2*, **Sci. Adv. 8**, eadd2392.

Xu, X. Z., Z. H. Zhang, J. C. Dong, D. Yi, J. J. Niu, M. H. Wu, L. Lin, R. K. Yin, M. Q. Li, J. Y. Zhou, S. X. Wang, J. L. Sun, X. J. Duan, P. Gao, Y. Jiang, X. S. Wu, H.

L. Peng, R. S. Ruoff, Z. F. Liu, D. P. Yu, E. G. Wang, F. Ding, and K. H. Liu, 2017, *Ultrafast epitaxial growth of metre-sized single-crystal graphene on industrial Cu foil*, **Sci. Bullet. 62**, 1074.

Xu, Y., Y. Yuan, H. F. Ding, L. Li, M. Guo, H. L. Tao, and M. Sun, 2014, *Voltage synchronization scheme and control strategy for 50 T flat-top pulsed magnetic field power system*, **IEEE Trans. Appl. Supercon. 24**, 3801305.

Xu, Z., W. G. Lu, W. L. Wang, C. Z. Gu, K. H. Liu, X. D. Bai, E. G. Wang, and H. J. Dai, 2008, *Converting metallic single-walled carbon nanotubes into semiconductors via boron-nitrogen co-doping*, **Adv. Mater. 20**, 3615.

Xue, M., T. Cao, D. Wang, Y. Wu, H. Yang, X. Dong, J. He, F. Li, and G. F. Chen, 2012, *Superconductivity above 30 K in alkali-metal-doped hydrocarbon*, **Sci. Rep. 2**, 389.

Xue, Q., A. J. Horsewill, M. R. Johnson, and H. P. Trommsdorff, 2004, *Isotope effects associated with tunneling and double proton transfer in the hydrogen bonds of benzoic acid*, **J. Chem. Phys. 120**, 11107.

Y

Yadav, A. K., C. T. Nelson, S. L. Hsu, Z. Hong, J. D. Clarkson, C. M. Schlepütz, A. R. Damodaran, P. Shafer, E. Arenholz, L. R. Dedon, D. Chen, A. Vishwanath, A. M. Minor, L. Q. Chen, J. F. Scott, L. W. Martin, and R. Ramesh, 2016, *Observation of polar vortices in oxide superlattices*, **Nature 530**, 198.

Yamanaka, T., and S. Morimoto, 1996, *Isotope effect on anharmonic thermal atomic vibration and kappa refinement of C-12 and C-13 diamond*, **Acta Crystallogr. B: Struct. Sci. 52**, 232.

Yan, J., Y. Zhang, P. Kim, and A. Pinczuk, 2007, *Electric field effect tuning of electron-phonon coupling in graphene*, **Phys. Rev. Lett. 98**, 166802.

Yan, L., F. W. Wang, and S. Meng, 2016, *Quantum mode selectivity of plasmon-induced water splitting on gold nanoparticles*, **ACS Nano 10**, 5452.

Yan, L., J. Y. Xu, F. W. Wang, and S. Meng, 2018, *Plasmon-induced ultrafast hydrogen production in liquid water*, **J. Phys. Chem. Lett. 9**, 63.

Yanagisawa, Y., K. Kajita, S. Iguchi, Y. Xu, M. Nawa, R. Piao, T. Takao, H. Nakagome, M. Hamada, T. Noguchi, G. Nishijima, S. Matsumoto, H. Suematsu, M. Takahashi, and H. Maeda, 2016, *27.6T generation using Bi-2223/REBCO superconducting coils*, **IEEE/CSC & ESAS Superconductivity News Forum** (Global Edition).

Yang, J., X. L. Zhu, J. P. F. Nunes, J. K. Yu, R. M. Parrish, T. J. A. Wolf, M. Centurion, M. Guhr, R. K. Li, Y. S. Liu, B. Moore, M. Niebuhr, S. Park, X. Z. Shen, S. Weathersby, T. Weinacht, T. J. Martinez, and X. J. Wang, 2020, *Simultaneous observation of nuclear and electronic dynamics by ultrafast electron diffraction*, **Science 368**, 885.

Yang, J., R. Dettori, J. P. F. Nunes, N. H. List, E. Biasin, M. Centurion, Z. Chen, A. A. Cordones, D. P. Deponte, T. F. Heinz, M. E. Kozina, K. Ledbetter, M. F. Lin, A. M. Lindenberg, M. Mo, A. Nilsson, X. Shen, T. J. A. Wolf, D. Donadio, K. J. Gaffney, T. J. Martinez, and X. Wang, 2021, *Direct observation of ultrafast hydrogen bond strengthening in liquid water*, **Nature 596**, 531.

Yang, J. J., S. Meng, L. F. Xu, and E. G. Wang, 2004, *Ice tessellation on a hydroxylated silica surface*, **Phys. Rev. Lett. 92**, 146102.

Yang, X. B., Y. Ding, and J. Ni, 2008, *Ab initio prediction of stable boron sheets and boron nanotubes: Structure, stability, and electronic properties*, **Phys. Rev. B 77**, 041402(R).

Yang, X. M., 2009, *The extent of Non-Born-Oppenheimer coupling in the reaction of* $Cl(^2p)$ *with* H_2, **Science 322**, 573.

Ye, Q.-J., Z.-Y. Liu, Y. Feng, P. Gao, and X.-Z. Li, 2018, *Ferroelectric problem beyond the conventional scaling law*, **Phys. Rev. Lett. 121**, 135702.

Yen, F., and T. Gao, 2015, *Dielectric anomaly in ice near 20 K: Evidence of macroscopic quantum phenomena*, **J. Phys. Chem. Lett. 6**, 2822.

Yoshimura, Y., S. T. Stewart, M. Somayazulu, H. K. Mao, and R. J. Hemley, 2006, *High-pressure x-ray diffraction and Raman spectroscopy of ice VIII*, **J. Chem. Phys. 124**, 024502.

You, P. W., C. Lian, D. Chen, J. Xu, C. Zhang, S. Meng, and E. G. Wang, 2021a, *Nonadiabatic dynamics of photocatalytic water splitting on a polymeric semiconductor*, **Nano Lett. 21**, 6449.

You, P. W., D. Chen, C. Lian, C. Zhang, and S. Meng, 2021b, *First-principles dynamics of photoexcited molecules and materials towards a quantum description*, **WIREs: Compu. Mol. Sci. 11**, e1492.

Yu, P., W. Luo, D. Yi, J. Zhang, M. Rossell, C.-H. Yang, L. You, Singh-Bhalla G., S. Yang, and Q. He, 2012, *Interface control of bulk ferroelectric polarization*. **Proc. Natl. Acad. Sci. U. S. A. 109**, 9710.

Yu, Z., 1985, *Study about mechanism to form interstellar hydrogen molecule by gas phase reaction*, **Chinese J. Atom. Mol. Phys. 2**, 25.

Yuan, K., R. N. Dixon, and X. Yang, 2011, *Photochemistry of the water molecule: Adiabatic versus nonadiabatic dynamics*, **Acc. Chem. Res. 44**, 369.

Z

Zacharias, M., C. E. Patrick, and F. Giustino, 2015, *Stochastic approach to phonon-assisted optical absorption*, **Phys. Rev. Lett. 115**, 177401.

Zacharias, M., and F. Giustino, 2016, *One-shot calculation of temperature-dependent optical spectra and phonon-induced bandgap renormalization*, **Phys. Rev. B 94**, 075125.

Zaric, S., G. N. Ostojic, J. Kono, J. Shaver, V. C. Moore, M. S. Strano, R. H. Hauge, R. E. Smalley, and X. Wei, 2004, *Optical signatures of the Aharonov-Bohm phase in single-walled carbon nanotubes*, **Science 304**, 1129.

Zeches, R., M. Rossell, J. Zhang, A. Hatt, Q. He, C.-H. Yang, A. Kumar, C. Wang, A. Melville, and C. Adamo, 2009, *A strain-driven morphotropic phase boundary in* $BiFeO_3$, **Science 326**, 977.

Zelevinsky, T., 2019, *Ultracold and unreactive fermionic molecules*, **Science 363**, 820.

Zewail, A. H., and J. M. Thomas, 2009, *4D electron microscopy: Imaging in space and time* (Imperial College Press, London).

Zha, C.-S., Z. Liu, and R. J. Hemley, 2012, *Synchrotron infrared measurements of dense hydrogen to 360 GPa*, **Phys. Rev. Lett. 108**, 146402.

Zhang, C. L., S. Y. Xu, C. M. Wang, Z. Q. Lin, Z. Z. Du, C. Guo, C. C. Lee, H. Lu, Y. Y. Feng, S. M. Huang, G. Q. Chang, C. H. Hsu, H. W. Liu, H. Lin, L. Li, C. Zhang, J. L. Zhang, X. C. Xie, T. Neupert, M. Z. Hasan, H. Z. Lu, J. F. Wang, and S. Jia, 2017, *Magnetic-tunnelling-induced Weyl node annihilation in TaP*, **Nat. Phys. 13**, 979.

Zhang, D. H., and J. Z. H. Zhang, 1994, *Full-dimensional Time-dependent treatment for diatom-diatom reactions: The* $H_2 + OH$ *reaction*, **J. Chem. Phys. 101**, 1146.

Zhang, G. Y., X. Jiang, and E. G. Wang, 2003, *Tubular graphite cones*, **Science 300**, 472.

Zhang, I. Y., and X. Xu, 2021, *On the top rung of Jacob's ladder of density functional theory: Toward resolving the dilemma of SIE and NCE*, **Wiley Interdiscip. Rev. Comput. Mol. Sci. 96**, 11.

Zhang, J., P. Chen, B. Yuan, W. Ji, Z. Cheng, and X. Qiu, 2013, *Real-space identification of intermolecular bonding with atomic force microscopy*, **Science 342**, 611.

Zhang, J., H. Hong, C. Lian, W. Ma, X. Xu, X. Zhou, H. Fu, K. H. Liu, and S. Meng, 2017, *Interlayer-state-coupling dependent ultrafast charge transfer in MoS_2/WS_2 bilayers*, **Adv. Sci. 4**, 1700086.

Zhang, J., C. Lian, M. X. Guan, W. Ma, H. X. Fu, H. Guo, and S. Meng, 2019, *Photoexcitation induced quantum dynamics of charge density wave and emergence of a collective mode in 1T-TaS2*, **Nano Lett. 19**, 6027.

Zhang, L., Q. Zheng, Y. Xie, Z. Lan, O. V. Prezhdo, W. A. Saidi, and J. Zhao, 2018, *Delocalized impurity phonon induced electron–hole recombination in doped semiconductors*, **Nano. Lett. 18**, 4592.

Zhang, L., W. Chu, Q. Zheng, A. V. Benderskii, O. V. Prezhdo, and J. Zhao, 2019, *Suppression of electron–hole recombination by intrinsic defects in 2D monoelemental material*, **J. Phys. Chem. Lett. 10**, 6151.

Zhang, L. J., Y. Xie, T. Cui, Y. Li, Z. He, Y. M. Ma, and G. T. Zou, 2006, *Pressure-induced enhancement of electron-phonon coupling in superconducting CaC_6 from first principles*, **Phys. Rev. B 74**, 184519.

Zhang, Q. F., 2010, *Full-quantum effect and path-integral molecular dynamics*, PhD dissertation, Graduate School of Chinese Academy of Sciences, Beijing, (in Chinese).

Zhang, Q. F., G. Wahnstrom, M. E. Bjorketun, S. W. Gao, and E. G. Wang, 2008, *Path integral treatment of proton transport processes in $BaZrO_3$*, **Phys. Rev. Lett. 101**, 215902.

Zhang, R., Y. Zhang, Z. C. Dong, S. Jiang, C. Zhang, L. G. Chen, L. Zhang, Y. Liao, J. Aizpurua, Y. Luo, J. L. Yang, and J. G. Hou, 2013, *Chemical mapping of a single molecule by plasmon-enhanced Raman scattering*, **Nature 498**, 82.

Zhang, W. T., G. D. Liu, L. Zhao, H. Y. Liu, J. Q. Meng, X. L. Dong, W. Lu, J. S. Wen, Z. J. Xu, G. D. Gu, T. Sasagawa, G. L. Wang, Y. Zhu, H. B. Zhang, Y. Zhou, X. Y. Wang, Z. X. Zhao, C. T. Chen, Z. Y. Xu, and X. J. Zhou, 2008, *Identification of a new form of electron coupling in the $Bi_2Sr_2CaCu_2O_8$ superconductor by laser-based angle-resolved photoemission spectroscopy*, **Phys. Rev. Lett. 100**, 107002.

Zhang, X. F., Q. J. Ye, H. J. Xiang, and X. Z. Li, 2020, *Quantum paraelectricity of $BaFe_{12}O_{19}$*, **Phys. Rev. B 101**, 104102.

Zhang, X. F., Q. J. Ye, and X. Z. Li, 2021, *Structural phase transition and Goldstone-like mode in hexagonal $BaMnO_3$*, **Phys. Rev. B 103**, 024101.

Zhang, Y., B. Yang, A. Ghafoor, Y. Zhang, Y. F. Zhang, R. P. Wang, J. L. Yang, Y. Luo, Z. C. Dong, and J. G. Hou, 2019, *Visually constructing the chemical structure of a single molecule by scanning Raman picoscopy*, **Natl. Sci. Rev. 6**, 1169.

Zhang, Y. B., Y. W. Tan, H. L. Stormer, and P. Kim, 2005, *Experimental observation of the quantum Hall effect and Berry's phase in graphene*, **Nature 438**, 201.

Zhao, C. L., Q. D. Wang, Z. P. Yao, J. L. Wang, B. Sanchez-Lengeling, F. X. Ding, X. Qi, Y. X. Li, Y. Lu, X. D. Bai, B. H. Li, H. Li, A. Aspuru-Giuzik, X. Huang, C. Delmas, M. Wagemaker, L. Chen, and Y. S. Hu, 2020, *Rational design of layered oxide materials for sodium-ion batteries*, **Science 370**, 708.

Zhao, H., Q. An, X. Ye, B. H. Yu, Q. H. Zhang, F. Sun, Q. Y. Zhang, F. Yang, J. D. Guo, and J. M. Zhao, 2021, *Second harmonic generation in AB-type $LaTiO_3/SrTiO_3$ superlattices*, **Nano Energy 82**, 105752.

Zhao, J. M., A. V. Bragas, D. J. Lockwood, and R. Merlin, 2004, *Magnon squeezing in an antiferromagnet: Reducing the spin noise below the standard quantum limit*, **Phys. Rev. Lett. 93**, 107203.

Zhao, J. M., A. V. Bragas, R. Merlin, and D. J. Lockwood, 2006, *Magnon squeezing in antiferromagnetic MnF_2 and FeF_2*, **Phys. Rev. B. 73**, 184434.

Zhao, R. J., P. W. You, and S. Meng, 2023, *Ring polymer molecular dynamics with electronic transitions*, **Phys. Rev. Lett. 130**, 166401.

Zhao, Y., T. Yamamoto, and W. H. Miller, 2004, *Path integral calculation of thermal rate constants within the quantum instanton approximation: Application to the H+CH_4 àH_2+CH_3 hydrogen abstraction reaction in full Cartesian space*, **J. Chem. Phys. 120**, 3100.

Zhao, Y. C., L. Du, W. Yang, C. Shen, J. Tang, X. Li, Y. Chu, J. Tian, K. Watanabe, T. Taniguchi, R. Yang, D. Shi, Z. Sun, and G. Zhang, 2020, *Observation of logarithmic Kohn anomaly in monolayer graphene*, **Phys. Rev. B 102**, 165415.

Zhao, Y. W., Y. Lu, H. P. Li, Y. B. Zhu, Y. Meng, N. Li, D. H. Wang, F. Jiang, F. N. Mo, C. B. Long, Y. Guo, X. L. Li, Z. D. Huang, Q. Li, J. C. Ho, J. Fan, M. L. Sui, F. R. Chen, W. G. Zhu, W. S. Liu, and C. Y. Zhi, 2022, *Few-layer bismuth selenide cathode for low-temperature quasi-solid-state aqueous zinc metal batteries*, **Nat. Commun. 13**, 752.

Zheng, J. J., S. Zhang, B. J. Lynch, J. C. Corchado, Y.-Y. Chuang, P. L. Fast, W.-P. Hu, Y.-P. Liu, G. C. Lynch, K. A. Nguyen, C. F. Jackels, A. F. Ramos, B. A. Ellingson, V. S. Melissas, J. Villà, I. Rossi, E. L. Coitiño, J. Pu, and T. V. Albu, 2010, *POLYRATE – version 2010-A* (University of Minnesota, Minneapolis, MN).

Zheng, Q. J., W. B. Chu, C. Y. Zhao, L. L. Zhang, H. L. Guo, Y. N. Wang, J. Xiang, and J. Zhao, 2019, *Ab initio nonadiabatic molecular dynamics investigations on the excited carriers in condensed matter systems*, **WIREs Comput. Mol. Sci. 9**, 1411.

Zhong, G. H., X. Chen, and H. Lin, 2015, *Prediction of superconductivity in potassium-doped benzene*, DOI: 10.48550/arXiv.1501.00240.

Zhong, G. H., C. L. Yang, X. J. Chen, and H. Q. Lin, 2018, *Superconductivity in solid benzene molecular crystal*, **J. Phys.: Condens. Matter. 30**, 245703.

Zhong, J. H., X. Jin, L. Y. Meng, X. Wang, H. S. Su, Z. L. Yang, C. T. Williams, and B. Ren, 2017, *Probing the electronic and catalytic properties of a bimetallic surface with 3 nm resolution,* **Nat. Nanotechnol. 12**, 132.

Zhong, W., D. Vanderbilt, and K. M. Rabe, 1995, *First-principles theory of ferroelectric phase transitions for perovskites: The case of $BaTiO_3$*, **Phys. Rev. B 52**, 6301.

Zhong, W., and D. Vanderbilt, 1996, *Effect of quantum fluctuations on structural phase transitions in $SrTiO_3$ and $BaTiO_3$*, **Phys. Rev. B 53**, 5047.

Zhong, W. L. (1996). *Ferroelectric Physics* (Science Press (in Chinese), Beijing).

Zhou, X. G., Y. Yao, Y. H. Matsuda, A. Ikeda, A. Matsuo, K. Kindo, and H. Tanaka, 2020, *Particle-hole symmetry breaking in a spin-dimer system $TlCuCl_3$ observed at 100 T*, **Phys. Rev. Lett. 125**, 267207.

Zhou, X. J., S. L. He, G. D. Liu, L. Zhao, L. Yu, and W. T. Zhang, 2018, *New developments in laser-based photoemission spectroscopy and its scientific applications: A key issues review*, **Rep. Prog. Phys. 81**, 062101.

Zhu, X., and J. J. Quinn, and G. Gumbs, 1990, *Excitonic insulator transition in a GaSb-AlSb-InAs quantum-well structure*, **Solid State Commun. 75**, 595.

Zhu, X. D., A. Lee, A. Wong, and U. Linke, 1992, *Surface diffusion of hydrogen on Ni(100): An experimental observation of quantum tunneling diffusion*, **Phys. Rev. Lett. 68**, 1862.

Ziman, J. M., 1960, *Electrons and phonons* (**Oxford University Press**, New York).

Zimmermann, T., and J. Vaníček, 2014, *Efficient on-the-fly ab initio semiclassical method for computing time-resolved nonadiabatic electronic spectra with surface hopping or Ehrenfest dynamics*, **J. Chem. Phys. 141**, 134102.

Zinchenko, K. S., F. Ardana-Lamas, I. Seidu, S. P. Neville, J. van der Veen, V. U. Lanfaloni, M. S. Schuurman, and H. J. Worner, 2021, *Sub-7-femtosecond conical-intersection dynamics probed at the carbon K-edge*, **Science 371**, 489.

Zollner, S, M. Cardona, and S. Gopalan, 1992, *Isotope and temperature shifts of direct and indirect band gaps in diamond-type semiconductors*, **Phys. Rev. B 45**, 3376.
Zuev, P., and R. S. Sheridan, 1994, *Tunneling in the C-H insertion of a singlet carbene: Tert-butylchlorocarbene*, **J. Am. Chem. Soc. 116**, 4123.
Zwanzig, R., 1965, *Time-correlation functions and transport coefficients in statistical mechanics*, **Annu. Rev. Phys. Chem. 16**, 67.

中文文献

曹端云, 2020, 界面水的结构与动力学研究, 北京大学.
陈基, 2014, 氢的原子核量子效应计算模拟研究, 北京大学.
丁哲, 石发展, 杜江峰, 2020, 基于金刚石量子传感的纳米磁成像及凝聚态物理应用, 物理**49**, 359.
高淼, 卢仲毅, 向涛, 2015, 通过金属化σ电子寻找高温超导体, 物理 **44**, 421.
郭见青, 2022, 轻元素液体结构和动力学的第一性原理计算研究, 北京大学.
郝文杰, 翟燕妮, 张倩瑜, 赵继民, 2021, 阿秒光源在材料领域的应用, 科学通报**66**, 856.
胡旭光, 蔡宇民, 李前树, 1991, 过渡态理论的发展（二）, 西安石油学院学报**6**, 55.
匡光力, 邵淑芳, 2018, 稳态强磁场技术及科学意义, 科技导报**36**, 93.
李宁, 2022, 低维氮化硼中声子与声子极化激元的电子能量损失谱研究, 北京大学.
陆果, 1997, 基础物理学, (高等教育出版社, 北京).
罗渝然, 1983, 过渡态理论的进展, 化学通报**10**, 8.
马杰, 2009, 水在石墨烯表面的范德瓦尔斯作用研究, 中国科学院研究生院.
孟胜, 王恩哥, 2014, 水基础科学理论与实验, (北京大学出版社, 北京).
宋凯, 徐扬, 宋林泽, 史强, 2016, 凝聚相质子转移反应的量子动力学理论, 中国科学: 化学**46**, 1039.
杨学明, 2009, 氯加氢化学反应中玻恩-奥本海默近似的适用性, 中国基础科学·研究进展.
张千帆, 2010, 全量子化效应与路径积分分子动力学研究, 中国科学院研究生院.
钟维烈, 1996, 铁电体物理学, 科学出版社.

Index

For EU product safety concerns, contact us at Calle de José Abascal, 56–1°,
28003 Madrid, Spain or eugpsr@cambridge.org.

www.ingramcontent.com/pod-product-compliance
Lightning Source LLC
LaVergne TN
LVHW062159200926
848484LV00006B/1287

* 9 7 8 1 0 0 9 5 6 3 9 1 8 *